T0183531

# Topics in Applied Physics

Volume 145

Topics in Applied Physics is a well-established series of review books, each of which presents a comprehensive survey of a selected topic within the domain of applied physics. Since 1973 it has served a broad readership across academia and industry, providing both newcomers and seasoned scholars easy but comprehensive access to the state of the art of a number of diverse research topics.

Edited and written by leading international scientists, each volume contains high-quality review contributions, extending from an introduction to the subject right up to the frontiers of contemporary research.

Topics in Applied Physics strives to provide its readership with a diverse and interdisciplinary collection of some of the most current topics across the full spectrum of applied physics research, including but not limited to:

- Quantum computation and information
- Photonics, optoelectronics and device physics
- Nanoscale science and technology
- Ultrafast physics
- Microscopy and advanced imaging
- Biomaterials and biophysics
- Liquids and soft matter
- Materials for energy
- Geophysics
- Computational physics and numerical methods
- Interdisciplinary physics and engineering

We welcome any suggestions for topics coming from the community of applied physicists, no matter what the field, and encourage prospective book editors to approach us with ideas. Potential authors who wish to submit a book proposal should contact Zach Evenson, Publishing Editor:

zachary.evenson@springer.com

Topics in Applied Physics is included in Web of Science (2020 Impact Factor: 0.643), and is indexed by Scopus.

More information about this series at https://link.springer.com/bookseries/560

Alkwin Slenczka · Jan Peter Toennies
Editors

# Molecules in Superfluid Helium Nanodroplets

Spectroscopy, Structure, and Dynamics

 Springer

*Editors*
Alkwin Slenczka
Institute of Physical and Theoretical
Chemistry
University of Regensburg
Regensburg, Germany

Jan Peter Toennies
MPI für Dynamik und Selbstorganisation
Göttingen, Germany

ISSN 0303-4216 ISSN 1437-0859 (electronic)
Topics in Applied Physics
ISBN 978-3-030-94898-6 ISBN 978-3-030-94896-2 (eBook)
https://doi.org/10.1007/978-3-030-94896-2

This Springer imprint is published by the registered company Springer Nature Switzerland AG
The registered company address is: Gewerbestrasse 11, 6330 Cham, Switzerland

### The Helium Poem

*This little atom helium*
*is cause of great delirium.*
*When one predicts what it will do,*
*one often finds, one has no clue.*

*At cryogenic temperature,*
*where other atoms cease to stir,*
*helium alone does not stagnate,*
*but forms a superfluid state.*

*In supersonic gas expansion,*
*droplets grow by aggregation,*
*And they cool down to just a smidge:*
*To form a supersonic fridge.*

*If in this enigmatic juice*
*a spinning molecule's let loose,*
*instead of slowing naturally,*
*it keeps on twirling merrily.*

*The secret behind this strange action*
*is the large superfluid fraction,*
*which to every one's surprise*
*have even clusters of smallest size.*

*Though superfluid helium*
*was studied ad absurdium,*
*by Landau, London, Yang and Lee,*
*They missed the droplets completely.*

*Now the droplets are the greatest,*
*being cold and gentle is the latest,*
*clusters grow and self organize*
*and form structures which tantalize.*

*Let's hope droplets of helium.*
*a most marvelous medium,*
*bring benefits to chemistry*
*for the rest of this century...*

*R. B. Doak, A. P. Graham, B. Holst, and J. P. Toennies*

*We dedicate this book to countless students, especially those not listed as coauthors. Without their optimism, enthusiasm, and hard work, many of the results in this book would not have been reported.*

# Preface

*Theory is a good thing but a good experiment lasts forever.*

*Pyotr Kapitza cited by Neville Mott in Nature 288, 627 (1980)*

Superfluid helium nanodroplets have in the last 25 years opened up a new era of high-resolution spectroscopy. This has been made possible by their sub-Kelvin temperatures and the gentle nature of the superfluid. As a matrix for spectroscopy, helium droplets have provided a new perspective on the physics and chemistry of atoms, molecules, both inorganic and organic, and even large biomolecules, also in the form of clusters and large aggregates. New information has also been obtained on the reactions and structures of radicals and ions. Moreover, much has also been revealed about the photon-excitation dynamics of molecules. At the same time, the molecular spectra have provided new insight into the microscopic many-body coherent quantum physics of the helium superfluid. For example, the very first evidence that small finite-sized objects with even less than 100 helium atoms can be superfluid comes from this new spectroscopy.

This is the first book that provides the reader with a coherent overview of the progress made in the last 25 years. A quarter of a century suffices to present a comprehensive description of the progress in understanding and the potential of using helium nanodroplets as "The ultimate spectroscopic matrix?" as Giacinto Scoles and Kevin Lehmann, two of the pioneers, proclaimed already in 1998. Compared to the 1990s when about 500 articles appeared under terms similar to the book's headline, presently more than 5000 articles appear every year. Additional evidence for the vitality of this area of research is the large number of 43 reviews that have appeared since 1998.

Spectroscopic matrices as molecular sample holders have a long tradition in chemistry. They were first introduced in the 1950s by George Pimentel and coworkers for stabilizing and studying unstable or reactive molecules by freezing and preserving them in a large excess of inert atoms or molecules condensed on a cold transparent substrate. This method has not only the advantage of synthetic variability but also the disadvantage of appreciable perturbations and the disabling of mobility like molecular rotation or reactive collision. In 1977, the seeded beam method was introduced by Richard Smalley, Lennard Wharton, and Don Levy as an alternative to matrix isolation. In this method, gas phase molecules to be investigated are cooled to low temperatures of several degrees Kelvin by coexpanding in an excess of an inert gas into a vacuum. At the end of the expansion, the molecules are unperturbed and are free to rotate and can even interact under single collision conditions with a crossed beam. The method is limited to small molecules to assure that the internal degrees of freedom are sufficiently cold.

The superfluid helium nanodroplet matrix unifies the advantages of both methods of synthetic flexibility and negligible perturbation enabling the molecules to move around and rotate, with the additional advantage of sub-Kelvin temperatures. Moreover, for helium droplets, the possible sizes of molecular dopant species are far larger. For example, large tailor-made molecular aggregates can be synthesised inside the droplets by consecutively adding the individual constituents to the droplets, which act as a kind of test tube. The secret behind these many advantages lies ultimately in the unique properties of helium. It is the only chemically completely inert element and the only element which remains liquid down to 0 K (at pressures below 25 bar) and below 2.17 K becomes the only naturally occurring superfluid.

In its eleven chapters the book provides a coherent overview of the many new advances that have been made in studying the physics and chemistry of a broad variety of atomic scaled systems doped into superfluid helium droplets. The eleven chapters are organized in the following way: The first two chapters serve to introduce the reader to the properties of pure helium droplets and to the extremely weak van der Waals interaction between two He atoms, which as described in the second chapter, leads to the largest of all small molecules. Then the next four chapters are devoted to the mass, infrared and optical spectroscopic investigations and chemical reactions of embedded foreign atoms, molecules, radicals and clusters. The following two chapters describe diffraction experiments to determine the structures of droplets and embedded large protein molecules. An overview of time dependent spectroscopy focusing on the short-term dynamics is provided in the next two chapters. Finally, the last chapter is devoted to the synthesize of new nanoscopic materials within nanodroplets.

Chapter 1 by one of the editors, J. Peter Toennies, starts with summarizing the history of superfluidity and droplet experiments. The different types of expansions used to produce droplets of different sizes and the velocities of droplets are surveyed. The theoretical and experimental studies of the physical properties of droplets, such as the total energies, excited state energies, densities, radial distribution and temperatures are also reviewed. Finally, the theoretical and experimental evidence for the microscopic superfluidity of droplets are summarized.

Chapter 2 by Maksim Kunitski is in a sense outside the area implied by the title of the book "Molecules in Helium" since his chapter is devoted to "Molecules of Helium". The remarkable aspect of the Coulomb Explosion Imaging (CEI) technique described in this chapter is that it enables imaging directly the Quantum Probability Density (QPD) of these exceptionally dispersed quantum molecules with remarkable precision. From the CEI measurements, the QPD could be determined over the entire quantum tunneling region from 14 Å out to 250 Å. From the exponential slope of the QPD, the extraordinarily weak binding energy of the helium dimer was finally precisely established to be $1.77 \pm 0.015$ $mK$ in excellent agreement with the latest quantum chemical calculations which included also non-adiabatic interactions. The structure of the very fascinating extremely quantum-fuzzy Efimov state of the helium trimer could be measured and the tetramer is on the horizon. The CEI technique is also used in Chap. 8 to follow the psec. motions of molecules in droplets.

Chapter 3 by Arne Schiller and colleagues describes the latest results of the Innsbruck group of Paul Scheier on using mass spectroscopy techniques to explore pure and doped neutral and ionic droplets. Since mass spectrometry was historically the first experimental technique for studying helium droplets, their chapter starts with an informatively compact overview of past experiments. Their chapter then deals with new studies of ion abundance distributions of embedded metal atoms and organic molecules. The last part reports on new exciting results in which ionized droplets are again ionized to form multiply charged ions.

Chapter 4 by Gary Douberly brings the reader up to date on the tremendous progress made in using infra-red spectroscopy as well as Stark and Zeeman spectroscopy for studying the rotational fine structure of vibrational modes in order to analyze the structure of organic molecules, and highly reactive organic radicals as, for example, some containing a divalent carbon atom. This chapter also reports on reactions involving highly reactive organic radicals.

In Chap. 5, Alkwin Slenczka and his colleagues report on electronic spectroscopy namely fluorescence excitation and dispersed emission of polycyclic aromatic hydrocarbons including smaller ones like polyacenes up to larger ones such as phthalocyanine or porphyrins. Moreover, the formation of van der Waals clusters of these molecules mainly with rare gas atoms is analyzed with electronic spectroscopy. Several examples of photolysis reactions and the influence of the superfluid environment are discussed in detail. Comparison of helium droplet data with corresponding gas phase experiments reveals not only advantages but also some open challenges in understanding microsolvation in helium droplets.

In Chap. 6, the group of Gert von Helden at the Fritz-Haber Institute in Berlin report on their pioneering experiments in which large ionized and mass selected organic biomolecules from an electrospray source are inserted into large droplets. With their method, they have extended helium droplet spectroscopy into the realm of large biomolecules with up to about 2000 Da. Highly resolved infra-red spectra of mono-, di-, tri-, and tetra-saccharides measured with an in-house free electron laser can even distinguish between conformers at selected sites in these large molecules. Among the tetra-saccharides studied are biologically active blood group antigens and human milk oligosaccharides. The experiments point to the great potential of droplet spectroscopy in pharmaceutical research.

Chapters 7 and 8 report on relatively recent progress in directly imaging very large pure and doped droplets and the structures of large aligned biomolecules. In Chap. 7, Thomas Möller and his colleagues report on the method of coherent diffraction imaging using X-rays or extreme ultra-violet light. These experiments have revealed that large droplets with up to $N \cong 10^{12}$ atoms and diameters of several hundred nanometers, previously thought to be spherical, have many different non-spherical shapes ranging from flat oblates to highly elongated prolate spheroids with aspect ratios up to 3.0. The images from xenon doped droplets show dopant traced superfluid vortex filaments and even Abrikosov vortex lattices. The shapes and vortex filaments reveal that the droplets are highly rotationally excited.

Chapter 8 by Wei Kong and colleagues describes their efforts to determine the structure of single oriented large biomolecules by electron diffraction. Most of what is known about the structure of large protein molecules is derived from X-ray diffraction from crystallites. Since many biomolecules cannot be crystallized, this experiment has the important goal of determining the structures of single biomolecules. In the ultracold superfluid environment of the droplets, the embedded molecules are easily aligned by an elliptically polarized laser field. The structures are then determined by diffracting electrons from different directions.

Chapters 9 and 10 are devoted to the time-resolved spectroscopy of molecules in helium nanodroplets. In Chap. 9, Henrik Stapelfeldt and his coworkers discuss the adiabatic and non-adiabatic alignment of molecules in the superfluid helium environment in comparison to the free molecules. They find that femtosecond or picosecond laser pulses can induce coherence in the rotation of embedded molecules that last for long times. Moreover, the Coulomb Explosion Imaging allows following the relaxation of laser pulse alignment of small linear species such as $I_2$, OCS, and $CS_2$ and in addition larger organic molecules and their clusters. Upon two-dimensional alignment, angular covariance maps of fragment pairs reveal deep insight into the structure of the solvated dopant system.

Chapter 10 from the group of Frank Stienkemeier also provides a picture of the dynamics of embedded molecules after they have been excited vibrationally or electronically via femtosecond pulse-probe spectroscopy. Separate subsections are devoted to time-resolved XUV spectroscopy and interatomic Coulombic decay processes and the dynamics of helium nano-plasmas. This chapter also introduces the reader to the technique of 2D spectroscopy, which enables observing the correlation between the absorption and emission processes in a single molecule. In their application, it is used to analyze $Rb_2$ and $Rb_3$ clusters on the surface of helium nanodroplets.

The last, but by no means less interesting, Chap. 11 by Florian Lackner describes the remarkable successes in synthesizing tailored nanoparticles. The chapter starts with an analysis of the mechanisms for nanoparticle growth. The discussion is illustrated with many examples of transmission electron microscopy, STM, and elemental images of core-shell metal, alloyed metal, and metal oxide and even metal core-transition metal oxide shell nanoparticles created in droplets and deposited on surfaces. The synthesis in helium droplet excels in the sub-10 nm size regime and has many potential applications in catalysis, plasmonics, magnetics, and photocatalysis.

Appendix A contains an annotated list of the 43 review articles devoted to the chemistry and physics of nanodroplets that have appeared since 1998. They provide a historical perspective and complement the chapters in this book.

This monograph was initially planned to appear as an ordinary volume in the series Topics in Applied Physics. Shortly before starting the printing process, it was unanimously decided by the authors to convert it to an open access book. For important advice with funding the open access fee, we are extremely thankful to Eberhard Bodenschatz (Max Planck Institut für Dynamik und Selbstorganisation, Göttingen): We wish to also thank him for a generous contribution. We are also indebted to the Max Planck Förderstiftung (https://www.maxplanckfoundation.org/) and a private sponsor for additional contributions which together made the conversion to an open access book possible.

The editors thank all 39 authors for their stellar contributions which will make this a long-lasting classical reference point for both practitioners and learners. As an open access book, we hope it will be widely used by students and scientists all around the globe. We the editors thank the authors for making this project a rewarding collaborative adventure with many minor ups and downs and for their patience.

Finally, the editors and authors are most grateful to Dr. Angela Lahee from Springer Nature for arranging that this book appears in the Springer Series Topics in Applied Physics. We the editors are thankful for her quick response and untiring patience in guiding us through our first venture as editors. Her congenial harmonious assistance made the project a pleasure for both of us.

Regensburg, Germany  
Göttingen, Germany

Alkwin Slenczka  
Jan Peter Toennies

# Contents

# Contributors

**Stephen Bradford** Chemistry Department, Oregon State University, Corvallis, OR, USA

**Lukas Bruder** Institute of Physics, University of Freiburg, Freiburg, Germany

**Adam S. Chatterley** Department of Chemistry, Aarhus University, Aarhus C, Denmark

**Lars Christiansen** Department of Chemistry, Aarhus University, Aarhus C, Denmark

**Gary E. Douberly** Department of Chemistry, University of Georgia, Athens, Georgia, USA

**Johannes Fischer** Institut für Physikalische und Theoretische Chemie, Universität Regensburg, Regensburg, Germany

**Yunteng He** Central Community College, Kearney, NE, USA

**Markus Koch** Institute of Experimental Physics, Graz University of Technology, Graz, Austria

**Wei Kong** Chemistry Department, Oregon State University, Corvallis, OR, USA

**Lorenz Kranabetter** Department of Chemistry, Aarhus University, Aarhus C, Denmark

**Maksim Kunitski** Institut für Kernphysik, Geothe-Universität Frankfurt am Main, Frankfurt am Main, Germany

**Florian Lackner** Institute of Experimental Physics, Graz University of Technology, Graz, Austria

**Felix Laimer** Institut für Ionenphysik und Angewandte Physik, Universität Innsbruck, Innsbruck, Austria

**Bruno Langbehn** Institut für Optik und Atomare Physik, Technische Universität Berlin, Berlin, Germany

**Lei Lei** Chemistry Department, Oregon State University, Corvallis, OR, USA

**Maike Lettow** Fritz-Haber-Institut der Max-Planck-Gesellschaft, Berlin, Germany

**Gerard Meijer** Fritz-Haber-Institut der Max-Planck-Gesellschaft, Berlin, Germany

**Eike Mucha** Fritz-Haber-Institut der Max-Planck-Gesellschaft, Berlin, Germany

**Marcel Mudrich** Institute of Physics and Astronomy, Aarhus University, Aarhus C, Denmark;
Indian Institute of Technology Madras, Chennai, India

**Alberto Viñas Muñoz** Department of Chemistry, Aarhus University, Aarhus C, Denmark

**Thomas Möller** Institut für Optik und Atomare Physik, Technische Universität Berlin, Berlin, Germany

**Jens H. Nielsen** Department of Physics and Astronomy, University of Aarhus, Aarhus C, Denmark

**Kevin Pagel** Fritz-Haber-Institut der Max-Planck-Gesellschaft, Berlin, Germany; Institut für Chemie und Biochemie, Freie Universität Berlin, Berlin, Germany

**Dominik Pentlehner** Department of Chemistry, Aarhus University, Aarhus C, Denmark;
Technische Hochschule Rosenheim, Campus Burghausen, Burghausen, Germany

**James D. Pickering** Department of Chemistry, Aarhus University, Aarhus C, Denmark

**Daniela Rupp** Laboratory for Solid State Physics, ETH Zürich, Zürich, Switzerland

**Arne Schiller** Institut für Ionenphysik und Angewandte Physik, Universität Innsbruck, Innsbruck, Austria

**Florian Schlaghaufer** Institut für Physikalische und Theoretische Chemie, Universität Regensburg, Regensburg, Germany

**Constant A. Schouder** Department of Chemistry, Aarhus University, Aarhus C, Denmark

**Benjamin Shepperson** Department of Chemistry, Aarhus University, Aarhus C, Denmark

**Alkwin Slenczka** Institut für Physikalische und Theoretische Chemie, Universität Regensburg, Regensburg, Germany

**Henrik Stapelfeldt** Department of Chemistry, Aarhus University, Aarhus C, Denmark

**Frank Stienkemeier** Institute of Physics, University of Freiburg, Freiburg, Germany

**Anders A. Søndergaard** Department of Physics and Astronomy, University of Aarhus, Aarhus C, Denmark

**Rico Mayro P. Tanyag** Institut für Optik und Atomare Physik, Technische Universität Berlin, Berlin, Germany

**Daniel Thomas** Fritz-Haber-Institut der Max-Planck-Gesellschaft, Berlin, Germany;
Department of Chemistry, University of Rhode Island, Kingston, RI, USA

**Lukas Tiefenthaler** Institut für Ionenphysik und Angewandte Physik, Universität Innsbruck, Innsbruck, Austria

**J. Peter Toennies** Max-Planck-Institut für Dynamik und Selbstorganisation, Göttingen, Germany

**Gert von Helden** Fritz-Haber-Institut der Max-Planck-Gesellschaft, Berlin, Germany

**Yuzhong Yao** Chemistry Department, Oregon State University, Corvallis, OR, USA

**Jie Zhang** Chemistry Department, Oregon State University, Corvallis, OR, USA

# Chapter 1
# Helium Nanodroplets: Formation, Physical Properties and Superfluidity

J. Peter Toennies

**Abstract** In this introductory chapter, we begin by informing the reader about the fascinating history of superfluidity in bulk liquid helium. This is followed by relating attempts in using liquid helium as a low temperature matrix for spectroscopy. After a brief review of the thermodynamic properties of helium in Sect. 1.2, the different types of free jet expansions used in experiments to produce clusters and nanodroplets of different sizes are described in Sect. 1.3. First it is shown how they depend on the nature and location in the phase diagram of the isentropes which determine the course of the expansion. Depending on the four regimes of isentropes, different number sizes and distributions are expected. Next in Sect. 1.4, the results of theoretical and, where available, experimental results on the total energies, excited states, radial density distributions, and temperatures of clusters and droplets are discussed. Finally, in Sect. 1.5 the theoretical and experimental evidence for the superfluidity of nanodroplets is briefly reviewed. For more information on the production and characteristics of nanodroplets, the reader is referred to the chapters in this book and to the reviews in Appendix.

## 1.1 History

### 1.1.1 History of Superfluidity in Helium

The first evidence for the existence of helium was a new spectral feature in the Fraunhofer spectrum of the sun discovered in 1868 by the French astronomer Jules Jansen and quickly confirmed in the same year by English astronomers. It was thought then to be another alkali atom and was named after the sun Helios with an ending of "um" as with all the then known alkali atoms. Helium is the second most abundant element in the universe after hydrogen. It is, of course, prevalent in the sun where it is produced by the nuclear fusion of hydrogen. On the heavy planets Saturn and

J. P. Toennies (✉)
Max-Planck-Institut für Dynamik und Selbstorganisation, Göttingen, Germany
e-mail: jtoenni@gwdg.de

A. Slenczka and J. P. Toennies (eds.), *Molecules in Superfluid Helium Nanodroplets*,
Topics in Applied Physics 145, https://doi.org/10.1007/978-3-030-94896-2_1

Jupiter it makes up 60–70% of their atmospheres. Because of the earth's light mass, helium easily escapes and is virtually not present in our atmosphere.

Helium was first liquified by Kamerlingh Onnes in 1908 with a six-stage cooling cascade in Leiden, Holland. Three years later, the ultralow temperatures enabled Kamerlingh to discover superconductivity in mercury at 4.17 K. Thirty years later, in 1938, the superfluidity of liquid helium was discovered at temperatures below 2.141 K by Pyotr Kapitza in Moscow [1] and simultaneously by Allen and Misener in Cambridge [2]. Both results were published back to back in Nature. The fascinating stories behind the race to discover superfluidity have been recently reviewed [3–6]. In the same year, László Tisza published a Nature article where he postulated that superfluid helium consists of two interpenetrating liquids called the superfluid and normal fractions (two fluid model) [7]. Also in 1938, Fritz London proposed that Bose–Einstein Condensation (BEC) [8] could explain the superfluidity of helium [9].

Superfluidity manifests itself through many different apparently unrelated phenomena, such as a vanishingly small viscosity, the fountain effect, the ability to creep out of a container defying gravity, and the extraordinary ability to conduct heat much more efficiently than even the purest metals. These many strange properties are all *macroscopic*. These properties inspired Fritz London in 1954 to proclaim that "superfluid helium, also called liquid helium II, is the only representative of a particular fourth state of aggregation" [10].

In 1941 when Nazi Germany invaded Russia, Lev Landau, then in Tbilisi, was the first to formulate a microscopic theory of superfluidity by assuming that the elementary excitations in a superfluid are dominated by highly coherent phonons at low energies [11]. Consequently, he postulated that the phonon dispersion curves in the superfluid are different than in an ordinary liquid. For one they are sharply defined, as in a solid, and further exhibit a maximum which he called a "maxon" and a minimum called a "roton" (Fig. 1.1). He also pointed out that the sharpness of the dispersion curve and the roton would allow the frictionless motion of a particle at velocities below 58 m/s. The sharp dispersion curves coupled with the two-fluid

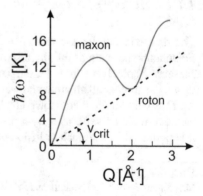

**Fig. 1.1** The dispersion curve of superfluid helium proposed by Landau. Particles moving in the superfluid at velocities less than the critical velocity of 58 m/s experience no resistance

model of Tisza [7] and the suggestion by London relating Bose–Einstein Condensation with superfluidity [9] provided a unifying framework within which many of the above macroscopic observations could be described in the following years.

In contrast, on a microscopic level, the physical mechanisms behind many of the unusual properties of both the quantum liquids $^3$He and $^4$He are yet not fully understood. Even today, the statement made in 1984 that the "connections between Bose condensation and superfluidity (in $^4$He) remain a deep and complex problem" [12] is still valid despite considerable progress in explaining many related phenomena in Bose–Einstein condensed gases [13, 14]. On the other hand, superfluid helium droplets and trapped Bose–Einstein condensates are both quantum many-body systems and have much in common [15].

Superfluidity and superconductivity are closely related [16]. Both are ubiquitous phenomena which occur in solids, in the nuclear matter in stars, as well as in nuclei [17]. and also in elementary particle physics, e.g. Higgs boson [18] The two volumes edited by Bennemann and Ketterson entitled "Novel Superfluids" list and discuss the many manifestations of superfluidity and superconductivity [19].

## 1.1.2  History of Helium as a Cryomatrix for Spectroscopy

The secret behind the advantages of helium as a cryomatrix for spectroscopy lies ultimately in the unique properties of the helium atom. The properties relevant to the use of droplets as a matrix are briefly summarized in the following four paragraphs:

1. The helium atom is the most inert of all atoms. No chemistry is known. In comparison with the simpler H atom, the He atom, with two electrons in a closed shell, is even smaller in size and has a three times smaller polarizability. The energy of the first excited state of 19.82 eV makes liquid helium transparent at wavelengths larger than 62.56 nm.

2. The inertness of the helium atom is responsible for the exceedingly weak van der Waals potential of the helium-4 dimer, which is just barely bound. As discussed in Chap. 2 by M. Kunitski in this book, the dimer has a mean internuclear distance of $45 \pm 2$ Å. This is about 3 times greater than its classically allowed maximum distance and is due to quantum tunneling. Consequently, it is the largest of all ground state diatomic molecules. Because of its low mass in comparison with other atoms and molecules, it has a large zero-point energy and therefore forms the weakest of van der Waals bonds. With alkali and alkaline earth atoms, the helium atom even has no bound state.

3. The extremely weak He–He bond also explains that it is the only element that, as discussed in the next section, at atmospheric pressure does not freeze at 0 K. Thus, droplets are definitely liquid. Due to its light mass, weak bonding, and bosonic characteristics, liquid $^4$He undergoes a second-order phase transition below 2.17 K, a type of Bose–Einstein Condensation called superfluidity.

4.  The extremely weak He–He bond also means that it has a very small heat of evaporation. In vacuum it evaporates very rapidly until the temperature is so low that further evaporation cannot occur. As discussed in Sect. 1.4.4, this property explains the low temperature of 0.38 K for $^4$He droplets and 0.15 K for $^3$He. Droplets are thus an ideal cryostat with a well-regulated constant temperature.

Most of the remarkable features listed above have been known for nearly a century through the extensive experiments carried out at the Physics Institute of the University of Leiden by Kamerlingh Onnes, his successor Willem Hendrik Keesom, and colleagues in the early half of the last century [20]. Only the superfluid nature of $^4$He is more recent. Thus, one may ask why hasn't helium been used as an ideal matrix earlier?

One of the first attempts was undertaken in 1964 by Jortner et al. who deposited $N_2$ and $O_2$ on the surface of liquid helium to produce colloidal aggregates. These were excited by a $^{210}P_0$ $\alpha$-source and the electronic emissions $N_2\left(A^3\sum_n^+ \rightarrow X^2\sum_g^+\right)$ and $O_2\left(C^3\sum_u^- \rightarrow X^3\sum_g^+\right)$ were observed [21]. The first attempt to inject atoms into the liquid made use of ion and electron emission from a hot filament. Later discharges were used as sources of ions and the ions were neutralized with electrons in situ. In 1973, Eugene Gordon et al. carried out a series of experiments in which a mixture of $N_2$: He (1: 1000) was passed through a discharge directed at the surface of liquid helium. Inside the liquid, a snow-like condensate was observed via the $\left(^2D \rightarrow {}^4S\right)$ emission of the N atom [22]. The group of Gordon in Chernogolovka also pioneered the growth of high aspect ratio metal nanowires inside bulk superfluid helium [23]. Metal nanoobjects grown in helium droplets are discussed in Chap. 11 by F. Lackner in this volume. Several groups used laser ablation of the metal inside the cryostat and investigated the atoms by laser-induced fluorescence [24, 25]. So far, only metal atoms and clusters have been investigated spectroscopically in the bulk liquid and solid. For a review, see [26]. Generally, however, the atomic lines of metals are considerably shifted and broadened as a result of the strong Pauli repulsion between the helium electrons and the unpaired electron of the electronically excited atom.

Bulk helium spectroscopic experiments have the advantage as a spectroscopic matrix over droplets of temperature and pressure flexibility. Consequently, the spectra inside the normal liquid can be compared with that inside the superfluid and solid, [27] further allowing investigations over a wide range of temperatures and pressures. Unfortunately, it still remains a challenge to stabilize closed-shell atoms and molecules in bulk liquid helium. Once inserted, they drift to the walls or coagulate to form large aggregates [28].

Parallel with and independent of the above developments there were a few experiments and theories exploring the physics of large helium droplets formed in gas expansions in vacuum. In 1961 Becker, Klingelhöfer and Lohse reported on the first investigation of the time-of-flight of droplets of helium formed in nozzle expansions [29]. These scientists were interested in possible applications to plasma physics and the experiments have been reviewed by E. W. Becker in 1986 [30]. In 1983 Stephens and King reported distinct magic numbers of the ion signals in the mass spectra of both $^3$He and $^4$He clusters [31]. The first reported theoretical prediction that finite-sized droplets like nuclei could be superfluid was published in 1978 [32]. In 1983 the nuclear theoretician V. R. Pandharipande and colleagues reported the first Variational Monte Carlo quantum calculations of the ground state properties of helium droplets as a function of the number of atoms up to N = 728 [33]. These pioneering calculations have been largely confirmed by many subsequent theories.

In 1986 our group reported mass spectroscopic investigations of $^4$He clusters beams [34]. Then, in 1990, Scheidemann, Toennies and Northby reported the observation of additional Ne cluster ions in the mass spectrum of the droplet beam after it had been scattered from a Ne atom nozzle beam [35]. A more complete description of some of the early mass spectrometer experiments, can be found in the chapter "Mass Spectroscopy of Pure and Doped Droplets" by A. Schiller, L. Tiefenthaler and F. Laimer in this volume.

Shortly afterwards in 1992 Goyal, Schutt and Scoles observed two unusually sharp infrared lines of a $CO_2$ laser in the spectrum of $SF_6$ molecules, which they thought were attached to the helium droplets [36]. Two years later, Fröchtenicht and Vilesov from our laboratory used a tunable diode laser and observed well-resolved rotational lines of the P- and R-branches centered around a sharp Q-branch of $SF_6$ in droplets of several thousand atoms [37, 38]. This experiement demonstrated for the first time that single molecules could be inserted into droplets and that the molecules were free to rotate. Since then rotationally resolved spectra have found a wide application as reported in Chap. 4 by Gary Douberly.

## 1.2  Thermodynamic Properties of Helium

As implied earlier helium is a unique element. It is also the best characterized of all the elements in all its phases. The book by Keller from 1969 [39] and the book edited by Bennemann and Ketterson [40] still provide the most comprehensive introduction to the many remarkable thermodynamical and physical properties of helium. The more recent book by Tilley and Tilley is also recommended as a modern source [16].

Figure 1.2 shows the phase diagram of helium in a double logarithmic plot. Helium is the only substance that remains liquid and does not solidify at 0 K under atmospheric pressure. To solidify helium, high pressures of about P = 29 bar at below 1.73 K are required. Its critical point is at a pressure of $P_{c.p.}$ = 2.27 bar and $T_{c.p.}$ = 5.2 K. At atmospheric pressure below about 4.2 K helium becomes a liquid, denoted He I. Then, on further cooling to 2.17 K it undergoes a second-order transition into the new liquid superfluid phase denoted He II (Fig. 1.2). In place of the triple point, where in all other substances the gas, liquid and solid states coexist, it has a unique point at which the gas, liquid He I and superfluid He II coexist, called the λ-point. Furthermore, instead of a phase line marking the transition from the liquid to solid, as in all other substances, it exhibits a λ-line, which marks the transition from the normal liquid He I to superfluid He II. The name derives from the sharp spike in the specific heat which approaches infinity at the superfluid transition. At the λ-line,

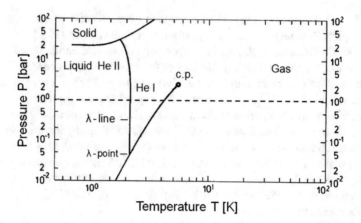

**Fig. 1.2** Pressure–temperature phase diagram of $^4$He in a double logarithmic diagram. The dashed horizontal line marks atmospheric pressure

helium undergoes many dramatic changes in its properties. For example, as a result of the superfluidity of He II the viscosity drops by 6 orders of magnitude to less than $10^{-11}$ poise and the heat conductivity jumps by almost 7 orders of magnitude.

Table 1.1 summarizes some of the thermodynamic properties of $^4$He and the rare $^3$He isotope. Because $^3$He is much lighter than $^4$He and is a fermion and not a boson as $^4$He, it has much different physical properties than the more abundant $^4$He. Below about 2.5 mK, $^3$He also becomes a superfluid by the pairing of two fermions to produce a type of boson. Because of the many different magnetic phases $^3$He is presently a hot topic in the low temperature community which regularly meet at the biannual conferences *Quantum Fluids and Solids* (QFS).

## 1.3 Formation and Characterization of Helium Nanodroplets

### 1.3.1 Production of Nanodroplets in Free Jet Expansions

Beams containing helium clusters and droplets are readily produced by expanding the high purity gas or the liquid at a stagnation source pressure $P_0$ between 1.0 and 80 bar and low temperatures $T_0$ between 3 and 40 K into vacuum. A thin-walled orifice with diameter $d = 5 - 10\,\mu$ is commonly used as the source [41, 42]. Since the flow is not guided, as compared to a Laval-shaped relatively long nozzle, the corresponding jet is termed "free jet". For preliminary collimation and to reduce the gas load on the vacuum system, the orifice is followed by a conical skimmer [43]. In the ensuing radial expansion of the gas, the particle density $n$ falls off with the inverse square of the distance $z$ measured from the orifice, $n \propto z^{-2}$. For a review on free jet expansions,

**Table 1.1** Thermodynamic and physical properties of $^4$He and $^3$He

| | $^4$He (boson) | $^3$He (fermion) (Fermi temp. 1.75 K) | Units |
|---|---|---|---|
| Relative abundance on earth | 5.2 | $7 \times 10^{-6}$ | ppm |
| Boiling point, $T_B$ | 4.2 | 3.191 | K |
| Binding energy per atom and heat of evaporation | 7.15 | 2.7 | K |
| Critical pressure $P_{c.p}$ | 2.2746 | 1.15 | bar |
| Critical temperature $T_{c.p}$ | 5.2014 | 3.324 | K |
| Critical enthalpy $h_{c.p}$ | 6.74 | – | kJ/kg |
| Critical entropy $s_{c.p}$ | 5.70 | – | kJ/(kg K) |
| Entropy at 2.0 K | 3.85 | 12.95 | J/mole K |
| Temperature of $\lambda$ point | 2.1773 | $2.5 \times 10^{-3}$ | K |
| Surface tension $\gamma$ at $T = 0\,\mathrm{K}$ | 0.3544 | 0.15 | dyn/cm |
| Compressibility at 0 K | 0.120 | 0.361 | cm$^3$/J |
| Viscosity at temperature of droplets | $\approx 0$ | 200 | micropoise |
| Liquid mass density | 0.146 | 0.0832 | g/cm$^3$ |
| Liquid particle density $n$ at 0 K | $2.18 \times 10^{22}$ | $1.64 \times 10^{22}$ | cm$^{-3}$ |
| Velocity of sound at 0 K | 239 | 183 | m/sec |
| Gas to liquid volume ratio (at normal conditions) | 662 | 866 | – |
| Thermal conductivity at $T_B$ | 0.26 | 0.20 | mW/cm$^2$ K |
| Conduction band (surface pot. Barrier) | 1.2 eV | – | eV |
| Radius of electron bubble in liquid helium | 18.91 (P = 0) | – | Å |
| Mean free path of quasiparticles at 0.38 K the temperature of droplets | $\approx 10^{-1}$ | $\approx 5 \times 10^{-7}$ | cm |
| Average kinetic energy per atom | $\approx 14$ | $\approx 10$ | K |

reference is made to the article by Miller [44]. Pulsed helium droplet sources were first successfully operated by Uzi Even and colleagues [45] and by Slipchenko et al. [46] and further improved by Pentlehner et al. [47] Pulsed sources of helium droplets at cryogenic temperatures were first developed by Ghazarian and colleagues [48]. Typically pulsed sources have two orders of magnitude larger orifices of about 0.5 mm dia. Since the short pulses of about 100 μs have a repeat frequency of about 50 Hz the load on the vacuum system is much the same as with a continuous source.

In both types of sources, changes in the thermodynamic state of helium as it flows through the orifice can be approximated by assuming an adiabatic isentropic process in which the gas or liquid is always in thermodynamic equilibrium. The gas or liquid can adjust adiabatically since it takes only less than about a hundred collisions ($\approx 10^{-12}$ s) to adapt to a new pressure and temperature. Thus, it is possible to describe the expansion by following the isentropes in the phase diagram shown in Fig. 1.2. The thermodynamic properties of helium and their dependence on pressure and temperature, as well as the isentropes, are available in tabular form from extensive calculations [49, 50] and measurements [51].

### 1.3.2    The 4 Regimes of Isentropic Expansions

The free jet expansions of helium depend on the nature of the corresponding isentropes. These can be divided into four flow regimes, which are denoted as Regimes I through IV. Figure 1.3 shows a number of isentropes as dashed colored lines starting from a fixed stagnation pressure of $P_0 = 20$ bar and for a range of source temperatures $T_0$ from 0 to 40 K. Regime I expansions are frequently referred to as supercritical since the temperature range lies above the critical temperature. Consequently, the

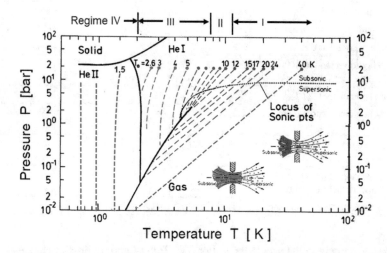

**Fig. 1.3** Pressure-temperature phase diagram for $^4He$ with isentropes (----) for free jet expansions starting from a stagnation pressure of $P_0 = 20\,bar$ and a range of temperatures $T_0$ [52]. As discussed in the text, qualitatively different behavior is expected for the four regimes indicated at the top of the diagram; Regime I:$T_0 \geq 12K$, Regime II : $T_0 \approx 8 - 12K$, Regime III : $T_0 \approx 2.8 - 8K$, and Regime IV :$< 2.2K$. The isentropes are based on data in [50, 51] for different temperatures $T_0$ and a single initial pressure of $P_0 = 20\,bar$ [52]. To obtain the state of the gas or liquid at the orifice, the velocity has been calculated from the change in enthalpy during the expansion (the lines of equal enthalpy are not shown). The sonic condition requires that the flow velocity equals the speed of sound at the narrowest point of the orifice and at this point the flow changes from subsonic to supersonic. Where this occurs is indicated by the dotted curves in Fig. 1.3 [52]. Reproduced from [52] with the permission of AIP Publishing

Regime III expansions below the critical point are termed subcritical. In Regime II expansions the isentropes pass to or near the critical point. Whereas in Regimes I, II and III normal He I is expanded, in Regime IV a jet of superfluid He II liquid is formed. Each regime has different operation conditions and leads to clusters or nanodroplets with different sizes and size distributions.

### 1.3.2.1 Isentropes: Regime I

In Regime I the source temperatures are well above the critical temperature. Accordingly, the isentropes are nearly those of an ideal gas and accordingly follow straight lines, and cross the phase line from the gas into the liquid phase. Early on in the expansion, the flow reaches the sonic condition and the expansion continues along the isentrope beyond the orifice. For an ideal gas, the changes in state $(T,P)$ for adiabatic isentropic flow can be calculated from the following equation

$$T^\gamma P^{1-\gamma} = T_0^\gamma P_0^{1-\gamma} \tag{3.1}$$

It follows then that for an ideal gas the temperature at some point downstream from the nozzle $z$ is given by

$$T(z) = T_0 \left( \frac{n(z)}{n_0} \right)^{\gamma-1}, \tag{3.2}$$

where $\gamma$ is the ratio of the specific heats ($\gamma = 1.67$ for atoms). The sharp drop in the density $n(z)$ beyond the orifice leads to a rapid decrease in the temperature. Stein has calculated the gas cooling rates of $H_2$, He, Ar, and $N_2$ as a function of orifice diameters at a source temperature of 100 K and higher [53]. The temperature drops continuously as long as the density is sufficient that collisions still occur in significant numbers to assure local equilibrium. Immediately, after crossing from the gas phase into the liquid He I or superfluid He II phase region the system will, before it can become liquid, continue on for some distance in a metastable supercooled gaseous state. At some point belonging to the *Wilson* line [54] condensation begins. The heat released will raise the temperature and bring the system trajectory back to the saturated vapor phase curve as the droplets continue to grow in size. This continues until collisions cease to occur (sudden freeze). Beyond the point of sudden freeze, the temperature of the droplets is determined by evaporation (see Sect. 1.4.4).

### 1.3.2.2 Isentropes: Regime II

In Regime II the source conditions correspond to isentropes that converge on or in the vicinity of the critical point. Table 1.2 lists the source temperatures $T_{CI}$ for several source pressures which correspond to critical isentropes passing exactly through the

**Table 1.2** Source temperatures for critical isentropes at four different source pressures

| $P_0$ [bar] | $T_{CI}$ [K] |
|-------------|--------------|
| 20          | 9.2          |
| 40          | 11.2         |
| 60          | 12.7         |
| 80          | 13.5         |

critical point. In expansions which converge at and near the critical point unusual behavior is expected. Here, the correlation length becomes macroscopically large, corresponding to large fluctuations in density. Light scattering experiments under similar steady state conditions reveal a critical opalescence which indicates the presence of large clusters under static conditions [55]. Also the speed of sound is minimized and the surface tension vanishes [56]. The transition from Regime I to III is, therefore, not sharply defined and occurs over a range of isentropes and source temperatures.

Regime II was accessed in the first experiment demonstrating that He clusters could capture foreign particles, in this case Ne atoms [35]. The capture probability was found to be largest for expansions in this regime.

### 1.3.2.3    Isentropes: Regime III

In Regime III at source pressures of 20 bar and temperatures less than about 8 K, below the range of Regime II, but higher than the temperature corresponding to the superfluid transition of helium, the isentropes deviate greatly from the straight lines of the ideal gas. This is consistent with the increasing fraction of liquid and the large increase in heat capacity at the transition from He I to superfluid He II as the isentrope approaches the λ-line. Thus, the isentropes all bend downwards, and cross the liquid–gas phase line from the liquid side. Since the sonic point is at the phase line, the liquid flashes and disintegrates into large clusters after leaving the orifice. Related processes occur in the fragmentation of brittle materials, laser ablation, nuclear collisions, and in the *big bang*. Holian and Grady among others have simulated this phase transition in a molecular dynamics simulation which provided an expression for the distribution of cluster sizes after the break-up of a liquid [57–59].

### 1.3.2.4    Isentropes: Regime IV

At even lower temperatures in superfluid He II, the isentropes are vertical lines. The liquid leaves the orifice initially as a flowing cylinder with approximately the orifice diameter [60]. After a few millimeters, capillary instabilities inherent in the passage through the orifice lead to a Rayleigh break-up into a sequence of large, equally-sized droplets about 1.89 times the nozzle diameter. Depending on the orifice diameter, droplets produced from liquid helium jet breakup could contain about $10^{10}$ atoms. Flashing has not been observed in these expansions [60].

## 1.3.3 Droplet Sizes and Size Distributions in Regimes I, II, III, and IV

Figure 1.4 presents an overview of droplet sizes obtained with 5 μ orifices as a function of source pressure and temperatures for each of the four regimes. Different nanodroplet sizes over a range of more than 10 orders of magnitude are obtained in the four different regimes. Correspondingly different methods have been found to be suitable for characterizing the sizes, size distributions and velocities. Some of the methods employed rely on the fact that the velocity of the helium cluster and droplet beams have a sharp velocity distribution.

The final cluster sizes are determined not only by growth processes, such as droplet coagulation, but also by subsequent cooling by evaporation with a correspondingly large loss of atoms. As discussed in Sect. 1.4.4, see also Fig. 1.14, the droplets lose about half of their atoms resulting from the extensive evaporation. The evaporation is very rapid and the internal temperatures drop with a time constant of about $10^{-11}$ – $10^{-9}$ s until the droplet gets so cold that the rate of evaporation becomes negligible. After travelling typical apparatus distances of one meter, the final droplet temperatures have been calculated to be $0.38 K \left(^4 He\right)$ [61, 62] and $0.15 K \left(^3 He\right)$ [61–63] (see Fig. 1.14).

### 1.3.3.1 Droplet Sizes Regime I

The initial growth from the gas phase of small helium clusters, which serve as nuclei for further growth, is accelerated due to a quantum resonance enhanced cross section resulting from the weak 1 mK binding energy of the dimer [65, 66]. For a detailed description of the helium dimer see the chapter by M. Kunitski, *Small helium clusters studied by Coulomb explosion imaging,* in this volume. The resonance has been calculated to increase the cross section from 30 $Å^2$ at room temperature to 259,000 $Å^2$ as $T \to 0 K$ [67]. The large increase in the quantum cross sections and resulting increase in collision frequency can also explain the narrow velocity distributions of $\frac{\Delta v}{v} \leq 0.01$ of helium atom beams at $T_0 = 80 - 300$ K [66, 68]. Under conditions where condensation does not take place, this narrow velocity distribution corresponds to an ambient temperature in the moving frame of the beam of only about $T_{amb} \cong 10^{-3}$ K. In mild cryogenic expansions ($T_0 = 30$ K, $P_0 = 5$ bar), the heat released in the coagulation to small clusters increases the velocity half width to only about $\frac{\Delta v}{v} = 0.015$ ($T_{amb} = 2 \times 10^{-3}$ K). In an expansion at higher source pressures ($T_0 = 30$ K, $P_0 = 80$ bar), the velocity half width increases further to $\frac{\Delta v}{v} = 0.06$ ($T_{amb}$

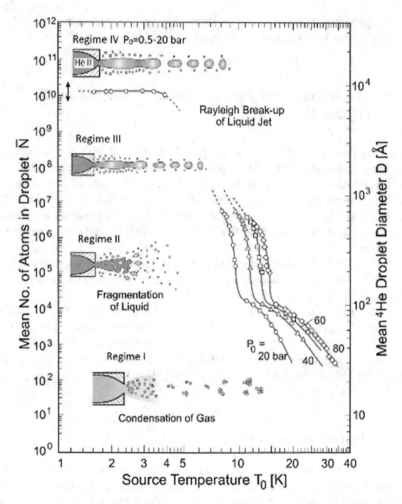

**Fig. 1.4** Overview of the average number sizes $\overline{N}$ and liquid droplet diameters D in Å of $^4$He clusters and droplets as a function of source temperature and pressures in the four expansion regimes (Fig. 1.2). In all regimes the sizes increase with decreasing source temperature. Small clusters and droplets are obtained in supercritical expansions (Regime I). Larger droplets are produced by expanding liquid helium in critical (Regime II) and subcritical (Regime III) expansions. Within Regime III at the lowest temperatures of about 3 K liquid He I leaves the orifice in a cylindrical column which breaks up into a series of nearly uniformly-sized droplets with about $10^{10}$–$10^{12}$ atoms. Similar phenomena are observed in the expansion of the superfluid He II liquid in Regime IV. Adapted from [64]

$= 3 \times 10^{-2}$ K) [69]. Compared to expansions with heavier rare gases, however, this resolution is still very good.

The growth processes leading to small clusters at $T_0 = 6$, 12 and 30 K, $P_0 \leq$ 1.5, 7.0 and 80 bar, respectively, have been analyzed with a detailed kinetic theoretical model which takes account of three body recombination and break-up processes [69]. The formation of small clusters $He_2$, $He_3$ and $He_4$ from a 5 $\mu$ diameter orifice starts already several nozzle diameters from the source and continue to grow up to a distance of about 220 nozzle diameters ($\approx$1mm) [69]. There the expansion undergoes a "sudden freeze" which implies that collisions essentially cease and the clusters continue their forward motion in vacuum without further encounters.

In Regime I, $^4$He dimers, trimers, small clusters, and droplets have been produced with sizes up to about $10^4$ atoms. The size distributions of small clusters up to about $N \approx 100$ have been measured with matter-wave diffraction from nanoscopic transmission gratings [69]. Since the diffraction pattern is made up of a superposition of many coherent de Broglie particle waves, which pass through the slots of the grating without interaction with the slits, the method is non-destructive. Matter-wave diffraction from nanoscopic transmission gratings is discussed in more detail in the chapter *Small Helium Clusters studied by Coulomb Explosion Imaging* by M. Kunitski in this volume.

In other matter-wave diffraction experiments from transmission nano-gratings at 1.25 bar and 6.7 K, the number size distributions peaked at about 10 atoms with an exponential fall-off extending up to cluster size of about 100 atoms [70, 71]. The results for helium clusters with $N$ atoms at $T_0 = 6.7$ K have been shown to be well fitted by the following distribution: $P(N) = AN^a e^{-bN}$ derived from a theory which accounts for all the recombination and break-up rate constants [72, 73].

For a review of experimental techniques to determine the sizes of larger clusters and droplets up to 1997 the reader is referred to the article by Knuth [74]. The size distributions of larger clusters and droplets of helium were first measured with mass spectroscopy [52, 75]. The ion intensity distributions are severely affected by the heat released (2.2 eV) in the recombination of the initially formed $He^+$ with a nearby He atom to form $He_2^+$, which is sufficient to evaporate about 3500 atoms [76]. For this reason, the method is not very suitable in Regime I. For the latest applications of mass spectroscopy see the chapter *Helium Droplet Mass Spectroscopy* by Schiller, Laimer and Tiefenthaler in this volume.

A more suitable and gentle method, which was pioneered by Gspann in 1981, consists of deflecting the helium droplet beam by collisions with a secondary beam of heavy atoms [77–79]. The method was later used in our group for studying the size distributions in Regime I by scattering from a secondary beam of heavy particles such as krypton or $SF_6$ [80, 81]. A careful analysis of the deflected droplets, taking account of the velocity spreads of both beams, indicated that the Kr atoms were fully captured by the droplets so that the entire momentum was imparted to the droplets. From the angular distribution of deflected droplets at angles as small as $10^{-3}$radians, the momentum distribution of the incident droplets could be measured. Moreover, since the velocity of the droplet beam is sufficiently sharp the droplet mass and number size distributions could be determined. The number sizes were found to be

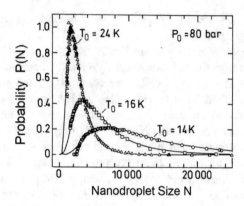

**Fig. 1.5** Droplet size distributions in Regime I for three different source temperatures and a source pressure of 80 bar (5-μ diameter orifice). The distributions were obtained from the angular distribution of the deflected droplets after collisions with a beam of $SF_6$ molecules [80]. The curves are least squares fits to a log-normal distribution. Reproduced from [80] with the permission of AIP Publishing

well fitted by a log-normal distribution shown in Fig. 1.5 and given in (3.3) [80].

$$P(N) = \frac{1}{\sqrt{2\pi}N\delta}exp\left[-\frac{(lnN - \mu)^2}{2\delta^2}\right],\tag{3.3}$$

where the mean number of atoms $\overline{N}$ is

$$\overline{N} = exp\left(\mu + \frac{\delta^2}{2}\right),\tag{3.4}$$

and the width of the distribution (FWHM) is

$$\Delta N_{1/2} = exp\left(\mu - \delta^2 + \delta\sqrt{2ln2}\right) - exp\left(\mu - \delta^2 - \delta\sqrt{2ln2}\right),\tag{3.5}$$

Typically, the FWHM is very similar in size as the mean $\overline{N}$. Best fit parameters $\mu$ and $\delta$ are tabulated in [81].

The mean sizes in Regime I over a wide range of temperatures and pressures have also been determined from the size dependent attenuation of the droplet beam by an electron beam [82] Based on these measurements and beam velocities, Knuth et al. have derived expressions for estimating the fraction of the beam which is condensed as liquid [83].

Mean sizes can also be estimated from kinetic and thermodynamic scaling parameters first introduced by Hagena for a wide variety of substances including metals [84]. Knuth found these scaling laws to be unsatisfactory for helium since they considered the clusters to solidify [74]. Knuth introduced kinetic and thermodynamic scaling parameters which took account of the fluid nature of helium clusters and droplets to reliably predict the number sizes of droplets produced in Regime I.

The sizes of droplets from pulsed sources operating at cryogenic temperatures are much larger than droplets from the continuous sources. To investigate the distributions in pulsed beams, Slipchenko et al. [46] doped the droplet bursts with the dye phthalocyanine. The droplet sizes were estimated from the suppression of the laser-induced fluorescence (LIF) signal by scattering from argon gas. The beam attenuation provided information on the collision cross section, which in turn depends on the number size. Slipchenko et al. reported mean droplet sizes as a function of temperature from $T_0 = 10$ to 26 K at $P_0 = 6$, 20 and 40 bar. In comparison with continuous droplet sources under the same conditions, they found an increase in droplet sizes between a factor 50–100 at comparable source temperatures. In Regime I Kuma and Azuma also reported a similar increase in sizes [85]. In a more recent experiment from the Vilesov group, the peak flux from a pulsed source was found to be a factor $10^3$ greater than with a continuous source [86].

Yang and Ellis observed that the heavier droplets with $10^5$ atoms from a pulsed source have significantly longer flight times than the lighter droplets with $4 \times 10^3$ atoms [87]. In several pulsed studies using dopants, which are ionized inside the droplet, the mass distribution of the host droplets could be measured from the intensity of the ions which passed a retarding potential grid [88]. For a similar recent experiment see the chapter by Zhang et al. entitled *Electron diffraction of molecules and clusters in superfluid helium* in this volume. In a very recent experiment from the same group, Pandey et al. measured the flight time of large droplets which had been doped with cations of the laser dye Rhodamine 6G from an electrospray source [89]. This technique opens up the possibility to select the sizes of droplets containing an ion.

### 1.3.3.2 Droplet Sizes: Regime II

In Regime II the source conditions correspond to isentropes that converge on the critical point. As mentioned earlier, since large fluctuations occur in the medium at the critical point the transition from Regime I to III is expected to occur over a range of temperatures near the critical isentrope. Gomez et al. carried out a careful study of the sizes in going from Regime I through Regime II to Regime III [90]. They found that there was only a small increase in in the droplet sizes in going from about 9 to 7 K at 20 bar and that a sharp increase by several orders of magnitude occurred below 7 K. As described in the next section, the same large increase could be found at about the same temperature with negatively charged droplets [91]. With pulsed beams, Verma and Vilesov observed a similar plateau in droplet sizes from about 9 to 7.5 K at source pressures of 5 and 10 bar [86].

Only a few studies have been carried at exactly or close to the critical expansion which at 20 bar is at $T_o = 9.2$ K. (Table 1.2) As mentioned earlier, a large enhancement of the pick-up-probability was observed in expansions in which the isentropes converged on the critical point [35].

Only one experiment has been reported on the properties of beams emerging from the source at conditions corresponding to the critical point ($T_0 = T_{c.p.} = 5.2$ K, and $P_0 = P_{c.p.} = 2.3$ bar) [92]. As the critical pressure was approached, the velocity of the cluster beam dropped from about 200 m/s to only 50 m/s, whereas the velocity of the atom beam component remained surprisingly unchanged. In a subsequent investigation, the droplets were found to have sizes of $N = 1.5 \times 10^9$ atoms and a velocity half width of 5% [93]. Under similar source conditions, $^3$He droplets were found to have slightly smaller sizes of $10^8$ atoms [93]. Since the droplets have velocities below the Landau critical velocity of 58 m/s, they were used to demonstrate the superfluid transmission of $He^4$ and $He^3$ atoms through the droplets [93].

### 1.3.3.3  Droplet Sizes: Regime III

In Regime III, liquid He I flows from the orifice and then once in vacuum cavitates and flashes into smaller droplets and clusters. In 1986, in one of the first experiments in our group, the mass spectrometer signal at critical point conditions jumped-up by several orders of magnitude [34]. As mentioned above, at temperatures well below the transition to Regime III, the droplet sizes increase by about five orders of magnitude.

The first size measurements in this region made use of mass spectroscopy [52]. Jiang et al. [94] tagged the droplets by attaching electrons to their surface with very small binding energies of between $4 \times 10^{-6}$ and $1.8 \times 10^{-4}$ eV [91] so that fragmentation was minimal. The negatively charged droplets were either retarded or deflected in an electric field to determine the mass distribution [94, 95]. At $P_0 = 20.7$ bar they observed an increase from $N = 4.2 \times 10^4$ at 10 K to $5.5 \times 10^5$ at 6.5 K. Similar measurements were later carried out by Henne and Toennies [91]. As seen in Fig. 1.6 the distribution of negative ions and positive ions are similar for large droplet sizes with $10^5$ or more atoms.

Figure 1.7 shows the distribution of sizes measured at 20 bar from the deflection of positive ions at 10 K and 8 K on both sides of the critical isotherms at 9.2 K [91]. At 10 K (Fig. 1.7a) in Regime I in addition to the log-normal distribution an additional long tail extending out to larger sizes was found. Since the tail is much less intense than the log-normal distribution it was overlooked in the collisional deflection experiments described in Sect. 1.3.3.1. At 8 K the distribution displays a pure exponential shape (Fig. 1.7b). Confirmation comes from measurements in which the size distribution of very large droplets with more than $10^9$ atoms was determined by analyzing the amplitude of $He_2^+$ ion pulses following ionization by electrons of 100 eV. At 5.4 K (20 bar) an exponential size distribution of droplets was determined from the amplitude distribution of the $He_2^+$ ion pulses [96]. A combined theoretical and experimental study shows that an exponential distribution is also consistent with theory and is given by

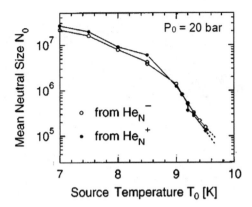

**Fig. 1.6** Mean number sizes of droplets in Regime III measured by deflecting the negatively and positively charged beams in an homogeneous electric field directed perpendicular to the beam direction. Despite the large loss of 3500 atoms upon ionization the cation sizes agree with anion sizes at sizes $N \geq 10^5$. Reproduced from [91] with permission of AIP Publishing

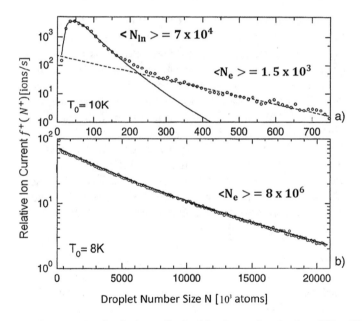

**Fig. 1.7** Typical number size distributions of ionized droplets produced at $P_0 = 20$ bar. The distribution of the positive ion current is measured by deflecting the charged droplets in an electric field. **a** The distribution in Regime I at 10 K shows in addition to the expected log-normal distribution $N_{ln}$ a much smaller exponential tail $N_e$ extending out to $10^6$ atoms, which was overlooked in earlier scattering deflection measurements. **b** In Regime III at 8 K a pure exponential distribution is observed with a most probable size two orders of magnitude larger. Reproduced from [97] with the permission of AIP Publishing

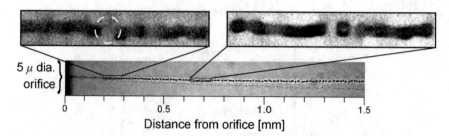

5 $\mu$ dia.
orifice

0                    0.5                        1.0                        1.5

Distance from orifice [mm]

**Fig. 1.8** Photographs of droplets in the early stages of a Regime III expansion at $P_0 = 20$ bars and $T_0 = 3.5$ K at distances from the orifice at the left. The dashed white oval at 0.5 mm indicates the point at which the liquid cylinder breaks up first into ligaments and then after 1.3 mm into discrete droplets. Adapted from [99]

$$P_N = \frac{e^{-\frac{N}{\overline{N}}}}{\overline{N}} \tag{3.6}$$

where $\overline{N}$ is the average number of atoms in the droplet [97]. This distribution is similar to that obtained in the theory of the fragmentation of brittle materials [98].

At very low temperatures of 3.5 K in Regime III, it was recently discovered by optical imaging that under these extreme conditions the liquid issues from the orifice as a liquid column before breaking up into larger entities [99]. As seen in Fig. 1.8. after a fraction of a mm from the orifice the liquid first breaks up into long ligaments and then, after about one millimeter further downstream, into small droplets. The droplets were measured to be 6.7–8.3 $\mu$ in diameter and larger than the source orifice of 5 $\mu$. This is consistent with the theory which goes back to a classic paper by Rayleigh (1879). The disturbances which lead to Rayleigh breakup have a length, which converted to the spherical droplets are estimated to have a diameter 1.89 times the nozzle diameter [100–102]. For a 5 $\mu$ nozzle, this translates to 9.45 $\mu$ droplets. These ultra large droplets have found applications as low-Z targets in high energy collisions in storage rings [103]. There, the interest is only in the alpha particle nucleus of the helium atom. A discussion of droplet sizes and size distributions and droplet properties in Regime III can be found in the review by Tanyag et al. [104].

These large droplets have been found to be filled with quantum vortices with interesting quantum properties. These are discussed in Chap. 7 entitled *X-ray Imaging of Helium Droplets* by Tanyag et al. in this volume. As discussed there, diffraction of X-rays provides detailed information on the sizes and even the shapes, which often deviate from spherical and can be either prolate or oblate. See also [105]

#### 1.3.3.4 Droplet Sizes: Regime IV

Source conditions in the liquid superfluid phase have been explored in only one experiment [60]. A 2 μ thin-walled orifice and also for comparison a convergent large aspect-ratio silica pipette with the same sized opening were used. The transition from the superfluid phase He II to the normal phase He I was explored at temperatures of $T_0 = 1.5$–2.6 K ($P_0 = 2.0 - 20$ bar). In both phases shortly after emerging from the orifice the intact beam disintegrated into a sequence of highly collimated droplets similar to those shown in Fig. 1.8. The liquid beam had a sharp velocity distribution of $\frac{\Delta v}{v} = 0.01$ both below and above the phase transition at velocities as low as 15 m/s [60]. Because of the high directionality a high brilliance of about $10^{22}$ atoms/s sr, which is 2–3 orders of magnitude more than possible when flashing occurs, is observed as in Regime III. At the second order superfluid λ-line phase transition, it appears that some turbulence was observed since the relative velocity spread increased from $1 \times 10^{-2}$ to $4 \times 10^{-2}$ and the beam doubled in the angular width. A similar and possibly related effect has been observed in the bulk liquid and claimed to be analogous to the formation of cosmic strings in cosmology [106, 107]. Presently, it is not clear how the droplet beams produced by expanding the superfluid differ from the beams from normal He I. Possibly, since the superfluid is a many body coherent system interesting interference effects might be observed in the scattering of two superfluid beams prior to break-up. Recently the break-up of jets of normal and superfluid liquid helium, issuing into the saturated vapor, where an entirely different behavior is expected, has been reported [108].

### 1.3.4 Velocities of Nanodroplets

In Regime I the velocities of small clusters with sizes up to about 6 atoms have been analyzed with a combination of matter wave diffraction and time-of-flight spectroscopy [69]. With increasing $P_0$ (0–80 bar at $T_0 = 30$ K) the velocities decrease from about 560 m/s by about 2% at 30 bar and then increase to 570 m/s at higher pressures. The increase is a result of the heat released in the condensation to clusters. At low pressures the velocities agree qualitatively with calculations based on the conservation of enthalpy, $v = \sqrt{\frac{2h_0}{m}}$, where $h_0$ is the enthalpy per atom at $P_0$ and $T_o$. As a result of condensation heating, the velocity resolution at $P_0 = 80$ bar and $T_0 = 30$ K decreased to $\frac{\Delta v}{v} \cong 6 \times 10^{-2}$ ($T_{amb} = 3 \times 10^{-2}$ K) [69] from the very sharp velocity resolution of beams free of clusters with $\frac{\Delta v}{v} \cong 10^{-2}$. Thus, in general beams of small clusters of helium have reasonable sharp velocities compared to cluster beams with heavier particles.

Figure 1.9 shows the velocities of nanodroplets as a function of the source temperature for 4 different source pressures encompassing Regimes I to III. As expected, the velocities increase linearly with the source temperature and do not reflect the sharp

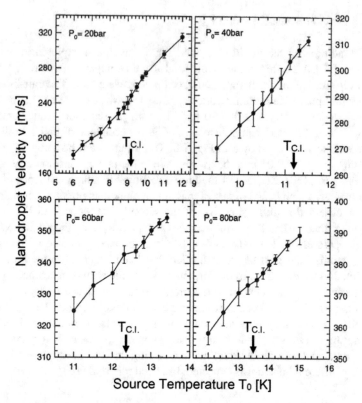

**Fig. 1.9** Experimental velocities of droplets as a function of source temperature for four different source pressures. The orifice used has a diameter of 5 μ [109]

rise in the droplet sizes following the transition region, Regime II. In the transition the linear increase shows only a small perturbation.

In Regime IV in expansions of He II very low velocities were achieved. With the pipette (see the previous section) in He II at $P_0 = 3$ bar the velocity was 30 m/s. At $P_0 = 0.5$ bar a velocity of 15 m/s was found [60]. Because of the low velocities in the apparatus great care had to be taken to avoid losing the beam by the downward deflection by gravity.

## 1.4 Physical Properties of Nanodroplets

The physical properties of pure helium clusters and droplets come mostly from theory. They have been calculated by many methods starting from a simple variational calculation in 1965 [110]. The methods used since then include the Green's Function Monte-Carlo method (GFMC), [33, 111] the Variational Monte-Carlo method (VMC), [33, 111] the Diffusion Monte-Carlo (DMC) method, [112, 113] the hypernetted-chain/Euler–Lagrange theory (HNC/EL) [114] and the Density Functional (DF) method using the Orsay-Trento (OT) finite range density functional [115]. All these methods are based on the assumption that the droplets have a thermodynamic temperature of 0 K. The Path Integral Monte-Carlo Method (PIMC) perfected by David Ceperley to simulate many of the temperature dependent properties of bulk helium [116] is one of the few temperature dependent methods.

### 1.4.1  Total Energies

Figure 1.10 shows an example of the many calculations of the total ground state energies per atom of $^4$He-nanodroplets as a function of the number of atoms for $N \geq 70$ from the HNC/EL and DMC calculations performed by Chin and Krotscheck [114]. The chemical potential given by $\mu(N) = E(N) - E(N-1)$ is also shown in Fig. 1.10 and was determined from the energy per atom. $\mu(N)$ is also a measure of the evaporation energy per atom. As shown in Fig. 1.10, the ground state energy and evaporation energy for clusters and droplets increases from about 2–3 to 7.21 K with increasing size.

The calculated energies are customarily fitted to a *mass formula* as a function of powers of $N^{-1/3}$. Equation (4.1) is a fit to the calculations of Chin and Krotscheck, where the coefficients are in K [114],

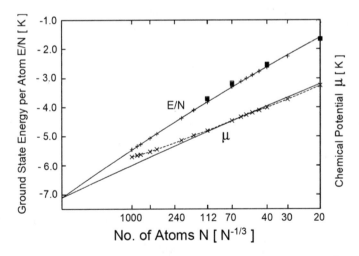

**Fig. 1.10** The top curve shows the results for the total ground state energy per atom from HNC/EL (+ symbols) and DMC ( black squares) calculations as a function of $N^{-1/3}$ [117]. The crosses (x) show the results for the chemical potential $\mu$ from a generalized Hartree function. The lower straight line is obtained by differentiating the mass formula fit to the upper curve. Reproduced from [117] with permission of AIP Publishing

$$\frac{E(N)}{N} = -7.21 + 17.71\,N^{-1/3} - 5.95N^{-2/3}, \tag{4.1}$$

The first term on r.h.s. in (4.1) is the evaporation energy in the bulk which is obtained by extrapolating to $N = \infty$. The coefficient of the second term $a_s$ on the r.h.s. in (4.1) accounts for the energy of the droplet surface and is calculated from $a_s = 4\pi r_0^2 \tau_4$, where $r_0$ is the bulk unit radius defined by $\frac{4}{3}\pi r_0^3 \rho_0 = 1$ and $\tau_4$ is the surface tension which is about $\tau_4 = 0.274 K \mathring{A}^{-2}$. The next coefficient is referred to as the curvature term. As discussed in the review by Barranco et al. [62] very similar results are also obtained with the following methods: GFMC MEDHE-2 [33, 111], DMC [114] OTDF [115, 118], VMC [33, 111].

### 1.4.2   Excited State Energies

The first attempts to account for the excited states of helium droplets were based on the liquid drop model which was introduced for dealing with the vibrations of nuclei [61, 118, 119]. The relevant normal modes of He clusters and droplets within the liquid drop model are *ripplons*, which are quantized capillary surface waves, and *phonons*, which are quantized bulk volume compressional waves. Krotscheck and colleagues have carried out both DMC and HNC/EL method calculations of the collective excitations of droplets with sizes from $N = 20$ up to 1000 atoms [113, 114, 117]. Instead of the usual angular momentum quantum number $l$ to account for the surface ripplon modes, these authors have introduced an effective wave number $k$ defined by

$$k = \sqrt{l(l+1)}/R \tag{4.2}$$

where $R$ is an equivalent hard sphere droplet radius given by $R = \sqrt{\frac{5}{3}}\,r_{ms}$ and $r_{ms}$ is the root mean square radius. The lowest compressional volume mode, with $l = 0$ in the limit of large $N(N > 300)$ can be approximated by

$$\hbar\omega_0 \cong 21N^{-1/3} \tag{4.3}$$

Their latest calculations for both types of excitations are displayed in a plot of the energy and the effective wave number $k$ in Fig. 1.11c for two particle sizes $N = 40$ and $N = 200$ [117]. The horizontal lines at $\hbar\omega_0 = 3.9K(N = 40)$ and $5.2K$ $(N = 200)$ mark the corresponding chemical potentials discussed in the previous section. The results contain both the actual quantum states and the virtual states lying above the chemical potential. In the $N = 200$ dispersion curve the remnant of the bulk maxon-roton dispersion curve of bulk He II (Fig. 1.1) can be clearly seen in the virtual states.

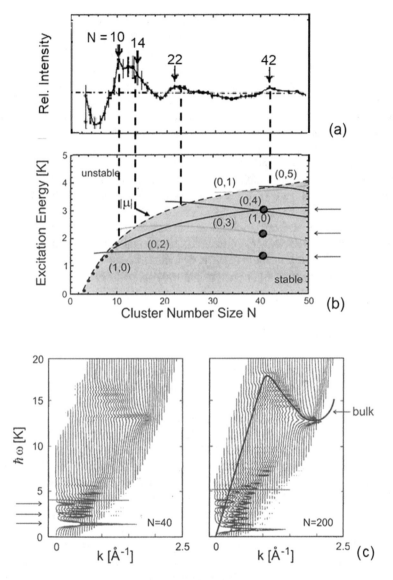

**Fig. 1.11  a** The experimental size distribution of small clusters corrected for the drop off in the average size distribution. Adapted from [71] **b** The grey area below the chemical potential (grey dotted curve) shows the region of real stable energies bounded by the chemical potential μ (dashed curve) from theory. Adapted from [71] The nearly horizontal line curves mark the ripplon *states (0,l)* and the only stable compressional state (1,0), indicated by the curve which extends from 1.5 K at N = 10 to 3.2 K at N = 50 [71]. The vertical dotted lines connect the maxima in (a) with the locations where the states cross the chemical potential. The arrows on the sides of part (b) mark the energies of the states (0,2), (0,3), and (0,4) for $N = 40$, indicated by the red dots, which are in agreement with the experiment in (**a**) and the theory of [71]. **c** Calculated density of states for a $N = 40$ and a $N = 200$ cluster. The arrows mark the energies of the same states in part (**b**) from the theory in [117]. Reproduced from [117] with permission of AIP Publishing

Experimental confirmation for some of these collective states has been found in matter wave diffraction measurements of helium clusters [70, 71]. As described in Sect. 1.3.3.1, the resolution could be increased to analyze the relative intensities of clusters with sizes up to about $N = 45$. Surprisingly, the intensities showed peaks at $N = 8, 14, 25$ and 41 [70, 71]. Because of the quantum liquid nature of helium clusters this observation was completely unexpected [120]. Calculations of the quantized collective excitations revealed that these magic numbers occur at exactly the threshold sizes at which an additional excited state became bound leading to a greater structural stability [71]. In Fig. 1.11a the grey area below the dashed line curve depicting the chemical potential shows the bound stable area. The nearly horizontal lines in the grey stable region mark the ripplon states $(0, l)$ and the only stable compressional state $(1, 0)$. The arrows in Fig. 1.11a for $(0, l) = (0, 2)$, $(0, 3)$, and $(0, 4)$ at $N = 40$ agree remarkably well with the calculations in part (b) of the same figure. As shown by the horizontal arrows in (c) which mark the same state-energies, highlighted by red dots in (b), the two theories are both in agreement and agree with the experiment in (a). The dynamics behind the magic numbers have recently been attributed to Auger inelastic processes inside the droplet [121].

## 1.4.3   Radial Distributions

Most of what is known about the particle number densities and radial distributions of droplets is also from theory. In most of the theories, a spherical shape and a temperature of 0 K are assumed. A typical result obtained using the OT-DF method is shown in Fig. 1.12a [62]. For all clusters with $N \geq 50$, the central density is about equal to the bulk atom density $\rho_B = 0.021\text{Å}^{-3}$. For smaller clusters, the central density increases from about $0.005\text{Å}^{-3}$ for $N = 3$ up to close to $\rho_B$ for $N \approx 20$[71]. Figure 1.12b compares the outer surface density fall-off region for $N = 50$ and higher. In each case, the surface region has been referenced to the radius at which each of the surface densities has fallen to $\rho_B/2$. Practically all the large clusters have the same outer shape. The small oscillations seen in the outer regions of radial distributions were found in one of the earliest calculations [33] and in several theories since. Evidence for the existence of oscillations is discussed at lengths in Dalfovo et al. [115]. At present, since the same structures appear in all the recent calculations they are considered to be significant. The authors of the 2006 review by Barranco et al. [62], where Fig. 1.12 is from, conjecture "One could think that such oscillations reflect an underlying quasi-solid structure".

The surface thickness is customarily defined as the difference between the densities at 10% and 90% of the internal density. The surface thickness of the calculations in Fig. 1.12b is 5.2 Å for all droplet sizes. It is about the same as the surface of films on polished silicon wafers and smaller than the surface thickness of the bulk liquid which has been measured with X-ray scattering to be about 9.2 Å [122].

The surface thickness and central densities have also been measured in a deflection scattering experiment discussed already in connection with droplet sizes and

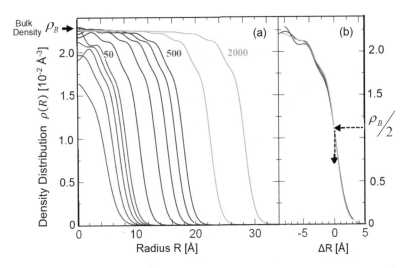

**Fig. 1.12 a** Radial density distributions $\rho(R)$ calculated with the OT-DF method for $^4He_N$ nanodroplets. The red curves correspond to steps of $N = 10$ up to 50. The blue curves correspond to steps of $N = 100$ up to 500 and the green curves are for $N \approx 1000$ and 2000. **b** The shape of the outer density profiles from 50, 100, 500 and 2000 are compared by shifting the radius to where the density is at $\rho_{B/2}$, where $\rho_B$ is the bulk density. Reproduced with permission from [62]. © copyright Springer Nature. All rights reserved

their distribution (see Sect. 1.3.3.1) In these deflection experiments the nanodroplets were deflected by free jet beams of Ar or Kr which crossed the droplet beam at an angle of 40 degrees [81]. As in the earlier size measurements [80], the mean number of atoms in the droplets was determined from the measured angular deflection pattern. Simultaneously, the attenuation of the droplet beam was also measured at a low angular resolution (implied by the large momentum of the droplets) which provides the "classical cross section". This is equivalent to the hard sphere area of the droplet. Assuming a spherical structure, an effective volume is determined. The density obtained from the mass and effective volume was found to be less than the bulk density. With a realistically assumed surface drop-off in density, the decrease compared to the bulk density could be attributed to the drop-off region. The results for 15 different sizes ranging from $N = 700$ to 13,000 resulted in an average surface thickness of 6.4 $\pm1.3$Å [81]. This result is in good agreement with five theories similar to the OT-DF method of Fig. 1.12 which gave values between 6 and 7 Å [81]. As shown in Fig. 1.13 the surface region occupies a significant volume of the droplets with less than about $10^4$ atoms.

Very large He droplets with up to $10^{10}$ atoms have been imaged individually with X-ray diffraction [105]. The images of single droplets reveal prolate and oblate as well as spherical shapes with typically average diameters of 300–2000 nm [123]. The shapes with large aspect ratios up to 3.0, indicate that the droplets are rotating with considerable angular momentum with a rotational frequency up to $\cong 10^7$ rad s$^{-1}$ [105, 123] The method and results are the subject of the chapter entitled "*X-ray and*

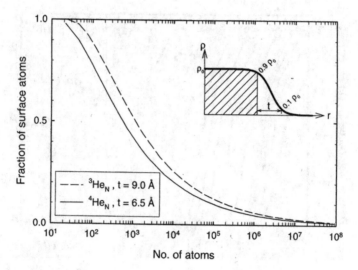

**Fig. 1.13** The fraction of atoms in the surface region, denoted by t in the inset, is shown as a function of the number of atoms in the nanodroplet

*XUV Imaging of Helium Nanodroplets"* by R. M. P Tanyag, B. Langbehn, D. Rupp, and T. Möller in this volume.

### 1.4.4 Internal Temperatures of Nanodroplets

Jürgen Gspann in 1982 was the first to predict that the temperature of helium droplets is less than 1 K [78]. He based his estimate on the correlation between the electron diffraction measured temperature of the heavier rare gas clusters and the potential depth of the corresponding dimers suggested by Farges et al. [124] Then in 1990, Brink and Stringari calculated the temperature from the rate of evaporation from the surface of droplets. For their theory, they estimated the state density of excited states, the energies and chemical potentials of the droplets [61]. They reported that after $10^{-3}$ s the temperature approached 0.3 K and also predicted the temperature of $^3$He droplets to be about 0.15 K. Similar results for $^3$He droplets have also been calculated by Guirao et al. [63].

The results of a more recent calculation, shown in Fig. 1.14 taken from the review by Barranco et al., illustrates the time dependent evaporation induced decrease in droplet sizes and temperatures [62]. Prior to the evaporation they assumed that the $^4$He droplet had grown to $10^3$ atoms and initially had a temperature of 4 K [62]. The evaporation is very rapid and the internal temperatures drop with a time constant of about $10^{-11}$–$10^{-9}$ s until the droplet is so cold that the rate of evaporation becomes negligible [61]. After travelling typical apparatus distances of one meter, corresponding to a flight time of about $10^{-3}$s, the final droplet temperatures have been calculated to

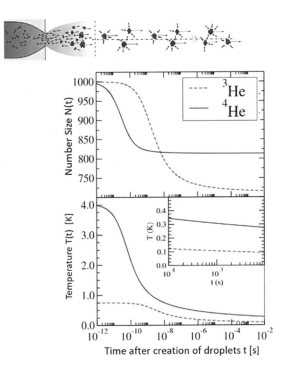

**Fig. 1.14** Calculated time evolution of the mean number sizes and temperatures of $^3$He and $^4$He droplets after they have grown to $10^3$ atoms. It is assumed that the $^4$He droplets have initially a temperature of 4.0 K and the $^3$He droplets 0.8 K. Concomitant with the large evaporative loss the temperatures decrease by about an order of magnitude to below 0.3 K ($^4$He) and 0.1 K ($^3$He) [62, 63]. Reproduced with permission from [62] © copyright Springer Nature. All rights reserved

be $0.38 K \left(^4 He\right)$ [61, 62] and $0.15 K \left(^3 He\right)$ [61–63] (see Fig. 1.14) in good agreement with experiment [38, 125–128].

Calculations by Tanyag et al. [104] for $N = 10^7$, $10^{10}$, and $10^{13}$ show that after $10^{-3}$ s the temperature versus time curves are virtually the same independent of the size. Interestingly they find that for these large droplets the reduction in sizes is always about 40% independent of the droplet size.

The first experimental evidence for the internal temperatures of droplets came from the Boltzmann distribution of rotational line intensities of the completely resolved spectra of SF$_6$ [37, 38]. Shortly afterwards in 1998, the rotational spectrum of linear carbonyl sulfide (O$^{32}$CS) molecule inside $^4$He droplets showed an even sharper well-resolved rotational fine structure [126, 128]. Figure 1.15 compares the spectrum of OCS measured in $^4$He droplets with that of the free molecules in a seeded beam [129] and with the spectrum inside non-superfluid $^3$He droplets [127]. In addition, the spectrum in the colder (0.15 K) inner $^4$He core of mixed $^4$He/$^3$He droplets is shown [126].

Since the OCS rotational line intensities closely follow a Boltzmann distribution the rotational temperature could be determined to be 0.37 K in excellent agreement with the earlier result obtained for SF$_6$ in $^4$He droplets and with the theory mentioned above. The agreement with theoretical predictions of the droplet temperatures from evaporation cooling indicates, moreover, that despite the weak coupling with the surrounding bath the dopant is thermally equilibrated in the vibrational and rotational

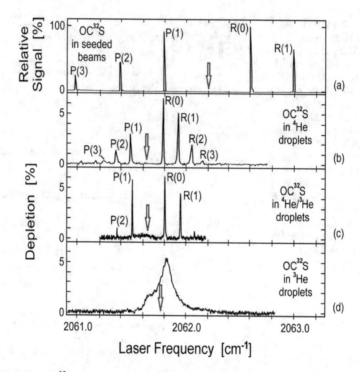

**Fig. 1.15 a** The $OC^{32}S$ rotational IR absorption spectrum of the free molecules in an Ar seeded beam [129] is compared with **b** the depletion IR spectrum in pure $^4He$ droplets ($\overline{N}_4 \approx 10^3$) [126, 128] and **c** with the depletion spectrum in the $^4He$ core of a mixed $^4He/^3He$ droplet ($\overline{N}_4 \approx 10^3$, $\overline{N}_3 \approx 10^4$) [126, 127] and (d) with the depletion spectrum in pure $^3He$ droplets ($\overline{N}_3 \approx 10^4$) [126, 127]. The P-branch corresponds to $j \rightarrow j - 1$ and R-branch to $j \rightarrow j + 1$ transitions and are labelled by the initial $j$ value. The corresponding transitions are coupled to the transition of the OCS asymmetric stretch vibration mode. Note the Q-branch is missing (position indicated by red arrows) both for the free molecule and in the mixed droplets as expected for a linear molecule. The reduced spacing of the lines in the $^4He$ droplet is due to the larger effective moment of inertia of the molecules with effectively several attached helium atoms. The smaller shift in the band origin in the $^3He$ droplets (Fig. 1.15d) can be explained by the lower density compared to $^4He$. Reproduced from [127] with the permission of AIP Publishing

degrees of freedom. The droplet temperature is much less than the temperature of 5 K in a seeded beam (Fig. 1.15a). In seeded beams the vibrational temperatures often lag behind the rotational temperature.

Several differences and similarities between the spectrum in $^4He$ droplets and that of the free molecule are noteworthy. For one, the line widths in the droplets are only slightly broadened compared to the free molecule. The well-resolved spectrum and the narrow lines indicate that the molecule rotates practically freely despite the dense liquid environment. Secondly, the vibrational band origins are only slightly red shifted by 0.557 cm$^{-1}$ ($^4He$) and 0.450 cm$^{-1}$ ($^3He$). The biggest effect seen in $^4He$ droplets, however, is the greatly reduced line spacing in the droplets by about a factor 2.8, indicating that the embedded molecule has a larger effective moment of inertia (MOI) by the same amount.

Several physical models have been proposed to explain the increase in the MOI and are discussed in [125, 128, 130]. Several theories have also been published; see for example [131]. Most theories assume that a small number of He atoms are attached to the molecule and rotate with it. The increase in the MOI has been observed in many molecules. The review by Callegari et al. contains a list with about 50 small molecules all of which have a greater MOI in helium droplets than in the gas phase [132]. The infrared spectroscopy of radicals, carbenes and ions is the subject of Chap. 4 in this volume by Gary Douberly.

Also of significance is that the Q-branch is missing in both the $^4$He and the mixed $^4$He/$^3$He droplet spectra (Fig. 1.15) as is the case in the free molecule. The Q-branch, which corresponds to transitions in which $j$ does not change, is forbidden for free linear rotor type molecules but is allowed for symmetric top molecules. Thus, the absence of a Q-branch in the droplet spectrum indicates that the helium environment has no apparent effect also on the symmetry of the rotating chromophore.

Mixed droplets produced by expanding mixtures of $^4$He and $^3$He have been found experimentally [126, 133] and theoretically [134, 135] to consist of an inner core of nearly pure $^4$He atoms and an outer shell of $^3$He atoms. The latter, being on the outside, serve for evaporative cooling and thereby determine the temperature of the molecules inside the central $^4$He core predicted to be about 0.1–0.15 K [61, 63] and experimentally determined at 0.15 K [126, 127, 133]. Pure $^3$He droplets have even a lower experimentally determined temperature of 0.07 K [127]. As seen in Fig. 1.15, the rotational line widths in the colder mixed droplets are less than in the pure $^4$He droplets and approach those of the free molecule.

At the time when these spectra were first measured, it was speculated that since bulk helium is superfluid below 2.17 K the droplets might be superfluid. Alternatively, it was argued that perhaps the low temperature and the inertness of He might account for the free rotation. This interpretation could be excluded by the broad practically structureless spectrum in the $^3$He droplets (Fig. 1.15d). Since $^3$He is lighter and has a weaker interaction with dopants the spectrum would be even sharper. Moreover since $^3$He is a fermion, superfluidity in the bulk occurs at much lower temperatures of $3 \times 10^{-3}$ K and can be excluded even at the low experimentally determined temperature of 0.07 K [127]. Thus, the structureless spectrum in the $^3$He droplets is evidence that the bosonic nature of the $^4$He droplets explains the sharp rotational features. The spectra in Fig. 1.15 and the earlier rotational resolved IR spectra of SF$_6$ [36, 37] provided the first evidence that the droplets might be superfluid. The phenomenon of free rotations has been designated as a manifestation of *molecular superfluidity* [126].

Confirmation that indeed the droplets are superfluid came shortly afterwards in 1989 from theory [136] and the experiments described in the next section.

## 1.5   Evidence for Superfluidity in Finite-Sized Helium Nanodroplets

Prior to the theory from 1989 [136] and the advent of the spectroscopic experiments in helium nanodroplets described above, it was not clear if finite-sized objects could support superfluidity. Up to this time all the evidence for superfluidity was based on the macroscopic properties of bulk He II, such as the viscous-less flow through narrow channels, the fountain effect, film flow and the enormous heat conductivity compared to He I [16]. The only confirmation that superfluidity might occur at the microscopic level came from a 1977 experiment in which the drift velocity of electrons was measured in the bulk at high pressures close to freezing [137]. In this experiment the electron drift velocity was found to satisfy the upper limit of about 58 m/s specified by Landau and thereby provided confirmation for Landau's prediction of frictionless motion in a superfluid (Fig. 1.1) [137].

The first theoretical evidence that small clusters of $^4He$ with 64 and 128 atoms were superfluid came from the 1989 Path-Integral Monte Carlo calculations of Sindzingre et al. [136]. The superfluidity was determined by simulating a slow rotation of the entire cluster and calculating the reduction of the moment of inertia of the cluster from its classical value. The transition was found to be smeared as expected for a finite-sized system. The superfluid fraction increased from about 50% at about 1.5 K and approached 100% below 1 K. The same effect was at the basis of the macroscopic experiment of Andronikashvili which provided the first evidence for the temperature dependence of the bulk superfluid fraction (two fluid model) below the 2.2 K $\lambda$-transition [141].

Previously in 1988 Lewart, Pandharipande and Pieper had calculated the Bose condensate fraction for a $^4$He droplet of 70 atoms [140]. Later, in related variational Monte Carlo calculations, Chen, McMahon and Whaley calculated the Bose fraction in smaller clusters with 7 and 40 atoms shown in Fig. 1.16 [139]. Further evidence for superfluidity came from Rama Krishna and Whaley in 1990 [119]. They calculated the dispersion curves for the $l = 0$ compression mode which in the spherical droplet is equivalent to the phonon dispersion curve of the bulk. In their $T = 0$ K calculations, they found for $N = 240$ cluster a visible equivalent of a roton which for $N = 70$ was weaker and disappeared at $N = 20$ [119]. See also Fig. 1.11 and the related discussion. For a more detailed theoretical discussion of the superfluidity of helium droplets, we call attention to the early review by Whaley [142] and the reviews by Dalfovo and Stringari [15] and by Szalewicz [143], which are also listed in Appendix A.

The first direct unequivocal experimental evidence that $^4$He droplets are indeed superfluid came in 1996 from electronic excitation spectra of the $S_1 \leftarrow S_0$ transition of the glyoxal $(C_2H_2O_2)$ molecule embedded in a 5,000 atom droplet [144]. In addition to a sharp zero phonon line (ZPL) at higher frequencies an unusual clearly peaked phonon wing (PW) was found to be well separated from the ZPL as shown in Fig. 1.17. Broad PWs are well known for species trapped in low-temperature classical matrices where they are found to be proportional to the density of the available phonon

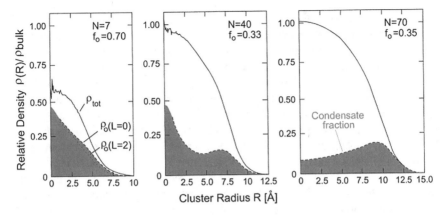

**Fig. 1.16** Variational Monte-Carlo calculations of the radial total density distribution $\rho_{tot}$ of small clusters of $^4$He. The blue area shows the radial distribution of the Bose–Einstein condensate $\rho_0$. $f_0$ is the overall cluster condensate fraction for angular momentum $l = 0$. Note that in the low density on the periphery the local condensate fraction approaches unity as in BEC of alkali gases [138]. The results for $N = 7$ and 40 are from Cheng et al. [139] and those for $N = 70$ are from [140]. Note also that in the bulk the condensate fraction is only about 10%. Adapted from [139] and [140]

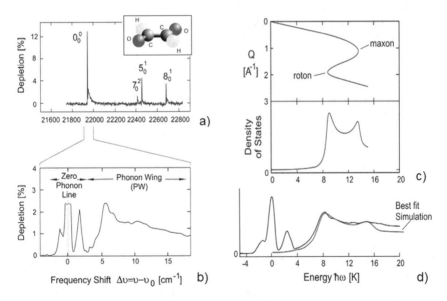

**Fig. 1.17** **a** Depletion spectrum of the band origin of the $S_1 \leftarrow S_0$ in glyoxal ($C_2H_2O_2$) in a $^4$He droplet with $\overline{N}_4 = 5 \times 10^3$ [147]. **b** The fine structure of the $0_0^0$ line reveals a rotationally resolved zero phonon line (ZPL) followed by a structured phonon wing (PW). **c** The density of states corresponding to the sharp dispersion curves predicted by Landau for superfluid $^4$He. **d** A simulation of the phonon wing (red curve) based on the position of the roton and maxon agrees with the shape of the PW

states. While usually the PW is broad and largely merged with the ZPL, the glyoxal PW is sharply peaked and well separated from the ZPL. More remarkable is that the exact shape reflects the Landau dispersion curve with a large maximum at 6 cm$^{-1}$ (8.4 K) equal to the energy of the roton and another peak at the energy of the maxon (Fig. 1.1). The distinct PW and the coincidence of its energy distribution with the density of states associated with the roton minimum provides direct evidence that the $^4$He droplets are indeed superfluid in the microscopic sense. Phonon wings have been found in other systems but few have been found to be so sharp as in glyoxal. For further details the reader is referred to Chap. 5 entitled *"Electronic Spectroscopy in Superfluid Helium Droplets"* by F. Schlaghäufer, J. Fischer and A. Slenczka.

The above interpretation is confirmed by the spectrum of glyoxal in pure non-superfluid $^3$He droplets [125, 145]. There the sharply peaked phonon wing is replaced by a gradually falling intensity starting at the ZPL and extending to higher energies. This is consistent with the fact that in $^3$He the density of states is smeared out and partly concealed by particle-hole pair excitations at the Fermi level [145, 146].

Another equally important experiment is the observation of the frictionless ejection of atoms from the droplets with a velocity of about 58 m/s as predicted by Landau. In the 2013 experiment of Brauer et al., $Ag\,^2P_{1/2}$ and five different molecules, including NO, the large organic molecules such as trimethylamine(TMA),1,4-diazabicyclo[2.2.2]octane(DABCO), and 1-azabicyclo [2.2.2]octane (ABCO) were electronically photo-excited by a pulsed laser [148]. Previously, it had been established that certain electronically excited atoms and molecules are efficiently ejected from droplets [149]. The velocities of the ejected particles were measured by Brauner et al. with a velocity map imaging spectrometer. In all cases, the most probable speeds were in the range of 40 – 60 m/s. With adequate corrections the velocity of ejected excited $Ag\,^2P_{1/2}$ atoms was found to be 56 m/s in excellent agreement with the Landau velocity [148]. A wide range of different sized droplets were investigated and the limiting velocities were all quite similar. A small dependence on the masses of the 5 particles investigated was roughly in accord with theory [148]. For a more detailed discussion of these photoejection experiments see Chap. 5 entitled *"Electronic Spectroscopy in Superfluid Helium Droplets "* by F. Schlaghäufer, J. Fischer and A. Slenczka.

An equally direct experimental evidence that very large droplets with between $10^8$–$10^{11}$ atoms are superfluid comes from the observation of quantum vortices in the diffraction images from pulsed coherent X-ray scattering [105]. The droplet shapes were found to be either oblate or prolate [150]. Ancillotto et al. on the basis of DFT calculations conclude that the shapes observed experimentally can only be attributed to the presence of quantum vortices inside the large droplets [151]. By doping the droplets with Xe atoms, which are known to be localized at the vortex cores and which are strong scatterers it was possible to directly observe the vortices in the diffraction image. The Fourier-transformed images revealed a regular array with up to 170 vortices [152]. This is considered as evidence that very large droplets, which are close to macroscopic are also superfluid. In an earlier famous experiment, similar regular vortex arrays were found in a rotating cylinder containing superfluid bulk helium [153]. A related vortex lattice was seen much earlier at the surface of superconductors and named after the Nobel Prize laureate Aleksei Abrikosov [154]. The method of X-ray imaging of superfluid droplets and the experimental results are dealt with in detail in the chapter entitled *"X-ray and XUV Imaging of Helium Nanodroplets"* by. R. M. P. Tanyag, B. Lengbehn, D. Rupp, and T. Möller in this volume.

There are several additional experiments that can only be fully explained when the superfluidity of the droplets is invoked. These include the field detachment of electrons from droplets [155] and the transmission of $^3$He atoms through large $^4$He droplets [93].

At the present time the superfluidity of helium droplets is well established. An interesting, but not fully solved question is "How many atoms are required for superfluidity?" or to put it differently "How does superfluidity manifest itself in the limit of few atoms?". Related experiments can be found in the [126] and [156, 157].

This introduction covers only a small selection of the vast literature on the properties of helium droplets and the methods that have been developed to create clusters with tailored made properties. Also, the various techniques which are used in the spectroscopic investigations have hardly been touched upon. For more information the reader is referred to the chapters in this book and the reviews in the Appendix.

**Acknowledgements** I am grateful to Alkwin Slenczka and his colleagues Florian Schlaghäufer and Johannes Fischer for a careful meticulous reading and for very many suggestions for small but important corrections. I also thank Rico Mayro P. Tanyag and Henrik Høj Kristensen for many essential corrections which have significantly improved an earlier version of the manuscript.

# References

1. P. Kapitza, Viscosity of liquid helium below the λ-point. Nature **141**, 74 (1938)
2. J.F. Allen, A.D. Misener, Flow of liquid helium II. Nature **141**, 75 (1938)
3. S. Balibar, The discovery of superfluidity. J. Low Temp. Phys. **146**(5–6), 441 (2007)
4. A. Griffin, *The Discovery of Superfluidty: A Chronology of Events in 1935–1938*. www.phy sics.utoronto.ca/~griffin (2006)
5. A. Griffin, New light on the intriguing history of superfluidity in liquid He-4. J. Phys.-Condens. Matter **21**(16), 164220 (2009)
6. P.E. Rubinin, The discovery of superfluidity in letters and documents. Usp. Fiz. Nauk **167**(12), 1349 (1997)
7. L. Tisza, Transport phenomena in helium II. Nature **141**, 913 (1938)
8. S. Bose, Plancks Gesetz und Lichtquantenhypothese. Z. Phys. **26**, 178 (1924); A. Einstein, Quantentheorie des einatomigen idealen Gases, Sitzungsberichte der preussischen Akdemie der Wissenschaften, Berlin 1924, p.261
9. F. London, The alpha-phenomenon of liquid helium and the Bose Einstein degeneracy. Nature **141**, 643 (1938)
10. F. London, Superfluids: volume II macroscopic theory of superfluid helium, in *Superfluids*, ed. by J. Wiley (Wiley, New York, 1954)
11. L. Landau, Theory of the Superfluidity of helium II. J. Phys. USSR **5**, 71 (1941)
12. H.B. Ghassib, G.V. Chester, He-4 N-mers and Bose-Einstein condensation in He-II. J. Chem. Phys. **81**(1), 585 (1984)

13. F. Dalfovo, S. Giorgini, L.P. Pitaevskii, S. Stringari, Theory of Bose-Einstein condensation in trapped gases. Rev. Mod. Phys. **71**(3), 463 (1999)
14. A. Griffin, D.W. Snoke, G. Stringari, *Bose-Einstein Condensation* (University Press, Cambridge, 1995)
15. F. Dalfovo, S. Stringari, Helium nanodroplets and trapped Bose-Einstein condensates as prototypes of finite quantum fluids. J. Chem. Phys. **115**(22), 10078 (2001)
16. D.R. Tilley, J. Tilley, Superfluidity and superconductivity. in *Graduate Student Series in Physics* (Adam Hilger, Bristol and New York, 1990)
17. C.J. Pethick, D.G. Ravenhall, Matter at large neutron excess and the physics of neutron-star crusts. Annu. Rev. Nucl. Part. Sci. **45**, 429 (1995)
18. K. Aoki, K. Sakakibara, I. Ichinose, T. Matsui, Magnetic order, Bose-Einstein condensation, and superfluidity in a bosonic t-J model of $CP^1$ spinons and doped Higgs holons. Phys. Rev. B **80**(14) (2009)
19. K.H. Bennemann, J.B. Ketterson (ed.) *Novel Superfluids*, vol. 2 (Oxford University Press, 2014)
20. W.H. Keesom, *Helium* (Elsevier, Amsterdam-London-New York, 1942)
21. J. Jortner, S.A. Rice, E.G. Wilson, L. Meyer, Energy transfer phenomena in liquid helium. Phys. Rev. Lett. **12**(15), 415 (1964)
22. E.B. Gordon, L.P. Mezhovde, O.F. Pugachev, Stabilization of nitrogen atoms in superfluid-helium. Jetp Lett. **19**(2), 63 (1974)
23. E.B. Gordon, Y. Okuda, Catalysis of impurities coalescence by quantized vortices in superfluid helium with nanofilament formation. Low Temp. Phys. **35**(3), 209 (2009)
24. B. Tabbert, H. Gunther, G.Z. Putlitz, Optical investigation of impurities in superfluid He-4. J. Low Temp. Phys. **109**(5–6), 653 (1997)
25. M. Takami, Comm. At. Mol. Phys. **32**, 219 (1996)
26. J.P. Toennies, A.F. Vilesov, Spectroscopy of atoms and molecules in liquid helium. Annu. Rev. Phys. Chem. **49**, 1 (1998)
27. V. Lebedev, P. Moroshkin, J.P. Toennies, A. Weis, Spectroscopy of the copper dimer in normal fluid, superfluid, and solid He-4. J. Chem. Phys. **133**(15), 154508 (2010)
28. I.F. Silvera, Ultimate fate of a gas of atomic-hydrogen in a liquid-helium chamber—recombination and burial. Phys. Rev. B **29**(7), 3899 (1984)
29. E.W. Becker, R. Klingelhofer, P. Lohse, Strahlen aus kondensiertem Helium im Hochvakuum. Zeitschrift für Naturforschung Part a-Astrophysik Physik und Physikalische Chemie **16**(11), 1259 (1961)
30. E.W. Becker, On the history of cluster beams. Zeitschrift für Physik D-Atoms Mol. Clust. **3**(2–3), 101 (1986)
31. P.W. Stephens, J.G. King, Experimental investigation of small helium clusters—magic numbers and the onset of condensation. Phys. Rev. Lett. **51**(17), 1538 (1983)
32. M. Rasetti, T. Regge, Helium droplets as analogs of heavy nuclei, in *Quantum Liquids*, ed. by F. Ruvalds, T. Regge (North Holland, Amsterdam, 1978)
33. V.R. Pandharipande, J.G. Zabolitzky, S.C. Pieper, R.B. Wiringa, U. Helmbrecht, Calculations of ground-state properties of liquid-He-4 droplets. Phys. Rev. Lett. **50**(21), 1676 (1983)
34. H. Buchenau, R. Götting, A. Scheidemann, J.P. Toennies, Experimental studies of condensation in helium nozzle beams, in *15th International Symposium on Rarefied Gas Dynaimcs, June 16–20, 1986* (B. G. Teubner, Grado (Italy), 1986)
35. A. Scheidemann, J.P. Toennies, J.A. Northby, Capture of neon atoms by He-4 clusters. Phys. Rev. Lett. **64**(16), 1899 (1990)
36. S. Goyal, D.L. Schutt, G. Scoles, Vibrational spectroscopy of sulfur hexafluoride attached to helium clusters. Phys. Rev. Lett. **69**(6), 933 (1992)
37. R. Fröchtenicht, J.P. Toennies, A. Vilesov, High-resolution infrared-spectroscopy of $SF_6$ embedded in He clusters. Chem. Phys. Lett. **229**(1–2), 1 (1994)
38. M. Hartmann, R.E. Miller, J.P. Toennies, A. Vilesov, Rotationally resolved spectroscopy of Sf6 in liquid helium clusters—a molecular probe of cluster temperature. Phys. Rev. Lett. **75**(8), 1566 (1995)

39. W.E. Keller, *Helium-3 and Helium-4* (Plenum Press, New York, 1969)
40. K.H. Bennemann, J.B. Ketterson, (eds.), *The Physics of Liquid and Solid Helium*, vol. 1 (Wiley, 1976)
41. J.R. Buckland, R.L. Folkerts, R.B. Balsod, W. Allison, A simple nozzle design for high speed ratio molecular beams. Meas. Sci. Technol. **8**(8), 933 (1997)
42. R.M. Cleaver, C.M. Lindsay, Detailed design and transport properties of a helium droplet nozzle from 5 to 50 K. Cryogenics **52**(7–9), 389 (2012)
43. W.R. Gentry, C.F. Giese, High-precision skimmers for supersonic molecular beams. Rev. Sci. Instrum. **46**(1), 104 (1975)
44. D.R. Miller, Free jet sources, in *Atomic and Molecular Beam Methods*, ed. by G. Scoles (Oxford University Press, New York, Oxford, 1988), p. 14
45. U. Even, J. Jortner, D. Noy, N. Lavie, C. Cossart-Magos, Cooling of large molecules below 1 K and He clusters formation. J. Chem. Phys. **112**(18), 8068 (2000)
46. M.N. Slipchenko, S. Kuma, T. Momose, A.F. Vilesov, Intense pulsed helium droplet beams. Rev. Sci. Instrum. **73**(10), 3600 (2002)
47. D. Pentlehner, R. Riechers, B. Dick, A. Slenczka, U. Even, N. Lavie, R. Brown, K. Luria, Rapidly pulsed helium droplet source. Rev. Sci. Instrum. **80**(4), 043302 (2009)
48. V. Ghazarian, J. Eloranta, V.A. Apkarian, Universal molecule injector in liquid helium: pulsed cryogenic doped helium droplet source. Rev. Sci. Instrum. **73**(10), 3606 (2002)
49. J.S. Brooks, R.J. Donnelly, Calculated thermodynamic properties of superfluid He-4. J. Phys. Chem. Ref. Data **6**(1), 51 (1977)
50. R.D. McCarty, Thermodynamic properties of helium 4. J. Phys. Chem. Rev. Data **2**(4) (1973)
51. R.J. Donnelly, C.F. Barenghi, The observed properties of liquid helium at the saturated vapor pressure. J. Phys. Chem. Ref. Data **27**(6), 1217 (1998)
52. H. Buchenau, E.L. Knuth, J. Northby, J.P. Toennies, C. Winkler, Mass spectra and time-of-flight distributions of helium cluster beams. J. Chem. Phys. **92**(11), 6875 (1990)
53. G.D. Stein, Cluster beam sources—predictions and limitations of the nucleation theory. Surf. Sci. **156**(Jun), 44 (1985)
54. G. Koppenwallner, C. Dankert, Homogeneous condensation in N2, Ar, and $H_2O$ free jets. J. Phys. Chem. **91**(10), 2482 (1987)
55. H.A. Cataldi, H.G. Drickamer, light scattering in the critical region .1. Ethylene. J. Chem. Phys. **18**(5), 650 (1950)
56. J.O. Hirschfelder, C.F. Curtiss, R.B. Bird, *Molecular Theory of Gases and Liquids* (Wiley, New York, Chapman & Hall, London, 1954)
57. B.L. Holian, D.E. Grady, Fragmentation by molecular-dynamics—the microscopic big-bang. Phys. Rev. Lett. **60**(14), 1355 (1988)
58. D.E. Grady, Particle size statistics in dynamic fragmentation. J. Appl. Phys. **68**(12), 6099 (1990)
59. T.E. Itina, Decomposition of rapidly expanding liquid: molecular dynamics study. Chem. Phys. Lett. **452**(1–3), 129 (2008)
60. R.E. Grisenti, J.P. Toennies, Cryogenic microjet source for orthotropic beams of ultralarge superfluid helium droplets. Phys. Rev. Lett. **90**(23), 234501 (2003)
61. D.M. Brink, S. Stringari, Density of states and evaporation rate of helium clusters. Zeitschrift für Physik D-Atoms Mol. Clust. **15**(3), 257 (1990)
62. M. Barranco, R. Guardiola, S. Hernandez, R. Mayol, J. Navarro, M. Pi, Helium nanodroplets: an overview. J. Low Temp. Phys. **142**(1–2), 1 (2006)
63. A. Guirao, M. Pi, M. Barranco, Finite size effects in the evaporation rate of He-3 clusters. Zeitschrift Fur Physik D-Atoms Mol. Clust. **21**(2), 185 (1991)
64. J.P. Toennies, Physics and chemistry of helium clusters and droplets, in *Helium Occurance: Applications and Biological Effects* (Nova Science Publishers, Hanppange N.Y. (USA), 2013)
65. R.E. Grisenti, W. Schöllkopf, J.P. Toennies, G.C. Hegerfeldt, T. Kohler, M. Stoll, Determination of the bond length and binding energy of the helium dimer by diffraction from a transmission grating. Phys. Rev. Lett. **85**(11), 2284 (2000)

66. J.P. Toennies, K. Winkelmann, Theoretical studies of highly expanded free jets—influence of quantum effects and a realistic intermolecular potential. J. Chem. Phys. **66**(9), 3965 (1977)

67. F. Luo, G. Kim, G.C. Mcbane, C.F. Giese, W.R. Gentry, Influence of retardation on the vibrational wave function and binding energy of the helium dimer. J. Chem. Phys. **98**(12), 9687 (1993)

68. G. Brusdeylins, H.-D. Meyer, J.P. Toennies, K. Winkelmann, Production of helium nozzle beams with very high speed ratios, in *10th Symposium of Rarefied Gas Dynamics*, ed. by J.L. Potter (AIAA, New York, Aspen, 1977)

69. L.W. Bruch, W. Schollkopf, J.P. Toennies, The formation of dimers and trimers in free jet He-4 cryogenic expansions. J. Chem. Phys. **117**(4), 1544 (2002)

70. R. Bruhl, R. Guardiola, A. Kalinin, O. Kornilov, J. Navarro, T. Savas, J.P. Toennies, Diffraction of neutral helium clusters: evidence for "magic numbers". Phys. Rev. Lett. **92**(18), 185301 (2004)

71. R. Guardiola, O. Kornilov, J. Navarro, J.P. Toennies, Magic numbers, excitation levels, and other properties of small neutral He-4 clusters (N <= 50). J. Chem. Phys. **124**(8), 084307 (2006)

72. J. Chaiken, J. Goodisman, O. Kornilov, J.P. Toennies, Application of scaling and kinetic equations to helium cluster size distributions: Homogeneous nucleation of a nearly ideal gas. J. Chem. Phys. **125**(7), 074305 (2006)

73. O. Kornilov, J.P. Toennies, Para-hydrogen and helium cluster size distributions in free jet expansions based on Smoluchowski theory with kernel scaling. J. Chem. Phys. **142**(7), 074303 (2015)

74. E.L. Knuth, Size correlations for condensation clusters produced in free-jet expansions. J. Chem. Phys. **107**(21), 9125 (1997)

75. B. Schilling, *Molekularstrahlexperimente mit Helium Clusters*. Max-Planck-Institut für Strömungsforschung (1993)

76. M. Lewerenz, B. Schilling, J.P. Toennies, Successive capture and coagulation of atoms and molecules to small clusters in large liquid helium clusters. J. Chem. Phys. **102**(20), 8191 (1995)

77. J. Gspann, Helium microdroplet transparency in heavy atom collisions. Physica B & C **108**(1–3), 1309 (1981)

78. J. Gspann, Electronic and atomic impacts on large clusters, in *Physics of Electronic and Atomic Collisions*, ed. by S. Datz (Noth Holland, Amsterdam, 1982), p. 79

79. J. Gspann, Cluster size separation by crossjet deflection. Berichte der Bunsen-Gesellschaft-Phys. Chem. Chem. Phys. **88**(3), 256 (1984)

80. M. Lewerenz, B. Schilling, J.P. Toennies, A new scattering deflection method for determining and selecting the sizes of large liquid clusters of He-4. Chem. Phys. Lett. **206**(1–4), 381 (1993)

81. J. Harms, J.P. Toennies, F. Dalfovo, Density of superfluid helium droplets. Phys. Rev. B **58**(6), 3341 (1998)

82. O. Kornilov, J.R. Toennies, The determination of the mean sizes of large He droplets by electron impact induced attenuation. Int. J. Mass Spectrom. **280**(1–3), 209 (2009)

83. E.L. Knuth, O. Kornilov, J.P. Toennies, Terminal liquid mass fractions and terminal mean droplet sizes in He free-jet expansions, in *27th International Symposium on Rarefied Gas Dynamics*. 2010. AIP Conference Proceedings

84. O.F. Hagena, Condensation in free jets—comparison of rare-gases and metals. Zeitschrift für Physik D-Atoms Mol. Clust. **4**(3), 291 (1987)

85. S. Kuma, T. Azuma, Pulsed beam of extremely large helium droplets. Cryogenics **88**, 78 (2017)

86. D. Verma, A.F. Vilesov, Pulsed helium droplet beams. Chem. Phys. Lett. **694**, 129 (2018)

87. S.F. Yang, A.M. Ellis, Selecting the size of helium nanodroplets using time-resolved probing of a pulsed helium droplet beam. Rev. Sci. Instrum. **79**(1), 016106 (2008)

88. M. Alghamdi, J. Zhang, A. Oswalt, J.J. Porter, R.A. Mehl, W. Kong, Doping of green fluorescent protein into superfluid helium droplets: size and velocity of doped droplets. J. Phys. Chem. A **121**(36), 6671 (2017)

89. R. Pandey, S. Tran, J. Zhang, Y.Z. Yao, W. Kong, Bimodal velocity and size distributions of pulsed superfluid helium droplet beams. J. Chem. Phys. **154**(13), 164220 (2021)
90. L.F. Gomez, E. Loginov, R. Sliter, A.F. Vilesov, Sizes of large He droplets. J. Chem. Phys. **135**(15), 154201 (2011)
91. U. Henne, J.P. Toennies, Electron capture by large helium droplets. J. Chem. Phys. **108**(22), 9327 (1998)
92. J. Harms, J.P. Toennies, E.L. Knuth, Droplets formed in helium free-jet expansions from states near the critical point. J. Chem. Phys. **106**(8), 3348 (1997)
93. J. Harms, J.P. Toennies, Observation of anomalously low momentum transfer in the low energy scattering of large He-4 droplets from He-4 and He-3 atoms. J. Low Temp. Phys. **113**(3–4), 501 (1998)
94. T. Jiang, C. Kim, J.A. Northby, Electron attachment to helium microdroplets—creation induced magic. Phys. Rev. Lett. **71**(5), 700 (1993)
95. T. Jiang, J.A. Northby, Fragmentation clusters formed in supercritical expansions of He-4. Phys. Rev. Lett. **68**(17), 2620 (1992)
96. R. Sliter, L.F. Gomez, J. Kwok, A. Vilesov, Sizes distributions of large He droplets. Chem. Phys. Lett. **600**, 29 (2014)
97. E.L. Knuth, U. Henne, Average size and size distribution of large droplets produced in a free-jet expansion of a liquid. J. Chem. Phys. **110**(5), 2664 (1999)
98. D.E. Grady, M.E. Kipp, Geometric statistics and dynamic fragmentation. J. Appl. Phys. **58**(3), 1210 (1985)
99. R.M.P. Tanyag, A.J. Feinberg, S.M.O. O'Connell, A.F. Vilesov, Disintegration of diminutive liquid helium jets in vacuum. J. Chem. Phys. **152**(23), 234306 (2020)
100. J. Eggers, Nonlinear dynamics and breakup of free-surface flows. Rev. Mod. Phys. **69**(3), 865 (1997)
101. J. Eggers, E. Villermaux, Physics of liquid jets. Rep. Prog. Phys. **71**(3), 036601 (2008)
102. A. Frohn, N. Roth, *Dynamics of Droplets* (Springer, 2000)
103. M. Kuhnel, N. Petridis, D.F.A. Winters, U. Popp, R. Dorner, T. Stohlker, R.E. Grisenti, Low-Z internal target from a cryogenically cooled liquid microjet source. Nucl. Instrum, Methods Phys. Res. Sect. A-Accel. Spectrom. Detect. Assoc. Equip. **602**(2), 311 (2009)
104. R.M.P. Tanyag, C.F. Jones, C. Bernando, S.M.O. O'Connell, D. Verma, A.F. Vilesov, Experiments with large superfluid helium nanodroplets, in *Cold Chemistry: Molecular Scattering and Reactivity Near Absolute Zero*, ed. by O.A.O. Dulieu (The Royal Society of Chemistry, London, 2018)
105. L.F. Gomez, K.R. Ferguson, J.P. Cryan, C. Bacellar, R.M.P. Tanyag, C. Jones, S. Schorb, D. Anielski, A. Belkacem, C. Bernando, R. Boll, J. Bozek, S. Carron, G. Chen, T. Delmas, L. Englert, S.W. Epp, B. Erk, L. Foucar, R. Hartmann, A. Hexemer, M. Huth, J. Kwok, S.R. Leone, J.H.S. Ma, F.R.N.C. Maia, E. Malmerberg, S. Marchesini, D.M. Neumark, B. Poon, J. Prell, D. Rolles, B. Rudek, A. Rudenko, M. Seifrid, K.R. Siefermann, F.P. Sturm, M. Swiggers, J. Ullrich, F. Weise, P. Zwart, C. Bostedt, O. Gessner, A.F. Vilesov, Shapes and vorticities of superfluid helium nanodroplets. Science **345**(6199), 906 (2014)
106. P.C. Hendry, N.S. Lawson, R.A.M. Lee, P.V.E. Mcclintock, C.D.H. Williams, Generation of defects in superfluid He-4 as an analog of the formation of cosmic strings. Nature **368**(6469), 315 (1994)
107. W.H. Zurek, U. Dorner, P. Zoller, Dynamics of a quantum phase transition. Phys. Rev. Lett. **95**(10), 105701 (2005)
108. N.B. Speirs, K.R. Langley, P. Taborek, S.T. Thoroddsen, Jet breakup in superfluid and normal liquid He-4. Phys. Rev. Fluids **5**(4) (2020)
109. B. Samelin, *Lebensdauer und Neutralisation metastabiler negativ geladener Helium-Mikrotropfen*. Max-Planck-Institut für Strömungsforschung (1998)
110. E.W. Schmid, J. Schwager, Y.C. Tang, R.C. Herndon, Binding energies of polyatomic 4he-molecules. Physica **31**(7), 1143 (1965)
111. V.R. Pandharipande, S.C. Pieper, R.B. Wiringa, Variational Monte-Carlo calculations of ground-states of liquid-He-4 and He-3 drops. Phys. Rev. B **34**(7), 4571 (1986)

112. S.A. Chin, E. Krotscheck, Microscopic calculation of collective excitations in He-4 clusters. Phys. Rev. Lett. **65**(21), 2658 (1990)
113. S.A. Chin, E. Krotscheck, Structure and collective excitations of He-4 clusters. Phys. Rev. B **45**(2), 852 (1992)
114. S.A. Chin, E. Krotscheck, Systematics of pure and doped He-4 clusters. Phys. Rev. B **52**(14), 10405 (1995)
115. F. Dalfovo, A. Lastri, L. Pricaupenko, S. Stringari, J. Treiner, Structural and dynamical properties of superfluid-helium—a density-functional approach. Phys. Rev. B **52**(2), 1193 (1995)
116. D.M. Ceperley, Path-integrals in the theory of condensed helium. Rev. Mod. Phys. **67**(2), 279 (1995)
117. E. Krotscheck, R. Zillich, Dynamics of He-4 droplets. J. Chem. Phys. **115**(22), 10161 (2001)
118. M. Casas, S. Stringari, Elementary excitations of He-4 clusters. J. Low Temp. Phys. **79**(3–4), 135 (1990)
119. M.V.R. Krishna, K.B. Whaley, Collective excitations of helium clusters. Phys. Rev. Lett. **64**(10), 1126 (1990)
120. R. Melzer, J.G. Zabolitzky, No magic numbers in neutral He-4 clusters. J. Phys. A-Math. Gen. **17**(11), L565 (1984)
121. E. Spreafico, G. Benedek, O. Kornilov, J.P. Toennies, Magic numbers in Boson $^4$He clusters: the auger evaporation mechanism. Molecules **26**, 10.3390 (2021)
122. L.B. Lurio, T.A. Rabedeau, P.S. Pershan, I.F. Silvera, M. Deutsch, S.D. Kosowsky, B.M. Ocko, Liquid-vapor density profile of helium—an X-ray study. Phys. Rev. Lett. **68**(17), 2628 (1992)
123. C. Bernando, R.M.P. Tanyag, C. Jones, C. Bacellar, M. Bucher, K.R. Ferguson, D. Rupp, M.P. Ziemkiewicz, L.F. Gomez, A.S. Chatterley, T. Gorkhover, M. Muller, J. Bozek, S. Carron, J. Kwok, S.L. Butler, T. Moller, C. Bostedt, O. Gessner, A.F. Vilesov, Shapes of rotating superfluid helium nanodroplets. Phys. Rev. B **95**(6), 064510 (2017)
124. J. Farges, M.F. Deferaudy, B. Raoult, G. Torchet, Structure and temperature of rare gas clusters in a supersonic expansion. Surf. Sci. **106**(1–3), 95 (1981)
125. S. Grebenev, M. Hartmann, A. Lindinger, N. Pörtner, B. Sartakov, J.P. Toennies, A.F. Vilesov, Spectroscopy of molecules in helium droplets. Physica B **280**(1–4), 65 (2000)
126. S. Grebenev, J.P. Toennies, A.F. Vilesov, Superfluidity within a small helium-4 cluster: the microscopic Andronikashvili experiment. Science **279**(5359), 2083 (1998)
127. B.G. Sartakov, J.P. Toennies, A.F. Vilesov, Infrared spectroscopy of carbonyl sulfide inside a pure He-3 droplet. J. Chem. Phys. **136**(13), 134316 (2012)
128. S. Grebenev, M. Hartmann, M. Havenith, B. Sartakov, J.P. Toennies, A.F. Vilesov, The rotational spectrum of single OCS molecules in liquid He-4 droplets. J. Chem. Phys. **112**(10), 4485 (2000)
129. A.R.W. McKellar, *Private Communication* (2002)
130. C. Callegari, A. Conjusteau, I. Reinhard, K.K. Lehmann, G. Scoles, F. Dalfovo, Superfluid hydrodynamic model for the enhanced moments of inertia of molecules in liquid He-4. Phys. Rev. Lett. **83**(24), 5058 (1999)
131. Y. Kwon, P. Huang, M.V. Patel, D. Blume, K.B. Whaley, Quantum solvation and molecular rotations in superfluid helium clusters. J. Chem. Phys. **113**(16), 6469 (2000)
132. C. Callegari, K.K. Lehmann, R. Schmied, G. Scoles, Helium nanodroplet isolation rovibrational spectroscopy: methods and recent results. J. Chem. Phys. **115**(22), 10090 (2001)
133. J. Harms, M. Hartmann, B. Sartakov, J.P. Toennies, A.F. Vilesov, High resolution infrared spectroscopy of single $SF_6$ molecules in helium droplets. II. The effect of small amounts of He-4 in large He-3 droplets. J. Chem. Phys. **110**(11), 5124 (1999)
134. M. Barranco, M. Pi, S.M. Gatica, E.S. Hernandez, J. Navarro, Structure and energetics of mixed He-4-He-3 drops. Phys. Rev. B **56**(14), 8997 (1997)
135. M. Pi, R. Mayol, M. Barranco, Structure of large He-3-He-4 mixed drops around a dopant molecule. Phys. Rev. Lett. **82**(15), 3093 (1999)

136. P. Sindzingre, M.L. Klein, D.M. Ceperley, Path-integral Monte-Carlo study of low-temperature He-4 clusters. Phys. Rev. Lett. **63**(15), 1601 (1989)
137. D.R. Allum, P.V.E. Mcclintock, A. Phillips, R.M. Bowley, Breakdown of superfluidity in liquid-He-4—experimental test of Landaus theory. Phil. Trans. R. Soc. A-Math. Phys. Eng. Sci. **284**(1320), 179 (1977)
138. A. Griffin, S. Stringari, Surface region of superfluid helium as an inhomogeneous Bose-condensed gas. Phys. Rev. Lett. **76**(2), 259 (1996)
139. E. Cheng, M.A. McMahon, K.B. Whaley, Current and condensate distributions in rotational excited states of quantum liquid clusters. J. Chem. Phys. **104**(7), 2669 (1996)
140. D.S. Lewart, V.R. Pandharipande, S.C. Pieper, Single-particle orbitals in liquid-helium drops. Phys. Rev. B **37**(10), 4950 (1988)
141. E.L. Andronikashvili, J. Phys. U.S.S.R. **10**(201), 948 (1946)
142. B.K. Whaley, Structure and dynamics of quantum clusters. Int. Rev. Phys. Chem. **13**(1), 41 (1994)
143. K. Szalewicz, Interplay between theory and experiment in investigations of molecules embedded in superfluid helium nanodroplets. Int. Rev. Phys. Chem. **27**(2), 273 (2008)
144. M. Hartmann, F. Mielke, J.P. Toennies, A.F. Vilesov, G. Benedek, Direct spectroscopic observation of elementary excitations in superfluid He droplets. Phys. Rev. Lett. **76**(24), 4560 (1996)
145. N. Poertner, J.P. Toennies, A.F. Vilesov, G. Benedek, V. Hizhnyakov, Anomalously sharp phonon excitations in He-3 droplets. EPL **88**(2), 26007 (2009)
146. G. Benedek, V. Hizhnyakov, J.P. Toennies, The response of a He-3 Fermi liquid droplet to vibronic excitation of an embedded glyoxal molecule. J. Phys. Chem. A **118**(33), 6574 (2014)
147. M. Hartmann, R.E. Miller, J.P. Toennies, A.F. Vilesov, High-resolution molecular spectroscopy of van der Waals clusters in liquid helium droplets. Science **272**(5268), 1631 (1996)
148. N.B. Brauer, S. Smolarek, E. Loginov, D. Mateo, A. Hernando, M. Pi, M. Barranco, W.J. Buma, M. Drabbels, Critical Landau velocity in helium nanodroplets. Phys. Rev. Lett. **111**(15), 153002 (2013)
149. F. Federmann, K. Hoffmann, N. Quaas, J.D. Close, Rydberg states of silver: Excitation dynamics of doped helium droplets. Phys. Rev. Lett. **83**(13), 2548 (1999)
150. B. Langbehn, K. Sander, Y. Ovcharenko, C. Peltz, A. Clark, M. Coreno, R. Cucini, M. Drabbels, P. Finetti, M. Di Fraia, L. Giannessi, C. Grazioli, D. Iablonskyi, A.C. LaForge, T. Nishiyama, V.O.A. de Lara, P. Piseri, O. Plekan, K. Ueda, J. Zimmermann, K.C. Prince, F. Stienkemeier, C. Callegari, T. Fennel, D. Rupp, T. Moller, Three-dimensional shapes of spinning helium nanodroplets. Phys. Rev. Lett. **121**(25), 255301 (2018)
151. F. Ancilotto, M. Barranco, M. Pi, Spinning superfluid He-4 nanodroplets. Phys. Rev. B **97**(18), 184515 (2018)
152. C.F. Jones, C. Bernando, R.M.P. Tanyag, C. Bacellar, K.R. Ferguson, L.F. Gomez, D. Anielski, A. Belkacem, R. Boll, J. Bozek, S. Carron, J. Cryan, L. Englert, S.W. Epp, B. Erk, L. Foucar, R. Hartmann, D.M. Neumark, D. Rolles, A. Rudenko, K.R. Siefermann, F. Weise, B. Rudek, F.P. Sturm, J. Ullrich, C. Bostedt, O. Gessner, A.F. Vilesov, Coupled motion of Xe clusters and quantum vortices in He nanodroplets. Phys. Rev. B **93**(18), 180510 (2016)
153. E.J. Yarmchuk, M.J.V. Gordon, R.E. Packard, Observation of stationary vortex arrays in rotating superfluid helium. Phys. Rev. Lett. **43**(3), 214 (1979)
154. A.A. Abrikosov, The magnetic properties of superconducting alloys. J. Phys. Chem. Solids **2**(3), 199 (1957)
155. M. Farnik, U. Henne, B. Samelin, J.P. Toennies, Differences in the detachment of electron bubbles from superfluid He-4 droplets versus nonsuperfluid He-3 droplets. Phys. Rev. Lett. **81**(18), 3892 (1998)
156. A.R.W. McKellar, Y.J. Xu, W. Jager, Spectroscopic exploration of atomic scale superfluidity in doped helium nanoclusters. Phys. Rev. Lett. **97**(18), 183401 (2006)
157. A.R.W. McKellar, Y.J. Xu, W. Jager, Spectroscopic studies of OCS-doped He-4 clusters with 9–72 helium atoms: observation of broad oscillations in the rotational moment of inertia. J. Phys. Chem. A **111**(31), 7329 (2007)

# Chapter 2
# Small Helium Clusters Studied by Coulomb Explosion Imaging

**Maksim Kunitski**

**Abstract** Small helium clusters consisting of two and three helium atoms are unique quantum systems in several aspects. The helium dimer has a single weakly bound state and is of huge spatial extent, such that most of its probability distribution resides outside the potential well in the classically forbidden tunnelling region. The helium trimer possesses only two vibrational states, one of which is of Efimov nature. In this chapter, we discuss application of the Coulomb explosion imaging technique for studying geometries and binding energies of these peculiar two- and three-body quantum systems. Irradiation of a helium cluster by a strong laser field allows tuning interactions between helium atoms. Such ultrashort interaction modification induces response dynamics in a cluster that is observed by combination of the imaging technique with the pump-probe approach.

## 2.1 Introduction

Helium, being the second most abundant element in the universe, is a unique system in terms of its macroscopic and microscopic properties [1]. It is the only known substance that does not have a solid phase at the lowest temperatures under normal pressure. Helium is the only liquid that becomes superfluid in its natural state. This bulk behaviour is partially determined by microscopic properties such as an atomic polarizability, which is exceptionally small for helium. The polarizability is responsible for an extremely weak van der Waals interaction between two helium atoms, which was a reason for a long standing debates about existence of the helium dimer, until it was experimentally observed in the early 1990s [2, 3]. The helium dimer is a very distinctive quantum system, not alike commonly known covalent molecules and other van der Waals clusters. $He_2$ has only one weakly bound state; higher vibrational and even rotational states are not supported by the He-He potential. A tiny binding energy is a reason for huge spatial extent with an average interatomic

M. Kunitski (✉)
Institut für Kernphysik, Geothe-Universität Frankfurt am Main, Max-von-Laue-Straße 1, 60438 Frankfurt am Main, Germany
e-mail: kunitski@atom.uni-frankfurt.de

© The Author(s) 2022                                                                                     41
A. Slenczka and J. P. Toennies (eds.), *Molecules in Superfluid Helium Nanodroplets*,
Topics in Applied Physics 145, https://doi.org/10.1007/978-3-030-94896-2_2

distance of 4.7 nm. Moreover, two helium atoms in the dimer can be most frequently found outside the potential well in the classically forbidden tunnelling region.

The combination of three helium atoms, the helium trimer, is extraordinary as well. Though the ground state of the trimer is spatially more compact than the single state of the helium dimer, it is still exceptionally defuse, such that almost all triangular shapes are equally probable. In addition, the helium trimer forms an Efimov state of extreme spatial extent under natural conditions.

In this chapter we discuss application of the laser-based Coulomb explosion imaging for determination of structures and binding energies of small helium clusters consisting of two and three helium atoms. We will show the images of the Efimov state and discuss how a strong laser field can modify the interaction between helium atoms and how the helium dimer reacts to such modification.

## 2.2  Experimental

Small helium clusters are imaged using the experimental setup that consists of two parts: cluster preparation and their detection (Fig. 2.1). As a cluster source we use the supersonic expansion of helium gas into vacuum (Sect. 2.2.1). Subsequently, the clusters of desired size are selected using mater wave diffraction on a transmission grating. For cluster detection we employ the laser-based Coulomb explosion imaging (Sect. 2.2.2). The momenta of ions after Coulomb explosion are measured by the COLTRIMS technique (Sect. 2.2.3). The initial cluster structures are deduced from the ion momenta using the reconstruction procedure based on the classical simulation of Coulomb explosion (Sect. 2.2.4).

### 2.2.1  Preparation of Small Helium Clusters

Helium clusters are produced by expanding gaseous helium into vacuum through a nozzle with a 5 μm orifice. During such free jet expansion, the cluster formation, i.e. nucleation, is governed by collision processes and depends on the temperature and pressure of the gas prior to expansion [4] (see the Chap. 1 by J. Peter Toennies in this volume). Effective cluster formation with yields up to 6% was found at low nozzle temperatures below 30 K [4]. In our experiments we used temperatures in the range of 8–12 K. The nozzle temperature is stabilized within better than ±0.01 K by a continuous flow cryogenic cryostat (Model RC110 UHV, Cryo Industries of America, Inc.). The pressure dependence of the cluster yield (in $s^{-1}$) at a nozzle temperature of 8 K is shown in Fig. 2.2.

As seen in Fig. 2.2 it is barely possible to find expansion conditions (i.e. back pressure) where clusters of a particular size are formed. Another complication in the cluster experiments is that helium monomers dominate in the molecular beam under all expansion conditions [4]. In order to select a single cluster size from the molecular

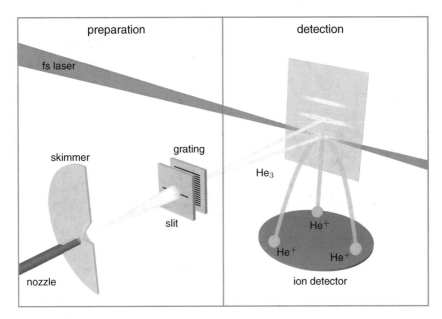

**Fig. 2.1** The experimental setup

beam we make use of matter wave diffraction. The technique was pioneered in the early 90s in the group of Prof. J. P. Toennies in Göttingen and is based on a transmission grating with a period of 100 nm [3, 5]. The clusters in the beam, having the same velocity, can be sorted by mass as their diffraction at the grating depends on the de Broglie wavelength $\lambda_{dB} = \frac{h}{mv}$ ($h$ is the Planck's constant, $m$ and $v$ are the mass and the velocity of a cluster, respectively). Clusters of different size are deflected to different angles that results in spatial separation of clusters at the detection site (Fig. 2.1).

## 2.2.2 Coulomb Explosion Imaging

Coulomb explosion imaging (CEI) was introduced in 1989 [7] as a relatively direct method for determining structure of small molecules (see [8] for the recent review). The main idea is to produce multiple charges in a molecule by, for instance, ionization, and let its constituents fly apart due to the mutual Coulomb repulsion. The momenta or kinetic energy that these charged molecular fragments (e.g. atomic ions) gain during this so-called Coulomb explosion depends on the initial distances between the parts in the neutral molecule. Thus, measuring momenta of charged fragments allows determining structural information about the molecule.

In the poof-of-the-principle experiment [7] the charges in the methane cation were produced by stripping off electrons when the cation was passing at high veloc-

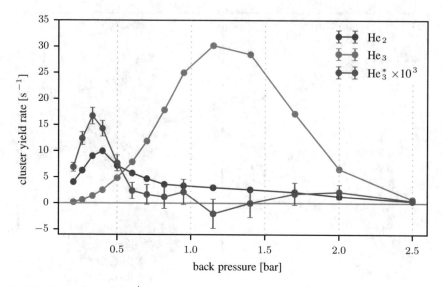

**Fig. 2.2** Dependence of the $^4$He cluster rates on the back pressure at a temperature of 8 K for a nozzle with a 5 μm orifice. The very low rate of the He$_3$ excited state (He$_3^*$, red) is scaled by a factor of $10^3$. The background caused by ground state structures has been subtracted from the excited state rate. The rates for the He$_3$ ground state and He$_2$ are shown in green and blue, respectively. The statistical error bars correspond to the standard deviation. The figure is adapted from [6], Copyright (2016) by American Association for the Advancement of Science

ity through a thin solid film. Alternatively, electrons can be removed from a molecule using single photon ionization with the subsequent multiple Auger-decay [9], collision with a charged projectile [10, 11] or strong field ionization in ultrashort laser pulses [12–15].

Coulomb explosion imaging has been widely used for obtaining structural information about van-der-Waals clusters, wave packet dynamics [16, 17], imaging of excited vibrational states [11] as well as determination of absolute configuration of chiral molecules [18, 19].

Within the classical description of the Coulomb explosion, so-called frozen nuclei reflection approximation (FNRA), the potential energy of $N$ singly charged ions located at distances $R_{ij}$ from each other is converted into a kinetic energy release (KER):

$$\text{KER}_N = \sum_{i \neq j} \frac{1}{R_{ij}}, i, j = 1, 2, .., N. \tag{2.1}$$

If the ionization process is instantaneous, distances $R_{ij}$ correspond to the structure of the neutral cluster. Therefore, by measuring the magnitudes and directions of the momenta (and correspondingly the KER) that the ions acquire during the Coulomb explosion, information on the geometrical structure of the cluster as well as its orientation in space can be obtained. Repeated measurements allow one to reconstruct the quantum mechanical structure distribution of the neutral cluster prior to the application of the ionization laser field. Note, that relation (2.1) is only valid for pure Coulomb repulsion, which in one dimension described by the $1/R$ potential. This is the case for weakly bound van-der-Waals clusters, which atomic orbitals are feebly overlapped.

Although, the FNRA has been successfully applied for obtaining ground state probability distributions of variety of small atomic and molecular clusters [14, 15, 20], including helium clusters [6, 21, 22], it was found to be inaccurate in imaging excited vibrational states of the $H_2^+$ cation [11]. The reason is two-fold: 1) the electronic and nuclear degrees of freedom are not fully decoupled in this system and 2) the local kinetic energy (kinetic energy at a given $R$) can be comparable to the $1/R$ potential energy.

In general, the following conditions should be fulfilled for accurate description of Coulomb explosion imaging by the frozen nuclei reflection approximation:

1. The ionization of the cluster, that transfers the neutral wave packet to the final repulsive ionic state should be faster than possible movements of the wave packet on the intermediate states.
2. The ionization probability should be independent of internuclear distances $R_{ij}$.
3. The recoil of electrons to the residual ion during ionization should be lower than the energy gained on the repulsive potential during Coulomb explosion.
4. There should be only one repulsive potential along which the cluster dissociates during Coulomb explosion.
5. The motion of the wave packet on the repulsive energy potential should be to a large extent "classical".

Coulomb explosion of small helium clusters in general meets all these requirements. The ionization of the cluster in a strong laser field happens sequentially via tunnelling ionization processes [23]:

$$He_N \rightarrow He_N^+ + e^- \rightarrow, ..., \rightarrow He_N^{(N-1)+} + (N-1)e^- \rightarrow He_N^{N+} + Ne^-.$$

Since tunnelling ionization is highly nonlinear process it mainly happens within a short time interval close to the field maximum. This interval is about 20 fs in case of 30 fs laser pulses with a peak intensity of about $10^{16}$ W/cm$^2$. On this time scale the ionization process can be treated as instantaneous, since intermediate ionic potentials of $He_2$ and $He_3$ are rather shallow [24, 25].

In general, the ionization probability of the ionization sequence shown above depends on the internuclear distance $R$. The reason for this is that the $R$-dependence of the ionic potentials determines the vertical ionization potential of the corresponding ionization step. The ionization potential, in turn, governs the ionization probability in a highly nonlinear manner [23]. It turns out, however, that the $R$-dependence is extremely weak in case of high laser intensities ($\sim 10^{16}$ W/cm$^2$) for which the single ionization is saturated.

Upon ionization the electrons "kick" the residual ions providing them with initial momenta prior to Coulomb explosion. According to our measurements, these initial momenta follow a Gaussian distribution centered around zero with a width of about few atomic units and point along the direction of the probe field polarization. In case of the dimer the FNRA, which does not account for the electron "kick", underestimates the probability distribution for large interatomic distances. Using simulation we estimated the corresponding correction and applied it to the FNRA data in order to get the probability distribution of the interatomic distance in the helium dimer (2.5). The recoil of electrons is also taken into account during structure reconstruction of the helium trimer (2.2.4).

### 2.2.3   COLTRIMS

The 3D momenta of the ions acquired during the Coulomb explosion are measured by cold target recoil ion momentum spectroscopy (COLTRIMS) [26]. In the COLTRIMS spectrometer a homogenous electric field (3–4 V/cm) guides ions onto a time and position sensitive micro-channel plate detector with hexagonal delay-line position readout [27] and an active area with a diameter of 80 mm (Fig. 2.1). The detector is typically placed at a distance of 40–50 mm away from the laser focus, which results in a $4\pi$ collection solid angle for atomic ions with an energy of up to 3 eV.

### 2.2.4   Structure Reconstruction from the Momentum Space

COLTRIMS allows to measure ion momenta after Coulomb explosion. In case of the dimer the interatomic distance can be deduced from these momenta using 2.1. In case of the trimer the structural reconstruction is not that straightforward. We devised a look-up table approach in which Coulomb explosion was simulated classically for many different trimer structures [6]. The obtained relations between structures and momenta were saved in a table. In order to reduce the dimensionality of the table we utilized two-dimensional representation proposed by Dalitz [28] for both structures and momenta (Fig. 2.3). Since the Dalitz representation encodes only the shape of a structure, we used the measured KER for deducing the absolute size of the trimer.

The Dalitz coordinates $x$, $y$ for both coordinate and momentum space are defined as follows:

$$x = \frac{\varepsilon_2 - \varepsilon_3}{\sqrt{3}}, y = \varepsilon_1 - \frac{1}{3}$$

**Fig. 2.3** Dalitz plot. Triangles correspond to structures of a trimer (for the coordinate space). Arrows represent the momentum vectors (for the momentum space)

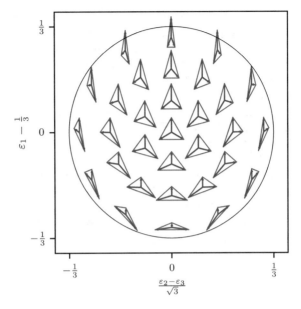

where

$$\varepsilon_i^{\text{coordinate}} = \frac{|\mathbf{r_i}|^2}{\sum\limits_{n=1}^{3} |\mathbf{r_n}|^2} \ , \ \varepsilon_i^{\text{momentum}} = \frac{|\mathbf{p_i}|^2}{\sum\limits_{n=1}^{3} |\mathbf{p_n}|^2}$$

with $\mathbf{r_i}$ being a position vector of $i$th atom of the trimer with respect to the center-of-mass and $\mathbf{p_i}$ being a momentum vector of $i$th atom after Coulomb explosion.

The coordinate and momentum Dalitz's spaces were binned by $1000 \times 1000$. For each bin the corresponding structure in coordinate space was numerically "exploded" several times with different randomly generated small initial momenta using Newton's equations of motion. These initial momenta is a result of the electron recoil during ionization. The three-dimensional distribution of the initial momenta was chosen in the way to match the experimental distribution of the singly charged helium ion.

## 2.3 Helium Dimer

The helium dimer is bound by a potential with a well depth of about 11 K. The well depth almost equals to the zero-point energy of the dimer, which was the reason for long debates about existence of the helium dimer. It turned out that the binding energy is below 2 mK. This tiny binding energy poses a challenge to the theory requiring very high accuracy in calculation of the interatomic potential. The accuracy of ab-initio methods was mediocre until 1990s. In those early days mainly empirical analytical expressions modelling the He-He potential have been proposed. The parameters in such expressions were optimized in order to reproduce the viral coefficients and other thermophysical properties of helium. The binding energies and the expectation values of the interatomic distance for some potentials are collected in Fig. 2.4. In the 1990s the accuracy of ab-initio methods based on the Born-Oppenheimer approximation was improved, which allowed developing better model potentials. One such potential is LM2M2; it has been frequently used for theoretical treatment of small helium clusters. In the 2000s it became possible to improve ab-initio methods further by including many corrections beyond the Born-Oppenheimer approximation. The most recent potential of Szalewicz and co-workers [2, 29, 30] accounts for adiabatic, relativistic, QED corrections as well as retardation. These corrections change the binding energy of the helium dimer by 6–10%. The biggest of these corrections is retardation (about 9%).

The tiny binding energy is responsible for a huge spatial extent of the dimer that spreads far beyond the potential well, such that about 80% of the probability distribution resides in the classically forbidden tunnelling region. Such few body systems have been termed "quantum halos". An expectation value of the interatomic distance of the helium dimer is predicted to be about 47 Å (Fig. 2.4).

Such fragile systems as helium dimer pose challenges also for experimentalists. During preparation of the dimer the temperature of environment should be lower than the binding energy, i.e. as low as 1 mK. Moreover, since the helium dimer does not possess any bound ro-vibrational states, it cannot be detected by standard spectroscopic tools. First experimental evidence for existence of the helium dimer was provided in 1993 by mass-spectrometry [37]. There, the helium gas at room temperature was expanded into the vacuum through the nozzle with an orifice of 150 $\mu$m. Subsequently, the supersonic gas expansion was collimated and ionized by electron impact ionization that allowed to detect $He_2^+$ ions. The main criticism of this experimental concept was that $He_2^+$ ions could originate from fragmentation of larger helium clusters. This issue was addressed by measuring dependence of the ion signal on the nozzle back pressure [37]. The pressure dependence turned out to be quadratic for the $He_2^+$ ion yield, which contradicts the scenario of large clusters fragmentation. Subsequent experiment in 1996 [47] utilized transmission through nanoscale sieves for measuring the mean internuclear distance of the cluster corresponding to the $He_2^+$ ion signal. The obtained value of $62\pm10$ Å was very close to the predicted expectation value of the interatomic distance in the helium dimer at that time (55 Å).

Another and more direct evidence of the existence of the helium dimer was obtained in 1994 using matter wave diffraction [3] in the group of Prof. J. P. Toennies. In experiments the helium beam produced under supersonic expansion was deflected by a tiny transmission grating with a period of 200 nm [48]. The different clusters were deflected at different angles due to the difference in the de Broglie wavelength. Analysis of many diffraction orders of He$_2$ with the theory that took into account

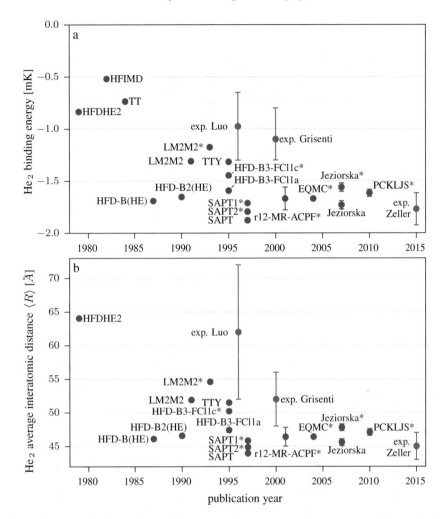

**Fig. 2.4** Predicted and experimental binding energies (**a**) and average interatomic distances (**b**) of the helium dimer. The following He-He potentials have been considered: HFDHE2 [31], HFIMD [32], TT [33], HFD-B(HE) [34],HFD-B2(HE) [35], LM2M2 [36], LM2M2* [37], TTY [38], HFD-B3-FCl1a [39], HFD-B3-FCl1c* [39], SAPT1* [40, 41], SAPT2* [40, 41], SAPT [40, 41], r12-MR-ACPF* [42], EQMC* [43], Jeziorska [44], Jeziorska* [44], PCKLJS* [30, 45]. Star symbol indicates inclusion of correction for retardation. The experimental values are labelled as "exp. Luo" [2], "exp. Grisenti" [46] and "exp. Zeller" [22]. Theoretical and experimental data is shown in blue and red, respectively. Panel (**a**) is adapted from [22], Copyright (2016) by National Academy of Sciences of the United States of America

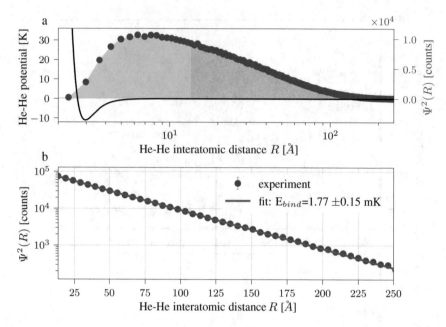

**Fig. 2.5** The measured square of the nuclear wave function of the helium dimer. The distribution is corrected for the electron recoil during ionization. **a** Green and pink colored areas under the distribution visualize the classically allowed and forbidden regions, respectively. The He-He interaction potential is shown in black. Note the logarithmic scale on the x-axis. **b** Fit of the experimental probability distribution by expression (2.2) is shown in red. Note the logarithmic scale on the y-axis. Statistical error bars are smaller than the radius of blue circles. Based partially on results reported in [22]

dispersive interaction with the grating bars allowed to estimate a mean internuclear distance and a binding energy of $52\pm4$ Å and $1.1+0.3/-0.2$ mK, respectively.

Using CEI we measured the square of the nuclear wave function of the helium dimer [22], which is shown in Fig. 2.5. This probability distribution consists of one experimental set, where Coulomb explosion was initiated by a strong laser field ($I_0 \approx 3 \times 10^{15}$ W · cm$^{-2}$, $\lambda = 780$ nm, $\Delta t_{FWHM} \approx 30$ fs). In this respect, the experimental data presented here is different from that in [22], which consists of two experimental sets: the region of short interatomic distances ($R < 14$ Å) were measured using a femtosecond laser, while the range of long interatomic distances ($R \geqslant 14$ Å) were measured at the free-electron laser facility in Hamburg (FLASH). As predicted, the distribution spreads far beyond the classically allowed region (shown in green in Fig. 2.5). The expectation value of the interatomic distance was found to be $45 \pm 2$ Å, which is in line with the most recent theoretical predictions (Fig. 2.4b).

Apart from being a benchmark system for theoretical methods, the helium dimer, as a quantum halo, is a perfect candidate for testing general predictions of quantum mechanics such as the exponential decay of the wave function in the tunnelling region. Indeed, as seen in Fig. 2.5b the decay of the measured squared wave function for $R > 25$ Å resembles closely the exponential one (note the logarithmic scale on the y-axis). Solving the Schrödinger equation for a particle below the barrier, one obtains the following expression for the squared wave function:

$$\Psi^2(R) \propto e^{-\frac{2}{\hbar}\sqrt{2\mu E_{bind}}R}. \tag{2.2}$$

Here $E_{bind}$ is the binding energy of the helium dimer, which corresponds to the barrier height, and $\mu = m_{He}/2$ is the reduced mass of the dimer. Fitting the experimental probability distribution in Fig. 2.5 with expression (2.2), we found the binding energy of the helium dimer to be $1.77 \pm 0.15$ mK. This experimental value is very close to the one obtained in [22] ($1.76 \pm 0.15$ mK) using interatomic distance distribution of the helium dimer measured at FLASH. Both experimental binding energies are in good agreement with the most recent theoretical calculations (Fig. 2.4a).

## 2.4 Helium Trimer

Presently it is known that the helium trimer, $^4He_3$, has two vibrational states: the ground state and the excited one of the Efimov nature [49]. Though the former state was observed experimentally back in 1994 using the matter wave diffraction technique [3], the excited Efimov state remained elusive for longer time and was detected only in 2014 [6], 37 years after its theoretical prediction [50].

In the following the size and structure of both states of $^4He_3$ as well as the single state of $^3He^4He_2$ will be discussed.

### 2.4.1 $^4He_3$: Ground State

The recent theory [51] predicts that the ground state of the helium trimer is more strongly bound ($E_{bind}$=131.84 mK) comparing to the helium dimer and, thus, has a more compact size with an average interatomic distance of 9.53 Å. This is in agreement within error bars with the experimentally obtained value of $11^{+4}_{-5}$ Å [52]. We measured the trimer using the CEI technique, where ionization was performed by a strong 30 fs laser field [21]. The trimer structures were reconstructed from the momentum space using the procedure described in 2.2.4. The obtained interatomic distance distribution (Fig. 2.6a) as well as the distribution of corner angles in a trimer (Fig. 2.6b) resemble closely theoretical ones, calculated with quantum Monte Carlo (QMC) [21, 53] and coupled-channel [6, 54] methods. In the Monte Carlo simulation the TTY helium-helium potential [38] was used, whereas the coupled-channel method utilized the most recent PCKLJS potential [30, 45]. Average interatomic distances estimated from the distributions in Fig. 2.6a are $9.3 \pm 1$ Å (experimental), 9.61 Å (QMC) and 9.53 Å (coupled-channel).

**Fig. 2.6** Interatomic distance and corner angle distributions of the helium trimer, $^4He_3$. The distributions obtained in the QMC (blue) and coupled-channel (green) simulations are almost identical. Based on results reported in [6, 21]

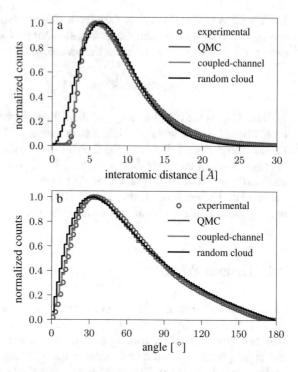

Though many theories agreed on the binding energy and the size of the helium trimer, for long time no consensus was achieved about its shape. The suggested typical structures were ranging from a nearly linear [55–58] to an equilateral triangle [59, 60]. Bressanini and co-workers pointed out that the question of typical structure in case of $^4He_3$ is ill-posed. Considering the two-dimensional angle distributions they found that more or less all structures are equally probable. The fact that some structures seem to be more favoured explained solely by the choice of the visualisation approach.

The idea of the not-well-defined structure of $He_3$ was additionally supported by the random cloud model proposed by Voigtsberger et al. [21]. In the model a trimer was constructed by picking three atoms randomly from the hypothetical three-dimensional cloud, defined by a spherically symmetric atom density. The density was considered to be constant within a sphere of a certain radius and decayed exponentially outside that sphere. The cloud was thus characterized by two parameters: a radius and a decay rate [21]. Here we have simplified the random cloud model by using only an exponential decay of the radial atom density of the cloud defined by the following function:

$$\rho(R) = e^{-aR}. \tag{2.3}$$

Fitting the interatomic distance distribution obtained from this random cloud model to the experimental one, the decay parameter $a = 0.525 \, \text{Å}^{-1}$ was determined. Hence, even a single parameter model is sufficient for fairly good reproduction of interatomic distance and angle distributions of the helium trimer in Fig. 2.6. In the region of interatomic distances $R < 4 \, \text{Å}$ the agreement is rather poor (Fig. 2.6a), since the simple model does not account for repulsion between two helium atoms at short distances.

Another way of visualising the trimer structure is to plot three atoms in the coordinate system defined by two principal axes of inertia, **a** (corresponding to the smallest moment of inertia) and **b** and having the center-of-mass of the cluster at the origin, as proposed by Nielsen et al. [59]. Using this representation (Fig. 2.7) one could argue that the shape of the helium trimer closely resembles the equilateral triangle. However, more or less the same two-dimensional distribution results from the random cloud model.

## 2.4.2  $^4He_3$: Excited Efimov State

In 1970 Vitaly Efimov predicted a peculiar quantum effect in a three-body system consisting of bosons [49]. Namely, at the limit of extremely weak interaction between two bosons, when a single state of a two-body system becomes unbound, infinite number of three-body bound states appear. It turned out, that an effective long range $1/R^2$ potential that arises between three particles under such conditions is responsible for such behavior. The effect is independent of the details of the underlying two-body interactions. In this respect, the Efimov effect is an universal phenomenon, which can be found in different fields of physics such as atomic [6, 61], nuclear [62], condensed matter [63] and high energy physics [64].

Seven years after Efimov's prediction it was suggested that the helium trimer could have an Efimov state [50]. However, not all calculations based on different realistic helium-helium potentials supported this conclusion [39]. Using most recent potential it was shown that $^4He_3$, indeed, has two states [51]. In theory one can investigate how binding energies of both states depend on the scattering length $a$ (relates to the depth of the two-body potential well) by artificially scaling the helium-helium potential. The corresponding so-called Efimov plot consists of three areas (Fig. 2.8). The top violet area ($E > 0$) belongs to three particle continuum. The area to the right of the blue line (binding energy of the dimer) corresponds to the three particle region where two particles are bound forming a dimer and the third one being free. To the left of the dimer binding energy curve lies the area where trimer bound states can exist. Two bound states of the trimer, one of which is the Efimov one (labelled as "1st ES"), correspond to the native scattering length of helium $a = 90.4 \, \text{Å}$. An ideal Efimov case with an infinite number of bound states would be at $a = \infty$. The size and the binding energy of such ideal Efimov states are scaled by factors 22.7 and $22.7^2$, respectively.

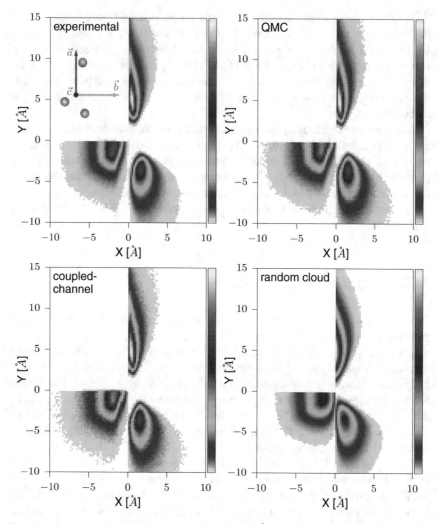

**Fig. 2.7** Structure of the ground state of the helium trimer, $^4He_3$. Three helium atoms are plotted in the coordinate system defined by the principal axes of inertia with the origin in the center-of-mass. Based on results reported in [6, 21]

In 2005 an experimental attempt was undertaken to detect the Efimov state of the helium trimer using matter wave diffraction at a 100 nm period transmission grating [52]. In the experiment the grating was rotated by an angle of 21°, which allowed to almost halve the slit width and, consequently, increase sensitivity of the method for the trimer detection. Since the effective slit width, which determines the diffraction pattern, depends furthermore on the cluster dimension, the authors were able to estimate the average trimer size (see Sect. 2.4.1). Comparing this value to the theoretical one it was concluded that the contribution of the large Efimov state was

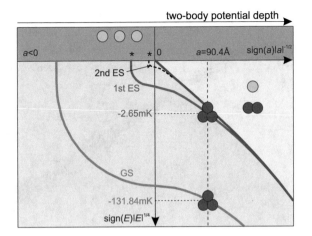

**Fig. 2.8** Theoretical dependence of the binding energy $E$ of two $^4$He$_3$ states: the ground state (GS) and the first excited state (1st ES) on the two-body scattering length $a$. The vertical dashed line corresponds to the naturally occurring $^4$He$_3$ with a scattering length of 90.4 Å. The area for $E > 0$ is the unbound three-particle continuum. The blue line corresponds to the binding energy of the helium dimer. To the right of this line the dimer-atom region that describes the dimer and a separate helium atom. To the left of the blue line the area where three helium atoms are bound. The figure is adapted from [6], Copyright (2016) by American Association for the Advancement of Science

below experimental sensitivity (6%). According to the theory of cluster formation, the concentration of the Efimov state under experimental conditions was estimated to be about 10%. The fact that the Efimov state was not detected led to doubting its existence.

One can see from Fig. 2.2 that the relative yield of the Efimov state (He$_3^*$, in red) with respect to the ground state (He$_3$, in green) is well below 1% under expansion conditions that are optimized for the ground state yield ($p \approx 1.2$ bar). This might be an explanation why the Efimov state was not detected in the experiment of Brühl et al. [52].

In 2014 we successfully observed the Efimov state of the helium trimer using laser-based CEI [6]. We used two approaches to reconstruct the pair distance distribution from the measured ion momenta after Coulomb explosion. In the first approach the filter in the momentum space was applied to cut the contribution of the ground state (Fig. 2.9 in black, for details see [6]), which was dominant in the molecular beam under all expansion conditions (Fig. 2.2). In the second reconstruction approach the ground state pair distance distribution was subtructed from that of the mixture of the ground and Efimov states (Fig. 2.9 in red). Both experimental distributions match the theoretical one (Fig. 2.9 in violet) very well for large interatomic distances ($R > 100$ Å). At small distances the resemblance is poor due to the remaining contamination of the ground state.

As seen from Fig. 2.9 the Efimov state of $^4$He$_3$ has a huge spatial extent, spreading beyond 300 Å. It thus also belongs to the family of quantum halos, i.e. most of

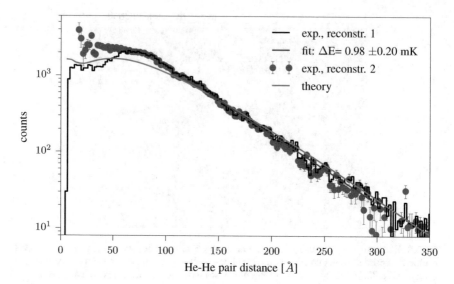

**Fig. 2.9** Pair distance distribution of the Efimov state of the helium trimer, $^4\text{He}_3$. Note logarithmic scale of the y axis. Two experimental distributions (labelled as "reconstr. 1" and "reconstr. 2") are obtained using different reconstruction approaches, see text for details. The fitted exponential decay according to expression (2.2) is shown in blue. Statistical error bars correspond to the standard deviation. The figure is adapted from [6], Copyright (2016) by American Association for the Advancement of Science

its probability distribution resides well outside the potential well in the classically forbidden region. Therefore the asymptotic part of the pair distance distribution shows an exponential decay, likewise the wave function of the helium dimer. Similarly, the expression (2.2) can be used for obtaining the binding energy of the Efimov state. In this case, however, the binding energy is defined not with respect to the dissociation continuum of three atoms, but relative to the binding energy of the helium dimer [51]. From the fit (Fig. 2.9, blue) this partial binding energy $\Delta E$ was found to be $0.98 \pm 0.20$ mK. Given the dimer binding energy of $1.77 \pm 0.15$ mK, the binding energy of the Efimov state with respect to the three-body dissociation continuum is estimated to be $2.75 \pm 0.25$ mK. This experimental value is in a good agreement with the theoretical value of 2.65 mK [6, 51].

Fig. 2.10 presents the first experimental image of the Efimov state [6]. For this representation the same coordinate system as in Fig. 2.7, namely, based on the principal axes of inertia, was used. The structure of the Efimov state looks substantially different to that of the ground state in Fig. 2.7. Whereas in the ground state all structures are equally probable (see Sect. 2.4.1), the excited Efimov state is dominated by configurations in which two atoms are close to each other with the third one being farther away.

Further insights in the structure of the Efimov state can be gained by considering the distribution of the shortest interatomic distance (Fig. 2.11). This distribution resembles very closely the interatomic distance distribution of the helium dimer

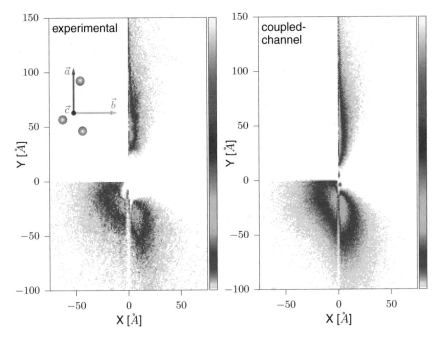

**Fig. 2.10** Structure of the Efimov state of the helium trimer, $^4He_3$. Three helium atoms are plotted in the coordinate system defined by the principal axes of inertia with the origin in the center-of-mass. The figure is adapted from [6], Copyright (2016) by American Association for the Advancement of Science

(Fig. 2.5), implying that the Efimov state consists of a dimer to which the third helium atom is weakly attached and located at even larger distance than two atoms in the dimer. This shape is justified by the position of the Efimov state in the Efimov plot (Fig. 2.8). Namely, it is located very close to the dimer-atom region.

### 2.4.3  $^3He^4He_2$

Substitution of one $^4He$ atom in the trimer by the lighter $^3He$ isotop decreases the binding energy of the ground state by about 8 times due to the increase in the zero-point vibrational energy [59, 65, 66]. Only one state in such heterotrimer remains bound [59]. It is so-called Tango state [62, 67], where only one pair of atoms out of three can form the two-body bound state. The existence of the cluster was confirmed in experiments with matter wave diffraction [68].

Voigtsberger et al. [21] measured the $^3He^4He_2$ trimer using laser-based CEI. The pair distance distributions are in fairly good agreement with the QMC theory by Dario Bressanini [21, 69]. It was found that the $^3He$-$^4He$ distances are longer than $^4He$-$^4He$ ones, and that size of the heterotrimer is larger than that of the $^4He_3$ ground state as

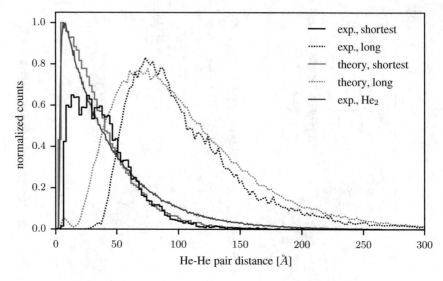

**Fig. 2.11** Distributions corresponding to the shortest and two longest (labelled as "long") inter-atomic pair distances of the Efimov state of $^4$He$_3$. The experimental distribution of He$_2$ (from Fig. 2.5) is shown in blue

expected from the lower binding energy (Fig. 2.12a). Along with angle distributions (Fig. 2.12b) one can conclude that $^3$He$^4$He$_2$ exists as an acute triangle having $^3$He atom at the corner with a smallest angle. As was however pointed out [69], the trimer is very defuse to talk about the well defined structure. The same conclusion can be drawn by considering plots in the coordinate system defined by the principal axes of inertia (Fig. 2.13).

## 2.5 Field-Induced Dynamics in the Helium Dimer

So far we have considered steady state structures and binding energies of small helium clusters consisting of two and three atoms. Ability of controlling interaction between helium atoms in a cluster would open up a door to series of experiments, where not only new exotic states can be created but also response dynamics of these unique quantum objects to an external disturbance can be investigated. In case of ultracold atomic gases such interaction control is achieved in the vicinity of Feshbach resonances through an application of a magnetic field [70, 71]. This is however not suitable for non-magnetic helium atoms. Nielsen and co-workers [72] suggested to use an external electric field to tune interaction between helium atoms and predicted appearance of bound states in the naturally unbound $^3$He$^4$He and $^4$He$^3$He$_2$. In addition, it has been suggested to use intense laser fields to modify rovibrational states of weakly bound molecules [73–76] as well as to turn the helium

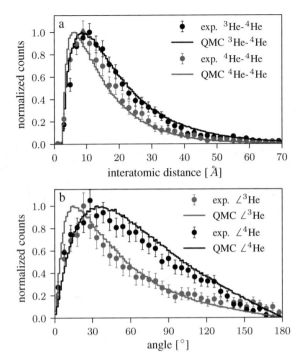

**Fig. 2.12** Interatomic distance and corner angle distributions of the helium heterotrimer, $^3$He$^4$He$_2$. Based on results reported in [21]

dimer into a "covalent"-like molecule that supports many bound states [75, 77]. These phenomena are unexplored experimentally due to challenges in preparation and detection of such fragile quantum states.

Another interesting aspect of a laser field interaction with molecules is a non-adiabatic alignment of molecules in space (see the Chap. 9 by Nielsen et al. in this volume). How this interaction manifests itself in quantum halos that do not support any bound rotational states? Would one expect to see angular anisotropy? Would the spatially extended system reacts as whole to a rotational "kick"? In order to answer these questions we applied a 310 fs laser pulse ($1.3 \times 10^{14}$ W $\cdot$ cm$^{-2}$) to the helium dimer and watched its response using CEI initiated by the delayed probe pulse (30 fs, ca. $10^{16}$ W $\cdot$ cm$^{-2}$) [78]. The pump laser pulse induces dipoles on helium atoms changing the native interaction potential. The overall potential becomes anisotropic: repulsive perpendicular to the laser polarization direction and about 3 times more attractive along the polarization direction than the native potential (Fig. 2.14). Note that the laser field modifies the interaction potential only in a very small region, namely, in the vicinity of the potential well ($R < 10$ Å).

The response of the helium dimer to the pump pulse is shown in Fig. 2.15 in terms of alignment parameter $\langle \cos^2\theta \rangle$. Prior to the pump pulse the angular distribution of the dimer axis in space is isotropic for all interatomic distances ($\langle \cos^2\theta \rangle = 1/3$).

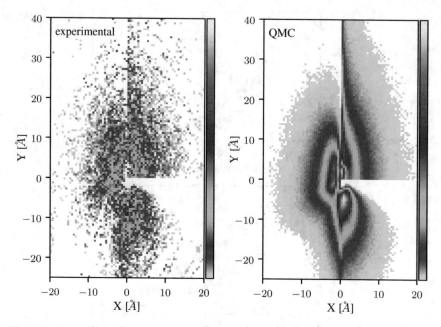

**Fig. 2.13** Structure distribution of the helium heterotrimer, $^3\text{He}^4\text{He}_2$. The coordinate system is defined by the principal axes of inertia with the origin in the center-of-mass

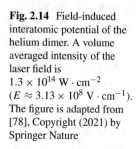

**Fig. 2.14** Field-induced interatomic potential of the helium dimer. A volume averaged intensity of the laser field is $1.3 \times 10^{14}\,\text{W} \cdot \text{cm}^{-2}$ ($E \approx 3.13 \times 10^8\,\text{V} \cdot \text{cm}^{-1}$). The figure is adapted from [78], Copyright (2021) by Springer Nature

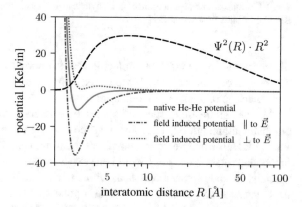

Right after arrival of the pump pulse the positive alignment ($\langle\cos^2\theta\rangle > 1/3$) emerges at short interatomic distances. Subsequently the alignment wave moves to larger interatomic distances and gets broader. This response is very different to what is known from typical alignment experiments, namely, periodic in time alignment signal, called recurrences [79]. The experimental observation is accurately reproduced by the parameter-free quantum simulation [78].

The field-induced dynamics of He$_2$ can also be visualised by changes in the probability density. These changes are shown in Fig. 2.16 for two orientations of the

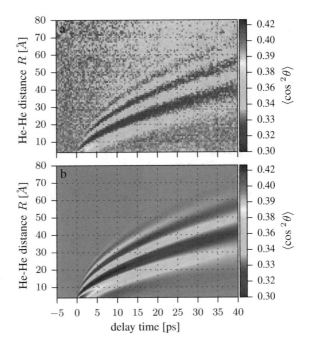

**Fig. 2.15** Temporal evolution of the field-induced alignment of $^4$He$_2$. **a** experiment, **b** theory. Expectation value of $\cos^2\theta$ is shown in color. $\theta$ is an angle between the dimer axis and the direction of laser polarization. The figure is adapted from [78], Copyright (2021) by Springer Nature

dimer axis with respect to the polarization direction of the laser field: parallel and perpendicular. Changes in the probability density resemble the outgoing alignment wave in Fig. 2.15. Remarkably, not only magnitude, but also the phase of the density wave is resolved. According to quantum mechanics, the phase of the wave packet is only accessible through interference with another wave packet or wave function. Note that two wave packets, propagating along the laser polarization (Fig. 2.16a) and perpendicular to it (Fig. 2.16b) are out of phase.

The picosecond response of the single state He$_2$ to an intense laser field (Figs. 2.15 and 2.16) can be understood as follows. Initially the laser field induces an isotropic effective potential in the short interatomic region of the dimer (Fig. 2.14). This potential results in transfer of some part of the ground state wave function at these distances to the unbound $J = 2$ rotational state, creating a dissociating wave packet. Subsequently, the wave packet moves along the repulsive $J = 2$ potential towards larger interatomic distances and spreads with time due to intrinsic dispersion of the matter wave. Its interference with a huge isotropic ground state wave function allows to measure the quantum phase of the wave packet. At large interatomic distances the $J = 2$ potential is very flat, implying that the dissociating wave packet moves without influence of any force. This is confirmed by the temporal evolution of the semi-classical phase of a free particle with a reduced mass of 2 amu (Fig. 2.16b, green dashed lines). Thus, the experiment shows not only how a halo state reacts to the non-adiabatic tuning of the two-body interaction, but also visualizes propagation of a freely moving quantum particle in space.

**Fig. 2.16** Temporal evolution of the field-induced changes in the probability distribution of $^4He_2$. **a** Dimer axis is within $\pm 40°$ to the laser polarization direction. **b** Dimer axis is within $90 \pm 40°$ to the laser polarization direction. The green dashed lines in panel b show the calculated constant-phase evolution of a free particle with a reduced mass of 2 amu. The figure is adapted from [78], Copyright (2021) by Springer Nature

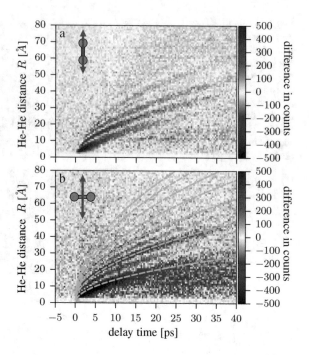

## 2.6 Conclusions

Coulomb explosion imaging is a powerful tool for retrieving probability distributions and binding energies of small helium clusters. Extension of the method towards larger clusters is feasible though the structure reconstruction seems to be rather challenging. A cluster consisting of four helium atoms, the helium tetramer, is particular interesting for the following reason. It has been shown [80, 81] that four-body systems with a huge two-body scattering length also show universal behaviour and have states that are connected to Efimov states in a trimer.

Combination of Coulomb explosion imaging with a control of interaction between helium atoms in a pump-probe manner, as demonstrated using the helium dimer, paving the way for studying field-induced dynamics in these peculiar quantum systems. One might envision application of such technique to the helium trimer in order to explore the birth and decay of an Efimov state in time.

**Acknowledgements** I would like to acknowledge Dario Bressanini for providing me with structures of $^4He_3$ and $^3He^4He_2$ trimers obtained from QMC simulations. I am very thankful to Dörte Blume for many years of theoretical support and successful collaboration. I would like to acknowledge Jörg Voigtsberger and Stefan Zeller for building the experimental setup and performing some experiments, presented in this chapter. I am very grateful to Reinhard Dörner for many years of extensive support, bright ideas and fruitful discussions. Moreover, I am thankful to Reinhard for proofreading. I am very much obliged to Till Jahnke, who together with Reinhard introduced me to the fascinating world of helium clusters. The experiments presented here would not be possible without high competence of Lothar Ph. H. Schmidt, Markus S. Schöffler, Achim Czasch and Anton Kalinin. I am thankful to the rest of the Frankfurt team for their permanent willingness to help and for the positive climate in the group.

# References

1. J.P. Toennies, Mol. Phys. **111**(12–13), 1879 (2013). https://doi.org/10.1080/00268976.2013.802039
2. F. Luo, G.C. McBane, G. Kim, C.F. Giese, W.R. Gentry, J. Chem. Phys. **98**(4), 3564 (1993). https://doi.org/10.1063/1.464079
3. W. Schöllkopf, J.P. Toennies, Science **266**(5189), 1345 (1994). https://doi.org/10.1126/science.266.5189.1345
4. L.W. Bruch, W. Schöllkopf, J.P. Toennies, J. Chem. Phys. **117**(4), 1544 (2002). https://doi.org/10.1063/1.1486442
5. T.A. Savas, S.N. Shah, M.L. Schattenburg, J.M. Carter, H.I. Smith, J. Vac. Sci. Technol. B **13**(6), 2732 (1995). https://doi.org/10.1116/1.588255
6. M. Kunitski, S. Zeller, J. Voigtsberger, A. Kalinin, L.P.H. Schmidt, M. Schöffler, A. Czasch, W. Schöllkopf, R.E. Grisenti, T. Jahnke, D. Blume, R. Dörner, Science **348**(6234), 551 (2015). https://doi.org/10.1126/science.aaa5601
7. Z. Vager, R. Naaman, E.P. Kanter, Science **244**(4903), 426 (1989). https://doi.org/10.1126/science.244.4903.426
8. T. Yatsuhashi, N. Nakashima, J. Photochem. Photobiol. C Photochem. Rev. **34**, 52 (2018). https://doi.org/10.1016/j.jphotochemrev.2017.12.001
9. M. Pitzer, G. Kastirke, M. Kunitski, T. Jahnke, T. Bauer, C. Goihl, F. Trinter, C. Schober, K. Henrichs, J. Becht, S. Zeller, H. Gassert, M. Waitz, A. Kuhlins, H. Sann, F. Sturm, F. Wiegandt, R. Wallauer, L.P.H. Schmidt, A.S. Johnson, M. Mazenauer, B. Spenger, S. Marquardt, S. Marquardt, H. Schmidt-Böcking, J. Stohner, R. Dörner, M. Schöffler, R. Berger, ChemPhysChem **17**(16), 2465 (2016). https://doi.org/10.1002/cphc.201501118
10. R.M. Wood, A.K. Edwards, M.F. Steuer, Phys. Rev. A **15**(4), 1433 (1977). https://doi.org/10.1103/PhysRevA.15.1433
11. L.P.H. Schmidt, T. Jahnke, A. Czasch, M. Schöffler, H. Schmidt-Böcking, R. Dörner, Phys. Rev. Lett. **108**(7), 073202 (2012). https://doi.org/10.1103/PhysRevLett.108.073202
12. M. Schmidt, D. Normand, C. Cornaggia, Phys. Rev. A **50**(6), 5037 (1994). https://doi.org/10.1103/PhysRevA.50.5037
13. S. Chelkowski, A.D. Bandrauk, J. Phys. B Atomic Mol. Opt. Phys. **28**(23), L723 (1995)
14. B. Ulrich, A. Vredenborg, A. Malakzadeh, L.P.H. Schmidt, T. Havermeier, M. Meckel, K. Cole, M. Smolarski, Z. Chang, T. Jahnke, R. Dörner, J. Phys. Chem. A **115**(25), 6936 (2011). https://doi.org/10.1021/jp1121245
15. J. Wu, M. Kunitski, L.P.H. Schmidt, T. Jahnke, R. Dörner, J. Chem. Phys. **137**(10), 104308 (2012). https://doi.org/10.1063/1.4750980

16. H. Stapelfeldt, E. Constant, P.B. Corkum, Phys. Rev. Lett. **74**(19), 3780 (1995). https://doi.org/10.1103/PhysRevLett.74.3780

17. H. Stapelfeldt, E. Constant, H. Sakai, P.B. Corkum, Phys. Rev. A **58**(1), 426 (1998). https://doi.org/10.1103/PhysRevA.58.426

18. M. Pitzer, M. Kunitski, A.S. Johnson, T. Jahnke, H. Sann, F. Sturm, L.P.H. Schmidt, H. Schmidt-Böcking, R. Dörner, J. Stohner, J. Kiedrowski, M. Reggelin, S. Marquardt, A. Schießer, R. Berger, M.S. Schöffler, Science **341**(6150), 1096 (2013). https://doi.org/10.1126/science.1240362

19. P. Herwig, K. Zawatzky, M. Grieser, O. Heber, B. Jordon-Thaden, C. Krantz, O. Novotný, R. Repnow, V. Schurig, D. Schwalm, Z. Vager, A. Wolf, O. Trapp, H. Kreckel, Science **342**(6162), 1084 (2013). https://doi.org/10.1126/science.1246549

20. A. Khan, T. Jahnke, S. Zeller, F. Trinter, M. Schöffler, L.P.H. Schmidt, R. Dörner, M. Kunitski, J. Phys. Chem. Lett. **11**(7), 2457 (2020). https://doi.org/10.1021/acs.jpclett.0c00702

21. J. Voigtsberger, S. Zeller, J. Becht, N. Neumann, F. Sturm, H.K. Kim, M. Waitz, F. Trinter, M. Kunitski, A. Kalinin, J. Wu, W. Schöllkopf, D. Bressanini, A. Czasch, J.B. Williams, K. Ullmann-Pfleger, L.P.H. Schmidt, M.S. Schöffler, R.E. Grisenti, T. Jahnke, R. Dörner, Nat. Commun. **5**, 5765 (2014)

22. S. Zeller, M. Kunitski, J. Voigtsberger, A. Kalinin, A. Schottelius, C. Schober, M. Waitz, H. Sann, A. Hartung, T. Bauer, M. Pitzer, F. Trinter, C. Goihl, C. Janke, M. Richter, G. Kastirke, M. Weller, A. Czasch, M. Kitzler, M. Braune, R.E. Grisenti, W. Schöllkopf, L.P.H. Schmidt, M.S. Schöffler, J.B. Williams, T. Jahnke, R. Dörner, Proc. Natl. Acad. Sci. **113**(51), 14651 (2016). https://doi.org/10.1073/pnas.1610688113

23. V.S. Popov, Physics-Uspekhi **47**(9), 855 (2004)

24. E. Scifoni, F. Gianturco, Eur. Phys. J. D Atomic Mol. Opt. Plasma Phys. **21**(3), 323 (2002). https://doi.org/10.1140/epjd/e2002-00211-3

25. J.C. Xie, S.K. Mishra, T. Kar, R.H. Xie, Chem. Phys. Lett. **605–606**, 137 (2014). https://doi.org/10.1016/j.cplett.2014.05.021

26. J. Ullrich, R. Moshammer, A. Dorn, R. Dörner, L.P.H. Schmidt, H. Schmidt-Böcking, Rep. Prog. Phys. **66**(9), 1463 (2003)

27. O. Jagutzki, A. Cerezo, A. Czasch, R. Dorner, M. Hattas, M. Huang, V. Mergel, U. Spillmann, K. Ullmann-Pfleger, T. Weber, H. Schmidt-Bocking, G. Smith, IEEE Trans. Nucl. Sci. **49**(5), 2477 (2002). https://doi.org/10.1109/TNS.2002.803889

28. R. Dalitz, Lond. Edinb. Dublin Philos. Mag. J. Sci. **44**(357), 1068 (1953). https://doi.org/10.1080/14786441008520365

29. M. Przybytek, W. Cencek, J. Komasa, G. Lach, B. Jeziorski, K. Szalewicz, Phys. Rev. Lett. **104**(18), 183003 (2010). https://doi.org/10.1103/PhysRevLett.104.183003

30. W. Cencek, M. Przybytek, J. Komasa, J.B. Mehl, B. Jeziorski, K. Szalewicz, J. Chem. Phys. **136**(22), 224303 (2012). https://doi.org/10.1063/1.4712218

31. R.A. Aziz, V.P.S. Nain, J.S. Carley, W.L. Taylor, G.T. McConville, J. Chem. Phys. **70**(9), 4330 (1979). https://doi.org/10.1063/1.438007

32. R. Feltgen, H. Kirst, K.A. Köhler, H. Pauly, F. Torello, J. Chem. Phys. **76**(5), 2360 (1982). https://doi.org/10.1063/1.443264

33. K.T. Tang, J.P. Toennies, J. Chem. Phys. **80**(8), 3726 (1984). https://doi.org/10.1063/1.447150

34. R.A. Aziz, F.R. McCourt, C.C. Wong, Mol. Phys. **61**(6), 1487 (1987). https://doi.org/10.1080/00268978700101941

35. R.A. Aziz, M.J. Slaman, Metrologia **27**(4), 211 (1990). https://doi.org/10.1088/0026-1394/27/4/005

36. R.A. Aziz, M.J. Slaman, J. Chem. Phys. **94**(12), 8047 (1991). https://doi.org/10.1063/1.460139

37. F. Luo, G. Kim, G.C. McBane, C.F. Giese, W.R. Gentry, J. Chem. Phys. **98**(12), 9687 (1993). https://doi.org/10.1063/1.464347

38. K.T. Tang, J.P. Toennies, C.L. Yiu, Phys. Rev. Lett. **74**(9), 1546 (1995). https://doi.org/10.1103/PhysRevLett.74.1546

39. A.R. Janzen, R.A. Aziz, J. Chem. Phys. **103**(22), 9626 (1995). https://doi.org/10.1063/1.469978

40. T. Korona, H.L. Williams, R. Bukowski, B. Jeziorski, K. Szalewicz, J. Chem. Phys. **106**(12), 5109 (1997). https://doi.org/10.1063/1.473556
41. A.R. Janzen, R.A. Aziz, J. Chem. Phys. **107**(3), 914 (1997). https://doi.org/10.1063/1.474444
42. R.J. Gdanitz, Mol. Phys. **99**(11), 923 (2001). https://doi.org/10.1080/00268970010020609
43. J.B. Anderson, J. Chem. Phys. **120**(20), 9886 (2004). https://doi.org/10.1063/1.1704638
44. M. Jeziorska, W. Cencek, K. Patkowski, B. Jeziorski, K. Szalewicz, J. Chem. Phys. **127**(12), 124303 (2007). https://doi.org/10.1063/1.2770721
45. M. Przybytek, W. Cencek, J. Komasa, G. Łach, B. Jeziorski, K. Szalewicz, Phys. Rev. Lett. **104**(18), 183003 (2010). https://doi.org/10.1103/PhysRevLett.104.183003
46. R.E. Grisenti, W. Schöllkopf, J.P. Toennies, G.C. Hegerfeldt, T. Köhler, M. Stoll, Phys. Rev. Lett. **85**(11), 2284 (2000). https://doi.org/10.1103/PhysRevLett.85.2284
47. F. Luo, C.F. Giese, W.R. Gentry, J. Chem. Phys. **104**(3), 1151 (1996). https://doi.org/10.1063/1.470771
48. D.W. Keith, M.L. Schattenburg, H.I. Smith, D.E. Pritchard, Phys. Rev. Lett. **61**(14), 1580 (1988). https://doi.org/10.1103/PhysRevLett.61.1580
49. V. Efimov, Phys. Lett. B **33**(8), 563 (1970). https://doi.org/10.1016/0370-2693(70)90349-7
50. T.K. Lim, S.K. Duffy, William C. Damert, Phys. Rev. Lett. **38**(7), 341 (1977). 10.1103/PhysRevLett.38.341
51. E. Hiyama, M. Kamimura, Phys. Rev. A **85**(6), 062505 (2012). https://doi.org/10.1103/PhysRevA.85.062505
52. R. Brühl, A. Kalinin, O. Kornilov, J.P. Toennies, G.C. Hegerfeldt, M. Stoll, Phys. Rev. Lett. **95**(6), 063002 (2005). https://doi.org/10.1103/PhysRevLett.95.063002
53. D. Bressanini, G. Morosi, J. Phys. Chem. A **115**(40), 10880 (2011). https://doi.org/10.1021/jp206612j
54. D. Blume, C.H. Greene, B.D. Esry, J. Chem. Phys. **113**(6), 2145 (2000). https://doi.org/10.1063/1.482027
55. M. Lewerenz, J. Chem. Phys. **106**(11), 4596 (1997). https://doi.org/10.1063/1.473501
56. T. González-Lezana, J. Rubayo-Soneira, S. Miret-Artés, F.A. Gianturco, G. Delgado-Barrio, P. Villarreal, J. Chem. Phys. **110**(18), 9000 (1999). https://doi.org/10.1063/1.478819
57. T. González-Lezana, J. Rubayo-Soneira, S. Miret-Artés, F.A. Gianturco, G. Delgado-Barrio, P. Villarreal, Phys. Rev. Lett. **82**(8), 1648 (1999). https://doi.org/10.1103/PhysRevLett.82.1648
58. D. Bressanini, M. Zavaglia, M. Mella, G. Morosi, J. Chem. Phys. **112**(2), 717 (2000). https://doi.org/10.1063/1.480604
59. E. Nielsen, D.V. Fedorov, A.S. Jensen, J. Phys. B Atomic Mol. Opt. Phys. **31**(18), 4085 (1998)
60. P. Barletta, A. Kievsky, Phys. Rev. A **64**(4), 042514 (2001). https://doi.org/10.1103/PhysRevA.64.042514
61. E. Braaten, H.W. Hammer, Ann. Phys. **322**(1), 120 (2007). https://doi.org/10.1016/j.aop.2006.10.011
62. A.S. Jensen, K. Riisager, D.V. Fedorov, E. Garrido, Rev. Mod. Phys. **76**(1), 215 (2004). https://doi.org/10.1103/RevModPhys.76.215
63. Y. Nishida, Y. Kato, C.D. Batista, Nat. Phys. **9**(2), 93 (2013). https://doi.org/10.1038/nphys2523
64. H.W. Hammer, L. Platter, Ann. Rev. Nucl. Part. Sci. **60**(1), 207 (2010). https://doi.org/10.1146/annurev.nucl.012809.104439
65. R. Guardiola, J. Navarro, Phys. Rev. A **68**(5), 055201 (2003). https://doi.org/10.1103/PhysRevA.68.055201
66. D. Bressanini, G. Morosi, Few-Body Syst. **34**(1), 131 (2004). https://doi.org/10.1007/s00601-004-0022-x
67. F. Robicheaux, Phys. Rev. A **60**(2), 1706 (1999). https://doi.org/10.1103/PhysRevA.60.1706
68. A. Kalinin, O. Kornilov, W. Schöllkopf, J.P. Toennies, Phys. Rev. Lett. **95**(11), 113402 (2005). https://doi.org/10.1103/PhysRevLett.95.113402
69. D. Bressanini, J. Phys. Chem. A **118**(33), 6521 (2014). https://doi.org/10.1021/jp503090f
70. C. Chin, R. Grimm, P. Julienne, E. Tiesinga, Rev. Mod. Phys. **82**(2), 1225 (2010). https://doi.org/10.1103/RevModPhys.82.1225

71. S. Kotochigova, Rep. Prog. Phys. **77**(9), 093901 (2014)
72. E. Nielsen, D.V. Fedorov, A.S. Jensen, Phys. Rev. Lett. **82**(14), 2844 (1999). https://doi.org/10.1103/PhysRevLett.82.2844
73. B. Friedrich, M. Gupta, D. Herschbach, Collect. Czech. Chem. Commun. **63**, 1089 (1998)
74. M. Lemeshko, B. Friedrich, Phys. Rev. Lett. **103**(5), 053003 (2009). https://doi.org/10.1103/PhysRevLett.103.053003
75. Q. Wei, S. Kais, T. Yasuike, D. Herschbach, Proc. Natl. Acad. Sci. **115**(39), E9058 (2018). https://doi.org/10.1073/pnas.1810102115
76. Q. Guan, D. Blume, Phys. Rev. A **99**(3), 033416 (2019). https://doi.org/10.1103/PhysRevA.99.033416
77. P. Balanarayan, N. Moiseyev, Phys. Rev. A **85**(3), 032516 (2012). https://doi.org/10.1103/PhysRevA.85.032516
78. M. Kunitski, Q. Guan, H. Maschkiwitz, J. Hahnenbruch, S. Eckart, S. Zeller, A. Kalinin, M. Schöffler, L.P.H. Schmidt, T. Jahnke, D. Blume, R. Dörner, Nat. Phys. **17**(2), 174 (2021). https://doi.org/10.1038/s41567-020-01081-3
79. H. Stapelfeldt, T. Seideman, Rev. Mod. Phys. **75**(2), 543 (2003). https://doi.org/10.1103/RevModPhys.75.543
80. H.W. Hammer, L. Platter, Eur. Phys. J. A **32**(1), 113 (2007). https://doi.org/10.1140/epja/i2006-10301-8
81. J. von Stecher, J.P. D/'Incao, C.H. Greene, Nat. Phys. **5**(6), 417 (2009). https://doi.org/10.1038/nphys1253

# Chapter 3
# Helium Droplet Mass Spectrometry

**Arne Schiller, Felix Laimer, and Lukas Tiefenthaler**

**Abstract** Mass spectrometry is of paramount importance in many studies of pristine and doped helium droplets. Here, we attempt to review the body of work that has been performed in this field. Special focus is given to experiments conducted by the group of Paul Scheier at the University of Innsbruck. We specifically highlight recent studies of highly charged helium droplets and the successive development of pickup into highly charged and mass selected droplets.

## 3.1 Foreword and Introduction

Mass spectrometry (MS) has been an invaluable tool in various fields of scientific research since the beginning of the 20th century. Many ground-breaking advances in atomic and molecular physics, chemistry and biology would have been impossible without this technique—research of helium nanodroplets (HNDs) being no exception. This chapter aims to give the reader an overview of mass spectrometric research utilizing HNDs (HND MS). We shall assume that the reader is familiar with, and therefore we will not discuss the basic principles, techniques and instrumentation of MS. We would, however, like to briefly discuss some important concepts for the interpretation of HND mass spectra that might not be as widely known—the experienced reader familiar with cluster and HND MS is encouraged to skip the following paragraph.

The size distribution of cluster ions as observed in a mass spectrum (i.e. the ion yield as a function of $m/z$) is influenced by various factors, such as the size distribution of the neutral precursor clusters, ionization cross section, ion transmission efficiency of the ion optics and the MS as well as the detector efficiency. While these factors are often difficult to determine, they can be expected to vary rather smoothly over the mass range. Hence, abrupt intensity variations of neighboring cluster ion peaks are unlikely caused by the experimental setup, but rather the intrinsic cluster properties

A. Schiller (✉) · F. Laimer · L. Tiefenthaler
Institut für Ionenphysik und Angewandte Physik, Universität Innsbruck, Technikerstr. 25, A-6020 Innsbruck, Austria
e-mail: Arne.Schiller@uibk.ac.at

© The Author(s) 2022
A. Slenczka and J. P. Toennies (eds.), *Molecules in Superfluid Helium Nanodroplets*,
Topics in Applied Physics 145, https://doi.org/10.1007/978-3-030-94896-2_3

which are known to be able to vary dramatically with the addition or removal of a single atom or molecule [1]. Cluster ion peaks with anomalously increased intensity compared to neighboring peaks are referred to as "magic number" clusters while those with anomalously decreased intensity are called "antimagic" [2–4], in analogy to the concept in nuclear physics [5–7]. The occurrence of magic cluster ions generally requires both a particularly stable cluster structure (neutral or ionic) and an energetic process inducing fragmentation of the emanating clusters or cluster ions. In principle, the variations could be produced by fragmentation during the dopant cluster formation process, however, HNDs are excellent at stabilizing even weakly bound structures via evaporative cooling. Hence, a smooth neutral cluster size distribution can be expected and fragmentation will typically be caused by excess energy transferred to the nascent cluster ion during the ionization process. This results in the depletion of weakly bound clusters and the corresponding relative enrichment of magic number cluster ions [8]. Many of the observed magic numbers are linked to shell closures [1] of geometric [9–13] or electronic nature, the latter being described by the jellium model [3, 14, 15]. Magic numbers are not only found in homogenous clusters, but also in progressions of heterogenous complexes where ions are solvated by different numbers of ligands such as rare gas atoms or small molecules. Again, these magic numbers can often be interpreted in terms of corresponding structures [16, 17] and may be used in the structure analysis of the solvated ion [18–21], for which rare gas atoms and especially He are desirable due to the minimal influence of the weakly bound ligands on the underlying structure. Nonetheless, the innermost ligands are often bound strongly enough to be heavily localized around the central ion in these complexes, which led to the term "snowballs" for ions complexed with He atoms [22–26].

Furthermore, we introduced a few simplifications in order to avoid repetition and for the sake of an easier reading experience:

(1)    While $^3$He is a unique and fascinating species in its own right, it is almost irrelevant in terms of natural abundance and HND MS. Thus, whenever "helium" or "He" is mentioned in this chapter, it refers to the dominant $^4$He isotope.

(2)    Electron ionization (EI) is by far the most common ionization method in HND MS. Thus, when no specific method of ionization is mentioned, EI was used in the discussed experiment.

(3)    Deuterium or deuterated compounds are sometimes used in HND MS instead of or complementary to naturally occurring hydrogen or its compounds, achieving mainly two things. For one, the larger spacing between attached D atoms or molecules can make the interpretation of mass spectra easier. In addition, ambiguities in chemical reactions may be resolved, for example one can determine whether a reaction involving hydrogen proceeds via a deuterated dopant or residual water molecules. Since hydrogen and deuterium atoms are chemically very similar, the results are usually transferrable between the two species. Thus, we simply refer to "hydrogen" or "H" in the corresponding discussion if deuterium or deuterated compounds are used in a complementary way and the results are very similar.

We structured our chapter as follows. First, we will present an overview of mass spectrometric work utilizing HNDs performed by groups all over the world. The second part of our chapter focuses on recent advances in mass spectrometry of pristine and doped HNDs by three experiments of our group. While we gave our best effort to include all work that has been performed in the field of HND MS and give credit to the contributors, this is obviously a difficult task due to limited time, space and the imperfect human nature. We want to apologize to any contributors whose name we did not mention or whose contribution we missed entirely. But without further ado, let's begin!

## 3.2 History of HND Mass Spectrometry

The first report of a mass spectrometric study of small helium clusters was made in 1975 by van Deursen and Reuss [27]. The authors attempted to answer the question whether a bound state of $He_2$ exists by observing $He_n^+$ up to $n = 13$. The experiment produced helium clusters in a supersonic nozzle expansion and used a magnetic sector analyzer instrument (MSA) for ion detection. Five years later, Gspann and Vollmar employed time-of-flight mass spectrometry (ToF-MS) to observe metastable excitations of large neutral or cationic $He_N$ clusters and determined their size to be in the range of $N = 10^6$–$10^8$ [28]. Gspann also observed the ejection of "charged miniclusters" [29] and confirmed the existence of large anionic $He_N$ ($N > 2\times10^6$) in 1991 using a similar experiment [30]. In 1983, Stephens and King used a free jet expansion cluster source and a quadrupole mass spectrometer (QMS) to record mass spectra of small $He_n^+$ and make the first report of magic numbers in small He cluster ions, finding $n = 7$, 10, 14, 23 and 30 to be anomalously abundant [31]. Naturally, these first studies attempted to understand HNDs and thus targeted pristine HNDs as well as small $He_n$ cluster (ions), however, this was about to change.

## 3.2.1  Pioneering Work by the Toennies Group (Göttingen)

Arguably the single most important, ground-breaking discovery in HND research was made in 1990 when Scheidemann and co-workers demonstrated the HND's ability to capture various foreign atoms and molecules (Ne, Ar, Kr, $H_2$, $O_2$, $H_2O$, $CH_4$ and $SF_6$) [32, 33]. A HND beam was crossed with beams of foreign atomic or molecular species in various configurations and fragments emerging from the doped droplets upon ionization were analyzed using a MSA. The mass spectra revealed that not only were the foreign gas-phase species captured by the HNDs, but they also coagulated and formed clusters of their own as large as $(H_2O)_{18}$, in or on—this was not clear at the time—the HNDs. These discoveries opened up a plethora of possibilities and applications such as HND isolation spectroscopy, growing dopant clusters and studying chemical reactions inside HNDs. The following years saw

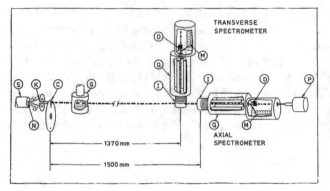

FIG. 1. Schematic diagram of the TOF and mass spectrometer apparatus. S: stagnation chamber; N: nozzle; K: skimmer; C: pseudorandom chopper; G: scattering gas; I: ionizers; Q: quadrupole spectrometers; D: deflection plates; M: electron multipliers; P: Pitot tube.

**Fig. 2.1** An early HND MS setup from the Toennies group in Göttingen. This basic combination of a HND source followed by a pickup chamber and a MS (in this case two) has been and still is utilized by groups all over the world. This experiment was designed to record mass spectra of small ionic fragments from doped or pristine HNDs and measure the velocity of the HND beam but could easily be modified to be used for laser depletion spectroscopy. Reproduced with permission from Ref. [34]. © copyright AIP Publishing.

an unprecedented growth of the field with numerous research teams around the globe picking up research based around HNDs, manifesting in a huge increase in publication numbers in the field.

The described foreign species pickup experiment came out of one of the first and most successful groups conducting extensive studies on pristine and doped HNDs, their properties, ionization mechanisms and applications, the research group for molecular interactions led by J. P. Toennies at the MPI for Fluid Dynamics (now the MPI for Dynamics and Self-Organization) in Göttingen, Germany.

Despite their pioneering work laying the foundation for applications of HNDs in spectroscopy, chemistry and cluster physics, much of the mass spectrometric work conducted in the Toennies group came rather early and was aimed at understanding the relevant characteristics of and processes in *pristine* HNDs (Fig. 2.1). Two early studies published by Buchenau and co-workers in 1990/91 characterized HNDs with source stagnation pressures $p_0$ between 8–20 bar and source temperatures $T_0$ ranging from 5–20 K by investigating a number of their basic properties [34, 36]. The experiments yielded time-of-flight (ToF) spectra, mass spectra of small $He_n^+$ ($n \leq 30$) cluster ions (Fig. 2.2a) and signal intensities as a function of the EI energy. By analyzing these measurements in combination with the He phase diagram, the authors classified different regimes for the formation of HNDs and the ionization/fragmentation processes leading to the ejection of small $He_n^+$ fragment ions. The photoionization (PI) of pure and doped helium droplets was studied by Fröchtenicht et al. in a comprehensive investigation using a synchrotron radiation source and a linear time-of-flight mass spectrometer (ToF-MS) [35]. The authors studied in detail the ionization processes of pristine and $SF_6$-doped HNDs with a wide range of droplet sizes ($<N> \sim 10^2$–$10^7$) at photon energies between 15 and 30 eV. The recorded mass spectra of small $He_n^+$ were found to be similar to those

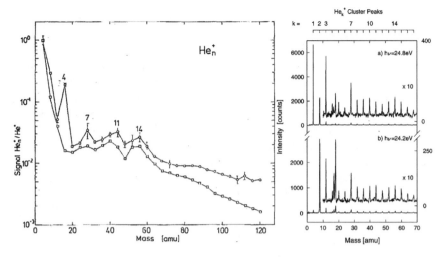

**Fig. 2.2  a** EI mass spectra of small $He_n^+$ from large droplets produced at $T_0 = 5K$, $p_0 = 20$ bar (circles, top trace), recorded with the setup shown in Fig. 2.1. The second mass spectrum (squares, bottom trace) was recorded using a similar setup but much smaller droplets produced at $T_0 = 4.2K$, $p_0 = 0.5$ bar. The most interesting feature is the dominant magic $He_4^+$ in the top mass spectrum which is entirely absent in the bottom one, indicating the difference in droplet size. **b** Mass spectrum showing $He_n^+$ fragments produced via photoionization of small HNDs ($<N> = 2100$). The close resemblance of the $He_n^+$ distributions produced via EI and PI indicates similarities in the respective ionization processes. Reproduced with permission from Refs. [31] (**a**) and [35] (**b**). © copyright AIP Publishing. All rights reserved

utilizing EI, with pronounced magic numbers at $n = 7$, 10 and 14 as previously observed in EI experiments (Fig. 2.2b) [31]. Ion signals of $He_n^+$ and impurities were also found below the ionization potential of free He atoms (<24.6 eV) and assigned to autoionization and Penning ionization processes at energies corresponding to excitations of free He atoms. On the other hand, no direct photoionization of embedded dopant $SF_6$ molecules was observed. These findings led the authors to conclude that both EI and PI proceed in a fairly similar manner in HNDs after being predominantly initiated by the excitation/ionization of a He atom. A set of studies by Lewerenz and co-workers as well as Bartelt and co-workers expanded on the findings of the pickup studies by growing clusters of the heavier rare gas atoms Ar, Kr and Xe, $H_2O$ and $SF_6$ molecules [37] as well as the metals Ag, In and Eu [38]. For the non-metals, the emanating cluster ions displayed a Poisson distribution, which is expected for pickup of gas-phase species and subsequent growth of neutral clusters [22, 39]. This indicates that fragmentation due to EI is largely suppressed by efficient cooling of the nascent cluster ions provided by the surrounding HND [37]. In contrast, the mass spectra of metal cluster ions produced in HNDs are fairly similar to those obtained from free cluster beams. The size distribution of silver cluster ions $Ag_n^+$ show the well-known odd-even oscillations [40] with a clear enhancement of the $n = 3$, 5, 7, 9 ions [38]. The indium mass spectrum features strong signals of both $In^+$ and $In_7^+$. Whereas the latter can simply be explained by an increased stability of $In_7^+$, the explanation for the enhanced monomer signal is a little more complex. Indium clusters between 8–20 atoms feature a similar or slightly higher ionization potential compared to the indium atom, thus fragmentation of these cluster is more likely to proceed by ejection of a charged monomer instead of a neutral one [41]. Additionally, the authors concluded that, in contrast to alkali metal atoms and clusters, the observed metal atoms and clusters reside inside the HNDs and should thus adopt the HNDs' temperature of ~380 mK [42], which would be about 2 orders of magnitude lower than previously observed for free metal clusters [38].

Another interesting topic tackled by the Toennies group are chemical reactions proceeding inside HNDs. In 1993, Scheidemann et al. utilized a MSA to record mass spectra of $SF_6$ molecules embedded in HNDs [43]. In contrast to EI of a beam of free $SF_6$ molecules, where all possible fragments $SF_n^+$ ($n = 0$–5) and $F^+$, but no intact $SF_6^+$ are detected [44–47], the HND mass spectrum essentially yields $SF_5^+$ as the only fragment and, seemingly, intact $SF_6^+$ ions. The authors concluded that the $SF_6$ molecules reside inside the HND, the dominant ionization mechanism is charge transfer from $He^+$ and the expected rich fragmentation of the nascent $SF_6^+$ ions is efficiently quenched by the HND. Later, the apparent observation of $SF_6^+$ was attributed to an impurity [48], likely $SF_5^+$ complexed with a water molecule. In the concluding remarks the authors hint at the potential application of HNDs of studying chemical reactions via the unique possibility of identifying "frozen" reaction intermediates [43]. In 2004, Farnik and Toennies followed up on this seminal idea with a comprehensive study of ion–molecule reactions, very fittingly describing HNDs as "flying nano-cryo-reactors" [49]. The authors utilized HNDs doped with $D_2$, $N_2$ and $CH_4$ in two successive pickup chambers and a MSA to record mass spectra of ions produced in initial charge transfer reactions with $He^+$ and secondary reactions. By monitoring the product intensities as a function of the pickup chamber pressure(s), reaction pathways were identified. The reaction products of the molecules with $He^+$ differ significantly from the gas-phase equivalents. Similar to the $SF_6$ experiment, dissociative charge transfer reactions are largely suppressed by the HND environment. For $CH_4$, the gas phase reaction produces all possible fragment ions except $C^+$ [50], but in the HND environment only $CH_3^+$ and $CH_4^+$ are detected, which are minor contributions to the total ion yield in the gas phase reaction. For $N_2$, both $N^+$ and $N_2^+$ are produced in the gas phase [50, 51], but again, the dissociation reactions appears to be suppressed in the HND, where $N_2^+$ is the sole reaction product. The dominating reaction products produced in the $D_2$–doped HNDs are $He_mD^+$ ($m < 20$), produced in a dissociative reaction of $He^+$ and $D_2$, and $D_3^+$ formed in a secondary reaction of $D_2$ with $D_2^+$, which is initially formed by non-dissociative charge transfer from $He^+$. Other secondary reaction products included $CH_5^+$, (from $CH_4^+ + CH_4$), but *not* $C_2H_5^+$ (from $CH_3^+ + CH_4$), which are both expected from gas-phase experiments [50, 52], as well as $CH_4D^+$ and $CH_3D_2^+$ (from reactions of $CH_4^+$ and $CH_3^+$ upon additional doping of $D_2$) and $N_2D^+$ (from $N_2^+ + D_2$ upon additional doping of $D_2$). The authors also observed molecular ions complexed with He, such as $He_mN_2^+$ ($m = 1, 2, 4$) and $He_mCH_3^+$ ($m = 1, 2$), which was unexpected at the time, since until then mostly atomic species were observed complexed with He [49].

## 3.2.2  Review of more recent research

The basic setup of a HND source, a pick-up region and a mass spectrometer proved to be a simple, but efficient way to investigate pure HNDs, dopant clusters, ionization mechanisms and chemical reactions inside HNDs that was soon applied and adapted by several groups to reach their individual scientific goals. Among the first of these groups were those of K. Janda at UC Irvine and V. Kresin at USC Los Angeles.

### 3.2.2.1  Janda Group (Irvine)

Janda and co-workers used a HND-pickup-QMS setup to study processes following the ionization of small HNDs ($<N> \sim$ 100-20000) in either pristine [53] or doped (NO, Ne, Ar or Xe) [48, 54–56] condition. The study utilizing pristine HNDs showed that the relative intensity of small $He_n^+$ fragments (up to $n \approx 135$) did not vary significantly with the average HND size in the range of 100–15,000 He atoms [53]. The authors concluded that small $He_n^+$ originating from larger HNDs are not produced in a thermal process by evaporating excess He, but rather an impulsive process following the formation of a $He_2^+$ core, leading to the ejection of small $He_n^+$ from the HND as the ionic core may drag along $n$–2 additional He atoms. By comparing the dopant and $He_2^+$ ion yields for different average HND sizes, Callicoatt et al. [48, 56] as well as Ruchti et al. [54, 55] determined the probability of charge transfer from $He^+$ to the dopant (cluster). It was clearly shown that the probability is highest for very small HNDs ($<N>$ on the order of a few hundred), but decreases gradually for larger HNDs ($<N>$ on the order of a few thousand). While this trend was universally observed, the exact probabilities ranged from a few per cent and unity, depending on the dopant species and dopant cluster size. The authors also estimated the average number of resonant charge hops by a positive $He^+$ hole before self-trapping and localizing as $He_2^+$. A relatively simple model yielded a value of 70 for HNDs doped with NO [56], considerably lower than previous estimates on the order of $10^4$ made by Scheidemann et al. [43]. A refined model taking into account polarization arrives at an even lower number of 3–4 hops before charge localization at $He_2^+$ or the dopant (Ar) [48], which is fairly close to the most recent estimate of ~10 hops before self-trapping as $He_2^+$ [57]. Three studies of HNDs ($<N>$ typically ~1000–3000 He atoms) doped with Ne, Ar and Xe investigated the production of small fragment ions produced upon EI by varying parameters such as the average HND size, dopant pickup pressure and electron energy [48, 54, 55]. The studies of Ne and Ar revealed qualitatively similar patterns where HNDs doped with single atoms produced $RgHe_n^+$ ($Rg$ = Ne, Ar) with evidence for a shell closure after $n$ = 12 for $ArHe_n^+$, but no bare ions. On the other hand, HNDs doped with two or more rare gas atoms primarily produced $Rg_2^+$ for the smaller and $RgHe_n^+$ as well as $Rg_2He_n^+$ for the larger droplet sizes studied [48, 54]. In contrast, bare $Xe^+$ is much more likely to be formed, both from droplets containing only a single Xe atom as well as more heavily doped HNDs with up to four Xe atoms, which is attributed to an electronically excited state of $Xe^+$ accessible

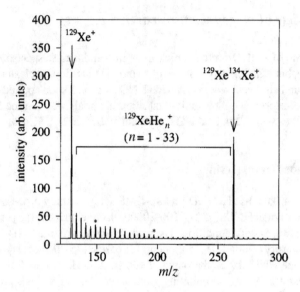

**Fig. 2.3** Simplified mass spectrum showing the progression of $XeHe_n^+$ from EI of Xe-doped HNDs. In contrast to $NeHe_n^+$ [54] and $ArHe_n^+$ [48], there is a higher yield of complexes with small n, however, there are no immediately obvious features. A possible magic number character of $n = 5$ and shell closure after $n = 17$ (marked with asterisks) are not discussed by the authors. Reproduced with permission from Ref. [55]. © copyright Royal Society of Chemistry. All rights reserved

by charge transfer from $He^+$, which does not exist for $Ne^+$ or $Ar^+$ [55]. While the most likely fragment from HNDs doped with two Xe atoms is $Xe_2^+$, analogous to Ne and Ar, small complexes $XeHe_n^+$ $(0 \leq n \leq 3)$ are more readily formed than for $ArHe_n^+$ and no shell closure is evident at $n = 12$ (a possible shell closure of $XeHe_n^+$ after $n = 17$ indicated in the mass spectrum (Fig. 2.3) is not discussed by the authors). Finally, $Xe_2He_n^+$ complexes are notably absent for all experimental conditions, which indicates a weaker caging effect of the relatively small HNDs ($<N>$ up to 3300) for $Xe_2^+$ compared to $Ne_2^+$ and $Ar_2^+$, which could be due to the production highly repulsive $Xe_2^+$ states that cannot be cooled by the HND.

### 3.2.2.2   Kresin Group (Los Angeles)

Kresin and co-workers introduced a similar HND-pickup-QMS setup to investigate alkali and alkaline earth metals as well as amino acids. Scheidemann, Vongehr and co-workers were able to show that alkali metal atoms and small clusters (Li and Na) are preferably ionized by excited He* via a Penning process, indicating these species, as He*, are located at or near the surface of HNDs (Fig. 2.4a) [58, 60]. This finding corroborated theoretical predictions made for alkali metal atoms (Li–Cs) [61–63] and dimers ($Li_2$ and $Na_2$) [64]. A few years later, Ren and Kresin were able to show that the alkali metal's neighbors in the periodic table, alkaline earth metals (Mg–Sr),

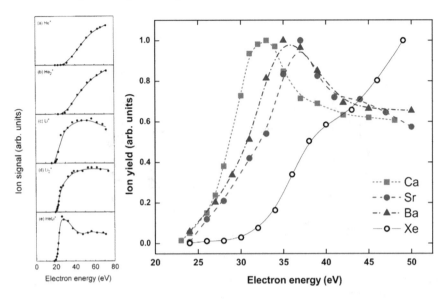

**Fig. 2.4 a** Ion yields as a function of electron energy for small He and Li complexes and **b** three earth alkali atoms as well as xenon grown in HNDs consisting of approximately 2000–5000 and 10000 atoms, respectively. The shape of the curves hint at the location of the neutral dopants: whereas the Xe ion yield closely resembles that of $He^+$ and $He_2^+$ and is consistent with an interior location, all other curves show distinct evidence of Penning ionization and thus a surface location. Reproduced with permission from Refs. [58] (**a**) and [59] (**b**). © copyright AIP Publishing (**a**) and American Physical Society (**b**). All rights reserved

are also located in surface "dimples" (Fig. 2.4b) [59]—this was previously heavily debated, especially for Mg [65–69]. Kresin's group also performed studies examining the possibilities of controlling amino acid fragmentation in HNDs [70, 71] as well as proton transfer reactions between amino acids in HNDs [72]. Ren and co-workers showed that by co-doping HNDs with water, fragmentation of embedded glycine and tryptophan (albeit to a lesser degree compared to glycine) molecules could be suppressed significantly, whereas the HND environment alone was unable to significantly reduce fragmentation compared to the gas phase [70]. The authors proposed a "charge-steering" effect (also see [73] which will be discussed later on) where charge transfer from $He^+$ is favored to occur on $(H_2O)_n$ due to attractive forces caused by the water's relatively large dipole moment, followed by ionization of the amino acid via proton transfer, much softer than direct charge transfer from $He^+$. Ren and Kresin conducted a follow-up study to support this hypothesis using glycine (complexes) as well as alkanes and alkanethiols [71]. They showed that fragmentation patterns of molecules with similar dipole moments as water were practically unaffected, whereas those with lower dipole moments were efficiently protected from fragmentation by the presence of water. Finally, Bellina and co-workers studied the proton transfer reaction in histidine-tryptophan complexes yielding protonated histidine. Additional mass spectra using methyl-tryptophan and indole instead of tryptophan together with

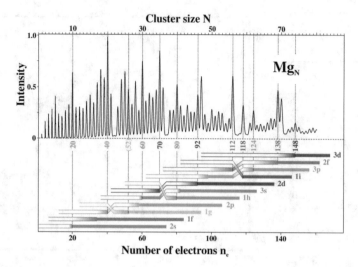

**Fig. 2.5** Smoothed mass spectrum of $Mg_N$ produced via PI of doped HNDs (top). Anomalously abundant ion signals indicate magic number clusters. The authors propose a model where electronic shells are filled by the $2N$ delocalized electrons of the cluster, as illustrated below the mass spectrum. Unexpected shell closures are explained in terms of level reorganization upon increased filling of some shells. Reproduced with permission from Ref. [74]. © copyright American Physical Society.

density functional theory (DFT) calculations revealed that protonation occurs from the indole to the imidazole side chain of tryptophan and histidine, respectively, at the site of the N–H• • •N bond between the two functional groups, which can be viewed as a model system for heterodimers between aromatic amino acids [72] ( (Figs. 2.5 and 2.6).

### 3.2.2.3 Meiwes-Broer and Tiggesbäumker Group (Rostock)

One of the first HND MS experiments utilizing a high-resolution time-of-flight mass spectrometer (HR-ToF-MS) was employed by the cluster and nanostructures group of K.-H. Meiwes-Broer and J. Tiggesbäumker at the University of Rostock, Germany. The capabilities of the apparatus originally constructed by the Toennies group were expanded greatly by adding different, versatile laser systems and a HR-ToF-MS ($m/\Delta m \approx 2000$). While the group mainly focuses on spectroscopic studies of various clusters, HND MS was and is frequently utilized in combined studies of metal clusters (for a review of studies of metal clusters in HNDs see [76]). Since the early 2000s, the group has conducted a number of detailed studies on magnesium clusters. Diederich, Döppner and co-workers observed a highly structured $Mg_n^+$ abundance distribution with strong magic and antimagic cluster ions in both EI and PI (for a review of PI studies of HNDs see [77]) mass spectra (Fig. 2.5) [74, 78]. While some magic numbers agree well with shell closing predictions of the jellium model, others clearly

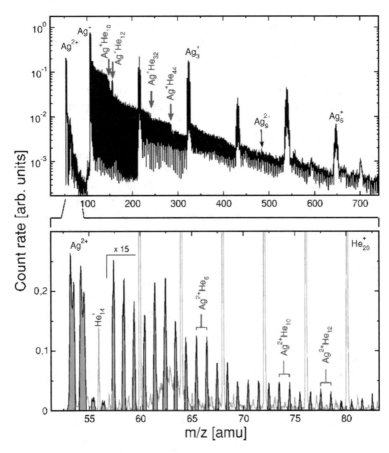

**Fig. 2.6** Mass spectra showing bare silver clusters $Ag^{2+}He_n$ and silver ion-snowball complexes produced via PI of doped HNDs. Shell closures at $n = 10, 12, 32$ and 44 are clear signs of icosahedral-shape solvation shells. The bottom excerpt mainly shows doubly charged silver $Ag^{2+}He_N$ displaying similar shell closures at $n = 6, 10$ and 12. Reproduced with permission from Ref. [75]. © copyright AIP Publishing. All rights reserved

do not. The authors attempt to explain the latter in terms of a level interchange model, where additional shell closings arise due to electron rearrangement caused by interchange of electronic levels. Furthermore, $He_nMg^+$ and $He_nMg^{2+}$ ions with up to $n \approx 150$ were observed upon irradiation with femtosecond laser pulses [78]. The authors additionally studied ion snowball formation dynamics using the pump-probe technique (also see discussion on [75] below. In a subsequent study, Diederich and co-workers revisited their level rearrangement model and expanded their work on $Mg_n^+$ clusters towards larger clusters ($n$ up to ~2500) as well as dications $Mg_n^{2+}$ ($5 \leq n \leq 50$) [79]. Note that the separation of overlapping isotopic patterns provided by the HR-ToF-MS is crucial for the identification of even-numbered $Mg_n^{2+}$. The authors find that the cluster abundance distributions of $Mg_n^+$ display the coexistence of electronic shell closures and geometrical packing schemes at $n \geq 92$ between and clear evidence of icosahedral packing for $n \geq 147$. Magic numbers of dicationic clusters such as $n = 30$ rather agree with the standard jellium electronic shell closures known e.g. from cationic sodium clusters [3], with only one level interchange at n = 41 (i.e. 80 electrons) observable within the limited range. Electronic shell effects are also found for smaller clusters of other divalent metals, namely cadmium and zinc. While some features agree with those of magnesium, the authors find the level interchange model inapplicable and clear signs of icosahedral packing such as an enhanced abundance of $n = 147$ missing for $Cd_n^+$ and $Zn_n^+$. In a number of follow-up studies, the group further exploited the capabilities of their femtosecond laser setup for studying dynamics using the pump-probe technique. Döppner and co-workers extended their study of ion-induced snowballs [78] towards He-solvated Mg and Ag [75], observing He snowballs $He_n Mg^{Z+}$ and $He_n Ag^{Z+}$ ($Z = 1, 2$) with up to $n \approx 150$ He atoms for the singly charged species upon femtosecond laser ionization with intensities of $10^{13}$–$10^{14}$ W/cm$^2$ (nanosecond PI as well as EI were found to produce similar results). The distributions of $He_n Mg^{Z+}$ appear relatively feature-poor with steps at $n = 4$ ($Z = 1$) and $n = 4, 8$ ($Z = 2$) as well as kinks at $n = 19$–$20$ ($Z = 1$) and $n = 10$–$11$ ($Z = 2$), the latter of which are interpreted as closures of the first, liquid-like solvation shells [80]. On the other hand, the distributions of $He_n Ag^{Z+}$ are much more structured and display indications of icosahedral packing schemes with pronounced steps at $n = 10, 12, 32$ and $44$ ($Z = 1$) as well as $n = 6, 10$ and $12$ ($Z = 2$). A subsequent study of $He_n Pb^{Z+}$ finds magic numbers of $n = 12$ and $17$ ($Z = 1$) as well as $n = 12$ ($Z = 2$), indicating a closure of the first shell at 17 and 12 He atoms, respectively [81]. In both cases, a higher charge state likely leads to a stronger interaction, resulting in smaller, more tightly bound structures. He-solvated ions are found to preferably form in HNDs predominantly doped with single metal atoms or small clusters [75]. Additionally, the authors study the dynamics of snowball formation and fragmentation of larger Mg and Ag clusters upon femtosecond laser ionization using the pump-probe technique. Increasing the pump-probe delay reveals a strong decrease in the yield of larger, multiply charged clusters with a minimum at ~30 ps, mirrored by a simultaneous increased production of $He_n Ag^+$. Both signals are found to fully recover at pump-probe delays of ~100 ps. The first part of the signal progression up to 30 ps can be understood in terms of intense heating and fragmentation of the initial cluster, also leading to an increased number of single atoms/ions or small clusters being present in the HND and thus promoting snowball formation. The recovery of signals clearly illustrates the cage effect of the He environment, allowing fragments to dissipate excess energy and eventually recombine to form once again larger clusters and fewer snowballs. Similar trends are observed for Mg clusters, although on shorter timescales with local extrema located around 7 ps. Three studies by Döppner and co-workers further used the pump-probe technique to perform an extensive investigation of plasmon enhanced ionization of metal clusters (Ag, also Cd and Pb in [81]) embedded in HNDs, previously studied in free metal clusters [82]. The authors studied in detail how to manipulate the ion yield and maximum charge state of highly charged metal ions via laser pulse width and field strength [83], delay between two identical pulses [84] as well as a combination of all the above, additionally introducing asymmetry to dual delayed pulses [81]. Narrower pulses, stronger laser fields [83] and shorter delays in between pulses [84] are found to promote production of more highly charged ions, as is cluster irradiation by an initial weaker pulse, followed by a stronger one [81]. The highest achievable charge states were $Z = 11$ (Ag) [83, 84] and $Z = 13$ (Cd) [81]. Expanding on these experiments, Truong and co-workers utilized an ultrafast pulse shaper [85] and a feedback algorithm in order to further substantially increase the yield of highly charged $Ag^{Z+}$ ions, producing charge states up to $Z = 20$ [86]. The authors found the ideal pulse shape to have a double pulse structure, where a weaker (~ 15% intensity) pre-pulse is followed by a main pulse around 140 fs later, in good qualitative agreement with computations based on the nanoplasma model by Ditmire and co-workers [87]. A spectroscopic study by Przystawik and co-workers found that in a HND environment, Mg does not form classical clusters, but rather agglomerate in loosely bound, metastable complexes ('foams'), each Mg atom separated by an estimated ~10 Å with a layer of He in between [88]. The proposed structure is in good agreement with a subsequent computational study by Hernando and co-workers [89] and found to collapse into hot, compact clusters upon laser excitation on a timescale of 20 ps [88]. This collapse was studied in detail by Göde and co-workers using femtosecond dual-pulse spectroscopy [90]. Mass spectra of Mg-doped HNDs subject to femtosecond multiphoton ionization (MPI) reveal $Mg_n^+$ ($n \leq 20$) with enhanced signals of $Mg_5^+$ and $Mg_{10}^+$ as well as $He_nMg^+$ snowballs, in accordance with previous studies [74, 81]and [81], respectively. By studying the response of the various ion signals to changes in delay and intensity ratio of the dual femtosecond pulses, the authors were able to unravel the dynamics of cluster, snowball and even electronically excited, neutral complex (exciplex) formation following the light-induced collapse of Mg foams in HNDs.

### 3.2.2.4  Stienkemeier and Mudrich Group (Freiburg)

Similar experiments to those of Meiwes-Broer and Tiggesbäumker were conducted since the early 2000s in Germany by F. Stienkemeier (University of Freiburg, previously University of Bielefeld), M. Mudrich (University of Freiburg, now University of Aarhus) and co-workers. While the group specializes in spectroscopic techniques, important mass spectrometric contributions involving alkali metals and PI of HNDs were also made. The basic setup consisted of a HND source (typically used to produce droplets with 5000–20,000 He atoms), femtosecond PI and QMS analysis. In an early investigation employing this setup, Schulz and co-workers studied complexes of up to 25 alkali atoms (Na and K), which were found to aggregate into highly spin-polarized, weakly bound van-der-Waals complexes instead of covalently bound or metallic clusters [92]. Upon PI of the alkali-doped HNDs, the authors observe the collapse and fragmentation of the system, evident in mass spectra featuring an exponential decrease of clusters sizes and displaying well-known characteristics such as pronounced odd-even oscillations and magic numbers (e.g. $n = 5, 9, 21$) corresponding to electronic shell closures. The authors observe a maximum cluster ion

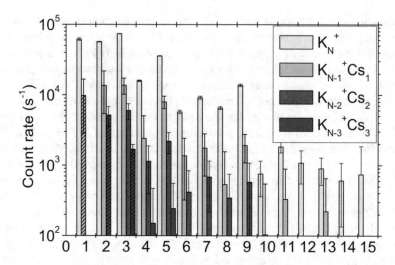

**Fig. 2.7** Cluster size distribution of mixed potassium-cesium clusters from PI of co-doped HNDs as a function of total number of atoms. It appears that, to an extent, Cs atoms can replace K atoms in $K_N^+$ and still preserve the characteristic features of the cluster size distribution. However, the ion yield quickly drops off upon addition of further Cs atoms. Reproduced with permission from Ref. [91]. © copyright AIP Publishing. All rights reserved

size of $n = 3$ and 5 for Rb and Cs, respectively, interpreting it as evidence for a lack of stable high-spin states. Droppelmann and co-workers expanded the investigation towards HNDs co-doped with two alkali metal species (lighter Na or K with heavier Rb or Cs) [91]. The authors observed pure clusters of both species as well as significantly less abundant mixed clusters upon PI of the doped HNDs, regardless of pickup order (Fig. 2.7). Species appear to be interchangeable to a degree, evident in the abundance/stability patterns of mixed clusters which appear to be governed by the *total* number of atoms (i.e. valence electrons), once more reproducing known odd-even oscillations and magic numbers such as $n = 5$ and 9. Incorporation of increasing numbers of heavier alkali metal atoms, however, tends to destabilize mixed clusters, resulting in quickly decreasing ion signals which is attributed to second-order spin orbit interaction. In two follow-up studies, Müller and co-workers shifted their focus towards helium snowballs around cationic alkali metal atoms and dimers (Na-Cs, $Na_2$ and $Cs_2$) [93] as well as reactions between alkali metal atoms (Na, Cs) and water [94]. The authors observed snowball formation with up to 3 and ~10 He atoms around light alkali cations (Na and K, respectively), with much further progressions of up to ~40 He atoms attached to the heavier $Rb^+$ and $Cs^+$ [93]. Additionally, weak signals of small snowballs around the dimers $Na_2^+$ and $Cs_2^+$ are detected. Local anomalies at $n = 4$ for $K^+He_n$ and $n = 12$ for $Cs^+He_n$, hint at especially stable, possibly ring-like (cf. $Mg^{2+}$ and $Ag^{2+}$ [83]) and icosahedral structures, respectively. Shell closures are indicated at $n = 14$ for $Rb^+He_n$ as well as $n = 16$ for $Cs^+He_n$. Generally, snowball formation is found to be favored around ion fragments produced from larger, multiply charged clusters rather than single atoms. In a separate study, reaction products of

**Fig. 2.8** Mass spectra demonstrating the isomer selection capabilities of the OSMS technique [73]. Different fragmentation patterns are clearly evident from the mass spectra for the selected HCN/HCCCN isomers. Reproduced with permission from Ref. [95]. © copyright AIP Publishing. All rights reserved

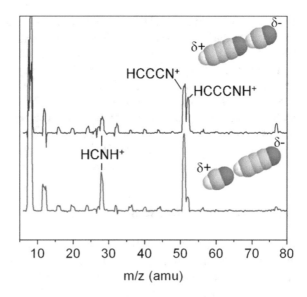

Na or Cs with $H_2O$ were observed upon PI of co-doped HNDs [94]. Whereas Na was found to form primarily weakly bound van-der-Waals complexes $Na_m(H_2O)_n$, mass spectra of HNDs co-doped with Cs and $H_2O$ revealed a variety of compounds indicating efficient chemical reactions between $Cs_m$ and $(H_2O)_n$ *prior* to PI.

Another group utilizing HND MS in the early 2000s consisted of W. Lewis, R. Miller and co-workers who performed detailed studies of pickup and ionization processes in HNDs, developing clever techniques for manipulating dopant species in HND MS experiments along the way. In an early experiment, Lewis and co-workers employed a combination of a QMS and a Threshold PhotoElectron PhotoIon COincidence (TPEPICO) setup to study and control the fragmentation of triphenylmethanol and quantify the energetics of HND cooling [96]. In a subsequent study, Lewis and co-workers expanded the capabilities of the basic QMS setup by introducing an IR laser. The authors demonstrated the possibility of selecting specific isomers of dopant complexes in neutral HNDs using a technique called "optically selected mass spectrometry" (OSMS) [73]. By irradiating the HND with an IR laser tuned to an isomer-specific vibrational transition, the dopant complex is heated, resulting in the evaporation of He atoms from the droplet and a corresponding reduction of the EI cross section which manifests as a depletion of the ion signal in the mass spectrum. Using this technique to study the charge transfer processes in HNDs doped with HCN, HCCH and HCCCN (Fig. 2.8), the authors show that the charge transfer probability from $He^+$ to a dopant molecule (or complex) is heavily dependent on the latter's dipole and higher electrostatic moment(s). The authors explore possibilities of the developed techniques such as isomer selective mass spectrometry and controlling fragmentation patterns of complexes by charge-steering [73] and expand on the described findings in two follow-up studies of non-thermal ion cooling [97] as well as ionization and fragmentation processes in HNDs [95]. A different method of forming ionic complexes in HNDs, avoiding the large amounts of energy transferred to a dopant during ionization via $He^+$ (or $He^*$) and concomitant fragmentation, was developed by Falconer et al. The authors doped HNDs doped with $Na^+$ ions and only then performed pick-up of neutral molecules, followed by desolvation to extract gas-phase analytes from the HNDs. The mass spectra revealed sodiated ion-molecule clusters $[Na \bullet M_n]^+$, where $M = H_2O$, HCN or $N_2$, and could be explained by pickup statistics, suggesting no fragmentation occurs [98]. Lewis et al. also developed a calorimetry technique able to determine the binding energies of moderately to strongly bound clusters such as $(H_2O)_n$ and $C_n$ by observing the threshold HND size necessary to observe a certain cluster ion using a ToF-MS setup [99, 100].

### 3.2.2.5 Ellis and Yang Group (Leicester)

Most experiments produce HNDs via *continuous* (cw) expansion of pressurized, cold He into vacuum. In fact, all HND sources were continuous until 2002, when the first *pulsed* HND source was reported by Slipchenko, Vilesov and co-workers [101]. Pulsed sources achieve a much higher HND flux, have distinct advantages in combination with elements such as pulsed lasers and generally reduce pump load and consumption of high-purity He [101–103]. Besides technical issues like additional heat load on the nozzle region due to the valve operation, it quickly became apparent that pulsed HND sources generally behave differently from cw sources, so that well-established knowledge such as scaling laws could not easily be transferred. The pulsed HND source of Slipchenko et al. could only produce HNDs within a narrow average size range between 20000 and 70000 He atoms. The design was adapted by A. Ellis, S. Yang and co-workers in an attempt to improve the performance of pulsed HND sources. By experimenting with different nozzle shapes, the authors were successful in both widening the accessible size range of HNDs produced and achieving a more predictable behavior with varying stagnation pressure and nozzle temperature [104]. The authors utilized ToF-MS in combination with $H_2O$ and toluene doping to determine the average HND sizes and extract a scaling law, which suggests that the effect of stagnation pressure on the produced HND sizes is negligible, albeit in a limit temperature and pressure range ($T = 10$–$16$ K, $p = 8$–$20$ bar). Later, Yang and Ellis extended their studies to show that HNDs produced in a pulsed source exhibit velocity dispersion according to their size [105], in contrast to cw sources where the velocity spread of differently sized HNDs was found to be uniform for a given set of source conditions [36, 106]. The authors highlight the possibility of probing differently sized HNDs by simply probing the droplet beam at different times instead of changing the source conditions, as in a cw source. Yang, Ellis and co-workers were very productive in utilizing their pulsed HND source and ToF-MS setup to study a number of topics such as ionization and fragmentation/dissociation dynamics, atomic and molecular clusters and ion-molecule reactions. A summary of these extensive and rich experimental studies is attempted below.

The ionization of small dopant molecules via $He^+$ charge transfer and subsequent fragmentation and dissociation products were investigated for several alcohols ($C_1$–$C_6$) and ethers [107], haloalkanes ($C_1$–$C_3$) [108] and diatomic molecules ($O_2$, CO and $N_2$) [109]. The authors found that for the small to medium-sized alcohols and ethers as well as the haloalkanes studied, the HND environment did alter the fragmentation patterns compared to the gas phase, but mostly quantitatively, enhancing certain channels like H-abstraction. Apart from the cyclic $C_5$- and $C_6$-alcohols, the parent ion remained a minor product, i.e. the HND was unable to prevent the excessive fragmentation. The authors concluded that while EI of doped HNDs could not be considered a soft ionization method suited for analytical mass spectrometry for the rather small molecules studied, it might still be worthwhile for larger species such as typical biomolecules [107, 108]. A similar, extensive study was carried out by Boatwright and co-workers on clusters of small molecules such as aliphatic alcohols ($C_1$–$C_3$), several halomethanes and inorganic triatomic molecules ($H_2O$, $SO_2$ and $CO_2$) embedded in HNDs [110]. Again, the mass spectra suggest that the EI-initiated chemistry of both clusters and at least one of the single molecules differs significantly from gas-phase studies and proceed via direct bond fission processes instead of ion-molecule chemistry. The findings further support the authors' conclusion that EI of HNDs doped with small molecules and their cluster cannot be considered a soft ionization route [110]. In contrast to the small molecules discussed so far, EI of HNDs doped with the diatomics $O_2$, CO and $N_2$ show a significant reduction of dissociation (fragmentation) compared to the gas phase reaction between the diatomic species and $He^+$ [109]. The authors consider two possible explanations: suppression of the dissociation channel in the ion-molecule reaction by the HND or acting as a reservoir, allowing for recombination of the products after initial dissociation. In order to determine which is the case, the energetics of the ion-molecule reactions and the corresponding amount of He evaporation to dissipate the energy difference were calculated. The authors conclude that while the dissociation reaction itself must be suppressed in the case of $O_2$ and CO, no conclusion can be drawn for $N_2$ from the calculated energetics [109]. Shepperson and co-workers investigated the formation of small $He_n^+$ cluster ions from HNDs with different sizes (average droplet size $<N>$ between 4000 and 90000) and dopants (pristine, $H_2O$ and Ar) [111]. The authors find that the $He_n^+$ / $He_2^+$ signal ratio increases with increasing HND size, reaching an asymptotic limit at around $<N> = 50000$. The authors conclude that larger HNDs favor the formation of $He_n^+$ ($n > 2$) from $He_2^+$ due to the larger number of collisions of the latter and surrounding He atoms on its way of leaving the droplet. The introduction of Ar mainly leads to the asymptotic value being reached at a significantly smaller $<N> = 10000$, explained by a potential energy gradient, steering the charge hopping of the initially formed $He^+$ (before formation of $He_2^+$) towards the impurity, which is most likely located close to the droplet center. The introduction of $H_2O$ also lowers $<N>$ where the asymptotic limit is approached, although not as clearly. More interestingly though, the limit is approached from the opposite side, i.e., with $H_2O$ as a dopant, the $He_n^+$ / $He_2^+$ signal ratio decreases with increasing $<N>$. The authors propose that the dipole moment creates as stronger gradient, which makes it likely that $He_2^+$ is formed in the vicinity of the impurity, additionally lowering the kinetic energy of the $He_2^+$ on its way to leaving the droplet, thus lowering the collision energy with He atoms, increasing the chance of additional He attaching to form $He_n^+$ ($n > 2$) [111].

The first studies on clusters formed inside HNDs by Yang, Ellis and co-workers were performed on small molecules M such as aliphatic alcohols ($C_1$–$C_5$) [112] and $H_2O$ [113], using the pulsed HND source and ToF-MS setup. In both cases, mass spectra differed significantly from gas phase experiments, displaying higher abundances of unfragmented cluster ions $M_n^+$, besides the dominant protonated $[M_nH]^+$ ions. Other important signals were due to dehydrogenated alcohol clusters $[(ROH)_n–H]^+$ and intact, He-tagged water clusters $[(H_2O)_nHe]^+$. The emergence of the unfragmented, dehydrogenated and He-tagged species are attributed to the efficient cooling/quenching effects of the HND environment. In a follow-up study, Liu et al. formed and studied core-shell clusters of water and several different co-dopant species (Ar, $O_2$, $N_2$, CO, $CO_2$, NO and $C_6D_6$), with similar results for both orders of doping [114]. Both binary and pure water clusters are observed in the mass spectra. The authors focused on pure water clusters and describe a "softening" effect on the charge transfer reaction for co-doping with non-polar molecules ($O_2$, $N_2$, $CO_2$, and $C_6D_6$). The softer ionization process manifested in a higher ratio of intact water cluster ion signals $(H_2O)_n^+$ compared to protonated $(H_2O)_nH^+$ for all cluster sizes. The effect strength increased in the order $N_2$, $O_2$, $CO_2$, $C_6D_6$ and is explained in terms of energy dissipation via evaporation of the co-dopant species. The different strength of the softening effect is caused by an interplay of the molecules' dipole polarizabilities (i.e. binding energy to an ionic core), vibrational degrees of freedom and ionization energies. Whereas argon is fairly similar to the pure HND environment, showing no significant softening effect, the influence of the polar molecules CO and NO is more complicated. While a weaker softening effect is observed for some water cluster sizes, in most cases, fragmentation of intact water clusters to protonated $(H_2O)_nH^+$ is actually enhanced by the presence of CO and NO. It is suggested $(H_2O)_nH^+$ production is enhanced by CO and NO readily accepting OH radicals to form the stable HOCO and HONO in secondary ion-molecule reactions, which is supported by a subsequent ab initio theoretical study of Shepperson and co-workers [111].

In addition to the previously discussed pulsed HND source and ToF-MS setup, the group of Yang and Ellis recently employed a different setup consisting of a classical $cw$ HND source and a QMS, using it to conduct two studies on metal clusters, namely bi-metallic core-shell nanoparticles [115, 116]. In the former study, two temperature scan series illustrate the influence of HND source and oven temperatures on the yield of $Ni_n^+$ cluster ions from EI of doped HNDs. While a certain size is required to ensure the pickup of multiple Ni atoms, charge transfer to the dopant cluster becomes less likely in larger droplets. Similarly, a certain dopant partial pressure is required for the capture of multiple Ni atoms, but HNDs are quickly evaporated by collisions if the pressure is too high. Additionally, a mass spectrum shows that mixed Au/Ag cluster cations with up to seven Au and six Ag atoms are formed (Fig. 2.9). Furthermore, various produced nanoparticles (Ag, Ni, Au, Ag/Au and Ni/Au) were also deposited on a substrate for ex situ analysis. Transmission electron microscope (TEM) images

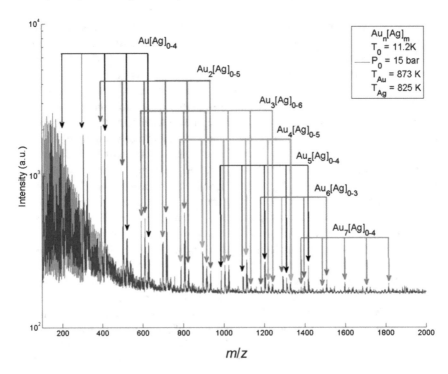

**Fig. 2.9** Mass spectrum demonstrating the formation of binary gold/silver clusters grown in HNDs ($<N> = 16000$). The largest detected cluster ion consists of seven Au and four Ag atoms. Weak odd-even oscillations as well as distinct magic number clusters (e.g. $AuAg_2^+$, $Au_2Ag_3^+$, $Au_3Ag_2^+$, …) can be found, however, these are not further analyzed by the authors since the main scope of the study was the synthesis, deposition and ex situ of larger nanoparticles. Reproduced with permission from Ref. [115]. © copyright Royal Society of Chemistry. All rights reserved

show that Ag-Au nanoparticles with a diameter of a few nm formed in HNDs adopt a crystalline structure, but do not allow conclusions about their structure. Evidence for a core-shell structure of Ni-Au nanoparticles is found via the absence of any Au 4f shift in X-Ray Photoelectron Spectroscopy (XPS) spectra characteristic for an Au/Ni alloy. A recent study by Spence and co-workers performed on the same setup demonstrates the formation of various aluminum cluster ions such as $Al_n^+$ ($n \leq 15$), $[(Al_2O)Al_n]^+$($n \leq 12$) as well as $Al^+He_n$ ($n \leq 17$) and $Al_2^+He_n$ ($n \leq 6$) [116]. These observations contrast a previous study by Krasnokutski and Huisken, where no formation of Al clusters inside HNDs was observed [117]. Spence et al. suggest that inadequate pickup conditions prevented the formation of clusters in the experiment of Krasnokutski and Huisken [116]. The cw apparatus was further used to investigate complexes of and ion-molecule reactions between organic molecules and noble metal atoms in two recent studies. In the first study, complexes of tetrapyridyl porphyrin (5,10,15,20-tetra(4-pyridyl)porphyrin, H2TPyP) and gold were formed and analyzed by Feng and co-workers [118]. The mass spectrum of H2TPyP-doped

HNDs shows excessive fragmentation of the molecules. Upon co-doping with Au, complexes of H2TPyP (fragments) and Au are detected and the H2TPyP fragmentation is generally reduced. The authors assume that Au atoms and small clusters attached to H2TPyP can dissipate excess energy from the charge transfer process and might additionally introduce a charge-steering/buffering effect where Au can act as a "buffer" in a sequential charge transfer (see earlier discussion about *co-doping of amino acids and water by Ren et al.* [70, 71]). The effect appears to be even stronger when Au is picked up after H2TPyP, attributed due to enhanced binding of Au (clusters) to multiple sites of H2TPyP instead of attachment of a previously formed Au cluster to a single site of H2TPyP. Additionally, the composition of fragments is altered, e.g. some protonated fragment channels and fragment "dimer" channels are clearly weakened. The second study by Sitorus and co-workers observed dissociative ion-molecule reactions in complexes of 1-pentanol and 1,9-decadiene, co-doped with gold or silver [119]. The mass spectra show a number of fragments as well as complexes of the organic molecule (fragments) and Au/Ag. However, in contrast to the previously discussed study, little to no softening effects on the fragmentation pattern of the organic molecules were observed upon co-doping with Au/Ag. An exception was the case of 1,9-decadiene and Au, where fragmentation was found to be efficiently reduced, especially for small fragments. The authors found that only for 1,9-decadiene and Au, the ionization potential of the metal atom was higher than for the organic molecules (Ag: 7.6 eV, Au: 9.2 eV [120], 1-pentanol: 10 eV [121] and *1,9-decadiene:* 8.6 eV [119]), concluding that this was a prerequisite for a softening effect to occur.

### 3.2.2.6   Ernst Group (Graz)

The group of W. Ernst at TU Graz is dedicated to the study of optical and catalytic properties of metal clusters grown in HNDs. Whereas analysis is mostly performed using spectroscopy and microscopy methods, the group employs a ToF-MS as well (Fig. 2.10). While its primary use is the in-situ monitoring of cluster growth, some mass spectrometric studies involving alkali and earth-alkali as well as other metal species have also been conducted. Theisen and co-workers studied the submersion of and snowball formation around alkali metal ions (Rb and Cs) in HNDs by resonant two-photon ionization (R2PI) involving initial excitation to selected, non-desorbing states [122, 123]. The authors observed small $Rb^+He_n$ snowballs (up to $n \sim 20$) as well as intense ion signals at heavier masses, which are attributed to large $Rb^+He_n$ snowballs ($n > 500$) and well-described by a log-normal distribution [122]. While observations made for Cs are generally similar, $Cs_2^+He_n$ snowballs are also detected. The authors tentatively assign possible shell closures of $Cs^+He_n$ around $n = 17$ and 50 while missing a noteworthily enhanced abundance of $Cs^+He_{12}$ [123]. A follow-up study by Theisen and co-workers covered in detail the ionization process for Rb and Cs at HNDs via R2PI and determined the (slight) lowering of ionization thresholds due to the HND environment [124]. Theisen and co-workers also expanded the studies of Schulz and co-workers on high-spin alkali clusters [92] towards larger oligomers

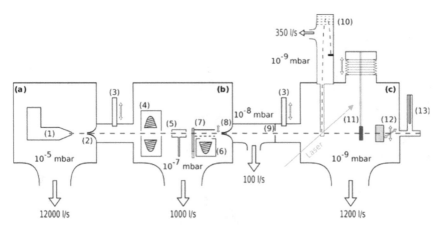

**Fig. 2.10** Setup used by the Graz group for the production, in situ analysis and deposition of metal clusters. The main parts of the experiment are the source chamber (**a**), doping chamber equipped for sequential doping from various sources (**b**) and analysis chamber (**c**) containing a ToF-MS (10) with possibilities for both PI and EI as well as a substrate holder (11) and a quartz microbalance (12). Reproduced with permission from Ref. [127]. © copyright AIP Publishing. All rights reserved

$Rb_n^+$ and $Cs_n^+$ with up to $n = 30$ and 21 atoms (in contrast to five and three in the previous study), respectively [125]. Whereas the authors observed these larger clusters using single-PI with photon energies of 3.3 and 4.1 eV, no ions larger than trimers were detected using multi-PI at 1.4 eV. Familiar magic numbers such as $n = 5$, 9 and 19 are found for both species, with additional anomalies at $n = 13$, 21 and 28 for $Rb_n^+$ [126] as well as $n = 15$ for $Cs_n^+$. Barring explicit evidence, the authors consider several models of cluster formation to argue that neutral alkali clusters should be able to form in both high and low spin states on the larger HNDs ($<N>$ ~20,000) used in the experiment. A cleverly constructed setup allowing for in-situ monitoring (EI/ToF-MS) of NPs *while* depositing them on a substrate for ex-situ analysis by methods such as electron microscopy was employed by the group in 2015. The setup was first used by Thaler and co-workers in an extensive study of metal (Ni, Cr or Au) nanoparticle (NP) formation of up to ~500 atoms (diameters up to 2 nm) in large HNDs ($N = 10^5$ to $>10^8$), with good agreement between in-situ and ex-situ analysis [127]. The presented mass spectra extending to $m/z$ ~20,000 show well-known features of metal clusters such as odd-even oscillations and magic numbers corresponding to highly geometric structures such as a pentagonal bi-pyramid ($Cr_7^+$) or an icosahedron ($Cr_{13}^+$). Magic numbers of Au clusters are also evident throughout the mass spectrum (Fig. 2.11), but were not further analyzed by the authors. A more detailed analysis of small $Au_n^+$ ($n \leq 9$) produced via R2PI can be found in [128] in the frame of a spectroscopic study. Messner and co-workers observe pronounced odd-even oscillations with even-numbered species practically absent, except for the dimer. Further mass spectrometric work can be found in studies that have a different general focus such as spectroscopy or characterization of an evaporation source. Mass spectra (EI/QMS) of neat and hydrated $Cr_m(H_2O)_p^+$ ($m \leq 9, p = 0, 1$), $Cu_n(H_2O)_p^+$ ($n$

**Fig. 2.11** Mass spectrum of large gold clusters extending beyond 100 Au atoms demonstrating the capabilities of the setup shown in Fig. 2.10. The inserts show the prominent odd-even oscillations of small cluster ions and the resolution towards high masses exceeding $m/z =$ 15000. Reproduced with permission from Ref. [127]. © copyright AIP Publishing. All rights reserved

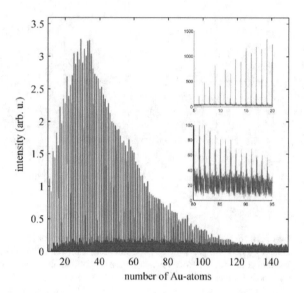

$\leq 7, p = 0\text{--}2)$ as well as the detection of small, mixed $CrCu_n^+$ $(n = 1\text{--}3)$ were reported by Ratschek and co-workers [129] as well as Lindebner and co-workers [130]. Krois and co-workers present a mass spectrum demonstrating the efficient production of RbSr molecules via (R)2PI of co-doped HNDs in their spectroscopic study of the molecule [131]. Complexes of Li such as $Li_2^+$, $LiHe_n^+$ $(n \leq 3)$ and $Li_nOH^+$ $(n = 1, 2)$ were observed by Lackner and co-workers in a laser spectroscopy study [132]. The detection of the latter species and the concurrent absence of $Li_n(H_2O)_p^+$ is interesting since Müller and co-workers observed the opposite in HNDs co-doped with Na and $H_2O$ [94].

### 3.2.2.7 Krasnokutski and Huisken Group (Jena)

One of the most intriguing applications of HNDs as a "nano-cryo-reactor" is the possibility to study astrochemically relevant processes in the laboratory. F. Huisken, S. Krasnokutski and co-workers successfully employed a HND source and a QMS to conduct several clever mass spectrometric studies of such processes in the last decade. Additional techniques such as calorimetry (via shrinking of HNDs due to evaporation) and chemiluminescence (CL) measurements were used to complement mass spectrometry. The first studies investigated reactions of astronomically relevant atomic species (Mg, Al, Si and Fe) with $O_2$, and in some cases $H_2O$ and $C_2H_2$ as well. Krasnokutski and Huisken performed an extensive study combining mass spectrometry and CL measurements, which found that single atoms of Mg react with $O_2$ in the HND environment, but the most likely product $MgO_2$ is expelled from the droplet, preventing detection by MS [135]. If on the other hand, Mg clusters were allowed to form prior to the reaction with $O_2$, a number of $Mg_xO_y^+$ $(x \leq 4, y$

**Fig. 2.12** Differential mass spectra from a study probing possible astrochemically relevant pathways of glycine formation in HNDs. The top figure shows the effect of adding atomic $^{13}C$ to HNDs doped with two different gas mixtures (top, gas 1: 3:2:1 mixture of $NH_3$, $H_2$ and $CO_2$, gas 2: ~2:1 mixture of $H_2$ and $CO_2$) The bottom figure shows the same mass spectrum for gas 1, overlaid with a gas-phase EI mass spectrum of glycine [133]. Reproduced with permission from Ref. [134]. © copyright American Astronomical Society.

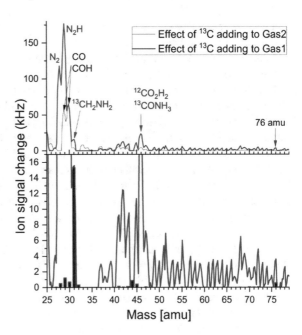

$\leq 2$) compounds were detected in the mass spectra. The authors unexpectedly find that these reactions occur faster than $5 \times 10^4$ s$^{-1}$ (with some ambiguity about the influence of the HND environment) suggesting a need to include this reaction in astrochemical modelling. Similar experiments were conducted on the reactions of Si with $O_2$ (and their respective clusters) as well as $H_2O$ [136]. In a calorimetric approach, Krasnokutski and Huisken utilized pressure measurements to monitor the evaporation of HNDs due to the energy released during the reaction of Si and $O_2$. The authors conclude that the entrance channel is barrierless and calculate the lower limit of the reaction rate to be $5 \times 10^{-14}$ cm$^3$ mol$^{-1}$ s$^{-1}$, indicating relevance in interstellar environments. Binary complexes $Si_xO_y^+$ ($x \leq 2$, $y \leq 3$) were detected in the mass spectra when Si clusters were allowed to form in larger HNDs ($N_{He} \geq 15{,}000$). While SiOH$^+$ are detected when droplets are doped with Si and (residual) $H_2O$, the authors attribute this reaction to occur after EI of the doped HND. A subsequent study employs similar methods to study reactions of Al with $O_2$ and $H_2O$ [117]. The authors found that Al atoms do not coagulate to form regular clusters in HNDs, a conclusion that has since been disputed (see [116] and previous discussion in this chapter). Despite this possible mishap, further findings are likely valid, such as the reaction of Al with $O_2$ to form AlO$_2$ occurring in HNDs, detection of $Al_xO_y^+$ ($x \leq 2$, $y \leq 3$) in mass spectra of larger HNDs and the nonreactivity of single Al with $H_2O$, whereas reactions occur if any of the species is allowed to form a cluster. Last in this series of studies is the exploration of iron reactivity with $O_2$, $H_2O$ and $C_2H_2$ [137]. Combining mass spectrometry, R2PI spectroscopy [138] and quantum chemical simulations, the authors find that Fe atoms undergo reactions with the reactant species, but only form weakly bound complexes without significantly altering the molecular geometries.

Since these complexes easily dissociate, weak bonding of Fe to interstellar ice and dust grains is also indicated.

In their most recent studies, Krasnokutski and co-workers employed a home-built atomic carbon source [139] to tackle two hot topics in astrochemistry—the chemical evolution of polycyclic aromatic hydrocarbons (PAHs) and the formation of glycine in astrophysical environments. PAHs are found to be a ubiquitous component of organic matter in space [140], suggested to be responsible for the IR emission features in the 3-to-15 $\mu$m range that dominate spectra of most galactic and extragalactic sources [140–143] and considered to be carriers of the diffuse interstellar bands (DIBs) [144, 145], a number of features in the UV-vis-IR-range first reported in 1922 by Heger [146], but remain largely unidentified to date except for a few bands recently attributed to $C_{60}^+$ [4, 147–150]. Motivated by the poorly understood chemical evolution of PAHs in astrophysical environments [143], Krasnokutski and co-workers investigated the reaction of C atoms with (deuterated) benzene [151] and PAHs (naphthalene, anthracene and coronene) [152] embedded in HNDs. Whereas the gas phase reaction $C_6D_6 + C$ dominantly yields $C_7D_5$ [153, 154], mass spectra recorded by Krasnokutski and Huisken suggest that in the HND environment ($N_{He} \geq 4000$), the reaction intermediate $C_7D_6$ is stabilized as the lone reaction product. Calorimetry measurements show that stabilization requires the dissipation of ~2.75 eV, considered feasible to occur on the surface of cold interstellar dust grains by the authors. Quantum chemical calculations reveal that the C atom is inserted into the aromatic ring, resulting in a seven-membered ring structure. Expanding their efforts to larger aromatic hydrocarbons, Krasnokutski and co-workers performed a similar study on naphthalene, anthracene and coronene [152]. Like the benzene study, mass spectra and calorimetry measurements show barrierless reactions resulting in the formation of C(PAH) in all three cases, however, the energy release detected via calorimetry is considerably less in the case of coronene. Quantum chemical calculations show that C atoms initially attach to CC bonds, but the binding energies for interior CC bonds (i.e. ones shared between two aromatic rings) are found to be < 1 eV, significantly lower than for peripheral CC bonds (i.e. belonging to form a single aromatic ring). This affects the subsequent reactions—whereas a C atom initially attaching to a peripheral bond leads to a ring opening and insertion into the carbon network to form a seven-membered ring structure in all examined PAHs, a ring opening is not possible for initial attachment to an interior CC bond. Consequently, the basic structure of the PAH is largely preserved, with a weakly bound out-of-plane carbon attached. The latter case is associated with a lower energy release, which, considering the higher ratio of interior versus peripheral CC bonds in larger PAHs such as coronene, explains the lower energy release evidenced by calorimetry measurements. The authors point out implications of their findings for the growth and destruction of PAHs in astrophysical environments. The described formation of seven-membered ring species from the reaction of "classical" PAHs) with atomic carbon could lead to a large number of "nonclassical" PAHs and furthermore facilitate formation and bottom-up growth of small PAHs by sequential reactions with carbon and hydrogen atoms. On the other hand, PAHs with a higher number ratio of interior to peripheral CC bindings, i.e. larger and more compact species are becoming increasingly chemically inert towards reactions with atomic carbon, limiting their growth.

The latest study discussed here covers glycine (Gly), the simplest of amino acids, essential building blocks of proteins and imperative for the emergence of life. While Gly was detected in comets [155, 156] and meteorites [157], there is no convincing observational evidence for the existence of Gly in the interstellar medium yet. Lacking observational evidence, possible astrochemical formation pathways of Gly are being assessed in experimental and theoretical efforts. Chemical surface reactions occurring on ice surfaces of interstellar dust particles are thought to facilitate the formation of a large number of organic species, including amino acids [158–160]. Krasnokutski, Jäger and Henning studied two possible low-temperature routes of Gly formation considered feasible to occur on such interstellar dust particles by means of HND MS, calorimetry and theoretical calculations [134]. Both $NH_3$ and $H_2$ were calorimetrically shown to perform barrierless and highly exoergic reactions upon addition of C atoms, forming $CH_2NH$ and $HCH$, respectively, precursors for the formation of Gly. The pathway $HCH + NH_3 + CO_2 \rightarrow$ Gly was investigated theoretically by calculating the potential energy surface of the three-body reaction, revealing a barrierless reaction pathway of Gly formation. Experimentally, the reaction was studied by adding C atoms to clusters of $H_2$, $NH_3$ and $CO_2$ grown in HNDs and analyzing the reaction products via EI of HNDs and MS (Fig. 2.12). While some evidence for the formation of Gly was found, the results were not entirely conclusive due to low ion signals and numerous interference mass peaks on the positions of the parent and typical Gly fragment ions. The identification of mass peaks could be improved by using a high-resolution ToF-MS instead—e.g. for the mass spectrometric separation of $NH_2{}^{13}CH_2COOH^+$ ($m/z$ 76.035) and $He19^+$ ($m/z$ 76.049), a mass resolution of $R = \frac{m}{\Delta m} \sim 5000$ is required.

### 3.2.3 Mass Spectrometry as a Complimentary Tool

Many groups conduct HND experiments that use MS as a complementary tool while relying on other techniques such as spectroscopy to achieve their scientific goals. Consequently, many authors have occasionally published MS studies or included valuable mass spectrometric work in studies with a different main focus. We attempt to summarize this body of work in the following section.

#### 3.2.3.1 Vilesov Group (Los Angeles)

Beginning in 2000, the group of A. Vilesov at the University of Southern California in Los Angeles has performed groundbreaking work around fundamental properties and applications of HNDs. Some notable examples are the development of pulsed HND sources (also see previous discussion about the work of Ellis, Yang and co-workers) [101, 103], spectroscopic studies of various dopant clusters [166–168] and

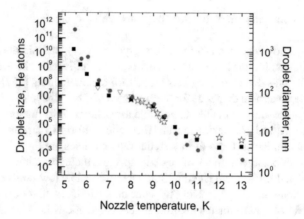

**Fig. 2.13** Average helium droplet sizes <N> as produced in a cw expansion (5μm nozzle diameter, $p_0 = 20$ bar) and measured with different methods. The results of the titration method used by Gomez et al. is shown for two collision gases as filled squares (He) and circles (Ar). Early deflection measurements are shown as open triangles [106, 161] and stars [162–164]. The agreement between the methods is generally good, especially for the experimentally relevant intermediate temperatures between 6 and 10K. Reproduced with permission from Ref. [165]. © copyright AIP Publishing. All rights reserved

studies of quantized vortices [169, 170]. The group has also developed and studied techniques for the determination of HND sizes that could be considered mass spectrometry in a broader sense. These serve as alternative or complementary methods to the common electrostatic deflection techniques which have difficulties with large HNDs due to their enormous mass and variety of charge state [171]. Gomez and co-workers measured the attenuation of a HND beam from a cw source upon introduction of He or Ar collision gas by monitoring the partial pressure in a downstream detection chamber and used their results for the determination of droplet sizes [165]. HND sizes determined by this titration method in the 7-to-10 K range (<N> ~ $10^5$– $10^7$) are in good agreement with previous deflection measurements (Fig. 2.13) [161, 162]. The method finds that HNDs in a previously uncharacterized size range of $10^7$–$10^{10}$ He atoms are produced in the cw expansion at low nozzle temperatures (5.6–7 K). Furthermore, the ion yield of $He_4^+$ exhibits an anomalously large increase (relative to other $He_n^+$) with HND size in the size range of <N> ~ $10^4$–$10^9$, as previously observed [34, 172] and attributed to formation and ejection of $He_4^+$ at the HND surface via collision of two metastable $He_2^*$ [34]. The authors propose that the yield of $He_4^+$ relative to $He_2^+$ can be used as a secondary method of determining HND sizes in the mentioned size range and apply it in the measurement of HND sizes produced in a pulsed source [165]. The process of $He_4^+$ formation in large HNDs was further investigated in a detailed study by Fine and co-workers by monitoring the ion yield of $He_4^+$ (relative to other $He_n^+$) from EI of HNDs, produced in a pulsed source at temperatures between 9 and 19K [173]. The authors used two mass spectrometers, an in-line QMS and an orthogonal-extraction ToF-MS. Surprisingly, the observed $He_4^+$ intensity progression with HND size was in poor agreement between both mass spectrometers initially, with the one measured by ToF-MS considerably lower. Only upon increasing the time width of the electron pulse used for ToF-MS, the $He_4^+$ intensity approached that measured by the QMS. The authors concluded that two processes are involved in the formation of $He_4^+$. One is fast and produces $He_4^+$ as part of a "regular" $He_n^+$ distribution. The other one, suspected to be responsible for the anomalous increase of $He_4^+$ intensity, is slower, occurring on timescales of the order of 10 μs, has an electron energy threshold of 40.5 ± 1.0 eV and selectively produces $He_4^+$ [172]. Fine and co-workers carefully elucidate possible formation mechanisms on the surface of large HNDs compatible with the observed energy threshold (40.5 ± 1.0 eV) and slow reaction times (~10 μs). The authors conclude that binary collisions of He* + He*, He* + $He_2$* and $He_2$* + $He_2$* (provided $He_2$* highly vibrationally excited) can all contribute to the selective formation of $He_4^+$, which consists of two $He_2^+$ cores in perpendicular orientation and a shared Rydberg electron [174].

### 3.2.3.2  Neumark Group (Berkeley)

The group of D. Neumark studies several different topics in chemical physics, one of which is spectroscopy and dynamics in HNDs. When first venturing into the field of HND research, three detailed studies on PI of pristine and doped droplets were conducted. Kim and co-workers utilized a synchrotron light source and a ToF-MS to study the PI characteristics of small HNDs ($N$ ~ 8000) doped with rare gases (Rg = Ne–Xe) at photon energies 10–30 eV [175]. Mass spectra reveal small cluster ions $Rg_m^+$ ($m$ ~ 1–3) and ion snowball complexes $He_nRg_m^+$ ($n, m \geq 1$) alongside the well-known distribution of $He_n^+$ including features such as a magic $He_{14}^+$ ion (Fig. 2.14), in good agreement with a previous PI [35] as well as numerous EI studies. Ion yield curves recorded as a function of photon energy indicate that ionization of rare gas atoms and clusters embedded in HNDs is always mediated by He atoms, analogous to EI. The two possible methods are excitation transfer with the dominant resonance at 21.6 eV (atomic $1s$–$2p$ transition) below the He ionization threshold at 24.6 eV, or charge transfer following direct He ionization above it. Further notable features are the absence of a bare $Ne^+$ or $Ar^+$ signal from doped droplets as well as the magic number character of $He_{12}Kr_2^+$ and $He_{12}Kr_3^+$. In a subsequent investigation, Peterka and co-workers studied in detail the PI and photofragmentation of $SF_6$ embedded in HNDs at photon energies of 21.8 and 25.5 eV [176]. Ionization was equally found to proceed via He as in the previous study of rare gas dopants. The recorded mass spectra were compatible with the EI study of Scheidemann and co-workers [43] in that $SF_5^+$ was the dominant fragment with $SF_3^+$ and $SF_4^+$ signals largely suppressed (less so in the 25.5 eV spectrum) compared to the gas phase. Additionally detected compounds were $He_nSF_5^+$ ($n \leq 25$), $(SF_6)_mSF_5^+$, $H_2O \bullet SF_5^+$ and $He_nSF_3^+$ (the latter only at 25.5 eV)—the suspected magic number ion $He_{12}SF_5^+$ displayed no local abundance anomaly. Due to the similarity of photoelectron spectra of $SF_6$ in HNDs and gas phase, the authors conclude that the dissociative states involved in the gas phase fragmentation are equally accessible in HNDs and consequently attribute the suppressed

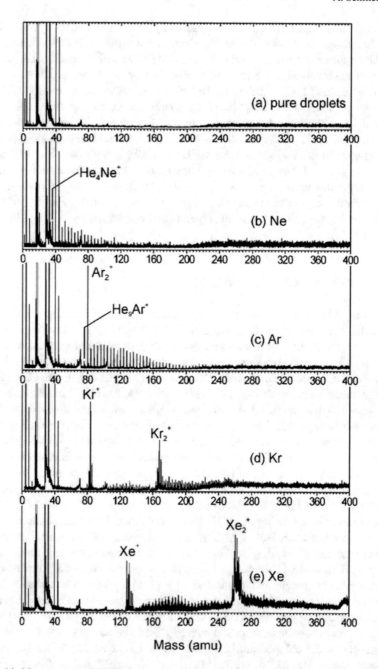

**Fig. 2.14** Mass spectra of HNDs in pure condition (**a**) and doped with rare gases Ne-Xe (**b**–**e**) subject to PI at photon energies of 21.6 eV. The mass spectra reveal the formation of $RgHe_n^+$ as well as $Rg_2He_n^+$ (Rg = Kr, Xe) complexes comparable to EI studies, but exhibit fewer features such as magic number or apparent shell closures. Reproduced with permission from Ref. [175]. © copyright AIP Publishing. All rights reserved

fragmentation to the rapid cooling of the hot nascent ion by the HND environment. Peterka and co-workers used the same setup to study the ionization dynamics of pristine HNDs at photon energies between 24.6 and 28 eV using photoelectron spectroscopy [177].

### 3.2.3.3 Von Helden Group (Berlin)

Many molecules of biological relevance, or biomolecules, are notoriously difficult to bring into the gas phase due to their large mass and/or low vapor pressure. A method capable of producing ions of such non-volatile compounds is electrospray ionization (ESI), which has developed into a tremendously successful and popular technique for analytical biochemists in the past decades [178–181]. The technique was employed by G. von Helden and his group to produce HNDs doped with large biomolecules such as amino acids, peptides and even proteins as large as ~12000 amu for spectroscopic investigations. In the experimental setup used by Bierau and co-workers, ions are produced in an ESI source, mass-selected in a QMS and finally transferred into either a ToF-MS for analysis or a hexapole ion trap for storage. After the trap is loaded, a beam of very large HNDs generated by a pulsed source is guided through the trap. Ions picked up by HNDs can leave the trap due to the HNDs' high kinetic energy [182]. The authors demonstrated the doping of HNDs with singly charged phenylalanine as well as the much larger protein cytochrome C (~12,000 amu) in multiple charge states and determine the mean HND sizes as $<N> \sim 10^{10}–10^{12}$ by using a deflection method. In a subsequent study, Filsinger and co-workers exchange the deflection setup for two different ToF-MSs to monitor the HND size distribution ($<N> \sim 10^{5}–10^{7}$) as well as ions expelled from the HNDs upon laser irradiation in a photoexcitation/spectroscopy study of hemin, an iron-containing porphyrin in HNDs [183]. The setup was additionally used by Flórez and co-workers in an IR spectroscopy study of the peptide leu-enkephalin in its protonated form as well as its complex with a crown ether [184].

### 3.2.3.4 Drabbels Group (Lausanne)

The group of M. Drabbels studies the spectroscopic and dynamic properties of nanoscale systems using a variety of methods with his group at EPF Lausanne. For the purpose of studying atomic and molecular species in a HND environment and the interactions between the two, Drabbels and co-workers employ a versatile setup consisting of a HND source (typical droplet sizes of $<N> \sim 10^{3}–10^{4}$) followed by a pickup section and a multifunctional analysis region [185–187]. The cleverly designed analysis contains a QMS as well as a flight tube equipped with a set of electric lenses that can be used in a multitude of ways, e.g. as a ToF-MS, for photoelectron spectroscopy (PES) and velocity map imaging (VMI). Various lasers are used to excite and/or ionize dopant species. The setup has been used in several combined studies of mostly spectroscopic nature, but including mass spectrometric work as well. In a study of the excited state dynamics of silver atoms in HNDs upon photoexcitation, Loginov and Drabbels that $AgHe_n$ exciplexes ($n = 1, 2$) are efficiently formed upon excitation of embedded Ag atoms to the $^2P_{3/2}$ state [187]. These exciplexes generally form more efficiently in small droplets and are found to become solvated in the HND as $AgHe_2$. Curiously, the natural isotope ratio of silver is not conserved in AgHe, which forms more efficiently with the $^{107}$Ag isotope. The authors suggest that tunneling is responsible for the effect which manifests in an abundance ratio of $^{107}$AgHe to $^{109}$AgHe that is ~20–50% above the natural abundance ratio, being higher in larger HNDs. Taking full advantage of the previously described setup's capabilities, Braun and Drabbels conducted an extensive, three-part study about photodissociation (PD) of alkyl iodides in HNDs, investigating in detail the kinetic energy transfer [185], solvation dynamics [188] and fragment recombination [189]. Dopant alkyl iodides $CH_3I$, $CF_3I$ and $C_2H_5I$ are dissociated by 266nm laser irradiation, followed after 50 ns by non-resonant 780nm femtosecond PI of the products. Whereas the authors find that He-solvated fragment complexes $He_nI$ (from PD of $CH_3I$ and $CF_3I$) and $He_nCH_3^+$ (only from $CH_3I$) were produced in the PD/PI of alkyl iodides with up to $n = 15$ or more attached He atoms, $CF_3$ and $C_2H_5$ were only detected as bare fragments [188]. Based on the fragment velocity characteristics, Braun and Drabbels concluded that He-solvated fragment complexes are formed in the interior of the HND in a dynamic process balancing the formation of solvation shells around the escaping fragment and the encountered flow of He atoms due to the relative motion on their way through the HND. The He solvation and desolvation of barium (ions) as a probe for surface-generated ions was investigated by Zhang and Drabbels utilizing PI of Ba attached to small HNDs ($<N>$ ~6000), analyzed via ToF-MS and excitation spectroscopy [186]. Judging from the recorded excitation spectra, which are practically identical to those of $Ba^+$ in bulk liquid helium [190] as well as the mass spectra, the authors concluded that $Ba^+$ becomes fully solvated in the HND upon PI before being ejected as bare $Ba^+$ or $Ba^+He_n$ following photoexcitation of two different $6p \longleftarrow 6s$ transitions (D1 and D2 lines). A follow-up study by Leal and co-workers in cooperation with the theoretical group around Barranco and Pi attempted to unravel in detail the desolvation dynamics of photoexcited $Ba^+$ [191]. The authors find that half of the expelled ions are fully desolvated $Ba^+$ while the rest are detected as $Ba^+He_n$ ($n \leq 25$). Both the He attachment distribution of $Ba^+He_n$ exciplexes as well as the velocity of the desolvated $Ba^+$ and $Ba^+He_n$ showed different characteristics for the different excited states, but was independent of droplet size or laser intensity. Whereas the performed time-dependent density functional calculations are able to reproduce the formation of $Ba^+He_n$ exciplexes and the measured excitation spectra very well, they fail to capture the experimentally observed desolvation of $Ba^+$ and $Ba^+He_n$.

### 3.2.3.5 Kong Group (Corvallis)

Electron diffraction is a widely used tool for structure analysis of crystals, surfaces and gas-phase molecules [192]. The group of W. Kong at Oregon State University is working towards the construction of a molecular goniometer, capable of collecting electron diffraction images of single, oriented macromolecules as well as complexes grown in HNDs. This would allow the structure determination of species that are difficult to access by conventional methods. The group relies on pulsed droplet sources and uses ToF-MS technology to monitor and adjust doping conditions, which is crucial for electron diffraction imaging of HNDs. He and co-workers conducted a study of methods for the characterization of a pulsed beam of pristine and halomethane-doped HNDs using two simple and cost-effective, home-built ToF-MSs utilizing EI and MPI [193]. In two subsequent studies, Alghamdi and co-workers investigated EI and MPI of large HNDs ($N \sim 10^8$) doped with aniline. The authors found that in both cases, the large HND environment suppressed the ionization mechanism via shielding of the neutral dopant against charge transfer from $He_2^+$ [194] and caging of photoelectrons allowing for recombination with photoionic (fragments) [195], respectively. A recent study demonstrated the feasibility of electron diffraction of smaller HNDs ($N = 800$–$1.4 \times 10^5$) doped with $CS_2$ [196].

### 3.2.3.6 Ichihashi Group (Tokyo)

The group of M. Ichihashi at Toyota Technological Institute conducts research on catalytic activity of charged metal nanoparticles. Odaka and Ichihashi used a recently constructed HND setup employing a pulsed HND source and a QMS in a study colliding size-selected cobalt clusters $Co_m^+$ ($m \le 5$) with HNDs ($N \sim 1600$) to produce $Co_m^+He_n$ ($m \le 5$, $n \le 21$) [197]. The authors found several abundance anomalies, indicating increased stability of the corresponding configurations, such as $Co_2^+He_n$ ($n = 2, 4, 6$ and $12$), $Co_3^+He_n$ ($n = 3, 6$), $Co_4^+He_n$ ($n = 4$), and $Co_5^+He_n$ ($n = 3, 6, 8$ and $10$). By observing the total $Co_m^+He_n$ yield as a function of the relative velocity, different collision regimes are identified. The ion yield was highest at low relative velocity ($\sim 100$ m/s) due to the attractive electrostatic van-der-Waals interactions between $Co_m^+$ and HND, whereas a steep decline in yield towards higher relative velocities indicated a transition to hard-sphere collisions.

## 3.3 Review of Recent Work by the Scheier Group (Innsbruck)

We will now review the work performed by the group led by P. Scheier at the University of Innsbruck, which the authors of this chapter are currently part of. Much of the more recent work has already been discussed in three comprehensive review article about collisions, ionization and reactions in HNDs [22], the solvation of ions in helium [198] as well as the chemistry and physics of dopants embedded in HNDs [199]. We will thus focus on the most recent work performed in our group, including our own. The three following chapters are dedicated to the three experimental HND MS setups currently active in the group.

First, we will discuss the "ClusToF" experiment, a classical setup consisting of a HND source, followed by a pick-up region, an EI source and a high resolution ToF-MS. The main focus are mass spectrometric studies of chemical reactions and cluster growth inside HNDs, with the additional capability of performing action spectroscopy on He-tagged ions.

The "Snowball" experiment was designed to investigate the behavior of large, pristine HNDs. Utilizing a tandem MS setup consisting of two electrostatic SF-MSs, each equipped with an EI source, Laimer and co-workers recently demonstrated the existence of highly charged, stable HNDs [171]. The setup can be modified to allow the synthesis of nanoparticles in these large, highly charged and $m/z$-selected HNDs, followed by deposition on a substrate for *ex situ* analysis.

The third experiment, "Toffy" combines the newly developed technique of pickup into highly charged, $m/z$-selected HNDs [200] with a high-resolution ToF-MS. Tiefenthaler and co-workers constructed a highly versatile experimental setup which was able to produce considerable scientific output in a fairly short period of time. While both ClusToF and Toffy are similar in their scientific scope, the latter is additionally capable of performing collision-induced dissociation studies and controlling the helium attachment to analytes in an unprecedented manner. A more detailed description of each experiment is given in the respective chapter.

### 3.3.1   Classical HND MS Experiments

Early HND MS experiments by the Innsbruck group between 2006 and 2009 utilized either a QMS [201] or SF-MS [202–205] for mass analysis. It quickly became clear, however, that none of the instruments were ideal for HND MS. Whereas the QMS lacked in both resolution and accessible mass range, the SF-MSs' low ion yields forced experimenters to slowly scan the mass range, which is problematic since experimental conditions changing over time may produce distorted mass spectra. These issues led to the construction of a new apparatus in 2010, the "ClusToF" experiment, now the longest running experiment in the Innsbruck group. It consists of an HND source (5$\mu$m nozzle diameter, typical stagnation pressure $p_0 \sim 20$ bar, nozzle temperature $T_0 \sim 10$K) followed by two differentially pumped pickup chambers, each equipped with an oven and gas inlet ports, and a third chamber housing a Nier-type (EI) ion source (Fig. 2.15). Ions created in this region are extracted via a weak electric field and focused by an electrostatic ion guide into an orthogonal extraction reflectron-ToF-MS. The ToF-MS can be mounted such that the extraction region is mounted in-line with the HND beam or perpendicular to it. A more detailed description of

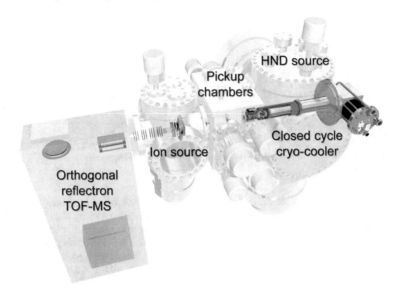

**Fig. 2.15** Schematic setup of the "ClusToF" experiment. The HND beam traverses two differentially pumped pickup chambers before entering an ion source chamber, follow by the ToF-MS. In the current version, the ToF-MS is rotated by 90°, mounted horizontally at the open flange of the ion source chamber facing the viewer, however, there were no significant differences found between the two alignments. Reproduced with permission from Ref. [22]. © copyright Elsevier. All rights reserved

the experiment (in-line configuration) can be found in [172]. The employed ToF-MS is able to overcome the deficiencies of the previously used MSs by being able to record full-range mass spectra ($m/z$ up to ~50 kDa) with a mass resolution of $R$ ~6000 while maintaining a reasonable ion yield. An example mass spectrum of pure $He_n^+$ fragments from larger droplets is shown in Fig. 2.16. The ion yield can be increased manifold if a high mass resolution is not required, but a higher ion yield is beneficial, e.g. for action spectroscopy measurements, which can be performed on the experiment as well by aligning a tunable OPO laser system with the ToF-MS extraction region [150, 206–208]. The ClusToF experiment has proven to be a reliable and versatile design, in large part responsible for the Innsbruck group's successful application of HND MS in the last decade. In the following section, we will present the most recent studies performed with this experimental setup.

### 3.3.1.1 Cationic and Protonated Clusters of Rare and Inert Gases

The capabilities of the high resolution ToF-MS setup were recently demonstrated when Gatchell and co-workers were able to resolve an issue that had been a topic of debate for more than 30 years. Magic numbers found in early mass spectrometric studies of cationic rare gas, but especially Ar cluster ions, associated with shell

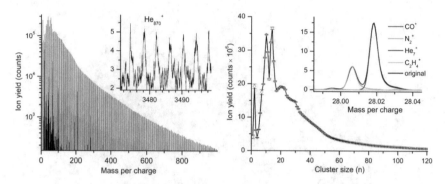

**Fig. 2.16** Mass spectrum of small $He_n^+$ from EI of pristine HNDs (**a**) and extracted cluster size distribution (**b**) as recorded with the ClusToF setup. Magic numbers typically observed for $He_n^+$ with this setup are as $n = (7)$, 10, 14, 23 and 30 which can be seen in the cluster size distribution (**b**). Both insets demonstrate the capabilities of the high resolution ToF-MS by displaying the mass resolution around $m/z = 3500$ (**a**) and the deconvolution of an isobaric mass peak at $m/z = 28$ (**b**). Reproduced with permission from Ref. [22]. © copyright Elsevier. All rights reserved

closures of icosahedral structures resulting from sphere-packing models [11, 209], were not always reproducible [13, 210–214]. In a recent study, Gatchell et al. studied HNDs co-doped with $H_2$ and Ar [215]. High-resolution mass spectra reveal both bare $Ar_n^+$ and protonated $Ar_nH^+$ cluster ion series which exhibit vastly different characteristics such as magic numbers. The bare $Ar_n^+$ series exhibits relatively few features (magic $n = 16$, 19, 23, 27, 81, 87, antimagic $n = 20$) that agree well with previously reported measurements [13, 210–214], but miss key features of sphere-packing models such as the first and second icosahedral shell closures at $n = 13$ and 55. In contrast, magic numbers found in the protonated $Ar_nH^+$ series are in essentially perfect agreement with previously mentioned early measurements [11] as well as the predictions of sphere-packing models (e.g. $n = 13$, 19, 26, 29, 32, 34, 49, 55, ...). In addition, the $Ar_7H^+$ ion is found the be magic, a feature that is rarely found in the literature [215]. Ab initio structure calculations performed on $Ar_n^+$ and $Ar_nH^+$ systems up to $n = 21$ are able to reproduce very well the magic numbers found in the small cluster ion series. The calculated structures and magic numbers (where available) also show good agreement with previous theoretical work on pure cationic [216] and protonated Ar clusters [217–219]. The studies of protonated rare gas clusters were extended towards Ne and Kr by Gatchell and co-workers [220] (Kr was co-doped with $D_2$) as well as towards He by Lundberg and co-workers [221]. In contrast to Ar, which is essentially monoisotopic, the richer isotopic patterns of Ne and especially Kr complicate the distinction between pure and protonated rare gas cluster series, preventing the authors from studying cluster sizes larger than $n$ ~ 30. Similar to Ar, the mass spectra of $Ne_n^+$ and $Ne_nH^+$ are found to be fairly different. Whereas the $Ne_n^+$ and $Ar_n^+$ series share a few magic numbers such as $n = 14$ and 21 (although both are much weaker in $Ar_n^+$), other features are completely different, the magic $n = 19$ and 27 of $Ar_n^+$ are both slightly antimagic in $Ne_n^+$ and the prominent antimagic $n = 20$ of $Ar_n^+$ does not present a noteworthy anomaly in $Ne_n^+$

at all. In contrast, the agreement between the protonated $Ne_nH^+$ and $Ar_nH^+$ series is much better. Some features become much more apparent upon calculating the second differences, defined as $\ln\left[\frac{2I_n}{(I_{n-1}+I_{n-1})}\right]$ (where $I_n$ denotes the yield of the $n^{th}$ ion in a series), as shown in Fig. 2.17b. It is now clear that all main features of $Ar_nH^+$ up to $n = 32$, including $n = 7$, are equally found in $Ne_nH^+$. Calculations reveal that $Ne_7H^+$ consists of a linear $Ne$-$H^+$-$Ne$ core, surrounded by $Ne$ atoms spanning a pentagon, which is oriented perpendicular to the axis of the ionic core and shares a common plane with the proton, a structure that is equally found for $Ar_7H^+$ [215, 217, 219].

Calculations performed by Gatchell and co-workers [215] also reveal the reason for the difference between shell closures at $Ar_{13}H^+$ and $Ar_{14}^+$ (instead of $Ar_{13}^+$ as would be expected from sphere packing schemes). The central ion present in smaller clusters is a linear $Ar_3^+$ molecule with reduced binding lengths between $Ar$ atoms due to the presence of the charge. This compression of binding length leaves room in the icosahedral structure for a 14th $Ar$ atom interacting relatively strongly with the ionic core, essentially forming an $Ar_4^+$ central ion, surrounded by two pentagonal rings of five $Ar$ atoms. In the protonated $Ar_{13}H^+$ structure the proton, carrying most of the charge, can squeeze in between two of the central three $Ar$ atoms without significantly distorting the icosahedral structure of the neutral cluster. Calculations of neutral, cationic and protonated $Ne_{19}$ systems show analogous effects [220]. Again, $Ne_{19}$ and $Ne_{19}H^+$ adopt similar, highly symmetric geometries with the proton squeezing in between two $Ne$ atoms in the center of the cluster. The central ion in $Ne_{19}^+$ is a contracted, covalently bound $Ne_2^+$, which again distorts the highly symmetric structure. Assuming this allows for the addition of two strongly interacting $Ne$ atoms along the axis of the central ion could explain the observed magic character of $Ne_{21}^+$.

In contrast to the lighter rare gas species, both the $Kr_n^+$ and $Kr_nH^+$ series share some key features of sphere packing, such as magic numbers at $n = 13$, 19 and 29. Other magic numbers found in all $Rg_nH^+$, such as $n = 17$ and 26 appear to be shifted down to $n = 16$ (cf. Ar series) and 25. It appears as the structures of $Kr_n^+$ are distorted less by the presence of the charge compared to the lighter rare gas species. This is supported by calculated structures of neutral, cationic and protonated $Kr_{19}^+$ [220], showing that indeed the charge-induced contraction of the $Kr$-$Kr$ binding length is relatively smaller than in the lighter rare gas species.

The study of protonated $He$ clusters by Lundberg et al. shows that the lightest rare gas presents a special case once again, with neither $He_n^+$ nor $He_nH^+$ showing good agreement with icosahedral packing structures [221]. Besides the strong $He_2^+$ ion signal, the most prominent magic numbers of small $He_n^+$ are $n = 10$, 14, 23 and 30, with the latter two being much less prominent, yet easy to identify upon careful analysis. Both $n = 14$ and 23 are found in other series as well, the former in bare ions such as $Ne_n^+$ and $Ar_n^+$, while the latter is seen in $Ne_nH^+$ and $Ar_nH^+$. The mass spectrum of $He_nH^+$ is fairly unique among the studied rare gas species. While one magic number, $n = 13$, is shared between $He_nH^+$ and all other protonated cluster ion series as well as $Kr_n^+$, none of the other magic numbers found, namely $n = 6$, 11, 22 (weak), 35 or 39 show enhanced abundance in any other series, except perhaps for $n = 22$, which has enhanced abundance in $Kr_n^+$, but is not considered magic since $Kr_{23}^+$ has similar intensity. A detailed theoretical study of $He_nH^+$ ($n \leq$ 18) by Császár and co-workers shows significant drops in the calculated successive electronic evaporation energies (i.e. the energy difference of the reaction $He_nH^+$ &#xF0E0; $He_{n-1}H^+ +$ He) after $n = 6$ and 13 at the HF, MP2 and CCSD(T) (only up to $n = 7$) levels of theory levels of theory [222]. The calculated successive electronic evaporation energies are found to be comparable between MP2 and CCSD(T), but consistently lower at the HF level of theory. The values are lying between 25 (HF) and 44 meV (MP2/CCSD(T)) for $n = 3$–6 (this is incorrectly given as ~250 meV in [221]), dropping to between 5 and 17 meV for $n = 8$–13, with $n = 7$ slightly higher between 12 and 27 meV. Surprisingly, none of the theoretical methods indicate an enhanced binding energy of $He_{11}^+$, the strongest anomaly found by Lundberg et al., quite the contrary, as the MP2 calculations even display a slight local minimum. After $n = 13$, the successive electronic evaporation energies are dropping towards 0 for HF and 6 meV for MP2 and CCSD(T). The binding energy of additional He atoms can be expected to converge to the HND "bulk" value of 0.6 meV per atom [22], which is also recognized as a possible onset for microscopic superfluidity by Császár et al. [222].

The findings of Gatchell, Lundberg and co-workers showed that the characteristics of cationic and protonated (as well as neutral) rare gas clusters, especially of the lighter species (He–Ar) can be fairly different, probably due to charge-induced distortion of the cluster geometries. Gatchell et al. [215] suggest that the discrepancies of magic numbers in early measurements of rare gas cluster ions, especially

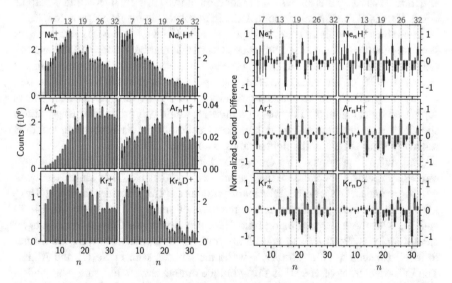

**Fig. 2.17** Size distributions (**a**) and normalized second differences (see text for an explanation) (**b**) of pure and protonated rare gas clusters with common local anomalies indicated above. Many magic numbers corresponding to (partial) shell fillings are found to be shared between the different rare gas species. Reproduced from [220]. Licensed under CC BY 4.0

Ar, were due to erroneous identification of unresolved, protonated $Rg_nH^+$ cluster ions where protonation was likely caused by water impurities. The studies further illustrate how sensitive cluster systems can be towards small changes such as the addition of a minor impurity.

While no further studies of the heaviest rare gases xenon and radon were conducted due to the increasingly difficult analysis of the very rich isotopic pattern of Xe and the radioactivity of Rn, respectively, Martini and co-workers studied a different van-der-Waals system, namely mixed cationic clusters of nitrogen and deuterium [223]. Again, the high resolution of the ToF-MS is crucial in the identification of the myriad ions such as $(N_2)_n^+$, $(N_2)_nD_m^+$, $He_nD_m^+$ and $D_m^+$. The main ion series identified is due to $(N_2)_nD^+$, which, in analogy to the discussed rare gas studies [215, 220, 221], shows very different characteristics from $(N_2)_n^+$. Whereas the lone notable feature in the pure cationic series is the enhanced intensity of $(N_2)_{19}^+$, an array of magic numbers, namely $n = 2, 6, 7, 12, 17$ is easily identified in the protonated $(N_2)_nD^+$ series. Structures calculated for the magic number ions $(N_2)_nD^+$ ($n = 6, 7, 17$) reveal a linear $[N_2–D–N_2]^+$ central ion with a relatively even charge distribution between the three constituents. The structure of $(N_2)_6H^+$ is found to be octahedral and slightly energetically favorable over the pentagonal bipyramid found for $(N_2)_7H^+$. Similar to the rare gas studies, the magic numbers $n = (2), 7, 12$ and $17$ can be explained in terms of packing molecules around the central $[N_2–D–N_2]^+$ ion in icosahedral (sub-)shells. The calculated structure of the magic $(N_2)_{19}^+$ is found to be similar, but the linear, central $(N_2)_2^+$ ion is contracted in comparison to $[N_2–D–N_2]^+$, allowing for two additional $N_2$ molecules placed along the central axis. By increasing the $D_2$ partial pressure, mixed $(N_2)_nD_m^+$ cluster ions were observed, predominantly with odd $m$. The data reveal a few indications that $N_2$ molecules can be partly replaced by $D_2$ in these mixed cluster ions. For one, the strong $n = 6$ anomaly in $(N_2)_nD^+$ seemingly shifts down to $n = 5$ in $(N_2)_nD_3^+$, but remains at $n = 5$ for $(N_2)_nD_5^+$ instead of further shifting down. Additionally, the progression of the $(N_2)_nD_m^+$ ($n = 0$–12) series shows a strong drop in intensity after $2n + m = 35$ (i.e., a total number of 35 atoms), except for $n = 0$ and 1. The interpretation of the authors is that any mixed cluster containing a total number of 17 molecules and at least two $N_2$ molecules forms a magic cluster ion. The minimum of two $N_2$ molecules suggests that a central $[N_2–D–N_2]^+$ forms the ionic core, surrounded by a structure similar to the one found for $(N_2)_{17}D^+$. These interpretations are supported by theoretical calculations. Structure optimizations of three possible $(N_2)_{16}D_2D^+$ isomers support this interpretation, since the structure of the $(N_2)_{17}D^+$ cluster ion upon the exchange of $N_2$ with $D_2$ is found to be largely preserved in all cases. However, while $N_2$ is found to prefer a linear orientation towards the charge center, $D_2$ aligns perpendicular to it. While no further structure optimizations of mixed $(N_2)_n(D_2)_mD^+$ cluster ions were attempted, potential energy surfaces of the interaction of a point charge (i.e., a proton) with $N_2$ and $D_2$ were calculated. The position with the lowest interaction energy is found to be along the molecular axis in the case of $N_2$ and perpendicular to it for $D_2$. Additionally, the interaction energy is found to be higher for $N_2$ compared to $D_2$, which explains why a $[N_2–D–N_2]^+$ ionic core is formed where possible. Furthermore, by considering the atomic van-der-Waals radii and the D–D bond length, the authors conclude that the spacing of molecules in the structural lattice remains very similar upon exchange between the two species, explaining the observed efficient mixing of $N_2$ and $D_2$ in mixed $(N_2)_n(D_2)_m D^+$ cluster ions.

### 3.3.1.2    Anionic Complexes of Hydrogen

It is no surprise that hydrogen is one of the most widely investigated chemical elements and perhaps the one that is best understood. Its simplicity, being comprised of only protons and electrons (as well as neutrons in the case of its isotopes deuterium and tritium), makes hydrogen appealing for theoretical studies. It is ubiquitous, both in the wider universe as well as a constituent of countless chemical compounds vital to life on earth. In the past decades, there has been growing interest in and rapid development of the possibilities of using hydrogen for energy storage due to the rising demand for renewable and sustainable energy sources in the face of global warming [224]. Hydrogen complexes are also of astrophysical relevance and have been considered [225, 226] as potential carriers of the diffuse interstellar bands (DIBs) [146, 227]. Here, we take a closer look at anionic complexes of hydrogen. Since $H_2^-$ is unstable except for highly rotationally excited metastable states [228], the simplest stable polyatomic anion is $H_3^-$. First theoretical investigations were published in 1937 [229] and suggested the existence of a stable $H^-(H_2)$ anionic complex. It took close to 40 years until the first reports of its detection were made [230, 231]. Since the first observations were disputed due to poor signal quality [232], it was only relatively recent that $H_3^-$ was [32, 33] shown to be stable using high-quality theoretical calculations [233, 234] and experimental detection in a dielectric discharge plasma experiment [235]. Motivated by the possible formation of anionic hydrogen clusters in space, the structure and stability of $H_n^-$ anions with odd $n$ between 3 and 13 was first investigated around 1980 at the Hartree-Fock level of theory [226, 236] and more recently in 2011 by DFT [237]. Despite the efforts, there was no consensus on the equilibrium geometries of $H_n^-$ anions or the stability of such systems.

In 2016, Renzler et al. reported the first direct observation of anionic hydrogen and deuterium clusters $H_n^-$ ($D_n^-$) with $n \geq 5$ formed by doping large HNDs (~$10^6$ He atoms) with molecular hydrogen and deuterium using the ClusToF setup [238]. The authors observed exclusively $H_n^-$ clusters with odd $n$ (Fig. 2.18) and concluded that these clusters consist of an anionic $H^-$ core, formed by dissociative electron attachment (DEA) to $H_2$, with two to well over 100 $H_2$ molecules attached, bound by ion-induced dipole interaction. No significant differences in the mass spectra were found between $H_n^-$ and clusters. The authors also obtained equilibrium geometries from DFT calculations for $H_n^-$ ($n = 3$, 5 and 7), finding that $H_5^-$ adopts a linear structure, in agreement with [226, 236] but in contrast to the bent structure found by [237]. Meanwhile, $H_7^-$ is predicted to have trigonal pyramid shape, agreeing with [237], but contradicting [226, 236]. Furthermore, clusters with $n = 25$, 65 and 89 (i.e. 12, 32 and 44 $H_2$ ($D_2$) molecules attached to the $H_n^-$ core) were found to be anomalously abundant, suggesting increased stability of these clusters, or magic

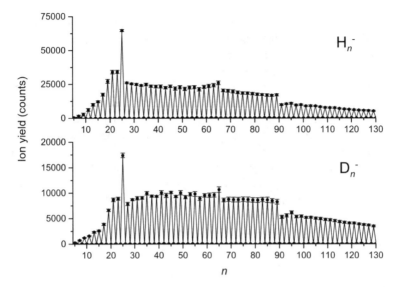

**Fig. 2.18** Size distributions of anionic hydrogen (deuterium) clusters from doped HNDs as measured with the ClusToF setup. Both species produce essentially identical features such as apparent shell closures at $n = 25$, 65 and 89, corresponding to 12, 32 and 44 $H_2$ molecules surrounding an $H^-$ ionic core. Reproduced with permission from Ref. [238]. © copyright American Physical Society. All rights reserved

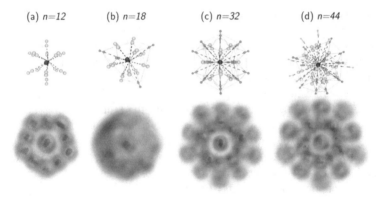

**Fig. 2.19** Classical (top) and quantum structures (bottom) of anionic hydrogen clusters $H^-(H_2)_n$ as calculated using PIMD. The strong symmetry and heavy localization of solvating $H_2$ molecules in the magic number clusters ($n = 12$, 32 and 44) is clearly visible in comparison with the non-magic $n = 18$ cluster. Reproduced with permission from Ref. [239]. © copyright AIP Publishing. All rights reserved

character. It was found that these clusters consist of the anionic core, surrounded by up to three solid-like solvation shells of 12, 20 and 12 molecules in an icosahedral-like geometry, a structure that was previously found for cations solvated in helium, such as theoretically in $Na^+He_n$ [240] and experimentally in $Ar^+He_n$ [241]. Calvo and Yurtsever modeled the structure of $H^-(H_2)_n$ and $D^-(D_2)_n$ clusters for a wide range of $n = 1$-54 using path integral molecular dynamics (PIMD) with a potential optimized for yielding accurate geometries [239]. Their findings clearly show the icosahedral-like shape of the magic cluster anions while also being able to qualitatively reproduce the geometries found by DFT for $n = 1$–6 (Fig. 2.19) [237]. In a separate follow-up study, Joshi et al. used dispersion-corrected DFT to determine the geometries of the experimentally found magic anionic hydrogen clusters, again confirming the icosahedral-like structure and increased stability [242]. The authors further calculated the binding energies and HOMO-LUMO gaps for systems of the form $X(H_2)_n$, for $n = 12, 32$ or $44$ and a variety of metal and non-metal anionic and neutral atomic impurities X. The findings suggest that many of the investigated $X(H_2)_n$ systems exhibit similar (icosahedral) geometries and binding energies as the corresponding anionic hydrogen clusters, predicting similar stability and feasibility of experimental detection. Another recent study by Mohammadi et al. attempted to more accurately predict structures and energies of small $H^-(H_2)_n$ clusters ($n = 1$–5) reported by previous publications. While the structures for $n = 1$–4 agree well with those found by [237, 239], the structures for $n = 5$ differ from those previously reported. The authors also report structural isomers for $n = 3$–5 and different binding energies and bond lengths compared to the previous DFT and PIMD studies. Huang et al. recently suggested that small anionic hydrogen clusters $H^-(H_2)_n$ clusters ($n = 1$–7) could be carriers of DIBs [225]. The authors identified 25 absorption bands between 440 and 620 nm which coincide with DIBs within the uncertainty (which is rather large at ~±5nm) of the calculated absorption wavelength. Since the energies involved in these transitions are much larger than the binding energies of such clusters, an excited cluster would be metastable and dissociate shortly after excitation. Thus, an effective production mechanism must exist for the clusters to contribute to any DIBs, which the authors propose to occur on interstellar dust grains.

In a follow-up study, Renzler et al. studied mixed oxygen and hydrogen (as well as deuterium) cluster anions produced by attachment of low-energy electrons (0–30 eV) to large helium droplets ($N \sim 5 \times 10^5$) co-doped with oxygen and hydrogen [243]. The experiment produced $H_nO_m^-$ anions with a wide variety of combinations of $n$ and $m$. The authors find that anions, even pure $O_m^-$, are predominantly produced at electron energies associated with resonances of electron attachment (EA) to $H_2$. In contrast to the pure hydrogen experiment, however, anions with even $n$ are not only readily observed, but in many cases more abundant than neighboring clusters with odd $n$. It appears that electron-induced chemistry between the two dopants leads to the production of different central anions in mixed $H_nO_m^-$ anion clusters, most notably $HO_2^-$ and $H_2O^-$. While the former can certainly be expected due to DEA to $H_2$ and subsequent reaction of $H^-$ with $O_2$, $H_2O^-$ is unstable with respect to autodetachment and thus its detection is surprising. The authors concluded that a rapid electron transfer mechanism from $H_2^-$ to $O_2$ must exist to initiate DEA to $O_2$, producing $O^-$ that subsequently reacts with $H_2$ to produce the $H_2O^-$ anion, which is stabilized in the presence of at least one other molecule. Surprisingly, $H_nO_m^-$ anions with both even $n$ and $m$ are also observed, leading the authors to conclude that the HND environment is able to efficiently quench both DEA processes and stabilize the resulting cluster anions. All the phenomena and processes described above are likely unique to mixed clusters formed in helium droplets.

### 3.3.1.3  Anionic Complexes of Nitrogen

Motivated by their possible application as non-polluting high-energy-density materials (HEDMs), polynitrogen compounds (i.e., compounds comprised primarily or exclusively of nitrogen) have received growing interesting in past decades [244–248]. While theoretical investigations led the way predicting structures of possible polynitrogen compounds, experimental verification generally proved difficult. Pure cationic nitrogen clusters have been widely produced and studied using various methods such as supersonic molecular jets [249–253], sputtering of solid $N_2$ [254, 255] and ionization of doped helium droplets [49]. Studies of pure anionic nitrogen clusters, however, are rare. Reports of nitrogen cluster anions $N_n^-$ with $n > 3$ were made by Tonuma et al., who observed $N_n^-$ ($n = 1$-$9, 11, 13$) upon sputtering of frozen $N_2$ with highly charged, energetic Ar ions from a linear particle accelerator [254], Vij et al., who isolated cyclic $N_5^-$ in an ESI-MS-MS experiment [256], Vostrikov and Dubov, who measured the cross sections for attachment of low energy electrons to large $(N_2)_n$ clusters [257] and Pangavhane et al. in 2011, who detected $N_n^-$ ($n = 6, 10$-$15$) upon laser desorption ionization of phosphorus nitride ($P_3N_5$) [258].

Weinberger et al. recently published a detailed study on the formation of nitrogen cluster anions $N_n^-$ with $n \geq 3$ upon EA to large helium droplets ($N \sim 10^5$) doped with $N_2$ [259]. The most prominent anion series observed was $N_m^-$ with $3 \leq m$ odd $<$ 140. Further anion series arising due to impurities are odd-numbered nitrogen clusters complexed with water and $C_2N_2^-$ complexed with intact nitrogen molecules. Evidence for the existence of even-numbered $N_m^-$ with $m > 2$ can be found, but the corresponding anion signals are roughly two orders of magnitude weaker than those of the nearest odd-numbered clusters, barely exceeding background noise levels. No evidence is found for $N^-$ or $N_2^-$, in accordance with the short lifetimes of $N_2^-$ (on the order of $10^{-15}$s [260]) and negative electron affinity of N ($-0.07$ eV [261]). The authors concluded that odd-numbered $N_m^-$ are comprised of a central $N_3^-$ azide anion solvated by $n = \frac{m-3}{2}$ $N_2$ molecules. The $N_3^-(N_2)_n$ series exhibits distinct anomalies in the ion intensity after $n = 4$ and $11$, which indicates increased stability of these compounds. A follow-up study by Calvo and Yurtsever employed a variety of quantum chemical methods to find the stable structures of odd-numbered nitrogen anion clusters [262]. The calculations confirm the assignment of an $N_3^-$ anionic core solvated by $N_2$ molecules and yield optimized geometries for $N_3^-(N_2)_n$ for $n = 1$-$9$. The first five $N_2$ molecules are found to arrange in parallel around the linear $N_3^-$ and form the first solvation shell. Calculations of binding and dissociation energies confirm the higher stability of $N_3^-$ $(N_2)_5$ expected from the calculated equilibrium geometries. While the general agreement between the two studies is good, this finding is at odds with the experimental observations of Weinberger et al., who find $N_3^-(N_2)_4$ to be particularly stable. Unfortunately, the pronounced enhanced stability of $N_3^-(N_2)_{11}$ could not yet be investigated theoretically due to the size of the system.

### 3.3.1.4    Solvation of Alkali Clusters in HNDs

Alkali atoms and clusters have been a popular study target in connection with HNDs, owing largely to their simple electronic structure and exceptional doping characteristics. Whereas an interior location is favorable for most other dopants, alkali atoms and small clusters are found to reside in dimples at the HND surface due to short range Pauli repulsion between their and the surrounding He atoms' $s$ valence electrons [62, 63, 263–266]. Cationic alkali atoms and clusters, however, are solvated due to the strong, charge-induced van der Waals interaction with the surrounding He atoms. This causes the closest He atoms to form a snowball structure with a high degree of localization around the ion. The solvation of alkali ion impurities and the structure of snowballs around them was studied in detail theoretically by Reatto et al. [80, 267, 268], Gianturco et al. [269–271] and other authors [272, 273] with a wide variety of methods. Experimental work can be found in a photoionization study of HNDs doped with Na–Cs [93] and an EI study of Na- and K-doped HNDs [274]. Lithium is often omitted from HND MS studies due to complications in the evaluation caused by numerous mass interferences. Recently, Rastogi et al. conducted a combined study of $Li^+$ solvation combining high-resolution MS and different theoretical methods [275]. The mass spectra revealed $Li_m^+$ ($m \leq 3$) and $He_nLi^+$ (extending to $n > 50$), with prominent local anomalies in the ion abundance of $He_nLi^+$ at $n = 2, 6,$ 8, 14 and 26, a weaker anomaly at $n = 4$ as well as pronounced minima at $n = 21, 24,$ 27 and 28. Theoretical calculations suggest particularly strongly bound structures of $He_nLi^+$, $n = 4, 6$ and 8, in decent agreement with the experimental observations and are in good agreement among each other, regardless of whether quantum or classical approaches are used. However, the calculations failed to reproduce the strong $n = 2,$ 14 anomalies and mostly indicate a structure of slightly increased stability at $n = 10$ which is not reflected in the mass spectra.

Interestingly, creating a positively charged alkali cluster is not the only way leading to its solvation. Using a classical model, Stark and Kresin predicted that, upon growing larger, the relative strength of attractive and repulsive forces between dopant alkali clusters and HNDs shifts, eventually tipping the scale in favor of the submersion of alkali clusters into the HND beyond a certain, element-specific, critical size $n_c$ [276]. The authors calculated $n_c = 23, 21, 78$ and $>100$, for Li, Na, K and Rb, respectively, while for cesium a failure of submersion is indicated by the exceedingly large $n_c$ value of 625. Shortly after, An der Lan et al. conducted studies utilizing EI MS of HNDs doped with sodium [277] and potassium [278] in an attempt to experimentally determine $n_c$. The authors followed the suggestion given by Stark and Kresin [276] to exploit the fact that dopants can be selectively ionized via Penning

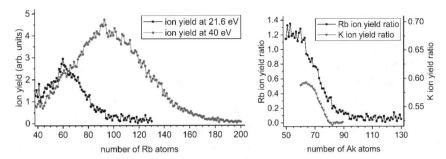

**Fig. 2.20** Cluster size distributions extracted of cationic Rb clusters grown in large HNDs at different electron energies (**a**) and ratio of ion yields recorded at 21.6 and 40 eV for Rb (**b**) [126]. The gradual submersion of Rb clusters into the HND is evident from the decrease in ion yield between 60 and 100 atoms in the 21.6 eV mass spectrum in (**a**). A gradual decline of ion yield ratio corresponding to a submersion into the HND is noticeable starting around 50 Rb atoms is evident in (**b**). Rubidium clusters larger than ~100 atoms are assumed to be fully submerged. Also shown is a similar transition for K clusters suggesting submersion into the HND around 80 K atoms [278]. Reproduced from [126]. Licensed under CC BY 4.0

ionization *if* they are located at or near the HND surface. By monitoring the cluster ion yield as a function of both cluster size and incident electron energy, An der Lan et al. were able to measure $n_{c, exp} = 21$ for Na and $n_{c, exp} \approx 80$ for K, in excellent agreement with the theoretical predictions. While the transition from an exterior to an interior location was observed sharply for Na clusters, it is more diffuse for K and was expected to be even more diffuse for rubidium [278].

Recently, Schiller et al. expanded the experimental evidence of alkali cluster submersion with a study of rubidium [126]. The authors recorded EI mass spectra of clusters with up to 200 Rb atoms grown on or in HNDs at electron energies between 8 and 160 eV (Fig. 2.20a). For determination of $n_c$, they followed the same principle proposed by Stark and Kresin and implemented by An der Lan et al. and indeed found a diffuse transition suggesting full submersion of Rb clusters larger than ~100 atoms (Fig. 2.20b). While this observation is still compatible with $n_c > 100$ predicted by Stark and Kresin (the exact calculation yielded $n_c = 131$ which was interpreted carefully by the authors), the agreement is worse than for the lighter alkali species Na and K. It will most likely be difficult to further extend the series of studies towards the heaviest (nonradioactive) and lightest alkali species in cesium and lithium, respectively. Since Stark and Kresin's model slightly overestimates the critical submersion size of Rb clusters, it is possible that this is also true and perhaps exaggerated for cesium. This could, in principle, warrant revised theoretical work and an attempt of experimental observation of Cs cluster submersion. Lithium, on the other hand, does not form large enough clusters on HNDs to allow for the experimental observation with the previously used method. A possible solution would be the production of larger, neutral Li clusters, e.g. in a gas aggregation source followed by HND pickup and analysis via the discussed or other methods.

### 3.3.1.5 Studies of Gold Complexes

Gold has always been precious to humans ever since its discovery. This is in part due to the chemical inertness of bulk gold under normal conditions, enabling pieces of art crafted from the noble metal to retain their beauty over long periods of time [279]. Gold has been a source of fascination, if not obsession for alchemists who attempted to transmute lead into gold. Gold nanoparticles (AuNPs) have been used for millennia, e.g. for their optical [280] and medical properties [281], long before the nature of the phenomena could be explained. The first publication starting to recognize AuNPs was Faraday's 1857 report, mentioning a "finely divided metallic state" [282]. Faraday's work laid the foundation for expanding the knowledge about nanoparticles via colloid science, especially during the first half of the 20th century [283]. It was the 1980s ground-breaking inventions of the scanning tunneling (STM) and atomic force microscopes (AFM), which allowed for the routine analysis and manipulation of nanoscale materials down to atomic precision, that helped colloid science to develop into today's ever-present field of nanotechnology. Gold became the metal of choice in many nanotechnology experiments and applications since it is not, like most other metals, readily covered by a passivating oxide layer [283]. Its widespread use in nanotechnology ultimately lead to the discovery that while bulk gold is inert, nanoscale gold actually exhibits significant catalytic activity, be it as ions [284] or neutral nanoparticles [285]. Ironically, while gold's inertness long prevented chemists from developing a real interest, it also promoted its widespread use in nanotechnology and, in turn, helped open up the field of gold catalysis. It has since become clear that in many cases gold is not only a good, but the best known catalyst [279, 283, 286–289]. Additionally, metal nanoparticles emerged as good candidates for hydrogen storage as part of a green energy revolution [279, 290291292293]. These discoveries have sparked great interest in the fundamental research of nanoscale gold and gold chemistry with the goal to better understand and further develop gold catalysis.

Gold clusters were mainly produced using sputtering [294–296] and laser evaporation methods [297, 298]. In a recent study, Martini et al. presented mass spectra of cationic and anionic gold clusters produced via EI/EA of gold-doped helium droplets [299]. The ion abundance distribution extracted from the cationic mass spectrum is roughly shaped like a log-normal distribution, caused by the statistics of picking up dopant atoms in helium droplets, superimposed with a pronounced odd-even oscillation up to $n = 21$, with odd-numbered $Au_n^+$ being significantly more abundant (Fig. 2.21a). Becker et al. performed CID on $Au_n^+$ ($2 \leq n \leq 23$) and found that clusters with $n \leq 15$ dissociate by dimer evaporation if $n$ is odd, while larger clusters dissociate by monomer evaporation [300], which was later supported by ion mobility measurements and theoretical calculations [301]. The observed odd-even oscillations were also found by Katakuse et al. in mass spectra of cationic gold clusters produced by sputtering a gold foil with 10 keV $Xe^+$ ions [294]. Although the general shape of the ion abundance distribution presented by Katakuse et al. is an exponential decay versus a log-normal distribution in the helium droplet experiment, the most prominent features are in excellent agreement—apart from the discussed

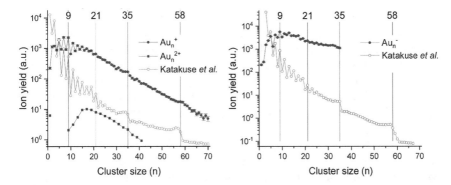

**Fig. 2.21** Cluster size distributions of $Au_n^{\pm}$ produced from EI of gold-doped HNDs (solid symbols) and by bombardment of a gold sheet with 10 keV $Xe^+$ (open symbols) [294]. Both methods produce similar results in terms of observed odd-even oscillations and shell closures. The fact that features are less apparent in the HND mass spectra can be attributed to the suppressed fragmentation via cooling of nascent ions by the HND environment. Reproduced with permission from Ref. [299]. © copyright Elsevier. All rights reserved

odd-even oscillations, both studies reveal local anomalies at $n = 21$, 35 and 58 as well as a local minimum at $n = 30$. Katakuse et al. applied a one-electron shell model used in modeling $Na_n^+$ systems [3], explaining the enhanced stability of odd-numbered (even-electron) $Au_n^+$ and the magic numbers, which mark electronic shell closures [294]. Apart from monocationic gold clusters, dicationic $Au_n^{2+}$ (only odd-numbered are detected due to interference from $Au_n^+$) are also detected with roughly two orders of magnitude lower abundance and an appearance size of $n = 9$. The appearance size agrees with previously reported ones [302], however, there are also reports of appearance sizes as low as $n = 2$–5 [303–305].

Besides the previously discussed cationic clusters, Martini et al. also report mass spectra of anionic gold clusters revealing $Au_n^-$ with up to $n = 35$ atoms (Fig. 2.21b). No dianionic $Au_n^{2-}$ could be detected. The ion abundance distribution also features a log-normal shape with similar, but much less pronounced, odd-even oscillations between 5 and 21 gold atoms [299]. Anionic mass spectra by Katakuse et al. extending to $n = 67$ reveal more pronounced odd-even oscillations between $n \sim 3$–25 as well as abrupt drops at $n = 35$, 58 [295], neither of which could be confirmed from the mass spectra recorded by Martini et al. since, unfortunately, the recorded mass spectrum ends at $n = 35$. Interesting trends are noted in the helium droplet experiment upon scanning the ion yield of $Au_n^{\pm}$ as a function of electron energy. For both $Au_n^{\pm}$ species, the maximum ion yield shifts to a different electron energy value with increasing cluster size, however, the shifts are opposite in cationic and anionic clusters. For $Au_n^+$, the maximum shifts to lower electron energies for larger gold clusters and vice versa for $Au_n^-$. Both can be explained in terms of the dopant cluster ionization processes. In helium droplets, dopant clusters are usually not directly ionized, but rather via an intermediate $He^+$, $He^{*-}$ or $He^*$ species. While the energy transferred by these species to the dopant cluster is independent of the incident electron energy, the energy available to dopant clusters from inelastic scattering of secondary electrons is increased. This energy is assumed to enhance fragmentation of larger clusters, resulting in a higher yield of small cluster (fragment) ions. In the case of anionic $Au_n^-$, the trend is reversed—higher electron energy leads to an increased yield of larger cluster anions. The authors propose that larger helium droplets produce larger clusters, but also require larger amounts of helium to be evaporated to produce gas-phase ions—the energy for this being provided by the incident or secondary electrons [299].

A key step to understanding gold catalysis on small clusters is determining their structure. It was shown that in the regime of small ($n = 1$–20) gold clusters, the addition or removal of a single atom can drastically alter the cluster's catalytic activity [306]. Methods employed for structure determination of small $Au_n^+$ are ion mobility spectroscopy of bare $Au_n^+$ [301, 307] or analysis of $Au_n^+$ using different probing techniques (e.g. spectroscopy or mass spectrometry) and tagging species, such as $CH_3OH$ [308], Ar [309, 310], $H_2$ [311] and CO [312, 313]. A key disadvantage of the tagging methods is that the interaction of the tagging species might alter the underlying gold cation structure (the "gold skeleton") that is being probed, making a weakly interacting tagging species such as He desirable. Two recent studies combined aspects of both structure determination of $Au_n^+$ and Au-Rg (Rg = He– Xe) chemistry in a series of helium droplet experiments. The first study examined the adsorption of He to $Au_n^+$ ($n \leq$ ~15) [21], while the second one focused on the adsorption of rare gases He through Xe on $Au^+$ [314]. In the latter study, large helium droplets ($n \sim 10^6$) were doped, first with the rare gas (except in the case of He), then with Au, followed by EI and analysis of the ejected ions via ToF-MS. The experimental findings are shown to be reproducible in three separate measurements with different experimental conditions and are complemented by MP2 calculations for $n =$ 1–6, aiming to resolve the structure of the complexes detected by mass spectrometry. By analyzing the ion abundance of $Au_nHe_m^+$ as a function of $m$ for a given $n$, magic numbers can be identified, indicating structures of enhanced stability. The abundance distribution of $AuHe_m^+$ reveals a pronounced maximum at $m = 12$, indicative of the formation of an icosahedral solvation shell, as observed for other atomic ions such as the previously discussed $Rg^+$ (Rg = Ar–Xe) [241], anionic hydrogen and deuterium clusters [238] or $Cu^+$ [315] and. The distribution of $AuHe_m^+$ bears close resemblance to that of $KrHe_m^+$, which, like $AuHe_m^+$, additionally displays a local anomaly at $m = 14$ that is not found for other $RgHe_m^+$ [241]. In principle, further solvation structures of dodecahedral and icosahedral geometry, so-called anti-Mackay layers [16, 17] can be formed. These are easily assigned via distinct magic numbers at $m = 32$, 44 in the mass spectra of $ArHe_m^+$ [241] as well as anionic hydrogen and deuterium clusters [238], but are not clearly seen in other $RgHe_m^+$ or $AuHe_m^+$. Analysis of the magic numbers obtained from the abundance distributions of $Au_nHe_m^+$ for $n \geq 2$ in connection with the MP2 calculations can be used to investigate the structure of the underlying cationic gold clusters. The gold skeleton structures for $n = 2$–6 obtained by MP2 calculations agree well with those presented by Schooss et al. [307]. The first He atoms attached to these structures are found to locate in the molecular plane, binding directly to gold atoms, thus maximizing the charge-induced dipole interaction. The binding energies are strongest for $Au_2He_m^+$ ($m = 1, 2$) and $Au_3He_m^+$ ($m = 1$–3) with 13 and 12 meV, respectively. Binding energies for the first He atoms in $Au_nHe_m^+$ ($n = 3$–6 , $m \leq n$) range between 0.9 and 6 meV. For gold skeletons where the charge is distributed unevenly between the gold atoms (e.g. $n = 4, 6$), binding energies vary by a factor of up to ~5 depending on the binding site, which is reflected by corresponding drops in the ion abundance distributions after the more strongly bound sites are occupied. No MP2 calculations were attempted for $n \geq 7$, but assuming the gold skeleton is identical to the predicted hexagonal $Au_7^+$ structure [301, 307], an observed anomaly in $Au_7He_m^+$ at $m = 8$ could be explained by assuming that six He atoms bind to the corner Au atoms with two additional He are bound to the central Au atom above and below the molecular plane. An enhanced ion abundance of $m = 1, 2$ may indicate that the latter are even more strongly bound than the He atoms located in the molecular plane. It is interesting to note that for $Au_5He_m^+$, the authors only show four He atoms bound to the corner atoms of the X-shaped $Au_5^+$, but the abundance distribution of $Au_5He_m^+$ indicates increased stability of the $m = 5$–8 ions. The additional four He atoms could be speculated to locate around the central Au atom which would be unoccupied in the presented $Au_5He_4^+$ structure [21]. The structure of $Au_n^+$ changes from planar or near-planar to a 3-dimensional geometry around $n = 8$. While most of the literature agrees that $Au_7^+$ is planar and $Au_9^+$ is 3-D, there is no conclusive evidence for $Au_8^+$. Most theoretical predictions conclude that the global minimum structures are 3-D conformers, however, planar isomers are found to be rather close in energy [301, 307, 316, 317]. Whereas ion mobility spectroscopy measurements show the best agreement with a 3-D $Au_8^+$ [301, 307], the abundance distributions of $Au_8He_m^+$ show an anomaly at 3-4 attached He atoms that cannot be reconciled with any of the predicted 3-D structures, but could be explained by a planar structure with He atoms bound to four equivalent corner Au atoms [316, 317]. It should be noted that this feature is not seen as clearly in one of the measurements, which indicates it might be dependent on experimental conditions. One should also keep in mind that in a helium droplet environment, nascent clusters and cluster ions can be trapped in states other than the global minimum configuration. It is also conceivable that a mix of structures is produced. For $n = 9$–14, the abundance distributions of $Au_nHe_m^+$ do not show any anomalies at small values of $m$, consistent with a predominance of 3-D structures.

The interest in gold chemistry led to the discovery of, amongst many other, complexes that curiously feature bonds between a noble metal and rare gas atoms. The first report of such a complex was made in 1977 by Kapur and Müller [318] who detected $NeAu^+$ using a magnetic sector instrument with a field evaporation/ionization source, however, not reproducibly so. In 1995, Pyykkö predicted the cationic gold-rare gas complexes $AuRg^+$ ($Rg$ = He-Xe) and $RgAuRg^+$ ($Rg$ = Ar-Xe) to be stable and calculated the bond dissociation energies (BDE) of $AuRg^+$ ranging from 0.02 eV for up to 0.9 eV and atomization energies of $RgAuRg^+$ ($Rg$ = Ar-Xe) ranging from 0.9 to 2.25 eV, with the heavier rare gases more strongly bound [319]. The author states that "one can expect a degree of covalent bonding between $Au^+$ and the heavier [noble] gases", referring to Ar–Xe. Three years later, the existence of $AuXe^+$ (and $AuXe_2^+$) was experimentally confirmed by Schröder et al. via reaction of $Au^+$ produced in a laser-desorption/laser-ionization source with $C_6F_6$ and Xe in an FT-ICR mass spectrometer [320]. The authors estimate 1.3 eV as the BDE of $AuXe^+$. Two years later, Seidel and Seppelt attempted to produce AuF, but the reaction unexpectedly produced $AuXe_4^{2+}$ as $AuXe_4^{2+}$-$(Sb_2F_{11}^-)_2$ crystals instead [321]. Relatively strong interactions between coinage metal atom clusters and rare gas atoms have also been found in neutral $Au_n$ and mixed $Au_nAg_m$ clusters complexed with a few Ar and Kr atoms [322–327] and cationic $CuNe_m^+$ and $CuAr_m^+$ complexes [315].

Martini et al. recently published a mass spectrometric study of cationic atomic gold ions complexed with all non-radioactive rare gases by analyzing the abundance distributions of $AuRg_m^+$ (Rg = He–Xe), supported by CCSD(T)/def2-TZVPP ab initio calculations [314]. Large helium droplets were doped, first with the ligand rare gas species (except in the case of He), then with gold, and subsequently submitted to EI and analysis via ToF-MS. The mass spectra reveal a local anomaly of $AuRg_2^+$ for all rare gas ligands except He. While this is consistent with theoretical studies by Li, Zhao and co-workers, predicting a drop in BDE after $m = 2$ for Rg = Ar–Xe, it is surprising for Ne, where $AuNe_3^+$ is expected to actually have a slightly higher BDE than $AuNe_2^+$ [328–332]. The bond nature of $AuRg_m^+$ ($m = 1, 2$) is predicted to exhibit significant covalent character for Rg = Kr and Xe, slight covalent character for Ar [319, 329, 333–335], but to be entirely physical for Ne [328, 333, 335]. Alkali metal ions embedded in rare gas clusters $A^+Rg_m$, can be used as an entirely physically bound model system, but no anomalies at $m = 2$ can are predicted for A = Li–K and Rg = Ar, Xe [336–339]. It is conceivable that the theoretical treatment of $AuNe_m^+$ is inaccurate, failing to reproduce the observed anomaly at $m = 2$, which might also infer that the character of binding in $AuNe_m^+$ is not entirely physical in the end.

The mass spectra reveal further magic numbers of $AuRg_m^+$, namely $m = 12$ for He and Ne as well as 6 and 9 for Ar. No further magic numbers after $m = 2$ are detected for either Kr or Xe. The magic numbers of He and Ne suggest the well-known icosahedral cage structure to be formed around the central $Au^+$ ion. Magic numbers at $m = 6, 9$ for Ar suggest that these complexes adopt a different packing scheme. A hard sphere model assuming pair-wise $Au^+$–Rg and Rg–Rg interaction is adapted from previous work [339, 340] and employed to predict the energetically most favorable packing scheme for the rare gas solvation shells around the central ion, with the ratio between $Au^+$–Rg and Rg–Rg bond lengths [335, 341, 342] as the only parameter. The model predicts close-packed (fcc or hcp crystal structure), icosahedral and octahedral structures for He, Ne and Ar, respectively, which leaves the magic numbers of these ligands in good agreement with shell closure predictions ($m = 12$ for He/Ne and 6 for Ar). The higher degree of covalent bonding in the Kr and Xe complexes likely prevents the simple model from making accurate predictions for the corresponding $AuRg_m^+$ [314]. Additional support for these findings is provided by molecular dynamics simulations performed by Martini et al. for $AuHe_m^+$ and $AuAr_m^+$ [314]. Three distinct solvation shell closures after $m = 12, 32$ and 44 are clearly observed in radial density function plots extracted from the simulation of $AuHe_{110}^+$, indicating formation of the previously mentioned anti-Mackay cage structure. Although the general shape of the distribution indicates a more liquid-like behavior, weak corresponding drops in the ion abundance of $AuHe_m^+$ at $m = 32, 44$ can be made out [21]. Simulations of $AuAr_{108}^+$ at 0 K reveal further distinct shell closures at $m = 10$, 22 and 34. These do not have corresponding local anomalies in the mass spectra, although 10 is close to the anomaly observed at $m = 9$. While the outer solvation structures are no longer distinct in simulations of $AuAr_{108}^+$ at a finite temperature of 10 K, the first and second shells remain distinct, with a total of 9 atoms in them, matching the observed magic number. The same $m = 9$ magic number is also found in dissociation energy calculations performed on much smaller systems, which additionally predict a drop after $m = 14$ that is not reflected in the mass spectra of $AuAr_m^+$. Furthermore, the simulation fails to reproduce the $m = 2$ anomaly since it does not correctly account for covalent bonding.

We will now take a look closer look at the chemistry of gold (clusters) with a reactant more common than rare gases, namely hydrogen. Early reports of bond formation between gold and hydrogen were made more than a century ago [343, 344], followed by numerous further studies [345, 346]. Lundberg et al. recently extended the study of gold chemistry towards the chemical bonding of hydrogen to gold clusters and their binding capacities for molecular hydrogen by investigating mixed $Au_nH_m^+$ ($n \leq 8, m \leq 20$) complexes via HND MS and MP2 calculations [347]. The presented mass spectra are in good agreement with previously discussed studies, showing the known odd-even oscillations of pure gold clusters as well as hydrogen attachment to clusters with 8 and fewer gold atoms. The ion abundance distributions as a function of $m$ for a given $n$ (Fig. 2.22a) display a rich structure including odd-even oscillations and intense magic numbers for $n \leq 5$. Several local anomalies are found in complexes with $n = 6$–8 gold atoms. For $n \leq 5$, the most abundant ions by far are $Au_nH_{n+3}^+$. The MP2 calculations reveal that, for $n = 2$–5, the structures of the odd-numbered magic number ions are essentially identical to previously discussed $Au_nHe_m^+$, with each He replaced by one $H_2$ molecule (Fig. 2.22b) [21]. In even-numbered (i.e. $n = 2, 4$) magic ions, one H atom is chemically bound between two gold atoms to form a structure reminiscent of the next-highest odd-numbered gold skeleton—these structures can also be considered as protonated gold clusters. Again, up to $n = 4$, the adsorption of $H_2$ is equivalent to He adsorption (in the even-numbered gold skeletons, H replacing Au does not offer a binding site for $H_2$). For $n = 6, 7$ there are "unoccupied" gold atom sites where no $H_2$ is adsorbed, whereas for magic ions in the $Au_nHe_m^+$ experiment, gold sites bind one or even two (in $Au_7He_8^+$) He atoms [21]. The calculations also yield binding energies for the first few (up to 5) attached $H_2$ molecules, which support the notion of enhanced stability of the magic number ions. For $n = 1,2$, the binding energies of the first two $H_2$ molecules range between 0.8 and 1.1 eV, with a steep drop to 0 eV or even slightly negative energy for the third $H_2$. In complexes with larger gold skeletons (i.e., $n = 3$-7), the binding energies gradually decrease to around 0.5 eV and the stepwise drops after a magic number ion become less pronounced.

In a follow-up study, Lundberg et al. showed that water molecules will efficiently replace $H_2$ in $Au_nH_m(H_2O)_p^+$ complexes, evident from a clear, systematic shift up

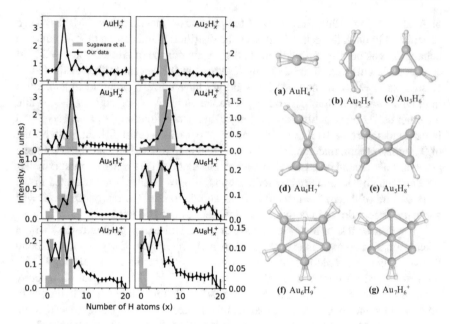

**Fig. 2.22** **a** Cluster size distributions of $Au_nH_x^+$ ($n = 1$–8) as a function of $x$, measured via EI of co-doped HNDs and **b** structures of particularly abundant cluster ions as calculated using MP2. The cluster size distributions reveal intense magic numbers and odd-even oscillations. Structures calculated for even numbers of gold atoms are reminiscent of the next highest odd-numbered gold skeleton where one gold atom is replaced by a hydrogen atom. Reproduced from [347]. Licensed under CC BY 4.0

to $n = 6$ in the magic numbers of the ion abundance series [348]. Binding energies obtained from MP2 calculations for attached $H_2O$ are roughly twice as large as for $H_2$ on average. While these binding energies are substantial, no significant changes in the underlying gold skeletons are found in the calculated geometries. The findings of the study indicate that water contaminations could be detrimental for the hydrogen storage capabilities of gold nanoparticles but could also be deliberately used to trigger the release of stored hydrogen by adding water to the system.

### 3.3.1.6 Studies of Imidazole

Imidazole ($C_3H_4N_2$, Im), is an aromatic, nitrogen-containing compound and ubiquitously found in nature as an important building block of various biomolecules [349]. Kuhn et al. recently studied the EI/EA-induced intra-cluster chemistry of Im clusters grown in HNDs by monitoring cationic and anionic fragments utilizing HND MS [350]. Anions were found to predominantly form via collision of the dopant cluster with He*−, evident from electron energy scans exhibiting a strong, well-known 22 eV resonance typical for anion formation in the HND environment [351–354]. The dominant anion series consists of dehydrogenated [$Im_n$–H]− ($n$ up to ~25), with several weaker series also observed consisting of intact $Im_n^-$, either pure or complexed with different combinations of CN and $C_2H_4$ fragments. The observed anions likely consist of an ionic $[Im–H]^-$ (if no CN is fragment is present) or $CN^-$ core based on the respective electron affinities of 2.67 (Im–H) and 3.8 eV (CN), solvated by $n-1$ neutral Im molecules. The existence of the more complex species with two fragment hints at complex reactions taking place in the Im cluster upon EA. Cationic species are formed via $He^+$ as evident from the onset of formation at the He ionization threshold (24.6 eV) and predominantly exist as protonated $Im_nH^+$. Other cationic series observed are intact $Im_nH^+$ and $[Im_nCH_3]^+$. The latter could consist of either $CH_3^+$ solvated by $n$ Im molecules or methylated $[ImCH_3]^+$ solvated by $n-1$ Im molecules. While both protonated and dehydrogenated species $[Im_n \pm H]^\pm$ were always more abundant than the corresponding pure $Im_n^\pm$, the ion abundance distributions of the latter decay less quickly towards larger cluster sizes in comparison, indicating that larger clusters are more stable against intra-cluster protonation and dehydrogenation reactions. Interestingly, local abundance anomalies hinting at magic number clusters are few and inconsistent. The only examples given are notable abundance drops in $[(Im_n–H)C_2H_4]^-$ after $n = 6$ and 11. While no further magic numbers are discussed by the authors, these can also be found, albeit less pronounced, in $Im_n^-$, perhaps along with $n = 8$. The cationic ion abundance distributions are notably smoother, with the only possible magic numbers at $n = 6$ and perhaps 10 in $Im_n^+$.

A different recent study investigated organometallic compounds formed by co-doping HNDs with gold and imidazole and analyzing the mixed $Au_mIm_n^+$ ($m \leq 6$, $n \leq 15$) complexes ejected upon EI with high-resolution ToF-MS (Fig. 2.23) [355]. The recorded mass spectra reveal that for every number of gold atoms, there is a number of Im molecules corresponding to a particularly abundant compound (for $m = 4$, both $n = 3$ and 4 are particularly abundant). It is tempting to assign these as magic number ions corresponding to particularly stable structures, however, DFT

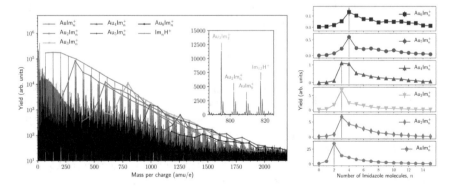

**Fig. 2.23 a** Mass spectrum of organometallic $Au_mIm_n^+$ with different ion series marked in color and **b** extracted cluster size distribution ($m = 1–6$) as a function of $n$, with vertical lines indicating the most abundant ions. While the pure $Im_nH^+$ series (gray) exhibits a rather smooth distribution, distinct maxima are found in the $Au_mIm_n^+$ series with $m = 1–6$. Reproduced from [355]. Licensed under CC BY 4.0

calculations performed for up to $m = 4$ gold atoms reveal that this is not universally accurate. Whereas the optimized structures of $AuIm_2^+$ and $Au_3Im_3^+$ are indeed found to be particularly stable compared to their $n \pm 1$ counterparts, the structures calculated for the anomalously abundant compounds $Au_2Im_3^+$, $Au_4Im_3^+$ and $Au_4Im_4^+$ were not (the binding energy of $Au_4Im_4^+$ is incorrectly given as 1.7 eV in Fig. 2.7 of [355]. The correct value of 1.26 eV is displayed in Fig. 2.8 of the same publication). The authors find a different explanation in that compounds with more Im molecules than the experimentally observed magic numbers generally fragment by loss of neutral Im, thus enriching these sizes. Compounds with a lower number of Im molecules tend to fragment by loss of neutral Au atoms, perhaps along with additional Im molecules, possibly further contributing to the strong ion signals of the particularly stable $AuIm_2^+$ and $Au_3Im_3^+$.

### 3.3.1.7 Studies of Buckminsterfullerene $C_{60}$

Fullerenes, the cage-shaped molecular allotropes of carbon have created tremendous interest in their unique physical and chemical properties since their discovery by Kroto et al. in 1985 [4], recognized in 1996 with the Nobel Prize in Chemistry. A short comprehensive review of fullerenes was published by Klupp et al. [356]. The most common fullerene, Buckminsterfullerene ($C_{60}$), has been widely investigated by a variety of techniques, including mass spectrometry of $C_{60}$-doped helium droplets. Notable findings include the first dianions detected in a helium droplet environment [357] and the independent confirmation [150, 206] of the assignment of $C_{60}^+$ as the first (and so far only) diffuse interstellar band (DIB) carrier, first assigned by Campbell, Walker and co-workers in their ion trap experiment [147, 148, 358]. Mauracher and co-workers give a detailed summary of $C_{60}$ experiments performed in helium droplets by the Innsbruck group in their comprehensive review from 2018 [22]. Besides the interaction of fullerenes with the helium environment, the experiments investigated in detail the adsorption of atoms and small molecules such as He, $H_2$, $N_2$, $O_2$, $H_2O$, $NH_3$, $CO_2$, $CH_4$ and $C_2H_4$, to positively or negatively charged $C_{60}$ and $C_{70}$ as well as fullerene aggregates (see [22, 359] and references therein). Additionally, complexes of $C_{60}$ with atomic carbon [360] as well as sodium and cesium [361, 362] were investigated. While former experiments aimed to understand the growth processes involved in the formation of fullerenes themselves and their role in extraterrestrial chemistry, research of metal-doped (especially with alkali and earth alkali species) fullerides is driven by hot topics such as potential applications as superconductors [363–366] and high capacity hydrogen storage materials [367–370]. A peculiar structure formed by reaction of $C_{60}$ with carbon atoms is the particularly stable $C_{60}=C=C_{60}$ "dumbbell" structure, which was first synthesized in the late 1990s [371, 372] and recently formed in helium droplets and detected as $[C_{60}=C=C_{60}]^\pm$ [360]. Goulart and co-workers recently observed small ionic complexes between gold and $C_{60}$ formed in helium droplets and found a similar, highly stable $[C_{60}AuC_{60}]^\pm$ dumbbell structure, both in anionic and cationic form (Fig. 2.24) [373]. The structure may be reminiscent of the linear $[XeAuXe]^+$,

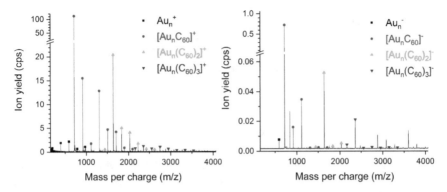

**Fig. 2.24  a** Cationic and **b** anionic mass spectra of HNDs co-doped with gold atoms and $C_{60}$ molecules. The most striking feature is the dramatically increased abundance of both cationic and anionic $(C_{60})_2$Au. These ions were found to adopt a highly stable dumbbell-shaped structure similar to $[C_{60}=C=C_{60}]^+$ [371372360]. Reproduced from [373]. Licensed under CC BY 4.0

which has also been found to be surprisingly stable, considering it consists of a noble metal and two rare gas atoms [319, 374]. A follow-up study by Martini and co-workers examined larger anionic and cationic clusters of up to ~10 $C_{60}$ molecules and up to ~20 metal atoms (either gold or copper) formed in helium droplets and subsequently ionized by electron impact or EA [375]. The observed ion intensities show several other local anomalies indicating ions of increased stability. As expected from the previously described experiment, $(C_{60})_2$Au$^\pm$ exhibit strong ion signals, which are interpreted as the previously discussed $[C_{60}$AuC$_{60}]^\pm$ dumbbell structure. Similarly, strong ion signals observed for $(C_{60})_2$Cu$^\pm$ may lead to the assumption of a similar structure, however, there is no theoretical evidence to support this conclusion and the observed ion abundances are not as dominant as in the Au series. Most $(C_{60})_m$M$_n^\pm$ (M = Cu or Au) ion series with $m \leq 3$ exhibit some form of odd-even oscillations, where ions with an odd number of metal atoms $n$ are generally found to be more stable ($n = 3$ is often prominent). A few exceptions where compounds show increased stability for even $n$ are $C_{60}$Au$_n^-$ ($n = 2, 4$), $(C_{60})_3$Au$_n^+$ ($n = 4, 6, 8$), $(C_{60})_3$Au$_n^-$ ($n = 4$, but also 1) and $(C_{60})_2$Cu$_n^+$ ($n = 4, 6, 10$, but also 1). Two more noteworthy observations can be made for Cu compounds. First, the ion intensity series of $(C_{60})_m$Cu$_n^\pm$ exhibits a sudden change from heavily structured distributions for $m \leq 2$ to a smooth ones for $m \geq 3$, with only $(C_{60})_3$Cu$_4^-$ showing slight magic character. A similar change is observed in the Au series as well, but for significantly bigger ensembles between $m \leq 4$ and $m \geq 5$). The reason for this sudden change in small $(C_{60})_m$Cu$_n^\pm$ compounds remains unclear. Second, di-anionic compounds $(C_{60})_m$Cu$_n^{2-}$ with $m = 2, 4, 6$ and odd $n$ can be observed in the mass spectra (compounds with even $n$ might also be present, but would be buried in the ion signals of singly charged species), significantly smaller than the smallest bare $C_{60}$ dianion observed in helium droplets, $(C_{60})_5^{2-}$ [357, 376]. Interestingly, no $(C_{60})_m$Cu$_n^{2-}$ with odd $m$ are observed, even though such ions should be discernible if the ion signal intensities were comparable with dianions of neighboring, even $m$. This circumstance hints at a pronounced odd-even effect in the formation or stability of $(C_{60})_m Cu_n^{2-}$, which unfortunately cannot be explained at this point. While the measured ion abundances for small $(C_{60})_m Au_n^{\pm}$ ($m \leq 2$, $n \leq 2$ and $m = 1$, $n = 3$) agree well with theoretical predictions [373], there are only few other theoretical investigations of small $(C_{60})_m M_n$ compounds that would help to explain the observations made.

### 3.3.1.8 Complexes of Formic Acid

Formic acid (HCOOH, abbreviated FA) is the simplest carboxylic acid and can be used as a model system for the larger, more complex RCOOH compounds (where R is one of many possible organic substituents). FA is ubiquitous in earth's atmosphere and plays an important role in atmospheric chemistry [377], it was the first organic acid detected in interstellar space [378], has further been detected in galactic hot molecular cores, regions of increased density and temperature within molecular clouds that feature rich chemistry [379] and was considered to be involved in the formation of glycine precursors, albeit with a negative result [380]. In terms of technical applications, FA has been demonstrated to be an excellent fuel for use in direct fuel cells, achieving superior performance compared to direct methanol fuel cells at ambient conditions [381, 382]. The extensive research devoted to finding suitable catalysts for direct formic acid fuel cells, a key problem in their commercial application, has found palladium-based catalysts as the best-performing candidate so far [382, 383]. Recent investigations have considered nanostructures such as hollow Pd-nanospheres [384], various core-shell nanoparticles [385] as well as Pd and La nanoparticles supported on modified $C_{60}$ [386–388], which show promise for the use in direct FA fuel cells. Interestingly, little research has been devoted to the study of pure complexes of FA and $C_{60}$.

Two recent studies by Mahmoodi-Darian et al. investigated the interaction of HNDs doped with FA [389] as well as FA and $C_{60}$ with low-energy electrons (Fig. 2.25) [390]. The most abundant anion cluster series in the purely FA-doped droplet experiment was dehydrogenated $[FA_n-H]^-$, whereas for cations, protonated $[FA_n+H]^+$ dominates the mass spectrum. Less abundant, but also observed are undissociated $[FA_n]^{\pm}$, deprotonated $[FA_n-H]^+$, and the more abundant ions complexed with one or more water molecules [389].

Whereas residual $H_2O$ is a very common impurity in mass spectra recorded in helium droplet experiments, the amount of water in the present mass spectra is much larger than in previous experiments with other hydrocarbon, $C_{60}$ or metal clusters on the same apparatus, but comparable to a study of methanol and ethanol clusters, where the large amounts of water observed were attributed to be produced an intra-cluster reaction [391]. While no such attempt has been made in the present study, Bernstein and co-workers added small amounts of water to the FA sample without observing a significant difference in soft x-ray ionization mass spectra of pure FA clusters, concluding that the presence of $(H_2O)[FA_n+H]^+$ can be explained by loss of a CO molecule from $[FA_{n+1}+H]^+$ [392]. While a contamination of either the vacuum chamber and/or the sample cannot be ruled out, it is likely that the

**Fig. 2.25** Size distributions of various mixed cluster ions composed of FA, water (W) and $C_{60}$. By far the most intense ion series belongs to $[FA_nH]^+$, which does not exhibit any noteworthy features, however. Series with added W and $C_{60}$ molecules reveal a richer structure. Interestingly, the magic character of $WFA_5$ and $W_2FA_6$ indicated in (**b**) appears to be preserved when adding one $C_{60}$ molecule, as evident in (**d**), (**e**). Reproduced with permission from Ref. [390]. © copyright Elsevier. All rights reserved

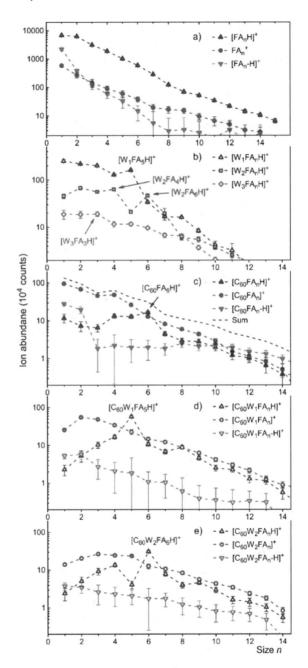

large amount of water in the mass spectra arises from intra-cluster ion-molecule reactions [389]. The findings are in good general agreement with a study utilizing sputtering of frozen FA films by energetic $^{252}$Cf fission fragments—the anion mass spectra are dominated by $[FA_n–H]^-$ with minor contributions of $(H_2O)_m[FA_n–H]^-$ ($m \leq 2$), similarly the cationic spectra feature mostly $[FA_n+H]^+$, with $(H_2O)_m[FA_n+H]^+$ ($m \leq 2$) also being detected [393, 394]. In this study, undissociated compounds or $[FA_n–H]^+$ could not be detected due to poor mass resolution. A study of low energy (1 eV) EA to FA clusters produced by supersonic expansion agrees well in observing primarily $[FA_n–H]^-$ and $[FA_n]^-$ in the mass spectra ranging up to $m/z$ 200 [395]. A notable difference compared to the helium droplet and sputtering experiments in the mass spectra is the absence of $(H_2O)[FA_n–H]^-$ for $n = 2$ and 3, accompanied by the appearance of peaks 13 and 21 u above $[FA_n]^-$ ($n = 2, 3$), whose composition remains unclear. In the helium droplet experiment, the anion series of $[FA_n]^-$ and $(H_2O)[FA_n–H]^-$ show abrupt onsets following $n = 2$ and $n = 5$, respectively [389]. The observed onset of $[FA_n]^-$ is in good agreement with the pure FA cluster experiment [395]. In addition, theoretical studies predict a substantial negative adiabatic electron affinity (AEA) of around -1 eV for both the formic acid monomer [396, 397] and dimer [396]. The negative electron affinities likely render the corresponding anions short-lived, whereas the trimer is found to have an AEA closer to 0 eV [396] and can thus be assumed to possess a longer lifetime, explaining the observed onset in the abundance distribution. While calculations are available to explain the sudden onset in $(H_2O)[FA_n–H]^-$, the sputtering experiments reveal a similar onset [393, 394]. Whereas most anionic and cationic cluster series in the helium droplet experiment show a rather smooth abundance distribution, some noteworthy magic numbers such as $n = 5$ for $[FA_n–H]^-$ and $(H_2O)[FA_n+H]^+$ as well as $n = 4, 6$ for $(H_2O)_2[FA_n+H]^+$ can be found. The magic number character of $[FA_5–H]^-$ is rather weak, but can be found in the sputtering experiments (potentially together with $n = 6$) [393, 394]. The magic number character of $(H_2O)[FA_5+H]^+$ is much stronger and universally found in other experiments utilizing fission fragment sputtering of frozen FA [393, 394], soft x-ray photoionization of pure FA clusters [392], a variable temperature and pressure electron ion source (though it should be noted that the distributions depend strongly on the source conditions) [398], and an electron impact study of mixed FA-water clusters [399]. Theoretical studies found that the composition of $(H_2O)[FA_n+H]^+$ probably changes from a central FA·H$^+$ ion to a central $H_3O^+$ between $n = 3$ and 4 with $(H_2O)[FA_5+H]^+$ adopting a structure where the central $H_3O^+$ is surrounded by a cyclic arrangement of 5 FA molecules [399, 400]. While the magic character of $(H_2O)_2[FA_n+H]^+$ ($n = 4, 6$) is also found in the soft x-ray photoionization [392] and variable temperature and pressure electron ion source [398] experiments, no theoretical work has been published on this ion. Magic numbers in the series of $[FA_n+H]^+$ were found in the soft x-ray experiment ($n = 5$) if a sufficient delay between laser pulse and ion extraction is introduced, allowing for slow decays of $[FA_n+H]^+$ into $[FA_{n-1}+H]^+$ and $[FA_{n-2}+H]^+$ to occur [392], as well as in the variable temperature and pressure ion source experiment, where the magic number varied between $n = 4$–6 depending on source conditions [398]. None of these additional magic numbers were found in the helium droplet and sputtering experiments or theoretical investigations [389, 399–402]. An interesting feature, unsurprisingly exclusive to the helium droplet experiment, is the detection of FA anions complexed with up to 15 helium atoms, namely $He_m[FA–H]^-$ and $He_m[FA_2–H]^-$ [389]. The abundance distribution of $He_m[FA–H]^-$ shows several local anomalies at $m = 5$, 7 and 11 helium atoms attached. A possible anomaly at $m = 9$, which would complete a weak odd-even oscillation, is unfortunately obscured by the dominating interference of $(H_2O)_2[FA–H]^-$. While the authors do not draw further conclusions from these observations, and no theoretical work has been devoted to FA anions complexed with helium, a comparison can be made with of acetic acid (AA) anions complexed with heliums observed in a previous study by da Silva et al. Attachment of up to 14 He atoms to $AA_n^-$ and $[AA_n–H]^-$ ($n \geq 2$) was observed, but no helium attachment to the (dehydrogenated or undissociated) AA monomer anions was detected [403]. The authors assumed that the attachment of He atoms is likely hindered by the excess charge in $AA^-$, but favored by the neutral AA moieties present in larger $AA_n^-$ clusters. This is contrasted by the FA study, where $He_m[FA–H]^-$ actually presents the strongest signal among the He attachment series. Since FA and AA are chemically similar and it can be very tricky to find the right conditions for helium attachment to any ions, it seems more plausible that experimental conditions are responsible for the lack of helium attachment to AA monomer anions.

In a follow up study, Mahmoodi-Darian et al. utilized EA and EI of large helium droplets co-doped with FA and $C_{60}$ to record mass spectra of mixed complexes of FA and $C_{60}$ (as well as $H_2O$) [390]. Again, the presence of large amounts of water in complexes involving FA can be noticed, however, the relative intensity of $C_{60}$ complexed with $H_2O$ is much lower, which strengthens the interpretation of $H_2O$ in FA complexes arising mainly from intra-cluster reactions. Some interesting features can be found in the abundance distributions of $[(C_{60})_p(H_2O)_mFA_n]^\pm$ complexes (along with their hydrogenated and dehydrogenated counterparts) when compared to complexes without $C_{60}$. Whereas $[FA_n+H]^+$ and $[FA_n–H]^-$ are the dominant series of bare FA clusters in the positive and negative mass spectra of pure FA clusters, the hydrogenation and dehydrogenation reactions are found to be strongly suppressed in the presence of $C_{60}$ for low numbers ($\lesssim 6$) of FA molecules. A comparison of the ionization energies (7.6 vs. 11.3 eV [133]) and electron affinities of $C_{60}$ and FA (2.68 [133] vs. -1.27 eV [396]) suggests that in both cases it is more favorable for the additional charge to reside on $C_{60}$ rather than FA, thus suppressing the autoprotonation and dehydrogenation reactions. The distributions of $[C_{60}FA_n]^+$ and $[C_{60}FA_n+H]^+$ show an enhanced abundance at $n = 4$ and $n = 4–6$, respectively, in contrast to the smooth, featureless distributions of the compounds without $C_{60}$. While $[C_{60}FA_6H]^+$ exhibits the strongest enhancement in these series, its magic character is lost upon further addition of $C_{60}$ molecules, whereas the anomaly at $n = 4$ persists for up to at least four $C_{60}$ molecules. The strong magic character found in $[(H_2O)FA_5H]^+$ is preserved, if not strengthened, upon addition of at least up to four $C_{60}$ molecules. It is not obvious, however, whether the presumed structure of a central $H_3O^+$ surrounded by a cyclic arrangement of five FA molecules [399, 400] changes when $C_{60}$ is added, since the proton affinity of $C_{60}$ (8.75 eV [404]) is significantly higher than that of FA and water (7.69 and 7.16 eV, respectively [133]). In the $[C_{60}(H_2O)_2FA_nH]^+$ series, the magic character of $[(H_2O)_2FA_6H]^+$ is not only clearly preserved, but further enhanced upon addition of $C_{60}$. The magic character of $[(H_2O)_2FA_4H]^+$ is probably preserved since after $n = 4$, a strong drop in the ion abundance can be observed, however, the local anomaly is more clearly seen in the bare $[(H_2O)_2FA_nH]^+$ series, perhaps due to different experimental conditions shaping the size distribution.

In the anionic mass spectra, the distributions of $[C_{60}FA_n]^-$ exhibits a local anomaly at $n = 4$, which is not reflected in the bare cluster $[FA_n]^-$ distribution. The local anomaly is not found in $[(C_{60})_2FA_n]^-$, but reappears in $[(C_{60})_3FA_n]^-$. The shape of the $[FA_n–H]^-$ distribution changes from a plateau-shaped maximum at $n = 3–5$ in the bare FA study to a maximum at $n = 3$ in the $C_{60}$ study, which likely coincides with the maximum of the envelope of the distribution and is thus not considered a local anomaly. Additionally, $[FA_5–H]^-$ no longer presents a local anomaly, illustrating the weakness of its magic character described earlier. The changes in the distribution shape are presumably caused by different experimental conditions. Interestingly, $n = 1$ represents the strongest ion signal in the $[C_{60}FA_n]^+$ series, but becomes strongly suppressed in $[(C_{60})_pFA_n]^+$ for $p = 2–4$ where the strongest signal is found at $n = 2$. Similar can be observed in the $[(C_{60})_pFA_n]^-$ series ($p = 1–3$) although $[(C_{60})_2FA]^-$ is not suppressed as strongly, and, to a lesser extent, in the $[(C_{60})_pFA_nH]^+$ and $[(C_{60})_pFA–H_n]^-$ series. In complexes of $C_{60}$ and ammonia, the signal of $[(C_{60})_p(NH_3)]^+$ is also found to be strongly suppressed for $n = 1$ and $p = 2, 3$ [405]. Certain small ionic complexes are also found to be absent in complexes of $C_{60}$ and either Cu or Au), although no consistent, simple pattern is observed for these compounds [375] (also see previous discussion on $C_{60}$-metal complexes).

### 3.3.1.9 Complexes of Adamantane and Water

Adamantane ($C_{10}H_{16}$, Ad) is basically a hydrogenated unit cell of the diamond crystal structure and the smallest of the diamondoids, a group of hydrocarbons sometimes referred to as nanodiamonds. They have intriguing physical and chemical properties as well as a wide array of applications in chemistry, material science and medicine [406–409]. Recently, Goulart et al. conducted the first study of Ad clusters synthesized in HNDs [410]. The authors observed $Ad_n^{z+}$ clusters with $n \leq 140$, $z \leq 3$ and the series for $z = 2$ and 3 starting at $n = 19$ and 52 molecules, respectively (Fig. 2.26). Several magic numbers are identified which correspond to particularly stable, geometric packing schemes and that mostly persist regardless of the charge state. Whereas the first magic numbers $n = 13$ and 19, only observed in $Ad_n^+$, suggest icosahedral packing schemes, further magic numbers associated with icosahedral packing like $n = 55$ are notably absent in the mass spectra. Instead, magic numbers of $n = 38, 52, 61, 68, 75$ and 79 are observed for all charge states (beyond their appearance size in the case of trications), which line up perfectly with predictions for face-centered cubic (fcc) packing structures [411]. Apparently, the

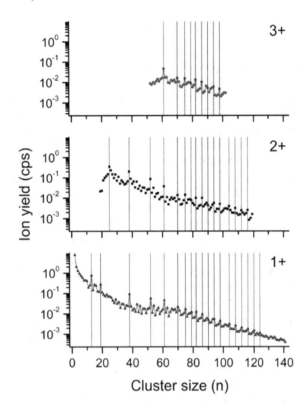

**Fig. 2.26** Ion abundance distributions of adamantane cluster ions $Ad_n^{z+}$ ($z = 1-3$) [410]. Interestingly, all magic numbers appear independent of the charge state. While small magic numbers ($n = 13$ and 19, only observable for $z = 1$) indicate cluster ions adopting icosahedral shape, the higher magic numbers of $n = 38$, 52, ... are in perfect agreement with predictions for fcc packing structures. Reproduced with permission from Ref. [411]. © copyright American Chemical Society. All rights reserved

preferred packing scheme of Ad clusters changes from icosahedral to fcc between $n = 19$ and 38. The study of Ad was extended with an investigation of HNDs co-doped with Ad and water by Kranabetter et al. [412]. The authors recorded mass spectra of both pure $Ad_n^+$ and protonated $(H_2O)_mH^+$ as well as mixed $(H_2O)_mAd_n^+$ clusters. Notable magic numbers observed in these experiments were again $n = 13$ and 19 for $Ad_n^+$ as well as $m = 4, 11, 21, 28$ and 30 for $(H_2O)_mH^+$, with $m = 21$ clearly showing the strongest magic character. These features were observed regardless of the doping order. The mixed cluster series displayed an interesting interplay between the magic numbers of both pure series. For one, the exceptional magic character of $(H_2O)_{21}^+$ is preserved when complexed with any number of Ad molecules between 6 and 19, growing relatively stronger with larger numbers of Ad molecules. Additionally, $(H_2O)_mAd_{12}^+$ and $(H_2O)_mAd_{18}^+$ are particularly abundant for $5 < m < 21$, suggesting that intermediate size water clusters are able to effectively replace a single Ad molecule in the icosahedral magic number structures of $Ad_{13}^+$ and $Ad_{19}^+$. The strongest magic character is observed when both these features are combined in $(H_2O)_{21}Ad_{12}^+$ and $(H_2O)_{21}Ad_{18}^+$. The authors also attempted, but were unable to find, evidence for the formation of clathrate hydrates [413–415] in which small numbers of Ad are trapped within a structured water cluster.

### 3.3.2  Multiply Charged Droplets

The consequences of electron impact ionization in pristine helium droplets have recently been studied using a tandem mass spectrometer setup that is able to resolve the charge state of ionized droplets [171, 416]. In the referenced setup, neutral helium droplets are produced through expansion of helium gas (Messer 99.9999% purity) into vacuum at 20 bar stagnation pressure through a 5 μm nozzle, cooled down to 4-10 K. After leaving the cluster source through a skimmer, the neutral droplets are ionized by electron impact before entering a 90° spherical electrostatic analyzer, where charged droplets can be selected by their mass-per-charge ratio. In the second stage of the tandem mass spectrometer, previously selected charged droplets can again be ionized and analyzed for their final mass-per-charge ratio in an identical electrostatic analyzer. By scanning the second electrostatic analyzer and detecting the charged droplets that pass the analyzer with a channel electron multiplier, mass-per-charge spectra for positively and negatively charged droplets can be obtained.

Positive charge carriers in helium droplets can be formed if an electron above the ionization threshold of helium [IE(He) = 24.6 eV] collides with a He atom and forms a $He^+$ ion [22]. Resonant charge hopping processes and subsequent formation of $He_2^+$ upon charge localization [57] will lead to the formation of an Atkins Snowball [25], an ionic core of $He_3^+$ [417] surrounded by a rigid shell of helium atoms. An electron that has a kinetic energy higher than 48 eV can also produce multiple positive charges in a droplet. Negative charge carriers in helium droplets can be introduced upon impact of electrons with energies below the ionization threshold of helium. At a kinetic electron energy of 22 eV excited $He^*$ can be formed which, contrary to ground state helium, can capture a slow electron and form highly mobile, heliophobic $He^{*-}$ [352]. The resonant formation of molecular $He_2^{*-}$ in helium droplets has also been observed [22], however, the production of the molecular anion seems to be far less efficient. At even lower electron energies of a few eV, which resembles the energy necessary for an electron to penetrate the surface of a helium droplet [418], slow electrons can form voids in the droplet and reside as electron bubbles [419].

Subjecting a precursor droplet to subsequent ionization leads to the appearance of product droplets detected at rational fractions of the precursor peak as shown in Fig. 2.27. The individual peaks originate from multiple additional charges added to the precursor droplets and therefore reducing their mass-per-charge ratio in the second ionizer. For instance, the peak at 1/2 in the anionic spectrum is produced by adding one additional charge to the singly charged precursor droplet, respectively adding two additional charges for the peak at 1/3. For all peaks in the shown cationic spectrum to appear, the initially negatively charged precursor must be positively charged multiple times while passing the second ionizer to bear a net positive charge. Peaks above the ratio of 1/2 can only be seen when the precursor droplet already carries multiple charges before being subjected to repeated ionization in the second ionizer. In Fig. 2.28a a charge state spectrum of such a positively multiply charged droplet is shown. To resolve the individual charge states in the second analyzer, the kinetic energy of the electrons in the second ionization stage is set as to introduce

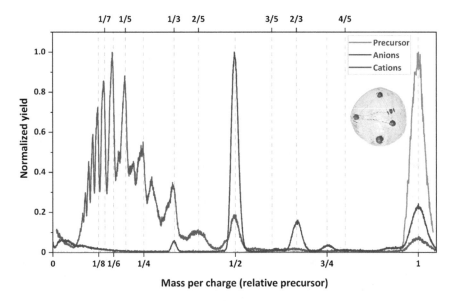

**Fig. 2.27** Mass-per-charge spectra for positively (red) and negatively (blue) charged droplets obtained by subjecting anionic precursor droplets (green) with a droplet size of $25 \times 10^6 \text{He}/z$ to subsequent electron impact ionization. Precursor droplets were produced at 6K by ionizing neutral droplets with electrons of 25.9 eV kinetic energy and 615 μA electron current in the first ionizer. Product droplets at rational fractions relative to the precursor result from ionizing selected precursor droplet with 167 μA electron current at 23.0 eV (anions) and 38.0 eV (cations)

negative charges into the droplets that recombine with already present positive charge centers and therefore decrease the net charge state of the droplet. As the charge state of a droplet decreases, the mass-per-charge ratio increases and therefore the droplet is detected at a higher mass-per-charge ratio than the precursor droplet.

As described above, negative charges can either be introduced into a droplet by capturing slow electrons that form electron bubbles or through the formation of $\text{He}^{*-}$. Electrons that have a kinetic energy of 22 eV or 44 eV can inelastically scatter with helium atoms, leading to excitation of a helium atom and capture of the free electron by the excited $\text{He}^{*}$. The corresponding resonances in the kinetic energy spectrum of incident electrons in the second ionizer is shown in Fig. 2.28b. While both curves of the charge reduction process show a resonance around 3 eV electron energy, which corresponds to the formation of an electron bubble, only the reduction to 3/2 has two distinguishable resonances at and above 22 eV, the latter resembling the formation of $\text{He}_2^{*-}$. In the region between 23 and 34 eV the formation of negative and positive charge carriers is possible, leading to a competition between both processes in the droplets. Charge reduction of multiply charged anions follows the same mechanisms as in cations. However, as observed autodetachment of negative charges in precursor droplets, shown in Fig. 2.29a, suggests, negative charges in helium droplets are not as stable as positive charges during the approximately 2 ms of flight time from the first ionizer to the detector.

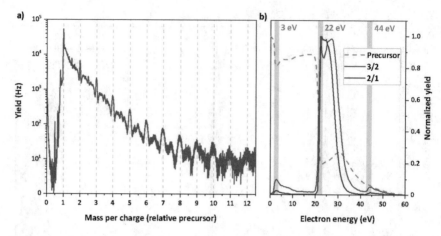

**Fig. 2.28 a** Charge state analysis of multiply charged cationic droplets with a droplet size of $1.89 \times 10^6 \mathrm{He}/z$, produced at 7 K with an electron current of 294 μA at 39.6 eV in the first ionizer. Using an electron current of 195 μA at 23 eV in the second ionizer to introduce negative charges into the droplets and decrease net positive charge shows droplets that had their charge state reduced from up to +12 to +1 (integer positions) and +17 to +2 (half-integer positions) respectively. **b** Energy spectra of a charge reduction process in positively charged precursor droplets resulting in product droplets with a relative mass-per-charge ratio of 2/1 and 3/2. The electron energy range where resonances of negative charge formation, i.e. electron bubbles and $\mathrm{He}^{*-}, \mathrm{He}_2^{*-}$ are observed are highlighted in grey

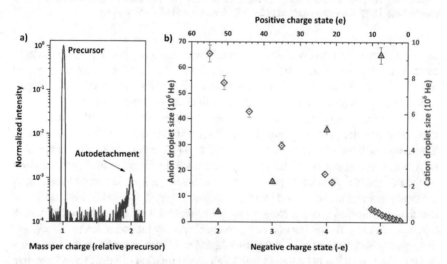

**Fig. 2.29 a** Autodetachment of negative charge carriers in anionic precursor droplets with a droplet size of $3.78 \times 10^6 \mathrm{He}/z$, produced at 7 K with an electron current of 685 μA at 28.5 eV in the first ionizer and second ionizer off. **b** Charge state of droplets versus the minimum droplet size at which the charge state can be observed for anionic (blue triangles) and cationic (red diamonds) droplets

When changing the charge state of precursor droplets, all observed patterns in resulting product droplets are strongly dependent on nozzle temperature, precursor size and ionization settings. Increasing electron current or the kinetic energy of electrons generally leads to a relative increase in intensity of lower mass-per-charge ratios in product droplets. An increase in precursor droplet size results in a higher number of achievable charge states upon secondary ionization. The possible share of already multiply charged droplets when selecting a droplet mass-per-charge ratio as precursor droplet is governed by the neutral droplet size distribution at a given temperature. For lower temperatures, more droplets double the size of the selected precursor droplet will be available, which, when being doubly charged upon ionization will appear at the same mass-per-charge ratio as the precursor droplet. The same principle holds for triply as well as more highly charged droplets.

The highest number of charges a droplet can carry for the ms timescale of the experiment is limited by the size of the droplet. In Fig. 2.29b the minimum droplet size for the observed charge states is shown both for positive and negative charges. While the threshold size for doubly charged cations is $(1.00 \pm 0.05) \times 10^5$ helium atoms, doubly charged anions can only be observed above a threshold size of $(3.95 \pm 0.20) \times 10^6$ atoms, at which a positively charged droplet can already hold more than 30 charges. This trend continues for further thresholds and illustrates a remarkable difference in charge stabilization mechanisms between negatively and positively charged droplets. As the threshold sizes for observable positively charged droplets scale in a linear fashion with the square of the droplet radius, it is likely that the positive charge carriers arrange near the surface of the droplets.

The experimental findings presented above show that helium droplets are capable of stabilizing multiple positive [171] and negative [416] charge carriers due to sufficiently large cohesive forces in helium droplets. In previous experiments, only doubly positively charged helium droplets [420] could be observed. The liquid droplet model [10] that takes possible multiply charging of helium droplets into account [421] underestimates the stability threshold sizes measured up to a factor of four [422]. Previous works with multiply charged Ne droplets also show a distinct discrepancy [423] in predicted threshold sizes by the liquid droplet model, which is explained by quantum effects.

### 3.3.2.1   Size Distributions

The size distributions of helium nano droplets produced by expansion of precooled helium through a pinhole into vacuum have been investigated under various conditions by means of titration [165] and deflection of charged droplets [161] during the last 30 years. Here we present recent results [424] that have been obtained by producing helium droplets in continuous supersonic expansion with 20 bar stagnation pressure, ionizing via electron impact and deflecting the charged droplets in an electrostatic sector analyzer (90° spherical, 0.07 m radius, 0.02 m electrode distance; 3.67° cylindrical, 5m radius, 0.01 m electrode distance, used for temperatures below 8 K).

As presented in the previous section "Multiply Charged Droplets", especially large droplets are able to stabilize multiple charge centers upon electron impact ionization. To minimize the influence of multiply charging on the droplet size distributions, the electron current is reduced to 50 nA, the lowest current that can be achieved while still guaranteeing stable electron emission from the ionizer filament. Even at this low electron current, due to the large geometrical cross sections of helium droplets produced at temperatures below 9 K, multiple electron hits per droplet are expected. Therefore, the electron energy of the ionizer was set to 22 eV to ensure the formation of negative droplet ions that have substantially larger critical droplet sizes for stabilizing multiple charges than cationic droplets.

To calibrate the results of electrostatic deflection measurements with regards to droplet size, the characterization of the beam velocity for each expansion condition is obligatory. Prior investigations by Buchenau [36] and Henne [425] showed that the velocity of individual droplets at a given temperature is dependent on the mass-per-charge ratio of the droplet. Therefore, droplet velocities were recorded for different selected mass-per-charge ratios. The velocities were obtained by time-of-flight measurements over the distance from the ionizer to the detector. The charged droplet beam was chopped by switching off the acceleration voltage of the electrons in the ionizer with a high-speed voltage switch. The switching pulse was used as trigger to start the time measurement. Figure 2.30a shows a set of droplet velocities for 5 K nozzle temperature. While the individual time-of-flight measurements with a peak width of only 5% justify the use of our electrostatic sector as a mass-per-charge selector, the overall velocity span in the range of 20% at a given expansion temperature underlines the importance of mass-per-charge selected velocity measurements for the evaluation of droplet size distributions in deflection measurements. The beam velocities at the mean size of the recorded log-normal shaped droplet size distributions are shown in Fig. 2.30b in comparison to experimental results reported in the literature. Possible sources for the discrepancies between our measurements and the literature could be a high fraction of fast small droplets that are suppressed by our sector geometry or a larger fraction of droplets in a high angular momentum state due to the used laser drilled nozzle that would lower the overall forward momentum of the droplet beam.

The resulting droplet size distributions shown in Fig. 2.31 follow a clear log-normal shape, as has been observed for small droplets in previous experiments [161]. Contrary to previous studies, we report a log-normal shaped distribution in the whole temperature range from 4 to 9 K and corresponding mean droplet sizes. The average sizes of these log-normal shaped distributions are compared with previous measurements by the group of Vilesov on neutral droplets [165] and deflection measurements by Henne and Toennies [161] in Fig. 2.32. In the temperature range of 6.5–9 K, our measurements are in reasonable agreement with previously reported average sizes. Below this range, we observe a plateauing of the mean droplet size, contrary to He titration measurements by the group of Vilesov [165]. The He titration method, in which the average He droplet size is deduced from the measured attenuation of a droplet beam colliding with collisional helium atoms, shows an exponential trend of increasing average size of neutral droplets, suggesting the existence of droplets

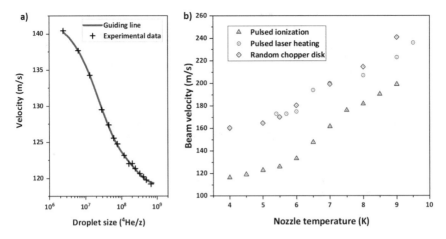

**Fig. 2.30  a** Droplet velocity distribution for different mass-per-charge values selected from a negatively charged droplet beam produced at 5 K. **b** Measured temperature dependent helium beam velocities determined by various experimental means. Time of flight velocity measurements by pulsing the helium beam via heating the nozzle with a pulsed laser by the group of Vilesov [165] are shown as green circles, random chopper disk measurements by Henne and Toennies [161] as red diamonds. Data obtained via pulsed ionization in our setup, as described in the text, are shown as blue triangles

having average diameters in the micrometer range. Such micrometer sized droplets have recently been identified optically [426] by the same group. A possible source for the deviation from titration measurements lies in the charged nature of the helium droplets used in our experiment. As discussed before, the possibility of a droplet carrying multiple charge centers can lead to a lower apparent droplet size than if the droplet would be singly charged. Assuming the geometrical cross section of a droplet resembling the ionization cross section, an exponential rise in droplet size for temperatures below 7 K would promote multiply charging if the ionization current, as in our case, is kept constant.

Another plausible cause for the deviation from the literature could be the pickup of dopant molecules, such as residual water [113] in the vacuum vessel. A charge transfer from He*⁻ to dopant species could create charge centers that are not able to trigger secondary electron emission due to a lack of sufficient ionization energy for overcoming the work function of the surface of the channel electron multiplier detector.

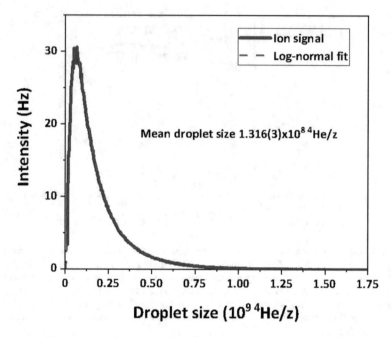

**Fig. 2.31** Distribution of helium droplet size recorded at 5 K with an electron current of 50 nA at 22 eV. A log-normal fit to the data gives a mean droplet size of $1.316(3) \times 10^8$ He atoms

### 3.3.3 Pickup with Charged HNDs

Experiments using helium droplets for pickup have been performed for many years now [427]. Helium droplets have proven to be very suitable for experiments at ultra-cold conditions as well as for spectroscopy studies [206, 427]. Furthermore, HNDs have been found to be a very sophisticated way of growing nanoparticles of a desired composition and size [428]. Measured size distributions of HNDs, produced via free jet expansion, exhibit a strong correlation between ionization conditions and mean droplet size. As discussed previously, Laimer et al. developed a new experimental setup to investigate the underlying mechanisms. The tandem mass spectrometry experiment was designed to study possible fragmentation channels of highly charged HNDs. To the surprise of the experimentalists, the HNDs were found to be stable up to very high charge states. If HNDs are charged beyond a certain maximum charge density, low mass singly charged ions are ejected, leading to a negligible reduction in the amount of helium of the residual HNDs not visible within the resolution of the experiment [171]. The knowledge about the stability of highly charged HNDs was pivotal in the process of the development of further experiments performing pickup into (highly-) charged HNDs. Pickup of dopants into multiply charged HNDs promises a higher yield of dopant cluster ions since every charge center located close to the surface of a HND acts as a seed for dopant cluster growth. Thus, multiple dopant

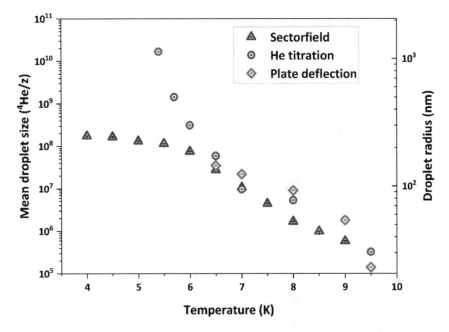

**Fig. 2.32** Mean helium droplet sizes for expansion temperatures that have been measured by different experimental methods and groups. Measurements produced with the presented experimental setup are shown as blue triangles, neutral He titration measurements by the group of Vilesov [165] as red circles and plate deflection measurements by Henne and Toennies [161] as green diamonds

clusters are formed in every HND as opposed to only one, as is the case for pickup into neutral HNDs.

In this chapter, an experimental setup is described where pickup of neutral dopants into stable, highly charged HNDs is realized. A rendered picture of the source used to form such HNDs is shown in Fig. 2.33, while a schematic of the apparatus is shown in Fig. 2.34. The focus of this work lies on the technical realization as well as on the characterization of the experiment. A more detailed description of this setup and its performance in comparison to conventional techniques was published recently [200].

### 3.3.3.1   Overview of the Experimental Setup

Neutral HNDs are produced via a continuous free jet expansion of pressurized and precooled helium through a pinhole nozzle (5 μm diameter) as described previously [172, 429]. After passing a skimmer, the droplets travel a distance of about 10 cm before being ionized in an electron impact ion source. A DC quadrupole is used to select HNDs of a specific mass-per-charge ratio and deflect these by 90° in either of two directions. On one side (left), a channel electron multiplier (CEM) is installed for monitoring the helium droplet beam. On the opposite side (right), the HNDs enter

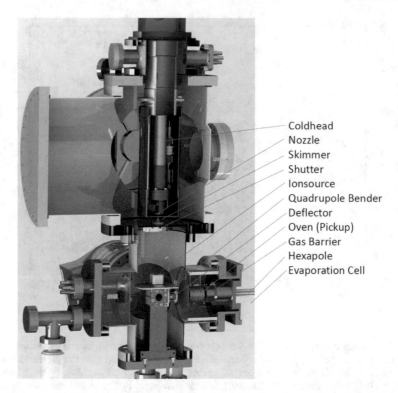

Coldhead
Nozzle
Skimmer
Shutter
Ionsource
Quadrupole Bender
Deflector
Oven (Pickup)
Gas Barrier
Hexapole
Evaporation Cell

**Fig. 2.33** A cutaway rendering of the experimental setup using charged pickup. Including the helium droplet source, ion source, quadrupole bender, and high temperature oven. Reproduced with permission from Ref. [200]. © copyright AIP Publishing. All rights reserved

a differentially pumped pickup region. In addition to the oven shown in Fig. 2.34, a second pickup chamber can be installed between the first oven and the gas barrier, housing a second oven as well as a gas inlet. This source for the production of highly charged, mass-selected and doped HNDs is connected to a modified QMS/ToF-MS tandem mass spectrometer (Q-Tof Ultima, Waters/Micromass). Figure 2.35 depicts a schematic of the entire experimental setup.

Following the pickup region, dopant ions are extracted by evaporating excess helium from the HNDs in collisions with a gas of variable pressure (typically He) in an evaporation cell equipped with a RF hexapole ion guide (a detailed description of this method can be found in 3.3.4). The dopant ions now enter a region housing a RF quadrupole mass filter followed by another RF hexapole, designed to perform collision induced dissociation (CID) measurements. The RF quadrupole mass filter is used to select precursor ions of a specific mass-to-charge ratio which are guided into the collision cell. Here, the ions perform collisions of controlled energy with a monoatomic gas (typically Ar). When no CID measurements are performed, the collision gas is removed and the quadrupole mass filter simply acts as an ion guide.

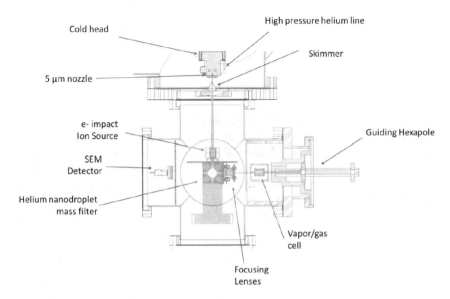

**Fig. 2.34** Schematic of the HND source used for charged pickup. The green lines represent simulated trajectories of helium droplets. Reproduced with permission from Ref. [200]. © copyright AIP Publishing.

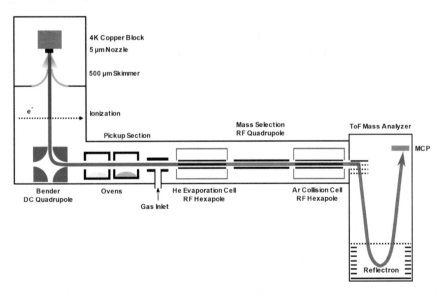

**Fig. 2.35** Schematic of the experimental setup including the QMS/ToF-MS tandem mass spectrometer. The blue line indicates the beam of neutral HNDs; the purple line represents the trajectory of the charged mass-selected HNDs and the red line indicates the path of the extracted dopant cluster ions

The final analysis is performed by an orthogonal extraction reflectron ToF-MS with a resolving power $m/\Delta m$ of up to 10,000.

### 3.3.3.2 Pickup and Growth of Dopant Clusters

Pickup of single neon atoms in HNDs was first reported by Scheidemann et al. [32]. Since then, HNDs have been used to study numerous different species, all of which have been picked up by neutral HNDs [22]. The growth of a dopant cluster within or at the surface of a neutral helium droplet depends almost exclusively on the pickup probability. Dopants are attracted to each other via van-der-Waals interaction which leads to the formation of dopant clusters within or at the surface of a HND [22].

Samples with high vapor pressures can be introduced into our machine via a stainless-steel gas line. For samples with lower vapor pressures, two different ovens were constructed. For organic samples, a CNC-machined molybdenum holder is constructed around a glass cup that holds the sample, which is ideal for fragile samples such as biomolecules since it is highly unreactive. Molybdenum was chosen for its thermal stability and to prevent surface charges due to its conductivity. The holder is heated via a tantalum filament able to generate temperatures of up to 800K. For samples with even lower vapor pressures, e.g. metals with high melting points such as gold, a CNC-machined aluminium nitride ceramic oven is used. It is heated via a tantalum-rhenium filament and can be used at temperatures of up to 1800 K. The disadvantages, compared to the molybdenum/glass oven, are that it is harder to clean, it is more delicate and one must face the issue of surface charges on the ceramic.

### 3.3.3.3 Ionization and Stability of the Dopant Clusters

In order to analyze dopants picked up by neutral HNDs via mass spectrometry, the doped HNDs have to be ionized. Based on the knowledge obtained from the formation and stability of highly-charged HNDs, we have to conclude that dopant cluster ions can only be released from large HNDs upon multiply-charging of the doped HNDs and subsequent asymmetric Coulomb explosion [171]. Although superfluid helium is one of the best matrices to dissipate excess energy, the ionization process via charge transfer from $He^+$ or $He_2^+$ is so exothermic (typically more than 10 eV) that the dopant clusters created are often fragmented.

This effect is illustrated in Fig. 2.36. A comparison between pickup into charged HNDs (Fig. 2.36a) and conventional pickup into neutral HNDs (Fig. 2.36b) demonstrates the effect of fragmentation of neutral gold clusters inside large HNDs upon charge transfer from $He^+$ or $He_2^+$. Both methods produce $Au_n^+$ distributions featuring similar log-normal envelopes, governed by Poissonian pickup statistics. However, whereas the distribution of $Au_n^+$ produced by pickup into charged HNDs is smooth, indicating little or no fragmentation (Fig. 2.36a), intense fragmentation of $Au_n^+$ upon EI of neutral gold-doped HNDs is indicated by intense odd-even oscillations

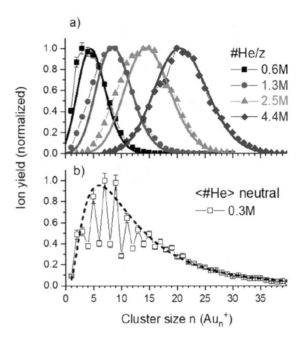

and pronounced drops in the ion yield after $n = 9$ and $21$, suggesting magic number character of these ions (Fig. 2.36b). The very same intensity anomalies have been observed for alkali [3, 14] or noble metal cluster ions [294] when formed via conventional techniques. High ion yields relative to neighboring cluster sizes are the result of an increased stability of a specific cluster ion, either due to electronic [3] or geometric [430] reasons. The stability of $Au_n^+$ and many small metal cluster ions is predominantly determined by their electronic structure. Spin pairing of the single $6s$ electrons of gold explains the odd-even oscillation while closure of the $1p$ and $1d$ orbitals for 8 and 20 electrons is responsible for the intensity drops at $n=9$ and $21$ (one extra gold atom is required to obtain a positively charged cluster). Therefore, what we observe in the experiments upon pickup into neutral HNDs, is not the size distribution of neutral dopant clusters but rather their charged fragments.

In the experiment using pickup into charged HNDs, no such oscillations are observed. This means that species created after pickup into charged HNDs grow by exclusively following the statistics of the pickup process, and do not fragment in the ionization process. This effect is especially beneficial when the species of interest is rather fragile. Performing pickup into charged HNDs enables the production of high yields of fragile ions such as even-numbered gold clusters or biomolecules which are prone to fragmentation when using conventional pickup into neutral HNDs followed by an ionization process.

### 3.3.3.4 Extracting Dopant Clusters from Charged HNDs: An Efficient Method to Obtain Helium-Tagged Ions

In conventional HND experiments, the pickup section of the experiment is usually followed by an ionization device, the most common being an EI or PI source. When a HND becomes highly charged, some low mass singly charged ions are ejected from the HND. These ejected ions are then accessible for mass spectrometric analysis.

In the case of pickup into charged HNDs, dopant cluster ions are ejected from the HND only if the droplet shrinks below the critical size for its charge state due to the evaporation of helium atoms. The evaporation of helium atoms is caused by pickup events and the release of binding energy during dopant cluster formation [171]. In the case of strongly bound gold clusters, every pickup and addition of a gold atom to a cluster will evaporate between 5000 and 8300 helium atoms [171, 431].

In order to extract the dopant cluster ions from the HNDs in the setup shown in Fig. 2.34, the following approach is chosen. The HNDs are continuously evaporated in a collision cell where the HNDs are heated by collisions with room temperature stagnant helium gas. In the process, the HNDs shrink until they are no longer stable with respect to their charge state. As a result, charged dopant cluster ions are ejected from the HNDs which often remain solvated by a small number of helium atoms, referred to as helium-decorated or helium-tagged ions [198]. From the helium decoration, one can obtain valuable information about the dopant (cluster) ions. The temperature of these ions can be estimated very conveniently since helium evaporates at temperatures above the binding energy of the weakest bound helium atom. Apart from that, observed shell closures contain information about the geometric structure of the dopant cluster [299]. Most importantly, He-decorated species are perfectly suitable for messenger-type action spectroscopy. When light at a certain wavelength is absorbed by a helium tagged ion, energy transfer to the adsorbed helium will lead to its evaporation. Measuring either the depletion of a helium tagged ion or the formation of its photoproduct while scanning the photon energy of a tunable laser provides an absorption spectrum of ions at very low temperatures [206, 432–434]. Hence, being able to produce large amounts of helium-decorated sample species is highly desirable.

The process of helium decoration in neutral pickup regimes where dopant ions are ejected from the HND is not easily controlled. The internal energy of the dopant cluster is often high at the time it is ejected and therefore most of the dopants are ejected with very little to no helium attached. As mentioned above, in a setup using pickup into charged HNDs, the ejection of dopant molecules is triggered by relatively slow evaporation of the HNDs in a collision cell. In the present setup, this evaporation cell is a 25 cm long guiding RF hexapole filled with stagnant helium gas at room temperature. When the HNDs collide with gaseous helium atoms, the collision energy leads to evaporation of helium from the droplet. The decreasing size of the HNDs lead to the ejection of cationic dopants, often solvated by up to a few hundred He atoms [25, 171, 435] whenever the droplet size shrinks below the critical size of the current charge state. The hexapole serves to collect and guide the small ejected ions since the ejection occurs in all directions. In further collisions of the ejected dopants with gaseous helium, the helium attached to the ions can be gradually removed. Thus, the amount of helium that is still attached to the dopants can be controlled by varying the pressure in the evaporation cell. Figure 2.37 shows a comparison of different helium decoration measurements made in a setup using pickup into neutral HNDs and the described apparatus.

The upper panel of Fig. 2.37 represents a measurement where helium-decorated fullerene $C_{60}$ cations were produced via pickup into charged HNDs. For the measurement, $C_{60}$ was evaporated in the high temperature ceramic oven. The mass spectrum shown in the lower panel represents a measurement of helium-decorated $C_{60}^+$ using electron ionization of neutral HNDs doped with $C_{60}$ in an action spectroscopy measurement of $C_{60}^+$ [206]. The comparison shows that the production efficiency of $C_{60}^+$ with one helium attached is increased by more than a factor of 1000 using pickup into charged HNDs. Please note that the intense peaks at $m/z=724$ and $m/z=725$ are predominantly due to isotopes of the bare fullerene cation containing four and five $^{13}C$. The high production efficiency of helium-tagged ions using charged pickup facilitates action spectroscopy of cold ions with a minimal matrix shift. Furthermore, the knowledge of the matrix shift as a function of the attached helium atoms provides information on the structure and size of the solvation layer. For a sufficiently large number of attached helium atoms, a constant matrix shift is observed.

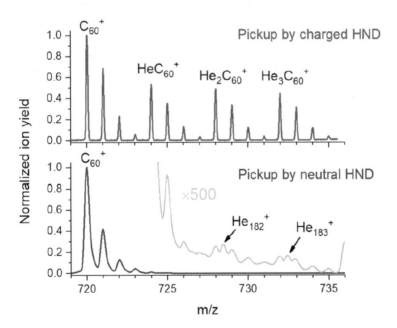

**Fig. 2.37** A comparison of helium-decorated fullerenes produced in the described experiment using pickup into charged HNDs (top panel) and an experiment using pickup into neutral HNDs (bottom panel). Reproduced with permission from Ref. [200]. © copyright AIP Publishing. All rights reserved

### 3.3.3.5    Tailoring the Size Distribution of Dopant Clusters Produced

By preselecting a mass-per-charge ratio of HNDs for the pickup into charged HNDs, the size distribution of dopant clusters produced can be manipulated in a very convenient way. A demonstration of this technique is shown in Fig. 2.38. Using this method to optimize the experimental parameters to produce the desired dopant cluster size distribution is especially interesting when sublimation sources are used where the partial pressure of the sample cannot be controlled easily. An example is the evaporation of sensitive amino acids which are destroyed by too high temperatures during sublimation of the sample.

Not only can the mean size of the dopant clusters be altered by the selection of different mass-per-charge ratios of HNDs (see Fig. 2.36), but also the width of the dopant cluster distribution, as can be seen in Fig. 2.38. As discussed earlier, a small mass-per-charge ratio of HNDs produces smaller dopant cluster ions. When the partial pressure of the sample in the pickup cell is increased, the dopant cluster sizes produced are increased as well. The width of the dopant cluster size distribution, however, does not increase proportionally.

Figure 2.38 demonstrates this effect using gold clusters as an example. The different partial pressures of gold in the pickup cell are realized by adjusting the oven temperature. The measurement shown in the left panel was taken at an oven heating power of 60 W, whereas the measurement shown in the right panel was taken at an oven power of 75 W.

For the measurements shown in Fig. 2.38, the nozzle temperature was held at 9.2 K for the left panel and at 9.6 K for the right panel. As shown in [171], a HND can only hold more than one charge when its size exceeds $10^5$ helium atoms. The plot in the right panel in Fig. 2.38 shows a measurement where pickup is performed predominantly with singly charged HNDs. The left panel shows a measurement with HNDs that are well above the threshold for multiply charging. In this panel, a Poissonian distribution with an expected value of $\lambda = 16$ is plotted with a red line together with the mass spectrum (black line) to clarify that the distribution follows the

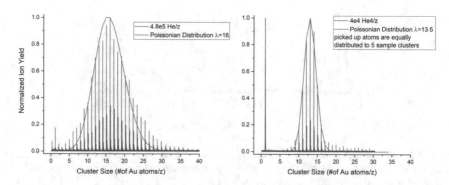

**Fig. 2.38** Demonstration of the altering of size distributions of gold clusters created by tuning the mass-per-charge ratio and the size of the HNDs used for pickup

pickup statistics. In the right panel a Poissonian distribution is shown (red line) that has an expected value of $\lambda=65$ and was divided by a factor of 5. The size distribution of singly charged gold clusters produced is five times narrower than expected for a "normal" pickup process. Our tentative explanation is that the gold cluster ions produced in this measurement are very close to the maximum size of gold clusters that can be produced with HNDs of this size. With every gold atom picked up by the HND, a substantial part of the droplet is evaporated, hence, the HND gradually shrinks. Since the partial pressure of gold in the pickup region is rather high, every HND that passes the cell is expected to have picked up the largest number of gold atoms possible for its size. The measured distribution is therefore not determined by the pickup probability but rather the maximum number of gold atoms that a HND can pick up before it evaporates almost entirely. The tail to the right side can be explained by pickup with HNDs bigger than 50,000 helium atoms per charge. A few of these bigger droplets also exit the DC quadrupole since it has limited resolving power. Droplets exceeding 50000 helium atoms per charge can be doubly charged, according to [171]. As soon as these HNDs pick up a few gold atoms, they drop below the stable size for holding two charges and eject a small, singly charged unit. These fragment ions can be detected as monomer ions $Au^+$. The remaining, singly charged HND now has a mass-to-charge ratio that is far bigger than the initially selected one of 40,000 helium atoms per charge, thus creating gold clusters that do not fit the main distribution, explaining the origin of the secondary, low-intensity size distribution of gold cluster ions observed between $n=17$ and $n=30$.

### 3.3.3.6  Proton Transfer Ionization in HNDs Pre-doped with $H_2$

Since our experiments are typically performed with HNDs containing more than $10^5$ helium atoms per cluster, ionization of the dopant occurs almost exclusively via charge transfer from helium. This is also the case for experiments utilizing pickup into neutral HNDs. For EI, the ionization probability is dependent on the geometrical cross section of the target helium droplet. The following equations represent the most common way for cation formation within a HND

$$e^- + He \rightarrow He^+ + 2e^- \tag{2.1}$$

$$He^+ + He \rightarrow He_2^+ \tag{2.2}$$

$$He_2^+ + M \rightarrow 2He + M^{*+} \tag{2.3}$$

Helium has the highest ionization energy of all elements at 24.6 eV [436]. Charge transfer from $He^+$ or $He_2^+$ to all other elements and molecules is therefore energetically possible. Ionization energies of molecules are typically around or below 10 eV, which results in a charge transfer from $He^+$ being exothermic by more than 14 eV in most cases. This excess energy is initially absorbed by the dopant in the form of electronic and ro-vibrational excitation. If this excitation energy is not sufficiently dissipated by the HND, it may lead to fragmentation of the dopant ion. Rovibrational excitation of dopants can be cooled very efficiently via evaporation of helium atoms due to the exceptionally high thermal conductivity of superfluid helium. However, electronic excitation into an antibonding state of a molecular dopant cannot be dissipated by the sourrounding helium. As a result, such excited dopant molecules are likely to be observed as fragments, even when picked up into a charged HND. To this end, a soft ionization process for doped HNDs was recently developed in our group utilizing the instrument shown in Figs. 2.33 and 2.34. During the time the multiply charged droplets require to travel from the ionization region to the pickup cell, all charge centers will localize as $He_2^+$ or $He_3^+$ ionic cores [417] surrounded by a dense, solid-like layer of He atoms [25]. As in the case of a conducting sphere, the charge centers will reside close to the surface of the HNDs and mutual repulsion will arrange them in some form of a two-dimensional Wigner crystal on the surface of the droplet. The ionization energy of $He_2^+$ [22] or $He_3^+$ [437] is still higher than the ionization energies of any dopant. Whereas the first dopant molecule colliding with an ionic core might fragment due to the excess energy, remaining excitation energy will be dissipated rapidly by the HND environment. Thus, the next dopant molecule which attaches to the fragmented molecular ion will not undergo fragmentation. This way, mixed clusters consisting of a fragment ion that is solvated by several intact molecules are formed. In order to produce intact dopant cluster ions, the fragmentation process needs to be suppressed.

Charge transfer from $He_2^+$ or $He_3^+$ transfers one electron from the first dopant molecule M to the $He_2^+$ or $He_3^+$ and leads to the formation of an excited cationic molecule $M^{+*}$. If the cation $M^+$ has an open electronic shell it can be highly reactive and unstable towards fragmentation. Adding a proton to M leads to a protonated cation $MH^+$ which has a closed electronic shell and thus produces a more stable cation. Proton transfer reaction is one of the softest ways of forming cations from neutral bio molecules [438]. Proton transfer to a molecule M from a hydrogen containing cation that has a lower proton affinity than M is a very efficient process. In gas phase, exothermic proton transfer often is associated with unwanted fragmentation of the proton acceptor molecule which requires proton donors with only slightly lower proton affinity than the reaction partner to minimize the transfer of excess energy. The enormous cooling power of HNDs, however, should be capable of dissipating excess energy originating from proton transfer processes, which is relatively slow process compared to an electron transfer reaction. If molecular hydrogen $H_2$ is picked up by charged HNDs, the following reactions will lead to the formation of $(H_2)_n H^+$ ions [171].

$$He_2^+ + H_2 \rightarrow 2He + H + H^+ \qquad (2.4)$$

$$H^+ + nH_2 \rightarrow (H_2)_n H^+ \qquad (2.5)$$

The small proton affinity of these ions makes them perfect proton donors for almost every molecular dopant. With these pre-doped, charged HNDs, pickup of bio molecules is performed. Thus, the charge transfer process that forms sample ions within the HNDs is no longer

$$M + He_2^+ \rightarrow M^+ + 2He + \Delta E \tag{2.6}$$

but rather a proton transfer process

$$M + (H_2)_n H^+ \rightarrow MH^+ + (H_2)_n + \Delta E' \tag{2.7}$$

where typically $\Delta E' < \Delta E$, reducing ionization-induced fragmentation. Figure 2.39 shows a comparison of two mass spectra of HNDs doped with the amino acid valine. Both mass spectra were obtained via pickup into charged HNDs utilizing the setup shown in Figs. 2.33 and 2.34. Proton transfer ionization by hydrogen pre-doping (lower diagrams) can be compared with charge transfer ionization from $He_2^+$ (upper diagrams).

For this measurement, D-valine was vaporized in a low temperature oven heated to 110 °C. In the upper left-hand panel, strong fragmentation of the biomolecule with a mass of 117 amu to its main fragment with 72 amu can be observed, corresponding to the loss of the carboxylic group COOH. As a comparison, in the lower left-hand panel, a measurement with hydrogen pre-doping at a pressure of 0.08 mPa and otherwise identical conditions as above is shown. The ionization process clearly appears to be softer indicated by the strongly suppressed COOH loss. A satellite peak that is 18 amu higher than the main ion series can be assigned to protonated valine clusters complexed with a water molecule. Since the ion signal is not distributed among the different fragmentation channels as is the case with charge transfer ionization from

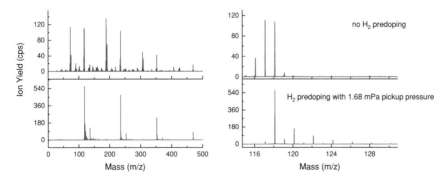

**Fig. 2.39** A comparison of two measurements with identical settings apart from the pre-doping with $H_2$ in the measurement shown in the lower panels. The residual pressure of helium in the hydrogen pickup region was 0.08 mPa. The pressures are corrected for the gas type, according to the manual

$He_2^+$, the signal intensity of the pure valine clusters has more than quadrupled for proton transfer ionization. It should be noted that from the mass spectra alone it is not clear whether the peaks corresponding to valine or protonated valine clusters are still intact or fragmented with the COOH neutral product still attached.

A closeup look at the monomer region is provided in the right-hand panels. In the top right panel, one can see that charge transfer from helium cations to valine, also produces protonated valine with similar intensity as the parent molecular cation. In this case, the likely proton donor is another a valine molecule. There are also peaks located one and two mass units below the parent cation, indicating hydrogen loss as a fragmentation channel, which can be explained by the large amount of excess energy available after charge transfer from $He_2^+$. In the case of $H_2$ pre-doping, protonated valine is by far the most abundant ion and no hydrogen loss is observed. Additionally, hydrogen tagging with up to ten $H_2$ molecules bound to protonated valine can be observed. The low binding energy of such taggants is a measure of the internal energy of the valine cluster ions. They also provide suitable messengers for action spectroscopy of cold ions [432, 433, 439–441].

## 3.4 Conclusion/Outlook

Many important discoveries and developments in HND research have been propelled by mass spectrometry since the emergence of the field. We attempted to review the work performed in the last decades by groups all over the world as well as our own group's most recent research in this chapter. However, while a lot of experimental techniques and components have been refined in various ways over the last decades, the vast majority of HND MS experiments never diverted from the tried-and-true basic principle of doping neutral helium droplets with an analyte, followed by ionization and analysis. Building on our recently gained knowledge about highly charged HNDs, further experiments showed that a simple re-ordering of these steps and ionizing HNDs *before* pickup opens up a plethora of exciting possibilities and applications, extending far beyond the field of HND MS. The ability to control the produced dopant cluster size distribution and efficiently produce helium-tagged ions are only two examples of what could be in store for the future.

## References

1. P. Jena, A.W. Castleman, in *Science and Technology of Atomic, Molecular, Condensed Matter & Biological Systems*, ed. by P. Jena, A.W. Castleman (Elsevier, 2010), pp. 1–36
2. O. Echt, K. Sattler, E. Recknagel, Phys. Rev. Lett. **47**, 1121 (1981)
3. W.D. Knight et al., Phys. Rev. Lett. **52**, 2141 (1984)
4. H.W. Kroto et al., Nature **318**, 162 (1985)
5. M.G. Mayer, Phys. Rev. **74**, 235 (1948)
6. O. Haxel, J.H.D. Jensen, H.E. Suess, Phys. Rev. **75**, 1766 (1949)

7. G. Audi, Int. J. Mass Spectrom. **251**, 85 (2006)
8. P.G. Lethbridge, A.J. Stace, J. Chem. Phys. **89**, 4062 (1988)
9. T.P. Martin et al., J. Phys. Chem. **95**, 6421 (1991)
10. M. Pellarin et al., Chem. Phys. Lett. **217**, 349 (1994)
11. I.A. Harris, R.S. Kidwell, J.A. Northby, Phys. Rev. Lett. **53**, 2390 (1984)
12. J.A. Northby, J. Chem. Phys. **87**, 6166 (1987)
13. W. Miehle et al., J. Chem. Phys. **91**, 5940 (1989)
14. W.A. de Heer, Rev. Mod. Phys. **65**, 611 (1993)
15. M. Brack, Rev. Mod. Phys. **65**, 677 (1993)
16. M. R. Hoare, in *Advances in Chemical Physics* (Wiley, 2007), pp. 49–135
17. J.P.K. Doye, D.J. Wales, R.S. Berry, J. Chem. Phys. **103**, 4234 (1995)
18. M.B. Knickelbein, W.J.C. Menezes, J. Phys. Chem. **96**, 6611 (1992)
19. E. Janssens *et al.*, Phys. Rev. Lett. **99**, 063401 (2007).
20. W. Huang, L.-S. Wang, Phys. Rev. Lett. **102**, 153401 (2009)
21. M. Goulart et al., Phys. Chem. Chem. Phys. **20**, 9554 (2018)
22. A. Mauracher et al., Phys. Rep. **751**, 1 (2018)
23. L. Meyer, F. Reif, Phys. Rev. **123**, 727 (1961)
24. L. Meyer, F. Reif, Phys. Rev. **110**, 279 (1958)
25. K.R. Atkins, Phys. Rev. **116**, 1339 (1959)
26. M.W. Cole, R.A. Bachman, Phys. Rev. B **15**, 1388 (1977)
27. A.P.J. van Deursen, J. Reuss, J. Chem. Phys. **63**, 4559 (1975)
28. J. Gspann, H. Vollmar, J. Chem. Phys. **73**, 1657 (1980)
29. J. Gspann, Surf. Sci. **106**, 219 (1981)
30. J. Gspann, Physica B **169**, 519 (1991)
31. P.W. Stephens, J.G. King, Phys. Rev. Lett. **51**, 1538 (1983)
32. A. Scheidemann, J.P. Toennies, J.A. Northby, Phys. Rev. Lett. **64**, 1899 (1990)
33. A. Scheidemann et al., Physica B **165–166**, 135 (1990)
34. H. Buchenau, J.P. Toennies, J.A. Northby, J. Chem. Phys. **95**, 8134 (1991)
35. R. Fröchtenicht et al., J. Chem. Phys. **104**, 2548 (1996)
36. H. Buchenau et al., J. Chem. Phys. **92**, 6875 (1990)
37. M. Lewerenz, B. Schilling, J.P. Toennies, J. Chem. Phys. **102**, 8191 (1995)
38. A. Bartelt et al., Phys. Rev. Lett. **77**, 3525 (1996)
39. S. Yang, A.M. Ellis, Chem. Soc. Rev. **42**, 472 (2012)
40. C. Jackschath, I. Rabin, W. Schulze, Z. Phys. D Atoms Mol. Clust. **22**, 517 (1992)
41. D. Rayane et al., J. Chem. Phys. **90**, 3295 (1989)
42. J.P. Toennies, A.F. Vilesov, K.B. Whaley, Phys. Today **54**, 31 (2001)
43. A. Scheidemann, B. Schilling, J.P. Toennies, J. Phys. Chem. **97**, 2128 (1993)
44. V.H. Dibeler, F.L. Mohler, J. Res. Natl. Bur. Stan. **40**, 25 (1948)
45. B.P. Pullen, J.A.D. Stockdale, Int. J. Mass Spectrom. Ion Phys. **21**, 35 (1976)
46. T. Stanski, B. Adamczyk, Int. J. Mass Spectrom. Ion Phys. **46**, 31 (1983)
47. D. Margreiter et al., Int. J. Mass Spectrom. Ion Process. **100**, 143 (1990)
48. B.E. Callicoatt et al., J. Chem. Phys. **108**, 9371 (1998)
49. M. Fárník and J. P. Toennies, J. Chem. Phys. **122**, 014307 (2004)
50. N.G. Adams, D. Smith, J. Phys. B: At. Mol. Phys. **9**, 1439 (1976)
51. B.R. Rowe et al., Chem. Phys. Lett. **113**, 403 (1985)
52. D. Smith, N.G. Adams, Int. J. Mass Spectrom. Ion Phys. **23**, 123 (1977)
53. B.E. Callicoatt et al., J. Chem. Phys. **109**, 10195 (1998)
54. T. Ruchti et al., J. Chem. Phys. **109**, 10679 (1998)
55. T. Ruchti, B.E. Callicoatt, K.C. Janda, Phys. Chem. Chem. Phys. **2**, 4075 (2000)
56. B.E. Callicoatt et al., J. Chem. Phys. **105**, 7872 (1996)
57. A. M. Ellis and S. Yang, Phys. Rev. A **76**, 032714 (2007).
58. A.A. Scheidemann, V.V. Kresin, H. Hess, J. Chem. Phys. **107**, 2839 (1997)
59. Y. Ren and V. V. Kresin, Phys. Rev. A **76**, 043204 (2007).
60. S. Vongehr et al., Chem. Phys. Lett. **353**, 89 (2002)

61. F. Dalfovo, Z. Phys. D Atoms Mol. Clust. **29**, 61 (1994)
62. F. Ancilotto, P.B. Lerner, M.W. Cole, J. Low Temp. Phys. **101**, 1123 (1995)
63. F. Ancilotto et al., Z. Physik B Condens. Matter **98**, 323 (1995)
64. P.B. Lerner, M.W. Cole, E. Cheng, J. Low Temp. Phys. **100**, 501 (1995)
65. F. Stienkemeier, F. Meier, H.O. Lutz, J. Chem. Phys. **107**, 10816 (1997)
66. J. Reho et al., J. Chem. Phys. **112**, 8409 (2000)
67. F. Stienkemeier, F. Meier, H.O. Lutz, Eur. Phys. J. D **9**, 313 (1999)
68. M. Mella, G. Calderoni, F. Cargnoni, J. Chem. Phys. **123**, 054328 (2005).
69. A. Hernando et al., J. Phys. Chem. A **111**, 7303 (2007)
70. Y. Ren, R. Moro, V.V. Kresin, Eur. Phys. J. D **43**, 109 (2007)
71. Y. Ren, V.V. Kresin, J. Chem. Phys. **128**, 074303 (2008).
72. B. Bellina, D.J. Merthe, V.V. Kresin, J. Chem. Phys. **142**, 114306 (2015)
73. W.K. Lewis et al., J. Am. Chem. Soc. **127**, 7235 (2005)
74. T. Diederich et al., Phys. Rev. Lett. **86**, 4807 (2001)
75. T. Döppner *et al.*, J. Chem. Phys. **126**, 244513 (2007)
76. J. Tiggesbäumker, F. Stienkemeier, Phys. Chem. Chem. Phys. **9**, 4748 (2007)
77. M. Mudrich, F. Stienkemeier, Int. Rev. Phys. Chem. **33**, 301 (2014)
78. T. Döppner et al., Eur. Phys. J. D **16**, 13 (2001)
79. Th. Diederich et al., Phys. Rev. A **72**, 023203 (2005)
80. M. Rossi et al., Phys. Rev. B **69**, 212510 (2004)
81. T. Döppner et al., Phys. Chem. Chem. Phys. **9**, 4639 (2007)
82. L. Köller et al., Phys. Rev. Lett. **82**, 3783 (1999)
83. T. Döppner et al., Eur. Phys. J. D **43**, 261 (2007)
84. T. Döppner et al., Phys. Rev. Lett. **94**, 013401 (2005)
85. F. Verluise et al., Opt. Lett., OL **25**, 575 (2000)
86. N.X. Truong et al., Phys. Rev. A **81**, 013201 (2010)
87. T. Ditmire et al., Phys. Rev. A **53**, 3379 (1996)
88. A. Przystawik et al., Phys. Rev. A **78**, 021202 (2008)
89. A. Hernando et al., Phys. Rev. B **78**, 184515 (2008)
90. S. Göde et al., New J. Phys. **15**, 015026 (2013)
91. G. Droppelmann et al., Eur. Phys. J. D **52**, 67 (2009)
92. C. P. Schulz et al., Phys. Rev. Lett. **92**, 013401 (2004)
93. S. Müller, M. Mudrich, F. Stienkemeier, J. Chem. Phys. **131**, 044319 (2009)
94. S. Müller et al., Phys. Rev. Lett. **102**, 183401 (2009)
95. W.K. Lewis, C.M. Lindsay, R.E. Miller, J. Chem. Phys. **129**, 201101 (2008)
96. W.K. Lewis et al., J. Am. Chem. Soc. **126**, 11283 (2004)
97. W. K. Lewis, R. J. Bemish, and R. E. Miller, J. Chem. Phys. **123**, 141103 (2005).
98. T. M. Falconer et al., Rev. Sci. Instrum. **81**, 054101 (2010)
99. W. K. Lewis et al., Rev. Sci. Instrum. **83**, 073109 (2012)
100. W. K. Lewis et al., Rev. Sci. Instrum. **85**, 094102 (2014)
101. M.N. Slipchenko et al., Rev. Sci. Instrum. **73**, 3600 (2002)
102. R. Katzy et al., Rev. Sci. Instrum. **87**, 013105 (2016)
103. D. Verma, A.F. Vilesov, Chem. Phys. Lett. **694**, 129 (2018)
104. S. Yang, S.M. Brereton, A.M. Ellis, Rev. Sci. Instrum. **76**, 104102 (2005)
105. S. Yang, A.M. Ellis, Rev. Sci. Instrum. **79**, 016106 (2008)
106. U. Henne, Untersuchung großer durch Elektronenstoß erzeugter negativer und positiver Helium-Clusterionen, PhD dissertation, University of Göttingen/Max-Planck-Institut für Strömungsforschung, 1996
107. S. Yang et al., Phys. Chem. Chem. Phys. **7**, 4082 (2005)
108. S. Yang et al., J. Phys. Chem. A **110**, 1791 (2006)
109. A.M. Ellis, S. Yang, Chin. J. Chem. Phys. **28**, 489 (2015)
110. A. Boatwright, J. Jeffs, A.J. Stace, J. Phys. Chem. A **111**, 7481 (2007)
111. B. Shepperson et al., J. Chem. Phys. **135**, 041101 (2011)
112. S. Yang, S.M. Brereton, A.M. Ellis, Int. J. Mass Spectrom. **253**, 79 (2006)

113. S. Yang et al., J. Chem. Phys. **127**, 134303 (2007)
114. J. Liu et al., Phys. Chem. Chem. Phys. **13**, 13920 (2011)
115. A. Boatwright et al., Faraday Discuss. **162**, 113 (2013)
116. D. Spence et al., Int. J. Mass Spectrom. **365–366**, 86 (2014)
117. S.A. Krasnokutski, F. Huisken, J. Phys. Chem. A **115**, 7120 (2011)
118. C. Feng et al., Phys. Chem. Chem. Phys. **17**, 16699 (2015)
119. B. Sitorus et al., AIP Conf. Proc. **2049**, 020066 (2018)
120. H.-P. Loock, L.M. Beaty, B. Simard, Phys. Rev. A **59**, 873 (1999)
121. F.S. Ashmore, A.R. Burgess, J. Chem. Soc. Faraday Trans. 2 **73**, 1247 (1977)
122. M. Theisen, F. Lackner, W.E. Ernst, Phys. Chem. Chem. Phys. **12**, 14861 (2010)
123. M. Theisen, F. Lackner, W.E. Ernst, J. Chem. Phys. **135**, 074306 (2011)
124. M. Theisen et al., J. Phys. Chem. Lett. **2**, 2778 (2011)
125. M. Theisen, F. Lackner, W.E. Ernst, J. Phys. Chem. A **115**, 7005 (2011)
126. A. Schiller et al., Eur. Phys. J. D **75**, 1 (2021)
127. P. Thaler et al., J. Chem. Phys. **143**, 134201 (2015)
128. R. Messner et al., J. Chem. Phys. **149**, 024305 (2018)
129. M. Ratschek, M. Koch, and W. E. Ernst, J. Chem. Phys. **136**, 104201 (2012).
130. F. Lindebner et al., Int. J. Mass Spectrom. **365–366**, 255 (2014)
131. G. Krois et al., Phys. Chem. Chem. Phys. **16**, 22373 (2014)
132. F. Lackner et al., J. Phys. Chem. A **117**, 11866 (2013)
133. P.J. Linstrom, W.G. Mallard, *NIST Chemistry WebBook, NIST Standard Reference Database Number 69* (National Institute of Standards and Technology, 2018)
134. S.A. Krasnokutski, C. Jäger, T. Henning, ApJ **889**, 67 (2020)
135. S.A. Krasnokutski, F. Huisken, J. Phys. Chem. A **114**, 7292 (2010)
136. S.A. Krasnokutski, F. Huisken, J. Phys. Chem. A **114**, 13045 (2010)
137. S.A. Krasnokutski, F. Huisken, J. Phys. Chem. A **118**, 2612 (2014)
138. S.A. Krasnokutski, F. Huisken, J. Chem. Phys. **142**, 084311 (2015)
139. S.A. Krasnokutski, F. Huisken, Appl. Phys. Lett. **105**, 113506 (2014)
140. F. Salama, Proc. Int. Astron. Union **4**, 357 (2008)
141. A. Leger, J.L. Puget, Astron. Astrophys. **137**, L5 (1984)
142. L.J. Allamandola, A.G.G.M. Tielens, J.R. Barker, Astrophys. J. Lett. **290**, L25 (1985)
143. A.G.G.M. Tielens, Annu. Rev. Astron. Astrophys. **46**, 289 (2008)
144. M.K. Crawford, A.G.G.M. Tielens, L.J. Allamandola, Astrophys. J. Lett. **293**, L45 (1985)
145. A. Leger, L. D'Hendecourt, Astron. Astrophys. **146**, 81 (1985)
146. M.L. Heger, Lick Observatory Bull. **10**, 146 (1922)
147. E.K. Campbell et al., Nature **523**, 322 (2015)
148. G.A.H. Walker et al., ApJL **812**, L8 (2015)
149. H. Linnartz et al., J. Mol. Spectrosc. **367**, 111243 (2020)
150. S. Spieler et al., ApJ **846**, 168 (2017)
151. S.A. Krasnokutski, F. Huisken, J. Chem. Phys. **141**, 214306 (2014)
152. S.A. Krasnokutski et al., ApJ **836**, 32 (2017)
153. R.I. Kaiser et al., J. Chem. Phys. **110**, 6091 (1999)
154. I. Hahndorf et al., J. Chem. Phys. **116**, 3248 (2002)
155. J.E. Elsila, D.P. Glavin, J.P. Dworkin, Meteorit. Planet. Sci. **44**, 1323 (2009)
156. K. Altwegg et al., Sci. Adv. **2**, e1600285 (2016)
157. A.K. Cobb, R.E. Pudritz, ApJ **783**, 140 (2014)
158. H. Linnartz, S. Ioppolo, G. Fedoseev, Int. Rev. Phys. Chem. **34**, 205 (2015)
159. G.M. Muñoz Caro et al., Nature **416**, 403 (2002)
160. E. Congiu et al., J. Chem. Phys. **137**, 054713 (2012)
161. U. Henne, J.P. Toennies, J. Chem. Phys. **108**, 9327 (1998)
162. M. Lewerenz, B. Schilling, J.P. Toennies, Chem. Phys. Lett. **206**, 381 (1993)
163. B. Schilling, Bericht 14/1993. Ph.D. dissertation, University of Göttingen/Max-Planck-Institut für Strömungsforschung, 1993

164. B. Samelin, Bericht 16/1998. Ph.D. dissertation, University of Göttingen/Max-Planck-Institut für Strömungsforschung, 1998
165. L.F. Gomez et al., J. Chem. Phys. **135**, 154201 (2011)
166. M.N. Slipchenko et al., J. Chem. Phys. **124**, 241101 (2006)
167. K. Kuyanov-Prozument, M.Y. Choi, A.F. Vilesov, J. Chem. Phys. **132**, 014304 (2010)
168. E. Loginov et al., Phys. Rev. Lett. **106**, 233401 (2011)
169. L.F. Gomez, E. Loginov, A.F. Vilesov, Phys. Rev. Lett. **108**, 155302 (2012)
170. L.F. Gomez et al., Science **345**, 906 (2014)
171. F. Laimer *et al.*, Phys. Rev. Lett. **123**, 165301 (2019).
172. H. Schöbel et al., Eur. Phys. J. D **63**, 209 (2011)
173. J. Fine et al., J. Chem. Phys. **148**, 044302 (2018)
174. P.J. Knowles, J.N. Murrell, Mol. Phys. **87**, 827 (1996)
175. J.H. Kim et al., J. Chem. Phys. **124**, 214301 (2006)
176. D.S. Peterka et al., J. Phys. Chem. B **110**, 19945 (2006)
177. D.S. Peterka et al., J. Phys. Chem. A **111**, 7449 (2007)
178. C.G. Edmonds, R.D. Smith, in *Methods in Enzymology* (Academic Press, 1990), pp. 412–431
179. J.B. Fenn et al., Mass Spectrom. Rev. **9**, 37 (1990)
180. C. Ho et al., Clin Biochem Rev **24**, 3 (2003)
181. J.J. Pitt, Clin Biochem Rev **30**, 19 (2009)
182. F. Bierau et al., Phys. Rev. Lett. **105**, 133402 (2010)
183. F. Filsinger et al., Phys. Chem. Chem. Phys. **14**, 13370 (2012)
184. A.I.G. Flórez et al., Phys. Chem. Chem. Phys. **17**, 21902 (2015)
185. A. Braun, M. Drabbels, J. Chem. Phys. **127**, 114303 (2007)
186. X. Zhang, M. Drabbels, J. Chem. Phys. **137**, 051102 (2012)
187. E. Loginov, M. Drabbels, J. Phys. Chem. A **111**, 7504 (2007)
188. A. Braun, M. Drabbels, J. Chem. Phys. **127**, 114304 (2007)
189. A. Braun, M. Drabbels, J. Chem. Phys. **127**, 114305 (2007)
190. H.J. Reyher et al., Phys. Lett. A **115**, 238 (1986)
191. A. *Leal et al.*, J. Chem. Phys. **144**, 094302 (2016)
192. B.K. Vainshtein, *Structure Analysis by Electron Diffraction* (Elsevier, 2013)
193. Y. He et al., Review of Scientific Instruments **86**, 084102 (2015)
194. M. Alghamdi, J. Zhang, W. Kong, J. Chem. Phys. **151**, 134307 (2019)
195. M. Alghamdi et al., Chemical Physics Letters **735**, 136752 (2019)
196. J. Zhang et al., J. Chem. Phys. **152**, 224306 (2020)
197. H. Odaka, M. Ichihashi, Eur. Phys. J. D **71**, 99 (2017)
198. T. González-Lezana et al., Int. Rev. Phys. Chem. **39**, 465 (2020)
199. S. Albertini, E. Gruber, F. Zappa, S. Krasnokutski, F. Laimer, P. Scheier, Chemistry and physics of dopants embedded in helium droplets. Mass Spectrom. Rev. (2021)
200. L. Tiefenthaler et al., Rev. Sci. Instrum. **91**, 033315 (2020)
201. S. Denifl et al., J. Chem. Phys. **124**, 054320 (2006)
202. K. Głuch et al., J. Chem. Phys. **120**, 2686 (2004)
203. S. Feil et al., Int. J. Mass Spectrom. **252**, 166 (2006)
204. F. Zappa et al., Eur. Phys. J. D **43**, 117 (2007)
205. S. Denifl *et al.*, Phys. Rev. Lett. **97**, 043201 (2006).
206. M. Kuhn et al., Nat. Commun. **7**, 13550 (2016)
207. M. Gatchell et al., Faraday Discuss. **217**, 276 (2019)
208. L. Kranabetter et al., Phys. Chem. Chem. Phys. **21**, 25362 (2019)
209. I.A. Harris et al., Chem. Phys. Lett. **130**, 316 (1986)
210. T.A. Milne, F.T. Greene, J. Chem. Phys. **47**, 4095 (1967)
211. A. Ding, J. Hesslich, Chem. Phys. Lett. **94**, 54 (1983)
212. P. Scheier, T.D. Märk, Int. J. Mass Spectrom. Ion Process. **76**, R11 (1987)
213. N.E. Levinger et al., J. Chem. Phys. **89**, 5654 (1988)
214. F.F. da Silva et al., Phys. Chem. Chem. Phys. **11**, 9791 (2009)
215. M. Gatchell et al., Phys. Rev. A **98**, 022519 (2018)

216. T. Ikegami, T. Kondow, S. Iwata, J. Chem. Phys. **98**, 3038 (1993)
217. K.T. Giju, S. Roszak, J. Leszczynski, J. Chem. Phys. **117**, 4803 (2002)
218. T. Ritschel, P.J. Kuntz, L. Zülicke, Eur. Phys. J. D **33**, 421 (2005)
219. D.C. McDonald et al., J. Chem. Phys. **145**, 231101 (2016)
220. M. Gatchell et al., J. Am. Soc. Mass Spectrom. **30**, 2632 (2019)
221. L. Lundberg et al., Molecules **25**, 1066 (2020)
222. A.G. Császár et al., Mol. Phys. **117**, 1559 (2019)
223. P. Martini et al., J. Chem. Phys. **152**, 014303 (2020)
224. F. Zhang et al., Int. J. Hydrogen Energy **41**, 14535 (2016)
225. L. Huang et al. arXiv:1912.11605 [astro-ph, physics:physics] (2019)
226. A.M. Sapse et al., Nature **278**, 332 (1979)
227. G.H. Herbig, Ann. Rev. Astron. Astrophys. **33**, 19 (1995)
228. B. Jordon-Thaden et al., Phys. Rev. Lett. **107**, 193003 (2011)
229. D. Stevenson, J. Hirschfelder, J. Chem. Phys. **5**, 933 (1937)
230. R.E. Hurley, Nucl. Inst. Methods **118**, 307 (1974)
231. W. Aberth, R. Schnitzer, M. Anbar, Phys. Rev. Lett. **34**, 1600 (1975)
232. Y.K. Bae, M.J. Coggiola, J.R. Peterson, Phys. Rev. A **29**, 2888 (1984)
233. J. Stärck, W. Meyer, Chem. Phys. **176**, 83 (1993)
234. M. Ayouz et al., J. Chem. Phys. **132**, 194309 (2010)
235. W. Wang et al., Chem. Phys. Lett. **377**, 512 (2003)
236. K. Hirao, S. Yamabe, Chem. Phys. **80**, 237 (1983)
237. L. Huang, C.F. Matta, L. Massa, J. Phys. Chem. A **115**, 12445 (2011)
238. M. Renzler et al., Phys. Rev. Lett. **117**, 273001 (2016)
239. F. Calvo, E. Yurtsever, J. Chem. Phys. **148**, 102305 (2017)
240. D.E. Galli, D.M. Ceperley, L. Reatto, J. Phys. Chem. A **115**, 7300 (2011)
241. P. Bartl et al., J. Phys. Chem. A **118**, 8050 (2014)
242. M. Joshi, A. Ghosh, T.K. Ghanty, J. Phys. Chem. C **121**, 15036 (2017)
243. M. Renzler et al., J. Chem. Phys. **147**, 194301 (2017)
244. P.C. Samartzis, A.M. Wodtke, Int. Rev. Phys. Chem. **25**, 527 (2006)
245. V.E. Zarko, Combust. Explos. Shock Waves **46**, 121 (2010)
246. B. Hirshberg, R.B. Gerber, A.I. Krylov, Nat. Chem. **6**, 52 (2014)
247. M.J. Greschner et al., J. Phys. Chem. A **120**, 2920 (2016)
248. S. Liu et al., Adv. Sci. **7**, 1902320 (2020)
249. H.T. Jonkman, J. Michl, J. Am. Chem. Soc. **103**, 733 (1981)
250. L. Friedman, R.J. Beuhler, J. Chem. Phys. **78**, 4669 (1983)
251. P. Scheier, A. Stamatovic, T.D. Märk, J. Chem. Phys. **88**, 4289 (1988)
252. T. Leisner et al., Z. Phys. D Atoms Mol. Clust. **12**, 283 (1989)
253. Y.K. Bae, P.C. Cosby, D.C. Lorents, Chem. Phys. Lett. **159**, 214 (1989)
254. T. Tonuma et al., Int. J. Mass Spectrom. Ion Process. **135**, 129 (1994)
255. F.A. Fernández-Lima et al., Chem. Phys. **340**, 127 (2007)
256. A. Vij et al., Angew. Chem. Int. Ed. **41**, 3051 (2002)
257. A.A. Vostrikov, DYu. Dubov, Tech. Phys. **51**, 1537 (2006)
258. S.D. Pangavhane et al., Rapid Commun. Mass Spectrom. **25**, 917 (2011)
259. N. Weinberger et al., J. Phys. Chem. C **121**, 10632 (2017)
260. J.S.-Y. Chao, M.F. Falcetta, K.D. Jordan, J. Chem. Phys. **93**, 1125 (1990)
261. J. Mazeau et al., J. Phys. B: At. Mol. Phys. **11**, L557 (1978)
262. E. Yurtsever, F. Calvo, J. Phys. Chem. A **123**, 202 (2019)
263. F.R. Brühl, R.A. Trasca, W.E. Ernst, J. Chem. Phys. **115**, 10220 (2001)
264. F. Stienkemeier et al., Z. Phys. D Atoms Mol. Clust. **38**, 253 (1996)
265. F. Ancilotto, G. DeToffol, F. Toigo, Phys. Rev. B **52**, 16125 (1995)
266. A. Leal et al., Phys. Rev. B **90**, 224518 (2014)
267. M. Buzzacchi, D.E. Galli, L. Reatto, Phys. Rev. B **64**, 094512 (2001)
268. D.E. Galli, M. Buzzacchi, L. Reatto, J. Chem. Phys. **115**, 10239 (2001)
269. E. Coccia *et al.*, J. Chem. Phys. **126**, 124319 (2007).

270. C. Di Paola et al., J. Chem. Theory Comput. **1**, 1045 (2005)
271. F. Sebastianelli et al., Comput. Mater. Sci. **35**, 261 (2006)
272. N. Issaoui et al., J. Chem. Phys. **141**, 174316 (2014)
273. S. Paolini, F. Ancilotto, F. Toigo, J. Chem. Phys. **126**, 124317 (2007)
274. L. An der Lan et al., Chem. A Eur. J. **18**, 4411 (2012)
275. M. Rastogi et al., Phys. Chem. Chem. Phys. **20**, 25569 (2018)
276. C. Stark, V.V. Kresin, Phys. Rev. B **81**, 085401 (2010)
277. L. An der Lan et al., J. Chem. Phys. **135**, 044309 (2011)
278. L. An der Lan et al., Phys. Rev. B **85**, 115414 (2012)
279. G.J. Hutchings, M. Haruta, Appl. Catal. A **291**, 2 (2005)
280. P. Mulvaney, MRS Bull. **26**, 1009 (2001)
281. C.L. Brown et al., Gold Bull. **40**, 245 (2007)
282. M. Faraday, Philos. Trans. R. Soc. Lond. **147**, 145 (1857)
283. G.J. Hutchings, M. Brust, H. Schmidbaur, Chem. Soc. Rev. **37**, 1759 (2008)
284. G.J. Hutchings, J. Catal. **96**, 292 (1985)
285. M. Haruta et al., Chem. Lett. **16**, 405 (1987)
286. M.-C. Daniel, D. Astruc, Chem. Rev. **104**, 293 (2004)
287. M. Haruta, Nature **437**, 1098 (2005)
288. A. Corma, H. Garcia, Chem. Soc. Rev. **37**, 2096 (2008)
289. G.C. Bond, D.T. Thompson, Catal. Rev. **41**, 319 (1999)
290. Y. Li, R.T. Yang, J. Phys. Chem. C **111**, 11086 (2007)
291. M. Yamauchi, H. Kobayashi, H. Kitagawa, ChemPhysChem **10**, 2566 (2009)
292. T. Hussain et al., Appl. Phys. Lett. **100**, 183902 (2012)
293. C.M. Ramos-Castillo et al., J. Phys. Chem. C **119**, 8402 (2015)
294. I. Katakuse et al., Int. J. Mass Spectrom. Ion Process. **67**, 229 (1985)
295. I. Katakuse et al., Int. J. Mass Spectrom. Ion Process. **74**, 33 (1986)
296. I. Rabin, C. Jackschath, W. Schulze, Z. Phys. D Atoms Mol. Clust. **19**, 153 (1991)
297. A. Herlert et al., J. Electron Spectrosc. Relat. Phenom. **106**, 179 (2000)
298. H. Sik Kim et al., Chemical Physics Letters **224**, 589 (1994)
299. P. Martini et al., Int. J. Mass Spectrom. **434**, 136 (2018)
300. St. Becker et al., Computational Materials Science **2**, 633 (1994)
301. S. Gilb et al., J. Chem. Phys. **116**, 4094 (2002)
302. W.A. Saunders, Phys. Rev. Lett. **64**, 3046 (1990)
303. W.A. Saunders, Phys. Rev. Lett. **62**, 1037 (1989)
304. W.A. Saunders, S. Fedrigo, Chem. Phys. Lett. **156**, 14 (1989)
305. J. Zieglera et al., Int. J. Mass Spectrom. **202**, 47 (2000)
306. A. Sanchez et al., J. Phys. Chem. A **103**, 9573 (1999)
307. D. Schooss et al., Philosophical transactions of the royal society a: mathematical. Phys. Eng. Sci. **368**, 1211 (2010)
308. R. Rousseau et al., Chem. Phys. Lett. **295**, 41 (1998)
309. A. Schweizer et al., J. Chem. Phys. **119**, 3699 (2003)
310. A. N. Gloess et al., J. Chem. Phys. **128**, 114312 (2008)
311. K. Sugawara, F. Sobott, A.B. Vakhtin, J. Chem. Phys. **118**, 7808 (2003)
312. M. Neumaier et al., J. Chem. Phys. **122**, 104702 (2005)
313. A. Fielicke et al., J. Am. Chem. Soc. **127**, 8416 (2005)
314. P. Martini et al., J. Phys. Chem. A **123**, 9505 (2019)
315. G.E. Froudakis et al., Chem. Phys. **280**, 43 (2002)
316. F. Remacle E.S. Kryachko, J. Chem. Phys. **122**, 044304 (2005)
317. E. M. Fernández et al., Phys. Rev. B **70**, 165403 (2004)
318. S. Kapur, E.W. Müller, Surf. Sci. **62**, 610 (1977)
319. P. Pyykkoe, J. Am. Chem. Soc. **117**, 2067 (1995)
320. D. Schröder et al., Inorg. Chem. **37**, 624 (1998)
321. S. Seidel, K. Seppelt, Science **290**, 117 (2000)
322. L. M. Ghiringhelli et al., New J. Phys. **15**, 083003 (2013)

323. A. Shayeghi et al., Angew. Chem. Int. Ed. **54**, 10675 (2015)
324. Z. Jamshidi, M.F. Far, A. Maghari, J. Phys. Chem. A **116**, 12510 (2012)
325. L.A. Mancera, D.M. Benoit, J. Phys. Chem. A **119**, 3075 (2015)
326. A. Ghosh, T.K. Ghanty, J. Phy. Chem. A **120**, 9998 (2016)
327. L.M. Ghiringhelli, S.V. Levchenko, Inorg. Chem. Commun. **55**, 153 (2015)
328. X. Li et al., Mol. Phys. **107**, 2531 (2009)
329. X.-Y. Li, X. Cao, Y. Zhao, J. Phys. B At. Mol. Opt. Phys. **42**, 065102 (2009)
330. L. Xinying, C. Xue, Z. Yongfang, Theor. Chem. Acc. **123**, 469 (2009)
331. L. Xinying, C. Xue, Z. Yongfang, Aust. J. Chem. **62**, 121 (2009)
332. P. Zhang et al., J. Mol. Struct. (Thoechem) **899**, 111 (2009)
333. S.J. Grabowski et al., Chem. A Eur. J. **22**, 11317 (2016)
334. L. Belpassi et al., J. Am. Chem. Soc. **130**, 1048 (2008)
335. W.H. Breckenridge, V.L. Ayles, T.G. Wright, J. Phys. Chem. A **112**, 4209 (2008)
336. M. Al-Ahmari et al., J Clust Sci **26**, 913 (2015)
337. M. Slama et al., Eur. Phys. J. D **70**, 242 (2016)
338. M. Slama et al., Mol. Phys. **115**, 757 (2017)
339. J. Hernández-Rojas, D.J. Wales, J. Chem. Phys. **119**, 7800 (2003)
340. D. Prekas, C. Lüder, M. Velegrakis, J. Chem. Phys. **108**, 4450 (1998)
341. A. Yousef *et al.*, J. Chem. Phys. **127**, 154309 (2007).
342. C. Kittel, *Introduction to Solid State Physics*, 8th edn. (Wiley, 2004)
343. J. Chem. Soc., Abstr. **112**, ii199 (1917)
344. E. Hulthèn, R.V. Zumstein, Phys. Rev. **28**, 13 (1926)
345. G.N. Khairallah, R.A.J. O'Hair, M.I. Bruce, Dalton Trans. 3699 (2006)
346. H. Schmidbaur, H.G. Raubenheimer, L. Dobrzańska, Chem. Soc. Rev. **43**, 345 (2013)
347. L. Lundberg et al., J. Am. Soc. Mass Spectrom. **30**, 1906 (2019)
348. L. Lundberg et al., Eur. Phys. J. D **74**, 102 (2020)
349. H. Rosemeyer, Chem. Biodivers. **1**, 361 (2004)
350. M. Kuhn et al., Eur. Phys. J. D **72**, 38 (2018)
351. D. *Huber et al.,* J. Chem. Phys. **125**, 084304 (2006)
352. A. Mauracher et al., J. Phys. Chem. Lett. **5**, 2444 (2014)
353. N. Weinberger et al., Eur. Phys. J. D **70**, 91 (2014)
354. E. Jabbour Al Maalouf et al., J. Phys. Chem. Lett. **8**, 2220 (2017)
355. M. Gatchell et al., Phys. Chem. Chem. Phys. **20**, 7739 (2018)
356. G. Klupp, S. Margadonna, K. Prassides, in *Reference Module in Materials Science and Materials Engineering* (Elsevier, 2016)
357. A. Mauracher et al., Angew. Chem. Int. Ed. **53**, 13794 (2014)
358. E.K. Campbell et al., ApJ **822**, 17 (2016)
359. O. Echt et al., ChemPlusChem **78**, 910 (2013)
360. S.A. Krasnokutski et al., J. Phys. Chem. Lett. **7**, 1440 (2016)
361. M. Harnisch et al., Eur. Phys. J. D **70**, 192 (2016)
362. M. Renzler et al., J. Phys. Chem. C **121**, 10817 (2017)
363. A.F. Hebard et al., Nature **350**, 600 (1991)
364. T.T.M. Palstra et al., Solid State Commun. **93**, 327 (1995)
365. A.Y. Ganin et al., Nat. Mater. **7**, 367 (2008)
366. A.Y. Ganin et al., Nature **466**, 221 (2010)
367. M. Yoon et al., Phys. Rev. Lett. **100**, 206806 (2008)
368. Q. Wang, P. Jena, J. Phys. Chem. Lett. **3**, 1084 (2012)
369. M. Robledo et al., RSC Adv. **6**, 27447 (2016)
370. A. Kaiser et al., Int. J. Hydrog. Energy **42**, 3078 (2017)
371. J. Osterodt, F. Vögtle, Chem. Commun. 547 (1996)
372. N. Dragoe et al., Chem. Commun. 85 (1999)
373. M. Goulart et al., J. Phys. Chem. Lett. **9**, 2703 (2018)
374. L. Xin-Ying, Cao xue, Phys. Rev. A **77**, 022508 (2008)
375. P. Martini et al., J. Phys. Chem. A **123**, 4599 (2019)

376. A. Mauracher et al., J. Chem. Phys. **142**, 104306 (2015)
377. P. Khare et al., Rev. Geophys. **37**, 227 (1999)
378. B. Zuckerman, J.A. Ball, C.A. Gottlieb, Astrophys. J. Lett. **163**, L41 (1971)
379. S.-Y. Liu, D.M. Mehringer, L.E. Snyder, ApJ **552**, 654 (2001)
380. P. Redondo, A. Largo, C. Barrientos, A&A **579**, A125 (2015)
381. C. Rice et al., J. Power Sourc. **111**, 83 (2002)
382. Y. Zhu, S.Y. Ha, R.I. Masel, J. Power Sourc. **130**, 8 (2004)
383. X. Wang et al., J. Power Sourc. **175**, 784 (2008)
384. Z. Bai et al., J. Phys. Chem. C **113**, 10568 (2009)
385. T. Gunji, F. Matsumoto, Inorganics **7**, 36 (2019)
386. Z. Bai et al., Int. J. Electrochem. Sci. **8**, 12 (2013)
387. F. Nitze et al., Electrochim. Acta **63**, 323 (2012)
388. M. Meksi et al., Chem. Lett. **44**, 1774 (2015)
389. M. Mahmoodi-Darian et al., J. Am. Soc. Mass Spectrom. **30**, 787 (2019)
390. M. Mahmoodi-Darian et al., Int. J. Mass Spectrom. **450**, 116293 (2020)
391. M. Goulart et al., Phys. Chem. Chem. Phys. **15**, 3577 (2013)
392. S. Heinbuch et al., J. Chem. Phys. **126**, 244301 (2007)
393. D.P.P. Andrade et al., J. Phys. Chem. C **112**, 11954 (2008)
394. D.P.P. Andrade et al., J. Electron Spectrosc. Relat. Phenom. **155**, 124 (2007)
395. I. Martin et al., Phys. Chem. Chem. Phys. **7**, 2212 (2005)
396. L. Ziemczonek, T. Wróblewski, Eur. Phys. J. Spec. Top. **144**, 251 (2007)
397. Y. Valadbeigi, H. Farrokhpour, Int. J. Quantum Chem. **113**, 1717 (2013)
398. W.Y. Feng, C. Lifshitz, J. Phys. Chem. **98**, 6075 (1994)
399. Y. Inokuchi, N. Nishi, J. Phys. Chem. A **106**, 4529 (2002)
400. V. Aviyente et al., Int. J. Mass Spectrom. Ion Processes **161**, 123 (1997)
401. R. Zhang, C. Lifshitz, J. Phys. Chem. **100**, 960 (1996)
402. L. Baptista et al., J. Phys. Chem. A **112**, 13382 (2008)
403. F.F. da Silva et al., Phys. Chem. Chem. Phys. **11**, 11631 (2009)
404. M. Šala et al., J. Phys. Chem. A **113**, 3223 (2009)
405. H. Schöbel et al., Phys. Chem. Chem. Phys. **13**, 1092 (2010)
406. H. Schwertfeger, P.R. Schreiner, Chem. unserer Zeit **44**, 248 (2010)
407. H. Schwertfeger, A.A. Fokin, P.R. Schreiner, Angew. Chem. Int. Ed. **47**, 1022 (2008)
408. K.-W. Yeung et al., Nanotechnol. Rev. **9**, 650 (2020)
409. A.M. Schrand, S.A.C. Hens, O.A. Shenderova, Crit. Rev. Solid State Mater. Sci. **34**, 18 (2009)
410. M. Goulart et al., J. Phys. Chem. C **121**, 10767 (2017)
411. R.M. Sehgal, D. Maroudas, Langmuir **31**, 11428 (2015)
412. L. Kranabetter et al., Phys. Chem. Chem. Phys. **20**, 21573 (2018)
413. B.A. Buffett, Annu. Rev. Earth Planet. Sci. **28**, 477 (2000)
414. Yu.A. Dyadin et al., J Struct Chem **40**, 645 (1999)
415. P. Englezos, Ind. Eng. Chem. Res. **32**, 1251 (1993)
416. F. Laimer et al., Chem. A Eur. J. **n/a**, (2020)
417. D. Mateo, J. Eloranta, J. Phys. Chem. A **118**, 6407 (2014)
418. T. Jiang, C. Kim, J.A. Northby, Phys. Rev. Lett. **71**, 700 (1993)
419. J.A. Northby, C. Kim, Physica B **194–196**, 1229 (1994)
420. M. Farník et al., 6 (1996)
421. W. Saunders, Phys. Rev. A **46**, 7028 (1992)
422. O. Echt et al., Phys. Rev. A **38**, 3236 (1988)
423. I. Mähr et al., Phys. Rev. Lett. **98**, 023401 (2007)
424. F. Laimer, F. Zappa, P. Scheier, Size and Velocity Distribution of Negatively Charged Helium Nanodroplets. J. Phys. Chem. A. **125**(35), 7662–7669 (2021)
425. U. Henne, Ph.D. Dissertation, available as Bericht 5/1996, Max-Planck-Instiut für Stömungs- forschung, Göttingen, Germany, 1996
426. R. M. P. Tanyag *et al.*, J. Chem. Phys. **152**, 234306 (2020).
427. M. Hartmann et al., Phys. Rev. Lett. **75**, 1566 (1995)

428. S. Yang et al., Nanoscale **5**, 11545 (2013)
429. J.P. Toennies, A.F. Vilesov, Angew. Chem. Int. Ed. **43**, 2622 (2004)
430. J. Mansikka-Aho, M. Manninen, E. Hammarén, Z. Phys. D Atoms Mol. Clust. **21**, 271 (1991)
431. Y. Dong, M. Springborg, J. Phys. Chem. C **111**, 12528 (2007)
432. J. Roithová et al., Faraday Discuss. **217**, 98 (2019)
433. A. Günther et al., J. Mol. Spectrosc. **332**, 8 (2017)
434. E.K. Campbell, J.P. Maier, ApJ **850**, 69 (2017)
435. P. Bartl et al., J Phys Chem A **118**, 8050 (2014)
436. A. Kramida, Y. Ralchenko, (1999)
437. S. Denifl et al., J Chem Phys **124**, 054320 (2006)
438. R.J. Beuhler et al., Biochemistry **13**, 5060 (1974)
439. M.P. Ziemkiewicz, D.M. Neumark, O. Gessner, Int. Rev. Phys. Chem. **34**, 239 (2015)
440. J.P. Maier, E.K. Campbell, Int. J. Mass Spectrom. **434**, 116 (2018)
441. M. Töpfer et al., Mol. Phys. **117**, 1481 (2019)

# Chapter 4
# Infrared Spectroscopy of Molecular Radicals and Carbenes in Helium Droplets

**Gary E. Douberly**

**Abstract** The helium droplet is an ideal environment to spectroscopically probe difficult to prepare molecular species, such as radicals, carbenes and ions. The quantum nature of helium at 0.4 K often results in molecular spectra that are sufficiently resolved to evoke an analysis of line shapes and fine-structure via rigorous "effective Hamiltonian" treatments. In this chapter, we will discuss general experimental methodologies and a few examples of successful attempts to efficiently dope helium droplets with organic molecular radicals or carbenes. In several cases, radical reactions have been carried out inside helium droplets *via* the sequential capture of reactive species, resulting in the kinetic trapping of reaction intermediates. Infrared laser spectroscopy has been used to probe the properties of these systems under either zero-field conditions or in the presence of externally applied, homogeneous electric or magnetic fields.

## 4.1 Infrared Spectroscopy of Molecular Radicals and Carbenes in Helium Droplets

The objective of our experimental research program is to isolate and stabilize transient intermediates and products of prototype gas-phase reactions relevant to both combustion and atmospheric chemistry. Helium Nanodroplet Isolation (HENDI) [1–12] is well-suited for this because liquid helium droplets have shown potential to freeze out high energy configurations of a reacting system, permitting infrared (IR) spectroscopic characterizations of reactive intermediates lying between the sequentially captured reactants and the various products associated with the potential energy surface. Hydrocarbon radical reactions with molecular oxygen or other small molecules relevant to combustion environments have been the focus of our recent work in this area (see Refs. [13–51] for our recent contributions applying this method). Here we describe a selection of the single and double [52–57] resonance IR laser spectroscopy

G. E. Douberly (✉)
Department of Chemistry, University of Georgia, Athens, Georgia 30602, USA
e-mail: douberly@uga.edu

155

A. Slenczka and J. P. Toennies (eds.), *Molecules in Superfluid Helium Nanodroplets*, Topics in Applied Physics 145, https://doi.org/10.1007/978-3-030-94896-2_4

techniques that are being used to probe the structural and dynamical properties of molecular radical and carbene systems solvated in helium droplets.

## 4.1.1 Experimental Methods

### 4.1.1.1 Droplet Production and Doping

A diagram of the HENDI apparatus is shown in Fig. 4.1. Helium droplets are formed ($10^{12}$ droplets per second) by the continuous expansion of He gas through a 5 micron diameter pin-hole nozzle (Fig. 4.1a). The average droplet size is controlled by changing the nozzle temperature, providing a dynamic range from $\sim 10^3$ atoms at 24 K to $\sim 10^6$ atoms at 8 K [58–61]. Upon leaving the high pressure region of the expansion, the droplets cool by evaporation to 0.4 K [62]. The droplet expansion is skimmed into a beam, which passes into the differentially pumped "pick-up" chamber (Fig. 4.1b). Here the droplets are doped by passing them through the vapor of the molecular species of interest (approximately $10^{10}$ molecules/cm$^3$ over a 1 cm path length). The internal energy of the captured molecule is rapidly removed by He evaporation, which returns the system to 0.4 K [63]. Each evaporating He atom reduces the internal energy of the system by 5 cm$^{-1}$ (0.014 kcal mol$^{-1}$) [63]. The density of molecules in the "pick-up" region can be varied such that each droplet captures one (or more) molecules on average. Molecular species of differing composition can be added to the same droplet by implementing multiple "pick-up" zones. An effusive pyrolysis source (Fig. 4.2) has been successfully used to fragment precursor molecules and dope He droplets with halogen atoms and molecular radicals or carbenes [17, 18, 20–23, 25, 27, 29, 30, 34, 35, 37–42, 46, 50, 51].

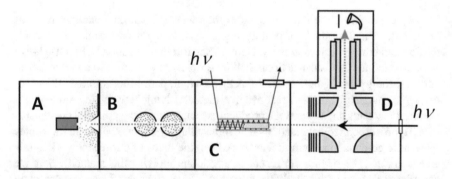

**Fig. 4.1** Schematic of the UGA HENDI spectrometer. The pyrolysis source for generating halogen atoms and molecular radicals or carbenes is load-locked into the vacuum chamber (section B). The laser excitation can be switched between a counter-propagating configuration to the laser-multipass/Stark/Zeeman cell configuration (section C). Detection of the droplet beam is achieved with a crossed-beam ionizer, quadrupole mass spectrometer (section D)

**Fig. 4.2** A schematic drawing of our SiC high temperature pyrolysis source. The copper electrodes are water-cooled and the length of the hot zone can be adjusted

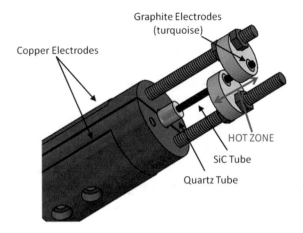

### 4.1.1.2    Droplet Detection *via* Mass Spectrometry

The droplet beam is detected with a quadrupole mass spectrometer (MS) (Fig. 4.1d). Electron impact ionization leads to the production of a $He^+$ cation within the droplet. The outcome of this ionization process is now well known to produce either a $He_n^+$ distribution or ions associated with the charge-transfer ionization and fragmentation of the molecular dopant [4, 6, 10]. The MS in Fig. 4.3a shows the $He_n^+$ ions associated with the electron impact ionization of a neat (dopant free) droplet beam. Figure 4.3b shows the MS of a droplet beam doped with *n*-butyl nitrite (one molecule is captured per droplet, on average; the energized molecular ion fragments predominately to $C_3H_3^+$, m/z=39). Thermal dissociation of *n*-butyl nitrite in a pyrolysis source leads to the production of the propargyl radical ($C_3H_3$), nitric acid (NO), and formaldehyde ($CH_2O$). The number density of molecules in the pyrolysate is sufficiently low such that droplets passing through it will capture either nothing or only one of the three fragment molecules. Multiple capture events by a single droplet occur with low probability due to the Poisson statistics associated with the capture process [4, 6]. Figure 4.3C shows the MS of the droplet beam after having passed through the pyrolysate associated with *n*-butyl nitrite pyrolysis. An intensity reduction is observed at m/z=39, along with intensity gains at m/z=29, 30, and 38. The peaks at m/z=29 and 30 are largely due to the ionization of droplets that have captured either NO or $CH_2O$. The peak at m/z=38 is a signature of the ionization and fragmentation of the He-solvated $C_3H_3$ radical.

### 4.1.1.3    Infrared Laser Spectroscopy

After traversing the pick-up zones and prior to entering the mass spectrometer, the beam of droplets is irradiated with the idler output from a continuous-wave optical parametric oscillator (OPO) [24]. Survey spectra are recorded with the laser beam

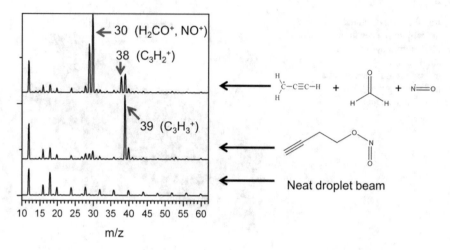

**Fig. 4.3** **a** MS of the neat droplet beam. **b** MS of droplets doped with single *n*-butyl nitrite molecules. **c** MS of droplets doped with single molecules, NO, formaldehyde, or the propargyl radical, $C_3H_3$

aligned counter-propagating to the droplet beam, whereas the laser is aligned into a two-mirror multipass cell for Zeeman, Stark or Polarization spectroscopy measurements. Vibrational excitation of He-solvated dopants leads to the evaporation of several hundred He atoms, which reduces both the geometric and ionization cross sections of the irradiated droplets. This photo-induced cross section reduction for electron impact ionization is measured as ion-signal depletion in selected mass channels. For example, the IR spectrum of *n*-butyl nitrite is measured as a depletion signal in mass channel m/z=39 (Fig. 4.4a; experimental conditions same as those in Fig. 4.3b), whereas the spectrum of the propargyl radical is measured in mass channel m/z=38 (Fig. 4.4b; experimental conditions same as those in Fig. 4.3c).

Along with the sharp acetylenic CH stretch band near 3333 cm$^{-1}$, the spectrum of *n*-butyl nitrite contains several bands below 3000 cm$^{-1}$ arising from the two $CH_2$ moieties. In comparison, the propargyl radical contains three sharp bands arising from the acetylenic CH stretch ($\nu_1$ at 3322 cm$^{-1}$), the symmetric $CH_2$ stretch ($\nu_2$ at 3039 cm$^{-1}$), and the antisymmetric $CH_2$ stretch ($\nu_8$ at 3130 cm$^{-1}$). The broader features in the precursor spectrum below 3000 cm$^{-1}$ are completely absent in the spectrum of the propargyl radical. For the acquisition of molecular radical spectra, as demonstrated in Fig. 4.4b, mass channels can usually be selected (judiciously) to discriminate against spectral features associated with droplets containing unpyrolyzed precursor molecules or other fragments in the pyrolysate.

The weakly perturbing, superfluid He solvent allows for highly resolved vibrational spectroscopy studies of these species, which can be compared directly to the predictions of quantum chemistry [4, 6]. Indeed, in the case of vibrational frequencies, when comparisons are available, the band origins of He-solvated molecules and molecular complexes differ little from those measured in the gas phase ($\sim$1 cm$^{-1}$

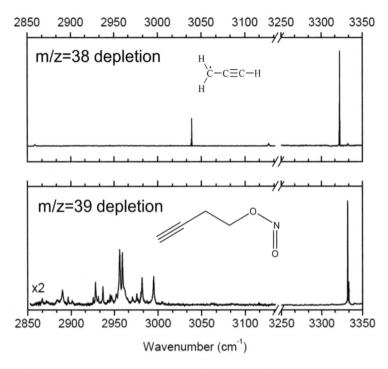

**Fig. 4.4**   Infrared spectra of **a** *n*-butyl nitrite and **b** the propargyl radical measured as ion-signal depletion in mass channels 39 and 38 u, respectively

or less) [6]. For example, the acetylenic CH stretch of the propargyl radical is red shifted by only 0.14 cm$^{-1}$ upon solvation in a helium droplet [64].

### 4.1.1.4   Spectra Exhibiting Rotational Fine Structure

For small molecules and molecular complexes assembled in He droplets, it is often the case that vibrational bands exhibit rotational fine structure. This fine structure results from the simultaneous change of vibrational and rotational quantum numbers upon vibrational excitation. The origin of resolved rotational fine structure has been discussed extensively in the helium droplet literature, [1, 3–6] and it may be thought of qualitatively as resulting from the fact that the rotational degrees of freedom of the embedded molecule are only weakly coupled to the helium bath; thereby, it is often the case that molecular rotations are sufficiently long-lived such that rovibrational bands are observed. This is readily apparent in the IR spectrum of the propargyl radical. Upon closer inspection of the $v_1$ and $v_2$ bands in Fig. 4.4b, the spectral patterns shown in Fig. 4.5 are revealed. For both of these rovibrational bands, the pattern of lines can be directly attributed to the orientation of the vibrational transition dipole moment in the molecular frame of reference.

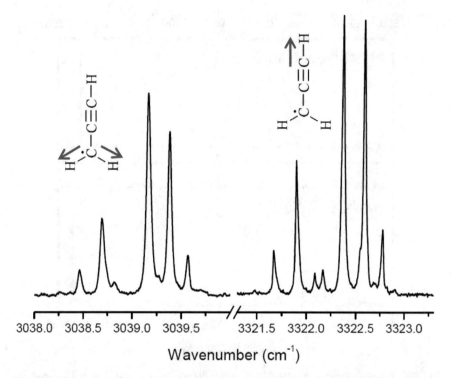

**Fig. 4.5** Higher resolution scans of the $\nu_1$ and $\nu_2$ acetylenic CH and symmetric CH$_2$ stretching bands for the propargyl radical. Both bands have $a_1$ symmetry and can be reproduced in simulations as $a$-type bands

Simulations of rotationally resolved spectra of helium-solvated molecules can be achieved by employing traditional effective Hamiltonian approaches, in which a gas-phase Hamiltonian is used with renormalized rotational constants. Moreover, Stark spectroscopy can be implemented and analyzed in the traditional sense, in which an external electric field perturbs the free-rotor behavior of the molecule or complex, and an additional term is appended to the zero-field effective Hamiltonian to account for the field-induced perturbation [41]. Our instrument is equipped with a laser multipass cell (Fig. 4.1c) that allows us to perform Stark spectroscopy measurements by applying a static electric field (0 to 80 kV/cm) to electrodes that surround the droplet beam/laser interaction region. The use of various Stark field strengths and laser polarization orientations (leading to different selection rules), allows us to accurately determine dipole moments of He-solvated species [21, 41, 44, 65, 66]. For example, the experimental (black) and simulated (red) Stark spectra of the linear OH-CO complex are shown in Fig. 4.6 [44]. The zero-field spectrum is shown along the bottom of the figure, and the Stark spectra recorded at three different field strengths provide the dipole moments of the complex in both the ground and excited OH stretch vibrational states.

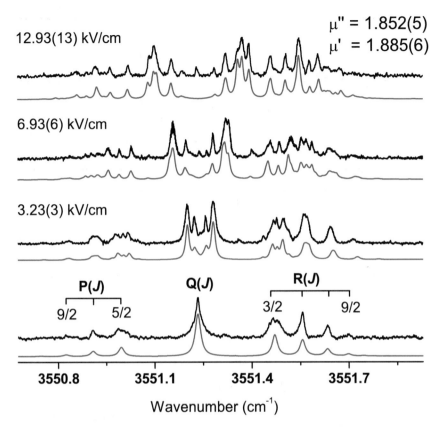

**Fig. 4.6** Rovibrational spectra of the OH stretch band of the linear OH-CO hydrogen bonded complex. Individual transitions are labeled above the zero-field spectrum (bottom). Infrared Stark spectra were obtained with a perpendicular laser polarization configuration and three separate static field strengths, revealing the magnitude of the permanent dipole moments in both the ground and excited vibrational states. The red traces are simulations using an effective Hamiltonian model. Reproduced with permission from Ref. [44]. © copyright *American Institute of Physics*. All rights reserved

Upon replacing the two stainless steel Stark electrodes with Neodymium rare-earth permanent magnets, the laser multipass cell can be used to record IR Zeeman spectra. Again, the analysis of such spectra is carried out by appending an additional term to the zero-field effective Hamiltonian. The Zeeman term is parameterized by the permanent magnetic field strength and the various g-factors associated with the interaction of the molecular magnetic moments with the external field. Experimental (black) and simulated (red) Zeeman spectra of the linear OH-CO complex are shown in Fig. 4.7. Here we find that the g-factor of the electron is unchanged from the gas-phase value, indicating the absence of any significant solvent-induced quenching of the electron's orbital angular momentum [44].

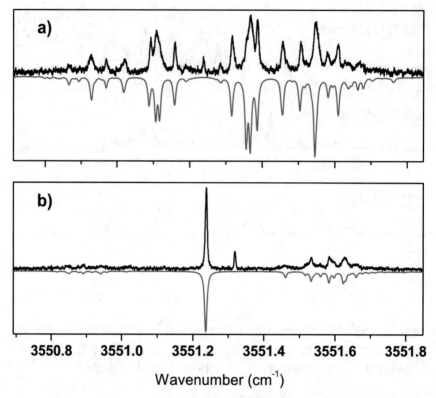

**Fig. 4.7** Infrared Zeeman spectra of the OH stretch band of the linear OH-CO hydrogen bonded complex. Zeeman spectra were obtained with both **a** perpendicular and **b** parallel laser polarization configurations. The red traces are simulations using an effective Hamiltonian model and a field strength of 0.425(2) Tesla. Reproduced with permission from Ref. [44]. © copyright *American Institute of Physics*. All rights reserved

### 4.1.1.5   Spectra Lacking Rotational Fine Structure

Often, the natural line width due to vibrational relaxation is broader than the rotational contour at 0.4 K ($\sim$1 cm$^{-1}$), precluding the determination of dipole moments *via* the aforementioned Stark measurements. This is a common feature for larger He-solvated systems that have a relatively high density of vibrational states, leading to more efficient coupling to the solvent and more rapid vibrational relaxation [6, 13, 14, 67, 68]. Nevertheless, by measuring the electric field dependence of the band intensity, it is possible to simultaneously obtain both the permanent electric dipole moment ($\mu_p$) and the vibrational transition moment angle (VTMA) [14, 67, 69] associated with each vibrational band [69, 70]. For any one normal mode vibration, the VTMA is defined as the angle $\mu_p$ makes with the transition dipole moment vector ($\mu_t$). Given a particular combination of VTMA and $\mu_p$ (obtained from *ab initio* calculations), this field dependence can be simulated and compared to the experiment [71–73]. Moreover, to make this comparison, the theoretical results do not require the scaling that often plagues the comparisons of experimental vibrational band origins to those obtained from *ab initio* harmonic frequency calculations [69]. Because dipole moments and VTMAs are accurately determined even at modest levels of *ab initio* theory, these *Polarization Spectroscopy* measurements provide key structural information that can be employed to assign vibrational spectra that contain contributions from multiple species or structural isomers [69]. An example of this is given in Fig. 4.8, where two closely spaced vibrational bands are attributed to two cyclic isomers of the $OH-(D_2O)_2$ trimer complex [38]. Being separated by only a few $cm^{-1}$, the two bands cannot be assigned to specific isomers on the basis of frequency computations alone. However, comparison of the experimental and computed VTMAs leads to a definitive assignment (see details in figure caption).

### 4.1.1.6 Organic Pyrolysis Precursors

Efficient doping of He droplets is essential for the acquisition of high-quality IR spectra, which is one longstanding goal of our research program. Although this is trivially achieved for stable, closed-shell systems, much of the He droplet spectroscopy being carried out in our research group requires the clean, continuous generation of carbenes, hydrocarbon radicals, the hydroxyl radical, or halogen atoms. We find that this is most efficiently achieved *via* flash vacuum pyrolysis of organic precursors; photolysis and RF discharge sources have been explored with less success. Initial studies of He-solvated radicals employed a rather simple low-pressure, continuous, effusive pyrolysis source composed of a radiatively heated quartz tube. A tantalum filament connected to two water cooled electrodes heats the tip of the quartz tube, and radicals are produced by pyrolysis as precursor molecules collide with the walls of the heated tube. The effusive beam of radicals crosses the path of the droplet beam, and the concentration of radicals in this "pick-up" zone is controlled with a fine metering valve that is located between the heated output region and the precursor reservoir. The resulting pressure in the pyrolysis region is near $2 \times 10^{-4}$ Torr (inside the quartz tube). Under these conditions, precursor molecules undergo only a few collisions with the walls of the heated tube; essentially no collisions occur in the gas phase, and the probability for radical recombination within the source is minimized. The maximum temperature achieved is $\sim$1400 K, which is the major drawback of this pyrolysis source. We have now expanded upon this original pyrolysis source design, increasing the upper temperature that can be achieved in our experiments. We incorporated a resistively heated silicon carbide (SiC) tube that can be heated up to $\sim$2100 K (Fig. 4.2). This new pyrolysis design was inspired by the pulsed pyrolysis sources originally reported by P. Chen and co-workers [74] and G. B. Ellison and co-workers [75]. At $\sim$2100 K, the range of precursor systems that can be pyrolyzed to create radicals is vastly expanded. A second-generation SiC pyrolysis source has now been designed to allow for the efficient use of solid organic precursors that have little vapor pressure at room temperature.

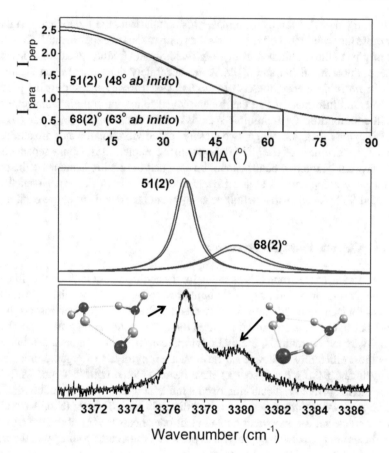

**Fig. 4.8** Vibrational transition moment angle analysis of the bands assigned to cyclic isomers of $OH(D_2O)_2$. The structures of the two cyclic trimers and the associated assignments are shown as insets. The middle frame contains the Lorentzians obtained from fitting the high-field spectra (31.0 kV/cm) obtained with parallel (red) and perpendicular (blue) polarization configurations. The Lorentzian areas are normalized to zero-field values obtained with an identical fitting procedure. The top frame shows the computed parallel to perpendicular intensity ratios expected at high-field versus VTMA for the ud/du (black) and uu/dd (red) cyclic trimers. Using the computed intensity ratio curve, the experimental intensity ratios are used to obtain semi-empirical VTMAs of 51(2) and 68(2)° for the 3377 and 3380 cm$^{-1}$ bands, respectively. These values compare favorably to the *ab initio* VTMAs computed for the OH stretch bands of the ud/du (48°) and uu/dd (63°) cyclic trimers. Reproduced with permission from Ref. [38]. © copyright *American Institute of Physics*. All rights reserved

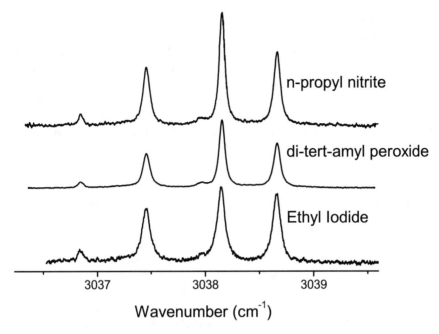

**Fig. 4.9** Ethyl (CH$_3$CH$_2$) zero-field rovibrational spectra ($\nu_1$ CH$_2$ symmetric stretch). The radical is produced from the pyrolytic dissociation of three different organic precursors

Our spectroscopic studies of smaller hydrocarbon radicals have mostly employed halogenated, peroxide, or nitrite precursors [20, 23, 25, 27]. For example, Fig. 4.9 shows three rovibrational spectra of the ethyl radical obtained with various precursors. Halogenated pyrolysis precursors (RI) perform the poorest, because thermal decomposition branches significantly to alkene+HI products. We observe increased branching to closed-shell products upon increasing the size of the hydrocarbon group. Nitrite precursors (RCH$_2$ONO) are easy to synthesize and perform well for generating somewhat larger hydrocarbon radicals (e.g. propyl radicals; see Fig. 4.10). The drawback to nitrite precursors is the simultaneous production of formaldehyde and NO (*i.e.* in addition to R$^\bullet$). Because droplet doping is a statistical process, when three fragments are produced upon pyrolysis, only 12% of the droplet ensemble is doped with R$^\bullet$ (as an upper limit). Although largely commercially unavailable and difficult to synthesize, R(CH$_3$)$_2$COOC(CH$_3$)$_2$R, peroxide precursors exhibit the best performance, because thermal decomposition leads to 2R$^\bullet$ + 2(CH$_3$)$_2$CO, resulting in 18% of droplets being doped by R$^\bullet$. Evidence for this can be directly observed in the signal to noise ratios in the ethyl radical spectra (Fig. 4.9). Diazo, diazirine, and diacyl compounds are also well-known to be efficient pyrolysis precursors for carbenes and radicals, [76] although the synthesis of these is more complex, and the resulting compounds can be too unstable.

With the new SiC based pyrolysis source, we are now in a position to test the efficacy of alternative pyrolysis precursors. For example, production of the vinyl radical was achieved *via* the pyrolysis of di-vinyl sulfone (DVS), [30] which was obtained from a commercial vendor. Sulfone precursors decompose to give $2R^\bullet + SO_2$, which results in 24% of droplets being doped with the radical, although the temperature necessary to achieve efficient pyrolysis is somewhat larger than is possible with the quartz pyrolysis source used in our previous work. Future collaboration with synthetic groups will be aimed at the synthesis of novel sulfone systems, which may serve as high quality pyrolysis precursors.

### 4.1.2  Infrared Spectroscopy of Hydrocarbon Radicals

Thermal decomposition of organic precursors in a continuous, effusive pyrolysis source allows for the helium nanodroplet isolation and spectroscopic interrogation of a variety of hydrocarbon radicals (see Section 4.1.1 for a detailed description of the HENDI methodology). Many of these initial studies involved small radicals that had been spectroscopically probed in the gas phase. Nevertheless, as summarized here, the low temperature afforded by He droplets allows for a characterization of these systems beyond what has so far been achieved in the gas phase. More recent studies of larger radical systems that have yet to be spectroscopically probed in the gas phase are encouraging (e.g. propyl radicals), [45] as it seems the only limitation to the HENDI method is the availability of suitable pyrolysis precursors.

#### 4.1.2.1  Helium-Mediated Tunneling Dynamics of the Vinyl Radical

The vinyl radical ($H_2C_\beta = C_\alpha H$) was trapped in liquid He droplets *via* the use of a di-vinyl sulfone pyrolysis precursor [30]. At 0.4 K, the entire population of nuclear spin isomers is cooled to either the $0_{00}^+$ (ortho) or $0_{00}^-$ (para) rotovibrational level. IR spectra in the fundamental CH stretch region revealed three bands that we assigned to the symmetric $CH_2$ ($\nu_3$), antisymmetric $CH_2$ ($\nu_2$) and lone $\alpha$–CH ($\nu_1$) stretch bands. The vinyl radical CH stretch band origins in He droplets differ from vibrational configuration interaction calculations [77] of J. Bowman and co-workers by $\sim$1, 2 and 10 cm$^{-1}$ for the $\nu_3$, $\nu_2$ and $\nu_1$ modes, respectively. Each band consists of a-type and b-type transitions from the $0_{00}$ level, and each of these is split by either the *difference in* or *sum of* the $v = 0$ and $v = 1$ tunneling splittings. Comparing the He droplet spectra to previous high-resolution spectroscopy of the $\nu_3$ band (D.J. Nesbitt and co-workers), [78, 79] we found that the $A' - B'$ rotational constant for this mode is reduced to 89% of its gas-phase value, and the tunneling splittings (ground and $\nu_3$ excited states) are both reduced by $\sim$20%. In addition, the relative intensities of the $\nu_3$ transitions indicate 4:4 spin statistics for ortho and para nuclear spin isomers, suggesting a facile interchange mechanism [80] for all three H atoms within the $\sim$1200 K pyrolysis source, prior to the pick-up and cooling of the hot vinyl radical by the He droplet. The $\sim$20% reduction in the ground and $\nu_3$ excited state tunneling splittings is due to two contributing effects from the He solvent. The He droplet can modify both the tunneling barrier and the effective reduced mass for motion along this coordinate. We have estimated that either an $\sim$40 cm$^{-1}$ increase in the effective barrier height or an $\sim$5% increase in the effective mass of the tunneling particles (both as upper limits) is sufficient to account for the observed $\sim$20% tunneling splitting reduction. Future theoretical work will be required to assess the extent to which each of these effects contribute to the overall modification of the vinyl radical tunneling dynamics upon solvation in liquid He.

### 4.1.2.2  Methyl, Ethyl, Propargyl, Allyl, and Propyl Radicals

The methyl ($CH_3$) and ethyl ($C_2H_5$) radicals were produced *via* the pyrolysis of peroxide precursors and isolated and spectroscopically characterized in He droplets [25, 27]. The five fundamental CH stretch bands of $C_2H_5$ near 3 $\mu$m were each observed within 1 cm$^{-1}$ of the band origins reported for the gas phase species (D.J. Nesbitt and co-workers) [81, 82]. The symmetric $CH_2$ stretching band ($\nu_1$) is rotationally resolved, revealing nuclear spin statistical weights predicted by $G_{12}$ permutation-inversion group theory. The ethyl radical's permanent electric dipole moment (0.28(2) D) was obtained *via* the Stark spectrum of the $\nu_1$ band. Three $a_1'$ overtone/combination bands were also observed, each having resolved rotational substructure. These were assigned to $2\nu_{12}$, $\nu_4+\nu_6$, and $2\nu_6$ through comparisons to anharmonic frequency computations at the CCSD(T)/cc-pVTZ level of theory and *via* an analysis of the rotational substructure observed for each band.

Rotationally resolved IR spectra were obtained for the propargyl ($C_3H_3$) and allyl radicals ($C_3H_5$) [20, 23]. In the IR spectrum of He-solvated allyl, we observed rovibrational bands near the band origins previously reported in high resolution gas-phase studies carried out by D.J. Nesbitt and co-workers [83] and R. Curl and co-workers [84–86]. In addition to the fundamental CH stretching modes, four other bands were assigned to the allyl radical using a consistent set of rotational constants. Indeed, in the gas-phase studies, it was noted that the CH stretch bands are heavily perturbed, but no explanation was given as to the nature of the perturbations. Isolating the radical in He droplets greatly decreases the number of populated rovibrational levels, and aided by anharmonic frequency computations and the resolved rotational substructure, we assigned the $\nu_1$ ($a_1$), $\nu_3$ ($a_1$), $\nu_{13}$ ($b_2$) fundamentals and the $\nu_{14}/(\nu_{15}+2\nu_{11})$ ($b_2$) and $\nu_2/(\nu_4+2\nu_{11})$ ($a_1$) Fermi dyads, in addition to an unassigned resonant polyad near the $\nu_1$ mode.

In our most recent work, [45] gas-phase *n*-propyl and *i*-propyl radicals ($C_3H_7$) were generated *via* pyrolysis of *n*-butyl nitrite and *i*-butyl nitrite, respectively. An Ar-matrix isolation study from the late 1970s represents the only previous molecular spectroscopy of these radicals [76, 87]. Several previously unreported bands were observed in the IR spectrum between 2800 and 3150 cm$^{-1}$ (Fig. 4.10). The CH stretching modes observed above 2960 cm$^{-1}$ are in excellent agreement with anharmonic frequencies computed using second-order vibrational perturbation

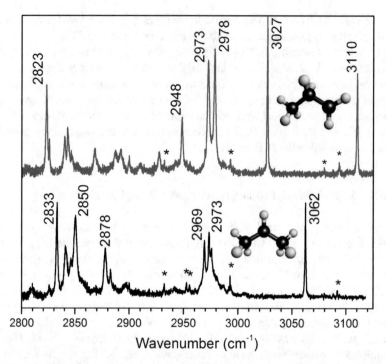

**Fig. 4.10** Comparison of the IR spectra of *n*-propyl (top, red) and *i*-propyl radicals (bottom, black). All vibrational bands are broadened beyond the rotational contour expected at 0.4 K. Residual propene absorptions are marked by *. Reproduced with permission from Ref. [45]. © copyright *American Institute of Physics*. All rights reserved

theory. However, between 2800 and 2960 cm$^{-1}$, the spectra of *n*- and *i*-propyl radicals become congested and difficult to assign due to the presence of multiple anharmonic resonances. Computations employing a local mode Hamiltonian reveal the origin of the spectral congestion to be strong coupling between the high frequency CH stretching modes and the lower frequency bending/scissoring motions. The most significant coupling is between stretches and bends localized on the same $CH_2/CH_3$ group. This work was carried out as a collaboration between the experiment/theory groups at the University of Georgia and Edwin L. Sibert at the University of Wisconsin-Madison.

### 4.1.2.3 Anharmonic Resonance Polyads in the Mid-IR Spectra of $^\bullet C_n H_{2n+1}$ Radicals: Vibrational Complexity in the CH Stretching Region

High-resolution, gas-phase, mid-IR spectra of alkyl radicals larger than ethyl are entirely missing from the spectroscopic literature. High-quality infrared spectra of *n*- and *i*-propyl radicals in the CH stretching region were recently obtained *via* the helium droplet isolation method [45]. In the limit of $3N - 6$ uncoupled oscillators, one expects seven CH stretch vibrations for both $n$- and $i$-propyl. However, the resolution achieved in the experiment reveals a vibrational complexity that demands a treatment beyond the harmonic approximation (see Fig. 4.11 for the $n$- propyl example, black trace). Second-order vibrational perturbation theory, VPT2, accurately predicts the high-frequency stretching vibrations localized on the radical site ($\alpha$-CH$_2$ for $n$-propyl). The CH stretch vibrations localized on carbon atoms adjacent to the radical center are red shifted, due to a hyperconjugative stabilization of the system and concomitant softening of the CH oscillators ($\beta$-CH$_2$ for $n$-propyl) [45]. The associated red shifts drive these CH stretch modes into resonance with the overtones and combination tones of CH$_n$ bending modes. This effect contributes substantially to the spectral complexity observed between 2800 to 3000 cm$^{-1}$. Clearly, VPT2 alone cannot account for the complexity that emerges in this lower frequency region (see Fig. 4.11, red trace).

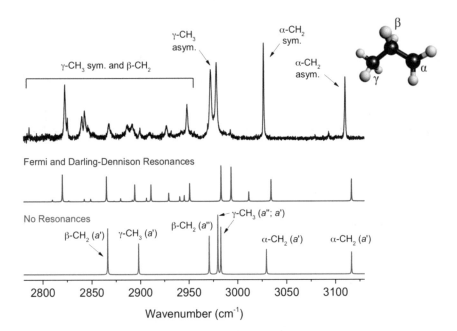

**Fig. 4.11** Comparison of the experimental $n$-propyl spectrum (top, black) to VPT2 simulated spectra. The bottom (red) trace represents a full VPT2 treatment with no resonance treatments whatsoever. The blue trace includes explicit treatment of Fermi and Darling-Dennison resonances. Labels correspond to the carbon atoms around which the vibrations are localized. The frequencies are the eigenvalues of a 22-dimensional effective Hamiltonian. Symmetry labels are included on the bottom trace; these are for the $C_s$ average structure (minimum energy structure on the zero-Kelvin enthalpic surface), although we note that the computation is carried out at the $C_1$ symmetry electronic global minimum structure. Reproduced with permission from Ref. [45]. © copyright *American Institute of Physics*. All rights reserved

The pervasive anharmonic coupling and intensity borrowing evident in the CH stretch region was modeled with two separate effective Hamiltonian approaches [45]. (1) The VPT2+K approach treats Fermi and Darling-Dennison resonances explicitly *via* the diagonalization of an effective Hamiltonian matrix (see Fig. 4.11, blue trace). The matrix contains deperturbed diagonal elements and off-diagonal coupling terms derived from quartic force fields computed at the CCSD(T)/ANO0 level of theory. The effective Hamiltonian is represented in a normal mode basis consisting of $CH_n$ stretching fundamentals and $CH_n$ bending overtones/combinations. (2) The local mode effective Hamiltonian approach employs a localization scheme that takes as input a harmonic frequency computation at the B3LYP/6-311++G(d,p) level of theory (see Fig. 4.12, blue trace). The localized basis states correspond to CH stretching fundamentals and overtones/combinations of HCH scissor modes. Refined harmonic scale factors and anharmonic coupling terms are taken from previous studies of closed shell hydrocarbon CH stretch spectra and are transferred to the local mode model without modification [88]. Both approaches generate Hamiltonian matrices that are

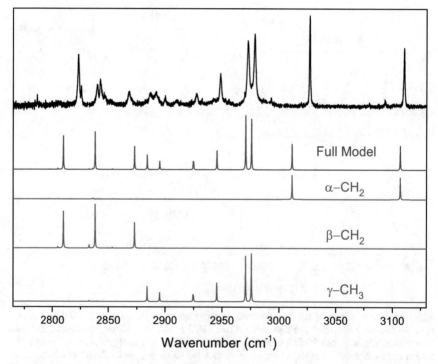

**Fig. 4.12** Dipole decomposition of local mode Hamiltonian simulations (*n*-propyl). The experimental spectrum, the full model local mode simulation, and the dipole decomposed simulations are shown as black, blue, and red traces, respectively. Reproduced with permission from Ref. [45]. © copyright *American Institute of Physics*. All rights reserved

22-dimensional for $n$-propyl. The computational cost of the local mode approach is far lower than the VPT2+K method, because it does not require a quartic force field as input.

Local mode predictions are generally in very good agreement with experiment, despite there being zero adjustable parameters in the model (see Fig. 4.12). The success of the local mode model indicates a rather robust transferability of the anharmonic coupling terms [88]. The presence of a radical center apparently does not significantly affect the cubic coupling between localized CH stretch and HCH scissor modes for the propyl radicals. On the other hand, the quadratic force field is strongly affected by the radical site. For example, the two $\alpha$-CH$_2$ stretches are shifted to higher energy and coupled more strongly by quadratic terms in the Hamiltonian. In contrast, the $\beta$-CH$_2$ stretches are largely decoupled from each other and shifted to lower energy. Both observations are consistent with the approximately $sp^2$ hybridization of the $\alpha$-CH$_2$ group and the hyperconjugative stabilization of the $\beta$-CH$_2$ group.

The choice of representation, local versus normal mode, appears to result in different convergence behavior. The coupling between basis states in the local mode model more accurately reflects the salient interactions responsible for the experimental spectral complexity (stretch-scissor coupling), and is therefore more easily converged. Because of its success in predicting the complexity associated with the propyl radical spectra, we expect the local mode model to accurately predict the CH stretch spectra of larger alkyl radical systems, and because of the low cost of such computations, this approach provides an excellent alternative to the more expensive VPT2+K method. The weakly interacting nature of superfluid helium allows for a direct comparison between experimental band origins and computed spectra using the local mode model. We propose to continue along this direction to explore the spectroscopy of larger alkyl radical systems that exhibit multiple conformations, such as the butyl radicals. Moreover, we plan to probe the CH stretch spectra of a series of cycloalkyl radicals. These studies will allow us to test/refine the local mode model and probe the anharmonic resonance polyads that emerge in the spectra of primary, secondary and tertiary alkyl and cycloalkyl radical systems. The spectra of these helium-solvated hydrocarbon radicals will provide a robust starting point for future high resolution gas-phase spectroscopic studies.

### 4.1.2.4  Infrared Spectroscopy of Cyclobutyl, Methylallyl, and Allylcarbinyl Radicals

Gas-phase cyclobutyl radical (•C$_4$H$_7$) was produced *via* pyrolysis of cyclobutyl-methyl nitrite (C$_4$H$_7$(CH$_2$)ONO) [47]. Other •C$_4$H$_7$ radicals, such as 1-methylallyl and allylcarbinyl, were similarly produced from nitrite precursors. For the cyclobutyl and 1-methylallyl radicals, anharmonic frequencies were predicted by VPT2+K simulations based upon a hybrid CCSD(T) force field with quadratic (cubic and quartic) force constants computed using the ANO1 (ANO0) basis set. A density functional theoretical method was used to compute the force field for the allylcarbinyl radical. For all •C$_4$H$_7$ radicals, resonance polyads in the 2800-3000 cm$^{-1}$ region appear as a result of anharmonic coupling between the CH stretching fundamentals and $CH_2$ bend overtones and combinations. VPT2+K simulations are generally good at predicting the spectral complexity in the CH stretch region for the cyclobutyl radical; however, the predictions are less satisfactory for the 1-methylallyl and allylcarbinyl radicals. Upon pyrolysis of the cyclobutylmethyl nitrite precursor to produce the cyclobutyl radical, an approximately two-fold increase in the source temperature leads to the appearance of spectral signatures that can be assigned to 1-methylallyl and 1,3-butadiene. On the basis of a previously reported $^\bullet C_4H_7$ potential energy surface, this result is interpreted as evidence for the unimolecular decomposition of the cyclobutyl radical *via* ring opening, prior to it being captured by helium droplets. On the $^\bullet C_4H_7$ potential surface, 1,3-butadiene is formed from cyclobutyl ring opening and H atom loss, and the 1-methylallyl radical is the most energetically stable intermediate along the decomposition pathway. The allylcarbinyl radical is a higher energy $^\bullet C_4H_7$ intermediate along the ring opening path, and the spectral signatures of this radical are not observed under the same conditions that produce 1-methylallyl and 1,3-butadiene from the unimolecular decomposition of cyclobutyl.

### 4.1.3    R· + $(^3\Sigma_g^-)O_2$ Chemistry in Helium Droplets

#### 4.1.3.1    Methyl Peroxy Radical

We have demonstrated that R· + $(^3\Sigma_g^-)O_2$ reactions can be carried out within the low temperature, He droplet environment. For example, the sequential addition of a methyl radical and molecular oxygen to He droplets leads to the barrierless reaction, $CH_3 + O_2 \rightarrow CH_3OO$ [17]. The reaction enthalpy is exothermic by $\sim$30 kcal mol$^{-1}$ and therefore requires the dissipation of $\sim$2000 He atoms to cool $CH_3OO$ to 0.4 K. The $CH_3OO$ radical remains in the droplet and is observed downstream with IR laser beam depletion spectroscopy. All three CH stretch bands are observed, and rotational fine structure is partially resolved for the $\nu_2$ totally symmetric CH stretch band, indicating complete internal cooling of the reaction product to the droplet temperature. Electron impact ionization of the droplets containing $CH_3OO$ results in the charge transfer reaction $He^+ + CH_3OO \rightarrow CH_3O_2^+ + He$, which is followed by the fragmentation of the $CH_3O_2^+$ ion. The major fragmentation channel is the production of $HCO^+$ and $H_2O$. The outcome of this work demonstrates that IR laser spectroscopy can be employed as a probe of the outcome of organic radical-radical reactions carried out in the dissipative environment of a He nanodroplet.

#### 4.1.3.2    Propargyl and Allyl Peroxy Radicals

IR spectroscopy was used to probe the outcome of the reaction between the propargyl radical ($C_3H_3$) and $(^3\Sigma_g^-)O_2$ within He droplets [23]. Helium droplets doped with a propargyl radical (generated *via* pyrolysis of 1-butyn-4-nitrite) were subsequently doped with an $O_2$ molecule. The reaction carried out at 0.4 K resulted in the exclusive formation of the acetylenic-*trans*-propargyl peroxy radical ($HC\equiv C-CH_2-OO^\bullet$). This work helped to elucidate the shape of the entrance channel on the ground-state potential energy surface, as it was unclear whether or not there exists a small barrier to formation of the peroxy species. The rapid cooling afforded by the He droplets motivates the conclusion that if a barrier does indeed exist, it is too small to kinetically stabilize a van der Waals complex between $C_3H_3$ and $O_2$. MRCI computations carried out in collaboration with Stephen Klippenstein and co-workers indicate that the reaction is barrierless for $O_2$ addition to the $-CH_2$ "tail" group, similar to alkyl + $O_2$ reactions. Apparently, $O_2$ addition to the $HC\equiv C-$ "head" group proceeds *via* a positive entrance channel barrier, consistent with the absence of *allenic* peroxy radicals in the He droplet IR spectra.

Five stable conformers were predicted for the allyl peroxy radical ($H_2C=CHCH_2-OO^\bullet$) [89]. A two-dimensional potential surface was computed for rotation about the CC–OO and CC–CO bonds, [20] revealing multiple isomerization barriers greater than $\sim 300$ cm$^{-1}$. Nevertheless, the C-H stretch IR spectrum can be assigned assuming the presence of a *single* conformer following the allyl + $O_2$ reaction within He droplets [20]. This is similar to the observation for the propargyl peroxy system, and from this we can infer a cooling mechanism for the vibrationally hot reaction products ($R-OO^\bullet$) that is consistent with both sets of data. The mechanism assumes that the more closely spaced torsional levels ($<100$ cm$^{-1}$) are relaxed more efficiently by the He solvent in comparison to the higher frequency vibrations, allowing the system to funnel into the lowest energy conformational minimum as it cascades down the ladder of excited stretching/bending levels.

## 4.1.4 Infrared Spectroscopy of Hydroxycarbenes

### 4.1.4.1 Hydroxymethlyene, Dihydroxycarbene, Hydroxymethoxycarbene

Hydroxymethylene ($H\ddot{C}OH$) and its $d_1$-isotopologue ($H\ddot{C}OD$) were isolated in He droplets following the pyrolysis of glyoxylic acid [32]. Transitions identified in the IR spectrum were assigned exclusively to the *trans*-conformation based on previously reported anharmonic frequency computations [90, 91]. For the OH(D) and CH stretches, *a*- and *b*-type transitions were observed, and when taken in conjunction with CCSD(T)/cc-pVTZ computations, lower limits to the vibrational band origins were determined. The relative intensities of the *a*- and *b*-type transitions provide the orientation of the transition dipole moment in the inertial frame. The He droplet data are in excellent agreement with anharmonic frequency computations carried out in collaboration with John F. Stanton, confirming strong anharmonic resonance interactions in the high-frequency stretch regions of the mid-IR. Moreover, the He droplet spectra confirm appreciable Ar-matrix shifts of the OH and OD stretches, which were previously postulated by Schreiner and co-workers [90].

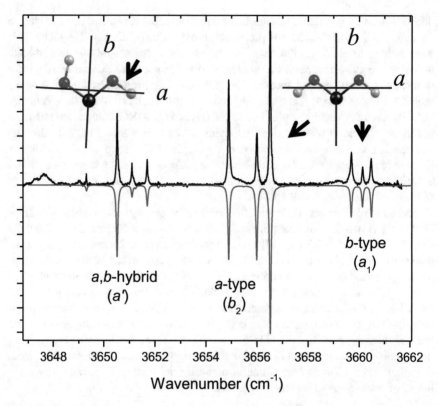

**Fig. 4.13** Rovibrational spectrum of *trans,trans-* and *cis,trans-*HOĊOH rotamers in the OH stretch region. A simulation (red) derived from an asymmetric top Hamiltonian is shown below the experimental (black) spectrum. Assignments are based on band-types and nuclear spin statistical weights. Pure *b-* and *a-*type bands are observed for the symmetric and antisymmetric OH stretching vibrations of the $C_{2v}$ *trans,trans-* rotamer, respectively. The *a,b-*hybrid band corresponds to the higher frequency OH stretch of the $C_s$ symmetry *cis,trans-* rotamer. Reproduced with permission from Ref. [35]. © copyright *American Institute of Physics*. All rights reserved

Dihydroxycarbene (HOĊOH) was produced *via* pyrolytic decomposition of oxalic acid, captured by He droplets, and probed with IR laser Stark spectroscopy [35]. Rovibrational bands in the OH stretch region were assigned to either *trans,trans-* or *trans,cis-* rotamers on the basis of symmetry type, nuclear spin statistical weights, and comparisons to electronic structure theory calculations (Fig. 4.13). The inertial components of the permanent electric dipole moments for these rotamers were determined with Stark spectroscopy. The dipole components for *trans,trans-* and *trans,cis-* rotamers are $(\mu_a, \mu_b) = (0.00, 0.68(6))$ and $(1.63(3), 1.50(5))$, respectively. The IR spectra lack evidence for the higher energy *cis,cis-* rotamer, which is consistent with a previously proposed pyrolytic decomposition mechanism of oxalic acid [92–95] and computations of HOĊOH torsional interconversion and tautomerization barriers [96].

Hydroxymethoxycarbene ($CH_3OCOH$) was similarly produced *via* monomethyl oxalate pyrolysis [36]. Two rotationally resolved *a,b*- hybrid bands in the OH-stretch region were assigned to *trans,trans-* and *cis,trans-* rotamers. Stark spectroscopy of the *trans,trans-* OH stretch band provided the *a*-axis inertial component of the dipole moment, namely $\mu_a = 0.62(7)$ D. The computed equilibrium dipole moment agrees with the expectation value determined from experiment, consistent with a semi-rigid $CH_3OCOH$ backbone computed *via* a potential energy scan at the B3LYP/cc-pVTZ level of theory, which reveals substantial conformer interconversion barriers of ~17 kcal mol$^{-1}$.

# References

1. K.K. Lehmann, G. Scoles, Science **279**, 2065 (1998)
2. J.P. Toennies, A.F. Vilesov, K.B. Whaley, Physics Today **54**, 31 (2001)
3. C. Callegari, K.K. Lehmann, R. Schmied, G. Scoles, J. Chem. Phys. **115**, 10090 (2001)
4. J.P. Toennies, A.F. Vilesov, Angew. Chem. Int. Ed. **43**, 2622 (2004)
5. F. Stienkemeier, K.K. Lehmann, J. Phys. B: At. Mol. Opt. Phys. **39**, R127 (2006)
6. M.Y. Choi, G.E. Douberly, T.M. Falconer, W.K. Lewis, C.M. Lindsay, J.M. Merritt, P.L. Stiles, R.E. Miller, Int. Rev. Phys. Chem. **25**, 15 (2006)
7. M. Barranco, R. Guardiola, S. Hernandez, R. Mayol, J. Navarro, M. Pi, J. Low Temp. Phys. **142**, 1 (2006)
8. J. Kupper, J.M. Merritt, Int. Rev. Phys. Chem. **26**, 249 (2007)
9. S.F. Yang, A.M. Ellis, Chem. Soc. Rev. **42**, 472 (2013)
10. A. Mauracher, O. Echt, A.M. Ellis, S. Yang, D.K. Bohme, J. Postler, A. Kaiser, S. Denifl, P. Scheier, Phys. Rep.-Rev. Sect. Phys. Lett. **751**, 1 (2018)
11. D. Verma, R.M.P. Tanyag, S.M.O. O'Connell, A.F. Vilesov, Adv. Phys.-X **4** (2019)
12. T. Gonzalez-Lezana, O. Echt, M. Gatchell, M. Bartolomei, J. Campos-Martinez, P. Scheier, Int. Rev. Phys. Chem. **39**, 465 (2020)
13. S.D. Flynn, D. Skvortsov, A.M. Morrison, T. Liang, M.Y. Choi, G.E. Douberly, A.F. Vilesov, J. Phys. Chem. Lett. **1**, 2233 (2010)
14. A.M. Morrison, S.D. Flynn, T. Liang, G.E. Douberly, J. Phys. Chem. A **114**, 8090 (2010)
15. T. Liang, S.D. Flynn, A.M. Morrison, G.E. Douberly, J. Phys. Chem. A **115**, 7437 (2011)
16. T. Liang, G.E. Douberly, Chem. Phys. Lett. **551**, 54 (2012)
17. A.M. Morrison, J. Agarwal, H.F. Schaefer, G.E. Douberly, J. Phys. Chem. A **116**, 5299 (2012)
18. P.L. Raston, T. Liang, G.E. Douberly, J. Chem. Phys. **137**, 184302 (2012)
19. L.F. Gomez, R. Sliter, D. Skvortsov, H. Hoshina, G.E. Douberly, A.F. Vilesov, J. Phys. Chem. A **117**, 13648 (2013)
20. C.M. Leavitt, C.P. Moradi, B.W. Acrey, G.E. Douberly, J. Chem. Phys. **139**, 234301 (2013)
21. T. Liang, D.B. Magers, P.L. Raston, W.D. Allen, G.E. Douberly, J. Phys. Chem. Lett. **4**, 3584 (2013)
22. T. Liang, P.L. Raston, G.E. Douberly, ChemPhysChem **14**, 764 (2013)
23. C.P. Moradi, A.M. Morrison, S.J. Klippenstein, C.F. Goldsmith, G.E. Douberly, J. Phys. Chem. A **117**, 13626 (2013)
24. A.M. Morrison, T. Liang, G.E. Douberly, Rev. Sci. Inst. **84**, 013102 (2013)
25. A.M. Morrison, P.L. Raston, G.E. Douberly, J. Phys. Chem. A **117**, 11640 (2013)
26. E.I. Obi, C.M. Leavitt, P.L. Raston, C.P. Moradi, S.D. Flynn, G.L. Vaghjiani, J.A. Boatz, S.D. Chambreau, G.E. Douberly, J. Phys. Chem. A **117**, 9047 (2013)

27. P.L. Raston, J. Agarwal, J.M. Turney, H.F. Schaefer, G.E. Douberly, J. Chem. Phys. **138**, 194303 (2013)
28. P.L. Raston, G.E. Douberly, J. Mol. Spec. **292**, 15 (2013)
29. P.L. Raston, T. Liang, G.E. Douberly, J. Phys. Chem. A **117**, 8103 (2013)
30. P.L. Raston, T. Liang, G.E. Douberly, J. Chem. Phys. **138**, 174302 (2013)
31. C.M. Leavitt, K.B. Moore, P.L. Raston, J. Agarwal, G.H. Moody, C.C. Shirley, H.F. Schaefer, G.E. Douberly, J. Phys. Chem. A **118**, 9692 (2014)
32. C.M. Leavitt, C.P. Moradi, J.F. Stanton, G.E. Douberly, J. Chem. Phys. **140**, 171102 (2014)
33. P.L. Raston, G.E. Douberly, W. Jager, J. Chem. Phys. **141**, 044301 (2014)
34. P.L. Raston, T. Liang, G.E. Douberly, Mol. Phys. **112**, 301 (2014)
35. B.M. Broderick, L. McCaslin, C.P. Moradi, J.F. Stanton, G.E. Douberly, J. Chem. Phys. **142**, 144309 (2015)
36. B.M. Broderick, C.P. Moradi, G.E. Douberly, Chem. Phys. Lett. **639**, 99 (2015)
37. G.E. Douberly, P.L. Raston, T. Liang, M.D. Marshall, J. Chem. Phys. **142**, 134306 (2015)
38. F.J. Hernandez, J.T. Brice, C.M. Leavitt, T. Liang, P.L. Raston, G.A. Pino, G.E. Douberly, J. Chem. Phys. **143**, 164304 (2015)
39. F.J. Hernandez, J.T. Brice, C.M. Leavitt, G.A. Pino, G.E. Douberly, J. Phys. Chem. A **119**, 8125 (2015)
40. C.P. Moradi, G.E. Douberly, J. Phys. Chem. A **119**, 12028 (2015)
41. C.P. Moradi, G.E. Douberly, J. Mol. Spec. **314**, 54 (2015)
42. C.P. Moradi, C. Xie, M. Kaufmann, H. Guo, G.E. Douberly, J. Chem. Phys. **144**, 164301 (2016)
43. M. Kuaffman, D. Leicht, M. Havenith, B.M. Broderick, G.E. Douberly, J. Phys. Chem. A **120**, 678 (2016)
44. J.T. Brice, T. Liang, P.L. Raston, A.B. McCoy, G.E. Douberly, J. Chem. Phys. **145**, 124310 (2016)
45. P.R. Franke, D.P. Tabor, C.P. Moradi, G.E. Douberly, J. Agarwal, H.F. Schaefer, E.L. Sibert, J. Chem. Phys. **145**, 224304 (2016)
46. P.L. Raston, E.I. Obi, G.E. Douberly, J. Phys. Chem. A **121**, 7597 (2017)
47. A.R. Brown, P.R. Franke, G.E. Douberly, J. Phys. Chem. A **121**, 7576 (2017)
48. J.T. Brice, P.R. Franke, G.E. Douberly, J. Phys. Chem. A **121**, 9466 (2017)
49. P.R. Franke, G.E. Douberly, J. Phys. Chem. A **122**, 148 (2018)
50. A.R. Brown, J.T. Brice, P.R. Franke, G.E. Douberly, J. Phys. Chem. A **123**, 3782 (2019)
51. P.R. Franke, J.T. Brice, C.P. Moradi, H.F. Schaefer III., G.E. Douberly, J. Phys. Chem. A **123**, 3558 (2019)
52. J.M. Merritt, G.E. Douberly, R.E. Miller, J. Chem. Phys. **121**, 1309 (2004)
53. G.E. Douberly, J.M. Merritt, R.E. Miller, Phys. Chem. Chem. Phys. **7**, 463 (2005)
54. G.E. Douberly, J.M. Merritt, R.E. Miller, J. Phys. Chem. A **111**, 7282 (2007)
55. G.E. Douberly, R.E. Miller, Chem. Phys. **361**, 118 (2009)
56. G.E. Douberly, P.L. Stiles, R.E. Miller, R. Schmied, K.K. Lehmann, J. Phys. Chem. A **114**, 3391 (2010)
57. P.L. Raston, G.E. Douberly, W. Jager, J. Chem. Phys. **141** (2014)
58. R. Sliter, L.F. Gomez, J. Kwok, A. Vilesov, Chem. Phys. Lett. **600**, 29 (2014)
59. L.F. Gomez, E. Loginov, R. Sliter, A.F. Vilesov, J. Chem. Phys. **135**, 154201 (2011)
60. E.L. Knuth, B. Schilling, J.P. Toennies. *International Symposium on Rarefied Gas Dynamics*, vol. 19 (Oxford University Press, Oxford, UK, 1995), pp. 270–276
61. M. Lewerenz, B. Schilling, J.P. Toennies, Chem. Phys. Lett. **206**, 381 (1993)
62. M. Hartmann, R.E. Miller, J.P. Toennies, A. Vilesov, Phys. Rev. Lett. **75**, 1566 (1995)
63. D.M. Brink, S. Stringari, Z. Phys, D Atom. Mol. Cl. **15**, 257 (1990)
64. J. Kupper, J.M. Merritt, R.E. Miller, J. Chem. Phys. **117**, 647 (2002)
65. P.L. Stiles, K. Nauta, R.E. Miller, Phys. Rev. Lett. **90**, 135301 (2003)
66. K. Nauta, R.E. Miller, Phys. Rev. Lett. **82**, 4480 (1999)
67. M.Y. Choi, F. Dong, R.E. Miller, Philos. T. Roy. Soc. A **363**, 393 (2005)
68. M.Y. Choi, R.E. Miller, J. Am. Chem. Soc. **128**, 7320 (2006)
69. F. Dong, R.E. Miller, Science **298**, 1227 (2002)

70. G.E. Douberly, R.E. Miller, J. Phys. Chem. B **107**, 4500 (2003)
71. W. Kong, J. Bulthuis, J. Phys. Chem. A **104**, 1055 (2000)
72. W. Kong, L.S. Pei, J. Zhang, Int. Rev. Phys. Chem. **28**, 33 (2009)
73. K.J. Franks, H.Z. Li, W. Kong, J. Chem. Phys. **110**, 11779 (1999)
74. D.W. Kohn, H. Clauberg, P. Chen, Rev. Sci. Inst. **63**, 4003 (1992)
75. X. Zhang, A.V. Friderichsen, S. Nandi, G.B. Ellison, D.E. David, J.T. McKinnon, T.G. Lindeman, D.C. Dayton, M.R. Nimlos, Rev. Sci. Inst. **74**, 3077 (2003)
76. J. Pacansky, D.E. Horne, G.P. Gardini, J. Bargon, J. Phys. Chem. **81**, 2149 (1977)
77. A.R. Sharma, B.J. Braams, S. Carter, B.C. Shepler, J.M. Bowman, J. Chem. Phys. **130**, 174301 (2009)
78. F. Dong, M. Roberts, D.J. Nesbitt, J. Chem. Phys. **128**, 044305 (2008)
79. D.J. Nesbitt, F. Dong, Phys. Chem. Chem. Phys. **10**, 2113 (2008)
80. A.R. Sharma, J.M. Bowman, D.J. Nesbitt, J. Chem. Phys. **136**, 034305 (2012)
81. S. Davis, D. Uy, D.J. Nesbitt, J. Chem. Phys. **112**, 1823 (2000)
82. T. Haber, A.C. Blair, D.J. Nesbitt, M.D. Schuder, J. Chem. Phys. **124**, 054316 (2006)
83. D. Uy, S. Davis, D.J. Nesbitt, J. Chem. Phys. **109**, 7793 (1998)
84. J.D. DeSain, R.I. Thompson, S.D. Sharma, R.F. Curl, J. Chem. Phys. **109**, 7803 (1998)
85. J.X. Han, Y.G. Utkin, H.B. Chen, N.T. Hunt, R.F. Curl, J. Chem. Phys. **116**, 6505 (2002)
86. J.D. DeSain, R.F. Curl, J. Mol. Spec. **196**, 324 (1999)
87. J. Pacansky, H. Coufal, J. Chem. Phys. **72**, 3298 (1980)
88. D.P. Tabor, D.M. Hewett, S. Bocklitz, J.A. Korn, A.J. Tomaine, A.K. Ghosh, T.S. Zwier, E.L. Sibert, J. Chem. Phys. **144**, 224310 (2016)
89. P.S. Thomas, T.A. Miller, Chem. Phys. Lett. **491**, 123 (2010)
90. P.R. Schreiner, H.P. Reisenauer, F.C. Pickard, A.C. Simmonett, W.D. Allen, A. Matyus, A.G. Csaszar, Nature **453**, 906 (2008)
91. L. Koziol, Y.M. Wang, B.J. Braams, J.M. Bowman, A.I. Krylov, J. Chem. Phys. **128**, 204310 (2008)
92. G. Lapidus, D. Barton, P.E. Yankwich, J. Phys. Chem. **70**, 407 (1966)
93. G. Lapidus, D. Barton, P.E. Yankwich, J. Phys. Chem. **70**, 1575 (1966)
94. G. Lapidus, D. Barton, P.E. Yankwich, J. Phys. Chem. **70**, 3135 (1966)
95. G. Lapidus, P.E. Yankwich, D. Barton, J. Phys. Chem. **68**, 1863 (1964)
96. P.R. Schreiner, H.P. Reisenauer, Angew. Chem. Int. Ed. **47**, 7071 (2008)

# Chapter 5
# Electronic Spectroscopy in Superfluid Helium Droplets

Florian Schlaghaufer, Johannes Fischer, and Alkwin Slenczka

**Abstract** Electronic spectroscopy has been instrumental in demonstrating the properties of helium droplets as a cryogenic matrix for molecules. The electronic spectrum of glyoxal, which was one of the first molecules investigated in helium droplets by means of electronic spectroscopy, showed two features that provided convincing evidence that the droplets were superfluid. These were free rotation and the distinct shape of the phonon side band which could be directly assigned to the characteristic dispersion curve of a superfluid. On closer examination, however, details such as increased moments of inertia and a spectral response on the droplet size distribution revealed unexpected features of microsolvation in the superfluid helium. In the course of studying many different molecules, it has become clear that electronic spectroscopy in helium droplets provides insight into the detailed effects of microsolvation. These in turn lead to numerous questions regarding the interaction with the superfluid which are discussed in this chapter. In addition, the influence of microsolvation in helium droplets on van der Waals clusters generated inside helium droplets are discussed. Finally, the effect of helium solvation on unimolecular or bimolecular elementary chemical reactions is evaluated in comparison with corresponding experiments in the gas phase. Particular focus of this article lies on the spectral features related to helium solvation which are not yet fully understood.

## 5.1 Introduction

This chapter is devoted to electronic spectroscopy of molecules in superfluid helium droplets. Compared to other spectroscopic techniques such as MW or IR spectroscopy that are discussed in Chap. 3 by Gary Douberly electronic excitation of molecules in the helium droplet environment shows two prominent features in the spectra. These are a phonon wing (PW) reflecting the excitation of the helium environment coupled to the electronic excitation of the dopant. The second feature is a spectral splitting

F. Schlaghaufer · J. Fischer · A. Slenczka (✉)
Institut für Physikalische und Theoretische Chemie, Universität Regensburg, 93053 Regensburg, Germany
e-mail: alkwin.slenczka@ur.de

A. Slenczka and J. P. Toennies (eds.), *Molecules in Superfluid Helium Nanodroplets*,
Topics in Applied Physics 145, https://doi.org/10.1007/978-3-030-94896-2_5

into multiplets that occurs at the pure molecular excitation called zero phonon line (ZPL). This splitting results from inhomogeneities in the solvation of the dopant inside the helium droplet [1–4] as shall be discussed below. Both features do not occur in IR or MW spectroscopy.

In the IR spectra the line resolved rotational bands prove that the molecules rotate freely inside the superfluid droplets. The helium environment has however a strong effect on the moments of inertia of the rotating dopant [1, 3–5]. As shown for the first time in the rotationally resolved IR spectrum of sulfur hexafluoride, the effective moments of inertia reveal a significant increase of the rotating mass in helium droplets [6]. In contrast to electronic excitation the coupling to phonons of the helium droplet has no noticeable effect on the rotations and/or the vibrations of the dopant. Molecular vibrations proceed mostly inside the dopant's electron cloud and, thereby, are shielded from the influence of the helium environment so that vibrational frequencies are almost identical to the corresponding gas phase values [5]. In contrast, the electron density distribution is that part of the dopant species which is in direct contact with the helium environment and upon electronic excitation is therefore coupled to the phonons of the helium environment. The shape of the dopant molecule as experienced by the helium environment is therefore defined by the outer electron density distribution. Besides an increase of the energy deposited into the dopant's electrons, electronic excitation is accompanied by a rearrangement of the electron density distribution.

The forces responsible for the dopant to helium interaction are dispersion forces and/or van der Waals forces which exceed dispersion forces among helium atoms. Thus, the dopant species attracts a layer of helium atoms [2–4]. Instead of an isolated and cold molecule doped into a non-viscous cryogenic environment one has to consider a dopant-helium solvation complex which rotates freely inside the superfluid droplet [3–5].

As the dopant enforces the formation of a non-superfluid helium solvation layer, the layer enforces modification of the electron density distribution of the dopant which leads to an energetic shift of electronic states compared to the isolated molecule. This modification is not simply dopant specific but in addition specific for different electronic states of the same dopant species. Depending on the relation of the corrugation of the dopant's electron density distribution to the size or the van der Waals radius of helium atoms, the solvation complex may exhibit different configurations which differ energetically. Thus, electronic spectra of molecules in helium droplets provide particularly insight into the solvation of molecules in helium and its influence on intramolecular dynamics. In order to make use of the twofold information namely about the dopant as well as about its solvation, it is necessary to decipher the helium induced spectral features. In other words, intramolecular and intermolecular contributions need to be disentangled which is quite a challenge.

Beyond molecular spectroscopy, the formation of a helium solvation complex has certainly some influence on steering, for example, the formation of weakly bound clusters. Since the cluster forming subunits are picked up consecutively, each subunit may attract a helium solvation layer prior to cluster formation and again helium solvation complexes approach each other instead of bare dopant units. Moreover, the ultra-low temperature conditions can promote cluster configurations which are absent at elevated temperatures. Thus, cluster formation in helium droplets provides a larger variety of metastable configurations and in addition those that incorporate helium atoms. So far it is an issue of microsolvation [1, 2, 6, 7].

Besides microsolvation of either single molecules or molecular compounds as for example van der Waals clusters, helium solvation can have a strong impact on molecular dynamics. Electronic excitation accompanying photophysical and photochemical processes are certainly affected by the low temperatures of the droplets. In combination with vanishing viscosity, superfluid helium droplets were expected to be an ideal host for the investigation of intramolecular and intermolecular photochemistry. We will have a critical view on this.

This article on electronic spectroscopy of molecules in helium droplets focusses on helium induced spectral features as revealed by comparison of helium droplet experiments with corresponding gas phase data. Out of the myriads of publications on electronic spectroscopy in helium droplets only selected studies will be discussed in order to highlight microsolvation expressed by helium induced spectral features. Besides several details readily explained by empirical conclusions or chemical intuition, there are many observations that elude empirical interpretation. The quest on modeling microsolvation in superfluid helium and its impact on chemical dynamics needs to consider all such peculiarities reported so far. Any progress in a quantitative understanding of these features is of fundamental importance for quantum chemical modeling.

## 5.2   Electronic Spectroscopy

In the past the following two techniques are most common in preparing samples of isolated molecules at low temperature. These are adiabatic expansion of a molecular gas into vacuum [8] or matrix isolation of molecules in solid crystals in many cases rare gas crystals or Spolskii matrices [9]. Molecular spectroscopy in superfluid helium droplets can be seen as a kind of hybrid of molecular beam and matrix isolation. Helium droplets are generated via expansion of helium and are provided as a beam of droplets propagating along a well-defined axis inside a vacuum machine. This is the molecular beam aspect. Afterwards, the molecule of interest is doped into or onto the helium droplet by a pick-up process. The solvated dopant resembles matrix isolation with the major difference or rather advantage that the helium matrix is superfluid instead of solid. The doped droplets propagate along a defined axis and, thus, resemble a transient sample similar as molecules do in a molecular beam. Doping of molecules into a helium droplet provides a temperature of only 0.38(1) K for all degrees of freedom of the dopant species except of spin states [1]. This temperature is much lower than can be reached by standard molecular beams. Since helium does not solidify upon cooling at pressures below 25 bar and instead undergoes a transition to a superfluid with vanishing viscosity below 2.17 K, it allows for free rotation of the dopant species. Molecular rotation subject to spectroscopic investigations is of high value for the analysis of the molecular structure. In a solid matrix the rotational degree of freedom is frozen.[1] In a standard molecular beam, however, rotation is cooled down to a temperature of roughly 1–10 K, however, often with non-thermal state population with a surplus at higher rotational states [8]. A Boltzmann ensemble at a temperature of 0.38 K represents perfect thermal conditions for making use of the rotational degree of freedom for structural analysis of molecular compounds as outlined in Chap. 3 of this monograph. A temperature of only 0.38 K warrants for eliminating vibrational hot bands entirely. Under these conditions an easy reading of vibrational modes from electronic spectra is warranted even selective for electronic states. At appropriate spectral resolution the rotational fine structure can also be resolved in electronic spectra of molecules in helium droplets as shall be discussed below.

Experimentally, electronic spectroscopy can be performed using different detection schemes which provide different information. The basic processes in electronic spectroscopy are absorption or emission of electromagnetic radiation in order to switch electronic states of atoms or molecules. Most of the spectroscopic data discussed in this article are based on these two fundamental processes, whereby absorption is recorded in two variants, namely depletion spectroscopy and fluorescence excitation spectroscopy [1]. Depletion spectroscopy makes use of energy dissipation from the excited dopant species into the helium droplet. The energy transfer initiates evaporative cooling whereby the droplet loses mass as well as volume. The reduction of mass can be monitored as a reduction of the energy flux into a bolometer placed on the droplet beam axis whereas the shrunk volume becomes effective in a reduced ionization cross section when using a quadrupole mass spectrometer as monitor detector. In both cases resonant absorption by the dopant generates a depletion of an intense signal, an effect which bore the term *depletion spectroscopy*. This technique allows for recording absorption spectra of highly diluted samples and is therefore the method of choice for recording IR spectra of molecules in helium droplets (cf. Chap. 3).

Instead of depletion spectroscopy, the fluorescence as response to electronic excitation of a molecule can be recorded. In general, it requires a radiative step on the decay path of the electronically excited molecule, a precondition which is not necessarily fulfilled. Thus, in contrast to absorption or depletion spectroscopy, resonances to non-radiating states, so-called dark states, are missing in fluorescence excitation spectra. The advantage of recording fluorescence is a zero-background signal which exceeds depletion in the sensitivity by orders of magnitude.

Starting always from the vibronic ground state, as is guaranteed by the droplet temperature of only 0.38 K, the frequency of a laser is tuned across the series of resonances when recording fluorescence excitation spectra or depletion spectra. A spectrum starts with a purely electronic transition, the so-called electronic band origin followed to the blue by vibronic transitions into the multitude of vibrational levels of electronically excited states. At appropriate spectral resolution the rotational band structure can additionally be resolved for each vibronic transition. Thus, at least

---

[1] Exceptions are the para-hydrogen matrix and methane as dopant [10].

the normal mode frequencies of electronically excited states if not in addition the moments of inertia of the dopant species are readily obtained when monitoring the dopant's fluorescence or the depletion signal as shown in the left half of Fig. 5.1.

Instead of recording the fluorescence in dependence on the excitation frequency, another variant of electronic spectroscopy records the fluorescence, however, upon excitation fixed at a certain resonance and dispersed by means of a grating spectrograph as shown in the right half of Fig. 5.1. Upon excitation at the electronic band origin, radiative transitions extend to the multitude of vibrational states of the electronic ground state. Dispersed emission spectra reveal information on the

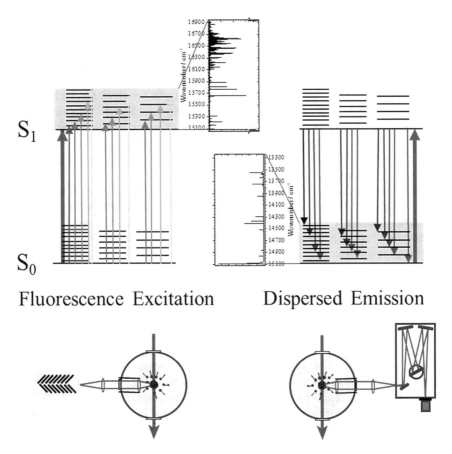

Fluorescence  Excitation          Dispersed  Emission

**Fig. 5.1** Schematic of electronic spectroscopy for a closed shell organic molecule indicating vibronic transitions. Left side: excitation spectroscopy starting with the electronic band origin. Left bottom: experimental setup for fluorescence detection for a view along the droplet beam axis. Right side: dispersed emission spectroscopy upon excitation at the electronic band origin. Right bottom: experimental setup for dispersed emission detection for a view along the droplet beam axis. Arrows resemble photons for excitation (up) or spontaneous emission (down)

normal mode frequencies of the electronic ground state as complement to the corresponding information on the electronically excited state from excitation spectra. Upon vibronic excitation or excitation to higher electronic states, the efficiency of dissipation of energy in excess to the electronic band origin into the helium droplet, the process depletion spectroscopy is based on, guarantees for radiative decay exclusively from the ground level of the first electronically excited state. As a consequence, dispersed emission spectra recorded for molecules in helium droplets are basically independent of the excitation frequency and start with the electronic band origin followed to the red by vibronic transitions as described above. Accordingly, the fluorescence excitation spectrum and dispersed emission spectra of a molecule in helium droplets should coincide in a single resonance line which represents the electronic band origin. A spectral gap instead of an overlap is indicative for dynamic processes in the electronically excited state.

Dissipation of excess excitation energy beyond the electronic band origin holds for rotational, vibrational, electronic, and in addition phonon energy. This process allows for easily identifying the resonance frequency of the electronic band origin of molecules in helium droplets by simply exciting into the quasi continuum of electronically excited states somewhere in the blue or near UV while recording the dispersed emission in the vicinity of the electronic band origin of the molecule. In general, the resonance with the maximum frequency in the corresponding dispersed emission spectrum represents the electronic band origin.

So far, selected variants of electronic spectroscopy were introduced which are relevant for what shall follow on electronic spectroscopy of molecules in superfluid helium droplets. Before going into the specific details, some general spectral features of helium solvation are mentioned, which are also present in electronic spectra of molecules isolated in solid state matrices. First, due to the weak nevertheless finite polarizability of the helium environment, electronic resonance frequencies are shifted compared to the isolated molecule either to the red or to the blue by rule of thumb about 1% of the transition frequency in the gas phase. Whether to the blue or to the red depends on the difference in the helium induced stabilization energy of the corresponding electronic states. Secondly, a ZPL is accompanied by a PW, representing the excitation of the helium environment coupled to the electronic excitation of the dopant species [11]. Third, the ZPL may exhibit a multiplet splitting [12]. In the case of a solid matrix this is an expression of different sites of the dopant within the solid. The correspondence to helium droplets will be discussed below. Finally, and in contrast to solid matrices, a ZPL exhibits a rotational band structure. In case of appropriate experimental conditions, molecular rotation can even be line resolved in electronic spectra of molecules in helium droplets.

As will be shown in the following, one of the strengths of electronic spectroscopy in helium droplets lies in obtaining vibrational frequencies of the dopant species specific for electronic states. Compared to the accuracy of corresponding theoretical values the helium induced shift of vibrational frequencies is rather small. Other experimental observables such as moments of inertia likewise rotational constants, electronic transition energies, and intramolecular dynamics induced by electronic excitation reveal the influence of helium solvation which becomes most evident in comparison to corresponding data from gas phase experiments. Thus, a key issue of electronic spectroscopy in helium droplets is the investigation of microsolvation which reveals information on both, the dopant as well as the helium droplet.

This chapter focusses on microsolvation of molecules and molecular compounds as well as its influence on molecular dynamics as revealed by electronic spectroscopy. This focus is highlighted by a comparative discussion of experiments made in helium droplets as well as under gas phase conditions. The influence of the helium environment is omnipresent for electronic spectra in helium droplets. We start with electronic spectroscopy of various dopant species in most cases closed shell organic molecules which are heliophilic and, therefore, reside fully solvated inside the helium droplet. Thereby, the focus will be on the spectral structure of the ZPL and the PW. Moreover, microsolvation of van der Waals clusters consisting of a single chromophore molecule and additional noble gas atoms generated inside helium droplets will be discussed. Finally, the influence of solvation on elementary chemical reactions involving electronic excitation or relaxation will be presented. Besides numerous helium induced spectral features which fit at least to empirical explanations, there are other helium induced spectral features which are counterintuitive to the current understanding of superfluid helium as host. Understanding the origin of such features is the challenge for future activities.

## 5.3   Electronic Spectra of Molecules in Helium Droplets

The discussion of electronic spectroscopy of molecules in superfluid helium droplets starts with an examination mainly of the ZPL and accompanying PW at the electronic band origin of various dopant species. It starts with glyoxal which behaves in many aspects as expected for a molecule solvated in a cryogenic superfluid. It will be continued with tetracene and related polycyclic aromatic hydrocarbons (PAH) before coming to larger organic compounds such as phthalocyanines and porphyrins. Along the line of dopant species, helium induced spectral features become more and more complex and require additional conceptions for an empirical explanation. Up to now, not all of the helium induced spectral features can be rationalized. Almost none of them can be simulated quantitatively.

### 5.3.1   Glyoxal in Superfluid Helium Droplets

The electronic absorption spectrum of glyoxal recorded by means of depletion spectroscopy is a textbook example for spectral features as expected for solvation in superfluid helium droplets. At appropriate spectral resolution the ZPL at the electronic origin appears as a line resolved rotational fine structure (cf. Fig. 5.2) [13]. While the asymmetric top character expressed by Ray's asymmetry parameter remained almost unaffected in helium droplets—an increase by only 1% for the ground state and 0.3%

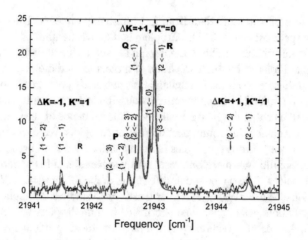

**Fig. 5.2** Rotationally resolved ZPL at the electronic band origin of glyoxal in superfluid helium droplets having on average 2600 atoms. The red line is a best fit calculated for a free asymmetric top, convoluted with a Lorentzian having a linewidth of dv = 0.035 cm$^{-1}$ (FWHM). The {j' ← j"} assignments of the transitions are given for the most prominent lines. (adapted from [13])

for the excited state—the moments of inertia of the $S_0$ state were found increased by factors of 2.87 (A), 2.22 (B), and 2.09 (C). Increased moments of inertia are well known mainly from rotationally resolved IR-spectra of molecules in helium droplets (cf. Chap. 3). The increase of the moments of inertia indicates increasing rotating mass. A countable number of helium atoms rigidly attached to the rotating molecule called solvation layer accounts for the increase quantitatively. Besides this increase of the mass, the observation of free rotation fulfills what is expected from a super-fluid solvent with vanishing viscosity. The cryogenic property of helium droplets is expressed by the intensity profile of the rotational fine structure which fits to a temperature of only 0.38(1) K. The ZPL at the electronic band origin is accompa-nied to the blue by a PW. For glyoxal the spectral shape of the PW fits perfectly to the spectrum of elementary excitations of superfluid helium (cf. Fig. 5.3) [14, 15]. In addition to the free rotation observed for the ZPL the spectral shape of the accompanying PW was a weighty argument for superfluidity and, thus, a milestone in molecular spectroscopy in helium droplets.

Remarkably, however, was the change of the moments of inertia upon electronic excitation of glyoxal in helium droplets [13]. This observation was readily explained by the helium attached to the dopant. As mentioned above, electronic excitation is accompanied by the change of the electron density distribution which by itself is of negligible effect on the moments of inertia for the bare molecule. However, in helium droplets the changing electron density has an impact on the attached helium atoms which follow the electron density distribution. Thus, the impact of electronic excitation on the moments of inertia is significantly amplified by the helium solva-tion layer of the dopant molecule. So far, the helium induced spectral features in

Fig. 5.3 **a** The dispersion curve of elementary excitations in bulk superfluid helium, **b** the corresponding density of states and **c** the density of states adopted to a small droplet (red) fitting almost perfectly to the experimental PW in the electronic spectrum of glyoxal in superfluid helium droplets (black) ($\overline{N} = 5500$). Adapted from [14]

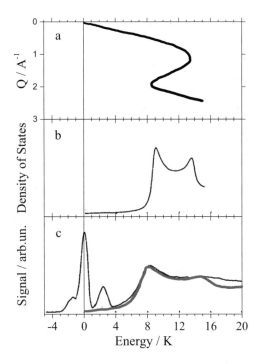

the electronic spectrum of glyoxal in helium droplets reflect expectations based on empirical understanding.

However, another observable which did not fit into the empirical modeling of solvation in superfluid helium was the dependence of the line shape within the rotational fine structure on the droplet size [13, 16]. For increasing droplet sizes beyond an average of 3000 helium atoms the line width increased significantly, certainly an indication for inhomogeneous line broadening. There are various line shape determining effects that scale with the average droplet size, namely, the width of the size distribution, the integral strength of dispersion forces of a polarizable environment, the density of states of the dopant inside the droplets—approximated by a particle in a spherical box, the spectral density of surface modes of the droplet, and last but not least vortices as the way a quantum fluid carries angular momentum, to name only the most evident ones. None of these effects succeeded in simulating the experimental observations on the line shapes of rotationally resolved electronic spectra of glyoxal. So far, the droplet size dependence of the line shape of glyoxal in helium droplets remains an open question. None of the helium induced spectral features were explained by specific properties of glyoxal as dopant species. Therefore, these features might be observable for other dopant species as well.

## 5.3.2    Tetracene in Superfluid Helium Droplets

Tetracene is among the first of organic molecules that have been studied in helium droplets by means of electronic spectroscopy [17]. Its electronic spectrum shown in Fig. 5.4 has undergone a very detailed examination. Neither does the ZPL reveal rotational fine structure nor does the PW reflect the spectral shape expected from superfluid helium [18]. Nevertheless, numerous unexpected spectral features provide information which is relevant for a deeper understanding of microsolvation in helium droplets.

A surprising feature in the electronic spectrum of tetracene was a ZPL that is split into a doublet without relation to a rotational fine structure (cf. Fig. 5.4). Pump probe experiments have proven for two independent systems represented by each peak of the doublet [18]. The observation of different dispersed emission spectra upon excitation at each of the two peaks confirmed the presence of two independent solvated systems [20, 21]. In most of the papers on tetracene in helium droplets the two peaks of the doublet are addressed as alpha (low frequency) and beta (high frequency) line which are separated by a gap of about 1 cm$^{-1}$ whereby the alpha line has roughly 1/3 of the peak intensity of the beta line. Empirical explanation for the doublet splitting by tetracene exhibiting two different configurations of a helium solvation complex was readily on hand and substantiated by quantum chemical modeling [17, 18, 20, 22]. The empirical modeling of helium solvation complexes with a countable number of helium atoms localized on the surface of tetracene in different configurations corresponds to the phenomenon of different sites of molecules in solid matrices. A first purely empirical discussion on possible configurations of solvation

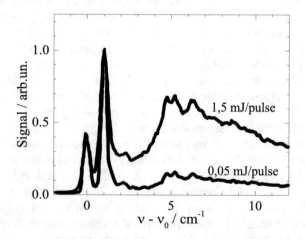

**Fig. 5.4** Fluorescence excitation spectrum of the electronic band origin of tetracene in helium droplets ($\overline{N} = 16{,}000$) with $v_0 = 22{,}293.4(5)$ cm$^{-1}$. The splitting of the ZPL is $\Delta v = 1.1(1)$ cm$^{-1}$ whereby the peak at 0 cm$^{-1}$ is the α-peak and the second and most intense is the β-peak. The phonon wing is enhanced by increasing the laser power from 0.05 to 1.5 mJ/pulse. Both spectra are saturation broadened. (adapted from [19])

complexes [18] was followed by path integral Monte Carlo (PIMC) simulations [22] of a helium solvation layer attached to tetracene embedded into up to 150 helium atoms. Finally, modeling of *quantum coherent, but strongly correlated, set of helium atoms* adsorbed in a linear arrangement on the quasi-planar molecular surface was capable of reproducing a doublet for the ZPL of tetracene [23]. The experimental spectrum chosen for fitting the doublet simulation was unfortunately a spectrum with poor spectral resolution which does not serve as appropriate experimental reference [24]. Furthermore, the model developed in Ref. [23] relates ZPL splitting to a linear arrangement of helium atoms and thereby serves in addition to explaining the missing of such splitting for a two-dimensional helium solvation layer as present for dopant species such as phthalocyanine and porphin. As a matter of fact, and in contrast to the earliest publication [12], the ZPL of porphin does show a triplet splitting as will be discussed below [25]. Moreover, the model developed in Ref. [23] has never been validated for other dopant species and in particular not for other linear PAH molecules shown in Fig. 5.5.

Further experimental information for the interpretation of helium induced spectral features and in particular for the doublet splitting at the ZPL of tetracene can be collected from comparison with related dopants. Among linear PAH species the electronic spectra of pentacene [11, 19], anthracene [26], naphthalene [27], and benzene [29, 28] have been reported. In contrast to the doublet of tetracene shown in Fig. 5.5b, a singly peaked ZPL was recorded for pentacene at the electronic band origin (cf. Fig. 5.5a) [19]. A different situation was found for anthracene. The ZPL at the electronic band origin in helium droplets was found split into a quartet with almost regular gaps of 1 $cm^{-1}$ (cf. Fig. 5.5c) which at the blue side merged into a broad and smoothly decreasing signal extending over tens of $cm^{-1}$. The latter part was assigned to the PW [26]. The characteristic phonon gap of superfluid helium was missing as were two maxima resembling the maxon and roton excitation of a superfluid. Dispersed emission spectra recorded upon excitation at each of the four peaks provided further details. Within the quartet the 1st and the 2nd peak exhibit identical emission spectra which differ from a second emission spectrum recorded upon excitation at the 3rd and 4th peak. Thus, within the quartet in the excitation the 1st and the 3rd line represent individual systems similar as the doublet of tetracene and are therefore assigned accordingly to $\alpha$ and $\beta$ line. The 2nd and 4th line in the quartet correlate with the 1st and 3rd, respectively, and are therefore assigned as $\alpha'$ and $\beta'$. In comparison to tetracene the spectral gap between the $\alpha$ and $\beta$ line has doubled and each of the two solvated systems comes with an additional line shifted by 1 $cm^{-1}$ to the blue. There are two empirical models explaining the $\alpha'$ and $\beta'$ line. Either, the electronic origin of each system is accompanied by a 1 $cm^{-1}$ van der Waals mode of the solvation complex whose energy dissipates prior to radiative decay. Alternatively, the entire quartet represents four different configurations of an anthracene helium solvation complex whereby the 2nd and the 4th peak represent configurations which are highly metastable in the electronically excited state, and, therefore, relax prior to radiative decay. Even though the series from pentacene to anthracene shows increasing multiplet splitting for decreasing size of the PAH species. A continuation for naphthalene and finally benzene is intuitively unlikely

**Fig. 5.5** Fluorescence excitation spectra of the electronic band origin of pentacene (**a**), tetracene (**b**), anthracene (**c**), benzene (**e**), and perylene (**f**), and depletion spectrum of the $8^0{}_1$ vibronic mode of naphthalene (**d**). ($\overline{N}$~20,000) for (**a**–**e**) and $\overline{N} = 8000$ for (**f**)). (adapted from **a**: [19], **b**: [19], **c**: [26], **d**: [27], **e**: [28], **f**: [21])

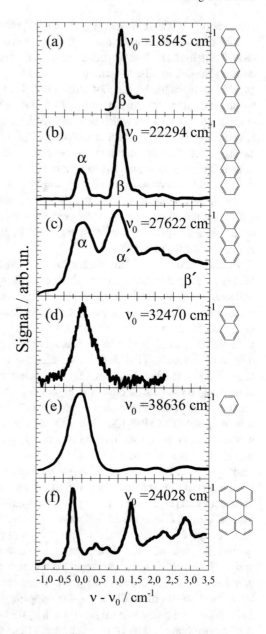

and was refuted by corresponding experiments [27, 29, 28] (cf. Fig. 5.5d, e). The electronic spectrum of naphthalene [27] in helium droplet was found singly peaked which is shown for a vibronic resonance recorded via depletion in Fig. 5.5d. Finally, the ZPL at the electronic band origin of benzene was also singly peaked [29, 28] (c.f. Fig. 5.5d).

Looking finally at perylene consisting of 5 benzene units similar to pentacene, however, in a two-dimensional arrangement, the ZPL at the electronic origin exhibits a rich fine structure with as much as 10 peak maxima within a spectral section of 6 cm$^{-1}$ [30] a part of which is shown in panel (f) of Fig. 5.5. This multiplet merges into a rather smooth and monotonously decreasing signal which extends over tens of cm$^{-1}$ assigned to the PW. The multiplet is dominated by a triplet of intense sharp peaks about 0.1 cm$^{-1}$ in width and with internal gaps of 1.65 cm$^{-1}$ and 1.50 cm$^{-1}$, respectively. The leading peak at the low frequency side of the mutiplet is not part of the dominating triplet and is quite low in intensity. Dispersed emission spectra recorded upon excitation at each of the peaks within the ZPL coincide perfectly among each other. Moreover, the leading peak in the dispersed emission coincides with the leading tiny peak of the multiplet in the excitation spectrum. Thus, the first tiny peak resembles the electronic band origin of perylene in helium droplets which contrasts to the assignment reported in the first paper on perylene in helium droplets [30] where the signal to noise limit did not allow for detecting the leading tiny peak.

Further important experimental details have surfaced within the ZPL doublet of tetracene [19]. Upon increased spectral resolution a substructure could be resolved for the more intense β line. In contrast, the spectral shape of the α line remained smooth, however, clearly asymmetric in shape. Within the substructure of the β line, 7 sharp peaks were resolved exhibiting irregular spectral separation among each other as shown in Fig. 5.6 as black line. An attempt to fitting this substructure by an asymmetric rotor with anisotropic angular momentum caused by the pick-up process did not provide convincing results [19, 31]. In case the model of an anisotropic rotor should explain the fine structure, an explanation for the missing of a corresponding fine structure for the alpha line would be needed. Very important in this context was the singly peaked ZPL of pentacene—shifted to the red of the tetracene resonance by roughly 3750 cm$^{-1}$ as to be expected for an additional carbon ring unit—which

**Fig. 5.6** High resolution electronic spectrum of the ZPL at the electronic band origin of tetracene (black) and pentacene (red) in superfluid helium droplets ($\overline{N} = 16{,}000$). Adapted from Ref. [19]

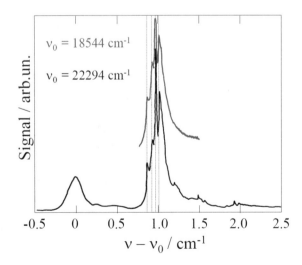

shows exactly the same fine structure as recorded for the β line of tetracene [19].
Consequently, this line of pentacene was assigned as β line. It is added in red to
the spectrum of tetracene in Fig. 5.6. The spectral identity of the beta line among
tetracene and pentacene is another argument against an assignment to rotational fine
structure as proposed in Refs. [19] and [31]. It is certainly helium induced and reveals
properties of microsolvation which need to be common to tetracene and pentacene.
Unfortunately, for anthracene, naphthalene, and benzene the spectral resolution did
not suffice to check for a fine structure as resolved for the β line of tetracene and for
pentacene.

An additional important detail came from the investigation of the droplet size
dependence of the alpha and beta line of tetracene and in particular of the fine structure
of the beta line [19]. As was discussed above already for glyoxal, even the limited
range for the average droplet size accessible under subcritical expansion conditions
can become effective on the line shape. For tetracene, the droplet size dependence has
been investigated for both, the subcritical and the supercritical regime of the droplet
source which was accomplished by recording fluorescence instead of depletion as
depicted in Fig. 5.7 [19]. The spectral response upon variation of the average droplet
size within the subcritical regime was a maximum solvent shift of about $100 \, \text{cm}^{-1}$ to
the red. Upon transition from sub- to supercritical expansion conditions this solvent
shift took a mild turn around to the blue while maintaining the internal gap of $1 \, \text{cm}^{-1}$
between the α and the β line (cf. Fig. 5.7 panel (d)). A closer look at the response
of the fine structure of the β line upon changing the average droplet size reveals the
following details. The fine structure does not shift upon variation of the droplet size

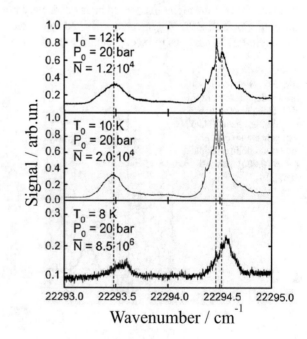

**Fig. 5.7** Fluorescence
excitation spectra with the α
and β line recorded for
increasing average droplet
size $\overline{N}$ obtained by
decreasing source
temperatures $T_0$ (from top to
bottom). The dashed vertical
lines mark the α line and the
two most pronounced peaks
in the fine structure of the β
line in the smallest droplets.
Adapted from Ref. [19]

as emphasized by dashed vertical lines in Fig. 5.7. Instead, for increasing droplets size the fine structure of the β line passes a maximum intensity and fades away for supercritical expansion conditions at the droplet source. From there on the spectral shape of the β line becomes smooth similar as found for the α line. Its spectral shape is still asymmetric and the peak shifts gradually to the blue with further increasing droplet size as does in parallel also the α line. It should be noted that this blue shift is in the order of the spectral width of the β line and, thus, tiny compared to the overall helium induced red shift of roughly $100 \, \mathrm{cm}^{-1}$. Not only the turnaround but in addition the curious disappearance of the fine structure in the β line reveals a fundamental difference for helium droplets generated under subcritical or supercritical expansion conditions. Thereby, one needs to consider, that in the vicinity of the transition from subcritical to supercritical expansion conditions a bimodal droplet size distribution is obtained which may be an expression of instabilities in the droplet source.

Finally, it should be noted that none of the PWs of all of these PAH compounds discussed above fits to the spectral shape of elementary excitations of superfluid helium as reported for glyoxal. Even though it reveals what is expected, the PW recorded for glyoxal is an exception in reflecting the spectral pattern of elementary excitations of superfluid helium as suggested in Fig. 5.3. Under the assumption of a non-superfluid helium solvation layer the PW may be dominated by excitations of the non-superfluid helium solvation layer which is evidenced by rather sharp spectral features already within the spectral section of the phonon gap of superfluid helium. It is rather surprising that the solvation layer of glyoxal evidenced by increased moments of inertia remains consealed within the PW.

In view of all the helium induced spectral features reported for the ZPL at the electronic band origin of tetracene in helium droplets an empirical understanding of microsolvation does not suffice for an explanation. The droplet size dependence exhibiting a turn-around in the solvent shift, the fine structure in the beta line which was identically resolved for pentacene, and, last but not least, the changing line shape upon switching from subcritical to supercritical expansion conditions, do not fit to empirical explanations of helium induced spectral features in electronic spectra.

### 5.3.3  Phthalocyanine in Superfluid Helium Droplets

Phthalocyanine belongs also to the first samples of organic molecules which were investigated by means of fluorescence excitation spectroscopy in superfluid helium nanodroplets [32]. At the electronic band origin which undergoes a helium induced shift of about $42 \, \mathrm{cm}^{-1}$ to the red, a sensitivity of the singly peaked ZPL on the droplet size distribution was immediately recognized. The accompanying PW peaks about $3.8 \, \mathrm{cm}^{-1}$ to the blue from the ZPL which is within the range of the phonon gap of superfluid helium. Within a spectral section of roughly $3 \, \mathrm{cm}^{-1}$, numerous side maxima are grouped around this most intense center peak of the PW. This spectral substructure indicates a quantized energy level structure as expected for a rather rigid helium solvation layer. Moreover, an investigation of the line shape of the PW at the electronic band origin, its missing response on variation of the droplet size distribution, and corresponding dispersed emission spectra provide strong evidence for the presence of a helium solvation layer and, thus, provide further insight to microsolvation in superfluid helium. Much of the information to be discussed could only be obtained under appropriate spectral resolution, whereby saturation broadening had to be avoided.

The spectral shape of the ZPL at the electronic band origin is asymmetric with a steep rise at the red edge and a tail extending to the blue towards the gas phase resonance frequency [32–34]. Under subcritical expansion conditions in the droplet source, the asymmetry of the ZPL expressed mainly by the spectral width of the tail to the blue decreases with increasing average droplet size and the peak position of the ZPL shifts to the red edge as shown in Fig. 5.8 for temperatures from 15 to 10 K. This behavior reminds of the investigation of electronic transition frequencies under the influence of a finite sized polarizable environment which could be simulated by the so-called excluded volume model [35]. Adapted to the effective size distribution of singly doped helium droplets, the asymmetric line shape of phthalocyanine in helium droplets and its development under variation of the droplet size distribution could be simulated quantitatively [34]. Moreover, the simulation procedure could be applied upside down to deduce the droplet size distribution from the line shape of the ZPL. In this model, the solvent induced shift of electronic transitions of a dopant species is the result of the solvent to solute dispersion interaction whose influence on the

**Fig. 5.8** Fluorescence excitation spectra of the ZPL at the electronic band origin of phthalocyanine in helium droplets recorded for increasing average droplet size (from bottom to top) as determined by the droplet source temperature indicated in each panel. Helium stagnation pressure was 20 bar. For further details see text. Adapted from [36]

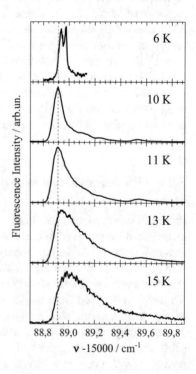

dopants energetics is of finite reach. Beyond a certain droplet size, the dopant species does not sense anymore the finite dimension of the droplet. For all droplets exceeding this limit, the dopant experiences bulk conditions. Beyond this limit inhomogeneous line broadening is expected to vanish even though there are still droplets with a broad size distribution.

Beyond the bulk limit, the line shape of the ZPL is expected to be dominated by the rotational fine structure of the solvated dopant species. In the case of phthalocyanine accessing the bulk-limit requires droplet sizes of at least $10^6$ helium atoms generated upon supercritical expansion conditions. Experimental results for the ZPL line shape reveal vanishing of the asymmetry upon approaching the transition from subcritical to supercritical expansion conditions in the helium droplet source [33, 34, 36]. Far beyond this limit a very sharp double peak structure is resolved (cf. Fig. 5.8 top panel) which was fitted by the envelope of the rotational band structure calculated for an oblate symmetric top rotor which is an approximative guess for phthalocyanine [37]. As obtained for numerous molecules in helium droplets, also for phthalocyanine the moments of inertia had to be increased by roughly a factor of three compared to the gas phase values. Unfortunately, the rotational fine structure resolved for phthalocyanine in helium droplets beyond the bulk limit is not line resolved [37]. Since the P, Q, and R-branch are merged, it is impossible to deduce the moments of inertia with high precision. Thus, the constants deduced in Ref. [37] represent a perfectly planar symmetric top rotor which is in contradiction not only to the asymmetric top rotor type of phthalocyanine but in addition to the non-planarity of a phthalocyanine helium solvation complex. Nevertheless, the consistency of the rotational band simulation with the experiment could be reached [37].

An important detail in the development of the line shape with the droplet size distribution is counterintuitive to the alleged consistency of helium droplets with the simple model of a polarizable environment. Besides the vanishing inhomogeneity observed for droplets generated under supercritical expansion conditions the ZPL experiences a tiny but clearly measurable shift to the blue [38] (cf. top panel Fig. 5.8) similar as discussed above for tetracene. Such a turn-around of the solvent shift is not expected from the excluded volume model [35]. In the case of tetracene the turnaround was also observed upon passing the limit from subcritical to supercritical expansion conditions in the droplet source so that the turn-around of phthalocyanine confirms fundamental differences in the properties of droplets generated under the two different expansion conditions. Instead, for glyoxal a blue shift was observed already within the subcritical regime of the droplet source. Nevertheless, the success of the excluded volume model applied to simulate the line shape of the ZPL of phthalocyanine in helium droplets generated under subcritical expansion conditions is quite convincing. So far, none of the spectroscopic signatures of phthalocyanine— neither in the ZPL nor in the PW – except of the turn-around of the solvent shift require an explanation based on features characteristic for a quantum fluid. This, however, should not be misunderstood as an indication against superfluidity of helium droplets. The effect of inhomogeneous line broadening as an expression of the droplet size distribution is expected to be ubiquitous throughout an electronic spectrum. Certainly, its strength depends on the van der Waals interaction which is not only dependent on the dopant species but may even be state-specific.

In contrast to the electronic band origin, the line shape at the ZPL of vibronic transitions of phthalocyanine in helium droplets was perfectly Lorentzian as shown in Fig. 5.9 [39]. For those transitions recorded in droplets with average size of 20,000 helium atoms, nothing reminds of the asymmetry due to inhomogeneous line broadening. The corresponding line width varies from peak to peak and can be transformed into the life time of the excited state. Obviously, for the ZPL of vibronic transitions of

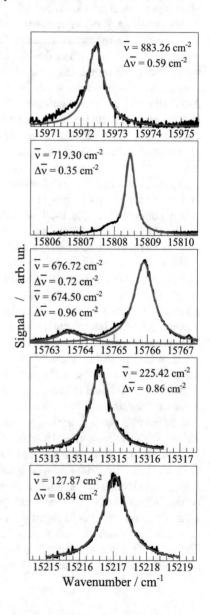

**Fig. 5.9** Fluorescence excitation spectra of ZPL at vibronic resonances of phthalocyanine in helium droplets ($\overline{N}$=20,000) (black line) fitted by Lorentzian line shapes (red line). Vibrational frequencies and Lorentzian line widths are added in $cm^{-1}$. Adapted from [39]

phthalocyanine the life time of the excited vibronic state is line shape determining and dominates over inhomogeneous line broadening. The average life time of vibronic levels of phthalocyanine in helium droplets was determined to about $15 \pm 8$ ps for vibrational modes up to $1000$ cm$^{-1}$. No correlation was found between life time and vibrational energy [39]. Beyond $1000$ cm$^{-1}$ the vibrational fine structure of phthalocyanine is obscured by contributions from a second electronically excited state which perturbs the line shapes. The radiative decay time of the first electronically excited state of phthalocyanine is in the order of 10 ns which is three orders of magnitude larger than the life times revealed by Lorentzian line shapes of the vibronically excited states in helium droplets as shown in Fig. 5.9. The missing correlation of the line widths or corresponding life times with vibrational energies speaks against direct dissipation of the vibrational energy into the helium droplet as the life time limiting process. More likely, the life time of the excited states is limited by internal vibrational redistribution (IVR) prior to energy dissipation into the helium droplet followed by radiative decay. Thus, the life times of vibrational states as deduced from the Lorentzian line widths reflect the mode specific IVR probabilities potentially modified by the helium environment. A convincing prove for bad coupling of high energy modes to the helium bath and the promotion of dissipation via IVR came from vibrationally excited hydrogen fluoride (HF) inside helium droplets [40]. Instead of recording a depletion as a result of evaporative cooling after dissipation of the almost $4000$ cm$^{-1}$ of rovibrational excitation the bolometer recorded an accretion from a rovibrationally excited HF molecule.

Valuable information on microsolvation of phthalocyanine in helium droplets and in particular on the presence and nature of the solvation layer came from dispersed emission spectra. As known already from matrix isolated molecules, radiative decay of electronically excited molecules in helium droplets originates only and exclusively from the ground level of the first electronically excited state independent of the initially excited level. Thus, radiative decay of vibronically excited dopant molecules is preceded by the dissipation of excitation energy in excess to the electronic band origin. Energy dissipation from the dopant species into the helium droplet reactivates evaporative cooling of the droplet which is the process depletion spectroscopy is based on. Coming back to the valuable information, for phthalocyanine in helium droplets dual emission was found. In addition to the expected emission spectrum coincident in its band origin with the excitation spectrum a second spectrum appeared [41]. The second emission spectrum was identical with respect to Franck–Condon factors (FCF) and vibrational frequencies to the first/expected whereas its helium induced red shift was increased by additional $10.8$ cm$^{-1}$. Its contribution to the integral emission intensity started with about 1.4% upon excitation at the electronic band origin, and increased monotonously with increasing excess excitation energy (cf. Fig. 5.10 from top to bottom). At about $1000$ cm$^{-1}$ of vibrational excess energy its contribution reached roughly 70% and for an excess energy of $15,000$ cm$^{-1}$ it contributed with 98% to the integral emission [39, 41]. Identical vibrational frequencies and FCF in the second spectrum reveals identical dopant species. The only difference is the solvent induced red shift of the electronic transition energy which has increased by roughly 25%. This type of dual emission is a strong indication for

**Fig. 5.10** Dispersed
emission spectra of
phthalocyanine in helium
droplets recorded upon
electronic excitation at the
electronic band origin (panel
(a)) and at vibronic
transitions as indicated in
panels (b) to (d). All spectra
are plotted twice in order to
emphasize the low intensity
vibronic transitions. Adapted
from [38] and [41]

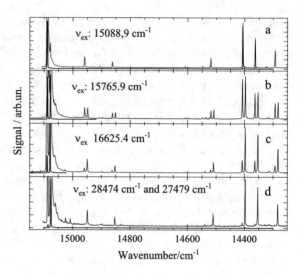

two configurations of a rigid helium solvation layer [41]. Prior to emission the decay
path including dissipation of vibrational excess excitation energy bifurcates, one of
which proceeds without and another with relaxation of the helium solvation layer of
the electronically excited dopant species. Thereby, the amount of excess excitation
energy drives the relaxation probability of the solvation layer. A four-level scheme
summarizes these experimental findings which is shown in Fig. 5.11e. For both electronic states of phthalocyanine involved in the observed electronic transition, the
helium solvation layer appears in two configurations. However, the global minimum
in the ground state becomes the local minimum in the electronically excited state
and vice versa. In both electronic states, relaxation into the corresponding global
minimum configuration of the helium solvation layer is possible. The relative intensity of the second emission spectrum images the relaxation probability into the global
minimum configuration of the helium solvation layer for the electronically excited
dopant. In contrast to the life time read from Lorentzian line widths discussed above,
the relaxation probability correlates with the excess excitation energy. Moreover, the
correlation between relaxation probability and excess excitation energy is indicative
for a barrier between the two helium solvation layer configurations in $S_1$.

A pump-probe spectrum shown in panel (c) of Fig. 5.11 reveals a similarly sharp
and asymmetric line shape for the excitation of the metastable solvation complex
(transition $|4> \rightarrow |3>$ in Fig. 5.11e) as for the stable configuration (transition $|1$
$> \rightarrow |2>$ in Fig. 5.11e) shown in panel (a) of Fig. 5.11. Panel (b) of Fig. 5.11
shows dispersed emission upon excitation with roughly 15,000 $cm^{-1}$ excess exci-
tation energy that has been the pump process for efficient populating of level $|3>$.
The integral intensity of the positive peak in panel (c) recorded under variation of
the delay time between pump and probe laser is shown in panel (d) of Fig. 5.11.
Since pump and probe laser operated in continuous wave mode a time delay was

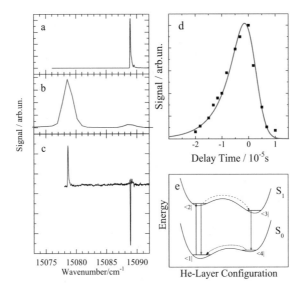

**Fig. 5.11** Left side: Fluorescence excitation spectrum (**a**), dispersed emission (**b**), and pump-probe spectrum (**c**) in the vicinity of the electronic band origin of phthalocyanine in helium droplets. Right side: pump-probe signal intensity (black dots) as function of the time delay between pump and probe laser and fit of exponential decay convoluted with Gaussian beam shape function (**d**). Four-level scheme explaining dual emission of phthalocyanine in helium droplets (**e**). For details see text. Adapted from [39]

accomplished by shifting the pump laser towards the droplet nozzle. After deconvolution of the gaussian overlap profile of pump and probe beam, a life time of 5.2 $\mu$s could be deduced [39]. Since the radiative life time of the electronically excited state in the order of 10 ns is negligible, the 5.2 $\mu$s reveal the life time of the metastable solvation complex of phthalocyanine in the electronic ground state. Compared to the pico-second time regime for thermalization of hot dopant species [2], the 5.2 $\mu$s life time is quite long and provides evidence for the barrier to the global minimum |1>. On the other hand, it is short enough to complete relaxation in the time between the pick-up and a spectroscopic investigation which explains the missing of corresponding signals in the fluorescence excitation spectrum. As mentioned above, the increase of the helium induced red shift of the electronic transition is surprisingly large. This difference relates to the difference in the electron density distribution of both electronic states of phthalocyanine as depicted schematically in Fig. 5.11e. The spectral response of dual emission is an experimental detail that reflects the charge density distributions and its change upon electronic excitation as sensitized by the helium environment. It serves as a bench mark for theoretical modeling of helium solvation. The observation of dual emission and the sharp peak signal of the metastable solvation complex as recorded in the pump probe experiment speaks for a rather rigid solvation complex most probably sandwich-like and, therefore, speaks also for a non-planar solvation complex. This is an important detail for the simulation of alleged rotational bands discussed in [37].

Additional evidence for the existence of two sandwich-like configurations of a solvation complexes for phthalocyanine in helium droplets came from path-integral-Monte-Carlo simulations (PIMC) [42]. Besides the global minimum configuration with a layer consisting of 24 helium atoms on each side of the planar dopant and both with almost perfect hexagonal structure, a metastable configuration was found with one of the two layers changed to a configuration rather commensurate to the corrugation of the phthalocyanine surface.

Similar multiplet splittings were reported for dispersed emission spectra of Mg-phthalocyanine, and a 1:1 cluster of phthalocyanine and argon in helium droplets [38]. Thereby, for Mg-phthalocyanine the second emission spectrum was shifted by $12.5\,\mathrm{cm}^{-1}$ to the red. Its relative intensity was 90% already without excess excitation energy and approached a value of 99% for an excess excitation energy of only $250\,\mathrm{cm}^{-1}$. According to a triplet splitting in the dispersed emission spectrum the Ar cluster of phthalocyanine exhibits even two additional metastable configurations. Upon increasing the excitation energy, the variation of the intensity profile within the triplet revealed a cascade of two consecutive relaxation steps. In the case of AlCl-phthalocyanine, both the excitation spectrum and the dispersed emission showed doublet features which, however, did not exhibit relaxation among each other [43, 44]. It speaks for an insurmountable barrier between the two systems. As the icing on the cake of the experimental evidence for the relaxation model of a solvation complex, a peak probability for relaxation was found upon excitation with less than $10\,\mathrm{cm}^{-1}$ excess excitation energy, but now injected directly into the mutually relaxing object via excitation of the PW [21]. In purely classical terms, one can conclude that shaking directly on the solvation layer drives the relaxation particularly efficiently even though the excess excitation energy is about a factor of 40 smaller than the smallest vibrational excess energy pumped into the dopant.

According to the empirical interpretation, the doublet in the emission of phthalo-cyanine is related to the doublet observed in the excitation of tetracene discussed above. Both are explained by the presence of different configurations of a helium solvation layer with the difference that tetracene does not allow for relaxation among the configuration. As seen for tetracene for the first time, multiplet splitting at a ZPL was observed for numerous other dopant species. These multiplet structures are generally understood as the presence of configurational variants of a helium solva-tion complex. The formation of a solvation complex accompanies helium solvation, however, with dopant specific expression concerning the number of configurational variants and the relaxation dynamics among them. Thereby, relaxation is certainly a response on the change of the electron density distribution upon electronic excitation.

## 5.3.4 Porphin in Superfluid Helium Droplets

Structurally related to phthalocyanines are porphyrin derivatives. Porphin was also among the first dopant species investigated by electronic spectroscopy in helium droplets [11, 32]. The ZPL at the electronic band origin of porphin in helium droplets exhibits a triplet consisting of a leading intense peak followed to the blue by two tiny peaks at an excess energy of 0.4 cm$^{-1}$ and 0.7 cm$^{-1}$, respectively [25]. The intensity profile within the triplet from red to blue was about 10:2:1. The PW is right in the middle of the phonon gap of superfluid helium and exhibits a fine structure qualitatively similar to that of phthalocyanine [11, 25]. Most probably, the ZPL triplet reveals three different configurations of a porphin helium solvation complex. The low oscillator strength and low fluorescence quantum yield of porphin prohibited an investigation of relaxation dynamics by means of dispersed emission spectra.

Similar as discussed for phthalocyanine, the line shape of the ZPL at the electronic origin of porphin is asymmetric, varies with the droplet size distribution, and, therefore, suffers inhomogeneous broadening [45]. However, the steep edge is at the blue side and a tail expands to the red which is inverted compared to the asymmetry found for phthalocyanine. The excluded volume model applied to log-normal droplet size distributions reveals a sharp edge marking the bulk limit of a solvent shift and a tail pointing towards the resonance of the isolated molecule. Accordingly, the inverted asymmetry is indicative for a helium induced solvent shift to the blue. And in fact, applying the excluded volume model with a blue shift the line shape and its dependence on the droplet size distribution can be simulated [45]. As closed shell organic molecule porphin is heliophilic in both electronic states. Whether the solvent shift is to the blue or to the red depends on the difference of the stabilization energy of the two electronic states involved in the transition. Thus, for heliophilic dopant species both a red or a blue shift is possible. What concerns porphin, the option of a helium induced blue shift as suggested by the inverted asymmetry in the line shape is refuted by the gas phase spectra from two independent sources revealing consistently a helium induced red shift of about 8 cm$^{-1}$ [46, 47].

Another remarkable feature in the droplet size dependence of the line shape was a sharp peak about 0.04 cm$^{-1}$ in width right on top of the inhomogeneous broadened ZPL [45]. It was observed for average droplet sizes below 10,000 helium atoms. Upon increasing of the droplets size, this peak remained fixed in frequency while the broad and asymmetric part of the ZPL shifted to the blue. Only the peak intensity did vary with the droplet size. This behavior reminds of the response of the fine structure resolved at the beta line of tetracene and on its response on the droplet size (cf. Sect. 5.3.2).

Porphin exhibits a helium induced solvation shift to the red without any doubt. Even though, the asymmetry in the line shape of the ZPL at the electronic band origin and its response to the variation of the droplet size distribution fits to the excluded volume model for a dopant system with a helium induced blue shift. A mistaken assignment of the spectrum to porphin instead to chlorin which is a side-product of the synthesis of porphyrins can safely be excluded. A turnaround of the solvent shift to the blue as discussed above for tetracene and phthalocyanine might be considered as the reason for an inverted asymmetric line shape. However, in this case the turn-around effect of porphin needs to surpass that for phthalocyanine or tetracene by orders of magnitude. Moreover, in contrast to tetracene and phthalocyanine the turnaround of the solvent shift should occur already within the subcritical expansion conditions and far from the transition to supercritical conditions in the droplet source. Thus, the blue shift observed for porphin is qualitatively rather similar to the blue shift reported for the rotationally resolved ZPL at the electronic band origin of glyoxal in helium droplets. Both are recorded within the subcritical expansion regime far from the transition to supercritical conditions at the helium droplet source.

### 5.3.5 Summary

Electronic spectra of glyoxal, tetracene and related PAH-compounds as well as phthalocyanine and porphin provide strong evidence for the presence of a helium solvation layer rigidly bound to the dopant species. Hence, spectroscopy of molecules doped into helium droplets deals with solvation complexes rather than the bare molecule. Solvation complexes differ from the bare molecule, most evident, in the moments of inertia. Depending on the size and shape of the molecules, several configurations of the solvation complex are to be expected which distinguish in the electronic transition energy as expressed by multiplet splitting at the ZPL and, case dependent, by a relaxation dynamic initiated via electronic excitation. In contrast, vibrational degrees of freedom remain almost unchanged by helium solvation and are insensitive to configurational variants of the helium solvation layer. In addition to the helium solvation layer, the entire droplet body has an influence on the electronic transition frequency of the dopant. In contrast to the solvation layer, which needs to be treated as a quantized multi-particle system of helium atoms attached to the dopant, the droplet body can be treated as a quasi-continuous and polarizable environment. The shift of electronic transition of a dopant depends on the thickness of the polarizable environment. Thus, a distribution of droplet sizes becomes effective in inhomogeneous line broadening. Inhomogeneous line broadening as an explanation for the asymmetry in the line shape of the ZPL works perfectly for phthalocyanine. However, it is put into question by the inverted asymmetry observed for porphin. While the droplet size dependence of the line shape is unquestionable, the excluded volume model does not suffice to explain line shapes in general. Most evident, the phenomenon of a turn-around in the solvent shift does not fit into the excluded volume model. While tetracene an phthalocyanine reveal a correlation between turn around and transition from subcritical to supercritical expansion conditions glyoxal and porphin exhibit turn around within the subcritical regime. Thus, the presence of a dopant helium solvation complex appears to be reasonable. However, several of the helium induced spectroscopic features are not understood.

## 5.3.6  Low Energy Torsional and Bending Modes
### in Electronic Spectra of Molecules in Helium Droplets

Dissipation of energy is a key feature accompanying solvation of molecules in helium droplets. Besides the practical benefits of depletion spectroscopy, it is the mechanism that cools molecules in helium droplets to a temperature of 0.38 K within picoseconds. As discussed above, the efficiency of energy dissipation is based on the coupling of the dopant's vibration to the helium droplet which decreases with increasing vibrational energy as exemplified by vibrationally excited HF molecules inside helium droplet [40]. As outlined above for phthalocyanine (cf. Sect. 5.3.3), the efficiency of energy dissipation may be mediated by IVR into low energy modes prior to energy transfer to the helium droplet. Ligands such as methyl-, buthyl-, ethyl-, phenyl-, cyano-moieties providing low energy torsional or bending modes or simply heavier atoms reducing vibrational frequencies may serve as acceptor modes for IVR and, thereby, promote dissipation of energy into the helium droplet. On the other hand, low temperature conditions in helium droplets are expected to be favorable for resolving vibrational progressions of such low energy modes. At 0.38 K these progressions are free of hot bands and reveal configurational changes accompanying electronic excitation. Electronic spectra of substituted derivatives of phthalocyanine and porphyrin, pyrromethene dyes, and anthracene have been recorded in helium droplets in order to study the influence of the substituents on the dopant to helium interaction.

As explained in Sect. 5.2. electronic spectra recorded as fluorescence excitation or dispersed emission of molecules in helium droplets reveal vibrational frequencies of electronically excited states and of the electronic ground state, respectively. The helium induced shift of vibrational frequencies is on the order of $\pm 1\%$. A more remarkable effect is line broadening as reported for low energy modes roughly below $300\,\mathrm{cm}^{-1}$ [32]. Low vibrational frequency correlates with large vibrational amplitude which may suffer damping by a solvent. Moreover, the presence of low energy modes might promote IVR and, thus, promotes energy dissipation from high energy modes into the helium droplet. In rather classical terms, the substituents can be seen as kind of antenna supporting the communication between dopant and helium droplet. This was investigated for several substituted derivatives by means of electronic spectroscopy.

### 5.3.6.1  Low Energy Torsional and Bending Modes of Phthalocyanine
#### and Porphin Derivatives in Helium Droplets

Among the phthalocyanine derivatives the electronic spectrum of 2,9,16,23-tetra-tert-butylphthalocyanine (TTBPc) has been examined [48]. At each of the four six membered rings a tert-buthyl-group substitutes one of the hydrogen atoms in the named position. Besides a vibrational fine structure attributable to the phthalocyanine core unit no additional low frequency progressions of torsional or bending modes of the tert-butyl-moieties appeared in the fluorescence excitation spectrum.

However, the ZPL at the electronic band origin appeared split into roughly 20 peaks within the first 3 cm$^{-1}$ each as narrow as 0.1 cm$^{-1}$. This large number reveals configurational variants not only with respect to the helium solvation layer. More likely, this compound by itself exhibits stereo isomers which at a temperature of only 0.38 K are spectrally well separated. However, no low energy progressions due to the substituents are observed in the fluorescence excitation spectrum. If at all, the influence of low energy modes of the substituents is a secondary effect observable in vibronic transitions involving normal modes of the phthalocyanine core unit. Below 400 cm$^{-1}$ the low energy modes of the core unit are significantly attenuated compared to the unsubstituted molecule. Moreover, the line widths of all vibronic resonances exceed that of the electronic band origin by at least an order of magnitude which was not the case for unsubstituted phthalocyanine. Both effects can be rationalized by IVR promoted by the presence of low energy modes of the substituents which is followed by energy dissipation into the helium droplet. The missing of progression of these low energy modes reveals almost identical configuration of the substituents in both electronic states. To our best knowledge, corresponding spectra from the gas phase are not available for this phthalocyanine derivative.

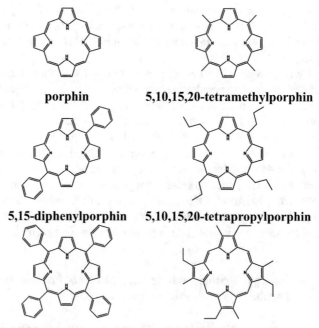

porphin               5,10,15,20-tetramethylporphin

5,15-diphenylporphin     5,10,15,20-tetrapropylporphin

5,10,15,20-tetraphenylporphin    2,7,12,17-tetraethyl-
3,8,13,18-tetramethylporphin

A larger variety of substituted derivatives has been investigated for porphin [25, 48]. These derivatives are specified by antenna like substituents of alkyl type such as methyl, ethyl, buthyl, or combinations of these moieties situated either at methine or pyrrole sites in the periphery of the planar porphin body. Among them are 5,15-diphenylporphin, 5,10,15,20-tetraphenylporphin, 5,10,15,20-tetramethylporphin, 5,10,15,20-tetrapropylporphin, and 2,7,12,17-tetraethyl-3,8,13,18-tetramethylporphin. All of these derivatives carry two or four substituents situated in such a way that inversion symmetry is maintained. To cut the story short, none of these derivatives show progressions of low energy modes as expected for torsional or bending modes of the corresponding substituents. Therefore, electronic excitation apparently maintains the steric arrangement of the substituents. Line broadening of high frequency vibronic transitions as a signature of promotion of IVR was rather insignificant [25]. In all cases, saturation broadening easily obtained by peak intensities of pulsed dye lasers was a serious side effect and gave an impression of broadened spectra as if they were dominated by the envelop of unresolved low energy progressions [25]. However, by using the moderate photon flux of a cw-dye laser the line widths shrank to an order of $0.1 \, cm^{-1}$ and all the spectrally broad signals vanished entirely. Thus, the broad spectral features can safely be assigned to PW [25]. Under these conditions, rich multiplet splitting of the ZPL at the electronic band origin was resolved. As in the case of the phthalocyanine derivative, the multiplet reveals not only variants of the solvation complex but in addition stereoisomers of the porphin derivative.

One of the porphin derivatives, namely tetraphenylporphin is an interesting example to demonstrate how saturation broadening misleads the interpretation of electronic spectra. Upon saturation broadening the electronic spectrum obtained after doping with a commercial sample of this molecules was dominated by PW which within the first $100 \, cm^{-1}$ exhibits a spectral substructure consisting of a kind of triple-peak feature which then merges into a broad and constantly decreasing signal to the blue [49] as depicted by the upper trace in Fig. 5.12. This feature repeats for two vibronic transitions within the spectral range shown in Fig. 5.12. Upon reducing the photon flux, most of the spectrally broad signal vanishes (cf. lower trace in Fig. 5.12). Now the signal peaks at the electronic band origin and at the two vibronic transitions. At a closer look (cf. inlay in Fig. 5.12), the electronic band origin reveals a ZPL with a single intense peak only $0.05 \, cm^{-1}$ in width [25, 48]. Furthermore, after a gap of about $1 \, cm^{-1}$, a series of similarly sharp peaks follows. Upon increasing the laser intensity, the unstructured PW grows in starting with the same $1 \, cm^{-1}$ gap to the leading peak. As discussed in Refs. [25, 48], the spectroscopic details speak for an assignment to a PW except of the leading sharp and intense peak. Furthermore, there are no contributions of low energy progressions of the four phenyl moieties which speaks for a rigid derivative with almost identical structure in both electronic states. This contrasts to the interpretation of the saturated spectrum as the signature of a floppy molecule [49]. Above all the spectra shown in Fig. 5.12 do not originate from tetraphenylporphin but instead from the corresponding chlorin derivative [25, 48] wel known as side product of the synthesis of porphin derivatives.

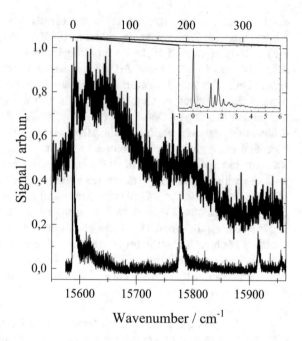

**Fig. 5.12** Fluorescence excitation spectrum of tetraphenylchlorin in helium droplets ($\overline{N}$=20,000) recorded under saturation (upper trace) and with reduced saturation (lower trace). The inset shows the ZPL at the electronic band origin without saturation. (Adapted from [25, 48], and [49])

### 5.3.6.2 Low Energy Torsional and Bending Modes of Anthracene Derivatives in Helium Droplets

Various derivatives of anthracene have been investigated by means of electronic spectroscopy in helium droplets. Among them are 9-phenylanthracene, 9-cyanoanthracene, 9-chloroanthracene, 9,10-dichloroanthracene, and three methylated derivatives, namely, 9-methylanthracene, 1-methylanthracene, and 2-methylanthracene [50, 51, 52, 53]. For some of these derivatives corresponding gas phase data have been published [54–63]. Within this series there are four singly substituted derivatives carrying substituents such as a chloro-, cyano-, methyl-, or phenyl-moiety in all cases at the 9-position. Moreover, there are tree derivatives singly substituted with a methyl group at position 9, 1, or 2. Among the singly methylated derivatives the electronic spectra of 1-methylanthracene and 9-methylanthracene do not show low energy torsional or bending progressions neither in the gas phase nor in helium droplets [50] (cf. Fig. 5.13b, c). Obviously, the steric configuration of the methyl substituent is identical in both electronic states. These two anthracene derivatives show rather similar vibronic structure as recorded for anthracene shown in panel (a) of Fig. 5.13. Similar observations were made for 9-cyanoanthracene, 9-chloroanthracene, and 9,10-dichloroanthracene.

In contrast, extended low energy progressions were observed for 2-methylanthracene in the gas phase and in helium droplets [50] (cf. Fig. 5.13d). At low temperatures these progressions reveal a twist of the methyl-substituent upon

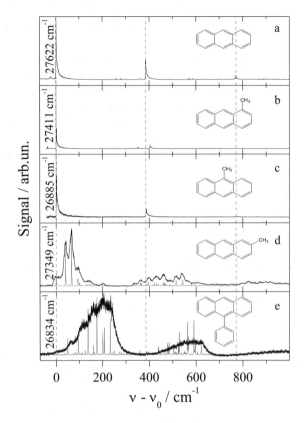

**Fig. 5.13** Fluorescence excitation spectra of anthracene derivatives in helium droplets ($\overline{N} =$ 20,000): **a** anthracene, **b** 1-methylanthracene, **c** 9-methylanthracene, **d** 2-methylanthracene, and **e** 9-phenylanthracene. The wavenumber scale is related to the corresponding origin given in each panel. In **d** and **e**, molecular beam spectra are added in red whose electronic band origin was shifted to coincide with the helium droplet experiment. Vertical lines mark the origin and two prominent vibronic transitions of non-substituted anthracene. Adapted from [50–52]

electronic excitation. Remarkably, however, in helium droplets significant line broadening throughout the low energy progressions for 2-methylanthracene was observed. A convolution of the gas phase spectrum with a line broadening function revealed almost perfect coincidence with the helium droplet spectrum. Such a coincidence proves for identical torsional frequencies in both experiments. Identical frequencies reveal identical torsional mass and, thus, no helium atoms attached to the methyl substituent. Moreover, the best fit was obtained for Lorentzian type line broadening an indication for helium induced reduction of the life time of the torsional mode. The low frequency torsion exhibits rather large amplitude motion which accomplishes efficient coupling to the helium environment. At this point it needs to be mentioned that line broadening was observed throughout the entire progression. We will come back to this issue below in the discussion of pyrromethene dye molecules. Similar line broadening was recorded for 9-phenylanthracene shown in panel (e) of Fig. 5.13 which was explained accordingly.

Slight modifications of the intensity pattern were observed in the helium droplet spectra. The process of dissipation of excess excitation energy into the helium droplet may become influential on the decay path of the excited dopant. As a consequence, the fluorescence quantum yield and, thus, the intensity recorded by means of fluorescence excitation might differ in helium droplets compared to gas phase [50, 53]. In the best case, dark states exhibiting a non-radiating relaxation for the isolated molecule can be bypassed to a fluorescent decay path. In the worst case, the bypassing may extinguish radiative decay.

### 5.3.6.3   Low Energy Torsional and Bending Modes of Pyrromethene Dyes in Helium Droplets

Dominant torsional and bending modes of aryl-, alkyl-, phenyl- and cyano-substituents in electronic spectra are known for borondipyrromethene dye molecules. Supersonic jet spectra revealed spectrally well resolved and extended low frequency progressions [64, 65]. Even the non-substituted 4-boro-3a,4a-diaza-s-indacene or borondipyrromethene (BDP) shows a progression of a flopping mode of the $BF_2$ unit. According to extended progressions of low energy modes recorded from cold samples, electronic excitation of these derivatives is accompanied by significant rearrangement of the steric configuration of the substituents. In addition to BDP the following derivatives have been investigated in the gas phase as well as in helium droplets by means of electronic spectroscopy: 8-phenylpyrromethene-difluoroborat (8-PhPM), 1,3,5,7,8-pentamethylpyrromethene-difluoroborat (PM546), 1,3,5,7,8-pentamethyl-2,6-diethylpyrromethene-difluoroborat     (PM567),     1,2,3,5,6,7-hexamethyl-8-cyanopyrromethene-difluoroborat (PM650) [53, 64, 65]. Gas phase spectra were recorded from a supersonic molecular jet which allowed to even resolve the rotational band contour at the electronic band origin. Thus, the rotational temperature could be determined to about 10 K. Upon doping into superfluid helium droplets, molecules are cooled down to 0.38 K for all internal degrees of freedom. Thus, hot bands are eliminated entirely and electronic spectra should exhibit reduced spectral density compared to the gas phase. This, however, was only the case for BDP where the band of the $BF_2$ flopping mode was missing. All the other derivatives undergo a counterintuitive development from spectrally resolved progressions of low energy modes in the gas phase to a kind of envelope of these progressions recorded in helium droplets as shown in the left column of Fig. 5.14 for three of the substituted pyrromethene derivatives. Similar as discussed above for 9-phenylanthracene and 2-methylanthracene, the helium droplet spectra resemble a convolution of the gas phase spectra with a line broadening function. As in the case of the anthracene derivatives, the torsional and bending modes show identical frequencies in both experiments. Thus, helium solvation maintains the mass involved in torsional or bending motion.

However, at a closer look depicted in the right column of panels in Fig. 5.14, one can recognize that for all of the pyrromethene derivatives the ZPL at the electronic

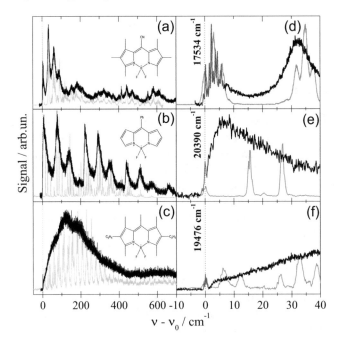

**Fig. 5.14** Fluorescence excitation spectra of three pyrromethene dye molecules (PM650 (**a**) (**d**), 8-PhPM (**b**) (**e**), and PM567 (**c**) (**f**)) recorded in helium droplets (black line) compared with corresponding spectra recorded in a molecular beam (grey line). Right column zooms into the electronic band origin for which a fine structure was resolved in helium droplets (red section). Adapted from [44]

band origin shows a spectrally well resolved multiplet with line widths in the order of 0.1 cm$^{-1}$. Only this particular ZPL at the electronic band origin is excluded from line broadening. In contrast to vibronic excitations the electronic band origin does not carry excess excitation energy and, therefore, no energy dissipation takes place. This observation is a strong indication for line broadening due to reduced excited state life time accomplished by highly efficient energy dissipation.

Coming back to 9-phenylanthracene and 2-methylanthracene, line broadening was also observed at the electronic band origin. Therefore, energy dissipation cannot be responsible for line broadening in the spectra of anthracene derivatives. Either intramolecular processes such as internal conversion (IC) or intersystem crossing (ISC) are involved and significantly promoted by the helium environment or the line broadening is an expression of severe perturbation of the closer helium environment induced by electronic excitation of the anthracene derivatives. For the two anthracene derivatives, the concomitant change of the electron density distribution enforces the substituent, either a methyl or a phenyl moiety, to undergo a twist of 60° and 30°, respectively [50, 66] as deduced from the Franck–Condon-pattern of the corresponding torsional progression. It is rather unlikely that such a change of the electron density distribution will not be sensitized in addition by the helium environment.

### 5.3.6.4 Summary

In summary, the substituted derivatives of phthalocyanine and porphin, anthracene, and pyrromethene have shown that helium solvation is of minor influence on the substituents and their intramolecular dynamics. Torsional and bending frequencies are almost identical as in the gas phase. Moreover, the spectral width of corresponding progressions which reflect the change of the equilibrium configuration of the steric arrangement of the substituents upon electronic excitation was also found unaltered in helium droplets. The major impact of the helium environment was line broadening. Life time broadening is certainly involved and induced by highly efficient energy dissipation particularly from low energy and large amplitude torsional and bending modes into the helium droplet. However, a perturbation of the entire solvation complex, namely, the helium solvation layer covering the dopant cannot be ruled out. To our best knowledge, there is no example where spectral resolution of low energy and large amplitude vibrational, torsional or bending modes has made profit from low temperatures in helium droplets with respect to spectral resolution.

## 5.4 Van Der Waals Clusters Generated in Helium Droplets

One of the exciting opportunities offered by helium droplets is the formation of clusters with well-defined stoichiometry. Although the droplet size as well as the doping process are subject to statistical distributions, individual spectroscopic signals can be assigned to clusters well defined with respect to the stoichiometry. A safe assignment of the cluster stoichiometry is particularly warranted for small clusters of less than say ten individual units which is the relevant range of sizes discussed in the following. For this range of size the number of configurational variants is an issue. The doping is accomplished via pick-up on the flight of the droplet beam through a compartment called pick-up unit providing the dopant species at a tunable particle density. Pick up via consecutive collisions of a droplet with a number of k dopant particles obeys Poisson statistics and, thus, depends on the droplet size, the dopant particle density, and the length of the path of the droplets through the pick-up unit. The underlying cluster stoichiometry can be deduced for each individual cluster signal by recording its intensity under variation of the particle density in the pick-up unit [67]. The number k of dopant particles is the fitted parameter to obtain an overlap of the experimental intensity profile with a Poisson distribution. In contrast to a standard mass spectrometer that measures the mass to charge ratio after ionizing the clusters, the procedure in helium droplets is entirely non-destructive. Thus, an unequivocal assignment of the nascent cluster size is obtained. Finally, the clusters generated in helium droplets are cooled down to the droplet temperature of 0.38(1)

K for all internal degrees of freedom whereby from other sources providing clusters in the gas phase internal temperatures up to the boiling point of the clusters are obtained. The sub 1 K temperature eliminates hot bands almost entirely. Moreover, metastable cluster configurations may be stabilized. From the very beginning, it was not surprising that electronic spectra of clusters generated in helium droplets show large numbers of cluster signals arising from cluster configurations that are stabilized only in helium droplets [17] as will be discussed in the following. Thereby, electronic spectroscopy is selective not only for different stoichiometry but in addition for configurational variants within a single cluster stoichiometry.

### 5.4.1  Van Der Waals Clusters of Tetracene with Argon Atoms

The formation of clusters inside a helium droplet by multiple doping via consecutive pick-up has first been demonstrated by the group of J. Peter Toennies via mass spectrometric detection [67]. Not much later, hetero clusters of tetracene and argon have been investigated by means of electronic spectroscopy [17]. In contrast to previous investigations of tetracene-argon clusters generated in a seeded beam expansion [68–72], the electronic spectra from clusters in helium droplets reveal spectrally much sharper lines. Vibrational and rotational hot bands as present in a supersonic jet experiment are eliminated almost entirely in helium droplets. For clusters in helium droplets consisting of tetracene and a single argon atom the most intense signal exhibits the largest argon induced red shift. Two additional signals were identified as configurational variants of the same cluster size exhibiting reduced red shift and intensity. The stoichiometry was determined unequivocally as described above. The assignment of the signals to different conformers was substantiated by pump-probe experiments. Further insight into the individual structure of these three clusters was not revealed by the experimental data.

In Ref. [17] a number of 21 signals of clusters consisting of a single tetracene molecule and up to almost five argon atoms have been recorded within a spectral range of 150 cm$^{-1}$ to the red of the electronic band origin of tetracene. The assignment of stoichiometric and of isomeric variants for clusters with more than one argon atom was based on empirical rules of additivity of argon induced red shifts as deduced from extensive gas phase studies of van der Waals clusters consisting of a single chromophore and a well-defined number of rare gas atoms [71, 72]. For these clusters, helium droplets provide a unique option for further analyzing the cluster structure which allows to validate these empirical rules. The option for cluster analysis goes beyond an assignment of the stoichiometry and is based on variation of the pick-up sequence of the chromophore and the rare gas atoms. Upon initial doping with rare gas atoms, the chromophore approaches a rigid rare gas cluster inside the helium droplet and most probably attaches to the cluster surface. The favored cluster configuration is single-sided in the attachment of rare gas with respect to the chromophore. Upon changing the pick-up sequence, the chromophore inside the helium droplet attracts one rare gas atom after the other. In case of a planar chromophore, the cluster configuration will be a double-sided attachment of the rare gas and single-sided configurations are less likely. Comparison of peak intensities for both pick-up sequences at identical conditions in the doping cell allows to identify single-sided and double-sided rare gas attachment to the chromophore.

Applying those methods in order to analyze the configuration of the tetracene-argon clusters in helium droplets, the limitations of the empirical additivity rule became obvious [73]. The advantage in terms of structural analysis as accomplished by the alternation of the pickup sequence is slightly relativized by the authors for good reason. In helium droplets the process of cluster formation proceeds under permanent dissipation of energy into the helium environment. Therefore, the path from separate components to the cluster is different from the path of cluster growth in a seeded beam expansion and so are the resulting cluster configurations.

The spectral shift of electronic resonances among configurational variants of a certain cluster size is on the order of frequencies expected for van der Waals modes. Therefore, the number of isomeric configurations does not necessarily correlate with the number of signals assigned to a particular cluster stoichiometry. For the 1:1 cluster of tetracene-argon the identification of band origins was accomplished by pump-probe experiments [17]. An alternative rather simple method to identify electronic band origins is provided by dispersed emission spectra. It is another uniqueness offered by helium droplets as host. In contrast to electronic band origins low energy modes carry excess excitation energy that dissipates prior to radiative decay which than occurs red shifted. Thus, a coincidence of the band origin in the dispersed emission spectrum with the excitation frequency identifies the corresponding resonance in excitation as electronic band origin which can be counted as one of the isomeric variants of the corresponding cluster size. Instead, a red shift in the electronic band origin of the dispersed emission with respect to the excitation frequency speaks against an assignment to a band origin. As a kind of drawback, one needs to keep in mind that besides dissipation of excess excitation energy a red shifted emission may reveal the relaxation of a particular helium solvation complex as observed for phthalocyanine in helium droplets (cf. Fig. 5.10). While emission coincident with the excitation proves for an electronic band origin of a particular cluster configuration a red shifted emission does not necessarily speak against it.

## 5.4.2   Van Der Waals Clusters of Anthracene with Argon Atoms

Among numerous other examples of clusters, those of a single anthracene molecule and a variable number of argon atoms have been investigated in helium droplets as well as in the gas phase. The initial motivation for the investigation of clusters consisting of a single anthracene molecule and increasing numbers of argon atoms was the interest in the transition from the isolated anthracene molecule to the fully solvated chromophore. Thereby, shell structures and specific interactions between chromophore and environment were expected [74, 75]. Stoichiometric analysis of such clusters generated in a seeded beam expansion proposed the presence of isomeric variants for all cluster sizes [57, 76–79]. A major breakthrough in the analysis of cluster configurations could be accomplished by rotationally resolved spectroscopy. Therefore, a smart approach was chosen by means of rotational coherence spectroscopy which allows to observe molecular rotation in the time domain. Thus, the limitations set by the spectral resolution in the frequency domain are eliminated by the time domain. As a result, for the clusters of anthracene with up to three argon atoms only a single configuration could be identified [80]. A final statement including clusters with up to 6 argon atoms came from an extensive study where clusters were interrogated by a broad set of experimental diagnostics based on mass-selective, fragmentation-free, two-color resonant two-photon ionization and laser induced fluorescence as well as a theoretical modeling of the ionization energy [81]. According to this study which refers also to almost all of the previous publications the clusters of anthracene with up to at least 6 argon atoms generated in a seeded supersonic beam expansion exhibit only a single configuration. The corresponding electronic band origins extrapolated to up to 8 argon atoms are depicted in Fig. 5.15 as black squares.

The investigation of anthracene argon clusters in superfluid helium droplets contrasts to the gas phase experiment [82]. By means of dispersed emission spectra, within the spectral range of 160 cm$^{-1}$ to the red of the electronic band origin of bare anthracene a number of 13 electronic band origins have been identified. By means of Poisson intensity profiles the band origins were assigned stoichiometrically. In Fig. 5.15, all red symbols mark electronic band origins sorted according to the cluster size given by the number of argon atoms. In helium droplets, a single argon atom finds two different sites whereas a maximum of five configurations were identified for two argon atoms attached to anthracene. Moreover, clusters with five and more argon atoms did not show spectrally resolved lines. Instead, spectrally broad signal was recorded. Therefore, these larger clusters could not be recorded

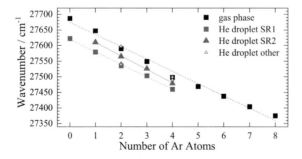

**Fig. 5.15** Frequency position of electronic band origin of anthracene and its clusters with 1 up to 8 argon atoms as revealed from gas phase experiments (black squares) and from a helium droplet experiment (red). Highlighted are two series from helium droplets as full red squares and full red triangles which develop with similar gradient as gas phase clusters. Adapted from Ref. [82]

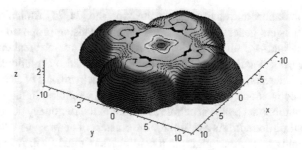

**Fig. 5.16** 3-dimensional plot of the minimum surface of the interaction potential of phthalocyanine in the electronic ground state and argon with phthalocyanine in the x–y plane. Axes are scaled in Å. The potential energy scales from pink/weak to red/strong. It is calculated from pair potentials parameters as given in Ref. [42]

individually and, thus, further stoichiometric assignment was not possible. By alternation of the pick-up sequence single-sided and double-sided cluster configurations could be assigned [82]. Clusters with two or four argon atoms exhibited single-sided and double-sided configurations while the cluster with three argon atoms appeared only double sided.

It is striking that the stoichiometric gradient of the electronic band origin for clusters exhibiting the larges red shift in helium droplets—marked by red squares and labeled as SR1 in Fig. 5.16.—is very similar to that reported for gas phase conditions—marked by black squares. The gradients of linear extrapolations amount to $-39(2)$ cm (black dashed line) and $-40(1)$ cm (red dashed line), respectively. Moreover, the assignment to single-sided and double-sided configurations within SR1 confirmed the results deduced from the gas phase studies. In the latter case, the structural assignment required theoretical input in addition to the experimental data. The configurations identified from rotational coherence spectroscopy for clusters with up to three argon atoms [80] were also consistent with the configurations reported in Ref. [81] and those from helium droplets [82]. Among additional isomeric variants identified in helium droplets a second series—highlighted in Fig. 5.15 by red full triangles (labeled as SR2)—exhibits a similar stoichiometric gradient of $-43(1)$ cm (red full line) as in the gas phase. According to the analysis performed in Ref. [82] these clusters of SR2 exhibit identical configuration with respect to single or double-sided occupation as those labeled SR1. The argon induced red shift of the cluster with a single argon atom from SR1 is similar to that in the gas phase which speaks for identical cluster configurations. The red shift of the corresponding cluster from SR2 in helium droplets reveals significant shielding of the argon induced solvent shift which continues similarly for the larger clusters of SR2. The reduction of the red shift in SR2 compared to SR1 is similar for each cluster size. The reduced argon induced red shift in combination with a similar stoichiometric gradient might be indicative for cluster configurations similar to those of SR1, whereby one argon atom is located in a more distant position most probably shielded by the helium solvation layer. It is evident that the effect of shielding of a single argon atom on the argon induced red shift decreases with increasing cluster size and vice versa as can be recognized from the linear extrapolation for SR2.

The number of theoretically proposed isomeric variants of the clusters of anthracene with up to three argon atoms listed for gas phase conditions in Ref. [80] contrasts to the single isomer identified experimentally. In contrast to both, the large number of isomeric variants identified in helium droplets which exceeds what is proposed in Ref. [80] speaks for an involvement of helium atoms as part of the solvated clusters. A deeper insight into the internal structure of those clusters requires detailed information on the mass distribution as revealed by a rotationally resolved spectroscopy similar as presented in Chaps. 3 or 8.

Additional peculiarities of cluster formation in helium droplets are revealed by electronic spectra for clusters of anthracene with five and more argon atoms. In contrast to corresponding gas phase spectra with spectrally isolated peaks the signals of those larger clusters in helium droplets merge into a broad and unstructured feature which does not allow to address individual clusters. Without helium droplets a broad signal might be an indication for intrinsic fluctionality of the clusters. However, inside helium droplets spectral broadening might also reveal an increased number of isomeric configurations possibly involving in addition a variable number of helium atoms. A recent theoretical investigation of those clusters embedded into 1000 helium atoms reported rigid cluster configurations for up to 9 argon atoms and, thus, excluded fluctionality [83]. Moreover, asymmetric cluster configurations with single-sided attachment of all argon atoms to the anthracene molecule were found favored. The latter disagrees with the experiment [81, 82], and it is not clear in how far the theoretical treatment allows to exclude cluster configurations involving helium atoms other than simply a solvation layer. A continuation of theoretical investigations for those clusters generated in helium droplets might start with confirming cluster configurations known from gas phase experiments and continue with investigation of their response to embedding into a helium bath. It should be continued by cluster formation inside helium droplets which then accounts for the possible involvement of helium atoms as part of the cluster.

## 5.4.3  Van Der Waals Clusters of Phthalocyanine with Argon Atoms

A cluster consisting of a single phthalocyanine molecule with a single argon atom was already addressed in Sect. 5.3.3 in the context of relaxation dynamics of helium solvated compounds induced by electronic excitation. The dispersed emission spectrum revealed two additional configurations accessible via relaxation of the electronically excited cluster [21].

Besides the cluster whose emission revealed additional metastable configurations which exhibit increased red shifts, the fluorescence excitation spectrum revealed additional configurations of this cluster with reduced red shift. While the first cluster exhibited an argon induced red shift of 15 cm$^{-1}$ identical to that reported from the gas phase [84], the argon induced red shift of a second cluster signal was only 1.3 cm$^{-1}$. Probably, a third isomer of this cluster size exists with an argon induced red shift of 3.6 cm$^{-1}$. Looking at the phthalocyanine to argon potential hypersurface deduced from pair potentials shown in Fig. 5.15 (such a potential model was also used for PIMC simulations on the solvation of phthalocyanine in helium droplets [42]) only a single global minimum about 3 Å above the center of mass of phthalocyanine was found. No additional local minima are present that could trap an argon atom even at low temperatures. Figure 5.16 shows a 3-dimensional plot representing the surface of the potential minimum for the electronic ground state. The color expresses the binding energy rising from pink to red. The red spot constitutes the global minimum with a binding energy of about 670 cm$^{-1}$. Besides the global minimum, additional cluster configurations require stabilization from the helium environment. The tiny red shift of two additional phthalocyanine-argon clusters is most probably induced by helium solvation. Moreover, instead of helium induced stabilization possibly in the periphery of the chromophore the argon atom might be shielded from phthalocyanine by the helium solvation layer.

According to a semiempirical model deduced from experimental studies of hetero clusters consisting of a single aromatic chromophore and variable numbers of rare gas atoms [72], the attachment of a rare gas atom, say argon atom to a chromophore inside helium droplets means replacing helium atoms by an argon atom [17]. According to Refs. [72], the argon induced red shift in helium droplets is reduced compared to argon attachment in the gas phase. In the case of the phthalocyanine-argon cluster the experimentally observed argon induced red shift amounts to 15 cm$^{-1}$ in the gas phase and in helium droplets. Obviously, sometimes empirical rules fail and the experiment asks for a solid theoretical model.

Besides relaxation dynamics of isomeric configurations among each other induced by electronic excitation as discussed in Sect. 5.3.3 for the most abundant phthalocyanine argon cluster, dispersed emission spectra revealed perceived dissociation [85]. Upon vibronic excitation of the most prominent cluster with only one argon atom not only the triple emission was observed that reveals a cascade of configurational relaxations. In addition, the emission spectrum showed a weak contribution of bare phthalocyanine [85]. According to the argon induced red shift of only 15 cm$^{-1}$, the first vibronic transition of the cluster suffices to exceed the electronic band origin of bare phthalocyanine. Thus, the balance of energy allows for dissociative relaxation of the vibronically excited cluster. However, according to the binding energy of argon at the global minimum, dissociation needs to pass a barrier of about 676 cm$^{-1}$ [83, 84]. In case the signal contribution from bare phthalocyanine should be due to dissociation of the cluster, the barrier needed a helium induced reduction to only 16% which is unlikely. In the meantime, we have repeated the experiment with the simple modification that doping with argon was eliminated. Upon excitation still at the frequencies of vibronic transitions of the argon cluster which means about 15 cm$^{-1}$ to the red of vibronic excitations of bare phthalocyanine, dispersed emission could be detected exclusively from bare phthalocyanine while the cluster signal had vanished. Even upon detuning the laser from the resonances of the argon cluster but still far from any resonance of bare phthalocyanine, the emission of phthalocyanine could be detected. Thus, the additional emission from bare phthalocyanine had nothing to do with the phthalocyanine-argon cluster and its dissociation. Instead, it came from excitation of bare phthalocyanine as accomplished by excitation at the high frequency tail of the PW. In addition to what was discussed for TPP in Fig. 5.12, this is another example for how easy one can be misled by helium induced spectral features.

Another curious dynamic process was observed in the dispersed emission of a cluster with two argon atoms in a single sided configuration. Upon vibronic excitation of this cluster, dual emission was observed. Besides the emission of the single-sided cluster that of a double-sided cluster of identical stoichiometry was detected. Without vibrational excess excitation this signal contribution was absent [85]. Such a vibronically induced isomerization from single-sided to double-sided cluster configuration suggests tunneling through the center of mass of phthalocyanine as isomerization coordinate. However, this is rather speculative and requires further experimental investigations.

### 5.4.4  Summary

In summary, the pick-up process is an ideal experimental method for multiple doping of helium droplets and thus, for designing clusters of well defined stoichiometry. In particular the generation of heterogeneous clusters by consecutive pick-up of the various components is a very favorable experimental technique. The identification of the number of isomeric variants is readily accomplished by means of dispersed emission spectra. Further structural information can be obtained from alternating the pick-up sequence of different cluster components. The low temperature conditions allow for elimination of hot bands and, thus, for unprecedented spectral clarity of cluster spectra. Thus, even tiny local minima in the configuration potential might be stabilized and detected. Besides stabilization of local minima in the configuration space of the cluster, additional configurations involving helium atoms cannot be excluded. An extreme scenario as proposed for anthracene-argon clusters is the attachment of cluster components to the helium solvation layer of another subunit. In this respect, the investigation of clusters generated inside helium droplets opens new perspectives beyond what is found under gas phase conditions.

## 5.5  Elementary Chemical Reactions in Helium Droplets

For the investigation of elementary chemical processes, helium droplets are known to serve as a cryogenic reactor which provides insight to low energy reaction paths otherwise not accessible. This is mainly due to highly efficient dissipation of excitation energy from the dopant system to the helium droplet prior to reactive encounter and throughout the reaction process. Chemistry induced by electronic excitation is certainly the key aspect of this section. Thereby, intramolecular dynamics is readily investigated by means of dispersed emission spectroscopy as discussed above in the context of heterogeneous clusters and solvation complexes in helium droplets. In the following three benchmark experiments will be discussed which address first a bimolecular elementary exchange reaction, namely, $N_2O + Ba \rightarrow BaO^* + N_2$, secondly, a photodissociation process, namely cleavage of iodine from $CH_3I$ and $CF_3I$, and, finally, excited state intramolecular proton transfer (ESIPT) for 3-hydroxyflavone. For all three reactions, the helium droplet experiment is contrasted with corresponding experiments in the gas phase.

### 5.5.1  Bimolecular Reaction of Barium with Nitrous Oxide

The first bimolecular chemical reaction investigated in helium droplets was the formation of barium oxide from the reaction of barium atoms with nitrous oxide, $Ba + N_2O \rightarrow BaO^* + N_2$ [86]. The interesting aspect of this reaction is the formation of an electronically excited product molecule which decays radiatively. The emitted chemiluminecence is readily detected by means of a grating spectrograph. The investigation of elementary chemical reactions under single collision condition allows to study this exchange reaction in its very details. The dispersed spectrum of the chemiluminescence provides insight into the internal energy distribution of the BaO* product molecules whereas tuning the collision energy provides insight into the entrance channel. The reaction of Ba with $N_2O$ has been investigated long before in the gas phase [87–91] and in addition on argon clusters [92, 93].

In the gas phase, crossed beam experiments have been performed in order to accomplish single collision conditions. The chemiluminescence of the BaO* upon reaction with nitrous oxide was found to cover almost the entire range of the visible spectrum and is shown in panel (a) of Fig. 5.17 for two different collision energies. Not much of a fine structure could be resolved and the maximum as well as the width of the emission was found to depend on the collision energy [90]. In contrast, upon replacing nitrous oxide by nitrogen dioxide the chemiluminescence spectrum was shifted slightly to the red and showed a clear fine structure which could be assigned to vibronic bands of the barium oxide product molecule [87]. The difference in the spectra reveal different energy distribution in the exit channel of both reaction systems. Besides the collision energy, the internal energy of the Ba educt has been varied. In addition to the ground state, electronically excited states of the barium atom [88] were studied for the reaction with nitrous oxide and in addition with nitrogen dioxide and also with ozone. In all cases electronic excitation of Ba became effective on the internal state distribution of the BaO* product molecule.

In a second experiment the same reaction system was studied on the surface of argon clusters [92, 93]. The purpose of these studies was to investigate the influence of solvents on elementary chemical processes. As a result, the chemiluminescence spectrum revealed two contributions shown in panel (b) of Fig. 5.17. One part of the spectrum was almost identical to the broad and unstructured spectrum recorded

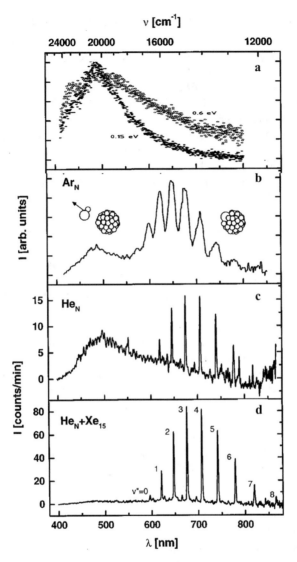

**Fig. 5.17** Chemiluminescence spectra of BaO* generated from the bimolecular reaction of Ba + $N_2O$ in the gas phase for two collision energies (**a**) (Ref. [90]), on argon clusters (**b**) (Refs. [92, 93]), on helium droplets ($\overline{N}\sim$20,000) (**c**) (Ref. [86]) and on $Xe_{15}$ clusters inside helium droplets ($\overline{N}\sim$20,000) (**d**) (Ref. [86]). **b** and **c** reveal signal from hot BaO* having left the cluster/droplet and from cold BaO* residing on or in the cluster/droplet as indicated by two cartoons in panel (**b**). In **d** BaO* escaped from the xenon cluster, however, radiates still inside the helium droplet

in the gas phase except of a slight spectral shift and much higher signal intensity. Superimposed to this a series of well separated peaks were recorded. Two reaction processes were proposed in order to explain the two signal contributions. In one case the reaction proceeds on the surface of the argon cluster so that the hot BaO* product molecule can leave the argon cluster prior to emission. In the second case the reaction happens inside the argon cluster. Thus, the internally hot BaO* product molecule cools down via energy dissipation into the argon cluster prior to radiative decay. In this case, hot bands are eliminated. The authors identify two major solvent effects. First, the effective reaction cross section was increased to the capture cross section of the Ba atom by the $N_2O$-doped argon cluster, which to a reasonable approximation corresponds to the geometric cross section of the argon cluster. Secondly, the collision energy in the entrance channel of the reaction was given by the mobility of the educt moieties on or in the argon cluster and, therefore, not a variable parameter as in the molecular beam experiment. Besides the latter issue providing rather indefinite energetic conditions, the approach on or in an argon cluster does not warrant for single collision conditions as in the molecular beam experiment. Finally, chemiluminescence from the reaction of barium in a single collision process with a cluster of the $N_2O$ instead of a single $N_2O$ molecule on an argon cluster was found to be rather quenched [91].

This bimolecular reactive experiment has been repeated in helium droplets [71, 86]. Thereby, helium droplets had first been doped with barium atoms and secondly with $N_2O$. Chemiluminescence has been collected by an optical fiber bundle and guided to a spectrograph. The corresponding spectrum shown in panel (c) of Fig. 5.17 was similar to that recorded for the reaction on argon clusters. Again, two contributions of chemiluminescence could be identified, one of which shows a broad signal across the entire visible spectral range and a second consisting of sharp peaks superimposed to the first. The former originates from hot product molecules and the latter from internally cold BaO*. The contribution from hot barium oxide reveals high escape probability from the helium droplet as accomplished by the heliophobic barium atoms shifting the reaction towards the surface of the droplet. Cold BaO* shows that product molecules remaining in contact with the helium droplet. One of the remarkable differences to the argon solvated reaction was a smaller solvent shift for the cold BaO* as to be expected for less polarizable helium as host. Furthermore, the vibronic lines of cold chemiluminescence were much narrower because of the lower temperature in helium droplets and possibly reduced inhomogeneous line broadening.

In a next step the experiment in helium droplets was modified in order to promote reaction inside the droplet and suppress reaction on the droplet surface. By additional doping of the helium droplets with on average 15 xenon atoms prior to doping of the reactants, the Ba atom was attracted by the xenon cluster into the helium droplet. Thus, the reactive encounter proceeds on the surface of the xenon cluster inside the helium droplet so that the escape probability of hot BaO* from the helium droplet prior to radiative decay became negligible. Under these conditions the signal from hot BaO* vanished almost entirely (cf. Fig. 5.17d). Despite additional doping with xenon the cold chemiluminescence spectrum was not shifted with respect to the corresponding signal without xenon doping. The missing spectral shift as expected for an attachment to the xenon cluster is indicative for detachment of the product from the xenon cluster inside the helium droplet. Similar as for argon clusters as host, helium droplets act as a catalyst by increasing the effective reactive cross section by orders of magnitude [86].

Along the path from the gas phase to the argon solvated and finally to the helium solvated version of this bimolecular reaction the focus has changed. While the gas phase experiment provides insight into an elementary chemical reaction process including energetic conditions in the entrance and exit channel under single collision conditions, argon or helium solvation reveals the influence of the solvent on the products in the exit channel. In fact, the effective reactive cross section is blown up to the pick-up cross section of the educts by the host cluster. However, single collision conditions are not warranted in helium droplets or in and on argon clusters. Any details of the reaction process other than the identification of BaO* as reaction product got lost and the information revealed from the chemiluminescence spectrum recorded from xenon doped helium droplets does not go far beyond mass spectrometric detection.

## 5.5.2 Photolysis of Iodomethane and Perfluorated Iodomethane in Helium Droplets

One of the most detailed investigations of chemistry inside superfluid helium droplets reports on the photolysis of iodomethane and its perfluorated isomer by means of velocity map ion imaging (VMII) [94–97]. One key issue of this investigation was the influence of helium droplets as a solvent on the energetics of a photoinduced unimolecular dissociation. In particular, possible expression of vanishing viscosity— a characteristic property of superfluidity—in the velocity distribution of photo fragments generated inside helium droplets was of interest. As in the case of the bimolecular chemiluminescent reaction discussed above, such an investigation profits from comparison with corresponding experimental data under gas phase conditions which are available for the chosen system [98]. Both, the gas phase experiment as well as the corresponding helium droplet experiment are benchmark studies which are making use of the capability of VMII in full depth. The authors of the gas phase experiment [98] are the pioneers in the development of VMII. The ideal photolysis process demonstrating the capability of VMII is photolysis with an abstraction of only a single atom. The atomic fragment may access only few energetically well separated electronic states, in the case of iodine two spin states. The other fragment, a molecular radical, appears with internal energy distributed over a variety of rovibrational states, if energetically accessible also in different electronic states. The elegance of VMII as experimental technique [99] lies in the wealth of information that can be obtained from the imaging of the three-dimensional velocity distribution of only the atomic photolysis product. According to conservation laws for energy and linear momentum the velocity map of the atomic fragment reveals kinetic energies of both photofragments as well as the internal energy distribution of the non-detected molecular fragment. The internal energy of the detected fragment is revealed by the VMII detection process. Finally, the spatial distribution of the fragment velocity reveals details on the intramolecular dissociation coordinate as well as on the timing of the dissociation process. Further details can be deduced from correlation of vector quantities obtained by polarization sensitive photolysis and detection schemes which, however, will not be subject of the experiments discussed in the following sections.

The gas phase experiment reported in Ref. [98] headed for the investigation of the dynamics of the photolysis of iodomethane. $CH_3I$ was prepared at low temperature accomplished by a seeded beam expansion. Thus, the energetic conditions are under control by the frequency of the photolysis laser. The entrance channel for the dissociation of $CH_3I \rightarrow CH_3 + I$ was accessed by electronic excitation into the so-called A-band of the $CH_3I$ molecule [98]. With the variation of the photon energy across the A-band the starting point for photolysis varies among three different electronically excited states which are assigned as $^1Q_1$, $^3Q_0$, and $^3Q_1$. At the Franck–Condon point the three levels scale with decreasing energy as listed. All three are repulsive with respect to the I-C coordinate. Among the three, $^1Q_1$ and $^3Q_1$ converge to $CH_3$ and I in the electronic ground state, whereas $^1Q_0$ converges to $I^*(^2P_{1/2})$. This latter state exhibits a conical intersection with the $^1Q_1$ state.

Velocity map ion images of the methyl radical shown in black and white in the left panel of Fig. 5.18 distinguish two exit channels for which the different velocities fit perfectly to the energy difference of the two spin states of the atomic iodine fragment. Besides other important details on the photolysis of $CH_3I$ such as energy disposal into kinetic and internal degrees of freedom as function of the photolysis energy, the anisotropy in the spatial distribution of the fragments revealed information on the probability of curve crossing at the conical intersection. Corresponding data for

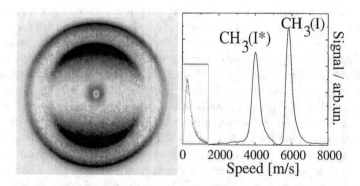

**Fig. 5.18** Left panel: Velocity map ion image of $CH_3$ radicals generated by photolysis of $CH_3I$. Map in black and white recorded from gas phase adapted from [98]. Center colored map recorded from superfluid helium droplets adapted from [95]. Right panel: Velocity distributions of $CH_3$ radicals from the gas phase resolves two channels in correlation with $I^*$ $(^2P_{1/2})$ and I $(^2P_{3/2})$. Inlay right panel: velocity distribution from the helium droplet experiment

the perfluorated analog of iodomethane allowed for comparison of the heavy-light product combination for $CH_3I$ with a rather equally weighted version for $CF_3I$. So far, from the gas phase experiment [96] only those results are reviewed which are relevant for comparison with the helium droplet experiment.

About a decade later, the photolysis of $CH_3I$ and $CF_3I$ was performed in superfluid helium droplets. In general, the influence of the solvent on the photolysis dynamics was of interest. Three consecutive papers [95–97] under the common title "Photodissociation of alkyl iodides in helium droplets" have been published each with an individual subtitle, namely, "Kinetic energy transfer", "Solvation dynamics", and "Recombination". In the first paper, the VMII of the fragments revealed substantial dissipation of kinetic energy from the photolysis products into the helium environment. The VMII of the $CH_3$ radical added in color to VMII from the gas phase experiment in Fig. 5.18 revealed a singly peaked velocity distribution (inset in right panel of Fig. 5.18) at on average less than 10% of the doubly peaked gas phase velocities plotted in the left panel of Fig. 5.18. For the heavier $CF_3$ radical the kinetic energy loss amounts to about 60% also without any signature of two channels as observed for the two electronic states of the iodine fragment in the gas phase. The heavier the product the less the energy loss and the larger the droplets the larger the energy loss of the escaping fragments. The angular distribution of the velocity map revealed anisotropy, however, substantially reduced for the light methyl radical fragment. Beyond an escape velocity of 600 m/s the anisotropy reached a constant average of about 70% of the gas phase value. Below 600 m/s the remaining anisotropy decreased with decreasing fragment velocity. Similar observations were reported for the heavier $CF_3$ radical whereby fragments with an escape velocity of more than 400 m/s reached an anisotropy almost identical to the gas phase experiment. Again, the heavier fragment is less perturbed by the helium environment. Due to numerous other experimental details in combination with a very careful analysis of the influence of the droplet size, the effect of helium solvation on the dynamics of the photolysis could be explained by a classical model of binary collisions of the fragments with individual helium atoms on their way out of the droplet. By means of quasi classical trajectory calculations the experimental velocity distributions could be reproduced quantitatively for both fragments of the photolysis of iodomethane and the perfluorated variant. Instead of characteristic features due to vanishing viscosity of superfluid helium, the influence of the helium environment on the photofragment dynamics revealed the image of a classical solvent.

The second article under the subtitle "Solvation dynamics" [96] addresses the solvation of the photolysis fragments along their path leaving the helium droplet. Solvation means the formation of clusters consisting of a photofragment and variable numbers of rigidly attached helium atoms. The corresponding cluster size distribution as deduced from mass selective VMII was recorded for clusters having left the helium droplet. Thereby it was found that the formation of such clusters took place along the escape path of the fragments. The dynamics of attachment and detachment by consecutive collisions along the escape path is responsible for the cluster size distribution. Since the fragment velocity ratio is inverse proportional to the mass ratio, the iodine fragment from $CH_3I$ is much slower than that from $CF_3I$. Slow fragment velocity was found to favor cluster formation. The dependence of the cluster size distribution on the escape velocity revealed a cluster growth along the escape path of the fragments through the droplet. Moreover, clusters generated as a result of photolysis could be excluded. In the case of the methyl radical, the protonated isomer did form clusters whereas the perfluorated isomer did not. The higher internal energy of the perfluorated fragment was made responsible for the missing of cluster formation.

The comparison of helium clusters of the iodine fragment generated either from the protonated or the perfluorated compound revealed further insight into the formation/solvation dynamics. For the slow iodine fragment the cluster size distribution was almost insensitive to the droplet size distribution. Instead, for the fast iodine fragment the cluster size distribution shifted to larger clusters with increasing droplet radius. For slow iodine fragments a dynamic equilibrium between attachment and detachment is reached whereas fast iodine fragments escape before reaching such an equilibrium.

Finally, the observation of drastically reduced fragment velocities raises the question on the possibility of recombination as discussed in Ref. [97]. Indeed, the authors could positively prove that one out of several overlapping signal contributions within the VMI images recorded for the nascent mass of $CH_3I$ shown in Fig. 5.19 was definitely the result of recombination inside the droplet. Besides the recombination signal marked in panel d) in Fig. 5.19 additional signals from background gas in the detection chamber (panel a)), an effusive beam emerging from the pick-up chamber (panel b)), and a contribution from escaped $IHe_4$ fragment clusters (panel c)) due to imperfect mass selection were detected. With increasing helium droplet size the recombination signal increased. Moreover, evidence was provided that the cut off in the helium cluster size distribution of solvated fragments was due to a vanishing escape probability beyond a certain cluster size. For the perfluorated variant no recombination could be observed. The velocities of the photo fragments from the perfluorated variant did not allow for sufficient deceleration to accomplish recombination. Therefore, corresponding VMII did not show a signal in correspondence to signal (d) (cf. Ref. [97]). In addition, signal (c) is missing since there were no fragment clusters close to the mass of the perfluorated iodomethane.

The investigation of photolysis of methyl iodide inside helium droplets is one of the most profound investigations of helium droplets acting as a solvent on a dopant system and its chemical dynamics. The wealth of experimental details reported in three consecutive papers [95–97] preceded by a letter [94] speaks for itself and goes far beyond what is reviewed above. However, in comparison with the gas phase experiment revealing the very details of the energetics in the entrance and the exit channel of the photolysis of methyl iodide and its perfluorated variant, not much is left over in the helium droplet experiment. Again, the continuous and highly efficient dissipation of internal and kinetic energy into the helium droplet alters the conditions for molecular dynamics substantially. The helium droplet experiment changes the focus from investigating properties of the dopant system to the influence of helium solvation on this system and its chemical dynamics. In this respect Refs. [94–97] are a benchmark project that provided deep insight into solvation in helium droplets.

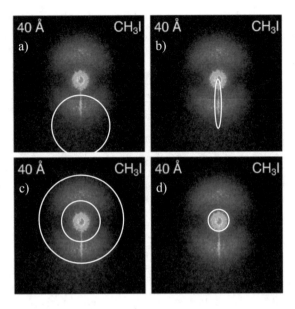

**Fig. 5.19** Velocity map image of non-solvated $CH_3I$ molecules from the detection volume of the ion imaging setup for average droplet radii of 40 Å. The time delay between UV dissociation and femtosecond ionization pulses was set to 50 ns. Note that the images contain signals from parent molecules present as background gas in the detection chamber (**a**) and of an effusive beam emerging from the doping chamber (**b**). The images of $CH_3I$ furthermore contain contributions from escaping $IHe_4$ fragments (**c**) whose mass differs by only 1 amu from the mass of $CH_3I$. The recombination signal (**d**) is readily observed for CH3I. Adapted from Ref. [97]

### 5.5.3   Excited State Intramolecular Proton Transfer (ESIPT) in Superfluid Helium Droplets

The third example studying molecular dynamics in helium droplets is ESIPT of 3-hydroxyflavone. ESIPT is a unimolecular process that is initiated by electronic excitation. According to Born–Oppenheimer approximation, the process of electronic excitation starts with a change of the electron density distribution and only afterwards the nuclei follow. In the case of ESIPT, the nuclear response is a proton transfer to another position. As depicted in Fig. 5.20 a hydrogen atom jumps from one oxygen to another oxygen. This rearrangement is enforced by the change in the electron density distribution accompanying electronic excitation which stabilizes the tautomer (T*) compared to the normal form (N*). As a consequence of ESIPT upon excitation at the electronic band origin of the normal form at about 351 nm, emission is recorded only in the green at and beyond 500 nm originating from the tautomer. In the electronic ground state, the energetic conditions between tautomer and normal form are inverted. The corresponding relaxation process is called back proton transfer (BPT). Under room temperature conditions this molecule is found exclusively in the normal form. Thus, electronic excitation at around 351 nm is the initiating step of

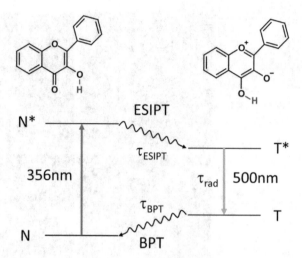

**Fig. 5.20** Jablonski diagram depicting ESIPT and BPT of 3-hydroxyflavone. Left side shows the normal form (N / N*) and right side shows the tautomeric form (T/T*)

a photocycle which is continued by ESIPT followed by radiative decay of the keto form and finished by BPT as depicted in Fig. 5.20. Within this photocycle the time constants for ESIPT and BPT are characteristic quantities. Moreover, the influence of solvents on ESIPT and BPT are important details in order to elucidate such chemical processes.

While ESIPT of 3-hydroxyflavone was known for long the particular dynamics was a matter of long-lasting investigations [100, 101]. The missing of any emission from the normal form (N) revealed that the rate constant for ESIPT exceeds the corresponding quantity for radiative decay. The rise time of tautomeric emission of 3-hydroxyflavone doped into an argon matrix revealed an upper limit in the time constant for ESIPT of 2 ps [102]. In those days, time resolved spectroscopy in the picosecond regime was a challenge. However, upon approaching the limits in the time domain, corresponding information can readily be obtained from the frequency domain. This has been done successfully by means of a spectrally highly resolved fluorescence excitation spectrum recorded from a seeded supersonic jet of 3-hydroxyflavone [103]. The line shape at the electronic band origin was almost perfectly reproduced by a Lorentzian type exhibiting a width of 4.1 cm$^{-1}$. After deconvolution of minor contributions of rotational bands and of the laser band width a purely Lorentzian contribution with a spectral width of 3.9 cm$^{-1}$ was determined which corresponds to an excited state life time of 1.4 ps. Since emission occurred exclusively from the tautomer the homogeneous line width reveals the life time of the electronically excited normal form that decays via ESIPT. Thus, the excited state life time corresponds to the rate constant for ESIPT which amounts to 740 GHz. A possible influence of the steric configuration of the phenyl moiety in the electronically excited system was a matter of discussion. A twist could be confirmed by the intensity pattern within a low energy vibronic progression of 45 cm$^{-1}$ in the excitation spectrum assigned to the phenyl torsion. Upon deuteration the emission was also exclusively of tautomeric origin. However, the line width in the electronic spectrum of the deuterated isomer was significantly smaller as compared to the protonated isomer which is indicative for a reduced ESIPT rate constant. Moreover, the line shape was dominantly of Gaussian type which did not allow for identifying a homogeneous contribution by means of deconvolution [104].

Further insight into the influence of the phenyl moiety on ESIPT has been obtained from corresponding experimental data from 3-hydroxychromone and from 2-(2-naphthyl)-3-hydroxychromone [105]. For all three compounds a Lorentzian line shape recorded at the electronic band origin revealed rate constants for ESIPT of 655 GHz for 3-hydroxyflavone, 145 GHZ for 2-(2-naphthyl)-3-hydroxychromone, and 1770 GHz for 3-hydroxychromone. Obviously, the two PAH substituents impede ESIPT the more the larger the substituent.

It was also known for long that a protic or polar solvent impedes ESIPT as revealed by ultraviolet emission from the solvated 3-hydroxyflavone [101]. The influence of water as a protic solvent was investigated for clusters consisting of a single 3-hydroxyflavone molecule and in addition one or two water molecules as generated in a seeded molecular beam experiment [105]. Dispersed emission revealed that ESIPT was blocked by adding water. Only a single water molecule sufficed to suppress ESIPT entirely. This result needs to be confronted with the observation of dual emission, namely, of normal and tautomeric origin obtained from 3-hydrxychromone derivatives dissolved in neat water [106]. The authors suggest to use the gradually changing emission of 3-hydroxychromone dyes as sensors for protic impurities and in particular of water in solutions. Another supersonic jet experiment on clusters of 3-hydroxyflavone with one or two water molecules revealed an obstacle to proton transfer only for the attachment of two water molecules, whereby reactivation of ESIPT could be accomplished upon sufficient excess excitation energy, indicative for a barrier in the proton transfer path [107, 108]. These results were obtained from mass selective IR/R2PI spectra and interpreted in combination with theoretical vibrational frequencies obtained by means of DFT and TDDFT calculations for the clusters of 3-hydroxyflavone with one or two water molecules. In summary, the experiments discussed above are benchmarks in featuring ESIPT of 3-hydroxyflavone as isolated molecule and under the influence of water on a molecular scale. While results for the isolated molecule are consistent the influence of water on ESIPT revealed inconsistencies among the different experiments.

For the investigation of ESIPT and in addition of further steps in the photocycle depicted in Fig. 5.20, helium droplets implemented as host can provide new details. In particular the influence of solvents such as water on a molecular scale can be realized even selective for isomeric variants of stoichiometrically selected complexes. Moreover, the efficient dissipation of rovibrational energy into the helium droplets allows for cooling of the electronically excited tautomer prior to radiative decay. With the expectation to record vibrationally resolved dispersed emission, the rate constant for BPT might be accessible from a line shape analysis of the electronic band origin similar as reported for ESIPT from the line shape in the excitation [102–105]. Last but not least, the influence of helium solvation on ESIPT is of interest. Thus, fluorescence excitation spectra and dispersed emission spectra of 3-hydroxyflavone and of its clusters with water have been recorded [109]. Dispersed emission spectra

**Fig. 5.21** Dispersed
emission spectrum of
3-hydroxyflavone in a
molecular beam upon
excitation at 356.12 nm
(black) (adapted from [105])
and in helium droplets ($\overline{N} =$
5500) upon excitation at
351 nm (red) (adapted from
[109])

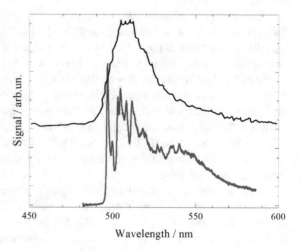

of 3-hydroxyflavone in helium droplets upon excitation at about 351 nm showed exclusively emission at and beyond 500 nm. This was a clear prove for unhindered ESIPT inside superfluid helium droplets as in the gas phase. In contrast to gas phase experiments, the dispersed emission of 3-hydroxyflavone in helium droplets revealed a vibrational fine structure as was expected from efficient cooling of the tautomer prior to radiative decay (cf. Fig. 5.21). This is similar to the cooling of electronically excited BaO* reaction products prior to radiative decay as discussed in Sect. 5.5.1. In the present cases the cooling allows to obtain information on the vibrational fine structure of the ground state of the metastable tautomer of 3-hydroxyflavone which otherwise is hardly accessible. Moreover, the dispersed emission of a rovibrationally cold tautomer allows for line shape analysis similar as done in Refs. [102–105] for the excitation spectrum. The line shape analysis at the electronic band origin revealed a Voigt profile with a Gaussian component of 27.2 cm$^{-1}$ in width and a Lorentzian component of 22.5 cm$^{-1}$ in width [109]. The latter represents the homogeneous contribution to the electronic transition of the tautomer which is determined by the life time of both electronic states involved in the transition. The contribution of the electronically excited state life time in the order of 10 ns [105] to the Lorentzian line width is negligible. Thus, the Lorentzian contribution can be attributed solely to homogeneous line broadening due to BPT. Accordingly, the rate constant for BPT is about 4.2 THz which corresponds to a time constant of 236 fs. An almost identical time constant was found from corresponding investigations in a Shpol'skii matrix [110]. So far, the influence of helium droplets on ESIPT in 3-hydroxyflavone could be classified as negligible. However, the cryogenic capability of helium droplets allowes to partly resolve the vibrational fine structure of the tautomer's electronic ground state and, thereby, deduce the rate constant for BPT.

As reported from gas phase experiments, the clusters of 3-hydroxyflavone with one or two water molecules suffer a red shift in the electronic band origin. Thus, 351 nm for excitation should excite both, bare 3-hydroxyflavone and in addition its clusters with water. Upon additional doping of the helium droplets with water post to 3-hydroxyflavone dispersed emission upon excitation at 351 nm showed spectrally well resolved cluster signals about $400 \, cm^{-1}$ further to the blue of bare 3-hydroxyflavone. According to the Poisson intensity profiles individual resonances could be assigned to clusters of 3-hydroxyflavone with one and with two water molecules as depicted in Fig. 5.22 as blue and green section, respectively. The red section is the electronic band origin of bare 3-hydroxyflavone. Only tautomeric emission was recorded for the clusters with one or two water molecules. The corresponding size distribution defined by the number of attached water molecules as revealed by a Poisson distribution is added as circles in the inlay using the same color code as in the spectrum. Upon increasing the pick-up probability to obtain the cluster size distribution plotted in black squares which corresponds to an average cluster size of four water molecules, a minor contribution of blue emission could be detected as a signature for hindered ESIPT. Surprisingly, however, this emission was spectrally broad without any kind of vibronic fine structure which contrasts to the emission of the tautomer whether from bare 3-hydroxyflavone or from the clusters with one or two water molecules. Under the given pick-up conditions only the largest clusters—probably with eight and more water molecules—might be responsible for hindered ESIPT. Instead of clarifying the inconsistency reported on the influence of water on ESIPT in 3-hydroxyflavone [105–108], the helium droplet experiment provides a third result [109] which is inconsistent to the gas phase data.

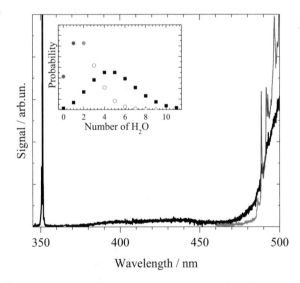

**Fig. 5.22** Dispersed emission spectra of 3-hydroxyflavone and its clusters with water in superfluid helium droplets ($\overline{N} = 5500$). The black spectrum was recorded for a cluster size distribution shown as black squares in the inset. Colored spectrum marks signal of bare 3-hydroxyflavone in red, of clusters with one water molecule in blue and with two water molecules in green. The corresponding cluster size distribution is added to the inset. (Adapted from [109])

**Fig. 5.23** Fluorescence
excitation spectrum of
3-hydroxyflavone in a
molecular beam (black)
(adapted from [105]) plotted
with an offset and in helium
droplets ($\overline{N}$=5500) (red)
(adapted from [109])

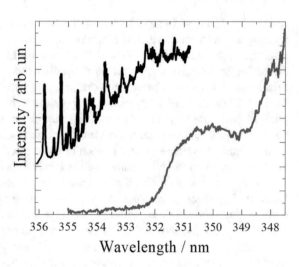

**Fig. 5.23** Fluorescence excitation spectrum of 3-hydroxyflavone in a molecular beam (black) (adapted from [105]) plotted with an offset and in helium droplets ($\overline{N}$=5500) (red) (adapted from [109])

To avoid speculations about the missing of vibrational fine structure in the blue emission of larger clusters with water, the investigation of the fluorescence excitation spectrum of 3-hydroxyflavone in helium droplets might provide explanations. However, instead of a vibrational fine structure as resolved in the gas phase experiment [102–104] shown as black line in Fig. 5.23, the spectrum from helium droplets showed two spectrally broad electronic bands starting at about 352 nm and 349 nm, respectively, as shown in the same figure in red. The helium induced blue shift is hard to quantify without a peak at the electronic band origin. A scenario responsible for vanishing of the vibrational fine structure is not evident. Upon increasing line widths in the gas phase spectrum, a factor of 10 suffice to hide the fine structure resolved in the gas phase. In this case, the factor of 10 in the line width can be interpreted as the helium induced increase of the ESIPT rate constant. Alternatively, a perturbation of the closer helium environment caused by electronic excitation of 3-hydroxyflavone might be a reason for line broadening. This scenario is similar to the explanation of line broadening for 9-phenylanthracene and 2-methylanthracene in helium droplets.

The investigation of ESIPT for 3-hydroxyflavone in helium droplets and in addition its response on attachment of water molecules revealed further insight into helium solvation. This experiment profited in the first place from highly efficient energy dissipation prior to radiative decay and in addition from stoichiometrically perfectly controlled cluster formation with water. The first advantage provided access to the BPT rate constant from line shape analysis in the cold emission of the tautomer. The latter gave access to the ESIPT under the influence of single water molecules. However, instead of clarifying in particular the effect of water on the ESIPT process, the helium droplet experiment was inconsistent to all of the previous results reported from gas phase experiments. As a matter of fact, the configuration of clusters generated in helium droplets might differ from those in the gas phase. However, to our best knowledge, there have never been reports on clusters in helium droplets that definitely excluded those configurations obtained in the gas phase. The stoichiometry of the clusters in helium droplets is unequivocal and so are all dispersed emission spectra. The problem of vanished vibrational fine structure in the excitation spectrum of the normal form and in addition in the corresponding dispersed emission from larger water clusters is an open issue which deserves further attention. In summary, helium solvation exhibits a significant influence of the helium environment on ESIPT in 3-hydroxyflavone and on its clusters with water.

### 5.5.4 Summary

The cryogenic capability of helium droplets has a significant influence on chemical processes whether bimolecular or unimolecular. Energy deposited in rovibrational degrees of freedom is instantaneously dissipated into the helium environment and, thus, the reaction path is forced to proceed without rovibrational contributions in the entrance channel as well as for reactive intermediates. Helium droplets as cryogenic reactor provide ideal conditions for the investigation of low temperature chemistry. The influence of solvents on a molecular scale in chemical processes is readily accessible under perfect control of the cluster stoichiometry. Nevertheless, one needs to consider that the attachment of solvent molecules—in the present case of water—to the reactive system—in the present case to 3-hydroxyflavone—allows for configurations that are not accessible without the helium environment.

## 5.6 Concluding Remarks on Electronic Spectroscopy of Molecules in Superfluid Helium Droplets

This article is far from a review on the title subject. Reviews on the title subject as well as on work related to helium droplets are listed in Appendix A of this book. Instead, this article aimed to highlight peculiar properties of helium droplets as cryogenic superfluid host for studying molecular systems and molecular dynamics by means of electronic spectroscopy. Several of the expectations that initiated and fueled the development of helium droplet sources and its application in molecular spectroscopy also described in Chap. 1 of this book were confirmed by corresponding experiments. Numerous rather surprising experimental observations could be explained by empirical conclusions and evidence-based arguments. However, some of the details of experimental observations were counterintuitive and thus do neither fit to theoretical nor empirical models. The experimental results presented in this article were selected with the idea to report on both, the expected and evident features of helium solvation as observed for glyoxal (Figs. 5.2 and 5.3) and in addition those which are not understood (cf. Figs. 5.5, 5.6, 5.7, 5.8, 5.16, and 5.23). Providing explanations for the latter is certainly a challenge and is vital for making use of the full capacity of helium droplets as cryogenic host. Among unsolved problems, the relation of line shapes and solvent shifts to the droplet size distribution stands out (Figs. 5.7, 5.8, and corresponding data for glyoxal from Ref. [16]). While the adaption of the excluded volume model to finite sized helium droplets is an unquestionable approach to handle the influence of a polarizable environment on the dopant species it does not suffice to describe all of what was reported experimentally on line shapes in helium droplets. As addressed also in Chap. 1, the source conditions such as nozzle temperature, stagnation pressure, and nozzle diameter are influential on the helium droplets. Whether from subcritical or supercritical expansion the change observed in the line shapes does not simply reflect what is expected from the accompanying change in the droplet size. Most curious was a turn-around in the helium solvation shift of electronic transitions as reported for tetracene (Fig. 5.7) and phthalocyanine at the transition from subcritical to supercritical droplet source conditions (Fig. 5.8 top panel). However, a kind of turn-around was also observed for glyoxal and in a different way for porphin, however, in both cases far from a transition in the droplet source conditions.

Besides sophisticated details revealed by the line shape and the overall line shift, the multiplet splitting observed at the ZPL of numerous dopant species deserves further investigations. Numerous experimental results similar as discussed above for a series of PAH species (cf. Fig. 5.5), for a series of derivatives of anthracene (Fig. 5.13), of pyrromethene dye molecules (Fig. 5.14), or of porphin can be taken as a guideline to develop an explanation possibly based on configurational variants of a helium solvation complex which needs to be manifested by theoretical modeling. It is the ultimate challenge for quantum chemical treatment of many particle systems steered by dispersion forces.

The phenomenon of multiplet splitting of the ZPL reveals insight into microsolvation. In addition, it is a key issue of studying van der Waals clusters and, in case of non-polar dopant species, a perfect example for cluster formation driven purely by dispersion interaction. In combination with electronic spectroscopy the investigation of heterogeneous clusters profits immensely from helium droplets as cryogenic host. In particular heterogeneous clusters consisting of a single chromophore molecule and a certain number of atoms, mostly rare gas atoms, or small molecules such as water, oxygen, nitrogen, or hydrogen, reveal information on microsolvation including details such as rigidity or shell structures on a molecular scale. Compared to alternative techniques of cluster generation as accomplished by seeded beam expansion, helium droplets as host provide unique advantages with respect to control of cluster stoichiometry and cluster temperature. However, one needs to keep in mind that the wealth of cluster configurations that are generated under support of the cryogenic environment and which can be addressed selectively by means of electronic spectroscopy might include species which involve helium atoms in addition as proposed for anthracene argon clusters (Fig. 5.16). Further experimental data and a deeper experimental insight into the cluster configuration is needed. Nevertheless, electronic spectra of van der Waals clusters generated in helium droplets are a valuable source of experimental data for the improvement of quantum chemical models dealing with dispersion interaction and with many particle systems that additionally allow to interpret and predict the splitting at the ZPL in electronic spectra.

Chemistry inside helium droplets deserves particular attention. Much of what happens under gas phase conditions is significantly modified by the helium environment as exemplified by the velocity map ion image of $CH_3$ fragment from the gas phase and from helium droplets shown in Fig. 5.18. The efficient dissipation of energy from the reaction system to the helium droplet which is active during the entire reaction process from the entrance channel to the exit channel has an immense impact on the reaction path. Exceeding the Landau velocity transforms the superfluid environment into a normal fluid with significant consequences on among others the kinetic energy distribution (cf. Fig. 5.18). As shown for the BaO* product molecule in Fig. 5.17 and for the tautomer of 3-hydroxyflavone in Fig. 5.21, reaction products as well as reactive intermediates are cooled prior to radiative decay. Thus, emission spectra are cleaned from hot bands and reveal vibrational fine structure of the electronic ground state which in the case of the metastable tautomer of 3-hydroxyflavone is otherwise not accessible. However, the genuine energy distribution as characterizing feature of molecular dynamics is entirely lost due to permanent dissipation of energy into the helium droplets. Low temperature chemistry and within this context the investigation of tunneling processes find unprecedented potential by using superfluid helium droplets as a cryogenic reactor.

Helium droplets and solvation of molecules in helium droplets bears numerous secrets that still need to be revealed. The exceptional sensitivity of electronic spectroscopy plays a key role in this endeavor. Quantitative understanding of helium droplets as nano-scaled quantum fluid and molecular solvation inside them is the ultimate goal. This is mandatory in order to make use of the full capacity of superfluid helium nanodroplets as cryogenic host for studying molecules, their dynamics, and fundamental chemical processes. Continuing work in this field warrants for gain of knowledge for all the various aspects offered by superfluid helium nanodroplets.

**Acknowledgements** We gratefully acknowledge critical comments from J. Peter Toennies. A.S. would like to acknowledge the tremendous contribution of his PhD students who have shared part of their life time to perform excellent experimental work in the Regensburg helium droplet laboratory. Among those are Rudolf Lehnig, Dominik Pentlehner, Ricarda Riechers, Eva-Maria Lottner née Wirths, Tobias Premke. Guest scientists who have joined the laboratory at different times are Alexander Vdovin, Joshua A. Sebree, Lars Christansen, and Jens H. Nielsen. Finally, A.S. acknowledges the continuous support the helium droplet laboratory received from Bernhard Dick and his successor Patrick Nürnberger as chairholders at the Regensburg Institute for Physical and Theoretical Chemistry.

# References

1. J.P. Toennies, A.F. Vilesov, Sectroscopy of atoms and molecules in liquid helium. Annu. Rev. Phys. Chem. **49**, 1–41 (1998)
2. F. Stienkemeier, A.F. Vilesov, Electronic spectroscopy in He droplets. J. Chem. Phys. **115**, 10119 (2001). https://doi.org/10.1063/1.1415433
3. J.P. Toennies, A.F. Vilesov, Suprafluide Heliumtröpfchen: außergewöhnlich kalte Nanomatrices für Moleküle und molekulare Komplexe. Angew. Chem. **116**, 2674–2702 (2004). https:// doi.org/10.1002/ange.200300611
4. J.P. Toennies, A.F. Vilesov, Superfluid helium droplets: a uniquely cold nanomatrix for molecules and molecular complexes. Angew. Chem. Int. Ed. **43**, 2622–2648 (2004). https:// doi.org/10.1002/anie.200300611
5. C. Callegari, K.K. Lehmann, R. Schmied, G. Scoles, Helium nanodroplet isolation rovibrational spectroscopy: methods and recent results. J. Chem. Phys. **115**, 10090 (2001). https:// doi.org/10.1063/1.1418746
6. M. Hartmann, R.E. Miller, J.P. Toennies, Rotaitionally resolved spectroscopy of $SF_6$ in liquid helium clusters: a molecular probe of cluster temperature. Phy. Rev. Lett. **75**, 1566–1569 (1995)
7. M.Y. Chio, G.E. Douberly, T.M. Falconer, W.K. Lewis, C.M. Lindsay, J.M. Merritt, P.L. Stiles, R.E. Miller, Infrared spectroscopy of helium nanodroplets: novel methods for physics and chemistry. Int. Rev. Phys. Chem. **25**, 15–75 (2006). https://doi.org/10.1080/014423506 00625092
8. G. Scoles (ed.) *Atomic and Molecular Beam Methods*, vol. 1 (Oxford University Press Inc, 1988) ISBN: 0195042808 (ISBN13: 9780195042801); vol. 2, (Oxford University Press Inc, 1988) ISBN: 0195042816 (ISBN13: 9780195042818)
9. I.R. Dunkin, *Matrix-Isolation Techniques—A Practical Approach* (Oxford University Press, Oxford, 1998). ISBN 0-19-855863-5
10. S. Tam, M.E. Fajardo, H. Katsuki, H. Hoshina, T. Wakabayashi, T. Momose, High resolution infrared absorption spectra of methane molecules isolated in solid parahydrogen matrices. J. Chem. Phys. **111**, 4191–4198 (1999). https://doi.org/10.1063/1.479717
11. M. Hartmann, A. Lindinger, J.P. Toennies, A.F. Vilesov, The phonon wings in the ($S_1 \leftarrow S_0$) spectra of tetracene, pentacene, porphin and phthalocyanine in liquid helium droplets. Phys. Chem. Chem. Phys. **4**, 4839–4844 (2002)
12. A. Lindinger, E. Lugovoj, J.P. Toennies, A.F. Vilesov, Splitting of the zero phonon lines of indole, 3-methyl indole, tryptamine and n-acetyl tryptophan amide in helium droplets Z. Phys. Chem. **215**, 401–416 (2001). https://doi.org/10.1524/zpch.2001.215.3.401
13. N. Pörtner, J.P. Toennies, A.F. Vilesov, The observation of large changes in the rotational constants of glyoxal in superfluid helium droplets upon electronic excitation. J. Chem. Phys. **117**, 6054–6060 (2002). https://doi.org/10.1063/1.1502643
14. M. Hartmann, J.P. Toennies, A.F. Vilesov, G. Benedek, Spectroscopic evidence for superfluidity in liquid He droplets. Czec. J. Phys. **46**, 2951–2956 (1996)
15. M. Hartmann, F. Mielke, J.P. Toennies, A.F. Vilesov, Direct spectroscopic observation od elementary excitations in superfluid he droplets. Phys. Rev. Lett. **76**, 4560–4563 (1996)
16. N. Pörtner, MPI für Strömungsforschung. Göttingen, Germany, Bericht 10 (200x)
17. M. Hartmann, A. Lindinger, J.P. Toennies, A.F. Vilesov, Laser-induced fluorescence spectroscopy of van der Waals complexes of tetracene–$Ar_N$ ($N \leq 5$) and pentacene–Ar within ultracold liquid He droplets. Chem. Phys. **239**, 139–149 (1998). https://doi.org/10.1016/S0301-0104(98)00250-X
18. M. Hartmann, A. Lindinger, J.P. Toennies, A.F. Vilesov, Hole-burning studies of the splitting in the ground and excited vibronic states of tetracene in helium droplets. J. Phys. Chem. A **105**, 6369–6377 (2001). https://doi.org/10.1021/jp003600t
19. N. Pörtner, J.P. Toennies, A. Vilesov, F. Stienkemeier, Anomalous fine structures of the $0^0_0$ band of tetracene in large He droplets and their dependence on droplet size. Mol. Phys. **110**, 1767–1780 (2012). https://doi.org/10.1080/00268976.2012.679633

20. D. Pentlehner, R. Riechers, B. Dick, A. Slenczka, U. Even, N. Lavie, R. Brown, K. Luria, Rapidly pulsed helium droplet source. Rev. Sci. Instrum. **80**, 043302 (2009). https://doi.org/10.1063/1.3117196

21. R. Lehnig, A. Slenczka, Spectroscopic investigation of the solvation of organic molecules in superfluid helium droplets. J. Chem. Phys. **122**, 244317 (2005). https://doi.org/10.1063/1.1946739

22. H.D. Whitley, J.L. DuBois, K.B. Whaley, Spectral shifts and helium configurations in $^4$He$_N$–tetracene clusters. J. Chem. Phys. **131**, 124514 (2009). https://doi.org/10.1063/1.3236386

23. H.D. Whitley, J.L. DuBois, K.B. Whaley, Theoretical analysis of the anomalous spectral splitting of tetracene in He droplets. J. Phys. Chem. A **115**, 7220–7233 (2011). https://doi.org/10.1021/jp2003003

24. S. Krasnokutski, G. Rouille, F. Huisken, Electronic spectroscopy of anthracene molecules trapped in helium nanodroplets. Chem. Phys. Lett. **406**, 386–392 (2005). https://doi.org/10.1016/j.cplett.2005.02.126

25. R. Riechers, D. Pentlehner, A. Slenczka, Microsolvation in superfluid helium droplets studied by the electronic spectra of six porphyrin derivatives and one chlorine compound. J. Chem. Phys. **138**, 244303 (2013). https://doi.org/10.1063/1.4811199

26. D. Pentlehner, A. Slenczka, Microsolvation of anthracene inside superfluid helium nanodroplets. Mol. Phys. **110**, 1933–1940 (2012). https://doi.org/10.1080/00268976.2012.695406

27. A. Lindinger, Ph. D. thesis, Universität Göttingen, Germany (1999)

28. R. Schmied, P. Carcabal, A.M. Dokter, V.P.A. Lonij, K.K. Lehmann, G. Scoles, UV spectra of benzene isotopomers and dimers in helium nanodroplets. J. Chem. Phys. **121**, 2701–2711 (2004). https://doi.org/10.1063/1.1767515

29. E. Loginov, A. Braun, M. Drabbels, A new sensitive detection scheme for helium nanodroplet isolation spectroscopy: application to benzene. Phys. Chem. Chem. Phys. **10**, 6107–6114 (2008). https://doi.org/10.1039/B807911A

30. P. Carcabal, R. Schmied, K.K. Lehmann, G. Scoles, Helium nanodroplet isolation spectroscopy of perylene and its complexes with oxygen. J. Chem. Phys. **120**, 6792–6793 (2004). https://doi.org/10.1063/1.1667462

31. N. Pörtner, A.F. Vilesov, M. Havenith, Spontaneous alignment of tetracene molecules in 4He droplets. Chem. Phys. Lett. **368**, 458–464 (2003). https://doi.org/10.1016/S0009-2614(02)01903-6

32. M. Hartmann, MPI für Strömungsforschung, Göttingen, Germany, Bericht **10** (1997)

33. A. Slenczka, B. Dick, M. Hartmann, J.P. Toennies, Inhomogeneous broadening of the zero phonon line of phthalocyanine in superfluid helium droplets. J. Chem. Phys. **115**, 10199 (2001). https://doi.org/10.1063/1.1409353

34. B. Dick, A. Slenczka, Inhomogeneous line shape theory of electronic transitions for molecules embedded in superfluid helium droplets. J. Chem. Phys. **115**, 10206 (2001). https://doi.org/10.1063/1.1409354

35. J. Jortner, Cluster size effects. Z. Phys. D **24**, 247–275 (1992). https://doi.org/10.1007/BF01425749

36. S. Fuchs, J. Fischer, A. Slenczka, M. Karra, B. Friedrich, Microsolvation of phthalocyanine molecules in superfluid helium nanodroplets as revealed by the optical line shape at electronic origin. J. Chem. Phys. **148**, 144301 (2018). https://doi.org/10.1063/1.5022006

37. R. Lehnig, M. Slipchenko, S. Kuma, T. Momose, B. Sartakov, A. Vilesov, Fine structure of the (S$_1$ ←S$_0$) band origins of phthalocyanine molecules in helium droplets. J. Chem. Phys. **121**, 9396–9405 (2004). https://doi.org/10.1063/1.1804945

38. R. Lehnig, A. Slenczka, Microsolvation of phthalocyanines in superfluid helium droplets. ChemPhysChem **5**, 1014–1019 (2004). https://doi.org/10.1002/cphc.200400022

39. D. Pentlehner, R. Riechers, A. Vdovin, G.M. Pötzl, A. Slenczka, Electronic spectroscopy of molecules in superfluid helium nanodroplets: an excellent sensor for intramolecular charge redistribution. J. Phys. Chem. A **115**, 7034–7043 (2011). https://doi.org/10.1021/jp112351u

40. K. Nauta, R.E. Miller, Metastable vibrationally excited HF (v=1) in helium nanodroplets. J. Chem. Phys. **113**, 9466–9469 (2000). https://doi.org/10.1063/1.1319965

41. R. Lehnig, A. Slenczka, Emission spectra of free base phthalocyanine in superfluid helium droplets. J. Chem. Phys. **118**, 8256–8260 (2003). https://doi.org/10.1063/1.1565313

42. H.D. Whitley, P. Huang, Y. Kwon, K.B. Whaley, Multiple solvation configurations around phthalocyanine in helium droplets. J. Chem. Phys. **123**, 054307 (2005). https://doi.org/10. 1063/1.1961532

43. A. Slenczka, Electronic spectroscopy of phthalocyanine and porphyrin derivatives in superfluid helium nanodroplets. Molecules **22**, 1244 (2017). https://doi.org/10.3390/molecules220 81244

44. T. Premke, E.-M. Wirths, D. Pentlehner, R. Riechers, R. Lehnig, A. Vdovin, A. Slenczka, Microsolvation of molecules insuperfluid helium nanodroplets revealed by means of electronic spectroscopy. Front. Chem. Phys. Chem. Chem. Phys. **2**, 51 (2014). https://doi.org/10.3389/ fchem.2014.00051

45. J. Fischer, S. Fuchs, A. Slenczka, M. Karra, V. Friedrich, Microsolvation of porphine molecules in superfluid helium nanodroplets as revealed by optical line shape at the electronic origin. J. Chem. Phys. **149**, 244306 (2018). https://doi.org/10.1063/1.5052615

46. U. Even, J. Jortner, Z. Berkovitch-Yellin, Electronic excitations of the free-base porphine–Ar van der Waals complex. Can. J. Chem. **63**, 2073 (1985). https://doi.org/10.1139/v85-342

47. F. Schlaghaufer, unpublished results (2020)

48. R.E.F.E. Riechers, High-resolution spectroscopy in superfluid helium droplets. Investigation of vibrational fine structures in electronic spectra of phthalocyanine and porphyrin derivatives, Dissertation, Universität Regensburg, Regensburg (2011)

49. C. Callegari, W.E. Ernst, Helium droplets as nanocryostats for molecular spectroscopy—from the vacuum ultraviolet to the microwave regime, 1551–1594, in *Handbook of High-resolution Spectroscopy*, ed. by M. Quack, F. Merkt (Wiley, Ltd., 2011). ISBN: 978-0-470-74959-3. http://onlinelibrary.wiley.com/book/https://doi.org/10.1002/9780470749593

50. D. Pentlehner, C. Greil, B. Dick, A. Slenczka, Line broadening in electronic spectra of anthracene derivatives inside superfluid helium nanodroplets. J. Chem. Phys. **133**, 1–9 114505 (2010). https://doi.org/10.1063/1.3479583

51. D. Pentlehner, A. Slenczka, Electronic spectroscopy of 9,10-dichloroanthracene inside helium droplets. J. Chem. Phys. **138**, 024313 (2013). https://doi.org/10.1063/1.4773894

52. D. Pentlehner, A. Slenczka, Helium induced fine structure in the electronic spectra of anthracene derivatives doped into superfluid helium nanodroplets. J. Chem. Phys. **142**, 014311 (2015). https://doi.org/10.1063/1.4904899

53. D. Pentlehner, Perturbations of electronic transitions of organic molecules in helium droplets generated with a new pulsed droplet source, Dissertation, Universität Regensburg, Regensburg (2010)

54. S. Hirayama, A comparative study of the fluorescence lifetimes of 9-cyanoanthracene in a bulb and supersonic free jet. J. Chem. Phys. **85**, 6867–6874 (1986). https://doi.org/10.1063/ 1.451424

55. A. Amirav, J. Jortner, S. Okajima, E.C. Lim, Manifestation of intramolecular vibrational energy redistribution on electronic relaxation in large molecules. Chem. Phys. Lett. **126**, 487–494 (1986). https://doi.org/10.1016/S0009-2614(86)80162-2

56. A. Amirav, Rotational and vibrational energy effect on energy-resolved emission of anthracene and 9-cyanoanthracene. Chem. Phys. **124**, 163–175 (1988). https://doi.org/10.1016/0301-010 4(88)87147-7

57. A. Amirav, C. Horwitz, J. Jortner, Optical selection studies of electronic relaxation from the S1 state of jet-cooled anthracene derivatives. J. Chem. Phys. **88**, 3092–3111 (1988). https:// doi.org/10.1063/1.453953

58. A. Amirav, J. Jortner, Laser-free absorption and fluorescence spectroscopy of large molecules in planar supersonic expansions. Chem. Phys. Lett. **94**, 545–548 (1983). https://doi.org/10. 1016/0009-2614(83)85052-0

59. A. Amirav, M. Sonnenschein, J. Jortner, Absolute fluorescence quantum yields of large molecules in supersonic expansions. Chem. Phys. **88**, 4214–4218 (1984). https://doi.org/10.1021/j150663a005

60. F. Tanaka, S. Yamashita, S. Hirayama, A. Adach, K. Shobatake, Fluorescence decays of 9,10-dichloroanthracene and its van der waals complexes with rare gas atoms in supersonic free jets. Chem. Phys. **131**, 435–442 (1989). https://doi.org/10.1016/0301-0104(89)80188-0

61. A. Penner, A. Amirav, V. Jortner, A. Nitzan, J. Gersten, Solvation effects on molecular pure radiative lifetime and absorption oscillator strength in clusters. J. Chem. Phys. **93**, 147–159 (1990). https://doi.org/10.1063/1.459613

62. N. Ben-Horin, D. Barhatt, U. Even, J. Jortner, Spectroscopic interrogation of herocluster isomerization. II. Spectroscopy of (9,10 dichloroanthracene)·(rare gas)$_n$ heteroclusters. J. Chem. Phys. **97**, 6011–6031 (1992). https://doi.org/10.1063/1.463712

63. A. Penner, A. Amirav, Vibrational predissociation of 9,10-dichloroanthracene—mixed and homo rare atom clusters. J. Chem. Phys. **99**, 9616–9629 (1993). https://doi.org/10.1063/1.465495

64. A. Stromeck-Faderl, D. Pentlehner, U. Kensy, B. Dick, High-resolution electronic spectroscopy of the BODIPY chromophore in supersonic beam and superfluid helium droplets. ChemPhysChem **12**, 1969–1980 (2011). https://doi.org/10.1002/cphc.201001076

65. A. Stromeck-Faderl, Hochauflösende Spektroskopie an isolierten Molekülen im Überschall-Düsenstrahl—Untersuchung von Pyrromethen-Farbstoffen, Dissertation, Universität Regensburg, Regensburg (2009)

66. D.W. Werst, W.R. Gentry, P.F. Barbara, The $S_0$ and $S_1$ torsional potentials of 9-Phenylanthracene. J. Phys. Chem. **89**, 729–732 (1985). https://doi.org/10.1021/j100251a001

67. M. Lewerenz, B. Schilling, J.P. Toennies, Successive capture and coagulation of atoms and molecules to small clusters in large liquid helium clusters. J. Chem. Phys. **102**, 8191–8207 (1995). https://doi.org/10.1063/1.469231

68. A. Amirav, U. Even, J. Jortner, Excited-state energetics and dynamics of pentacene-rare gas complexes. J. Phys. Chem. **85**, 309–312 (1981). https://doi.org/10.1021/j150604a002

69. S. Leutwyler, J. Jortner, The adsorption of rare-gas atoms on microsurfaces of large aromatic molecules. J. Phys. Chem. **91**, 5558–5568 (1987). https://doi.org/10.1021/j100306a014

70. W.M. Van Herpen, W.L. Meerts, A. Dymanus, Rotationally resolved laser spectroscopy of tetracene and its van der Waals complexes with inert gas atoms. J. Chem. Phys. **87**, 182–190 (1987). https://doi.org/10.1063/1.453613

71. N. Ben-Horin, U. Even, J. Jortner, S. Leutwyler, Spectroscopy and nuclear dynamics of tetracene–rare-gas heteroclusters. J. Chem. Phys. **97**, 5296–5315 (1992). https://doi.org/10.1063/1.463790

72. E. Shalev, N. Ben-Horin, U. Even, J. Jortner, Electronic spectral shifts of aromatic molecule–rare-gas heteroclusters. J. Chem. Phys. **95**, 3147–3166 (1991). https://doi.org/10.1063/1.460872

73. N. Pörtner, A.F. Vilesov, M. Havenith, The formation of heterogeneous van der Waals complexes in helium droplets. Chem. Phys. Lett. **343**, 281–288 (2001). https://doi.org/10.1016/S0009-2614(01)00648-0

74. T.R. Hays, V. Henke, V. Selzle, E.W. Schlag, Anthracene-argon complexes in a supersonic jet; spectra and lifetimes. Chem. Phys. Lett. **77**, 19–24 (1981). https://doi.org/10.1016/0009-2614(81)85591-1

75. A. Amirav, U. Even, J. Jortner, Electronic-vibrational excitations of aromatic molecules in large argon clusters. J. Phys. Chem. **86**, 3345–3358 (1982). https://doi.org/10.1021/j100214a017

76. W.F. Henke, W. Yu, H.L. Selzle, E.W. Schlag, D. Wutz, S.H. Lin, Theoretical study of electronic spectral shifts of van der Waals complexes. Chem. Phys. **97**, 205–215 (1985). https://doi.org/10.1016/0301-0104(85)87032-4

77. A. Heikal, L. Bañares, D.H. Semmes, A.H. Zewail, Real-time dynamics of vibrational predissociation in anthracene-Ar$_n$ (n = 1, 2, 3). Chem. Phys. **156**, 231–250 (1991). https://doi.org/10.1016/0301-0104(91)80092-V

78. T. Chakraborty, E.C. Lim, Study of van der Waals clusters of anthracene by laser-induced fluorescence in a supersonic jet: evidence for two structurally different dimers. J. Phys. Chem. **97**, 11151–11153 (1993). https://doi.org/10.1021/j100145a004

79. M.C.R. Cockett, K. Kimura, A study of anthracene–Arn (n=0–5) in the ground cationic state by laser threshold photoelectron spectroscopy: Selective ionization of complex isomers formed in the free jet expansion. J. Chem. Phys. **100**, 3429–3441 (1994). https://doi.org/10.1063/1.466386

80. S.M. Ohline, J. Romascan, P.M. Felker, Rotational coherence spectroscopy of aromatic-(Ar)n clusters: geometries of anthracene-(Ar)$_n$, 9,10-dichloroanthracene-Ar, and tetracene-Ar. J. Phys. Chem. **99**, 7311–7319 (1995). https://doi.org/10.1021/j100019a014

81. D. Uridat, V. Brenner, I. Dimicoli, J. Le Calvé, P. Millié, M. Mons, F. Piuzzi, Existence of two internal energy distributions in jet-formed van der Waals heteroclusters: example of the anthracene–argon$_n$ system. Chem. Phys. **239**, 151–175 (1998). https://doi.org/10.1016/S0301-0104(98)00269-9

82. E.-M. Lottner, A. Slenczka, Anthracene—argon clusters generated in superfluid helium nanodroplets: new aspects on cluster formation and microsolvation. J. Phys. Chem. A **124**, 311–321 (2020). https://doi.org/10.1021/acs.jpca.9b04138

83. F. Calvo, E. Yurtsever, The metastable structures of anthracene-argon clusters inside helium nanodroplets. Theor. Chem. Acc. **140**, 21 (2021). https://doi.org/10.1007/s00214-021-02721-4

84. S.H. Cho, M. Yoon, S.K. Kim, Spectroscopy and energy disposal dynamics of phthalocyanine–Arn (n=1,2) complexes generated by hyperthermal pulsed nozzle source. Chem. Phys. Lett. **326**, 65–72 (2000). https://doi.org/10.1016/S0009-2614(00)00777-6

85. R. Lehnig, J.A. Sebree, A. Slenczka, Structure and dynamics of phthalocyanine-Argon$_n$ (n=1-4) complexes studied in helium nanodroplets. J. Phys. Chem. A **111**, 7576–7584 (2007). https://doi.org/10.1021/jp0708493

86. E. Lugovoj, J.P. Toennies, A. Vilesov, Manipulating and enhancing chemical reactions in helium droplets. J. Chem. Phys. **112**, 8217–8220 (2000). https://doi.org/10.1063/1.481426

87. C. Ottinger, R.N. Zare, Crossed beam chemiluminescence. Chem. Phys. Lett. **5**, 243-248 (1970). https://doi.org/10.1016/0009-2614(70)85016-3

88. J.W. Cox, P.J. Dagdigian, Singlecollision chemiluminescence study of the Ba(1 S,3 D)+NO$_2$, N$_2$O, O$_3$ reactions. J. Chem. Phys. **79**, 5351–5259 (1983). https://doi.org/10.1063/1.445698

89. C.D. Jonah, R.N. Zare, C. Ottinger, Crossed-beam chemiluminescence studies of some group IIa metal oxides. J. Chem. Phys. **56**, 263–274 (1972). https://doi.org/10.1063/1.1676857

90. C. Alcaraz, P. de Pujo, J. Cuvellier, J.M. Mestdagh, Collision energy dependence of the chemiluminescent reaction: Ba+N$_2$O→BaO+N$_2$. J. Chem. Phys. **89**, 1945–1949 (1988). https://doi.org/10.1063/1.455092

91. J.P. Visticot, J.M. Mestdagh, C. Alcaraz, J. Cuvellier, J. Berlande, Reaction of barium atoms with N$_2$O clusters. J. Chem. Phys. **88**, 3081–3085 (1998). https://doi.org/10.1063/1.453951

92. A. Lallement, J. Cuvellier, J.M. Mestdagh, P. Meynadier, P. de Pujo, O. Sublemontier, J.P. Visticot, J. Berlande, X. Biquard, Reaction between (N$_2$O, Ar) binary clusters and barium atoms. Chem. Phys. Lett. **189**, 182–188 (1992). https://doi.org/10.1016/0009-2614(92)85120-Y

93. J.M. Mestdagh, V. Gaveau, C. Gee, O. Sublemontier, J.P. Visticot, Cluster isolated chemical reactions. Int. Rev. Phys. Chem. **16**, 215–247 (1997). https://doi.org/10.1080/014423597230280

94. A. Braun, V. Drabbels, Imaging the translational dynamics of CF$_3$ in liquid helium droplets. Phys. Rev. Lett. **93**, 253401 (2004). https://doi.org/10.1103/PhysRevLett.93.253401

95. A. Braun, M. Drabbels, Photodissociation of alkyl iodides in helium nanodroplets. I. Kinetic energy transfer. J. Chem. Phys. **127**, 114303 (2007). https://doi.org/10.1063/1.2767261

96. A. Braun, M. Drabbels, Photodissociation of alkyl iodides in helium nanodroplets. II. Solvation dynamics. J. Chem. Phys. **127**, 114304 (2007). https://doi.org/10.1063/1.2767262

97. A. Braun, M. Drabbels, Photodissociation of alkyl iodides in helium nanodroplets. III. Recombination. J. Chem. Phys. **127**, 114305 (2007). https://doi.org/10.1063/1.2767263

98. A.T.J.B. Eppink, D.H. Parker, Methyl iodide A-band decomposition study by photofragment velocity imaging. J. Chem. Phys. **109**, 4758–4767 (1998). https://doi.org/10.1063/1.477087

99. A.T.J.B. Eppink, D.H. Parker, Velocity map imaging of ions and electrons using electrostatic lenses: application in photoelectron and photofragment ion imaging of molecular oxygen. Rev. Sci. Instr. **68**, 3477–3485 (1997). https://doi.org/10.1063/1.1148310

100. P.K. Sengupta, M. Kasha, Excited state proton-transfer spectroscopy of 3-hydroxyflavone and quercetin. Chem. Phys. Lett. **68**, 382–385 (1979). https://doi.org/10.1016/0009-2614(79)872 21-8

101. D. McMorrow, M. Kasha, Intramolecular excited-state proton transfer in 3-hydroxyflavone. Hydrogen-bonding solvent perturbations. J. Phys. Chem. **88**, 2235–2243 (1984). https://doi.org/10.1021/j150655a012

102. B. Dick, N.P. Emsting, Excited-state intramolecular proton transfer in 3-hydroxylflavone isolated in solid argon: fluoroescence and fluorescence-excitation spectra and tautomer fluorescence rise time. J. Phys. Chem. **91**, 4261–4265 (1987). https://doi.org/10.1021/j10030 0a012

103. N.P. Ernsting, B. Dick, Fluorescence excitation of isolated, jet-cooled 3-hydroxyflavone: the rate of excited state intramolecular proton transfer from homogeneous linewidths. Chem. Phys. **136**, 181–186 (1989). https://doi.org/10.1016/0301-0104(89)80045-X

104. A. Mühlpfordt, T. Bultmann, N.P. Ernsting, B. Dick, Excited-state intramolecular proton transfer in jet-cooled 3-hydroxyflavone. Deuteration studies, vibronic double-resonance experiments, and semiempirical (AM1) calculations of potential-energy surfaces. Chem. Phys. **181**, 447–460 (1994). https://doi.org/10.1016/0301-0104(93)E0448-5

105. A. Ito, Y. Fujiwara, M. Itoh, Intramolecular excited-state proton transfer in jet-cooled 2-substituted 3-hydroxychromones and their water clusters. J. Chem. Phys. **96**, 7474–7482 (1992). https://doi.org/10.1063/1.462398

106. A.S. Klymchenko, A.P. Demchenko, 3-Hydroxychromonedyes exhibiting excited-state intramolecular proton transfer in water with efficient two-band fluorescence. New J. Chem. **28**, 687–692 (2004). https://doi.org/10.1039/B316149H

107. K. Bartl, A. Funk, M. Gerhards, IR/UV spectroscopy on jet cooled 3-hydroxyflavone $(H_2O)_n$ (n=1,2) clusters along proton transfer coordinates in the electronic ground and excited states. J. Chem. Phys. **129**, 234306 (2008).https://doi.org/10.1063/1.3037023

108. K. Bartl, A. Funk, K. Schwing, H. Fricke, G. Kock, H.-D. Martin, M. Gerhards, IR spectroscopy applied subsequent to a proton transfer reaction in the excited state of isolated 3-hydroxyflavone and 2-(2-naphthyl)-3-hydroxychromone. Phys. Chem. Chem. Phys. **11**, 1173–1179 (2009). https://doi.org/10.1039/B813425A

109. R. Lehnig, D. Pentlehner, A. Vdovin, B. Dick, A. Slenczka, Photochemistry of 3-hydroxyflavone inside superfluid helium nanodroplets. J. Chem. Phys. **131**, 194307 (2009).https://doi.org/10.1063/1.3262707

110. A.N. Bader, F. Ariese, C. Gooijer, Proton transfer in 3-hydroxyflavone studied by high-resolution 10 K laser-excited Shpol'skii spectroscopy. J. Phys. Chem. A **106**, 2844–2848 (2002). https://doi.org/10.1021/jp013840o

111. R. Lehnig, A. Slenczka, Quantum solvation of phthalocyanine in superfluid helium droplets. J. Chem. Phys. **120**, 5064–5066 (2004). https://doi.org/10.1063/1.1647536

# Chapter 6
# Spectroscopy of Small and Large Biomolecular Ions in Helium-Nanodroplets

**Eike Mucha, Daniel Thomas, Maike Lettow, Gerard Meijer, Kevin Pagel, and Gert von Helden**

**Abstract** A vast number of experiments have now shown that helium nanodroplets are an exemplary cryogenic matrix for spectroscopic investigations. The experimental techniques are well established and involve in most cases the pickup of evaporated neutral species by helium droplets. These techniques have been extended within our research group to enable nanodroplet pickup of anions or cations stored in an ion trap. By using electrospray ionization (ESI) in combination with modern mass spectrometric methods to supply ions to the trap, an immense variety of mass-to-charge selected species can be doped into the droplets and spectroscopically investigated. We have combined this droplet doping methodology with IR action spectroscopy to investigate anions and cations ranging in size from a few atoms to proteins that consist of thousands of atoms. Herein, we show examples of small complexes of fluoride anions ($F^-$) with $CO_2$ and $H_2O$ and carbohydrate molecules. In the case of the small complexes, novel compounds could be identified, and quantum chemistry can in some instances quantitatively explain the results. For biologically relevant complex carbohydrate molecules, the IR spectra are highly diagnostic and allow the differentiation of species that would be difficult or impossible to identify by more conventional methods.

E. Mucha · D. Thomas · M. Lettow · G. Meijer · K. Pagel · G. von Helden (✉)
Fritz-Haber-Institut der Max-Planck-Gesellschaft, Faradayweg 4-6, 14105 Berlin, Germany
e-mail: helden@fhi-berlin.mpg.de

D. Thomas
Department of Chemistry, University of Rhode Island, Kingston, RI 02881, USA

K. Pagel
Institut für Chemie und Biochemie, Freie Universität Berlin, Arnimallee 22, 14195 Berlin, Germany

© The Author(s) 2022
A. Slenczka and J. P. Toennies (eds.), *Molecules in Superfluid Helium Nanodroplets*,
Topics in Applied Physics 145, https://doi.org/10.1007/978-3-030-94896-2_6

## 6.1   Introduction

Experimental probes based on light-matter interactions have proven to be among the most effective means to elucidate molecular structure and dynamics. In early experiments, continuous-wave light sources were used to acquire information on the energy levels and structures of molecules in stationary states. Although knowing the equilibrium structures of molecules is crucial, many important processes such as excited state decay or chemical reactions involve electronic and nuclear dynamics as well as energy flow between various degrees of freedom. To investigate such processes, pulsed lasers with temporal widths reaching sub-femtosecond timescales have been used to investigate non-stationary states and to follow dynamical processes in real time. In spite of these significant advances, the complexity of many molecular systems, especially those of substantial size, precludes facile identification of even fundamental properties such as the three-dimensional molecular structure. Moreover, the dynamics of such systems remain to a great extent *terra incognita*.

Biological molecules are of special interest; the elucidation of their structure and dynamics is a prerequisite for an understanding of biological function and of biological processes in general. Numerous biological processes are carried out by proteins and carbohydrates (also frequently called glycans, sugars or oligosaccharides), often in their combination as in glycosylated proteins. These classes of biomolecules can form large and complex three-dimensional structures, and their analysis poses a formidable challenge. Over the last decades, much progress has been made, in particular in the analysis of proteins. A key role has been played by mass spectrometry (MS), as it has unparalleled sensitivity reaching down to the femtogram range. MS is therefore widely used for analytical purposes or as a selective detection method in basic research studies. Additionally, MS can be used in conjunction with dissociation experiments to gain structural information. In the case of peptides and proteins, this approach is the method of choice to determine primary structure (i.e. the sequence of amino acids). MS itself is, however, insensitive to the higher order three-dimensional structure of biomolecules and also reaches its limitations in the structural analysis of glycans.

To deduce higher order structures, condensed-phase spectroscopic or scattering techniques are often employed. However, these have a limited sensitivity and therefore require high sample densities, which are frequently not available. X-ray crystallography requires uniform crystals to achieve the high resolution required to distinguish between possible conformations, and important structural information can often remain hidden. Likewise, structures from multidimensional NMR experiments are restricted to solution, are often solvent dependent, and typically give only averaged conformational data.

### 6.1.1   Infrared Spectroscopy

An ideal companion for MS is optical spectroscopy and in particular infrared (IR) spectroscopy, as it can be highly sensitive and provide structural information far beyond what is available from MS alone. As a standalone technique, IR spectroscopy has a long history and is one of the most heavily utilized experimental methodologies for the analysis of molecular structure. In the region between 2800 and 4000 cm$^{-1}$, vibrations arising from the stretching motion of hydrogen atoms bound to heavy atoms are found. Upon deuteration, these modes shift to the 2000–2800 cm$^{-1}$ range. Located between 500 and 2000 cm$^{-1}$ is the so-called molecular fingerprint region, where the IR-active heavy-atom stretching and many bending modes are found. In the far IR (or THz) region below 500 cm$^{-1}$, primarily modes involving very heavy atoms and large-amplitude modes of groups of atoms are present. The exact position of IR bands depends on the details of chemical bonding and the local environment of the IR oscillators. IR spectra can therefore be used to obtain detailed information about the structure and the covalent and non-covalent interactions in molecules.

### 6.1.2   Action Spectroscopy

Most traditional condensed-phase IR spectroscopy experiments are performed by measuring the attenuation that light experiences after passing through a sample. Although this approach is also possible in the gas phase for molecules that have sufficient vapor pressure, it is not generally applicable to many interesting species, including molecular ions, because their attainable densities are many orders of magnitude too low for a measurable light attenuation. For such systems, a different spectroscopic approach can be taken by measuring not *what the molecules do to the light* (as in the aforementioned absorption spectroscopy) but rather *what the light does to the molecules*, thus employing one of several schemes of *action spectroscopy*. Whereas for regular photon absorption measurements, the figure of merit is given by the product $n * l * \sigma$, with $n$ being the number density of the molecules, $l$ the optical path-length and $\sigma$ the absorption cross section, the figure of merit for action spectroscopy is given by $F * \sigma$, with $F$ being the photon fluence. Thus, whereas a high number of molecules/cm$^2$ ($n * l$) is the requirement in regular absorption measurements, a high number of photons/cm$^2$ is necessary in action spectroscopy.

Examples of action spectroscopy include direct measurements of an excited state population via monitoring photon emission or employing techniques to cause photon-induced dissociation or ionization. In a related method, one can also monitor a change in mass of a molecule, complex or cluster that is directly or indirectly caused by the absorption of photons. Using UV light, this change can be caused by direct bond breaking, and when using IR light, the absorbed energy can be redistributed causing thermal dissociation.

### 6.1.3   IR Multiple Photon Dissociation (IRMPD) Action Spectroscopy

IR multiple photon dissociation (IRMPD) spectroscopy of gas-phase ions [1–3] is at present a widely used form of action spectroscopy [4–6]. In IRMPD, the species of interest are exposed to intense IR radiation. When the frequency of the light is resonant with an IR-active transition of the molecule, the sequential absorption of multiple photons can take place. The internal energy of the molecule can then increase to an extent that dissociation takes place. Monitoring the dissociation yield as a function of IR frequency gives the IRMPD spectrum. Over the years, IRMPD has proven to be a versatile and successful technique to record IR spectra of a wide range of species. Examples include biological molecules ranging from amino acids [7] to peptide aggregates and proteins [3, 8, 9], fragmentation products of gas-phase ions [10], species of astrophysical interest [11] and many other molecules and clusters [5, 6].

When considering IRMPD, it is important to distinguish between the sequential absorption of *multiple photons* and a *multiphoton* absorption process within one mode. In the former scenario, which is the relevant one for IRMPD, the timescale between successive absorption events is long compared to the timescale of internal vibrational redistribution (IVR) of the energy. A consequence is that vibrational modes are statistically populated, and hence the ground state of the light-absorbing mode is usually populated. Direct mode anharmonicities therefore play only a minor role. However, the frequencies of the light-absorbing modes are affected by cross-anharmonicities with other low-frequency modes that are statistically populated, giving rise to a dynamical broadening and frequency shift of the absorbing mode. These effects are qualitatively understood, can in some cases be quantitatively modeled [4], and give rise to shifts, broadening and differences between IRMPD and linear IR absorption spectra. These differences can be large when the species involved have very high fragmentation barriers, contain modes with very high anharmonicity, or isomerize during the excitation process. Molecules that are complex with many close-lying IR-active modes often have broad and unstructured room-temperature linear IR absorption spectra, and often multiple conformers are present. Performing IRMPD on such a system yields spectra that are often very congested with a correspondingly low information content. Examples of where IRMPD results in broad spectra include systems where a proton is shared between two functional groups [12]. In these cases, the motion of the proton occurs in an extremely anharmonic potential, causing couplings and spectral congestion, which can be resolved using alternative spectroscopic methods [13].

Many of the complications in IRMPD spectroscopy arise from the IRMPD excitation process itself, and the initial ion temperature (usually room temperature) as such is only a minor issue. Significantly improved spectra can be obtained by performing spectroscopy on ions that are cold and, at the same time, using methods where the absorption of a single, or maybe very few, photons is monitored. One possibility for cooling is to make use of a molecular beam expansion. Ions can then be directly generated in the source [14] or via resonant photoionization of a molecule-buffer gas cluster [15], and IR absorption is monitored via recording of the IR-induced dissociation. Methods where mass-to-charge selected ions are trapped in a cold ion trap and irradiated by IR light are more versatile, however, and are performed in a number of laboratories [16–19]. Detection of IR absorption can then occur by monitoring IR-induced fragmentation, either of the ion itself or of a weakly bound ion-messenger complex. Typical messengers are rare gas atoms or small molecules such as $H_2$ or $N_2$. For some species, alternatively, the depletion (or enhancement) in yield of a subsequent UV-photodissociation step can be monitored. Using these techniques, many ionic species have been investigated and characterized. However, in many instances, even lower temperatures are desirable, and in addition, the above schemes are difficult to apply to large molecules and ions. In these cases, embedding of the analyte in liquid helium is potentially advantageous and offers exciting new opportunities.

### 6.1.4  Action Spectroscopy Using Helium Nanodroplets

At low temperatures, helium shows behavior unlike any other element or material. Due to strong quantum behavior, bulk liquid helium becomes superfluid at 2.16°K, and helium is the only element that remains liquid down to the absolute zero of temperature. Particles in superfluid helium can move almost freely without any friction. Because helium provides an isothermal environment that interacts only weakly with dopants and is transparent throughout most of the electromagnetic spectrum, it can be regarded as the ultimate cryogenic matrix for spectroscopic investigations.

Experimentally, however, bulk liquid helium is not well suited to "dissolve" molecules, as these would rather aggregate or stick to the container walls. A breakthrough occurred some twenty years ago, when initial experiments were performed in which small, thermally evaporated gas-phase molecules were captured by helium droplets produced via supersonic expansion of low-temperature helium gas into vacuum [20–22]. Using this pickup technique, atoms [23], small molecules [24–27] and biomolecules [28–30] as well as large species such as $C_{60}$ [31] have successfully been doped into helium droplets. It was shown that, due to the weak interactions in liquid helium, molecules embedded in helium nanodroplets can rotate almost freely, and their optical spectra show narrow linewidths [24, 27], which evidenced that the helium droplets are indeed superfluid [32]. An analysis of the rotational population distribution revealed the droplet temperature to be only 0.4°K [27]. It has also been observed that in vibrational spectra, line positions are only very weakly affected by the presence of the helium environment [33]. In contrast, electronic spectra of molecules in helium droplets show a rather large influence of the helium on the line positions and general appearance of the spectra, which can be used to extract information on the interaction between the dopant and the helium surrounding [28, 34].

Using laser vaporization, less volatile materials such as refractory metal atoms can be evaporated and picked up by helium droplets [35]. Both techniques, thermal evaporation as well as laser vaporization, allowed for the growth of clusters inside helium droplets [36, 37]. These clusters occur in a distribution of sizes that is governed by Poisson statistics. Helium droplets are also uniquely suited for studies of the dynamics of dopants inside or on the surface of the droplets [38]. Using pulsed lasers, molecules can be aligned [39], or brought to rotation [40], or their wavepacket dynamics studied [41]. For small molecules, the rotational spectra and dynamics were recently described quantitatively with the help of a new quasi-particle, the angulon [42]. Further, helium droplets have been discussed and used as matrices for electron or X-Ray diffraction experiments [43–45]. Helium nanodroplets can also accommodate charged species. See [46] for a recent review. In such experiments, charged doped droplets can be produced by electron- or photoionization of doped neutral clusters [47], by the pickup of ions by helium droplets [48, 49] or by pickup of neutral species by charged droplets [50].

## 6.2 Experiments on Ions in Helium Nanodroplets

In the following, we describe an experimental setup that has been developed at the Fritz-Haber-Institut (FHI) over the last ten years to measure IR spectra of mass-to-charge selected ions in helium droplets. As the ion-doped droplets carry a charge, their mass-to-charge ratio can be measured using standard techniques. In the first experiments, large droplets were used to capture the mass-to-charge selected protein cytochrome c (molecular weight ~12,000 amu) in various charge states, and the droplet size distributions were determined [48]. Later, the first UV/VIS spectroscopic experiments were performed on the hemin ion [51]. After the FHI free-electron laser (FHI FEL) went online, IR action spectroscopy experiments were performed on small molecules [52–54] and clusters [55, 56], peptides and proteins [57, 58], DNA fragments [59], carbohydrates [60–64], reactive intermediates [65–68] and lipids [69, 70]. Here, we show some examples of IR spectroscopy of small anionic complexes as well as of carbohydrate ions in helium droplets.

### 6.2.1 Pickup of Mass-to-Charge Selected Ions in Helium Droplets

In most helium droplet experiments, neutral atoms or molecules of interest are brought into the gas phase, picked up by helium droplets, and probed spectroscopically. In typical experiments where a single atom or molecule per droplet is to be picked up, the sample is present in the pickup region at pressures on the order of $10^{-6}$–$10^{-3}$ mbar, corresponding to densities of ~$10^{10}$–$10^{13}$ molecules/cm$^3$. This density is limited by the corresponding vapor pressure, which, if needed, can be increased by heating the sample. Of course, all molecules will transform or decompose at a certain temperature, and for the vast majority of (biologically relevant) molecules, this temperature is much below the hypothetical temperature at which they would have a sufficient vapor pressure to enable helium droplet experiments. Under some circumstances, laser desorption might be a solution, as the heating rate caused by the laser is very high, and (thermal) desorption can occur before decomposition. However, also in this case, at least partial decomposition will occur, and experiments that do not involve mass separation or identification will yield ambiguous results.

A solution is to marry helium droplet methods with modern mass spectrometric techniques that enable the introduction of biological molecules into the gas phase as charged species and to perform mass-to-charge selection. Electrospray ionization (ESI) can be used to bring species ranging from small molecules to large proteins, or even entire viruses, as singly or multiply charged species into the gas phase. Usually, the charge stems from a lack or excess of protons or from complex formation with, for example, alkali cations ($Na^+$, $K^+$ etc.) or halide anions ($I^-$, $Br^-$, $Cl^-$ or $F^-$), and the observed charge or charge distribution often reflects the one that the molecule possesses in the condensed phase. Mass spectrometry can then be used to select and isolate an ion in an individual mass-to-charge state, which can then be picked up by a helium droplet.

In the instrument developed at the FHI in Berlin, the pickup of mass-to-charge-selected ions occurs in an ion trap. Conceptually, this is quite analogous to the pickup of neutral molecules in a pickup cell. The general scheme in shown in Fig. 6.1. Shown in (a) is the conventional approach used in many laboratories to capture neutral molecules in a pickup chamber. Depending on the molecule density, the length of the pickup chamber and the size of the helium droplets, a number of molecules can be picked up by the droplets according to a Poisson distribution. The doped droplets can then be interrogated further downstream. The analogous approach for ion pickup is shown in (Fig. 6.1b). In this method, the pickup cell is replaced by a linear ion trap. Mass-to-charge selected ions can be loaded into the trap up to the space charge limit ($\sim 10^6$ ions/cm$^3$). The ions remain stable in the trap for an extended period of time. The droplets traverse the trap and can pick up an ion, and because the kinetic energy of the doped droplet is higher than the trapping potential, the doped droplet can escape the trap and be interrogated further downstream.

An overview of the instrument is shown in Fig. 6.2. The front-end mass spectrometer is a modified commercial mass spectrometer (Waters Q-TOF Ultima). It is equipped with an ESI source (either regular or nano-ESI). The spray usually occurs from an aqueous solution of the molecules of interest into atmospheric pressure. Ions are transferred via several stages of differential pumping into high vacuum. Mass selection is performed with a quadrupole mass spectrometer (mass-to-charge range up to 3000 m/z). In the commercial instrument, this would be followed by a collision cell to induce fragmentation and a high-resolution time-of-flight (TOF) mass spectrometer to analyze the fragments. In our modified instrument, the collision cell is not used, and a quadrupole bender is inserted just in front of the TOF mass spectrometer. This bender allows the ion beam to be sent straight through to be analyzed in the TOF mass spectrometer, or to be deflected 90 degrees and injected into an ion trap. The ion trap is a linear hexapole trap of length 30 cm, and the six

**a)**

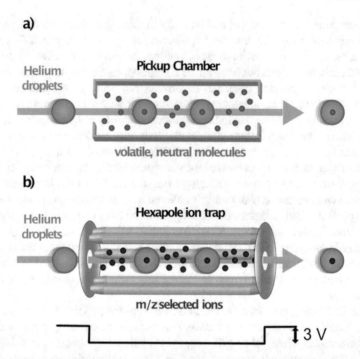

**Fig. 6.1 a** Pickup of neutral molecules by helium nanodroplets using a pickup chamber in which the molecule of interest is evaporated. This approach is only feasible when the molecule of interest has sufficient vapor pressure. **b** Pickup of ions from an ion trap. The trap can be filled with mass-to-charge selected ions from a mass spectrometer. Longitudinal trapping is achieved by a small trapping potential (1–3 V). The kinetic energy of a doped droplet is more than enough to overcome this trapping potential

rods (5 mm diameter) are mounted with an inscribed circle of 9.1 mm diameter. Alternating phase radio frequency voltage (~1 MHz, 100–300 V p–p) is applied to the six rods, providing for an effective confining potential in the transverse direction. In the longitudinal direction, electrodes at both ends of the trap provide for a shallow (1–3 V) trapping potential. The ions are injected slightly above the trapping potential, and a short pulse of helium gas (~$10^{-3}$ mbar) is supplied into the trap to allow for removal of the excess kinetic energy such that the ions remain stable in the trap for at least several minutes. The trap temperature can be varied in the range ~80–400 K, and the ions quickly thermalize to that temperature.

The trap is filled up to the space charge limit (on the order of $10^6$–$10^7$ ions/cm$^3$), which takes, depending on the ion current from the mass spectrometer, from less than one to several seconds. The trap is then traversed by helium droplets.

A pulsed, cryogenic Even-Lavie valve (EL-C-C-2013, Uzi Even & Nachum Lavie, Tel Aviv, Israel) is used to generate helium droplets. The valve-body can be cooled down to 6°K using a closed-cycle helium cryocooler (RDK 408D2, Sumitomo Heavy Industries Ltd., Tokyo, Japan). Combining a resistive heating element with a diode

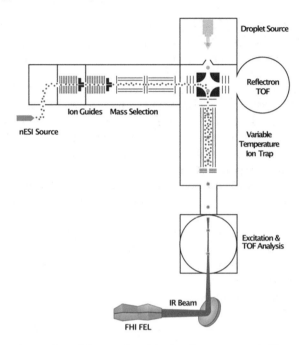

**Fig. 6.2** Schematic overview of the experimental setup. Ions are generated by nano-electrospray ionization (nESI) and transferred into high vacuum by two ion guides. The ions can be monitored by a reflectron time-of-flight (TOF) mass spectrometer. Once the ions of interest are isolated by the quadrupole mass filter and the ion current is optimized, they are injected into the hexapole ion trap. The ion trap is traversed by a beam of helium nanodroplets that can pick up trapped ions and thermalize them to 0.4 K. The helium droplets transport the embedded ions to the detection region where they are irradiated by an IR laser beam produced by the FHI free-electron laser (FHI-FEL) [71]

temperature sensor allows control of the valve temperature, which is typically kept between 15 and 25°K. Controlling the temperature is used to change the size of the helium droplets, and the size distribution can be well described by a log-normal distribution. In the present setup, the mean value of the size distribution can be shifted from around $10^4$ helium atoms at $25^{\circ K}$ to around $10^6$ helium atoms at 15°K. For experiments presented in this work, the valve was operated at a temperature of 21 or 23°K, a stagnation pressure of 70 bar, and a typical opening time of approximately 10 μs. The droplet source vacuum chamber is pumped by a turbomolecular pump with a pumping speed of 2400 l/s (Turbo-V 3°K-T, Agilent Technologies Italia, Italy) maintaining a pressure of <$10^{-5}$ mbar during operation. The central part of the molecular beam is transmitted towards the hexapole ion trap using a skimmer (Model 50.8, Beam Dynamics, Jacksonville, FL, USA) that has an aperture diameter of 4 mm and is placed around 15 cm away from the nozzle.

In the here presented experiment, the droplets have an average size of ~$5*10^4$ helium atoms (and therefore a much larger mass than the ions in the trap) and move at a velocity of ~500 m/s. With $E_{kin} = 1/2 \ mv^2$, the droplets have an average kinetic energy of ~520 eV. When a droplet collides with an ion, its capture can occur, and the then-doped droplet will move with almost unchanged velocity, kinetic energy and direction, compared to the droplet before the collision. As the kinetic energy of the doped droplet is much higher than the longitudinal trapping potential of the ion trap (~3 V), the doped droplets can exit the trap and travel towards the laser interaction region. It is important to note that the ions are otherwise stable in the trap, and the only relevant ion loss channel for the trap is the transport inside a helium droplet. This selective ion confinement makes the method extremely efficient.

Using the present setup, we demonstrated that small ions such as a protonated amino acid as well as large ions such as the protein cytochrome c (molecular weight ~12,000 amu) in charge states ranging from +6 to +14 can be efficiently incorporated into helium droplets [48, 51, 57, 58]. Measuring the current from doped droplets with a calibrated amplifier gives peak currents of up to ~20 pA, which implies about $10^4$ ion-doped droplets per pulse. Those intensities are high enough to enable many spectroscopic experiments, as ion-detection schemes can be employed that have sensitivities down to individual ions. Based on the ion density and the droplet sizes, it can be estimated that the probability that a helium droplet picks up an ion is about 2%, ensuring that multiple ion pick-up by a single droplet is negligible.

Because the doped droplets carry charge, they can be manipulated using electric fields. To determine the size distributions, the doped droplets can be accelerated using a static electric field after exiting the trap. As lighter droplets will arrive at earlier times at either a detector or in the laser interaction region, size distributions of the doped droplets can be measured, allowing experiments on droplets with defined sizes to be performed.

## 6.2.2 The FHI Free-Electron Laser

Several tens of centimeters downstream of the ion trap, the doped droplet beam is colinearly overlapped with the IR beam of the FHI FEL [71]. To date, a wide range of the electromagnetic spectrum can be covered by various commercially available benchtop laser systems. However, a single laser system often only provides a narrow spectral range with low-intensity radiation. If, on the other hand, high intensity radiation over a broad spectral range is required, an FEL can be the best option. This principle is implemented in the FHI-FEL [71], which is highly tunable and provides intense IR radiation in the range of 200–3500 cm$^{-1}$, thus covering the complete molecular fingerprint region as well as much of the light atom stretching region.

The concept of an FEL was proposed in 1971 [72], and the first realization was reported in 1976 [73]. In an IR FEL, a relativistic beam of electrons, produced by an accelerator, is injected into a resonator consisting of two high-reflectivity mirrors at each end of an undulator. The magnetic field in the undulator is perpendicular to the direction of the electron beam and periodically changes polarity a (large) number of times along its length. This causes a periodic deflection, a 'wiggling' motion, of the electrons while traversing the undulator. The transverse motion is quite analogous to the oscillatory motion of electrons in a stationary dipole antenna and hence will result in the emission of radiation with a frequency equal to the oscillation frequency. This oscillation frequency is given by the ratio of the velocity of the electrons to the path length travelled by the electrons per period of the undulator. This path length is larger than the period ($\lambda_u$) of the magnetic field by a factor $(1 + K^2)$ due to the transverse motion of the electrons induced by the magnetic field; the dimensionless factor $K$ is a measure of the strength of this magnetic field. The overall motion of the electrons in the undulator resembles the motion of oscillating electrons in a dipole antenna moving close to the speed of light. This high velocity results in a strong Doppler shift: the frequency of the radiation emitted in the forward direction as measured in the laboratory frame is typically up-shifted by a factor $\sim\gamma^2$, where $\gamma$, the Lorentz factor, is a measure of the electron energy in units of its rest mass. This radiation, referred to as spontaneous emission, is usually very weak. This is a consequence of the fact that the electrons are typically spread out over an interval that is much larger than the radiation wavelength and will therefore not emit coherently. However, on successive round trips in the resonator, this weak radiation will be amplified by fresh electrons, until saturation sets in at a power level that is typically $10^7$–$10^8$ times that of the spontaneous emission.

A schematic overview of the FHI-FEL and its main components is shown in Fig. 6.3. The electron beam used to generate the laser beam is produced by an electron gun emitting pulses of free electrons into vacuum. The time structure of the FEL is characterized by a macro- and a micro-repetition rate that also determines the time structure of the laser radiation. The macro-pulse repetition rate is usually set to 10 Hz, and each macro-pulse is 10 μs long and consists of $10^4$ micro-pulses generated at a repetition rate of 1 GHz. After traversing a buncher cavity that compresses the bunch-length, the electrons are accelerated by two linear accelerators (LINACs). The first LINAC accelerates the electrons to a fixed energy of around 20 meV and the second LINAC is used to vary the final electron energy between 18 and 45 meV. Next, the electron beam is directed through a U-shaped bend by dipole and quadrupole magnets before entering the laser cavity. Inside the cavity resides a 2 m long undulator containing 50 periods of oppositely poled NdFeB permanent magnets. The first electron bunch passing through the undulator emits an initial incoherent IR pulse. This optical pulse is reflected by the FEL cavity mirrors and passes through the undulator again. The length of the laser cavity (5.4 m) is set such that the cavity round-trip time of the optical pulse is synchronized with the electron bunches entering the cavity. As a consequence, both pulses travel along the magnetic fields of the undulator, and the electromagnetic field of the radiation can interact with each electron bunch. This interaction leads to micro-bunching of the electrons and the emission of coherent radiation with high intensity gains. The laser output characteristics can be precisely controlled by detuning (shortening) the optical cavity by a multiple of the laser wavelength. A small detuning, such as 1 $\lambda$, generates short IR pulses with high peak power. Because the spectrum of the radiation is Fourier-limited, these short pulses have a broader spectral width. In this work, the FEL was typically operated

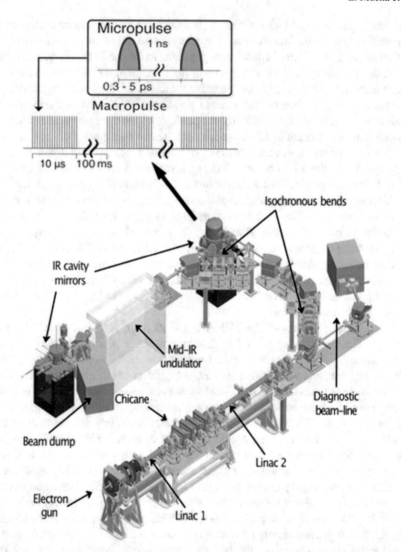

**Fig. 6.3** Schematic overview of the FHI-FEL [71]. An electron gun releases bunches of electrons into vacuum that are accelerated to relativistic velocities by two linear accelerators (LINACs). Inside the laser cavity, a periodic array of strong magnets forces the relativistic electrons on an oscillatory wiggling motion emitting monochromatic radiation. Constructive interference between the electromagnetic field of the radiation and the electrons amplifies the emission of coherent photons

at a detuning of 3–5 λ, which increases the pulse length to a few picoseconds and decreases the spectral width to around 0.2–0.5% (FWHM).

A hole in one of the cavity mirrors is used to transmit a fraction of the radiation to the user experiments and diagnostic elements. During a beam shift (typically one day), the electron energy is kept constant, and the laser wavelength can be tuned by changing the gap between the undulator magnets. The resulting laser radiation has a macro-pulse length of 10 μs and a macro-pulse energy of up to 120 mJ, maintaining a bandwidth of around 0.5% (FWHM). Changing the electron energy allows the FHI-FEL to produce photons with a wavelength between 3and 60 μm.

## 6.2.3   IR Excitation of Ions in Helium Droplets

Action spectroscopy can be performed on chromophores embedded in helium droplets. When the doped droplets are irradiated with laser light and when the dopant ion absorbs a photon, an energy relaxation cascade will follow, leading to warm-up of the droplet and finally to the evaporation of helium atoms. The pathways and timescales are generally not known and are topics in our research group. For almost all well-behaved molecules, however, these timescales will be much faster than the timescale at which the next micropulse arrives (1 ns).[1] The following micropulse will then encounter a dopant ion that is cooled down to the droplet equilibrium temperature of 0.4°K and embedded in a slightly smaller droplet. After successive absorption events, bare ions, void of any helium solvation shell, are observed. Some experiments indicate that bare ion generation is the result of the ions being ejected from the droplets [47]; the mechanism for such a process is, however, unclear. Another possibility is that the droplet is completely evaporated. Which of the two processes occurs might depend on the type of ion, the excitation process and other parameters. In any case, the resulting unsolvated ions can be detected in a TOF mass spectrometer. A scheme of the process is shown in Fig. 6.4.

Figure 6.5 shows mass spectra obtained after irradiating helium droplets that are doped with the protonated pentapeptide Tyr-Gly-Gly-Phe-Leu (YGGFL; $C_{28}H_{38}N_5O_7^+$) at 1704 cm$^{-1}$ and different laser macropulse energies. In the mass spectra in Fig. 6.5, essentially only the protonated YGGFL at m/z ≈ 556 is observed. At lower mass-to-charge, little to no signal from fragmentation of the peptide is observed. At higher mass-to-charge, no signal of adducts or the remainder of a helium solvation shell can be seen. The signal of the protonated YGGFL increases with increasing laser energy, however not linearly.

Figure 6.6 shows the dependence of the signal intensity on the laser energy for three different droplet size distributions that were selected by accelerating the droplet beam in an electric potential (30 V) and timing the FEL such that the IR light beam

---

[1] *This will be the case for essentially all molecules consisting of more than two atoms. An exception are diatomic molecules with only high frequency vibrational modes, such as HF, for which very long vibrational lifetimes in helium droplets have been observed* [74].

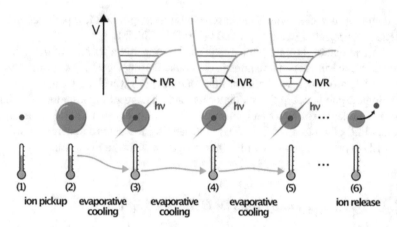

**Fig. 6.4** Schematic diagram of the mechanisms involved in IR spectroscopy of ions using helium droplets. The ions inside the hexapole ion trap (1) have an initial energy that is rapidly dissipated after pickup by helium droplets (2) that maintain an equilibrium temperature of 0.4°K by evaporative cooling. The absorption of a resonant photon (3) leads to vibrational excitation of the ion. The absorbed energy is quickly dissipated to the bath of vibrational degrees of freedom by intramolecular vibrational redistribution (IVR). Before the next laser micro-pulse arrives, the ion will return to its vibrational ground state by dissipating its energy to the helium droplet. Evaporation of helium atoms allows the helium droplet to re-thermalize to its equilibrium temperature. This process can repeat itself many times, and after the successive absorption of multiple resonant photons (4) + (5) the bare ion is released from the helium droplet (6). The number of released ions is then monitored as a marker for photon absorption

intersects with the desired droplet size distribution. The corresponding distributions are shown on the left in Fig. 6.3 (a, c, e). They result from simulations, based on experimental data on the time profiles of the entire distribution, measured further downstream in the instrument. The corresponding energy dependence curves on the right (b, d, f) show a clear nonlinear behavior. Whereas in the case of the distributions of the smaller droplets (a, c), signal is already observed for small laser energies, a substantial amount of laser energy is needed for onset of signal in the case of distribution (e). As the absorption of light will occur in different FEL micropulses in a sequential fashion, the number of absorbed photons is directly proportional to the laser energy. If the absorption cross section and the spatial characteristics of the laser beam would be known accurately, the x-axis could be converted from laser energy to number of photons absorbed. Measuring the ion signal at a specific laser energy setting could thus tell us directly how many photons on average are absorbed and therefore serve as an *if, and if so, how many* marker for photon absorption. By monitoring the bare ion yield as a function of wavelength, an IR-action spectrum can be obtained. Importantly, although multiple photons are absorbed, all absorption events will occur from the vibrational ground state of the ion at 0.4°K. Anharmonicities or cross anharmonicities that broaden spectra in IRMPD spectroscopy will therefore not play a role, and narrow absorption bands are expected.

**Fig. 6.5** Mass spectra obtained after irradiating helium droplets doped with the protonated pentapeptide YGGFL at 1704 cm$^{-1}$ at three different laser energy settings. In all mass spectra, essentially only the intact parent ion at m/z ≈ 556 is observed. The signal intensity increases non-linearly with increasing laser energy

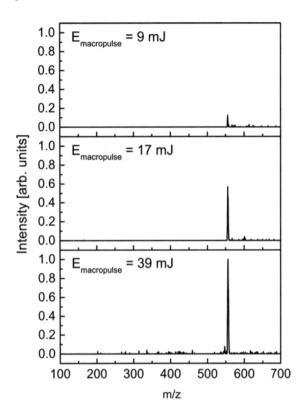

As examples, Fig. 6.7 shows experimental spectra of three different species in helium droplets. Shown in (a) is a spectrum of an anionic complex in which a proton holds together two formic acid anions $(HCO_2^-)_2 \cdot H^+$ [52]. Theory predicts that the equilibrium position of the proton is exactly halfway between the two formate ions. Clearly, the spectrum shows several sharp peaks. All of them are very narrow, and their width is determined by the spectral width of the FEL. An analysis of the spectrum shows that all lines in the spectrum below 1200 cm$^{-1}$ stem from motion of the shared proton coupled to deformation and torsional modes of the complex. Interestingly, the potential in which this proton moves is extremely anharmonic and more closely resembling a particle in a box potential than a harmonic oscillator. Maybe surprisingly, this does not lead to significant broadening. This shared-proton complex and the resulting spectrum can therefore serve as an interesting test case for experiment and theory of our understanding of anharmonic interactions [52].

Shown in (b) is the spectrum of a glycan, a naturally occurring tetrasaccharide, the blood group antigen Lewis b. The type and connectivity of the carbohydrate subunits is shown according to a symbol nomenclature [75]. This nomenclature and the spectroscopy of some carbohydrates will be discussed in more detail in forthcoming pages of this contribution. Considering the size and complexity of the molecule, a surprisingly well-resolved spectrum is obtained. Shown in (c) is the

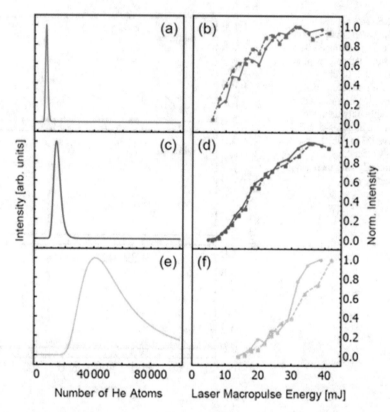

**Fig. 6.6** Droplet size distributions (**a, c, e**) and signal intensity as a function of laser energy (**b, d, f**). Clearly, especially for the case of the larger droplets, the signal depends non-linearly on the laser energy and shows a threshold energy. Small droplet sizes are useful for measuring absorption spectra, while larger droplets can be used as a marker for how many photons were absorbed in, for example, IR-IR pump probe experiments

IR spectrum of a non-covalently bound complex consisting of eight amino acids (serine) and one extra proton. This complex has received a great deal of attention, as its high abundance in mass spectrometry experiments indicates a high stability, and further because this complex has a strong preference for homochirality [76, 77]. Its IR spectrum between 600 and 1800 cm$^{-1}$ shows a multitude of sharp resolved bands that are characteristic for the complex and that led, together with other experiments and computation, to a structural assignment [55].

**Fig. 6.7** IR spectra of three species in helium droplets. **a** two formate anions bound by a proton, **b** a naturally occurring tetrasaccharide and **c** a complex containing eight serine molecules and a proton

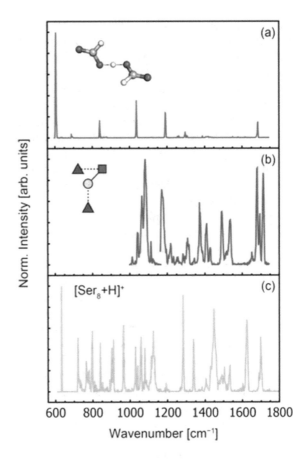

## 6.3   Spectroscopy of Ions in Helium Droplets: Results on Small Anionic Complexes and Carbohydrates

### 6.3.1   *Fluoride-$CO_2$-$H_2O$ Chemistry*

The halide ions $I^-$, $Br^-$, $Cl^-$ or $F^-$ are ubiquitous in nature, and their chemistry is of utmost importance not only in biology, but also in diverse fields such as geochemistry, atmospheric chemistry, and industrial chemistry. Important aspects of their properties include the ability to form strong hydrogen bonds and to participate in nucleophilic substitution reactions. The fluoride ion plays a special role among the halide ions. The high proton affinity and small ionic radius of fluoride promotes the formation of strong ionic hydrogen bonds in the complexation with protic molecules. Furthermore, the fluoride can also act as a potent nucleophile and for example undergo an exothermic reaction with carbon dioxide to yield fluoroformate, $FCO_2^-$, forming a covalent bond between F and C. This is of special interest, as the chemistry of $CO_2$

and possible ways to remove it from the gas phase are presently very active areas of research. Fluoroformate has been investigated by various experimental techniques, and McMahon and co-workers first generated gas-phase $FCO_2^-$ by fluoride ion transfer to study its thermochemistry by ion cyclotron resonance spectroscopy [78, 79]. The experimental findings gave a bond dissociation enthalpy ($DH_{298}$) of $133 \pm 8$ kJ/mol, significantly weaker than typical C–F bonds ($DH_{298} > 400$ kJ/mol) [80]. Infrared spectra of $FCO_2^-$ trapped in an argon matrix were measured, but the presence of multiple species within the matrix hindered band assignment [81]. Later, the tetramethylammonium salt of $FCO_2^-$ was isolated and characterized by infrared spectroscopy and solid-state NMR [82]. In addition, photoelectron spectroscopy was used to measure the electronic transitions between $FCO_2^-$ and the fluoroformyloxyl radical, $FCO_2$· [83]. Furthermore, $FCO_2^-$ as well as other halide ion—$CO_2$ complexes have been investigated by ab initio calculations [84–86].

We have explored the chemistry of $F^-$ with $CO_2$ and $H_2O$ by applying mass spectrometry coupled to infrared spectroscopy of ions in liquid helium droplets and ab-initio molecular dynamics simulations [53, 54]. In the experiment, a few $\mu$L of an aqueous solution of sodium fluoride (NaF) at a concentration of 1 mM/L is placed into a nESI capillary. The solution is sprayed into atmosphere to which additional $CO_2$ gas is added to increase the $CO_2$ level to a few % (the exact amount has not been quantified), and the instrument is set to transmit and detect anions. To aid in the assignment of molecular composition and to facilitate the assignment of IR spectra, deuteration experiments are also performed. To do so, the NaF is dissolved in $D_2O$, and the $CO_2$ is bubbled through $D_2O$ before being brought to the ESI region. Ions are then transferred into the instrument, and mass spectra are recorded using the on-axis TOF mass spectrometer.

Figure 6.8 shows a mass spectrum of the ions obtained. The spectrum is surprisingly simple, consisting of three dominant peaks. At the lowest m/z = 61, a peak corresponding to the bicarbonate anion is observed, and at m/z = 63, the $FCO_2^-$ ion is found. Interestingly, the most intense peak in the mass spectrum occurs 18 m/z

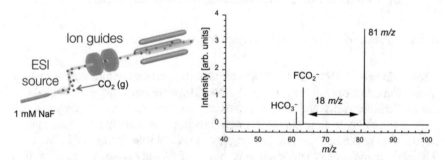

**Fig. 6.8** A 1 mM aqueous NaF solution is sprayed via nESI into air, to which some $CO_2$ gas is added. Anions are then transferred into the vacuum of the instrument and mass spectra are recorded (right side). The mass spectrum consists of three intense peaks, which can be assigned to $HCO_3^-$, $FCO_2^-$ and the adduct or reaction product of $FCO_2^- + H_2O$. The ions can then be captured in helium droplets and investigate via IR spectroscopy

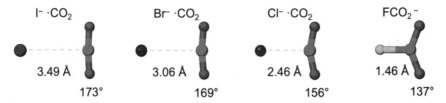

**Fig. 6.9** Calculated structures of halide ion—$CO_2$ complexes at the MP2/def2-TZVPP level. The initial structures for $X = I^- - Cl^-$ are taken from [86]

higher than (presumably) $FCO_2^-$, indicating the addition of one water molecule to $FCO_2^-$. The observation of the strong $FCO_2^-$ was somewhat surprising; the fact that $FCO_2^-$ is a stable ion was known, however, its facile production in the ESI process had not been reported. When replacing NaF by NaCl, no signal corresponding to $ClCO_2^-$ is observed, nor is signal indicating an ion 18 m/z larger observed.

Figure 6.9 shows structures calculated at the MP2/def2-TZVPP level for $X \cdot CO_2$ complexes ($X = I^-, Br^-, Cl^-$ and $F^-$) ($X = I^- - Cl^-$ from [86]; $X = F^-$ this work). It can be seen that for $I^- - Cl^-$, the halide-carbon distance remains larger than is common for a covalent bond, however also that the O–C–O angle decreases, indicative of at least some electron donation from the halide ion into antibonding O–C–O orbitals. The situation is different for $FCO_2^-$, for which both the structure (Fig. 6.9) as well as energetics [78, 79] indicate a stronger covalent character of the F–C bond and significant electron donation into the $CO_2$ moiety.

The $FCO_2^-$ ion can be incorporated into helium droplets, and the IR spectrum can be recorded, as shown in Fig. 6.10, middle trace. For comparison, the IR-spectrum of the well-known formate ion $HCO_2^-$ in helium droplets has been measured in the 1200–1800 cm$^{-1}$ range (top trace). This spectrum shows two very narrow peaks at 1317 and 1623 cm$^{-1}$ stemming from the symmetric and antisymmetric O–C–O stretching motion in $HCO_2^-$, respectively. These measurements compare very well to literature values of 1314 and 1622 cm$^{-1}$ for those modes [87]. It therefore seems that the helium environment does not induce a strong shift (at least no more than estimated 1–5 cm$^{-1}$) in line positions. Another striking observation concerns the width of the bands. In the spectra of both the formate and fluorofomate anions, the bands are very narrow, exhibiting full width at half maximum (fwhm) widths between 2 and 6 cm$^{-1}$. This is in some instances narrower than the spectral width of the FHI-FEL (typical fwhm of $\sim$0.5% of the photon energy). Most likely, this is caused by the non-linear nature of the dependence of the observed bare ion signal on laser fluence.

Shown in the middle trace of Fig. 6.10 is the helium droplet IR spectrum of the fluoroformate anion, and a total of six bands can be observed. For some of them, facile assignment based on previous experiments is possible [81, 82]. At 548 cm$^{-1}$, a band that can be assigned to result from C–F stretching motion is observed. This value is significantly red-shifted from the value of a typical C–F bond (>1000 cm$^{-1}$) due to the weakness of the C–F bond in $FCO_2^-$. The transition at 747 cm$^{-1}$ can be assigned to result from O–C–O bending motion ($\delta(OCO)$). At around 1270 cm$^{-1}$, only one

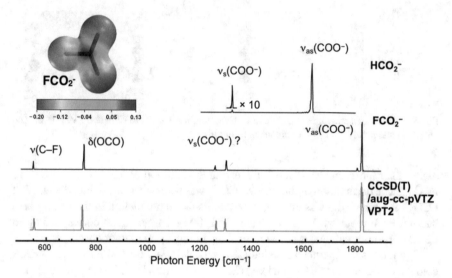

**Fig. 6.10** Vibrational spectrum of (**a**) formate and (**b**) fluoroformate obtained using the helium nanodroplet method. Shown in (**c**) is the theoretical vibrational spectrum of fluoroformate with anharmonic corrections calculated utilizing the VPT2 method at the CCSD(T)/aug-cc-pVTZ level of theory. Excellent agreement is observed between the experimental and theoretical vibrational spectra of fluoroformate, most notably for the Fermi resonance between $\nu_s(COO^-)$ and $\nu(C–F)$ + $\delta(OCO)$. However, theory overestimates the degree of vibrational wave function mixing in the Fermi resonance between $\nu_{as}(COO^-)$ and $\delta(C–F)$ + $\nu_s(COO^-)$. The top right corner shows the calculated electrostatic potential around the $FCO_2^-$ anion. It can be seen that the negative charge is distributed among the F- and O-atoms, with a slightly higher negative charge on the O-atoms

band resulting from symmetric O–C–O stretching motion ($\nu_s(COO^-)$) is expected, whereas the experiment shows two bands, one at 1251 cm$^{-1}$ and one at 1294 cm$^{-1}$. Similarly, only one band, stemming from antisymmetric O–C–O stretching motion ($\nu_{as}(COO^-)$) is expected to the blue, whereas the experiment shows two bands, a weak band at 1799 cm$^{-1}$ and a stronger band at 1816 cm$^{-1}$.

A complete assignment of the spectrum can be performed with the help of quantum chemistry. The ion is small enough for the "gold standard" method in quantum chemistry, coupled cluster with single, double and perturbative triple excitations (CCSD(T)) in combination with the rather large aug-cc-pVTZ basis set, to be applied. Calculations of vibrational frequencies are in most cases performed using the harmonic approximation. Here, we go beyond that and include anharmonicities, overtones, combination bands and Fermi resonances using the VPT2 method [88–90], as implemented within CFOUR [91]. Shown in the lower trace of Fig. 6.10 is the result of those calculations. The calculated line positions and intensities are convoluted with an 0.5% (full width at half maximum) gaussian function. No scaling of the horizontal axis has been applied. It can be observed that the calculations reproduce the experiment extremely well. The results from theory can also be used to elucidate the nature of the two doublets around 1270 and 1800 cm$^{-1}$. It turns out that the doublet near 1270 cm$^{-1}$ is caused by a Fermi resonance between the combination mode $\nu$(C–F) + $\delta$(OCO) with $\nu_s$(COO$^-$). Near 1800 cm$^{-1}$, theory also predicts two bands, however with much less splitting and almost equal intensity which stems from a Fermi interaction of $\nu_s$(COO$^-$) + $\delta$(C–F) with $\nu_{as}$(COO$^-$). The prediction for this Fermi interaction is less accurate, most likely as the theoretical description of the $\delta$(C–F) mode is poorer [53].

An interesting question is the nature of the ion that occurs at m/z = 81 in Fig. 6.8 and gives rise to the most intense peak. Typically, the intensity is directly related to the stability and further, as the ion is generated in in the ESI process in moist air, it must be largely unreactive towards at least oxygen, water or $CO_2$. The species with m/z = 81 is exactly 18 m/z higher in mass than $FCO_2^-$, indicating that it is either a adduct or reaction product of $FCO_2^-$ with $H_2O$.

To elucidate the structure of the m/z = 81 ion, we recorded its IR spectrum in helium droplets and performed calculations of possible $FCO_2^-$ + $H_2O$ adducts or reaction products [54]. The results are shown in Fig. 6.11, with the experimental spectrum shown in the top trace. Candidate structures were optimized using the MP2 method using the aug-cc-pVTZ basis set, anharmonic IR spectra were calculating using the GVPT2 method [92, 93] as implemented in Gaussian 16 [94]. When comparing the experimental spectrum to the results of the calculations, it is clear

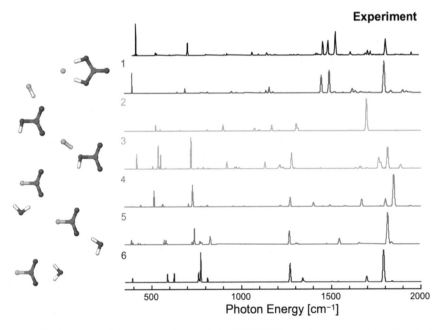

**Fig. 6.11** Experimental infrared spectrum of the [$H_2CO_3F$]$^-$ species compared to theoretical spectra of candidate structures. The experimental spectrum (black) was collected by ion infrared action spectroscopy in helium nanodroplets. Theoretical spectra for candidate structures **1–6** were calculated at the MP2/aug-cc-pVTZ level of theory with anharmonic corrections from the GVPT2 method. The experimental spectrum matches best to the spectrum calculated for structure **1**

that the best match of the experimental spectrum is to the spectrum of structure **1**—a complex of carbonic acid with $F^-$ [54].

The generation of a stable complex between carbonic acid and fluoride is highly intriguing. Carbonic acid is central to many chemical processes, yet it is thermodynamically unstable, undergoing decomposition to yield carbon dioxide and water. Although the barrier to dissociation is large for an isolated carbonic acid molecule, the dissociation reaction is catalyzed by complexation with many abundant molecular species, most notably water, ammonia, formic and acetic acid, and other carbonic acid molecules, which renders experimental characterization very challenging. The observed $H_2CO_3 + F^-$ complex is extremely stable and observed as the most abundant ion produced in the electrospray process under the employed conditions, providing a robust chemical trap for the normally elusive carbonic acid molecule. Complexation with anions and isolation in helium droplets was proven to be a successful method to further examine the structure and properties of this elusive molecule.

### 6.3.2 Carbohydrates

Among the four major classes of biomolecules in mammals—DNA, RNA, proteins and carbohydrates—carbohydrates remain the most poorly characterized. Carbohydrates are synonymously termed glycans, oligo- or polysaccharides or also in a broader sense simply sugars. Historically, carbohydrates were exclusively linked to functions such as energy storage, bioscaffolds, or cellular decorations in the absence of identified biological functions. In our everyday life, carbohydrates play a role as sources of nutrition, but carbohydrates are in fact a manifold class of biomolecules. The following pages present an introduction to some of the important functions of carbohydrates relevant to mammalian biology.

In biochemistry and medicine, the sequencing of the human genome was a breakthrough to the elucidation of all template-driven processes that derive from the identified genes. RNA and protein research advanced at incomparable speed. Glycosylation patterns, however, cannot be directly linked to the template of the human genome. Furthermore, many glycosylation patterns are not rigid but highly dynamic over time, which is obvious in an evolutionary sense, but accounts for physiological variations as well as pathological variations. The diverse roles of carbohydrates span from inflammation via cell adhesion to molecular recognition and cell–cell interactions [95–99]. To name an example of outstanding interest throughout the past year, the dense glycosylation on the outer shell of the spike proteins of coronaviruses shields the viruses from antibody recognition, enabling them to evade the host's immune response. In addition, some coronavirus glycosylation sites have a structural role relevant for host cell entry [100–103].

Formerly, carbohydrates, or the hydrates of carbons, were represented by the empirical formula $C_n(H_2O)_m$. As known today, many naturally occurring carbohydrates are exceptions from the empirical formula, for example deoxyhexoses or

*N*-acetylhexoses, the latter of which is depicted in Fig. 6.12. The chemical representation in Fig. 6.12 is typically used in the wider field of organic chemistry and known as a skeletal formula. Carbon and hydrogen atoms are only implicitly depicted or excluded for clarity reasons. All heteroatoms and the hydrogen atoms attached to these are explicitly depicted. Using wedged bonds, the chair conformation classically used for six-membered rings is illustrated in a three-dimensional way. To minimize steric repulsion between functional groups, monosaccharides in heterocyclic ring forms are not planar but puckered [104]. Besides the chair conformation, the families of conformations are boat, envelope, half-chair and skew. For simplicity, the chair conformation is most often used in illustrations.

The most basic units in carbohydrates are monosaccharides. Generally, monosaccharides can be present in an open chain form or in two different heterocyclic ring forms. Five-membered rings are called furanoses and the usually preferred six-membered rings pyranoses, see Fig. 6.12. Monosaccharides are grouped by the number of carbon atoms into tetroses, pentoses, hexoses or heptoses, containing four, five, six and seven carbon atoms, respectively. These groups are further divided by the functional groups present in open chain form, such as aldoses with an aldehyde or ketoses with a ketone functional group. Each group comprises a number of possible monosaccharides, isomers, that differ in the orientation of the hydroxyl groups, or in mentioned exceptions the orientation of other functional groups. Monosaccharides that differ in only one stereocenter are called epimers. The carbon atoms in a monosaccharide are labelled numerically, i.e. in the heterocyclic ring form counting clockwise from the ring oxygen beginning with the carbon atom at C1 position. The D- or L configuration is defined by the orientation of the hydroxyl group at the stereocenter furthest from the highest oxidized carbon atom in open chain form, or in the heterocyclic ring form at the carbon atom in the ring furthest from C1. The stereocenter at C1, termed anomeric center, is created with the ring-closing reaction. The two possible configurations are termed alpha (α) and beta (β) anomers, according to the relative orientation at the anomeric center and the highest, chiral carbon atom in the ring, the anomeric reference atom. In aqueous solution, α- and β-anomers can interconvert into each other in a process called mutarotation.

Going to larger carbohydrates, disaccharides are composed of two monosaccharides and oligo- or polysaccharides are polymers of larger numbers of monosaccharides. Monosaccharides, which define the composition of carbohydrates, are connected covalently via glycosidic linkages. In a condensation reaction, the glycosidic linkage is formed between the hydroxyl group at the anomeric center and one

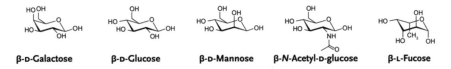

β-D-Galactose          β-D-Glucose          β-D-Mannose          β-N-Acetyl-D-glucose          β-L-Fucose

**Fig. 6.12** Examples for prominent monosaccharides in mammals. All monosaccharides are aldohexoses adopting a pyranose ring structure

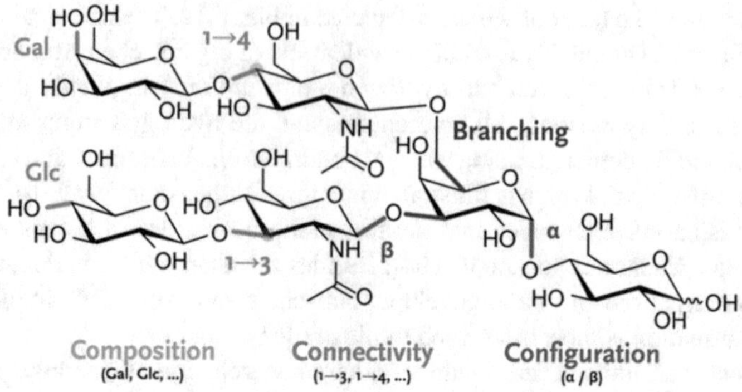

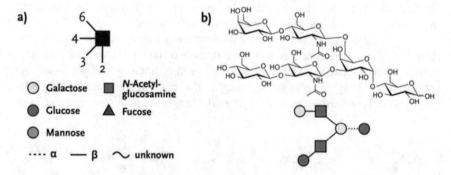

**Fig. 6.13** A generic glycan illustrating the structural details of glycans: branching, composition, connectivity and configuration. The composition of a glycan is defined by the monosaccharides present. The connectivity, in locked α- or β-configuration, indicates which hydroxyl group are involved in the glycosidic linkage. Branching is possible in larger carbohydrates

of the hydroxyl groups of the other monosaccharide, generating 1 → 2, 1 → 3, 1 → 4 or 1 → 6 linkages, as illustrated in Fig. 6.13. The formation of a glycosidic bond locks the anomeric center in either the α- or β-configuration. In contrast to DNA, RNA or proteins, oligo- and polysaccharides are not only linear molecules but can be branched, leading to significant challenges in glycan structural analysis.

Many representations of carbohydrates, such as the Fischer or Haworth projection, evolved to ease the reading of the chemical structures of carbohydrates. For example, for monosaccharides in their open chain form, the Fischer projection is typically used. The symbol nomenclature for glycans (SNFG) is the most commonly used structural representation in the field of carbohydrate chemistry [75, 105]. Figure 6.14b shows a carbohydrate composed of six monosaccharide units in its skeletal formula and

**Fig. 6.14 a** The symbol nomenclature for glycans (SNFG) [75, 105] is commonly used to illustrate glycan structures. Monosaccharides are represented in colored symbols. A glycosidic linkage is either a dashed or solid line indicating the configuration. The angle of the connecting line shows the connectivity. **b** A generic glycan structure in chemical representation and translated to SNFG

translated to the SNFG nomenclature. The composition of monosaccharides is visualized using colored symbols of different shapes, the connectivity is depicted with the angle of the connecting line at the symbol, and the configuration is represented with either a dashed or a solid line. The standardized use of symbols for monosaccharide building blocks is especially helpful in the depiction of larger structures.

The structural diversity and complexity of carbohydrates is thus very large, yet minute changes can alter the biological function. To address the question if, and in how far IR spectroscopy is sensitive to variations in carbohydrate structure, vibrational spectra of glycans in helium nanodroplets were recorded. Glycan ions are brought into the gas phase by nano-electrospray ionization (nESI). These ions are mass-to-charge selected and accumulated inside the ion trap. Helium nanodroplets with an average size of $10^5$ helium atoms traverse the trap and can pick up ions and rapidly cool them to the equilibrium temperature of the droplet. Inside the detection chamber, the embedded ions are investigated using laser radiation produced by the FHI-FEL. The subsequent absorption of resonant photons can lead to the release of the bare ion from the droplet, which is used as a marker for photon absorption. Plotting the yield of released ions as a function of laser wavelength yields a highly reproducible IR spectrum. Each spectrum shown here consists of at least two independent scans recorded with a wavenumber step-size of $\Delta v = 2$ cm$^{-1}$. Although the yield of released ions scales non-linearly with laser energy, a linear correction has been performed by dividing the signal intensity by the laser energy. As a result, relative intensities can be distorted to some extent. The spectra shown here were recorded between 950 and 1700 cm$^{-1}$ with two different photon fluxes using a variable focusing mirror. A softer focus (lower photon fluence) was used between 950 and 1200 cm$^{-1}$. To access the absorption bands between 1150 and 1700 cm$^{-1}$, a separate spectrum with a tighter focus (higher photon fluence) was recorded.

To assess the potential of cryogenic ion spectroscopy to resolve the minute structural details present in complex carbohydrates, a series of well-defined amino-alkyl linked carbohydrates as well as natural samples were analyzed. The analyzed samples consist of monosaccharides **1–5**, disaccharides **6–7**, trisaccharides **8–13**, and naturally occurring tetrasaccharides **14–17**, as shown in Fig. 6.15.

### 6.3.3  Mono- and Disaccharides

As discussed in the preceding section, different monosaccharides often share the same mass and only differ in the stereochemistry at single carbon atoms. Using cryogenic ion spectroscopy in helium droplets, IR spectra of the protonated aminoalkyl-linked monosaccharides were recorded (Fig. 6.16). Because the primary amine of the linker has the largest basicity, it will most likely accept a proton upon ionization and provide a localized charge.

The IR spectrum of α-mannose (**1**) shows a variety of absorption bands. Around 1150 cm$^{-1}$, two narrow and well-resolved absorption bands are found. Preliminary quantum chemical calculations indicate that the most intense bands in this

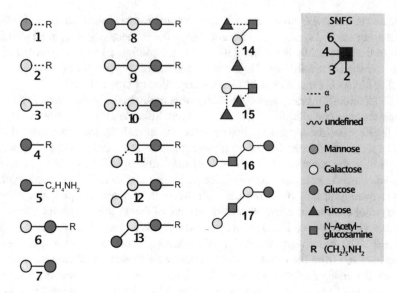

**Fig. 6.15** Schematic SNFG representation of the here investigated molecules. The sets consist of monosaccharides **1–5**, disaccharides **6** and **7**, trisaccharides **8–13**, and biologically relevant tetrasaccharides **14–17**

region (1000–1150 cm$^{-1}$) arise from C–O-stretching vibrations with strong transition dipole moments. Between 1150 and 1400 cm$^{-1}$, additional bands that likely originate from O–H-bend modes are present. Above 1400 cm$^{-1}$, a series of partially resolved bands are obtained. According to calculations, these bands correspond to the primary ammonium group of the linker. Interestingly, only three NH$_3{}^+$-bending modes are expected in this region. The presence of multiple, partially unresolved bands indicates that α-mannose adopts multiple coexisting conformers with different IR fingerprints in the gas phase. Moving to the isomeric α-galactose (**2**), the stereoconfiguration at C2 and C4 is inverted. The IR spectrum also shows a variety of resolved absorption bands, but the peak positions and intensities render it distinct from the spectrum obtained for α-mannose. Again, the presence of multiple absorption bands above 1400 cm$^{-1}$ indicates multiple coexisting conformers. The IR spectrum of the epimeric β-galactose (**3**) is characterized by several highly resolved absorption bands between 950 and 1400 cm$^{-1}$ and broader absorptions above 1400 cm$^{-1}$. Interestingly, some bands are as narrow as the bandwidth of the IR laser radiation (FWHM around 4 cm$^{-1}$). Also, the IR spectrum of β-glucose (**4**) exhibits a number of highly resolved absorption bands up to 1150 cm$^{-1}$. Between 1150 and 1500 cm$^{-1}$, only a few weaker absorptions are present. To examine the influence of the attached aminopentyl linker, β-glucose with a shorter aminoethyl linker (**5**) was investigated. A direct comparison shows that the two spectra are distinct. The spectrum of **5** exhibits fewer absorption bands than the one obtained for **4**, and the peak positions do not coincide. In contrast to the vibrational spectra of monosaccharides **1–4**, only one strong and three weak resolved bands appear above 1400 cm$^{-1}$. This

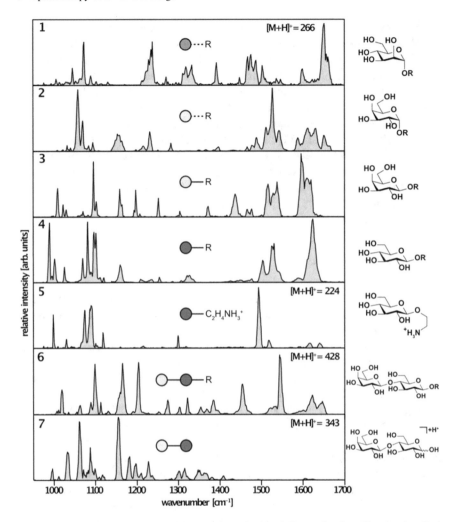

**Fig. 6.16** IR spectra of aminoalkyl-linked monosaccharides **1–7** as well as free (**7**) and aminoalkyl-linked lactose (**6**) investigated as [M + H]⁺ ions

could indicate that the gas-phase structure of **5** mainly adopts a single conformer, which could be caused by the reduced flexibility of the shorter linker.

Upon moving to disaccharides, which consist of two monosaccharide building blocks joined by a glycosidic bond, the structural complexity increases significantly. The glycosidic bond can be formed at different hydroxy groups (connectivity isomers) with either an α- or β-configuration. The increasing system size and the torsional angle of the glycosidic bond further increase the conformational complexity. Although an additional building block is present, the IR spectrum of β-lactose (**6**) shows a similar appearance to the monosaccharides described above. The variety of highly resolved bands provide a unique diagnostic pattern. To further investigate the influence of the aminoalkyl linker, a spectrum of protonated free lactose (**7**) was recorded. Here, the reducing end does not contain any linker, and the anomeric center can adopt either an α- or β-configuration. In addition, the location of the charge is uncertain, and different protomers may coexist. Nonetheless, the IR spectrum of **7** exhibits multiple well-resolved absorption bands between 950 and 1400 cm$^{-1}$. There are no obvious similarities between the vibrational signatures of both lactose variants, which indicates that their gas-phase structures are widely different. These structural differences most likely result from interactions between the charged linker and hydroxy groups of the disaccharide, or from a distinct charge distribution in **7**. The absence of characteristic absorption bands above 1400 cm$^{-1}$ confirms the assumption that this region is mainly governed by vibrational transitions of the charged aminoalkyl linker.

### 6.3.4 Trisaccharides

Next, a set of six isomeric trisaccharides was used to benchmark the method. The structures of trisaccharides **8–13** share the same reducing-end β-lactose core-motif and an aminopentyl linker, as shown in Fig. 6.17. The terminal building block was systematically varied to generate isomers that differ in composition, connectivity or configuration. For example, the glycan pairs **8/9** and **12/13** are compositional isomers and only differ in the identity of the terminal building block (Glc vs. Gal). Glycan pairs **9/10** and **11/12** are configurational isomers and differ in the stereoconfiguration of the glycosidic bond. Finally, glycan pairs **8/13**, **9/12** and **10/11** are connectivity isomers that only differ in the position of the glycosidic bond (1 → 3 vs. 1 → 4).

These isomeric trisaccharides are extremely difficult to distinguish using established LC–MS techniques. Recent studies [106] used ion mobility-mass spectrometry to approach this analytical challenge and measured the arrival-time distributions and collision cross sections of trisaccharides **8–13**. In negative ion mode, connectivity and configurational isomers were efficiently separated with baseline separation, allowing an unambiguous identification. The remarkable baseline separation also enabled a quantitative analysis of coexisting isomers inside a mixture. However, compositional isomers consistently showed very similar arrival time distributions and could therefore not be distinguished using IM-MS.

The IR spectra of trisaccharides **8–13** are shown in Fig. 6.18. In general, each spectrum exhibits a large number of highly resolved absorption bands and no significant spectral congestion. The configurational isomers **9/10** and **11/12** share the same connectivity and configuration, but differ in the configuration of the terminal glycosidic bond (α/β). The distinctness of the absorption patterns allows facile differentiation of these conformational isomers. Similar results are obtained for connectivity isomers **8/13**, **9/12** and **10/11**, which exhibit either a 1 → 3 or 1 → 4 terminal glycosidic bond. Again, the highly resolved optical signatures reveal differences that allow an unambiguous identification of each connectivity isomer. Most striking are the observations for compositional isomers **8/9** and **12/13**, which share the same

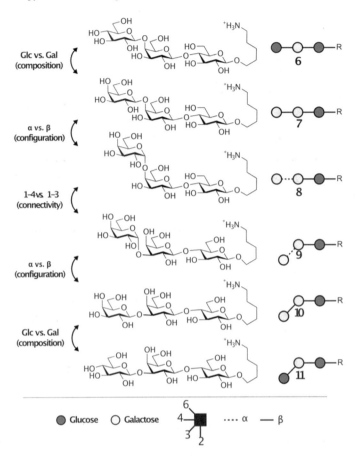

**Fig. 6.17** The synthetic trisaccharides **8–13** only differ in the connectivity (1 → 3 vs. 1 → 4), configuration (α vs. β), or composition (Glc vs. Gal) of the terminal building block. These isomeric oligosaccharides are extremely difficult to distinguish using established methods and therefore serve as a benchmark for this experimental approach

connectivity and configuration and only differ in the identity of the terminal building block (Glc vs. Gal). In other words, the trisaccharide structures within each isomeric pair only differ in the stereochemical orientation of a *single* hydroxy group. Surprisingly, these minute structural variations lead to substantial differences in the IR signatures, especially above 1300 cm$^{-1}$. For example, a characteristic high-intensity transition is observed at 1450 cm$^{-1}$ for trisaccharide **9**, but is absent in the spectrum of the corresponding compositional isomer **8**. Also, trisaccharide **12** features a strong absorption band around 1310 cm$^{-1}$ that is absent in the spectrum of the trisaccharide **13**. Taken together, each of the six trisaccharide isomers exhibits a variety of resolved absorption bands that lead to a unique IR signature and allow their unambiguous identification. Similarly to the previously discussed mono- and disaccharides, it is interesting to note that some IR spectra feature more than three absorption bands

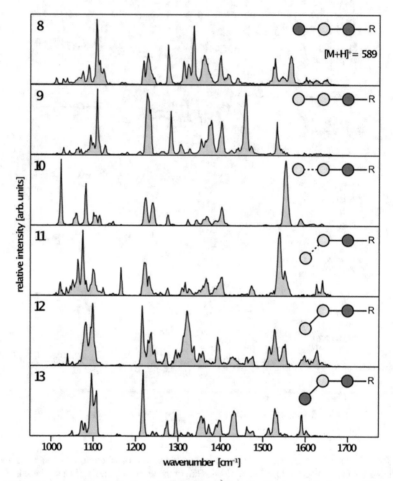

**Fig. 6.18** IR spectra of the isomeric trisaccharides **8–13**. Despite the marginal structural differences between these species, each trisaccharide can be readily distinguished by its unique absorption pattern that contains a variety of resolved bands

expected for the charged aminopentyl linker. For example, trisaccharide **12** features five resolved bands, which indicates the presence of multiple coexisting conformers. Other trisaccharides such as **10** show a much cleaner spectrum with a few absorption bands that are as narrow as the bandwidth of the laser radiation.

### 6.3.5 Naturally Occurring Tetrasaccharides

To extend the scope of this method from synthetic standards to naturally occurring glycans, the blood-group antigens Lewis b (Le^b, **14**) and Lewis y (Le^y, **15**) were investigated as sodium adducts. Although it is in principle possible to generate protonated species of these samples, the high salt concentrations found in many samples typically lead to strong signals of sodium adducts. The isomeric tetrasaccharides each consist of an $N$-acetylglucosamine, a galactose and two fucose building blocks and differ in their glycosidic linkages. Similarly to free lactose **7**, the absence of a linker leads to an undefined anomeric center that can adopt either an $\alpha$- or $\beta$-configuration. Although the number of expected vibrational transitions for molecules of this size is large, the corresponding IR spectra (Fig. 6.19) exhibit a remarkably small number of well-resolved absorption bands that allow an unambiguous discrimination between the two isomeric species. Especially the Le$^y$ tetrasaccharide shows an extremely clean IR spectrum with a total of eight features. The transitions around 1500 m$^{-1}$ and 1680 cm$^{-1}$ likely stem from the amide group of the GlcNAc building block and are assigned as amide II and amide I bands, respectively. A single amide group is expected to give only one amide II and one amide I band. Here, however,

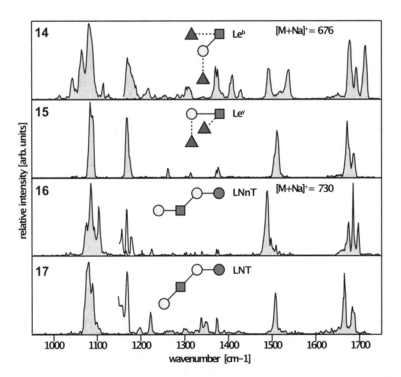

**Fig. 6.19** As representatives of biologically relevant glycans, the blood-group antigens Le$^b$ (**14**) and Le$^y$ (**15**), as well as the milk-sugar tetrasaccharides lacto-$N$-neotetraose (**16**) and lacto-$N$-tetraose (**17**) were investigated as [M + Na]$^+$ ions. The characteristic absorption patterns allow unambiguous discrimination between the corresponding isomers. The discontinuity in intensity around 1150 cm$^{-1}$ results from differences in photon fluence in the irradiation region prior to ion detection

multiple bands are observed, which indicates that multiple coexisting conformers with different amide band positions coexist in the gas phase.

As another example for naturally occurring glycans, lacto-*N*-neotetraose (**16**) and lacto-*N*-tetraose (**17**) were investigated. These isomeric tetrasaccharides belong to the group of human milk oligosaccharides and only differ in the connectivity of the terminal galactose building block. The corresponding IR spectra of the sodium adducts shown in Fig. 6.19 also exhibit distinct and well-resolved absorption features that allow their discrimination. Again, multiple bands in the amide II and amide I region indicate the presence of coexisting conformers.

The highly resolved IR spectra allow an unambiguous identification of complex isomeric glycans. But what are the underlying structures that lead to those absorption patterns? In general, theoretical methods are widely used to compute structural candidates and calculate their corresponding IR spectra. These theoretical IR spectra are then compared to experimental results in order to extract the structural information that is encoded in the vibrational fingerprint, and identify the underlying structure. For the glycans presented here, this approach turned out to be challenging. There is a threefold problem: (1) the molecules are extremely flexible, have many rotatable bonds and ring-pucker possibilities, giving rise to a very large conformational space; (2) the molecules are large, and accurate calculations of the structure and energies of individual conformers are very expensive; and (3) calculations of IR spectra beyond the harmonic approximations are presently out of the question for molecules of that size. It turns out that problem (1) is the most severe. For other systems such as peptides and proteins, a pre-screening of the conformational space using a less costly method, such as a simple empirical force field, is often performed. In order to be able to do so, these force fields need to have a certain minimum accuracy to give results that are meaningful for our experiments. For peptides and proteins, such force fields have been developed over several decades and area readily available. For glycans, however, no sufficiently accurate force fields are presently available. A firm conformational assignment of the here investigated molecules therefore requires improvements in the theoretical description.

Nonetheless, cryogenic vibrational spectroscopy is a valuable addition to the structural analysis toolbox for glycans. The low temperature environment of superfluid helium nanodroplets enables the acquisition of highly resolved absorption spectra: a true spectral fingerprint that is unique for each glycan. Even minute structural differences such as the stereochemical orientation of a single hydroxy group within trisaccharides lead to spectral differences that allow an unambiguous identification which is otherwise difficult or impossible to do. The method in general therefore offers possibilities for commercial application in the glycosciences. However, the sophisticated experimental setup used in this work involving a FEL will arguably not find a commercial application. This will be left to simpler implementations using cryogenic ion trap and commercially available benchtop laser systems.

## 6.4 Conclusions

Helium nanodroplets are indeed the ultimate cryogenic matrix for optical spectroscopy experiments. The method can not only be applied to volatile neutral species, but also to a wide variety of anionic or cationic gas-phase ions. When used in combination with mass spectrometric techniques, unprecedented selectivity and sensitivity can be obtained, allowing for the investigation of small ionic complexes, larger biological molecules and even entire proteins, containing thousands of atoms.

The IR spectra shown here are highly diagnostic for the species investigated. For small ions, rigorous theory can quantitatively explain the observations. For larger species, IR spectroscopy in helium nanodroplets can be used as an orthogonal technique to mass spectrometry and ion mobility spectrometry to fingerprint and distinguish molecular isomers or conformers which are indistinguishable by traditional bioanalytic methods. Present and future applications therefore range from fundamental studies on the structure and dynamics of molecules to real world applications in analytical chemistry.

**Acknowledgements** The authors thank the FHI-FEL staff, particularly Sandy Gewinner and Wieland Schöllkopf, for laser operation. D.A.T. gratefully acknowledges the support of the Alexander von Humboldt Foundation.

## References

1. L.I. Yeh, M. Okumura, J.D. Myers, J.M. Price, Y.T. Lee, Vibrational spectroscopy of the hydrated hydronium cluster ions $H_3O^+ \cdot (H_2O)_n$ (n=1, 2, 3). J. Chem. Phys. **91**(12), 7319–7330 (1989)
2. H. Oh, K. Breuker, S.K. Sze, Y. Ge, B.K. Carpenter, F.W. McLafferty, Secondary and tertiary structures of gaseous protein ions characterized by electron capture dissociation mass spectrometry and photofragment spectroscopy. Proc. Natl. Acad. Sci. USA **99**(25), 15863–15868 (2002). https://doi.org/10.1073/pnas.212643599
3. J. Oomens, N. Polfer, D.T. Moore, L. van der Meer, A.G. Marshall, J.R. Eyler, G. Meijer, G. von Helden, Charge-state resolved mid-infrared spectroscopy of a gas-phase protein. Phys. Chem. Chem. Phys. **7**(7), 1345–1348 (2005). https://doi.org/10.1039/b502322j
4. J. Oomens, B.G. Sartakov, G. Meijer, G. von Helden, Gas-phase infrared multiple photon dissociation spectroscopy of mass-selected molecular ions. Int. J. Mass. Spectrom. **254**(1–2), 1–19 (2006). https://doi.org/10.1016/j.ijms.2006.05.009
5. N.C. Polfer, J. Oomens, Vibrational spectroscopy of bare and solvated ionic complexes of biological relevance. Mass. Spectrom. Rev. **28**(3), 468–494 (2009). https://doi.org/10.1002/mas.20215
6. N.C. Polfer, Infrared multiple photon dissociation spectroscopy of trapped ions. Chem. Soc. Rev. **40**(5), 2211–2221 (2011). https://doi.org/10.1039/c0cs00171f
7. N.C. Polfer, B. Paizs, L.C. Snoek, I. Compagnon, S. Suhai, G. Meijer, G. von Helden, J. Oomens, Infrared fingerprint spectroscopy and theoretical studies of potassium ion tagged amino acids and peptides in the gas phase. J. Am. Chem. Soc. **127**(23), 8571–8579 (2005). https://doi.org/10.1021/ja050858u

8. H.B. Oh, C. Lin, H.Y. Hwang, H. Zhai, K. Breuker, V. Zabrouskov, B.K. Carpenter, F.W. McLafferty, Infrared photodissociation spectroscopy of electrosprayed ions in a Fourier transform mass spectrometer. J. Am. Chem. Soc. **127**(11), 4076–4083 (2005). https://doi.org/10. 1021/ja040136n

9. J. Seo, W. Hoffmann, S. Warnke, X. Huang, S. Gewinner, W. Schöllkopf, M.T. Bowers, G. von Helden, K. Pagel, An infrared spectroscopy approach to follow beta-sheet formation in peptide amyloid assemblies. Nat. Chem. **9**(1), 39–44 (2017). https://doi.org/10.1038/nchem. 2615

10. N.C. Polfer, J. Oomens, S. Suhai, B. Paizs, Infrared spectroscopy and theoretical studies on gas-phase protonated leu-enkephalin and its fragments: direct experimental evidence for the mobile proton. J. Am. Chem. Soc. **129**(18), 5887–5897 (2007). https://doi.org/10.1021/ja0 68014d

11. J. Oomens, B.G. Sartakov, A.G.G.M. Tielens, G. Meijer, G. von Helden, Gas-phase infrared spectrum of the coronene cation. Astrophys. J. **560**(1), L99–L103 (2001). https://doi.org/10. 1086/324170

12. D.T. Moore, J. Oomens, L. van der Meer, G. von Helden, G. Meijer, J. Valle, A.G. Marshall, J.R. Eyler, Probing the vibrations of shared, $OH^+O$-bound protons in the gas phase. ChemPhysChem **5**(5), 740–743 (2004). https://doi.org/10.1002/cphc.200400062

13. J.A. Fournier, C.J. Johnson, C.T. Wolke, G.H. Weddle, A.B. Wolk, M.A. Johnson, Vibrational spectral signature of the proton defect in the three-dimensional $H^+(H_2O)_{21}$ cluster. Science **344**(6187), 1009–1012 (2014). https://doi.org/10.1126/science.1253788

14. M.A. Duncan, Frontiers in the spectroscopy of mass-selected molecular ions. Int. J. Mass. Spectrom. **200**(1–3), 545–569 (2000). https://doi.org/10.1016/S1387-3806(00)00366-3

15. H. Piest, G. von Helden, G. Meijer, Infrared spectroscopy of jet-cooled neutral and ionized aniline-Ar. J. Chem. Phys. **110**(4), 2010–2015 (1999). https://doi.org/10.1063/1.477866

16. K.R. Asmis, N. Brummer, C. Kaposta, G. Santambrogio, G. von Helden, G. Meijer, K. Rademann, L. Wöste, Mass-selected infrared photodissociation spectroscopy of $V_4O_{10}^+$. Phys. Chem. Chem. Phys. **4**(7), 1101–1104 (2002). https://doi.org/10.1039/b111056j

17. T.R. Rizzo, J.A. Stearns, O.V. Boyarkin, Spectroscopic studies of cold, gas-phase biomolecular ions. Int. Rev. Phys. Chem. **28**(3), 481–515 (2009). https://doi.org/10.1080/014423509030 69931

18. A.B. Wolk, C.M. Leavitt, E. Garand, M.A. Johnson, Cryogenic ion chemistry and spectroscopy. Acc. Chem. Res. **47**(1), 202–210 (2014). https://doi.org/10.1021/ar400125a

19. J.G. Redwine, Z.A. Davis, N.L. Burke, R.A. Oglesbee, S.A. McLuckey, T.S. Zwier, A novel ion trap based tandem mass spectrometer for the spectroscopic study of cold gas phase polyatomic ions. Int. J. Mass. Spectrom. **348**, 9–14 (2013). https://doi.org/10.1016/j.ijms.2013.04.002

20. J.P. Toennies, A.F. Vilesov, Superfluid helium droplets: a uniquely cold nanomatrix for molecules and molecular complexes. Angew. Chem-Int. Edit. **43**(20), 2622–2648 (2004). https://doi.org/10.1002/anie.200300611

21. F. Stienkemeier, K.K. Lehmann, Spectroscopy and dynamics in helium nanodroplets. J. Phys. B-At. Mol. Opt. Phys. **39**(8), R127–R166 (2006). https://doi.org/10.1088/0953-4075/ 39/8/R01

22. M.Y. Choi, G.E. Douberly, T.M. Falconer, W.K. Lewis, C.M. Lindsay, J.M. Merritt, P.L. Stiles, R.E. Miller, Infrared spectroscopy of helium nanodroplets: novel methods for physics and chemistry. Int. Rev. Phys. Chem. **25**(1–2), 15–75 (2006). https://doi.org/10.1080/014423 50600625092

23. F. Stienkemeier, J. Higgins, W. Ernst, G. Scoles, Laser spectroscopy of alkali-doped helium clusters. Phys. Rev. Lett. **74**(18), 3592–3595 (1995). https://doi.org/10.1103/PhysRevLett.74. 3592

24. S. Goyal, D.L. Schutt, G. Scoles, Vibrational spectroscopy of sulfur-hexafluoride attached to helium clusters. Phys. Rev. Lett. **69**(6), 933–936 (1992). https://doi.org/10.1103/PhysRe vLett.69.933

25. R. Fröchtenicht, J.P. Toennies, A. Vilesov, High-resolution infrared-spectroscopy of $SF_6$ embedded in He clusters. Chem. Phys. Lett. **229**(1–2), 1–7 (1994). https://doi.org/10.1016/ 0009-2614(94)01026-9

26. S. Grebenev, M. Hartmann, M. Havenith, B. Sartakov, J.P. Toennies, A.F. Vilesov, The rotational spectrum of single OCS molecules in liquid $^4$He droplets. J. Chem. Phys. **112**(10), 4485–4495 (2000). https://doi.org/10.1021/cr00031a009
27. M. Hartmann, R. Miller, J. Toennies, A. Vilesov, Rotationally resolved spectroscopy of SF$_6$ in liquid-helium clusters—a molecular probe of cluster temperature. Phys. Rev. Lett. **75**(8), 1566–1569 (1995). https://doi.org/10.1103/PhysRevLett.75.1566
28. A. Lindinger, J. Toennies, A. Vilesov, High resolution vibronic spectra of the amino acids tryptophan and tyrosine in 0.38 K cold helium droplets. J. Chem. Phys. **110**(3), 1429–1436 (1999). https://doi.org/10.1063/1.478018
29. S. Denifl, F. Zappa, I. Mahr, J. Lecointre, M. Probst, T.D. Mark, P. Scheier, Mass spectrometric investigation of anions formed upon free electron attachment to nucleobase molecules and clusters embedded in superfluid helium droplets. Phys. Rev. Lett. **97**(4), 043201 (2006). https://doi.org/10.1103/PhysRevLett.97.043201
30. S. Smolarek, A.M. Rijs, W.J. Buma, M. Drabbels, Absorption spectroscopy of adenine, 9-methyladenine, and 2-aminopurine in helium nanodroplets. Phys. Chem. Chem. Phys. **12**(48), 15600–15606 (2010). https://doi.org/10.1039/c0cp00746c
31. J. Close, F. Federmann, K. Hoffmann, N. Quaas, Absorption spectroscopy of C$_{60}$ molecules isolated in helium droplets. Chem. Phys. Lett. **276**, 393–398 (1997)
32. S. Grebenev, J.P. Toennies, A.F. Vilesov, Superfluidity within a small helium-4 cluster: the microscopic andronikashvili experiment. Science **279**(5359), 2083–2086 (1998). https://doi.org/10.1126/science.279.5359.2083
33. K. Nauta, R.E. Miller, Solvent mediated vibrational relaxation: superfluid helium droplet spectroscopy of HCN dimer. J. Chem. Phys. **111**(8), 3426–3433 (1999). https://doi.org/10.1063/1.479627
34. A. Slenczka, B. Dick, M. Hartmann, J. Toennies, Inhomogeneous broadening of the zero phonon line of phthalocyanine in superfluid helium droplets. J. Chem. Phys. **115**(22), 10199–10205 (2001). https://doi.org/10.1063/1.1409353
35. P. Claas, S.O. Mende, F. Stienkemeier, Characterization of laser ablation as a means for doping helium nanodroplets. Rev. Sci. Instrum. **74**(9), 4071–4076 (2003). https://doi.org/10.1063/1.1602943
36. M. Hartmann, R. Miller, J. Toennies, A. Vilesov, High-resolution molecular spectroscopy of van der waals clusters in liquid helium droplets. Science **272**, 1631–1634 (1996). https://doi.org/10.1126/science.272.5268.1631
37. T. Döppner, T. Fennel, T. Diederich, J. Tiggesbäumker, K.H. Meiwes-Broer KH, Controlling the Coulomb explosion of silver clusters by femtosecond dual-pulse laser excitation. Phys. Rev. Lett. **94**(1), 013401(2005) https://doi.org/10.1103/PhysRevA.61.033201
38. M.P. Ziemkiewicz, D.M. Neumark, O. Gessner, Ultrafast electronic dynamics in helium nanodroplets. Int. Rev. Phys. Chem. **34**(2), 239–267 (2015). https://doi.org/10.1080/0144235x.2015.1051353
39. D. Pentlehner, J.H. Nielsen, A. Slenczka, K. Mølmer, H. Stapelfeldt, Impulsive laser induced alignment of molecules dissolved in helium nanodroplets. Phys. Rev. Lett. **110**(9), 093002 (2013). https://doi.org/10.1103/PhysRevLett.86.4447
40. B. Shepperson, A.A. Søndergaard, L. Christiansen, J. Kaczmarczyk, R.E. Zillich, M. Lemeshko, H. Stapelfeldt, Laser-induced rotation of iodine molecules in helium nanodroplets: revivals and breaking free. Phys. Rev. Lett. **118**(20), 203203 (2017). https://doi.org/10.1103/PhysRevLett.118.203203
41. M. Schlesinger, M. Mudrich, F. Stienkemeier, W.T. Strunz, Dissipative vibrational wave packet dynamics of alkali dimers attached to helium nanodroplets. Chem. Phys. Lett. **490**(4–6), 245–248 (2010). https://doi.org/10.1016/j.cplett.2010.03.060
42. M. Lemeshko, Quasiparticle approach to molecules interacting with quantum solvents. Phys. Rev. Lett. **118**(9), 095301 (2017). https://doi.org/10.1103/PhysRevLett.118.095301
43. O. Gessner, A.F. Vilesov, Imaging quantum vortices in superfluid helium droplets, in *Annual Review of Physical Chemistry,* vol. 70, ed. by M.A. Johnson, T.J. Martinez, (2019), pp. 173–198. https://doi.org/10.1146/annurev-physchem-042018-052744

44. L.F. Gomez, K.R. Ferguson, J.P. Cryan, C. Bacellar, R.M.P. Tanyag, C. Jones, S. Schorb, D. Anielski, A. Belkacem, C. Bernando, R. Boll, J. Bozek, S. Carron, G. Chen, T. Delmas, L. Englert, S.W. Epp, B. Erk, L. Foucar, R. Hartmann, A. Hexemer, M. Huth, J. Kwok, S.R. Leone, J.H.S. Ma, F. Maia, E. Malmerberg, S. Marchesini, D.M. Neumark, B. Poon, J. Prell, D. Rolles, B. Rudek, A. Rudenko, M. Seifrid, K.R. Siefermann, F.P. Sturm, M. Swiggers, J. Ullrich, F. Weise, P. Zwart, C. Bostedt, O. Gessner, A.F. Vilesov, Shapes and vorticities of superfluid helium nanodroplets. Science **345**(6199), 906–909 (2014). https://doi.org/10.1126/science.1252395

45. Y.T. He, J. Zhang, W. Kong, Electron diffraction of CBr4 in superfluid helium droplets: a step towards single molecule diffraction. J. Chem. Phys. **145**(3) (2016). https://doi.org/10.1063/1.4958931

46. T. Gonzalez-Lezana, O. Echt, M. Gatchell, M. Bartolomei, J. Campos-Martinez, P. Scheier, Solvation of ions in helium. Int. Rev. Phys. Chem. **39**(4), 465–516 (2020). https://doi.org/10.1080/0144235x.2020.1794585

47. S. Smolarek, N.B. Brauer, W.J. Buma, M. Drabbels, IR spectroscopy of molecular ions by nonthermal ion ejection from helium nanodroplets. J. Am. Chem. Soc. **132**(40), 14086–14091 (2010). https://doi.org/10.1021/ja1034655

48. F. Bierau, P. Kupser, G. Meijer, G. von Helden, Catching proteins in liquid helium droplets. Phys. Rev. Lett. **105**(13):133402(2010) https://doi.org/10.1103/PhysRevLett.105.133402

49. M. Alghamdi, J. Zhang, A. Oswalt, J.J. Porter, R.A. Mehl, W. Kong, Doping of green fluorescent protein into superfluid helium droplets: size and velocity of doped droplets. J. Phys. Chem. A. **121**(36), 6671–6678 (2017). https://doi.org/10.1021/acs.jpca.7b05718

50. L. Tiefenthaler, J. Ameixa, P. Martini, S. Albertini, L. Ballauf, M. Zankl, M. Goulart, F. Laimer, K. von Haeften, F. Zappa, P. Scheier, An intense source for cold cluster ions of a specific composition. Rev. Sci. Instrum. **91**(3) (2020) https://doi.org/10.1063/1.5133112

51. F. Filsinger, D.S. Ahn, G. Meijer, G. von Helden, Photoexcitation of mass/charge selected hemin$^+$, caught in helium nanodroplets. Phys. Chem. Chem. Phys. **14**(38), 13370–13377 (2012). https://doi.org/10.1039/c2cp42071f

52. D.A. Thomas, M. Marianski, E. Mucha, G. Meijer, M.A. Johnson, G. von Helden, Ground-state structure of the proton-bound formate dimer by cold-ion infrared action spectroscopy. Angew. Chem-Int. Edit. **57**(33), 10615–10619 (2018). https://doi.org/10.1002/anie.201805436

53. D.A. Thomas, E. Mucha, S. Gewinner, W. Schollkopf, G. Meijer, G. von Helden, Vibrational spectroscopy of fluoroformate, $FCO_2^-$, trapped in helium nanodroplets. J. Phys. Chem. Lett. **9**(9), 2305–2310 (2018). https://doi.org/10.1021/acs.jpclett.8b00664

54. D.A. Thomas, E. Mucha, M. Lettow, G. Meijer, M. Rossi, G. von Helden, Characterization of a trans-trans carbonic acid-fluoride complex by infrared action spectroscopy in helium nanodroplets. J Am Chem Soc **141**(14), 5815–5823 (2019). https://doi.org/10.1021/jacs.8b13542

55. V. Scutelnic, M.A.S. Perez, M. Marianski, S. Warnke, A. Gregor, U. Rothlisberger, M.T. Bowers, C. Baldauf, G. von Helden, T.R. Rizzo, J. Seo, The Structure of the protonated serine octamer. J. Am. Chem. Soc. **140**(24), 7554–7560 (2018). https://doi.org/10.1021/jacs.8b02118

56. M. Marianski, J. Seo, E. Mucha, D.A. Thomas, S. Jung, R. Schlogl, G. Meijer, A. Trunschke, G. von Helden, Structural characterization of molybdenum oxide nanoclusters using ion mobility spectrometry mass spectrometry and infrared action spectroscopy. J. Phys. Chem. C **123**(13), 7845–7853 (2019). https://doi.org/10.1021/acs.jpcc.8b06985

57. A.I. González Flórez, D.-S. Ahn, S. Gewinner, W. Schöllkopf, G. von Helden, IR spectroscopy of protonated leu-enkephalin and its 18-crown-6 complex embedded in helium droplets. Phys. Chem. Chem. Phys. **17**(34), 21902–21911 (2015). https://doi.org/10.1039/C5CP02172C

58. A.I. González Flórez, E. Mucha, D.-S. Ahn, S. Gewinner, W. Schöllkopf, K. Pagel, G. von Helden, Charge-induced unzipping of isolated proteins to a defined secondary structure. Angew. Chem. Int. Ed. **55**(10), 3295–3299 (2016). https://doi.org/10.1002/ange.201510983

59. D.A. Thomas, R. Chang, E. Mucha, M. Lettow, K. Greis, S. Gewinner, W. Schollkopf, G. Meijer, G. von Helden, Probing the conformational landscape and thermochemistry of DNA dinucleotide anions via helium nanodroplet infrared action spectroscopy. Phys. Chem. Chem. Phys. **22**(33), 18400–18413 (2020). https://doi.org/10.1039/d0cp02482a

60. E. Mucha, A.I. Gonzalez Florez, M. Marianski, D.A. Thomas, W. Hoffmann, W.B. Struwe, H.S. Hahm, S. Gewinner, W. Schollkopf, P.H. Seeberger, G. von Helden, K. Pagel, Glycan Fingerprinting via cold-ion infrared spectroscopy. Angew. Chem.-Int. Edit. **56**(37), 11248–11251 (2017). https://doi.org/10.1002/anie.201702896

61. E. Mucha, M. Lettow, M. Marianski, D.A. Thomas, W.B. Struwe, D.J. Harvey, G. Meijer, P.H. Seeberger, G. von Helden, K. Pagel, Fucose migration in intact protonated glycan ions: a universal phenomenon in mass spectrometry. Angew. Chem-Int. Edit. **57**(25), 7440–7443 (2018). https://doi.org/10.1002/anie.201801418

62. M. Lettow, E. Mucha, C. Manz, D.A. Thomas, M. Marianski, G. Meijer, G. von Helden, K. Pagel, The role of the mobile proton in fucose migration. Anal. Bioanal. Chem. **411**(19), 4637–4645 (2019). https://doi.org/10.1007/s00216-019-01657-w

63. M. Lettow, M. Grabarics, K. Greis, E. Mucha, D.A. Thomas, P. Chopra, G.J. Boons, R. Karlsson, J.E. Turnbull, G. Meijer, R.L. Miller, G. von Helden, K. Pagel, Cryogenic infrared spectroscopy reveals structural modularity in the vibrational fingerprints of heparan sulfate diastereomers. Anal. Chem. **92**(15), 10228–10232 (2020). https://doi.org/10.1021/acs.ana lchem.0c02048

64. M. Lettow, M. Grabarics, E. Mucha, D.A. Thomas, L. Polewski, J. Freyse, J. Rademann, G. Meijer, G. von Helden, K. Pagel, IR action spectroscopy of glycosaminoglycan oligosaccharides. Anal. Bioanal. Chem. **412**(3), 533–537 (2020). https://doi.org/10.1007/s00216-019-02327-7

65. E. Mucha, M. Marianski, F.F. Xu, D.A. Thomas, G. Meijer, G. von Helden, P.H. Seeberger, K. Pagel, Unravelling the structure of glycosyl cations via cold-ion infrared spectroscopy. Nat. Commun. **9**(1), 4174 (2018). https://doi.org/10.1038/s41467-018-06764-3

66. K. Greis, C. Kirschbaum, S. Leichnitz, S. Gewinner, W. Schollkopf, G. von Helden, G. Meijer, P.H. Seeberger, K. Pagel, Direct experimental characterization of the ferrier glycosyl cation in the gas phase. Org. Lett. **22**(22), 8916–8919 (2020). https://doi.org/10.1021/acs.orglett.0c0 3301

67. K. Greis, E. Mucha, M. Lettow, D.A. Thomas, C. Kirschbaum, S. Moon, A. Pardo-Vargas, G. von Helden, G. Meijer, K. Gilmore, P.H. Seeberger, K. Pagel, The impact of leaving group anomericity on the structure of glycosyl cations of protected galactosides. ChemPhysChem. **21**(17), 1905–1907 (2020). https://doi.org/10.1002/cphc.202000473

68. M. Marianski, E. Mucha, K. Greis, S. Moon, A. Pardo, C. Kirschbaum, D.A. Thomas, G. Meijer, G. von Helden, K. Gilmore, P.H. Seeberger, K. Pagel, remote participation during glycosylation reactions of galactose building blocks: direct evidence from cryogenic vibrational spectroscopy. Angew. Chem.-Int. Ed. **59**(15), 6166–6171 (2020). https://doi.org/10. 1002/anie.201916245

69. C. Kirschbaum, E.M. Saied, K. Greis, E. Mucha, S. Gewinner, W. Schollkopf, G. Meijer, G. von Helden, B.L.J. Poad, S.J. Blanksby, C. Arenz, K. Pagel, Resolving sphingolipid isomers using cryogenic infrared spectroscopy. Angew. Chem-Int. Edit. **59**(32), 13638–13642 (2020). https://doi.org/10.1002/anie.202002459

70. C. Kirschbaum, K. Greis, E. Mucha, L. Kain, S.L. Deng, A. Zappe, S. Gewinner, W. Schöllkopf, G. von Helden, G. Meijer, P.B. Savage, M. Marianski, L. Teyton, K. Pagel, Unravelling the structural complexity of glycolipids with cryogenic infrared spectroscopy. Nat. Comm. **12**(1) (2021) https://doi.org/10.1038/s41467-021-21480-1

71. W. Schöllkopf, S. Gewinner, H. Junkes, A. Paarmann, G. von Helden, H. Bluem, A.M.M. Todd, The new IR and THz FEL Facility at the Fritz Haber Institute in Berlin, in *Paper presented at the SPIE Optics + Optoelectronics* (2015)

72. J.M.J. Madey, Stimulated emission of bremsstrahlung in a periodic magnetic field. J. Appl. Phys. **42**(5), 1906 (1971). https://doi.org/10.1063/1.1660466

73. L.R. Elias, W.M. Fairbank, J.M.J. Madey, H.A. Schwettman, T.I. Smith, Observation of stimulated emission of radiation by relativistic electrons in a spatially periodic transverse magnetic-field. Phys. Rev. Lett. **36**(13), 717–720 (1976). https://doi.org/10.1103/PhysRe vLett.36.717

74. K. Nauta, R. Miller, Metastable vibrationally excited HF ($v$=1) in helium nanodroplets. J. Chem. Phys. **113**(21), 9466–9469 (2000)

75. A. Varki, R.D. Cummings, M. Aebi, N.H. Packer, P.H. Seeberger, J.D. Esko, P. Stanley, G. Hart, A. Darvill, T. Kinoshita, J.J. Prestegard, R.L. Schnaar, H.H. Freeze, J.D. Marth, C.R. Bertozzi, M.E. Etzler, M. Frank, J.F. Vliegenthart, T. Lutteke, S. Perez, E. Bolton, P. Rudd, J. Paulson, M. Kanehisa, P. Toukach, K.F. Aoki-Kinoshita, A. Dell, H. Narimatsu, W. York, N. Taniguchi, S. Kornfeld, Symbol nomenclature for graphical representations of glycans. Glycobiology **25**(12), 1323–1324 (2015). https://doi.org/10.1093/glycob/cwv091

76. R.G. Cooks, D. Zhang, K.J. Koch, F.C. Gozzo, M.N. Eberlin, Chiroselective self-directed octamerization of serine: implications for homochirogenesis. Anal. Chem. **73**(15), 3646–3655 (2001). https://doi.org/10.1021/ac010284l

77. S.C. Nanita, R.G. Cooks, Serine octamers: cluster formation, reactions, and implications for biomolecule homochirality. Angew. Chem-Int. Edit. **45**(4), 554–569 (2006). https://doi.org/ 10.1002/anie.200501328

78. T.B. McMahon, C.J. Northcott, The fluoroformate ion, $FCO_2{}^-$. An ion cyclotron resonance study of the gas phase lewis acidity of carbon dioxide and related isoelectronic species. Can. J. Chem. **56**(8), 1069–1074 (1978). https://doi.org/10.1139/v78-181

79. J.W. Larson, T.B. McMahon, Fluoride and chloride affinities of main group oxides, fluorides, oxofluorides, and alkyls. Quantitative scales of Lewis acidities from ion cyclotron resonance halide-exchange equilibria. J. Am. Chem. Soc. **107**(4), 766–773 (1985). https://doi.org/10. 1021/ja00290a005

80. S.J. Blanksby, G.B. Ellison, Bond dissociation energies of organic molecules. Acc. Chem. Res. **36**(4), 255–263 (2003). https://doi.org/10.1021/ar020230d

81. B.S. Ault, Matrix isolation investigation of the fluoroformate anion. Inorg. Chem. **21**(2), 756–759 (1982). https://doi.org/10.1021/ic00132a056

82. X. Zhang, U. Gross, K. Seppelt, Fluorocarbonate, $[FCO_2]^-$: preparation and structure. Angew. Chem. Int. Ed. **34**(17), 1858–1860 (1995). https://doi.org/10.1002/anie.199518581

83. D.W. Arnold, S.E. Bradforth, E.H. Kim, D.M. Neumark, Study of halogen-carbon dioxide clusters and the fluoroformyloxyl radical by photodetachment of $X^-(CO_2)$ (X=I, Cl, Br) and $F_CO_2-$. J. Chem. Phys. **102**(9), 3493–3509 (1995). https://doi.org/10.1063/1.468575

84. L.J. Murphy, K.N. Robertson, S.G. Harroun, C.L. Brosseau, U. Werner-Zwanziger, J. Moilanen, H.M. Tuononen, J.A.C. Clyburne, A simple complex on the verge of breakdown: isolation of the elusive cyanoformate ion. Science **344**(6179), 75–78 (2014). https://doi.org/ 10.1126/science.1250808

85. M. Torrent-Sucarrat, A.J.C. Varandas, Carbon dioxide capture and release by anions with solvent-dependent behaviour: a theoretical study. Chem-Eur. J. **22**(39), 14056–14063 (2016). https://doi.org/10.1002/chem.201602538

86. J.M. Weber, H. Schneider, Infrared spectra of $X^-\cdot CO_2\cdot Ar$ cluster anions (X=Cl, Br, I). J. Chem. Phys. **120**(21), 10056–10061 (2004). https://doi.org/10.1063/1.1736633

87. H.K. Gerardi, A.F. DeBlase, X. Su, K.D. Jordan, A.B. McCoy, M.A. Johnson, Unraveling the anomalous solvatochromic response of the formate ion vibrational spectrum: an infrared, Ar-tagging study of the $HCO_2{}^-$, $DCO_2{}^-$, and $HCO_2{}^-\cdot H_2O$ ions. J. Phys. Chem. Lett. **2**(19), 2437–2441 (2011). https://doi.org/10.1021/jz200937v

88. W. Schneider, W. Thiel, Anharmonic force fields from analytic second derivatives: method and application to methyl bromide. Chem. Phys. Lett. **157**(4), 367–373 (1989). https://doi. org/10.1016/0009-2614(89)87263-X

89. J.F. Stanton, J. Gauss, Analytic second derivatives in high-order many-body perturbation and coupled-cluster theories: computational considerations and applications. Int. Rev. Phys. Chem. **19**(1), 61–95 (2000). https://doi.org/10.1080/014423500229864

90. J. Vázquez, J.F. Stanton, Simple(r) algebraic equation for transition moments of funda-
    mental transitions in vibrational second-order perturbation theory. Mol. Phys. **104**(3), 377–388
    (2006). https://doi.org/10.1080/00268970500290367
91. J.F. Stanton, J. Gauss, L. Cheng, M.E. Harding, D.A. Matthews,P.G. Szalay, PG CFOUR,
    coupled-cluster techniques for computational chemistry.
92. V. Barone, Anharmonic vibrational properties by a fully automated second-order perturbative
    approach. J. Chem. Phys. **122**(1):014108 (2005) https://doi.org/10.1063/1.1824881
93. V. Barone, J. Bloino, C.A. Guido, F. Lipparini, A fully automated implementation of VPT2
    infrared intensities. Chem. Phys. Lett. **496**(1), 157–161 (2010). https://doi.org/10.1016/j.cpl
    ett.2010.07.012
94. M.J. Frisch, G.W. Trucks, H.B. Schlegel, G.E. Scuseria, M.A. Robb, J.R. Cheeseman, G.
    Scalmani, V. Barone, G.A. Petersson, H. Nakatsuji, X. Li, M. Caricato, A.V. Marenich, J.
    Bloino, B.G. Janesko, R. Gomperts, B. Mennucci, H.P. Hratchian, J.V. Ortiz, A.F. Izmaylov,
    J.L. Sonnenberg, Williams, F. Ding, F. Lipparini, F. Egidi, J. Goings, B. Peng, A. Petrone,
    T. Henderson, D. Ranasinghe, V.G. Zakrzewski, J. Gao, N. Rega, G. Zheng, W. Liang, M.
    Hada, M. Ehara, K. Toyota, R. Fukuda, J. Hasegawa, M. Ishida, T. Nakajima, Y. Honda, O.
    Kitao, H. Nakai, T. Vreven, K. Throssell, J.A. Montgomery Jr, J.E. Peralta, F. Ogliaro, M.J.
    Bearpark, J.J. Heyd, E.N. Brothers, K.N. Kudin, V.N. Staroverov, T.A. Keith, R. Kobayashi,
    J. Normand, K. Raghavachari, A.P. Rendell, J.C. Burant, S.S. Iyengar, J. Tomasi, M. Cossi,
    J.M. Millam, M. Klene, C. Adamo, R. Cammi, J.W. Ochterski, R.L. Martin, K. Morokuma,
    O. Farkas, J.B. Foresman, D.J. Fox, Gaussian 16 Rev. C.01. Wallingford, CT (2016)
95. A. Varki, Biological roles of oligosaccharides: all of the theories are correct. Glycobiology
    **3**(2), 97–130 (1993)
96. A. Varki, Biological roles of glycans. Glycobiology **27**(1), 3–49 (2016). https://doi.org/10.
    1093/glycob/cww086
97. R.A. Dwek, Glycobiology: toward understanding the function of sugars. Chem. Rev. **96**(2),
    683–720 (1996). https://doi.org/10.1021/cr940283b
98. K.W. Moremen, M. Tiemeyer, A.V. Nairn, Vertebrate protein glycosylation: diversity,
    synthesis and function. Nat. Rev. Mol. Cell Biol. **13**(7), 448–462 (2012). https://doi.org/
    10.1038/nrm3383
99. M.M. Fuster, J.D. Esko, The sweet and sour of cancer: glycans as novel therapeutic targets.
    Nat. Rev. Cancer **5**(7), 526–542 (2005). https://doi.org/10.1038/nrc1649
100. O.C. Grant, D. Montgomery, K. Ito, R.J. Woods, Analysis of the SARS-CoV-2 spike protein
     glycan shield reveals implications for immune recognition. Sci. Rep. **10**(1), 14991 (2020).
     https://doi.org/10.1038/s41598-020-71748-7
101. L. Casalino, Z. Gaieb, J.A. Goldsmith, C.K. Hjorth, A.C. Dommer, A.M. Harbison, C.A.
     Fogarty, E.P. Barros, B.C. Taylor, J.S. McLellan, E. Fadda, R.E. Amaro, Beyond shielding:
     the roles of glycans in the SARS-CoV-2 spike protein. ACS Cent. Sci. **6**(10), 1722–1734
     (2020). https://doi.org/10.1021/acscentsci.0c01056
102. Y. Watanabe, J.D. Allen, D. Wrapp, J.S. McLellan, M. Crispin, Site-specific glycan analysis
     of the SARS-CoV-2 spike. Science **369**(6501), 330–333 (2020). https://doi.org/10.1126/sci
     ence.abb9983
103. A. Shajahan, N.T. Supekar, A.S. Gleinich, P. Azadi, Deducing the N- and O-glycosylation
     profile of the spike protein of novel coronavirus SARS-CoV-2. Glycobiology **30**(12), 981–988
     (2020). https://doi.org/10.1093/glycob/cwaa042
104. D. Cremer, J.A. Pople, General definition of ring puckering coordinates. J. Am. Chem. Soc.
     **97**(6), 1354–1358 (1975). https://doi.org/10.1021/ja00839a011
105. Neelamegham S, Aoki-Kinoshita K, Bolton E, Frank M, Lisacek F, Lütteke T, O'Boyle N,
     Packer NH, Stanley P, Toukach P, Varki A, Woods RJ, Updates to the symbol nomenclature
     for glycans guidelines. Glycobiology **29**(9), 620–624 (2019)https://doi.org/10.1093/glycob/
     cwz045
106. J. Hofmann, H.S. Hahm, P.H. Seeberger, K. Pagel, Identification of carbohydrate anomers
     using ion mobility-mass spectrometry. Nature **526**(7572), 241–244 (2015). https://doi.org/
     10.1038/nature15388

# Chapter 7
# X-Ray and XUV Imaging of Helium Nanodroplets

Rico Mayro P. Tanyag, Bruno Langbehn, Thomas Möller, and Daniela Rupp

**Abstract** X-ray and extreme ultraviolet (XUV) coherent diffractive imaging (CDI) have the advantage of producing high resolution images with current spatial resolution of tens of nanometers and temporal resolution of tens of femtoseconds. Modern developments in the production of coherent, ultra-bright, and ultra-short X-ray and XUV pulses have even enabled lensless, single-shot imaging of individual, transient, non-periodic objects. The data collected in this technique are diffraction images, which are intensity distributions of the scattered photons from the object. Superfluid helium droplets are ideal systems to study with CDI, since each droplet is unique on its own. It is also not immediately apparent what shapes the droplets would take or what structures are formed by dopant particles inside the droplet. In this chapter, we review the current state of research on helium droplets using CDI, particularly, the study of droplet shape deformation, the in-situ configurations of dopant nanostructures, and their dynamics after being excited by an intense laser pulse. Since CDI is a rather new technique for helium nanodroplet research, we also give a short introduction on this method and on the different light sources available for X-ray and XUV experiments.

R. M. P. Tanyag (✉) · B. Langbehn · T. Möller
Institut für Optik und Atomare Physik, Technische Universität Berlin, Hardenbergstraße 36, 10623 Berlin, Germany
e-mail: tanyag@physik.tu-berlin.de

B. Langbehn
e-mail: bruno.langbehn@physik.tu-berlin.de

T. Möller
e-mail: thomas.moeller@physik.tu-berlin.de

D. Rupp
Laboratory for Solid State Physics, ETH Zürich, John-von-Neumann-Weg 9, 8093 Zürich, Switzerland
e-mail: ruppda@phys.ethz.ch

© The Author(s) 2022
A. Slenczka and J. P. Toennies (eds.), *Molecules in Superfluid Helium Nanodroplets*,
Topics in Applied Physics 145, https://doi.org/10.1007/978-3-030-94896-2_7

## 7.1  Introduction

X-ray and extreme ultraviolet (XUV) radiations are high energy forms of electromagnetic radiation with photon energies ranging from tens of electron volts (eV) to tens of kilo-electron volts (keV). Absorption resonances and binding energies of many atoms fall within this range, allowing the identification of the elements in a material, their oxidation states, and even their chemical environment [1–5]. The short wavelengths of X-ray and XUV, from a fraction to a few tens of nanometers, are also used for the structural determination of objects with high spatial resolution and chemical specificity [1–5]. The XUV regime from tens of electron volts to ~250 eV is commonly used in studying electronic structures, especially chemical bonding in molecules, using photoelectron spectroscopy. The X-ray regime, on the other hand, is divided into soft (low-energy, ~250 eV to ~10 keV) and hard (high-energy, ~10 to ~50 keV) X-rays. The latter is frequently employed in X-ray diffraction crystallography since the photons in this region have wavelengths comparable to interatomic distances; while the former is often used for core-level spectroscopy of many organic compounds because the K-absorption edges of carbon (284 eV), nitrogen (410 eV), and oxygen (543 eV) lie within this region. The photon range between the carbon K-edge and the oxygen K-edge is called the "water window" because carbon predominantly absorbs the photons there, whereas oxygen (or water) remains transparent. This window is particularly advantageous for improving contrast in the X-ray microscopy of many organic compounds and cellular components in their native aqueous environment.

The generation of X-rays and X-ray imaging evolve together [6]. Immediately after Röntgen's discovery in 1895, it was demonstrated that X-ray radiation penetrates through many materials and reveals the denser internal structure. There is a good contrast between different elements in an object because of absorption, which scales as a function of the atomic number almost to the fourth power [1, 2]. These rather unusual properties of X-rays lead to some of their earliest and dramatic applications in casting shadow images of bones and hidden objects; applications that are continuously used in X-ray radiography for making medical and dental diagnoses, and in scanning for dangerous and prohibited objects in security checkpoints. Moreover, through the principles of diffraction, X-rays are used for structure determination of crystals and quasicrystals, including biologically important molecular systems such as the ribosomes and the DNA. In fact, most structures reported in molecular structure databanks are collected from X-ray crystallography [7]. Another way of imaging structures is with the use of electrons, see the work by Zhang et al. in Chap. 8 of this volume [8].

While X-ray crystallography is nowadays a routine procedure for structure determination, its application remains limited to systems that can be crystallized. Many objects in the micro- and nanoscales, which are of interest in materials, physical, and biological sciences, especially in the so called "soft materials", (i) do not form reproducible structures, (ii) are often non-periodic, (iii) cannot be supported, (iv) are either difficult to crystallize or cannot be crystallized at all, and (v) are often too large to be imaged as a whole. X-ray microscopy addresses some of these issues [9–11]. However, one key challenge in X-ray microscopy comes from the fact that materials used for X-ray optics have refractive indices remarkably close to unity and do not significantly refract X-rays. The use of a zone plate is one option to focus an X-ray beam and to produce a real space X-ray image. These plates, which are also known as Fresnel zone plates, are designed with alternating opaque and transparent concentric rings and require bright and coherent light sources, usually from third-generation synchrotron sources. The resolution, however, is limited by the size of the outer most rings of the zone plates, which is of the order of a few tens of nanometers [12, 13]. Another approach to structure determination of non-periodic objects is with lensless coherent diffractive imaging (CDI) [14–18]. In this technique, the object of interest is directly illuminated by spatially coherent X-rays or XUV focused to a few microns. The data collected with CDI are the intensity distributions of the scattered photons recorded on a two-dimensional detector. Since only the intensities of the scattered photons are recorded, phase information is lost, and iterative transform algorithms are required to numerically calculate the density of the object giving the diffraction image [14, 19–23]. The X-ray or XUV laser must be sufficiently bright, and spatially and temporally coherent [2, 24]. Additionally, the focus of the light beam must also be larger than the size of the object, as this oversamples the information needed for determining the missing phase, see Sect. 7.2.3. Necessary parameters for CDI have been reached with recent technological developments of fourth generation XUV and X-ray light sources, which are also known as Free-Electron Lasers (FEL). XUV diffractive imaging with FELs was first demonstrated at the Free-Electron Laser in Hamburg (FLASH) facility in 2006 [25], and in the X-ray regime at the Linac Coherent Light Source (LCLS) at Stanford Linear Accelerator Laboratory in 2011 [26, 27]. Since then, single-shot diffraction and especially CDI has been applied to single viruses [27], soot particles [28], xenon clusters [29, 30], and silver nanocrystals [31], among others, with a nanometer resolution. Several reviews have been published recently on CDI with X-rays, especially focusing on biological samples [15, 24, 32–34].

The earliest X-ray CDI measurement with superfluid helium droplets was performed at LCLS with the aim of visualizing quantized vortices in finite and isolated quantum fluids [35]. Quantum vortices are the macroscopic manifestation of superfluidity, and they have been observed and studied in bulk superfluid helium and Bose–Einstein condensates [36–39]. Up until the first experiment at LCLS, traces of quantum vortices were only inferred from the electron micrographs of cluster deposits from metal-doped helium droplets on thin carbon films [40–44], see also the work by Lackner in Chap. 11 of this volume [45]. In contrast, spectroscopic signatures of quantum vortices in helium droplets are lacking [46, 47]. The first results published from the LCLS experiment have uncovered fresh facets of helium droplets and have reinvigorated research questions that can also be addressed by CDI, such as:

1.  How do superfluid helium droplets rotate? What causes the droplet's shape to
    distort? Is there a correlation between the size of a droplet and its shape? And
    how do droplets produced from molecular beams acquire rotation? [35, 48–52]
2.  What factors control the formation of dopant nanostructures inside the droplet?
    Is it possible to control their aggregation? [53–57]
3.  What are the structural changes occurring in a pure or doped droplet after it
    has been subjected to intense light pulse such as near-infrared (NIR) radiation?
    [58–61].

The CDI technique is not limited to X-ray FELs. It can also be applied to other
light sources producing spatially coherent radiation including visible lasers, intense
light pulses in the XUV radiation from FELs, such as FLASH in Germany and the
seeded Free Electron Laser Radiation for Multidisciplinary Investigations (FERMI)
in Trieste, Italy, and lab-based High Harmonic Generation (HHG) sources, which
are becoming widely available in many laboratories [17]. Experiments performed in
the XUV regime using either seeded FELs or HHG sources have used wide-angle
scattering approach to determine the three-dimensional shape of the helium droplet
[49, 50].

In this chapter, we review the current progress of research and discoveries in
coherent X-ray and XUV imaging with helium droplets. Since the application of
imaging is rather recent in the arsenal of techniques available for helium nanodroplet
science, we begin with a short introduction in Sects. 7.2 and 7.3 on single-shot, lensless coherent diffractive imaging; on how the structure of the pure and doped droplets
are determined from their corresponding diffraction image; and on the general experimental setup for imaging. In Sect. 7.4, we proceed in discussing the results on the
sizes and shapes of helium droplets and what the shapes of the droplets tell us about
their state of spin. In Sect. 7.5, we discuss results where numerical reconstructions
show the positions of dopant clusters, which in some cases reflect the configuration
and distribution of quantum vortices. We also consider the possibility of controlling the growth of dopant nanostructures by using different kinds of dopants, such
as xenon, silver, acetonitrile, and iodomethane. In Sect. 7.6, we introduce experimental results on imaging doped helium droplets after excitation with an intense
near infrared pulse. Finally, we present a brief outlook on further opportunities for
studying helium droplets with CDI in Sect. 7.7.

## 7.2  Imaging

### 7.2.1  Lens-Based and Lensless Imaging

Images map the spatial information of the scattered light from an object into an
image plane. Figure 7.1 shows sketches of two types of imaging systems with and
without the use of lenses. Here, the source of illumination comes from a distant
light source and is already considered a plane wave when it reaches the object.

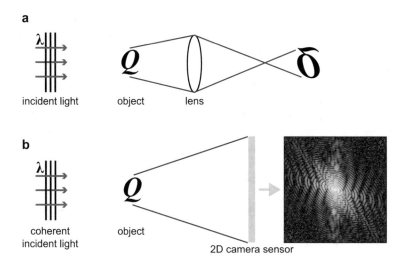

**Fig. 7.1** Image formation geometry **a** with and **b** without a lens. In both cases, the light source is far away from the object such that it can already be considered as a plane wave when it reaches the object

Lenses are an integral part in almost all optical imaging systems, such as our eyes, microscopes, and telescopes [62]. In lens-based imaging systems, such as in Fig. 7.1a, the lens collects the scattered light and transforms it into a distribution of intensities representing the object onto an image plane (usually a screen, a photographic plate, or a two-dimensional photosensor) located at some distance behind the lens [62, 63]. Mathematically, lenses can be seen as performing Fourier transforms of the collected scattered light to reproduce the object into the image plane. Based on Abbe's theory of image formation, the size of the lens defines the acceptance angles in collecting all the scattered light [64]. This limitation, along with lens aberrations, reduces the resolution of the image. In addition, depending on the geometry of the imaging setup, the image usually has a different magnification as that of the original object [62].

While lenses used in optical imaging are well-understood, similar components for X-ray and XUV imaging and X-ray microscopy remain in continuous development along with advances in nanofabrication [9, 10, 65]. The penetration depth of X-rays is large and the refractive index of most materials in the X-ray regime is very close to one. XUV radiation, on the other hand, is strongly absorbed in almost any material, rendering transmission optics almost unfeasible. The design, therefore, of X-ray and XUV optics is markedly different than that used in optical imaging. One of the key developments in X-ray focusing is the Kirkpatrick-Baez (KB) mirrors, which are a pair of ellipsoidal-shaped mirrors placed orthogonal with respect to each other. The incoming radiation is focused to a line by the first mirror, while the second focuses this line to a point [2]. These mirrors take advantage of the strong reflection of X-rays at glancing incidence angles with respect to the curved surface of the KB mirrors. Another kind of optics used with X-rays is the Fresnel zone plates. The current resolution achieved by these plates is limited to a few tens of nanometers [10, 65]. Overviews of X-ray and XUV microscopy methods and applications can be found in a number of reviews [1, 2, 9–11, 65].

In the absence of a lens, the type of image formed depends on the distance of the camera sensor with respect to the object [1, 62, 63]. From an optical point of view, an object consists of infinitesimal volume elements, where each element acts as a point source of spherical waves. After illumination, a path length difference develops among these scattered waves as they propagate towards a particular section of the detector. The path length difference is approximately given as $a^2 \cdot z^{-1}$, where $a$ is the typical object dimension, and $z$ is the distance between the object and the detector. If the detector is very close to the object, only a projection of the object is created on it; any visible contrast can be accounted from the different absorption properties of the materials that make up the object. This type of imaging is called *contact regime* and is commonly employed in X-ray radiography. As the detector is brought a bit farther, the effect of the path length difference becomes important. Fringes are observed in addition to the projection of the object on the detector. If the path difference is comparable to the wavelength of light, $\lambda$, i.e., $\lambda \approx a^2 \cdot z^{-1}$, the imaging is in the *near field* or *Fresnel regime*. When the detector is brought even farther, such that the arriving waves at the detector are approximately planar, the imaging is classified as *Fraunhofer* or *far-field regime*. In this case, the "image" on the detector no longer bears resemblance to the object. Rather, the image is a diffraction pattern corresponding to the Fourier transform of the density of the object. Positioning the detector even farther from the object does not change the diffraction pattern but only scales the size of it. The categorization of these different regimes is based on the Fresnel number, $f$, which is given as [1, 17]:

$$f = \frac{a^2}{\lambda \cdot z}. \tag{7.1}$$

If $f \gg 1$, imaging falls under the contact regime; Fresnel diffraction if $f \approx 1$; and Fraunhofer diffraction if $f \ll 1$. For a droplet with a diameter of 1000 nm, a wavelength of 0.826 nm (1.5 keV), and an object-detector distance of 565 mm, $f = 2.1 \times 10^{-3}$. With the same droplet diameter but for a wavelength of 72 nm (17 eV), which is within the short wavelength margin produced from an HHG source, and a detector distance of 50 mm, $f = 2.8 \times 10^{-4}$. Hence, for all intents and the purposes of X-ray and XUV imaging described in this chapter, the measured diffractions are classified under Fraunhofer or far-field diffraction.

In Fig. 7.1b, the diffraction image is registered within the field of view of the sensor. Because of the limited response time of the sensor, only the intensities, i.e., the squares of the scattering amplitudes, are measured, and the phase information of the scattered wave is lost. In order to recover the phase and to completely reconstruct the object from its diffraction image, numerical methods are applied that iterate between the Fourier space and the real space. These numerical methods analogously perform the same role as a lens in lens-based imaging. The resolution of lensless imaging at small scattering angles theoretically depends on the maximum collection angle subtended by the sensor, and by the wavelength of light used for imaging. Simply, the theoretical resolution is defined as [17] (Figs. 7.2 and 7.3):

$$r_t = \frac{\lambda}{2 \cdot \theta_{max}} \approx \frac{\lambda \cdot z}{N \cdot \Delta r},$$

(7.2)

where $\lambda$ is the wavelength of light; $\theta_{max}$ is the field of view as defined by the distance between the object and the detector, $z$; and the length along one side of the detector, $N \cdot \Delta r$, where $N$ is the number of pixels along one axis and $\Delta r$ is the size of one pixel, see Fig. 7.4. For imaging with X-rays at 0.826 nm using a 1024 × 1024 pixels detector, where each square pixel has a size of 75 μm, and located 565 mm from the object, $r_t \approx 6$ nm. Similarly, the depth of field is given by [17]:

$$r_l = \frac{2 \cdot \lambda \cdot z^2}{(N \cdot \Delta r)^2}.$$

(7.3)

For the same conditions as above, $r_l \approx 90$ nm.

## 7.2.2  Coherent Light Sources

A light source is considered coherent when there is a perfect correlation between the emanating complex field amplitudes, i.e., if the electric field is known at one point in space then the electric field at another point can be predicted based on their separation in space and time [2, 62, 63]. On the other hand, if a source produces light at various frequencies with no phase relationship whatsoever, then the source is incoherent. An image formed from *coherent illumination* is a result of the square of the linear superposition of individual exit waves from the different scattering points of the object, whereas an image formed from an *incoherent illumination* is due to the summation of individual intensities from the scattering points. Further comparisons between coherent and incoherent imaging are given in standard textbooks in Optics, in particular Fourier Optics [62, 63].

Real sources are only partially coherent since they are not perfectly monochromatic and do not propagate in a perfectly defined direction [1, 2]. Nevertheless, a region of coherence can be defined where the electromagnetic waves from a real light source remain in phase. Along the direction of propagation, the length of coherence, $l_L$, delimits the region from the source until the waves become out of phase. Mathematically, $l_L$ is defined as:

$$l_L = \frac{\lambda^2}{2 \cdot \Delta \lambda},$$

(7.4)

where, $\Delta\lambda$ is the spectral bandwidth of a source. The longitudinal coherence length is also known as temporal coherence. Transverse to the direction of propagation, the coherence length or spatial coherence, $l_T$, indicates the uniformity of the wavefront as it moves from a source and is defined as:

$$l_T = \frac{\lambda \cdot z}{2 \cdot s_d}, \tag{7.5}$$

where, $s_d$ is the typical spot size of a light source. Examples of spatially coherent light sources include optical lasers and FELs. However, incoherent light sources such as many synchrotrons, the sun, or arc lamps can also be made coherent through an introduction of a pinhole in between the object and the light source [2]. The setup of many X-ray imaging experiments done at synchrotrons is often based on spatial filtering using pinholes [9–11].

Developments in X-ray and XUV light sources usher new experimental techniques in imaging. These light sources produce coherent, intense, sub-100 fs pulses, which are characteristics conducive for single-shot CDI of nanometer-sized samples in free flight. The list of these new light sources includes: (i) X-rays from self-amplified spontaneous emission (SASE) FELs; and (ii) XUV radiation from SASE and seeded FELs, and lab-based HHG sources. All these new light sources can be used for CDI of large helium nanodroplets.

The concept of X-ray FELs builds on the technology of synchrotrons [1, 2]. Figure 7.2a shows the decade by decade development of X-ray sources from rotating anodes to FELs. Different X-ray sources are characterized by a quantity called brilliance, which is defined as the number of photons a source produces per second, the spectral bandwidth of a source at 0.1%, the focus size of the beam, and the beam divergence in milliradians. FELs deliver X-ray pulses with more than six orders

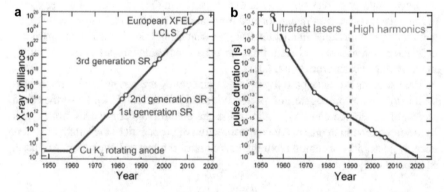

**Fig. 7.2** **a** The X-ray peak brilliance of light sources has improved by 20 orders of magnitude in last sixty years, see text for the definition of brilliance. **b** Decade by decade improvement in the pulse durations of light sources (Adapted with permission from Ref. [33]. © copyright <American Association for the Advancement of Science> All rights reserved.)

of magnitude peak brilliance than synchrotron sources, such as ESRF (European Synchrotron Radiation Facility) in Grenoble, France, and Spring-8 in Hyogo Prefecture in Japan. The coherent emission of X-rays is driven by the SASE process, where electron bunches from an injector gun are first accelerated through a linear accelerator (linac) hundreds of meters in length. Once the electron bunches travel close to the speed of light, they enter a series of undulators composed of a periodic series of alternating magnetic dipolar fields. As the electrons wiggle in the undulator, they radiate electromagnetic energy. Due to the high quality of the electron bunch or more precisely its small volume and narrow velocity spread, the electrons coactively interact with this emitted radiation, accelerating some of the electrons in the bunch and decelerating some depending on their position with respect to the crest of the electric field of the emitted radiation. This process creates electron microbunches that wiggle concurrently with the radiation and results into the exponential amplification of the radiation intensity within the SASE process. The photon energy produced can be tuned by adjusting the kinetic energy of the electrons and the vertical distance between the two sheets of the undulator, see Refs. [66, 67] for reviews on the physics of X-ray Free-Electron Lasers.

Due to the statistical fluctuations within the microbunches, which basically start from noise, the spectral bandwidth in the SASE process remains rather broad. This effect limits the application of the SASE pulse in studying few-femtosecond and attosecond electron dynamics in many atomic and molecular systems. Control over the spectral bandwidth and phase of FELs can be achieved by seeding the FELs with intense laser pulses or through self-seeding by introducing magnetic chicanes or dispersive elements along the trajectory of the electron bunch. The seed field arrives synchronously with the electron bunch at the beginning of the undulator, and the electric field of the seed laser dictates the microbunching process, significantly reducing the bandwidth of the FEL [68–70]. The FERMI FEL is one of these seeded FELs and can produce phase-coherent radiation in the XUV and soft X-ray regimes down to wavelengths as short as 4 nm [71].

FELs are large facilities. Access to these facilities is contingent upon a successful appropriation of a beamtime after an experimental proposal has been peer-reviewed. In short, experiments at FELs are limited by the allocation of available beamtimes. Alternatively, lab-based XUV sources are being developed from the nonlinear generation of high harmonics from an infrared (IR) drive laser, such as a femtosecond Ti:Sapphire laser centered at around 800 nm. Intense XUV pulses can be created via high harmonic generation (HHG) in gas using loose focusing geometries [72]. Atoms of a noble gas are ionized by the electric field of the loosely focused IR laser. The electrons are subsequently accelerated and thrown back onto the parent ion in the slowly changing IR field. This process results in recombination and emission of sub-femtosecond XUV bursts [2, 73, 74]. As the process is repeated twice for each IR cycle, the interaction produces odd harmonics of the driving laser [2, 73, 74]. In order to create high pulse energies, phase matching has to be achieved in a large volume of the focal area and requires optimizing conditions in a multiparameter space of IR intensity, gas pressure, and position relative to the focus [75]. The HHG can produce XUV pulse energies up to a few microjoules with pulse lengths as short as a few hundreds of attoseconds, see Fig. 7.2b. Short pulses open a new pathway for studying electron dynamics, such as electron density motions [76]. In terms of XUV imaging, tabletop HHG sources have been used in ptychographic reconstruction of patterned titanium on a silicon substrate [77, 78], ultrafast charge and spin dynamics in electronics [79], thermal transport in nanoscale systems [80], and wide-angle scattering of superfluid helium droplets [49]. Using HHG pulses for single-pulse, single-particle CDI, especially for dynamics studies, is very promising but still in its infancy.

## 7.2.3  Coherent Diffractive Imaging

Coherent diffractive imaging (CDI) is a lensless imaging technique where numerical methods are employed to reconstruct the object from its diffraction image, which represents the moduli squared of the complex scattering amplitudes [14, 15]. Details in the diffraction pattern play a crucial role in determining the location, amplitude, and spatial features of the different components in an object. For instance, speckles in a diffraction pattern denote specific spatial frequency that reflects the location or arrangement of substructures in the object. To regain the phase lost during the measuring process, iterative transform algorithms (ITA) are used to iteratively calculate Fourier transforms between the Fourier space and real space with the application of constraints in each space [15, 33, 81, 82]. These constraints are applied in order to minimize the error between the measured diffraction image from the calculated one.

Phasing algorithms usually do not converge to a unique solution due to ambiguities from unknown overall object boundary ("support"), signal noise, and missing data due to detector limitations. ITAs such as error-reduction (ER) [19] and hybrid input–output (HIO) [20, 21] were developed to bridge between the experimental data and the mathematical paradigm of oversampling theorem [21, 22]. The scattering phases can be retrieved from oversampling the recorded diffraction images, i.e., the volume of the object has to be smaller than the coherent volume of the light beam [81, 82]. Oversampling is essential in the convergence of ITAs. The phase retrieval is usually guided by self-consistency arguments, such as a good agreement between the Fourier transformation of the obtained densities and the measured diffraction amplitudes, along with the application of various physical constraints to minimize the sampled phase space and to prevent the algorithms from trapping in a local minimum that normally results into centro-symmetric image reconstructions. Trapping is often associated with the incorrect determination of the object boundary, which in some cases is unknown a priori. One pathway, called the "shrink-wrap" technique, progressively determines the object's support during reconstruction iterations [83]. The calculated object density or the output of a phasing algorithm is always represented by an average of hundreds of independent reconstruction runs in which each run consists of thousands of iterations [84–86]. Such a procedure is computationally expensive and incompatible with real-time data analysis [87]. The common practice of performing large numbers of reconstruction runs, where "acceptable" runs are averaged and "failed" runs are discarded, may also contribute to reconstruction ambiguity and reduce the image resolution defined by the wavelength of light and the geometry of the imaging setup, see Eq. (7.2). In Sect. 7.3.3, we discuss how helium droplets with embedded nanostructures promote the convergence of an iterative transform algorithm, since the overall object dimension, associated with the size and shape of the droplet, is easily determined from the diffraction pattern [53].

Although ITAs determine the missing phase information through iterative computation, another approach closely related to CDI is to encode phase information directly into the diffraction image through holographic imaging technique. Here, a point or an object with known dimensions placed in close proximity to the object of interest serves as a reference scatterer and interferes with the exit waves from the object [88–91]. Single-shot Fourier transform holography (FTH) has been successfully used in determining the structure of single viruses [91]. While FTH may provide a quick retrieval process, it is yet to be applied with helium droplets.

## 7.2.4  Small-Angle and Wide-Angle Scattering

Figure 7.3 shows a comparison between small- and wide-angle scattering regimes. Although there is no clear boundary separating them, small-angle scattering is taken to be $\theta_{max} < 5°$, while wide-angle scattering is certainly reached for angles larger than $10°$. $\theta_{max}$ is determined from the angle subtended by the edge of the detector at a certain distance from the interaction point. In Born's approximation, the scattering amplitude for a certain transferred momentum vector, $\vec{q}$, can be calculated from the two-dimensional Fourier transform of the projected density onto the plane defined by the normal vector $\vec{n}_p = 0.5\left(\vec{k}_{in} + \vec{k}_{out}\right)$. For small scattering angles, see Fig. 7.3a, the normal $\vec{n}_p$ vector is, to a good approximation, always parallel to the axis of direction of the main light beam. In this case, the scattering amplitude for all momentum vectors can be calculated in one step as the two-dimensional Fourier transform of the projected density onto the plane defined by $\vec{k}_{in}$. In other words, only information on the two-dimensional projection of the object is contained in the whole diffraction image. One visible signature of small angle scattering is that the diffraction image is point symmetric since the two-dimensional Fourier transform is a point symmetric operation, see Figs. 7.3a and 7.5.

One aim of X-ray and XUV imaging, however, is to determine the full three-dimensional structure of an object. In order to realize this goal, imaging techniques were developed in which multiple diffraction images are collected from different orientations of the same object (or its replica), or, similar to tomography, from different cross sections of the object [1, 18]. The application of these two techniques is feasible if the orientation of the object can be controlled, as a common practice in X-ray crystallography, or if the object itself is reproducible. Many objects of interest in the micro- and nanoscales, on the other hand, are non-reproducible,

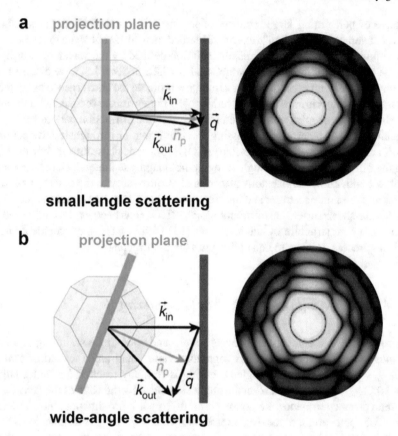

**a** projection plane

**small-angle scattering**

**b** projection plane

**wide-angle scattering**

**Fig. 7.3** Comparison between **a** small and **b** wide-angle scattering for a truncated octahedron (a common shape for silver clusters). The $\vec{k}_{in}$ and $\vec{k}_{out}$ are the wave vectors of the incoming and scattered light, respectively. For a certain transferred momentum vector $\vec{q}$, the diffracted intensity in Born's approximation can be calculated by the Fourier transform of the object's density projected on a plane perpendicular to the sum vector of $\vec{k}_{in}$ and $\vec{k}_{out}$ (displayed as *blue rectangle*). The normal vector of this projection plane is denoted as $\vec{n}_p$. In **a**, the transferred momentum vector $\vec{q}$ is always small compared to incoming and scattered wave vectors and the projection plane is therefore approximately parallel to the detector (represented as a *red rectangle*). In **b**, the length of the transferred momentum vector $\vec{q}$ is comparable to the wave vectors and large scattering angles are reached. Correspondingly, the projection planes are tilted to the detection plane. The diffraction patterns up to the same $\vec{q}$ vector differ in the outer features (Adapted with permission from Ref. [31], licensed under CC-BY)

and some only exist transiently. Therefore, only single-shot diffraction images can be collected for these transient objects. Some three-dimensional information can be retrieved when two particles are inside the focal volume of the light beam. This occurrence gives rise to concentric, off-axis diffraction rings [29] or in a holographic manner, the observation of Newton Rings [92]. Additionally, it also remains possible to encode some three-dimensional structural information from a single-shot coherent diffraction image of an isolated nanoparticle by collecting scattering information at wide scattering angles. In this case, different projection planes at large angles away from the main beam axis are recorded on the plane of the detector, see Fig. 7.3b [31, 93, 94].

Wide angle scattering is usually limited to rather long wavelengths in the extreme ultraviolet (XUV) regime, since the scattering intensity decreases dramatically with scattering angle due to Porod's law, and the scattering angle is proportional to the wavelength [1]. Diffraction images collected at wide angles are no longer necessarily point symmetric, see Figs. 7.3b, 7.8 and 7.9, and are no longer connected to the projected density by a simple two-dimensional Fourier transform. Additionally, these characteristics make data analysis more complicated as compared to those used for small angle scattering, where various phase retrieval methods are already available. In Sect. 7.3.4, we explain how the shape of an object can still be obtained from its wide-angle diffraction image through the forward fitting of a guess shape followed by an adaptive algorithm iteratively adjusting itself until the calculated diffraction pattern agrees well with that of the experimental data [31].

The choice between X-ray or XUV radiations for a particular experiment needs to be carefully adapted to the problem at hand. While scattering cross-sections are typically higher in the XUV regime than in the X-ray region and the wide-angle diffraction connected to XUV can provide three-dimensional information on the shape and orientation of the object being imaged, the non-trivial shape retrieval and the reduced spatial resolution both have to be taken into account as a clear trade-off.

## 7.3   Coherent Diffractive Imaging with Helium Droplets

### 7.3.1   Experimental Setup for X-Ray and XUV Imaging

The pioneering X-ray imaging of superfluid helium droplets was performed using the soft X-ray FEL at LCLS [35]. Since then, imaging experiments have been extended in the extreme ultraviolet regime using a seeded FEL [50], a lab-based HHG laser [49], and a multi-colored seeded FEL [95]. A general overview of the experimental setup of X-ray and XUV imaging of helium droplets is given in this section. Figure 7.4 shows an experimental layout for both static and dynamic imaging of pure and doped helium droplets, and Table 7.1 gives a list of contemporary imaging experiments. Images from static imaging are collected with one single pulse of the light beam and represent the instantaneous state of the droplet. These images describe the size and shape of a droplet, and, in the case of small scattering measurements of doped droplets, the different configurations of dopant nanoclusters assembled inside a droplet. On the other hand, the data collected with dynamic imaging represent the state of the droplet after it has been excited usually by an intense near-infrared pulse. These images build a picture on how the excited state of the droplet evolves through time.

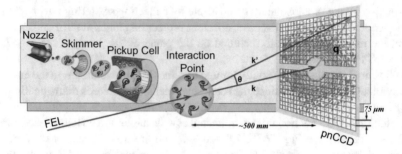

**Fig. 7.4** General schematic for imaging experiments. Helium droplets are produced either from a continuous or a pulsed *nozzle*. Depending on the experiment, the droplet can capture different types of dopants as it travels through the *pickup cell*. The droplet is imaged at the *interaction point* and the scattered photons are detected by a photon sensitive *detector* at some distance away from the interaction point. The droplet and the light beam are perpendicular with respect to each other. Note that the pixel size of the detector is exaggerated (Adapted with permission from Ref. [53], licensed under CC-BY 3.0)

**Table 7.1** Summary of X-ray and XUV imaging experiments with helium droplets, including the nozzle used, facility where experiments were performed, and the different photon energies, droplet isotopes, dopants, and detectors used

| Nozzle | Facility | Photon energy | Droplet Sample | Dopants Used | Detector used | Scattering angle |
|---|---|---|---|---|---|---|
| Continuous nozzle | LCLS-AMO | 1.5 keV | $^4$He, $^3$He | Xe | pnCCD | Small angle ~ 4° |
| Even-Lavie | FERMI-LDM | 19–39 eV | $^4$He | Xe | Microchannel plate | Wide angle |
| Even-Lavie | Max Born Institute | 17–27 eV | $^4$He | -- | Microchannel plate | Wide angle |
| Parker valve | European XFEL-SQS | 1 keV | $^4$He | Xe, Ag, CH$_3$CN, CH$_3$I | pnCCD | Small angle ~ 5° |

AMO—Atomic, molecular, and optical science instrument
LDM—Low density matter beamline
SQS—Small quantum systems instrument

Droplets of various sizes are generated from a nozzle cooled from ~14 to 3.5 K, spanning the gas condensation, liquid fragmentation, and jet disintegration regimes of droplet production [96], see also the work by Toennies in Chap. 1 of this volume [97]. While these helium droplets are created at nozzle temperatures above the superfluid transition, the droplets reach a temperature of ~0.4 K through cooling by evaporation at very short distances from the nozzle [96, 98–101]. This temperature is below the superfluid transition at 2.17 K. Currently, the types of nozzles used for the imaging experiments include the continuous nozzle based on the Göttingen design [101, 102], Even-Lavie valve [103, 104], and Parker valve [105–107]. The continuous source employs a pinhole nozzle with a nominal diameter of 5 μm, commonly used as an aperture for electron microscopes. The Even-Lavie valve is a commercially available cryogenic pulsed nozzle and is usually shipped with a trumpet-shaped nozzle having a throat diameter of 100 μm and an opening half-angle of 20°. The Parker valve is also commercially available; although, the nozzle plate can easily be replaced, for example, by a conical nozzle with a waist-diameter of 150–200 microns and an opening angle of 3–4°. The design of these conical nozzles are optimized for the generation of different types of large rare-gas clusters, which are samples used for studying the interaction between intense X-ray/XUV pulses and clusters [108]. Each nozzle is operated at different nozzle temperatures and stagnation pressures to produce droplet sizes at least on the order of ~100 nm in radius.

Helium droplets are known to capture different kinds of dopants [96, 109, 110]. These dopants may be introduced through a pick-up cell for gas and liquid samples, or through a ceramic oven for solid samples. The current list of dopants that have been used in imaging experiments includes xenon, silver, acetonitrile, and iodomethane. How these different dopant materials are assembled in the droplet can contribute to further understanding the mechanisms involved in the assembly of nanomaterials inside a superfluid droplet, see Sect. 7.5.

At the interaction point, the droplets, either pure or doped, meet the X-ray or the XUV pulse for single-shot imaging. For time-resolved dynamics imaging experiments, the pump laser arrives at the interaction point before or after the X-ray or XUV pulse. The scattered photons are collected either by a pnCCD (positive–negative charged-coupled device) sensor or a triumvirate of a microchannel plate, a phosphor screen, and a commercial camera. The pnCCD detector is a large area detector consisting of about one million pixels, where each pixel is sensitive to single photon and can detect linearly as many as a few hundred photons [111–113]. This detector consists of the two half plates ($512 \times 1024$ pixels) above and below the main axis of the FEL beam. In addition, the position of the detector with respect to the interaction point may be varied, which would change the maximum half scattering angles the detector could collect, from ~4 to 50° [111, 113]. In some instruments, two pairs of detectors are used with one being much closer to the interaction region than the other [111]. This configuration allows the simultaneous recording of scattering images at wide-angles and small-angles. The operation of the pnCCD detector is, however, rather complicated and expensive, requiring its own vacuum chamber and cooling system. A more affordable detector is a microchannel plate in tandem with a phosphor screen [29, 108]. The detected photons are converted to electrons and amplified by the microchannel plate, which consists of a two-dimensional array of microchannels acting as electron multipliers. These electrons then hit the phosphor screen. In this setup, a hole is introduced in the detector assembly since the intensity of the undeflected portion of the light beam is enough to damage it [29, 108]. Furthermore, a mirror, also with a hole, is placed at ~45° with respect to the plane of the phosphor screen and redirects the diffraction image on the phosphor screen to a camera sensor located outside vacuum [29, 108]. Unlike the pnCCD detector, however, the microchannel plate triumvirate does not give a linear response of the detected photons. As a consequence, the diffraction image cannot be easily processed with iterative transform algorithms for numerical image reconstructions. Further discussion on the treatment of data with the microchannel plate detector is presented in Sect. 7.3.4.

One important experimental parameter in performing imaging experiments is the hit rate, $HR$. Although the true hit rate is only known a posteriori of an experiment, it is possible to estimate the hit rate a priori. Two common definitions of hit rate in single-shot imaging are: (i) the number of images collected at a certain amount of time, with units in hits per hour; and (ii) the total number of images with respect to the total number of laser pulses at a given time. The latter definition is normally expressed in percentage. The hit rate indicates the probability of detecting a droplet in the focal volume of the laser pulse, $V_{\text{focal}}$, which is a function of the spot size of the FEL and its Rayleigh length [101]. From the first definition:

$$HR = \frac{F_D}{v_D \cdot d_{nozzle-IP}^2} \cdot V_{focal} \cdot R_{rr}, \tag{7.6}$$

in which the first factor gives the average number density of the helium droplets in the focal volume, $F_D$ is the flux of helium droplets, $v_D$ is the droplet velocity, $d_{nozzle-IP}$ is the distance between the nozzle and the interaction point, and $R_{rr}$ is the repetition rate of the light source. The second definition is similar to the definition of a duty factor and is based on the number of images collected as a function of the number of pulses from the light source:

$$\%HR = \frac{F_D}{v_D \cdot d_{nozzle-IP}^2} \cdot V_{focal} \cdot \frac{100}{t_{acq} \cdot R_{rr}}, \tag{7.7}$$

where $t_{acq}$ is the total time duration of measurement, and the factor 100 is for conversion to percentage.

The hit rate is affected by the overlap between the droplet beam and the laser beam, and the droplet density at the interaction point. Due to the stochastic nature of the SASE process and high harmonic generation, pulse intensities can vary from shot to shot, consequently varying the total number of detected photons from droplets of the same size. In addition, the X-ray or the XUV beam presumably has a Gaussian profile. The position of the droplet with respect to the beam axis will vary the intensity distribution in the diffraction image, i.e., with all things being the same, a droplet at the center of the Gaussian beam is expected to give a more intense diffraction image than those droplets imaged at the periphery of the light beam.

### 7.3.2 Diffraction Imaging of Helium Nanodroplets

The interaction between radiation and matter is important to image formation, and the optical response of matter strongly depends on the wavelength of light [1, 2,

114]. Changing the wavelength also changes the number of scattered or absorbed photons, assuming of course the intensity of light remains the same. In addition, for an object consisting of different materials, such as a doped helium droplet, variations in the refractive index or densities give contrast to imaging due to photoabsorption, which can induce structural damage and can contribute in reducing imaging resolution. Photoabsorption processes further lead to a cascade of ionization processes, including emission of photoelectrons, Auger electrons, and fluorescence [1, 2, 114–116]. Similarly, for helium droplets, the fraction of absorbed photons induces structural changes due to photoionization, where some of the photoionized electrons will escape and lead to charging of the droplet. The electrons that remained trapped in the Coulomb potential of a multiply-ionized droplet will further cause secondary ionization of the helium atoms that may result into the complete ionization of all the atoms in the droplet, creating a droplet nanoplasma. However, the details of these processes remain to be elucidated [117–128]. Experimentally, the disintegration of the droplet manifests itself through bursts of $He_n{}^+$ ions, which can be detected with a time-of-flight mass spectrometer. Some of the dynamics induced by the photoabsorption of intense infrared radiation can be studied with CDI, see Sect. 7.6. Although, photoabsorption causes structural changes in the droplet, the pulse length of X-ray and XUV light sources, on the order of a few hundreds of femtoseconds, is considered fast enough to capture the instantaneous state of the droplet before the onset of structural changes [14, 129, 130]. Imaging with the use of ultrashort light pulses is referred to as *diffraction before destruction* in single particle imaging. It must be noted, however, that electronic changes arising from photoionization and plasma formation in large xenon clusters occur on a sub-femtosecond time scale and may drastically influence the diffraction response [131, 132]. An example of accounting for the number of scattered and absorbed X-ray photons for a pure helium droplet is described in Ref. [101]. The power radiated by the free and bound electrons experiencing acceleration due to an incident electromagnetic field redirects radiation in a wide range of angles [2]. Due to both scattering and absorption, the intensity of the incident radiation is attenuated in the forward direction. The number of scattered and absorbed photons from an object can then be accounted from the scattering and absorption cross-sections of the elements constituting the object and the incident photon flux.

The collective interaction of X-ray or XUV with condensed matter can be described by the refractive index, $n$. In the X-ray regime, $n$ differs only by a small amount from unity [2, 133]. In the XUV region, many materials, including helium, exhibit strong electronic resonances that correspond to large deviations from unity. The refractive index is commonly written as:

$$n(\lambda) = 1 - \delta + i\beta = 1 - \frac{n \cdot r_e \cdot \lambda^2}{2\pi} \cdot \left( f_1^0(\lambda) - i \cdot f_2^0(\lambda) \right), \qquad (7.8)$$

in which $\delta$ and $\beta$ account for the phase variation and absorption of propagating waves, respectively. The term $r_e = 2.82 \times 10^{-6}$ nm is the classical electron radius, and $f_1^0(\lambda)$ and $f_2^0(\lambda)$ are the atomic scattering factors of an element. For helium at $\lambda = 0.826$ nm ($h\nu = 1.5$ keV), $f_{1,He}^0(\lambda) = 2.0$ and $f_{2,He}^0(\lambda) = 2.4 \times 10^{-3}$ [133]. For XUV, $\delta$ and $\beta$ can be obtained from Ref. [134]. Equation (7.8) has both refractive and absorptive components. The real part of the refractive index describes how the phase velocity of the X-ray/XUV wavefront changes due to the oscillations of the free and bound electrons; whereas the imaginary part corresponds to the amount of light absorbed during propagation. For a pure spheroidal droplet, these effects produce a diffraction pattern consisting of concentric rings. For a doped droplet, the dopants have different refractive indices and contribute to a phase shift with respect to that of helium. In coherent diffractive imaging, the exit waves from the helium and from the dopants interfere and thus modify the concentric ring patterns produced by a pure droplet alone. Figure 7.5 shows examples of diffraction patterns obtained from a pure spheroidal droplet and from a doped droplet [35, 53]. The detector is placed 565 mm from the interaction point and the wavelength of light used for imaging is ~0.826 nm. In the diffraction from a doped droplet, concentric rings are observed close to the center of the diffraction and specular patterns far from the center. These specular patterns correspond to the structure of the dopants inside the helium droplet. The dopant structures are naturally smaller than that of the droplet. A reconstruction algorithm in solving the structures of dopant clusters inside a helium droplet is presented in Sect. 7.3.3.

Using the droplet's diffraction pattern at small scattering angles, one can quickly estimate the droplet radius at a particular azimuthal angle, $\vartheta$, on the plane of the detector by determining the distance between the maxima of the rings, $\Delta N$:

$$R_{D,\vartheta} = \frac{\lambda \cdot z}{\Delta N \cdot \Delta r}. \tag{7.9}$$

For example, $\Delta N$ for the rings in the diffraction images shown in Fig. 7.5 is about 20 pixels for both the pure and doped droplets. Substituting this number in Eq. (7.9) and with $\Delta r = 75$ μm give a droplet radius of about 300 nm or about $2.5 \times 10^9$ helium atoms. For the diffraction images from doped droplets, the distance between the maxima of the concentric rings gives the size of the droplet after it has been doped. If the number and identity of the dopants are known, one can also estimate the initial size of the droplet before doping [101].

Following a more rigorous analysis from a collection of diffraction images of pure and doped droplets, it is also possible to determine droplet size distribution, which is historically presented in numbers of atoms per droplet. For superfluid helium-4 droplets produced using 5 μm pinhole nozzle at a stagnation pressure of 20 bars and nozzle temperature of 5 K, which fall under the liquid fragmentation regime of droplet production, both pure and doped droplets follow an exponential size distribution with its steepness increasing as more dopants are added [57]. Droplets produced using an Even-Lavie pulsed nozzle at a stagnation pressure of 80 bars and a nozzle temperature of 5.4 K also follow an exponential size distribution with an average droplet size of $6 \times 10^9$ or a radius of 400 nm [50]. These results agree with earlier measurements that determined an exponential size distribution for droplets produced from liquid fragmentation [96, 97, 135]. Similarly, helium-3 droplets, which remain as classical viscous droplets under experimental conditions, produced using a 5 μm nozzle at a stagnation pressure of 20 bars and at nozzle temperatures less than 5 K are likewise found to follow an exponential size distribution [52].

### 7.3.3   Dopant Clusters Image Reconstruction

There are two main interests in the static coherent diffractive imaging with helium droplets; the overall shape of the droplet, which indicates the rotational state of the droplet, and the assembly of dopant nanostructures inside the droplet, such as the dopant aggregation along the length of a vortex. In Sect. 7.2.4, we mention that methods are available to determine the two-dimensional or three-dimensional shape of the droplet, depending on whether diffraction was recorded at small or wide scattering angles. For the second interest, iterative numerical methods have to be utilized in order to reconstruct these nanostructures from the diffraction images of doped droplets. At the moment, these numerical methods are only applied at small scattering angles, where the idea that the diffraction image is simply the Fourier transform of the object density, including the nanostructures inside the droplet, holds. Hence, this section is only concerned with data from small-angle scattering. One critical prerequisite for the convergence of iterative numerical methods to a meaningful solution is the determination of the overall extent or dimension of the object being imaged. In the jargon of iterative transform algorithms, the overall dimension of the object is called the *support*, which could be the full dimension of a nanocrystal, a complex biological protein, a mimivirus, or a helium droplet. In many instances,

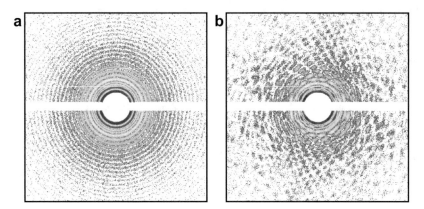

**Fig. 7.5** Examples of diffraction images obtained from **a** a pure droplet (adapted with permission from Ref. [35]. © copyright <American Association for the Advancement of Science> All rights reserved.), and from **b** a xenon-doped droplet (adapted with permission from Ref. [53], licensed under CC-BY 3.0) at small angle scattering. Only the central section of the diffraction is shown here

the support is unknown a priori due to varying sizes and shapes of the object being imaged. An incorrect determination of the support leads to incorrect numerical reconstructions. Many techniques have been developed that address the determination of the support, such as a technique called "shrinkwrap" algorithm [82, 83]. Additionally, support determination methods usually require ingenious integration of another set of algorithms that runs simultaneously with the iterative transform algorithms [83, 136–138].

The support in X-ray imaging of nanostructures in a helium droplet, on the other hand, is already defined by the dimensions of the droplet, which may also be taken as a reference scatterer aiding the convergence of an algorithm that solve for the missing phase information. Diffraction patterns from pure droplets have concentric circular or elliptical patterns ascribed to spherical or oblate pseudo-spheroidal droplet shapes, respectively, see Figs. 7.5 and 7.11 [35, 48]. The details of the droplet size and shape determination from the diffraction scattering patterns are described elsewhere [35, 48]. In addition to defining the support, the helium droplets also serve as vehicles to deliver and localize the dopant cluster structures in the laser focus. This technique of using the helium droplet as a support is referred to as droplet coherent diffractive imaging (DCDI). The primary goal of which is on the retrieval of the location and shapes of these dopant nanostructures from diffraction images of doped droplets.

The DCDI algorithm is based on the well-known error-reduction (ER) algorithm [21]. Figure 7.6 shows a flow diagram of the DCDI algorithm and the description follows from Ref. [53]. The algorithm is initiated by using the droplet density determined from the concentric ring patterns close to the center of the diffraction. After Fourier transform of this droplet density, the modulus of the scattering amplitude at each pixel is replaced by the square root of the measured intensity, $I_{Meas}$, whereas the initial phase, $\phi$, is retained. The intermediate scattering amplitude is called $G'$. The inverse Fourier transform of $G'$ gives an approximate solution, $\rho'$. However, as can be seen in Fig. 7.5, some of the intensity information are missing due to the central detector hole, the gap between the detector plates, and some arrays of damaged pixels. In this case, the algorithm sets some constraints in the real space, such that the missing pieces of information are ignored and that the approximate solution should not exceed the boundary defined by the droplet. This adjusted $\rho'$ then serves as a new input density, $\rho$, for the DCDI algorithm. In comparison to other methods, which require thousands of iterations and multiple initial guess inputs, DCDI converges to a meaningful solution within less than 100 iterations, as demonstrated in Fig. 7.7. Intermediate solutions at different stages of iterations are also shown. Aside from image reconstructions, results from DCDI algorithm can also be used to determine numerical values pertaining to the number of dopants in the droplet and the initial size of the droplet [53].

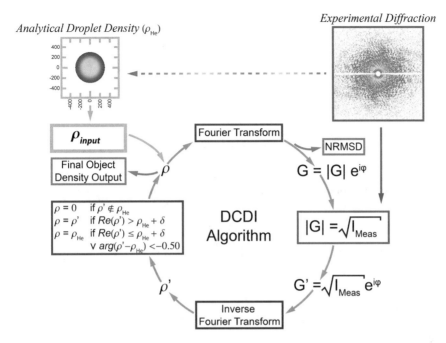

**Fig. 7.6** Schematic of droplet coherent diffractive imaging (DCDI). The algorithm is initiated using a preset helium droplet density, $\rho_{input}$. The series of Fourier and inverse Fourier transforms between the object- and reciprocal-space with iterative reinforcement of constraints in both spaces rapidly converge to yield the density of xenon clusters inside the droplet (Adapted with permission from Ref. [53], licensed under CC-BY 3.0)

## 7.3.4  *Forward Simulation and Machine Learning*

While X-ray small-angle scattering images can be analysed using iterative transform algorithms, the analysis of wide-angle diffraction images is made more complicated by the large index of refraction, both real and imaginary parts, at longer XUV wavelengths, where photoabsorption and phase shift along the direction of light beam propagation cannot be neglected. This phase shift also depends on the scattering angle, often exceeding several tens of degrees in wide-angle scattering, and can impede in the determination of the object structure. Even though obtaining structural information from wide-angle scattering images is nontrivial, these images can still be simulated using ideas developed in tomography in which full three-dimensional structure of an object, such as bones or internal organs, is reconstructed by combining different two-dimensional X-ray image projections of this object [139].

Currently, no rigorous algorithm is known that efficiently determines the object's structure form its wide-angle diffraction without prior information [140]. This problem has so far been approached via forward fitting methods, which simulate the diffraction image from a well-defined model shape [30, 31, 49, 50, 94, 141]. The simulated diffraction is then compared with that of the experimental, and shape

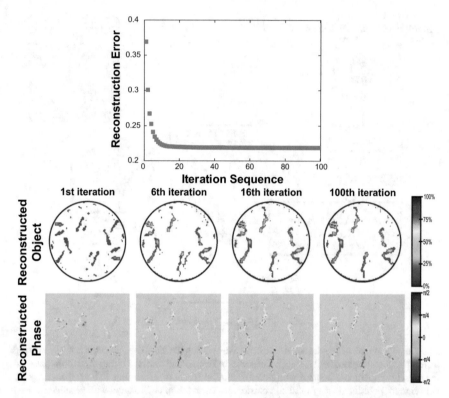

**Fig. 7.7** DCDI convergence and evolution of densities and phases within a 100-iteration run. The plot shows the rapid decrease in the reconstruction error as a function of the number of iterations. The *middle row* shows the calculated density modulus in a linear intensity scale. The *black circles* represent the droplet boundary enclosing the clusters. The *bottom row* corresponds to complex density phases at the same iteration as the droplets above it. Initially, DCDI finds a center-symmetric cluster configuration, but as the number of iterations increases, the enantiomer having positive imaginary density becomes dominant (Adapted with permission from Ref. [53], licensed under CC-BY 3.0)

parameters are refined to optimize the match between the diffraction images. Suitable pre-defined model shapes are chosen based on some known physical properties of the object being imaged, from general considerations on observed symmetries in individual diffraction patterns, or from the whole variety of observed patterns in the complete data set. In addition, the size, orientation, and other parameters of the model shape, such as eccentricity of an ellipsoid, are parametrized and used for fitting the simulation to the measured pattern. On the other end, the choice of the simulation method must (i) correctly reproduce the wide-angle features, (ii) account for (at least approximately) the optical properties of the particle, and (iii) be fast enough for a reasonable computational time in fitting many patterns. One simulation method for retrieving the object's structure from wide-angle scattering image is called the multi-slice Fourier transfom (MSFT), which has been previously applied in Refs. [31, 49,

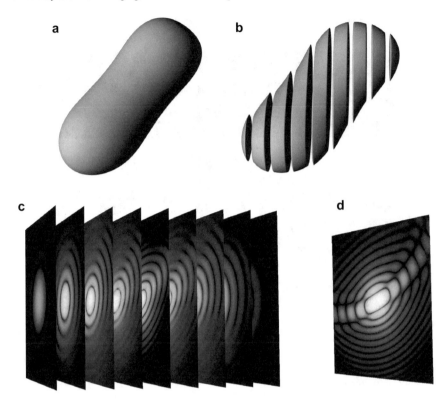

**Fig. 7.8** Schematic representation of the multi-slice Fourier transform (MSFT) approach: **a** three-dimensional rendering of the sample; **b** visualization of the spatial domain slicing following the multi-slice approach; **c** amplitudes of the fields scattered from each slice in **b**; and **d** square of the phase-corrected sum of all the wavefield amplitudes in **c**, which represents the final output of the simulation (This figure is a courtesy to the authors by Alessandro Colombo)

50, 141, 142]. MSFT was originally developed for electron diffraction [143] and has been previously applied to soft X-ray diffraction of supported particles [139]. Figure 7.8 shows a schematic of the MSFT approach. Here, the object is first divided into a stack of two-dimensional slices, whose normal vectors are oriented parallel to the incident photon beam, see Fig. 7.8b. Diffraction fields are then calculated from each slice, which corresponds to a two-dimensional distribution of refractive indices, see Fig. 7.8c. The final scattering intensity corresponds to the modulus square of the phase-corrected sum of the two-dimensional Fourier transforms from these different slices, see Fig. 7.8d. A fast implementation of the MSFT simulation code can be found in Ref. [141]. Material properties are approximately accounted by reducing the scattering amplitude as a function of propagation through the material according to Beer–Lambert's law. Multiple scattering events due to the rescattering of light inside the sample cannot be accounted correctly within the MSFT method. These events, however, only become significant when the real part of the index

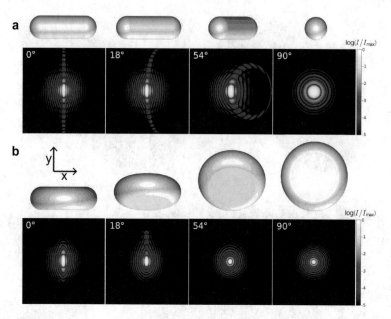

**Fig. 7.9** Scattering patterns calculated with MSFT for different rotation angles and droplet shapes: **a** pill-shaped, **b** wheel-shaped. The long axis is set to 950 nm, the short to 300 nm. The pill-shaped droplet is rotated around the *y-axis*, while the wheel-shaped droplet around the *x-axis*. In these simulations, the light beam propagates towards the image plane in **a** and out of the image plane in **b** ( Modified from Ref. [144], licensed under CC-BY 4.0)

of refraction is exceptionally large. For the analysis of helium nanodroplets using near resonant wavelengths, an approximation for the phase slip was also included in the MSFT simulations by considering an effective complex refractive index [50]. For the case of silver clusters imaged with non-resonant XUV wavelengths, the MSFT approach without an effective phase slip was benchmarked against full finite difference time domain (FDTD) simulations using tabulated optical constants, and a reasonable agreement between these two methods was obtained [31]. Although there are differences between the results of the MSFT and FDTD simulations, these differences are subtle.

XUV wide-angle diffraction imaging has the advantage of clearly distinguishing between shapes with similar two-dimensional projections. Therefore, experiments in the wide-angle regime substantially contribute to the discussion of rotation and shapes of helium droplets [49, 50]. As visualized in Fig. 7.9, wide-angle diffraction patterns from a pill-shaped droplet and a wheel-shaped droplet differ in a characteristic way. The scattering patterns for wheel and pill shaped droplet calculated with MSFT are displayed for different orientations to the incoming XUV pulse [144]. Since both shapes may result in an identical outline and a very similar two-dimensional projection, they are difficult to distinguish in the small-angle scattering regime. In contrast, there are noticeable deviations from the point symmetry of a diffraction image collected at wide scattering angles, especially those obtained for a tilted pill-shaped droplet. If the oblate particle's symmetry axis is neither oriented along the optical axis nor perpendicular to it, the diffraction patterns exhibit straight streaks to only one side. At 90° tilt angle between the symmetry axis and the optical axis or when one of the particle's symmetry axes is aligned along the optical axis, the two-dimensional projections are similar. However, the intensity distributions are clearly different and decay much faster for wheel-type than for pill-type shapes, compare Fig. 7.9a, b.

So far, forward fitting has only been applicable to small data sets with rather simple model shapes. However, huge data sets up to several million scattering patterns can be acquired in single-shot diffraction imaging during a single beamtime, emphasizing the need for rigorous and rapid reconstruction methods on data collected from both wide- and small-angle scattering regimes. Developments in X-ray and XUV light sources are proceeding toward high repetition rates as well. For instance, the European XFEL can run up to 4.5 MHz [145]; within an hour, it is possible to collect ~1.6 × 10^{10} diffraction images [146–148]. This situation is rather similar to other fields of big data science, such as in particle physics, where powerful analytical methods and algorithms are needed to extract significant information [140, 146–150]. Machine learning and neural networks are some contemporary approaches in managing these huge data. Some tools are available in structural biology [151–153], whereas the adaption of neural networks to coherent diffraction imaging of individual nanoparticles is still incipient [50, 142].

Being trained on a large augmented data set of simulated scattering patterns, neural networks can be of great help in extracting structural information and can especially account for image artefacts, such as noise, center hole, and the limited size of the detector [140]. In a recent wide-angle study on helium droplet shapes [50, 142], a supervised approach was used to exclude the existence of patterns from wheel-shaped droplet images, see Fig. 7.9b, from a large data set, see Sect. 7.4.1. In general, deep neural networks consist of many hierarchically structured nonlinear functions, referred to as layers of the neural network, which enable the network to learn intricate structures in the very high dimensional input space [154]. Having many layers between input and output levels, deep neural networks are well suited for extracting structural parameters as this procedure is equivalent to retrieving a small number of parameters from high-dimensional spaces [154]. A convolutional neural network is trained (*supervised training*) starting with a few thousand scattering patterns classified manually into a number of different classes [142]. The manual classification is a tedious process and could take weeks to sort through thousands of diffraction images. After training, a much larger data set can subsequently be analysed by the network and the scattering patterns are classified according to the learned classes, see Fig. 7.10. To optimize the classification accuracy of a given convolutional neural network, a priori knowledge about diffraction images can be used. For example, a logarithmically scaling activation function was introduced that boosted classification accuracy by about 2–4% as it accounts for the exponentially decaying intensity observed in diffraction images [142]. The network finally provides as output the classification of the scattering patterns independent of particle/droplet size, e.g., it gives statistics on how many spherical, oblate, prolate droplets are present

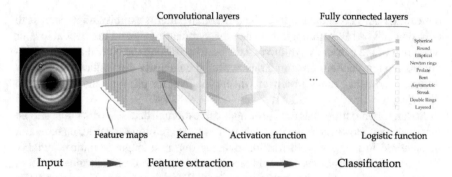

**Fig. 7.10** Schematic diagram of a convolutional neural network. First, sets of representative diffraction patterns are manually categorized. These patterns are then used as inputs to train the program. After which, the code runs through all the collected hits and sorts them into the manually classified categories. Finally, statistics from the automatically sorted files are produced, such as size and shape distributions, and possible three-dimensional reconstructions (Reused with permission from Ref. [142], licensed under CC-BY 4.0)

in a droplet beam [142]. These results provide a first step towards using deep learning techniques for a direct and fast determination of the object's shape from a diffraction pattern. Another key goal would be a routine from the unsupervised learning [155] paradigm for online analysis during an experiment, where manual pre-classification is no longer needed to sort through a large dataset.

## 7.4 Imaging Pure Helium Droplets

An isolated liquid droplet in equilibrium and held together by surface tension forces will adapt a spherical shape to minimize its surface area [156–160]. Once the droplet starts to rotate, its shape gets deformed, and capillary waves may be created on its surface [161]. For axisymmetric droplets, the shape deformation is given by their aspect ratio, $AR$, which is defined as the ratio between the droplet's half major and half minor axes. A spherical droplet has an $AR = 1$. The droplet becomes more oblate spheroidal as the droplet spins faster, or as the value of $AR$ increases from one [156–160]. Studies on the shapes and stabilities of liquid droplets have a long history, where a liquid drop serves as a model system in various length scales, from the shapes of self-gravitating astronomical and cosmological bodies, the fission of atomic nuclei, and the shapes of spinning uniformly-charged bodies [156–164]. The coupling between surface oscillations and rotation has also been considered in studying the stability of spinning droplets [165, 166].

Classical droplets are viscous, and their shape deformation may be described from the equilibrium shapes of rotating rigid bodies. In a rigid body rotation (RBR), its azimuthal speed, $v_{RBR}$, is a function of its angular velocity and the distance $r$ from the axis of rotation, while its vorticity, $\nabla \times \vec{v}$, where $\vec{v}$ is the velocity vector, is twice the angular velocity. In contrast, the viscosity of superfluid helium is negligible and superfluid flow is irrotational, i.e., $\nabla \times \vec{v} = 0$ [38, 167]. In a closed two-dimensional path of arbitrary shape in the superfluid, however, a phase defect may be present in order for the superfluid's wavefunction to be the same at the beginning and end of the closed path [168]. This also implies that the circulation around the closed path must either be zero or a multiple of the quantum of circulation, $\kappa = h/m_{He} = 9.97 \times 10^{-8}$ m$^2$ s$^{-1}$, where $h$ is the Planck's constant, and $m_{He}$ is the mass of the helium atom. The phase defect is known as a quantum vortex and the fluid's azimuthal speed, $v_{vort}$, around the vortex is given by [38, 168]:

$$v_{vort} = \frac{q \cdot \kappa}{2\pi \cdot r}, \tag{7.10}$$

where $q$ is a whole number multiplying the quantum of circulation, and $r$ is centred on the vortex core. Each of these vortices has a quantized circulation, hence the name. For superfluids at high angular momentum, it is energetically favourable to evenly distribute the angular momentum to many vortices than to a single vortex possessing all of the angular momentum [168]. The rotation of superfluid helium droplets depends on the presence of quantum vortices, in addition to surface shape oscillations [51, 169–171]. A collection of these vortices significantly contributes to the total angular momentum of the droplet [51, 171]. Early studies of magnetically levitated, charged, millimeter-sized superfluid helium droplets are interested in the decay of surface oscillations and the possible nucleation of quantum vortices [172–176]. In this section, we discuss the shapes and sizes of helium droplets in the size range of $10^7$ up to $10^{12}$ atoms, which are currently the size range of what can be measured with CDI, and how the shape of superfluid droplets is related to the droplet stability curve, which was derived from studies of classical viscous droplets.

### 7.4.1 Shapes of Pure Helium Droplets

Imaging the sizes and shapes of individual nanometer-sized superfluid helium droplets has only recently become possible with CDI. As discussed in Sect. 7.4.2, the diffraction images can be used to determine the state of rotation of the droplet. Figure 7.11 shows some characteristic single-shot diffraction images of helium droplets produced from a 5 μm pinhole orifice at a stagnation pressure of 20 bars and a nozzle temperature of around 5 K [35, 48]. These images were obtained at small scattering angles, where only the two-dimensional projection of the object is recorded onto the detector plane, and only the half-major axis and an upper bound of the half-minor axis of the droplet could be obtained from the image. Moreover, small angle diffraction cannot distinguish between a prolate and an oblate spheroidal droplet since both will generally result into similar patterns in the diffraction image, see Sect. 7.3.4. For small scattering angles, a handy equation in quickly determining the radius of the droplet at a particular azimuthal angle based on the distances between

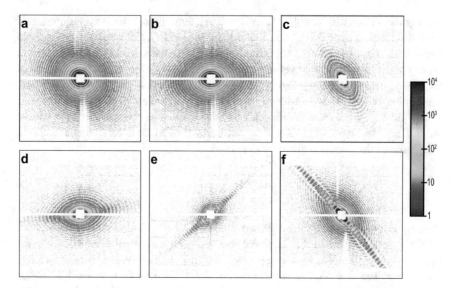

**Fig. 7.11** Diffraction images from helium droplets obtained at small scattering angles using soft X-ray FEL. The logarithmic intensity color scale reflects the number of photons per pixel and is shown on the *right*. Images **a–c** exemplify patterns corresponding to spheroidal droplets, whereas images **d–f** show streaks, which are features indicating high deformity of droplet shape (Modified with permission from Ref. [48]. © copyright <American Physical Society> All rights reserved.)

the maxima of the concentric rings in the diffraction image is given in Eq. (7.9). For a long time, helium droplets were thought to be spherical and not rotating [96]. The diffraction images in Fig. 7.11 show droplets with increasing $AR$, from about 1–2. These images revealed that the droplets are spinning tremendously fast as indicated by large shape deformations in the diffraction patterns. It was initially proposed that rotating helium droplets remain axially symmetric [35]. As will be shown later, further studies reveal that at high angular momentum the superfluid droplet may adapt a two-lobed shape [48–50], similar to what was observed in classical droplets [159, 160].

Figure 7.12 shows diffraction images of droplets, with an average droplet radius of ~400 nm, obtained from wide scattering angles, at most 30°, in the XUV regime at 19–24 eV [50]. The figure also shows the corresponding three-dimensional models of the droplets along with calculated diffraction images. In this experiment, the droplets were produced at ~5.4 K and 80 bars with a trumpet shaped nozzle, where a large dataset with a total of 38,500 bright scattering patterns was recorded at FERMI FEL. The vast majority (92.9%) of the bright scattering images exhibit concentric rings, see Fig. 7.12a, while the remaining images show various pronounced deformations of the rings. Some collected diffraction images at wide scattering angles, see Fig. 7.12b–e, show deviation from point symmetry, which is key to determining three-dimensional structure, see Sect. 7.3.4. Based on characteristic features in the diffraction image, five shape groups were identified. The respective relative abundance for each group

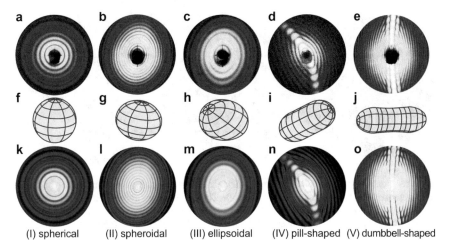

(I) spherical       (II) spheroidal      (III) ellipsoidal      (IV) pill-shaped  (V) dumbbell-shaped

**Fig. 7.12** Diffraction images from helium droplets obtained at wide scattering angles using XUV as the light source. Panels **a–e** show the experimental data and show implied evolution of spinning helium droplets. Panels **f–j** show simulated three-dimensional droplet sizes and shapes. Panels **k–o** show respective calculated diffraction images for the droplets in **f–j**. The data have been classified into five groups (I)–(V), with a transition from **f** spherical to **g** oblate and **h–j** prolate shapes (Reused with permission from Ref. [50], licensed under CC-BY 4.0)

was estimated using a neural network for automated image reconstruction [50, 142]. These groups are: (i) spherical (concentric circles, 92.9%), (ii) spheroidal (elliptical patterns or one-sided asymmetry, 5.6%), (iii) ellipsoidal (bent patterns, 0.8%), (iv) pill-shaped (streaked patterns, 0.6%), and (v) dumbbell-shaped (streaks with side maxima or pronounced side minima, less than 0.1%).

In addition to using an XUV FEL, wide-angle scattering has also been demonstrated using a lab-based HHG laser [49], transfering a very powerful imaging technique from large X-ray facilities to laboratories at research institutes and universities. Moreover, HHG sources can also be very intense and have the potential in producing very short XUV pulses up to the attosecond pulse duration [73, 74]. Figure 7.13a shows a scattering image obtained using a HHG laser consisting of the 11th until the 17th odd harmonics of the 792 nm IR seed laser. Figure 7.13 also shows simulations (panels b–c), which aid in the unique identification of the droplet shape. The optical axis of the extreme ultraviolet beam is directed into the image plane, while the tilt angle between the symmetry axis of the droplet and the optical axis is ~35°. From the simulations, the diffraction image is from a pill-shaped droplet with semi-minor radii of $b = c = 370$ nm and a semi-major radius of $a = 950$ nm. Figure 7.13c illustrates the origin of bent streaks occurring when a tilted pill-shaped structure diffracts the light. The constructive interference is analogous to the specular reflection at the surface of a macroscopic pill. Two bundles of constructively interfering rays are explicitly sketched. Note that the different ray colours in the figure do not refer to wavelengths but are applied to facilitate distinction in the specular reflection. Despite having a blurred pattern, due to multiple HHG harmonics involved in imaging, and

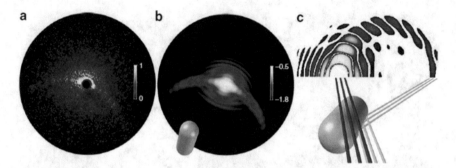

**Fig. 7.13** Scattering pattern obtained from a HHG laser and simulations of the scattering pattern. **a** Measured image and **b** matching simulation result. The droplet shape and orientation are visualized in *yellow*. **c** Illustration of the origin of bent streaks occurring when a tilted pill-shaped structure diffracts light (Adapted with permission from Ref. [49], licensed under CC-BY)

weaker scattering intensity as compared to images obtained using FELs, the droplet shape is still successfully retrieved from the HHG diffraction data.

### 7.4.2 Droplet Stability Curve

The shape of a droplet undergoing rigid body rotation is maintained by the balance between surface tension and centrifugal forces [158]. For classical viscous droplets, dimensionless parameters are introduced in order to facilitate comparison between different experiments and theories for droplets of different sizes and composition. The red curve in Fig. 7.14a shows the evolution of droplet shapes as given by the classical droplet stability curve, which is described in terms of reduced angular momentum, $\Lambda$, and reduced angular velocity, $\Omega$ [156, 158, 159]:

$$\Lambda = \frac{L}{\left(8 \cdot \sigma \cdot \rho \cdot R_D^7\right)^{1/2}}, \tag{7.11}$$

$$\Omega = \omega \cdot \left(\frac{\rho \cdot R_D^3}{8 \cdot \sigma}\right)^{1/2}, \tag{7.12}$$

where $L$ and $\omega$ are the angular momentum and angular velocity of the droplet, respectively. The surface tension of the liquid droplet is given by $\sigma$, the density by $\rho$, and its droplet radius by $R_D$. Droplets with different sizes but with the same values of $\Lambda$ and $\Omega$ belong to the same class of droplet shapes, which evolve from being spherical to oblate axisymmetric when the value of $\Lambda$ is increased from zero [156, 158, 159]. The curve starts to bifurcate at $\Lambda = 1.2$ and $\Omega = 0.56$. The upper branch, $\Omega > 0.56$, adapts a toroidal shape and is unstable to quadrupolar shape deformations, while the lower branch, $1.2 < \Lambda < 1.5$, is stable and represents prolate triaxial droplets. At

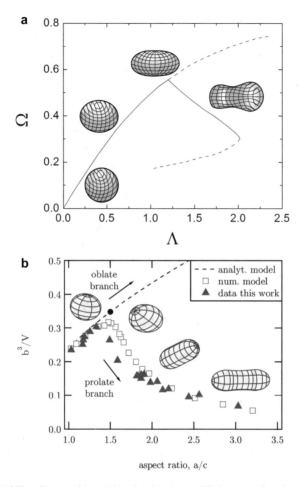

**Fig. 7.14 a** Stability diagram for rotating droplets in equilibrium as a function of the *reduced angular velocity*, $\Omega$, and the *reduced angular momentum*, $\Lambda$, see Eqs. (7.11) and (7.12). The upper branch of the *solid red line* corresponds to oblate axisymmetric shapes, whereas the lower branch to prolate two-lobed shapes. The bifurcation point is located at $\Lambda = 1.2$, $\Omega = 0.56$ with $AR = 1.48$. As for rigid bodies, they will show no distortions and would follow a straight line. The *dashed red lines* indicate the unstable portion of the droplet stability curve (Adapted with permission from Ref. [48]. © copyright <American Physical Society> All rights reserved.). **b** Ratios of the principal semiaxis lengths, a, b, and c, and V the volume of the droplet. The *dashed line* is from the analytical model of Chandrasekhar [156]. The *squares* are from numerical models for classical droplet shapes [163]. The *triangles* are data obtained from wide-angle scattering imaging of helium droplets using an XUV light source (Reused with permission from Ref. [50], licensed under CC-BY 4.0)

$\Lambda \approx 1.5$, the droplet starts to be highly deformed, first by forming a two-lobed droplet at $\Lambda > 1.5$, before undergoing fission at $\Lambda \approx 2$. Figure 7.14a also shows sketches of the droplet shapes at different parts of the stability curve. The solid red lines indicate the stable evolution of droplet shapes, from spherical to oblate spheroidal until it transforms into prolate triaxial. The dashed lines represent the unstable portion of the droplet stability curve.

For stable structures, $\Lambda$ and $\Omega$ exhibit a clear relation to the principal axes of the droplets, $a$, $b$, and $c$. The droplet radii at each of the principal axes are experimentally measurable quantities and represent the distances from the droplet's center of mass to the droplet's surface along each respective principal axes. Following convention, $a$ and $c$ are taken as the largest and smallest radii, respectively. For oblate droplets, $a = b \neq c$, while for prolate triaxial droplets, $a \neq b \neq c$. In small-angle scattering measurements, one of the principal axes in axisymmetric droplets is assumed to be the same as that of the major half-axis. On the other hand, wide-angle scattering measurements allow the determination of all these principal axes. Figure 7.14b shows the plot of $b^3/V$, where $V$ is the volume of the droplet, versus the aspect ratio $a/c$. At $AR = 1$, $b^3/V = R_D/V = 3/4\pi \approx 0.24$. The open square symbols in Fig. 7.14b were derived from classical models of rotating drops [163].

What is surprising with the results shown in Fig. 7.14 is that the evolution of shapes for large, oblate-spheroidal, superfluid helium nanodroplets seems to follow a similar behaviour as that of spinning classical viscous liquid drops. The diagram in Fig. 7.14b is another dimensionless representation of droplets that is used in classifying the shapes of spinning droplets. The numerical model, represented as purple squares, shows how the shape of classical spinning droplets evolves from axisymmetric oblate spheroid, $a/c < 1.5$, to triaxial prolate shapes, $a/c > 1.5$. The red triangles represent the different shape classes of superfluid helium droplets determined from diffraction images collected at wide scattering angles [50]. Aside from the points around $a/c = 1.5$, there seems to be a good agreement between the different shape classes of superfluid helium droplets with that of the classical droplets.

Quantum fluids, such as Bose–Einstein condensates and superfluid helium, are inviscid due to their negligible viscosity [38, 168]. The fluid is constrained to *rotate* with a quantized circulation around topological point defects, which are manifested as quantum vortices and store the most amount of the fluid's angular momentum [38, 168]. Due to the presence of a vortex array, a meniscus is formed in a rotating superfluid cylinder that would have otherwise been absent without the vortices [177]. It can be said that in the limit of multiple vortices, superfluids behave like classical viscous liquids [168, 177]. Other elementary excitations, such as phonons, rotons, and ripplons can also store angular momentum [47, 101]. However, due to the low temperature of the droplet at 0.4 K, the contribution of these excitations is negligible [39, 47]. A large, axisymmetric nanodroplet containing a large number of vortices, which arrange in a triangular lattice, rotates similarly to a classical droplet due to the velocity flow fields produced by each vortex [51, 169, 171, 178]. On the other hand, the shape of small helium droplets, $N_{He} < 20,000$, containing a few number of vortices deviate from the classical behaviour, as determined from density functional theory (DFT) calculations [171, 179, 180]. Capillary waves on the surface of the droplets also contribute to the angular momentum of the droplet, however, their contribution only becomes important for prolate-shaped droplets [50, 51, 171, 181]. Finally, the rather strong agreement between the different kinds of drops with theoretical models suggest a unified theory of the dynamics of fluid masses [161]. However, a caveat must be kept in mind that the rotational behavior of superfluid helium droplets is generally contingent upon the size of the droplet and the number of vortices in it [182].

## 7.4.3  Non-superfluid Helium Droplets

The results shown in earlier sections have demonstrated that superfluid helium-4 droplets produced from free jet expansion conditions are spinning considerably fast, as manifested by the shape distortion of these droplets. The droplet's shape distortion is ascribed to the presence of quantum vortices, which account for most of the angular momentum of the droplet [171]. The nucleation of these vortices, however, remains unknown and requires further experimental studies. One hypothesis speculates that the interaction of the droplets at the dense region around the proximity of the nozzle exit significantly contributes to the rotational state of superfluid helium-4 droplets, and thus nucleates the vortices in the droplet [40, 48]. To test this hypothesis, Ref. [52] produced non-superfluid helium-3 droplets using the same experimental setup as that used for the production of helium-4 droplets. The first creation of a beam of helium-3 droplets was done in late 1970's in Karlsruhe, Germany [183, 184]. Helium-3 is chemically equivalent to helium-4, and the strength of the van der Waals bonding is virtually identical. However, the smaller mass of helium-3 means that its zero-point energy is greater than that of helium-4 [185]. Therefore, in order to achieve the same average size as that of the helium-4 droplets, lower nozzle temperature is needed. While pairs of helium-3 atoms also undergo superfluid transition at a very low temperature of ~2.7 mK [185, 186], in molecular beam apparatus, helium-3 only cools down to ~0.15 K and, thus, remains non-superfluid [98]. It is interesting to note that seven different kinds of quantum vortices are determined in bulk superfluid helium-3 [187].

Figure 7.15 shows diffraction images of non-superfluid helium-3 droplets collected at small-angle scattering and demonstrates that the shapes of helium-3 droplets mostly correspond to oblate spheroids falling within the classical droplet stability curve shown in Fig. 7.14 [52]. Furthermore, these results agree with the shapes of superfluid helium-4 droplets, which were also produced and imaged under similar experimental conditions, i.e., same 5 $\mu$m diameter pinhole nozzle was used and the droplets were produced within the liquid fragmentation regime. Both isotopes were found to have the same average reduced angular velocities, $\Omega$, and reduced angular momenta, $\Lambda$, quantities which describe the rotational state of a droplet. One

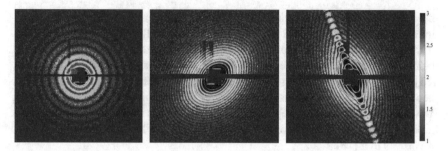

**Fig. 7.15** X-ray coherent diffractive imaging of pure $^3$He droplets at different rotational states. Only the central section of the diffraction is shown here (Adapted with permission from Ref. [52].

particular importance of this experiment is that it corroborates that similar mechanisms are involved in the shape distortion of the droplets, and that the origin of the droplet's vorticity may be due very well to the interaction of the droplets with the walls of the nozzle orifice and with each other close to the region of the nozzle exit [52]. It is also possible that quantized vortices are nucleated during the merging and coagulation of smaller droplet into bigger droplets [181, 188]. DFT calculations in a semi-classical approach for a $N_{He3} = 1500$ droplet also found that the shapes of spinning helium-3 droplets agree with that of classical droplets, although, the derived stability curve for helium-3 droplets has a slight deviation [189].

## 7.5 Imaging Dopant Cluster Structures in a Superfluid Helium Droplet

The unusual properties of superfluid helium droplets make them conducive media for investigating the formation and growth of out-of-equilibrium nanostructures in self-contained and isolated droplets, see also Chap. 11 in this volume [45, 55, 190]. This viability is due to the droplets' superfluidity, their very cold ambient temperatures of ~0.4 K, and the possibility to control the size and composition of embedded dopants [96, 109]. Dopants are captured by the droplets within pickup cells positioned along the droplets' flight path, see Fig. 7.4. Once a dopant is captured, it quickly thermalizes to the droplet temperature and is decelerated until it moves inside without friction [109]. When several dopants are successively captured, they coalesce stochastically with the influence of long-range van der Waals interaction allowing the dopants to form far-from-equilibrium nanostructures [190]. While some dopant materials may form compact clusters at one or several sites in the droplets [191], some polar molecules can arrange in long linear chains [192]. Other studies have shown a core–shell structure of a multicomponent doped droplet [193–195] or indicated to the formation of foam structures [196, 197]. Superfluid droplets are also used as a weakly interacting matrix for the study and control of orientation and alignment of embedded molecules [198–200]. Up to now, almost all of these very special structures could only be inferred from spectroscopic measurements on ensembles of many different droplet sizes. In the presence of quantum vortices, these vortices dominate structure formation processes [40–43, 53, 54, 178]. Imaging these nanostructures in situ can give us unprecedented insights into the processes underlying their formation.

## 7.5.1   *Vortex Structures in Superfluid Helium Droplets*

Probably, the most peculiar and clear manifestation of superfluidity is with the presence of quantum vortices in the droplet [39, 46, 170, 178, 201]. Traces of these vortices were first discovered from silver-doped helium droplets deposited on a carbon film for Transmission Electron Microscopy (TEM) imaging [40–43]. Silver atoms are attracted to the cores of the vortices. Some examples of these silver nanostructures are shown in Fig. 7.16. Further experiments showed that the traces of the vortex line remain observable even though room temperature induces structural changes in the deposits on the carbon film [43]. While the deposition and TEM imaging gave an idea on the traces of vortices inside the droplet, this type of ex situ imaging technique didn't provide enough information on the native configuration of the vortices in the droplet. Gessner and Vilesov [178] wrote a recent review on imaging quantum vortices in superfluid helium droplets.

In X-ray imaging experiments, the droplet needs to be doped first in order to visualize the in situ structures and positions of quantum vortices in the droplet, since bare quantum vortices have a diameter of roughly 2 Å [38] and have no noticeable contribution to the diffraction intensity. The dopants serve as contrast agents for imaging. The DCDI algorithm described in Sect. 7.3.3 solves for the positions and morphologies of the dopant structures inside the droplet, while the size and shape of the droplets can be determined from known analytical equations of light scattering by small particles [35, 202]. Almost all of the static images of doped helium droplets have so far used xenon as the dopant, partly due to its ease of handling. In Sect. 7.5.3, diffraction images of doped droplets from different dopant materials are presented.

Figure 7.17 shows how xenon clusters are arrange symmetrically in small droplets, <200 nm in diameter. Foreign particles approaching a quantized vortex core are attracted to it due to the pressure gradient experienced by the particles [38, 203]. The arrangement of these xenon clusters reflects the positions and configurations of quantum vortices in the droplet. However, compared to the symmetric positions of vortices observed in the bulk [36], or to the positions of bare vortices calculated for a nanocylinder [204] and nanodroplet [180], the positions of the xenon-traced vortices are farther away from the center of the droplet [54, 205]. This deviation can be accounted from the conservation of angular momentum. When the vortex core is filled with a dopant, the dopant rotates with the vortex core, which contributes to the total angular momentum. As a consequence, the angular momentum of the vortex must decrease resulting into the vortex slowing down and moving farther away from the center [54, 180, 205, 206]. The obtained symmetric positions of the vortices is in

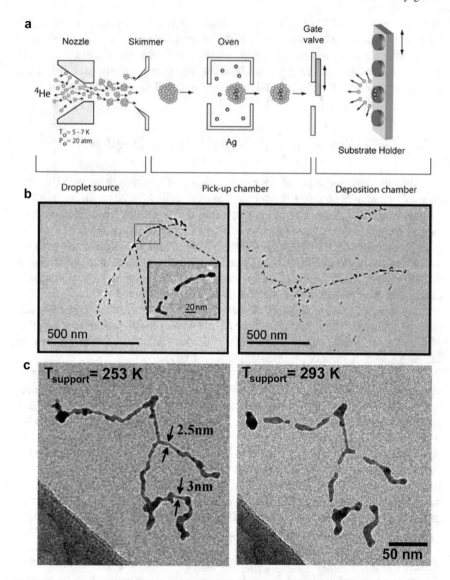

**Fig. 7.16** Traces of quantum vortices in superfluid helium nanodroplets. Panel **a** shows the schematic of the experiment. The droplet is first doped with silver atoms before being deposited on a carbon film for TEM imaging. The droplet evaporated as it collides with the carbon film (Adapted from Ref. [40]. © copyright <American Physical Society> All rights reserved.) **b** Typical silver traces obtained in 1000 nm helium droplets (from Ref. [40]. © copyright <American Physical Society> All rights reserved.). The *inset* shows an enlarged track segment. **c** Structure evolution of a silver nanowire at increasing carbon film temperature at 253 and 293 K (from Ref. [43], licensed under CC-BY 3.0)

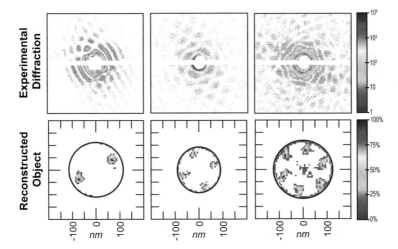

**Fig. 7.17** Symmetric vortex structures for droplets with radius from 80 to ~100 nm. The *upper row* shows single-shot diffraction images of xenon-doped droplets. The *bottom row* shows corresponding DCDI reconstructions, where the *black circle* represents the shape and boundary of the droplet. The diffraction images only show the central part of the pnCCD detector (Adapted with permission from Ref. [54]. © copyright <American Physical Society> All rights reserved.)

agreement with that expected for vortices having the same strength and with the same value as that of the quantum of circulation, $\kappa$. The kinematics of dopant positions is studied in Ref. [205].

In larger droplets, >200 nm in diameter, with moderate number of dopant atoms, $>10^6$, the configurations of the vortices also appear symmetric. However, one starts to see that the shape of the xenon-traced vortex filaments is not straight but wavy, see Fig. 7.18. In this category, the shapes of the vortex cores are given by the shape of the filaments due to particle trapping of the xenon atoms along the length of the vortex core. Aside from the undulations along a vortex length, vortices far from the center of the droplet are also curved, which is expected from calculations [47, 170, 180]. In Fig. 7.18, the appearance of vortex filaments indicate that the droplets are imaged with their rotational axis almost perpendicular to the propagation direction of the X-ray beam. It is energetically favourable to have the vortices aligned parallel to the rotational axis [38, 207]. Therefore, if the droplet is spheroidal, $1 < AR < 1.5$, the vortices are aligned parallel to the semi-minor axis of the droplet. Vortices along the major axis are not favourable. Similar observations of vortices aligning parallel to the axis of rotation have been noted in rotating Bose–Einstein condensates [201, 208–210]. On the other hand, if droplets are imaged with their rotational axis parallel to the direction of the X-ray beam propagation, then the filaments would appear as dots and would have a triangular lattice, as seen in Fig. 7.17 [54]. For example, the droplet in the first column of Fig. 7.18 is consistent with an approximately hexagonal pattern of C-shaped filaments imaged at some angle with respect to the symmetry axis [53]. The vortices in the droplets are expected to be curved as they must terminate

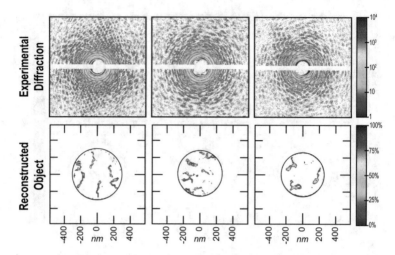

**Fig. 7.18** Xenon-traced quantum vortices for larger droplets, >200 nm in diameter. The *upper row* shows single-shot diffraction images of xenon-doped droplets. The *bottom row* shows corresponding DCDI reconstructions, where the *black circle* represents the shape and boundary of the droplet. The diffraction images only show the central part of the pnCCD detector (Adapted with permission from Ref. [53], licensed under CC-BY 3.0)

perpendicular to the surface [47, 170, 206]. In addition, the vortices in Fig. 7.18 are still well-separated from the other vortices in the droplet.

The waviness of the dopant-traced vortex filaments is a tell-tale sign of the dynamics involved as the dopant particles are trapped by the vortex lines, see Fig. 7.18. After being captured by the droplet, the dopant would walk randomly as it thermalizes. The droplet's surface acts as some sort of a net that prevents dopant particles from escaping. The droplet surface also bounces the dopant until it finds its most stable position [179, 206]. The process of penetration and translation of dopant particles in the droplet creates surface excitations. This process is also enough to create undulations along the vortex line as simulated using semi-classical matter wave theory [211] and DFT [206]. Distortions along a vortex line are classically known as Kelvin waves [38]. The dynamical picture is, of course, more complicated. In a superfluid helium droplet, the vortex lines are anchored perpendicular to the surface of the droplet [47, 170]. When a dopant particle, such as xenon and argon, approaches the vortex core, it is not captured right away. Instead, as demonstrated by the DFT numerical simulations, a dopant particle first orbits around the vortex core before being captured [206]. The dopant particle penetration and its orbiting motion would twist the vortex core. Upon vortex capture, the particle can cause bends and kinks along the vortex lines, which further create Kelvin waves [203, 212, 213]. The descriptions above are for a particle approaching a vortex core. Furthermore, since equilibrium positions of vortices can be determined from the reconstructed images, one can also infer the coupling between the dopant atoms and the vortices, i.e., from the images one can make a conjecture on the dynamics inside the superfluid, such as excitation of waves along the length of the vortex as the dopant approaches it. Further studies are of course needed to understand the undulations along the vortex filaments in the presence of multiple dopants and multiple vortices in a droplet [205].

## 7.5.2   Vortex Lattices and Angular Momentum Determination

Diffraction patterns obtained from very large droplets with droplet radius ranging from 500 to 1000 nm often contain a series of high intensity spots far from the center of the diffraction. These bright spots are known in X-ray crystallography as Bragg spots and indicate a crystal lattice structure in the object being imaged. For the xenon-doped droplets, the Bragg spots originate from the interference caused by the xenon-traced vortex lattice. Figure 7.19 gives some examples of diffraction images containing Bragg spots that either lie on a line crossing the image center, see Fig. 7.19a, or form an equilateral triangular pattern, see Fig. 7.19b. The diffraction pattern in Fig. 7.19b provides a direct measure of the vortex density, $n_V = 4.5 \times 10^{13}$ m$^{-2}$ [35]. The angular velocity of a rotating droplet can be calculated using [38, 214]:

$$\omega = \frac{1}{2} \cdot \kappa \cdot n_V, \tag{7.13}$$

where, $\kappa = h/m_{He}$ is again the quantum of circulation. For the droplet considered in Fig. 7.19b, $\omega = 2.2 \times 10^6$ s$^{-1}$, and for a droplet radius of $R_D = 1100$ nm, the total number of vortices is 170 [35].

Vortex configurations inside the droplet also give access to the rotational state of the droplet since both the structures of the vortices and the droplet shapes are imaged

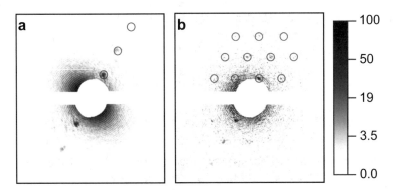

**Fig. 7.19** Diffraction images showing vortex lattice in the droplet. The *red circles* indicate the positions of the Bragg peaks (Reused with permission from Ref. [35]. © copyright <American Association for the Advancement of Science> All rights reserved.)

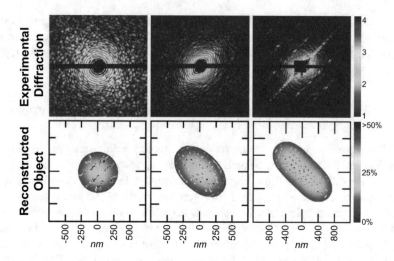

**Fig. 7.20** Diffraction patterns from xenon-doped droplets with various droplet shapes. *Upper row, left to right*: axisymmetric, near spherical droplet shape; triaxial pseudo ellipsoidal; and capsule shaped. The *lower row* shows the numerically reconstructed droplets using the DCDI algorithm. The diffraction images only show the central part of the pnCCD detector (Adapted with permission from Ref. [51]. © copyright <American Physical Society> All rights reserved.)

at the same time. From the positions of the vortices, one can estimate the angular velocity of the droplet due to the presence of quantum vortices using Eq. (7.13), while the droplet shape gives an overall account of the droplet's angular momentum. From these pieces of information, one can determine the total angular momentum state of the droplet and how they are distributed between different global excitations in the droplet, such as shape oscillations due to capillary surfaces waves and quantum vortices [51]. Figure 7.20 shows the evolution of vortex arrangements with changing droplet shapes. In axisymmetric, oblate superfluid droplets, the quantum vortices are arranged in a triangular lattice and solely contribute to the angular momentum of the droplet [51]. On the other hand, quantum vortices and capillary waves both contribute to the angular momentum of the droplet, especially for triaxial droplets [51, 171, 182].

### 7.5.3 Controlling Structures Formed in Helium Droplets

Droplets with quantum vortices are usually produced from the fragmentation of liquid helium close to the nozzle exit [178]. It is conjectured that vortices are nucleated due to the interaction of the droplets with each other and with the walls of the nozzle [40, 48, 52]. When dopants are introduced, the presence of these vortices dominate structure formation, and, consequently, the shapes of the dopants resemble that of a vortex core/line or a vortex lattice. One approach in reducing this type of vorticity acquisition is to produce large nanodroplets from the condensation of cold helium gas with the use of a conical nozzle. The design of conical nozzles promotes the condensation process and thus creates large clusters/droplets from the gas phase [215]. For a fixed stagnation pressure and nozzle temperature, the size of the cluster is increased by changing the effective diameter of the nozzle, $d_{eff}$, which can be estimated using [215]:

$$d_{eff} = \frac{0.72 \cdot d_{throat}}{\tan \varphi}. \tag{7.14}$$

The variables $d_{throat}$ and $\varphi$ correspond to the throat diameter and the opening half-angle of the nozzle, respectively. In a recent experiment at the European XFEL, helium nanodroplets were produced using a conical nozzle with $d_{throat} = 150$ μm, $\varphi = 3°$, and $d_{eff} \approx 2$ mm [55]. This configuration allows the generation of helium nanodroplets, >100 nm in diameter, from the gas phase. The increase in $d_{eff}$ also leads to an increase in the mass flow from the nozzle, and similarly the gas load on vacuum pumps. In order to avoid using large pumps, pulsed valves are employed.

Almost all diffraction images from pure droplets produced using a conical nozzle [55] exhibit the same concentric ring pattern as that shown in Fig. 7.5a. This observation indicates that an overwhelming majority of the droplets is spherical in shape. In contrast, some of the droplets produced at the liquid fragmentation regime from previous experiments at LCLS in the USA [35, 48], at FERMI FEL in Italy [50], and a HHG laser [49] showed extreme shape distortions, e.g. pill shapes or dumbbell shapes [178]. Theoretical work supports the idea that the shape of these distorted droplets is controlled by the presence of quantum vortices [171, 180].

In the absence of vortices, it may be possible to control the formation of dopant nanostructures by using different kinds of dopant materials. Figure 7.21 shows diffraction images for differently doped droplets: xenon, silver, acetonitrile, and iodomethane. The intermolecular interactions of these dopants include van der Waals, metallic, and dipole–dipole. The observed diffraction patterns show distinct features that were not previously observed, see Sect. 7.5.1. The diffraction images collected from atomic clusters suggest the presence of one to two cluster cores in the droplet, while that from molecular clusters suggest a complicated network of dopant clusters [55]. These results, however, are preliminary, further studies are needed in order to fully explore and understand factors influencing dopant nanostructure formation inside superfluid helium droplet using coherent diffractive imaging. Another possible means of investigating the formation of nanostructures inside a quantum fluid is through the use of non-superfluid helium-3 droplets, where xenon clusters were found to aggregation along the equatorial plane of the droplet [56].

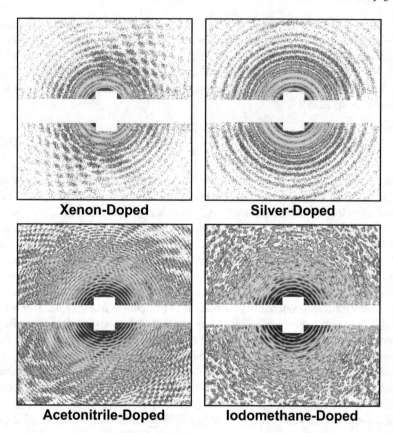

**Fig. 7.21** Examples of diffraction images from superfluid helium droplets containing different dopant materials. These images also demonstrate the possibility of CDI for superfluid helium droplet containing different dopant materials, opening opportunities for studying nanostructure formation inside the droplet (Modified with permission from Ref. [55])

## 7.6 Imaging Dynamical Processes in Helium Droplets

In an era of powerful lasers with intensities exceeding $10^{12}$ W cm$^{-2}$, the electric field strength of these lasers can become comparable to the Coulombic-binding fields in matter [216]. Atoms and molecules no longer simply absorb single or multiple photons [114, 217]. Instead, at these high-power densities, nonlinear ionization processes may occur. Moreover, molecules may start to get aligned along the direction of the laser polarization, and structural deformations can happen in addition to Coulomb explosion. At even higher intensities, plasmas may form that could lead to X-ray emission by relaxation of high-energy electrons in the system [114, 217]. In many studies, the energy and momentum distributions of ions and electrons upon strong-field ionization are detected using time-of-flight spectrometers or with velocity map imaging. These data are used to deduce the underlying dynamics in the system. On the other hand, with the development of ultra-bright and coherent light sources, these dynamical processes may also be directly imaged on the system. In clusters, structural changes are determined from the series of diffraction images at different delays after the introduction of an intense near-infrared pulse on the system [108, 218, 219]. One particularly intriguing result is the discovery that in a xenon cluster the surface first becomes less dense while keeping the core of the cluster relatively intact. In time, the blurred surface expands while the core shrinks [108, 218, 219]. Similarly, microscopic particle-in-cell or Mic-PIC calculations predict surface softening process for a 50 nm diameter hydrogen nanoplasma [220]. This result goes against earlier ideas that the cluster uniformly expands after excitation [221, 222]. The intensity of the FEL pulses needed for imaging clusters or droplets is so very high that fast ionization and, eventually, complete cluster destruction take place. In most cases, however, scattering images are collected before these events happen.

The CDI setup for investigating the mechanisms following strong field excitation is similar to that shown schematically in Fig. 7.4. The NIR pulse excites the droplet and initiates dynamical processes, while the X-ray or XUV pulse images the state of the droplet at a particular time delay after excitation. The NIR pulse is naturally synchronized at the same focal volume as that of the imaging pulse. Additionally, time-of-flight spectrometers for ions and electrons are placed perpendicular to the direction of both the light beam and the droplet beam. The combination of CDI with electrons/ions distribution measurements can address key questions such as: (i) how samples get damaged after interacting with the intense laser pulse; (ii) to what extent do samples get damaged; and (iii) how an excited system dissipates its available energy. Since most light sources now have pulse durations of a few femtoseconds, it is also possible to image these processes at their natural timescales [108, 114, 116]. It can be said that new XUV and X-ray light sources offer many exciting opportunities in imaging processes occurring in intense-light matter interaction, including helium nanodroplets.

At low laser intensities, liquid helium is transparent from far-infrared to vacuum ultraviolet radiation because of the large ionization potential of helium at ~25 eV or ~50 nm [96, 223]. This property quickened high resolution spectroscopy of many atoms and molecules inside a superfluid helium droplet [96, 224]. Theoretical studies, however, have predicted that ionization can be induced in a pure droplet with intense NIR pulses (800 nm, ~$10^{17}$ W/cm$^2$) [120] or with a moderately intense NIR pulse (800 nm, ~$10^{14}$ W/cm$^2$) if the droplets are doped [117]. Experimental studies for these ionization processes have so far been inferred from the energy and momentum distributions of ejected electrons and ions with time-of-flight spectrometers or with velocity map imaging [122, 124].

The first dynamics CDI was performed on pure, sub-micron sized, helium droplets [59]. The authors of this experiment reported an anisotropic surface softening on the droplet, in addition to a similar anisotropic shrinking of the plasma core [59]. The process of charging and ion ejection has similarly been studied by recording X-ray images at different time delays after excitation of pure helium droplets with an NIR pulse [61]. In the case of helium nanodroplets doped with rare gas atoms, such as xenon, which has a lower ionization potential than helium, irradiation with a moderate

NIR laser pulse was theoretically predicted to lead to the formation of a nanoplasma around the dopant cluster core, as shown in Fig. 7.22a [117]. The dopants provide seed electrons, which in turn will ionize the helium environment in an avalanche-like process [225, 226]. Hence, the dynamics following irradiation with the intense light field might be linked to the position of the dopants in a droplet. For example, the distribution of the $He^{2+}$ ions indicates whether the dopant is located at the surface, just below the surface, or very close to the center of the droplet [122].

In an experiment at the FERMI FEL, the light-induced dynamics in xenon doped helium nanodroplets were studied [60]. The doped droplet is first irradiated with a moderately intense NIR pulse (785 nm, ~8 × $10^{13}$ W/cm$^2$) followed by an XUV

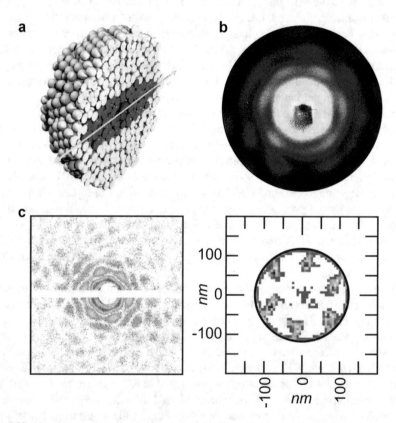

**Fig. 7.22** Nanoplasma dynamics in xenon doped helium nanodroplets and connection to vortex array. **a** Theory predicts that after irradiation with an NIR pulse a nanoplasma forms around the dopant cluster core. (Adapted with permission from Ref. [117]. © copyright <American Physical Society> All rights reserved.) **b** Wide-angle diffraction pattern of a xenon doped droplet recorded 20 ps after NIR irradiation. (Adapted with permission from Ref. [60], licensed under CC-BY 4.0) **c** Small-angle diffraction pattern and corresponding reconstruction of a xenon doped droplet revealing the vortex structure. (Adapted with permission from Ref. [54]. © copyright <American Physical Society> All rights reserved.) The similarity of the diffraction patterns in **b** and **c** points at a connection between the dynamics and the dopant cluster positions

pulse that can be delayed up to 800 ps. The diffraction images were recorded at wide scattering angles up to ~30° and exhibit different features as compared to the ones obtained from static diffraction images [60]. For example, a scattering image taken 20 ps after irradiation with the NIR pulse is shown in Fig. 7.22b. The pattern exhibits intensity maxima in a hexagonal configuration, indicating some kind of density fluctuation at multiple sites in the droplet with an intriguingly symmetric arrangement. In comparison, Fig. 7.22c shows a similar diffraction pattern from a xenon-doped droplet taken with a single X-ray pulse from LCLS and its corresponding DCDI reconstruction, which reveals a hexagonal pattern of xenon clusters [54]. Since the dopant atoms are attracted to the cores of the vortices, the hexagonal pattern is attributed to the vortex structure in the droplet, see also Sect. 7.5.1. However, the underlying processes leading to the observed diffraction patterns in Fig. 7.22b, c are completely different, as the XUV wavelength is not sensitive on the xenon filaments themselves. Changes in the diffraction pattern of xenon doped helium nanodroplets are only visible picoseconds after NIR irradiation. While data analysis is still ongoing, the similarity of the experimental patterns in Fig. 7.22b, c suggests that light-induced dynamics is connected to the position of the dopants in a superfluid helium droplet. The ignition of a nanoplasma at multiple sites and the nature of the dynamics in the droplet should be addressed in future studies.

## 7.7  Summary and Outlook

Recent developments of new light sources, such as X-ray and XUV Free-Electron Lasers and intense High Harmonic Generation sources, are enabling technologies that push the boundaries in studying the structure of matter to unprecedented resolution. These light sources are very intense and have pulse durations on the order of a few tens of femtoseconds down to the attosecond regime. These extraordinary characteristics are paving avenues to new experimental possibilities, such as single-shot imaging of non-periodic and transient systems, and new investigations on how matter interacts with intense light pulses on the timescale of electronic motions.

This chapter reviews developments in single-shot imaging of helium nanodroplets. In Sect. 7.2, the topic of lensless coherent diffractive imaging with the use of these new light sources is introduced. Scattering data are collected at small and wide angles. For small-angle scattering with X-rays, high spatial resolution can be achieved, although, only the two-dimensional density projection of the object is accessible. In contrast, wide-angle scattering with XUV wavelengths can provide three-dimensional information of the object but with reduced resolution. Both scattering techniques are complementary, and the choice on which technique to use depends on the goal of an experiment. In Sect. 7.3, the response of helium droplets with X-ray and XUV radiation is discussed, as well as strategies for solving the structures inside the droplet, and on how to simulate the wide-angle diffraction image of a three-dimensional droplet. In addition, since single-particle imaging can potentially collect millions of images in an hour, schemes for machine learning supported analysis that may handle huge amount of data during the experiment were also presented.

As for the experimental results presented in Sects. 7.4, 7.5 and 7.6, there are two types of studies that have been performed so far, static imaging and dynamic imaging. In the former, images of droplets are collected where the main research interests lie in determining the shapes of helium droplets and the structures of dopant clusters assembled inside them. The results from these experiments are rather surprising. For instance, the presence of multiple quantized vortices in superfluid helium-4 nanodroplets made them behave like classical droplets and generally follow the same droplet stability curve for classical droplets. In addition, two main types of droplets were identified, oblate spheroidal, which is axisymmetric, and prolate triaxial. The native configurations of xenon-traced vortices in the droplet were also seen for the first time. The arrangements of these vortices were even used in identifying factors that contribute to the angular momentum of the droplet. In dynamic imaging, the interest is on the behaviour of the helium droplets after they have been irradiated with an intense near-infrared pulse. For doped droplets, the diffraction images seem to suggest that the location of the dopants determine where the growth of nanoplasma is initiated in the droplet.

Our understanding of the many peculiar facets of helium nanodroplets has been advanced through new possibilities connected to imaging individual droplets with high spatio-temporal resolution. As is often the case with new fields of research, further questions are asked as new discoveries are made. One aspect that needs to be explored is the controlled rotation of the droplet. The experiments that have been done so far used a molecular beam apparatus with no control on the rotation of the droplet. In fact, the collection of highly deformed droplets is fortuitous. A more controlled evolution of the droplet rotation may help in exploring how its shape evolves as it spins faster and faster. This kind of experiment would also visualize the formation of a pill-shaped helium droplet and what conditions would favour its formation. Such controlled experiments would clearly define the rotational state of the droplet, along with the effect of capillary waves on the droplet shape. In addition, one might also be able to determine the nucleation of quantum vortices in the droplet. Another aspect that can be explored is with the production of helium droplets from the fragmentation of a liquid jet [227–229, 231]. Droplets produced this way have a narrow size distribution, and the droplet size can be controlled by changing the size of the nozzle diameter. The generation of droplets on demand, with known size and repetition, will aid in synchronizing the imaging pulse with the arrival of the droplet at the interaction point, creating the possibility of having 100% hit rate. Similarly, delivery of biological samples at the focus of a light source is continuously being developed in order to maximize the use of every single pulse [230, 232].

We also foresee a demand for further innovations in imaging the dopant nanostructures inside the droplet. For instance, X-ray imaging may be similarly extended to particle tracking methods commonly used in fluid mechanics experiments, where the trajectory of a particle can be traced in the droplet. Of course, this idea is still fraught with difficulties considering structural damages that may be incurred by the droplet and the dopant particles from photoabsorption. The possibility of X-ray particle tracking can help in visualizing how a dopant particle gets trapped by the vortex, how multiple dopants are distributed in a droplet containing many vortices, how the particle induces undulations along the vortex line, and how vortices connect and reconnect. The last two topics are related to quantum turbulence [233–235]. Another possible development is with three-dimensional imaging of the clusters inside the droplet. So far, only two-dimensional projections of xenon-traced vortices are reported.

One key technical challenge, which is now a limiting factor in CDI experiments and will become even more serious with the development of high repetition rate FELs, is connected to data management and processing. The strong need for fast and robust computational tools may be answered partially by the ongoing employment and adaption of machine learning techniques into the toolboxes of CDI. Finally, dynamic studies are currently receiving a strong push by the latest advances of FEL and HHG sources towards producing high intensity attosecond pulses. Thanks to its simple electronic structure, helium is an ideal model system for exploring ultrafast electron dynamics.

Single-shot coherent diffractive imaging is inaugurating new avenues of research in superfluid helium droplet science. The first experimental results were published in 2014, and many more experiments are being carried out by a growing number of research groups in pace with increasing availability of X-ray and XUV lasers around the world. At the time of this writing, there are roughly 13 experimental publications [35, 48–57, 59, 61], two reviews [101, 178], and more under way. We expect fruitful investigations and exciting discoveries using CDI technique in the years to come.

**Acknowledgements** This book chapter is made possible by the support provided by the Bundesministerium für Bildung und Forschung (BMBF) via grant No. 05K16KT3 within the BMBF Forschungsschwerpunkt Freie-Elektronen-Laser FSP-302, by the Leibniz-Gemeinschaft via grant No. SAW/2017/MBI4, by NCCR MUST of SNF, SNF grant 200021E_193642/1, and by the Deutsche Forschungsgemeinschaft (DFG) grants Mo 719/13 and Mo 719/14. The authors would also like to express their gratitude to Charles Bernando, Alessandro Colombo, Oliver Gessner, Luis F. Gomez, Curtis F. Jones, Katharina Kolatzki, Sean Marcus O'Connell-Lopez, Anatoli Ulmer, Deepak Verma, Andrey Vilesov, and Julian Zimmermann for constructive discussions and for providing us figures from their publications. In addition, we are grateful for Oliver Gessner and Andrey Vilesov for helping us in improving the content of our manuscript.

# References

1. J. Als-Nielsen, D. McMorrow, *Elements of modern X-ray physics*, 2nd edn. (John Wiley & Sons, Ltd., West Sussex, United Kingdom, 2011)
2. D. Attwood, A.E. Sakdinawat, *X-rays and extreme ultraviolet radiation: Principles and applications* (Cambridge University Press, Cambridge United Kingdom, 2017)
3. C. Bostedt, S. Boutet, D.M. Fritz, Z.R. Huang, H.J. Lee, H.T. Lemke, A. Robert, W.F. Schlotter, J.J. Turner, G.J. Williams, Linac coherent light source: The first five years. Rev. Mod. Phys. **88**(1), 015007 (2016)
4. J.A. van Bokhoven, C. Lamberti, *X-ray absorption and X-ray emission spectroscopy: Theory and applications* (Wiley, United Kingdom, 2016)

5. L. Young, K. Ueda, M. Gühr, P.H. Bucksbaum, M. Simon, S. Mukamel, N. Rohringer, K.C. Prince, C. Masciovecchio, M. Meyer, A. Rudenko, D. Rolles, C. Bostedt, M. Fuchs, D.A. Reis, R. Santra, H. Kapteyn, M. Murnane, H. Ibrahim, F. Légaré, M. Vrakking, M. Isinger, D. Kroon, M. Gisselbrecht, A. L'Huillier, H.J. Wörner, S.R. Leone, Roadmap of ultrafast X-ray atomic and molecular physics. J. Phys. B Atomic Mol. Opt. Phys. **51**(3), 032003 (2018)
6. H.H. Seliger, Wilhelm Conrad Röntgen and the glimmer of light. Phys. Today **48**(11), 25–31 (1995)
7. J.C. Cole, S. Wiggin, F. Stanzione, New insights and innovation from a million crystal structures in the Cambridge structural database. Struc. Dyn. **6**(5), 054301 (2019)
8. J. Zhang, Y. He, L. Lei, Y. Yao, S. Bradford, W. Kong, Electron diffraction of molecules and clusters in superfluid helium droplets, in *Molecules in helium nanodroplets*, ed. by A. Slenczka, J.P. Toennies (Springer, Berlin, 2021)
9. J. Kirz, C. Jacobsen, M. Howells, Soft-X-ray microscopes and their biological applications. Q. Rev. Biophys. **28**(1), 33–130 (1995)
10. A. Sakdinawat, D. Attwood, Nanoscale X-ray imaging. Nat. Photonics **4**(12), 840–848 (2010)
11. R. Falcone, C. Jacobsen, J. Kirz, S. Marchesini, D. Shapiro, J. Spence, New directions in X-ray microscopy. Contemp. Phys. **52**(4), 293–318 (2011)
12. W.L. Chao, B.D. Harteneck, J.A. Liddle, E.H. Anderson, D.T. Attwood, Soft X-ray microscopy at a spatial resolution better than 15 nm. Nature **435**(7046), 1210–1213 (2005)
13. T. Salditt, M. Osterhoff, X-ray focusing and optics, in *Nanoscale photonic imaging*. ed. by T. Salditt, A. Egner, D.R. Luke (Springer International Publishing, Cham, 2020), pp. 71–124
14. S. Marchesini, H.N. Chapman, S.P. Hau-Riege, R.A. London, A. Szöke, H. He, M.R. Howells, H. Padmore, R. Rosen, J.C.H. Spence, U. Weierstall, Coherent X-ray diffractive imaging: Applications and limitations. Opt. Express **11**(19), 2344–2353 (2003)
15. H.N. Chapman, K.A. Nugent, Coherent lensless X-ray imaging. Nat. Photonics **4**(12), 833–839 (2010)
16. A. Barty, J. Küpper, H.N. Chapman, Molecular imaging using X-ray free-electron lasers. Annu. Rev. Phys. Chem. **64**(1), 415–435 (2013)
17. W. Boutu, B. Carré, H. Merdji, Coherent diffractive imaging, in *Attosecond and XUV physics: Ultrafast dynamics and spectroscopy*. ed. by T. Schultz, M. Vrakking (Wiley-VCH, Singapore, 2014), pp. 557–597
18. R.A. Kirian, H.N. Chapman, Imaging of objects by coherent diffraction of X-ray free-electron laser pulses, in *Synchrotron light sources and free-electron lasers: Accelerator physics, instrumentation and science applications*. ed. by E. Jaeschke, S. Khan, J.R. Schneider, J.B. Hastings (Springer International Publishing, Cham, 2015), pp. 1–55
19. R.W. Gerchberg, W.O. Saxton, A practical algorithm for the determination of phase from image and diffraction plane pictures. Optik **35**, 237–246 (1972)
20. J.R. Fienup, Reconstruction of an object from the modulus of its Fourier transform. Opt. Lett. **3**(1), 27–29 (1978)
21. J.R. Fienup, Phase retrieval algorithms—a comparison. Appl. Opt. **21**(15), 2758–2769 (1982)
22. S. Marchesini, Invited article: A unified evaluation of iterative projection algorithms for phase retrieval. Rev. Sci. Instrum. **78**(4), 049901 (2007)
23. K.A. Nugent, Coherent methods in the X-ray sciences. Adv. Phys. **59**(1), 1–99 (2010)
24. A. Aquila, A. Barty, C. Bostedt, S. Boutet, G. Carini, D. de Ponte, P. Drell, S. Doniach, K.H. Downing, T. Earnest, H. Elmlund, V. Elser, M. Guhr, J. Hajdu, J. Hastings, S.P. Hau-Riege, Z. Huang, E.E. Lattman, F.R.N.C. Maia, S. Marchesini, A. Ourmazd, C. Pellegrini, R. Santra, I. Schlichting, C. Schroer, J.C.H. Spence, I.A. Vartanyants, S. Wakatsuki, W.I. Weis, G.J. Williams, The linac coherent light source single particle imaging road map. Struct. Dyn. **2**(4), 041701 (2015)
25. H.N. Chapman, A. Barty, M.J. Bogan, S. Boutet, M. Frank, S.P. Hau-Riege, S. Marchesini, B.W. Woods, S. Bajt, H. Benner, R.A. London, E. Plonjes, M. Kuhlmann, R. Treusch, S. Dusterer, T. Tschentscher, J.R. Schneider, E. Spiller, T. Möller, C. Bostedt, M. Hoener, D.A. Shapiro, K.O. Hodgson, D. Van der Spoel, F. Burmeister, M. Bergh, C. Caleman, G. Huldt, M.M. Seibert, F.R.N.C. Maia, R.W. Lee, A. Szöke, N. Timneanu, J. Hajdu, Femtosecond diffractive imaging with a soft-X-ray free-electron laser. Nat. Phys. **2**(12), 839–843 (2006)

26. H.N. Chapman, P. Fromme, A. Barty, T.A. White, R.A. Kirian, A. Aquila, M.S. Hunter, J. Schulz, D.P. DePonte, U. Weierstall, R.B. Doak, F.R.N.C. Maia, A.V. Martin, I. Schlichting, L. Lomb, N. Coppola, R.L. Shoeman, S.W. Epp, R. Hartmann, D. Rolles, A. Rudenko, L. Foucar, N. Kimmel, G. Weidenspointner, P. Holl, M.N. Liang, M. Barthelmess, C. Caleman, S. Boutet, M.J. Bogan, J. Krzywinski, C. Bostedt, S. Bajt, L. Gumprecht, B. Rudek, B. Erk, C. Schmidt, A. Homke, C. Reich, D. Pietschner, L. Struder, G. Hauser, H. Gorke, J. Ullrich, S. Herrmann, G. Schaller, F. Schopper, H. Soltau, K.U. Kuhnel, M. Messerschmidt, J.D. Bozek, S.P. Hau-Riege, M. Frank, C.Y. Hampton, R.G. Sierra, D. Starodub, G.J. Williams, J. Hajdu, N. Timneanu, M.M. Seibert, J. Andreasson, A. Rocker, O. Jonsson, M. Svenda, S. Stern, K. Nass, R. Andritschke, C.D. Schroter, F. Krasniqi, M. Bott, K.E. Schmidt, X.Y. Wang, I. Grotjohann, J.M. Holton, T.R.M. Barends, R. Neutze, S. Marchesini, R. Fromme, S. Schorb, D. Rupp, M. Adolph, T. Gorkhover, I. Andersson, H. Hirsemann, G. Potdevin, H. Graafsma, B. Nilsson, J.C.H. Spence, Femtosecond X-ray protein nanocrystallography. Nature **470**(7332), 73–77 (2011)

27. M.M. Seibert, T. Ekeberg, F.R.N.C. Maia, M. Svenda, J. Andreasson, O. Jonsson, D. Odic, B. Iwan, A. Rocker, D. Westphal, M. Hantke, D.P. DePonte, A. Barty, J. Schulz, L. Gumprecht, N. Coppola, A. Aquila, M.N. Liang, T.A. White, A. Martin, C. Caleman, S. Stern, C. Abergel, V. Seltzer, J.M. Claverie, C. Bostedt, J.D. Bozek, S. Boutet, A.A. Miahnahri, M. Messerschmidt, J. Krzywinski, G. Williams, K.O. Hodgson, M.J. Bogan, C.Y. Hampton, R.G. Sierra, D. Starodub, I. Andersson, S. Bajt, M. Barthelmess, J.C.H. Spence, P. Fromme, U. Weierstall, R. Kirian, M. Hunter, R.B. Doak, S. Marchesini, S.P. Hau-Riege, M. Frank, R.L. Shoeman, L. Lomb, S.W. Epp, R. Hartmann, D. Rolles, A. Rudenko, C. Schmidt, L. Foucar, N. Kimmel, P. Holl, B. Rudek, B. Erk, A. Homke, C. Reich, D. Pietschner, G. Weidenspointner, L. Struder, G. Hauser, H. Gorke, J. Ullrich, I. Schlichting, S. Herrmann, G. Schaller, F. Schopper, H. Soltau, K.U. Kuhnel, R. Andritschke, C.D. Schroter, F. Krasniqi, M. Bott, S. Schorb, D. Rupp, M. Adolph, T. Gorkhover, H. Hirsemann, G. Potdevin, H. Graafsma, B. Nilsson, H.N. Chapman, J. Hajdu, Single mimivirus particles intercepted and imaged with an X-ray laser. Nature **470**(7332), 78–81 (2011)

28. N.D. Loh, C.Y. Hampton, A.V. Martin, D. Starodub, R.G. Sierra, A. Barty, A. Aquila, J. Schulz, L. Lomb, J. Steinbrener, R.L. Shoeman, S. Kassemeyer, C. Bostedt, J. Bozek, S.W. Epp, B. Erk, R. Hartmann, D. Rolles, A. Rudenko, B. Rudek, L. Foucar, N. Kimmel, G. Weidenspointner, G. Hauser, P. Holl, E. Pedersoli, M. Liang, M.S. Hunter, L. Gumprecht, N. Coppola, C. Wunderer, H. Graafsma, F.R.N.C. Maia, T. Ekeberg, M. Hantke, H. Fleckenstein, H. Hirsemann, K. Nass, T.A. White, H.J. Tobias, G.R. Farquar, W.H. Benner, S.P. Hau-Riege, C. Reich, A. Hartmann, H. Soltau, S. Marchesini, S. Bajt, M. Barthelmess, P. Bucksbaum, K.O. Hodgson, L. Strüder, J. Ullrich, M. Frank, I. Schlichting, H.N. Chapman, M.J. Bogan, Fractal morphology, imaging and mass spectrometry of single aerosol particles in flight. Nature **486**(7404), 513–517 (2012)

29. C. Bostedt, M. Adolph, E. Eremina, M. Hoener, D. Rupp, S. Schorb, H. Thomas, A.R.B. de Castro, T. Möller, Clusters in intense FLASH pulses: Ultrafast ionization dynamics and electron emission studied with spectroscopic and scattering techniques. J. Phys. B Atomic Mol. Opt. Phys. **43**(19), 194011 (2010)

30. D. Rupp, M. Adolph, L. Flückiger, T. Gorkhover, J.P. Müller, M. Müller, M. Sauppe, D. Wolter, S. Schorb, R. Treusch, C. Bostedt, T. Möller, Generation and structure of extremely large clusters in pulsed jets. J. Chem. Phys. **141**(4), 044306 (2014)

31. I. Barke, H. Hartmann, D. Rupp, L. Flükiger, M. Sauppe, M. Adolph, S. Schorb, C. Bostedt, R. Treusch, C. Peltz, S. Bartling, T. Fennel, K.H. Meiwes-Broer, T. Möller, The 3D-architecture of individual free silver nanoparticles captured by X-ray scattering. Nat. Commun. **6**, 6187 (2015)

32. I. Robinson, R. Harder, Coherent X-ray diffraction imaging of strain at the nanoscale. Nat. Mater. **8**(4), 291–298 (2009)

33. J.W. Miao, T. Ishikawa, I.K. Robinson, M.M. Murnane, Beyond crystallography: Diffractive imaging using coherent X-ray light sources. Science **348**(6234), 530–535 (2015)

34. H.N. Chapman, X-ray free-electron lasers for the structure and dynamics of macromolecules. Annu. Rev. Biochem. **88**(1), 35–58 (2019)
35. L.F. Gomez, K.R. Ferguson, J.P. Cryan, C. Bacellar, R.M.P. Tanyag, C. Jones, S. Schorb, D. Anielski, A. Belkacem, C. Bernando, R. Boll, J. Bozek, S. Carron, G. Chen, T. Delmas, L. Englert, S.W. Epp, B. Erk, L. Foucar, R. Hartmann, A. Hexemer, M. Huth, J. Kwok, S.R. Leone, J.H.S. Ma, F.R.N.C. Maia, E. Malmerberg, S. Marchesini, D.M. Neumark, B. Poon, J. Prell, D. Rolles, B. Rudek, A. Rudenko, M. Seifrid, K.R. Siefermann, F.P. Sturm, M. Swiggers, J. Ullrich, F. Weise, P. Zwart, C. Bostedt, O. Gessner, A.F. Vilesov, Shapes and vorticities of superfluid helium nanodroplets. Science **345**(6199), 906–909 (2014)
36. E.J. Yarmchuk, M.J.V. Gordon, R.E. Packard, Observation of stationary vortex arrays in rotating superfluid-helium. Phys. Rev. Lett. **43**(3), 214–217 (1979)
37. J.R. Abo-Shaeer, C. Raman, J.M. Vogels, W. Ketterle, Observation of vortex lattices in Bose-Einstein condensates. Science **292**(5516), 476–479 (2001)
38. R.J. Donnelly, *Quantized vortices in helium II*, vol. 3, ed. by A.M. Goldman, P.V.E. McKlintock, M. Springford (Cambridge University Press, Cambridge, 1991)
39. F. Dalfovo, S. Stringari, Helium nanodroplets and trapped Bose-Einstein condensates as prototypes of finite quantum fluids. J. Chem. Phys. **115**(22), 10078–10089 (2001)
40. L.F. Gomez, E. Loginov, A.F. Vilesov, Traces of vortices in superfluid helium droplets. Phys. Rev. Lett. **108**(15), 155302 (2012)
41. E. Latimer, D. Spence, C. Feng, A. Boatwright, A.M. Ellis, S.F. Yang, Preparation of ultrathin nanowires using superfluid helium droplets. Nano Lett. **14**(5), 2902–2906 (2014)
42. D. Spence, E. Latimer, C. Feng, A. Boatwright, A.M. Ellis, S.F. Yang, Vortex-induced aggregation in superfluid helium droplets. Phys. Chem. Chem. Phys. **16**(15), 6903–6906 (2014)
43. A. Volk, D. Knez, P. Thaler, A.W. Hauser, W. Grogger, F. Hofer, W.E. Ernst, Thermal instabilities and Rayleigh breakup of ultrathin silver nanowires grown in helium nanodroplets. Phys. Chem. Chem. Phys. **17**(38), 24570–24575 (2015)
44. M. Schnedlitz, M. Lasserus, D. Knez, A.W. Hauser, F. Hofer, W.E. Ernst, Thermally induced breakup of metallic nanowires: Experiment and theory. Phys. Chem. Chem. Phys. **19**(14), 9402–9408 (2017)
45. F. Lackner, *Synthesis of metallic nanoparticles in helium droplets*, in *Molecules in helium nanodroplets*, ed. by A. Slenczka, J.P. Toennies (Springer, Berlin, 2021)
46. J.D. Close, F. Federmann, K. Hoffmann, N. Quaas, Helium droplets: A nanoscale cryostat for high resolution spectroscopy and studies of quantized vorticity. J. Low Temp. Phys. **111**(3–4), 661–676 (1998)
47. K.K. Lehmann, R. Schmied, Energetics and possible formation and decay mechanisms of vortices in helium nanodroplets. Phys. Rev. B Condens. Matter **68**(22), 224520 (2003)
48. C. Bernando, R.M.P. Tanyag, C. Jones, C. Bacellar, M. Bucher, K.R. Ferguson, D. Rupp, M.P. Ziemkiewicz, L.F. Gomez, A.S. Chatterley, T. Gorkhover, M. Muller, J. Bozek, S. Carron, J. Kwok, S.L. Butler, T. Möller, C. Bostedt, O. Gessner, A.F. Vilesov, Shapes of rotating superfluid helium nano-droplets. Phys. Rev. B Condens. Matter **95**(6), 064510 (2017)
49. D. Rupp, N. Monserud, B. Langbehn, M. Sauppe, J. Zimmermann, Y. Ovcharenko, T. Möller, F. Frassetto, L. Poletto, A. Trabattoni, F. Calegari, M. Nisoli, K. Sander, C. Peltz, M.J. Vrakking, T. Fennel, A. Rouzee, Coherent diffractive imaging of single helium nanodroplets with a high harmonic generation source. Nat. Commun. **8**, 493 (2017)
50. B. Langbehn, K. Sander, Y. Ovcharenko, C. Peltz, A. Clark, M. Coreno, R. Cucini, M. Drabbels, P. Finetti, M. Di Fraia, L. Giannessi, C. Grazioli, D. Iablonskyi, A.C. LaForge, T. Nishiyama, V. Oliver Álvarez de Lara, P. Piseri, O. Plekan, K. Ueda, J. Zimmermann, K.C. Prince, F. Stienkemeier, C. Callegari, T. Fennel, D. Rupp, T. Möller, Three-dimensional shapes of spinning helium nanodroplets. Phys. Rev. Lett. **121**(25), 255301 (2018)
51. S.M.O. O'Connell, R.M.P. Tanyag, D. Verma, C. Bernando, W.Q. Pang, C. Bacellar, C.A. Saladrigas, J. Mahl, B.W. Toulson, Y. Kumagai, P. Walter, F. Ancilotto, M. Barranco, M. Pi, C. Bostedt, O. Gessner, A.F. Vilesov, Angular momentum in rotating superfluid droplets. Phys. Rev. Lett. **124**(21), 215301 (2020)

52. D. Verma, S.M.O. O'Connell, A.J. Feinberg, S. Erukala, R.M.P. Tanyag, C. Bernando, W.W. Pang, C.A. Saladrigas, B.W. Toulson, M. Borgwardt, N. Shivaram, M.F. Lin, A. Al Haddad, W. Jager, C. Bostedt, P. Walter, O. Gessner, A.F. Vilesov, Shapes of rotating normal fluid He-3 versus superfluid He-4 droplets in molecular beams. Phys. Rev. B **102**(1), 014504 (2020)

53. R.M.P. Tanyag, C. Bernando, C.F. Jones, C. Bacellar, K.R. Ferguson, D. Anielski, R. Boll, S. Carron, J.P. Cryan, L. Englert, S.W. Epp, B. Erk, L. Foucar, L.F. Gomez, R. Hartmann, D.M. Neumark, D. Rolles, B. Rudek, A. Rudenko, K.R. Siefermann, J. Ullrich, F. Weise, C. Bostedt, O. Gessner, A.F. Vilesov, Communication: X-ray coherent diffractive imaging by immersion in nanodroplets. Struct. Dyn. **2**(5), 051102 (2015)

54. C.F. Jones, C. Bernando, R.M.P. Tanyag, C. Bacellar, K.R. Ferguson, L.F. Gomez, D. Anielski, A. Belkacem, R. Boll, J. Bozek, S. Carron, J. Cryan, L. Englert, S.W. Epp, B. Erk, L. Foucar, R. Hartmann, D.M. Neumark, D. Rolles, A. Rudenko, K.R. Siefermann, F. Weise, B. Rudek, F.P. Sturm, J. Ullrich, C. Bostedt, O. Gessner, A.F. Vilesov, Coupled motion of Xe clusters and quantum vortices in He nanodroplets. Phys. Rev. B Condens. Matter **93**(18), 180510 (2016)

55. R.M.P. Tanyag, D. Rupp, A. Ulmer, A. Heilrath, B. Senfftleben, B. Kruse, L. Seiffert, S.M.O. O'Connell, K. Kolatzki, B. Langbehn, A. Hoffmann, T. Baumann, R. Boll, A. Chatterley, A. de Fanis, B. Erk, S. Erukala, A.J. Feinberg, T. Fennel, P. Grychtol, R. Hartmann, S. Hauf, M. Ilchen, M. Izquierdo, B. Krebs, M. Kuster, T. Mazza, K.H. Meiwes-Broer, J. Montaño, G. Noffz, D. Rivas, D. Schlosser, F. Seel, H. Stapelfeldt, L. Strüder, J. Tiggesbäumker, H. Yousef, M. Zabel, P. Ziolkowski, A. Vilesov, M. Meyer, Y. Ovcharenko, T. Möller, Taking snapshots of nanostructures in superfluid helium droplets, in *European XFEL Annual Report 2019* (2020), pp. 20–21

56. A.J. Feinberg, D. Verma, S.M. O'Connell-Lopez, S. Erukala, R.M.P. Tanyag, W. Pang, C.A. Saladrigas, B.W. Toulson, M. Borgwardt, N. Shivaram, M.-F. Lin, A.A. Haddad, W. Jäger, C. Bostedt, P. Walter, O. Gessner, A.F. Vilesov, Aggregation of solutes in bosonic versus fermionic quantum fluids. Sci. Adv. **7**(50) (2021)

57. R.M.P. Tanyag, C. Bacellar, W. Pang, C. Bernando, L.F. Gomez, C.F. Jones, K.R. Ferguson, J. Kwok, D. Anielski, A. Belkacem, R. Boll, J. Bozek, S. Carron, G. Chen, T. Delmas, L. Englert, S.W. Epp, B. Erk, L. Foucar, R. Hartmann, A. Hexemer, M. Huth, S.R. Leone, J.H. Ma, S. Marchesini, D.M. Neumark, B.K. Poon, J. Prell, D. Rolles, B. Rudek, A. Rudenko, M. Seifrid, M. Swiggers, J. Ullrich, F. Weise, P. Zwart, C. Bostedt, O. Gessner, A.F. Vilesov, Sizes of pure and doped helium droplets from single shot X-ray imaging. J. Chem. Phys. **156**(4), 041102 (2022)

58. C. Bacellar, Ultrafast dynamics in helium nanodroplets probed by XUV spectroscopy and X-ray imaging. Ph.D. thesis, University of California, Berkeley, 2017

59. C. Bacellar, A.S. Chatterley, F. Lackner, C.D. Pemmaraju, R.M.P. Tanyag, C. Bernando, D. Verma, S. O'Connell, M. Bucher, K.R. Ferguson, T. Gorkhover, N.C. Ryan, G. Coslovich, D. Ray, T. Osipov, D.M. Neumark, C. Bostedt, A.F. Vilesov, O. Gessner, Evaporation of an anisotropic nanoplasma. EPJ Web Conf. **205**, 06006 (2019)

60. B. Langbehn, Imaging the shapes and dynamics of superfluid helium nanodroplets. Ph.D. thesis, Technical University of Berlin, 2021

61. C.A. Saladrigas, A.J. Feinberg, M.P. Ziemkiewicz, C. Bacellar, M. Bucher, C. Bernando, S. Carron, A.S. Chatterley, F.-J. Decker, K.R. Ferguson, L. Gomez, T. Gorkhover, N.A. Helvy, C.F. Jones, J.J. Kwok, A. Lutman, D. Rupp, R.M.P. Tanyag, T. Möller, D.M. Neumark, C. Bostedt, A.F. Vilesov, O. Gessner, Charging and ion ejection dynamics of large helium nanodroplets exposed to intense femtosecond soft X-ray pulses. Eur. Phys. J. Spec. Top. **230**(23), 4011–4023 (2021)

62. E. Hecht, *Optics*, 5th edn. (Pearson Education Limited, Essex, UK, 2017)

63. J.W. Goodman, *Introduction to Fourier optics* (The McGraw-Hill Companies Inc., New York, New York, 1996)

64. B.R. Masters, Abbe's theory of image formation in the microscope, in *Superresolution optical microscopy: The quest for enhanced resolution and contrast*. ed. by B.R. Masters (Springer International Publishing, Cham, 2020), pp. 65–108

65. A.T. Macrander, X. Huang, Synchrotron X-ray optics. Annu. Rev. Mater. Res. **47**(1), 135–152 (2017)
66. B.W.J. McNeil, N.R. Thompson, X-ray free-electron lasers. Nat. Photonics **4**(12), 814–821 (2010)
67. C. Pellegrini, A. Marinelli, S. Reiche, The physics of X-ray free-electron lasers. Rev. Mod. Phys. **88**(1), 015006 (2016)
68. E. Allaria, R. Appio, L. Badano, W.A. Barletta, S. Bassanese, S.G. Biedron, A. Borga, E. Busetto, D. Castronovo, P. Cinquegrana, S. Cleva, D. Cocco, M. Cornacchia, P. Craievich, I. Cudin, G. D'Auria, M. Dal Forno, M.B. Danailov, R. De Monte, G. De Ninno, P. Delgiusto, A. Demidovich, S. Di Mitri, B. Diviacco, A. Fabris, R. Fabris, W. Fawley, M. Ferianis, E. Ferrari, S. Ferry, L. Froehlich, P. Furlan, G. Gaio, F. Gelmetti, L. Giannessi, M. Giannini, R. Gobessi, R. Ivanov, E. Karantzoulis, M. Lonza, A. Lutman, B. Mahieu, M. Milloch, S.V. Milton, M. Musardo, I. Nikolov, S. Noe, F. Parmigiani, G. Penco, M. Petronio, L. Pivetta, M. Predonzani, F. Rossi, L. Rumiz, A. Salom, C. Scafuri, C. Serpico, P. Sigalotti, S. Spampinati, C. Spezzani, M. Svandrlik, C. Svetina, S. Tazzari, M. Trovo, R. Umer, A. Vascotto, M. Veronese, R. Visintini, M. Zaccaria, D. Zangrando, M. Zangrando, Highly coherent and stable pulses from the FERMI seeded free-electron laser in the extreme ultraviolet. Nat. Photonics **6**(10), 699–704 (2012)
69. E. Allaria, D. Castronovo, P. Cinquegrana, P. Craievich, M. Dal Forno, M.B. Danailov, G. D'Auria, A. Demidovich, G. De Ninno, S. Di Mitri, B. Diviacco, W.M. Fawley, M. Ferianis, E. Ferrari, L. Froehlich, G. Gaio, D. Gauthier, L. Giannessi, R. Ivanov, B. Mahieu, N. Mahne, I. Nikolov, F. Parmigiani, G. Penco, L. Raimondi, C. Scafuri, C. Serpico, P. Sigalotti, S. Spampinati, C. Spezzani, M. Svandrlik, C. Svetina, M. Trovo, M. Veronese, D. Zangrando, M. Zangrando, Two-stage seeded soft-X-ray free-electron laser. Nat. Photonics **7**(11), 913–918 (2013)
70. O.Y. Gorobtsov, G. Mercurio, F. Capotondi, P. Skopintsev, S. Lazarev, I.A. Zaluzhnyy, M.B. Danailov, M. Dell'Angela, M. Manfredda, E. Pedersoli, L. Giannessi, M. Kiskinova, K.C. Prince, W. Wurth, I.A. Vartanyants, Seeded X-ray free-electron laser generating radiation with laser statistical properties. Nat. Commun. **9**(1), 4498 (2018)
71. E. Allaria, L. Badano, S. Bassanese, F. Capotondi, D. Castronovo, P. Cinquegrana, M.B. Danailov, G. D'Auria, A. Demidovich, R. De Monte, G. De Ninno, S. Di Mitri, B. Diviacco, W.M. Fawley, M. Ferianis, E. Ferrari, G. Gaio, D. Gauthier, L. Giannessi, F. Iazzourene, G. Kurdi, N. Mahne, I. Nikolov, F. Parmigiani, G. Penco, L. Raimondi, P. Rebernik, F. Rossi, E. Roussel, C. Scafuri, C. Serpico, P. Sigalotti, C. Spezzani, M. Svandrlik, C. Svetina, M. Trovo, M. Veronese, D. Zangrando, M. Zangrando, The FERMI free-electron lasers. J. Synchrotron Radiat. **22**(3), 485–491 (2015)
72. P. Rudawski, C.M. Heyl, F. Brizuela, J. Schwenke, A. Persson, E. Mansten, R. Rakowski, L. Rading, F. Campi, B. Kim, P. Johnsson, A. L'Huillier, A high-flux high-order harmonic source. Rev. Sci. Instrum. **84**(7), 073103 (2013)
73. F. Krausz, M. Ivanov, Attosecond physics. Rev. Mod. Phys. **81**(1), 163–234 (2009)
74. T. Popmintchev, M.C. Chen, P. Arpin, M.M. Murnane, H.C. Kapteyn, The attosecond nonlinear optics of bright coherent X-ray generation. Nat. Photonics **4**(12), 822–832 (2010)
75. C.M. Heyl, C.L. Arnold, A. Couairon, A. L'Huillier, Introduction to macroscopic power scaling principles for high-order harmonic generation. J. Phys. B Atomic Mol. Opt. Phys. **50**(1), 013001 (2016)
76. M. Nisoli, P. Decleva, F. Calegari, A. Palacios, F. Martín, Attosecond electron dynamics in molecules. Chem. Rev. **117**(16), 10760–10825 (2017)
77. M.D. Seaberg, D.E. Adams, E.L. Townsend, D.A. Raymondson, W.F. Schlotter, Y.W. Liu, C.S. Menoni, L. Rong, C.C. Chen, J.W. Miao, H.C. Kapteyn, M.M. Murnane, Ultrahigh 22 nm resolution coherent diffractive imaging using a desktop 13 nm high harmonic source. Opt. Express **19**(23), 22470–22479 (2011)
78. M.D. Seaberg, B. Zhang, D.F. Gardner, E.R. Shanblatt, M.M. Murnane, H.C. Kapteyn, D.E. Adams, Tabletop nanometer extreme ultraviolet imaging in an extended reflection mode using coherent fresnel ptychography. Optica **1**(1), 39–44 (2014)

79. D. Rudolf, C. La-O-Vorakiat, M. Battiato, R. Adam, J.M. Shaw, E. Turgut, P. Maldonado, S. Mathias, P. Grychtol, H.T. Nembach, T.J. Silva, M. Aeschlimann, H.C. Kapteyn, M.M. Murnane, C.M. Schneider, P.M. Oppeneer, Ultrafast magnetization enhancement in metallic multilayers driven by superdiffusive spin current. Nat. Commun. **3**(1), 1037 (2012)
80. K.M. Hoogeboom-Pot, J.N. Hernandez-Charpak, X. Gu, T.D. Frazer, E.H. Anderson, W. Chao, R.W. Falcone, R. Yang, M.M. Murnane, H.C. Kapteyn, D. Nardi, A new regime of nanoscale thermal transport: Collective diffusion increases dissipation efficiency. Proc. Natl. Acad. Sci. **112**(16), 4846–4851 (2015)
81. J. Miao, D. Sayre, H.N. Chapman, Phase retrieval from the magnitude of the Fourier transforms of nonperiodic objects. J. Opt. Soc. Am. A **15**(6), 1662–1669 (1998)
82. H.N. Chapman, A. Barty, S. Marchesini, A. Noy, S.R. Hau-Riege, C. Cui, M.R. Howells, R. Rosen, H. He, J.C.H. Spence, U. Weierstall, T. Beetz, C. Jacobsen, D. Shapiro, High-resolution ab initio three-dimensional X-ray diffraction microscopy. J. Opt. Soc. Am. A-Opt. Image Sci. Vis. **23**(5), 1179–1200 (2006)
83. S. Marchesini, H. He, H.N. Chapman, S.P. Hau-Riege, A. Noy, M.R. Howells, U. Weierstall, J.C.H. Spence, X-ray image reconstruction from a diffraction pattern alone. Phys. Rev. B. **68**(14), 140101 (2003)
84. M.F. Hantke, D. Hasse, F.R.N.C. Maia, T. Ekeberg, K. John, M. Svenda, N.D. Loh, A.V. Martin, N. Timneanu, D.S.D. Larsson, G. van der Schot, G.H. Carlsson, M. Ingelman, J. Andreasson, D. Westphal, M.N. Liang, F. Stellato, D.P. DePonte, R. Hartmann, N. Kimmel, R.A. Kirian, M.M. Seibert, K. Muhlig, S. Schorb, K. Ferguson, C. Bostedt, S. Carron, J.D. Bozek, D. Rolles, A. Rudenko, S. Epp, H.N. Chapman, A. Barty, J. Hajdu, I. Andersson, High-throughput imaging of heterogeneous cell organelles with an X-ray laser. Nat. Photonics **8**(12), 943–949 (2014)
85. T. Kimura, Y. Joti, A. Shibuya, C.Y. Song, S. Kim, K. Tono, M. Yabashi, M. Tamakoshi, T. Moriya, T. Oshima, T. Ishikawa, Y. Bessho, Y. Nishino, Imaging live cell in micro-liquid enclosure by X-ray laser diffraction. Nat. Commun. **5**, 3052 (2014)
86. G. van der Schot, M. Svenda, F.R.N.C. Maia, M. Hantke, D.P. DePonte, M.M. Seibert, A. Aquila, J. Schulz, R. Kirian, M. Liang, F. Stellato, B. Iwan, J. Andreasson, N. Timneanu, D. Westphal, N.F. Almeida, D. Odic, D. Hasse, G.H. Carlsson, D.S.D. Larsson, A. Barty, A.V. Martin, S. Schorb, C. Bostedt, J.D. Bozek, D. Rolles, A. Rudenko, S. Epp, L. Foucar, B. Rudek, R. Hartmann, N. Kimmel, P. Holl, L. Englert, N.T.D. Loh, H.N. Chapman, I. Andersson, J. Hajdu, T. Ekeberg, Imaging single cells in a beam of live cyanobacteria with an X-ray laser. Nat. Commun. **6**, 5704 (2015)
87. H.J. Park, N.D. Loh, R.G. Sierra, C.Y. Hampton, D. Starodub, A.V. Martin, A. Barty, A. Aquila, J. Schulz, J. Steinbrener, R.L. Shoeman, L. Lomb, S. Kassemeyer, C. Bostedt, J. Bozek, S.W. Epp, B. Erk, R. Hartmann, D. Rolles, A. Rudenko, B. Rudek, L. Foucar, N. Kimmel, G. Weidenspointner, G. Hauser, P. Holl, E. Pedersoli, M.N. Liang, M.S. Hunter, L. Gumprecht, N. Coppola, C. Wunderer, H. Graafsma, F.R.N.C. Maia, T. Ekeberg, M. Hantke, H. Fleckenstein, H. Hirsemann, K. Nass, H.J. Tobias, G.R. Farquar, W.H. Benner, S. Hau-Riege, C. Reich, A. Hartmann, H. Soltau, S. Marchesini, S. Bajt, M. Barthelmess, L. Strüder, J. Ullrich, P. Bucksbaum, M. Frank, I. Schlichting, H.N. Chapman, M.J. Bogan, V. Elser, Toward unsupervised single-shot diffractive imaging of heterogeneous particles using X-ray free-electron lasers. Opt. Express **21**(23), 28729–28742 (2013)
88. W.F. Schlotter, R. Rick, K. Chen, A. Scherz, J. Stöhr, J. Lüning, S. Eisebitt, C. Günther, W. Eberhardt, O. Hellwig, I. McNulty, Multiple reference Fourier transform holography with soft X-rays. Appl. Phys. Lett. **89**(16), 163112 (2006)
89. S.G. Podorov, K.M. Pavlov, D.M. Paganin, A non-iterative reconstruction method for direct and unambiguous coherent diffractive imaging. Opt. Express **15**(16), 9954–9962 (2007)
90. B. Pfau, S. Eisebitt, X-ray holography, in *Synchrotron light sources and free-electron lasers: Accelerator physics, instrumentation and science applications*. ed. by E.J. Jaeschke, S. Khan, J.R. Schneider, J.B. Hastings (Springer International Publishing, Cham, 2016), pp. 1093–1133
91. T. Gorkhover, A. Ulmer, K. Ferguson, M. Bucher, F.R.N.C. Maia, J. Bielecki, T. Ekeberg, M.F. Hantke, B.J. Daurer, C. Nettelblad, J. Andreasson, A. Barty, P. Bruza, S. Carron, D.

Hasse, J. Krzywinski, D.S.D. Larsson, A. Morgan, K. Mühlig, M. Müller, K. Okamoto, A. Pietrini, D. Rupp, M. Sauppe, G. van der Schot, M. Seibert, J.A. Sellberg, M. Svenda, M. Swiggers, N. Timneanu, D. Westphal, G. Williams, A. Zani, H.N. Chapman, G. Faigel, T. Möller, J. Hajdu, C. Bostedt, Femtosecond X-ray Fourier holography imaging of free-flying nanoparticles. Nat. Photonics **12**(3), 150–153 (2018)

92. T. Gorkhover, S. Schorb, R. Coffee, M. Adolph, L. Foucar, D. Rupp, A. Aquila, J.D. Bozek, S.W. Epp, B. Erk, L. Gumprecht, L. Holmegaard, A. Hartmann, R. Hartmann, G. Hauser, P. Holl, A. Hömke, P. Johnsson, N. Kimmel, K.-U. Kühnel, M. Messerschmidt, C. Reich, A. Rouzée, B. Rudek, C. Schmidt, J. Schulz, H. Soltau, S. Stern, G. Weidenspointner, B. White, J. Küpper, L. Strüder, I. Schlichting, J. Ullrich, D. Rolles, A. Rudenko, T. Möller, C. Bostedt, Femtosecond and nanometre visualization of structural dynamics in superheated nanoparticles. Nat. Photonics **10**(2), 93–97 (2016)

93. K.S. Raines, S. Salha, R.L. Sandberg, H. Jiang, J.A. Rodríguez, B.P. Fahimian, H.C. Kapteyn, J. Du, J. Miao, Three-dimensional structure determination from a single view. Nature **463**(7278), 214–217 (2010)

94. D. Rupp, M. Adolph, T. Gorkhover, S. Schorb, D. Wolter, R. Hartmann, N. Kimmel, C. Reich, T. Feigl, A.R.B. de Castro, R. Treusch, L. Strüder, T. Möller, C. Bostedt, Identification of twinned gas phase clusters by single-shot scattering with intense soft X-ray pulses. New J. Phys. **14**(5), 055016 (2012)

95. L. Hecht, Zweifarben-Streubildaufnahme von Helium-Nanotröpfchen: Planung, Durchführung und erste Ergebnisse. Master Thesis, Technical University Berlin, 2018

96. J.P. Toennies, A.F. Vilesov, Superfluid helium droplets: A uniquely cold nanomatrix for molecules and molecular complexes. Angew. Chem. Int. Ed. **43**(20), 2622–2648 (2004)

97. J.P. Toennies, Helium nanodroplets: Formation, physical properties, and superfluidity, in *Molecules in helium nanodroplets*, ed. by A. Slenczka, J.P. Toennies (Springer, Berlin, 2021)

98. D.M. Brink, S. Stringari, Density of states and evaporation rate of helium clusters. Zeitschrift Fur Physik D-Atoms Molecules and Clusters **15**(3), 257–263 (1990)

99. K.B. Whaley, Structure and dynamics of quantum clusters. Int. Rev. Phys. Chem. **13**(1), 41–84 (1994)

100. M. Barranco, R. Guardiola, S. Hernandez, R. Mayol, J. Navarro, M. Pi, Helium nanodroplets: An overview. J. Low Temp. Phys. **142**(1–2), 1–81 (2006)

101. R.M.P. Tanyag, C.F. Jones, C. Bernando, S.M.O. O'Connell, D. Verma, A.F. Vilesov, Experiments with large superfluid helium nanodroplets, in *Cold chemistry: Molecular scattering and reactivity near absolute zero*. ed. by O. Dulieu, A. Osterwalder (The Royal Society of Chemistry, London, UK, 2018), pp. 389–443

102. H. Buchenau, R. Goetting, Scheidemann A., J.P. Toennies, Experimental studies of condensation in helium nozzle beams, in *Conference Proceedings of the 15th International Symposium on Rarefied Gas Dynamics*, ed. by V. Boffi, C. Cercignani (1986), pp. 197–207

103. U. Even, J. Jortner, D. Noy, N. Lavie, C. Cossart-Magos, Cooling of large molecules below 1 K and He clusters formation. J. Chem. Phys. **112**(18), 8068–8071 (2000)

104. D. Pentlehner, R. Riechers, B. Dick, A. Slenczka, U. Even, N. Lavie, R. Brown, K. Luria, Rapidly pulsed helium droplet source. Rev. Sci. Instrum. **80**(4), 043302 (2009)

105. M.N. Slipchenko, S. Kuma, T. Momose, A.F. Vilesov, Intense pulsed helium droplet beams. Rev. Sci. Instrum. **73**(10), 3600–3605 (2002)

106. S. Kuma, T. Azuma, Pulsed beam of extremely large helium droplets. Cryogenics **88**, 78–80 (2017)

107. D. Verma, A.F. Vilesov, Pulsed helium droplet beams. Chem. Phys. Lett. **694**, 129–134 (2018)

108. C. Bostedt, T. Gorkhover, D. Rupp, T. Möller, Clusters and nanocrystals, in *Synchrotron light sources and free-electron lasers: Accelerator physics, instrumentation and science applications*. ed. by E.J. Jaeschke, S. Khan, J.R. Schneider, J.B. Hastings (Springer International Publishing, Cham, 2020), pp. 1525–1573

109. M. Lewerenz, B. Schilling, J.P. Toennies, Successive capture and coagulation of atoms and molecules to small clusters in large liquid-helium clusters. J. Chem. Phys. **102**(20), 8191–8207 (1995)

110. J. Tiggesbäumker, F. Stienkemeier, Formation and properties of metal clusters isolated in helium droplets. Phys. Chem. Chem. Phys. **9**(34), 4748–4770 (2007)
111. L. Strüder, S. Eppa, D. Rolles, R. Hartmann, P. Holl, G. Lutz, H. Soltau, R. Eckart, C. Reich, K. Heinzinger, C. Thamm, A. Rudenko, F. Krasniqi, K.U. Kuhnel, C. Bauer, C.D. Schroter, R. Moshammer, S. Techert, D. Miessner, M. Porro, O. Halker, N. Meidinger, N. Kimmel, R. Andritschke, F. Schopper, G. Weidenspointner, A. Ziegler, D. Pietschner, S. Herrmann, U. Pietsch, A. Walenta, W. Leitenberger, C. Bostedt, T. Möller, D. Rupp, M. Adolph, H. Graafsma, H. Hirsemann, K. Gartner, R. Richter, L. Foucar, R.L. Shoeman, I. Schlichting, J. Ullrich, Large-format, high-speed, X-ray pnCCDs combined with electron and ion imaging spectrometers in a multipurpose chamber for experiments at 4th generation light sources. Nucl. Instrum. Methods Phys. Res. Sect. A Accel. Spectrometers Detect. Assoc. Equip. **614**(3), 483–496 (2010)
112. H. Ryll, M. Simson, R. Hartmann, P. Holl, M. Huth, S. Ihle, Y. Kondo, P. Kotula, A. Liebel, K. Müller-Caspary, A. Rosenauer, R. Sagawa, J. Schmidt, H. Soltau, L. Strüder, A pnCCD-based, fast direct single electron imaging camera for TEM and STEM. J. Instrum. **11**(04), P04006–P04006 (2016)
113. M. Kuster, K. Ahmed, K.-E. Ballak, C. Danilevski, M. Ekmedzic, B. Fernandes, P. Gessler, R. Hartmann, S. Hauf, P. Holl, M. Meyer, J. Montano, A. Munnich, Y. Ovcharenko, N. Rennhack, T. Ruter, D. Rupp, D. Schlosser, K. Setoodehnia, R. Schmitt, L. Struder, R.M.P. Tanyag, A. Ulmer, H. Yousef, The 1-megapixel pnCCD detector for the small quantum systems instrument at the european XFEL: System and operation aspects. J. Synchrotron Radiat. **28**(2), 576–587 (2021)
114. S. Hau-Riege, *High-intensity X-rays—interaction with matter: Processes in plasmas, clusters, molecules, and solids* (Wiley-VCH Verlag, Weinheim, Germany, 2011)
115. U. Saalmann, C. Siedschlag, J.M. Rost, Mechanisms of cluster ionization in strong laser pulses. J. Phys. B Atomic Mol. Opt. Phys. **39**(4), R39–R77 (2006)
116. T. Fennel, K.H. Meiwes-Broer, J. Tiggesbäumker, P.G. Reinhard, P.M. Dinh, E. Suraud, Laser-driven nonlinear cluster dynamics. Rev. Mod. Phys. **82**(2), 1793–1842 (2010)
117. A. Mikaberidze, U. Saalmann, J.M. Rost, Laser-driven nanoplasmas in doped helium droplets: Local ignition and anisotropic growth. Phys. Rev. Lett. **102**(12), 128102 (2009)
118. S.R. Krishnan, L. Fechner, M. Kremer, V. Sharma, B. Fischer, N. Camus, J. Jha, M. Krishnamurthy, T. Pfeifer, R. Moshammer, J. Ullrich, F. Stienkemeier, M. Mudrich, A. Mikaberidze, U. Saalmann, J.M. Rost, Dopant-induced ignition of helium nanodroplets in intense few-cycle laser pulses. Phys. Rev. Lett. **107**(17), 173402 (2011)
119. S.R. Krishnan, C. Peltz, L. Fechner, V. Sharma, M. Kremer, B. Fischer, N. Camus, T. Pfeifer, J. Jha, M. Krishnamurthy, C.D. Schröter, J. Ullrich, F. Stienkemeier, R. Moshammer, T. Fennel, M. Mudrich, Evolution of dopant-induced helium nanoplasmas. New J. Phys. **14**(7), 075016 (2012)
120. T.V. Liseykina, D. Bauer, Plasma-formation dynamics in intense laser-droplet interaction. Phys. Rev. Lett. **110**(14), 145003 (2013)
121. A.C. LaForge, M. Drabbels, N.B. Brauer, M. Coreno, M. Devetta, M. Di Fraia, P. Finetti, C. Grazioli, R. Katzy, V. Lyamayev, T. Mazza, M. Mudrich, P. O'Keeffe, Y. Ovcharenko, P. Piseri, O. Plekan, K.C. Prince, R. Richter, S. Stranges, C. Callegari, T. Möller, F. Stienkemeier, Collective autoionization in multiply-excited systems: A novel ionization process observed in helium nanodroplets. Sci. Rep. **4**, 3621 (2014)
122. M. Mudrich, F. Stienkemeier, Photoionisaton of pure and doped helium nanodroplets. Int. Rev. Phys. Chem. **33**(3), 301–339 (2014)
123. Y. Ovcharenko, V. Lyamayev, R. Katzy, M. Devetta, A. LaForge, P. O'Keeffe, O. Plekan, P. Finetti, M. Di Fraia, M. Mudrich, M. Krikunova, P. Piseri, M. Coreno, N.B. Brauer, T. Mazza, S. Stranges, C. Grazioli, R. Richter, K.C. Prince, M. Drabbels, C. Callegari, F. Stienkemeier, T. Möller, Novel collective autoionization process observed in electron spectra of He clusters. Phys. Rev. Lett. **112**(7), 073401 (2014)
124. M.P. Ziemkiewicz, D.M. Neumark, O. Gessner, Ultrafast electronic dynamics in helium nanodroplets. Int. Rev. Phys. Chem. **34**(2), 239–267 (2015)

125. M. Kelbg, M. Zabel, B. Krebs, L. Kazak, K.H. Meiwes-Broer, J. Tiggesbäumker, Auger emission from the coulomb explosion of helium nanoplasmas. J. Chem. Phys. **150**(20), 204302 (2019)
126. M. Kelbg, M. Zabel, B. Krebs, L. Kazak, K.H. Meiwes-Broer, J. Tiggesbäumker, Temporal development of a laser-induced helium nanoplasma measured through Auger emission and above-threshold ionization. Phys. Rev. Lett. **125**(9), 093202 (2020)
127. M. Mudrich, A.C. LaForge, A. Ciavardini, P. O'Keeffe, C. Callegari, M. Coreno, A. Demidovich, M. Devetta, M.D. Fraia, M. Drabbels, P. Finetti, O. Gessner, C. Grazioli, A. Hernando, D.M. Neumark, Y. Ovcharenko, P. Piseri, O. Plekan, K.C. Prince, R. Richter, M.P. Ziemkiewicz, T. Möller, J. Eloranta, M. Pi, M. Barranco, F. Stienkemeier, Ultrafast relaxation of photoexcited superfluid He nanodroplets. Nat. Commun. **11**(1), 112 (2020)
128. Y. Ovcharenko, A.C. LaForge, B. Langbehn, O. Plekan, R. Cucini, P. Finetti, P. O'Keeffe, D. Iablonskyi, T. Nishiyama, K. Ueda, P. Piseri, M. Di Fraia, R. Richter, M. Coreno, C. Callegari, K.C. Prince, F. Stienkemeier, T. Möller, M. Mudrich, Autoionization dynamics of helium nanodroplets resonantly excited by intense XUV laser pulses. New J. Phys. **22**(8), 083043 (2020)
129. R. Neutze, R. Wouts, D. van der Spoel, E. Weckert, J. Hajdu, Potential for biomolecular imaging with femtosecond X-ray pulses. Nature **406**(6797), 752–757 (2000)
130. H.N. Chapman, C. Caleman, N. Timneanu, Diffraction before destruction. Philos. Trans. R. Soc. B Biol. Sci. **369**(1647), 20130313 (2014)
131. C. Bostedt, E. Eremina, D. Rupp, M. Adolph, H. Thomas, M. Hoener, A.R.B. de Castro, J. Tiggesbäumker, K.H. Meiwes-Broer, T. Laarmann, H. Wabnitz, E. Plönjes, R. Treusch, J.R. Schneider, T. Möller, Ultrafast X-ray scattering of xenon nanoparticles: Imaging transient states of matter. Phys. Rev. Lett. **108**(9), 093401 (2012)
132. D. Rupp, L. Flückiger, M. Adolph, A. Colombo, T. Gorkhover, M. Harmand, M. Krikunova, J.P. Müller, T. Oelze, Y. Ovcharenko, M. Richter, M. Sauppe, S. Schorb, R. Treusch, D. Wolter, C. Bostedt, T. Möller, Imaging plasma formation in isolated nanoparticles with ultrafast resonant scattering. Struct. Dyn. **7**(3), 034303 (2020)
133. B.L. Henke, E.M. Gullikson, J.C. Davis, X-ray interactions—photoabsorption, scattering, transmission, and reflection at E = 50–30,000 eV, Z = 1–92. At. Data Nucl. Data Tables **54**(2), 181–342 (1993)
134. A.A. Lucas, J.P. Vigneron, S.E. Donnelly, J.C. Rife, Theoretical interpretation of the vacuum ultraviolet reflectance of liquid helium and of the absorption spectra of helium microbubbles in aluminum. Phys. Rev. B **28**(5), 2485–2496 (1983)
135. E.L. Knuth, U. Henne, Average size and size distribution of large droplets produced in a free-jet expansion of a liquid. J. Chem. Phys. **110**(5), 2664–2668 (1999)
136. J.R. Fienup, C.C. Wackerman, Phase-retrieval stagnation problems and solutions. J. Opt. Soc. Am. A **3**(11), 1897–1907 (1986)
137. J.R. Fienup, Reconstruction of a complex-valued object from the modulus of its Fourier transform using a support constraint. J. Opt. Soc. Am. A **4**(1), 118–123 (1987)
138. T.R. Crimmins, J.R. Fienup, B.J. Thelen, Improved bounds on object support from autocorrelation support and application to phase retrieval. J. Opt. Soc. Am. A **7**(1), 3–13 (1990)
139. A. Hare, G. Morrison, Near-field soft X-ray diffraction modelled by the multislice method. J. Mod. Opt. **41**(1), 31–48 (1994)
140. T. Stielow, R. Schmidt, C. Peltz, T. Fennel, S. Scheel, Fast reconstruction of single-shot wide-angle diffraction images through deep learning. Mach. Learn. Sci. Technol. **1**(4), 045007 (2020)
141. A. Colombo, J. Zimmermann, B. Langbehn, T. Möller, C. Peltz, K. Sander, B. Kruse, P. Tummler, I. Barke, D. Rupp, T. Fennel, The Scatman: An approximate method for fast wide-angle scattering simulations. Preprint at https://arxiv.org/abs/2202.03411 (2022)
142. J. Zimmermann, B. Langbehn, R. Cucini, M. Di Fraia, P. Finetti, A.C. LaForge, T. Nishiyama, Y. Ovcharenko, P. Piseri, O. Plekan, K.C. Prince, F. Stienkemeier, K. Ueda, C. Callegari, T. Möller, D. Rupp, Deep neural networks for classifying complex features in diffraction images. Phys. Rev. E **99**(6), 063309 (2019)

143. J.M. Cowley, A.F. Moodie, The scattering of electrons by atoms and crystals. I. A new theoretical approach. Acta Crystallogr. **10**(10), 609–619 (1957)
144. K. Sander, Reconstruction methods for single shot diffraction imaging of free nanostructures with ultrashort XUV laser pulses. Ph.D. thesis, Universität Rostock, 2019
145. S. Decking, et.al, A MHz-repetition-rate hard X-ray free-electron laser driven by a superconducting linear accelerator. Nat. Photonics **14**(6), 391–397 (2020)
146. S.A. Bobkov, A.B. Teslyuk, R.P. Kurta, O.Y. Gorobtsov, O.M. Yefanov, V.A. Ilyin, R.A. Senin, I.A. Vartanyants, Sorting algorithms for single-particle imaging experiments at X-ray free-electron lasers. J. Synchrotron Radiat. **22**(6), 1345–1352 (2015)
147. E. Sobolev, S. Zolotarev, K. Giewekemeyer, J. Bielecki, K. Okamoto, H.K.N. Reddy, J. Andreasson, K. Ayyer, I. Barak, S. Bari, A. Barty, R. Bean, S. Bobkov, H.N. Chapman, G. Chojnowski, B.J. Daurer, K. Dörner, T. Ekeberg, L. Flückiger, O. Galzitskaya, L. Gelisio, S. Hauf, B.G. Hogue, D.A. Horke, A. Hosseinizadeh, V. Ilyin, C. Jung, C. Kim, Y. Kim, R.A. Kirian, H. Kirkwood, O. Kulyk, J. Küpper, R. Letrun, N.D. Loh, K. Lorenzen, M. Messerschmidt, K. Mühlig, A. Ourmazd, N. Raab, A.V. Rode, M. Rose, A. Round, T. Sato, R. Schubert, P. Schwander, J.A. Sellberg, M. Sikorski, A. Silenzi, C. Song, J.C.H. Spence, S. Stern, J. Sztuk-Dambietz, A. Teslyuk, N. Timneanu, M. Trebbin, C. Uetrecht, B. Weinhausen, G.J. Williams, P.L. Xavier, C. Xu, I.A. Vartanyants, V.S. Lamzin, A. Mancuso, F.R.N.C. Maia, Megahertz single-particle imaging at the european XFEL. Commun. Phys. **3**(1), 97 (2020)
148. K. Ayyer, P.L. Xavier, J. Bielecki, Z. Shen, B.J. Daurer, A.K. Samanta, S. Awel, R. Bean, A. Barty, M. Bergemann, T. Ekeberg, A.D. Estillore, H. Fangohr, K. Giewekemeyer, M.S. Hunter, M. Karnevskiy, R.A. Kirian, H. Kirkwood, Y. Kim, J. Koliyadu, H. Lange, R. Letrun, J. Lübke, T. Michelat, A.J. Morgan, N. Roth, T. Sato, M. Sikorski, F. Schulz, J.C.H. Spence, P. Vagovic, T. Wollweber, L. Worbs, O. Yefanov, Y. Zhuang, F.R.N.C. Maia, D.A. Horke, J. Küpper, N.D. Loh, A.P. Mancuso, H.N. Chapman, 3D diffractive imaging of nanoparticle ensembles using an X-ray laser. Optica **8**(1), 15–23 (2021)
149. G. Carleo, I. Cirac, K. Cranmer, L. Daudet, M. Schuld, N. Tishby, L. Vogt-Maranto, L. Zdeborová, Machine learning and the physical sciences. Rev. Mod. Phys. **91**(4), 045002 (2019)
150. E.R. Cruz-Chú, A. Hosseinizadeh, G. Mashayekhi, R. Fung, A. Ourmazd, P. Schwander, Selecting XFEL single-particle snapshots by geometric machine learning. Struct. Dyn. **8**(1), 014701 (2021)
151. T. Ekeberg, M. Svenda, C. Abergel, F.R. Maia, V. Seltzer, J.-M. Claverie, M. Hantke, O. Jönsson, C. Nettelblad, G. van der Schot, Three-dimensional reconstruction of the giant mimivirus particle with an X-ray free-electron laser. Phys. Rev. Lett. **114**(9), 098102 (2015)
152. H. Liu, J.C. Spence, XFEL data analysis for structural biology. Quant Biol **4**(3), 159–176 (2016)
153. B.J. Daurer, K. Okamoto, J. Bielecki, F.R. Maia, K. Mühlig, M.M. Seibert, M.F. Hantke, C. Nettelblad, W.H. Benner, M. Svenda, Experimental strategies for imaging bioparticles with femtosecond hard X-ray pulses. IUCrJ **4**(3), 251–262 (2017)
154. Y. LeCun, Y. Bengio, G. Hinton, Deep learning. Nature **521**(7553), 436–444 (2015)
155. C.H. Yoon, P. Schwander, C. Abergel, I. Andersson, J. Andreasson, A. Aquila, S. Bajt, M. Barthelmess, A. Barty, M.J. Bogan, Unsupervised classification of single-particle X-ray diffraction snapshots by spectral clustering. Opt. Express **19**(17), 16542–16549 (2011)
156. S. Chandrasekhar, The stability of a rotating liquid drop. Proc. R. Soc. London Ser. A Math. Phys. Sci. **286**(1404), 1–26 (1965)
157. S. Chandrasekhar, Ellipsoidal figures of equilibrium—an historical account. Commun. Pure Appl. Math. **20**(2), 251–265 (1967)
158. S. Cohen, F. Plasil, W.J. Swiatecki, Equilibrium configurations of rotating charged or gravitating liquid masses with surface tension. II. Ann. Phys. **82**(2), 557–596 (1974)
159. R.A. Brown, L.E. Scriven, The shape and stability of rotating liquid-drops. Proc. R. Soc. London Ser. A Math. Phys. Eng. Sci. **371**(1746), 331–357 (1980)

160. R.J.A. Hill, L. Eaves, Nonaxisymmetric shapes of a magnetically levitated and spinning water droplet. Phys. Rev. Lett. **101**(23), 234501, (2008)

161. T.G. Wang, Equilibrium shapes of rotating spheroids and drop shape oscillations. in *Advances in applied mechanics*, ed. by J.W. Hutchinson, T.Y. Wu (Elsevier, Amsterdam, 1988), pp. 1–62

162. N. Bohr, J.A. Wheeler, The mechanism of nuclear fission. Phys. Rev. **56**(5), 426–450 (1939)

163. K.A. Baldwin, S.L. Butler, R.J.A. Hill, Artificial tektites: An experimental technique for capturing the shapes of spinning drops. Sci. Rep. **5**, 7660 (2015)

164. L. Liao, R.J.A. Hill, Shapes and fissility of highly charged and rapidly rotating levitated liquid drops. Phys. Rev. Lett. **119**(11), 114501 (2017)

165. F.H. Busse, Oscillations of a rotating liquid drop. J. Fluid Mech. **142**, 1–8 (1984)

166. N. Ashgriz, M. Movassat, Oscillation of droplets and bubbles, in *Handbook of atomization and sprays: Theory and applications*, ed. by N. Ashgriz (Springer US, Boston, MA, 2011), pp. 125–144

167. R.P. Feynman, Chapter II application of quantum mechanics to liquid helium, in *Progress in low temperature physics*, ed. by C.J. Gorter (Elsevier, Amsterdam, 1955), pp. 17–53

168. C. Barenghi, N.G. Parker, *A primer on quantum fluids* (Springer, Switzerland, 2016)

169. G.M. Seidel, H.J. Maris, Morphology of superfluid drops with angular-momentum. Phys. B **194**, 577–578 (1994)

170. G.H. Bauer, R.J. Donnelly, W.F. Vinen, Vortex configurations in a freely rotating superfluid drop. J. Low Temp. Phys. **98**(1–2), 47–65 (1995)

171. F. Ancilotto, M. Barranco, M. Pi, Spinning superfluid He-4 nanodroplets. Phys. Rev B **97**(18), 184515 (2018)

172. M.A. Weilert, D.L. Whitaker, H.J. Maris, G.M. Seidel, Laser levitation of superfluid helium. J. Low Temp. Phys. **98**(1), 17–35 (1995)

173. J.J. Niemela, Electrostatic charging and levitation of helium II drops. J. Low Temp. Phys. **109**(5), 709–732 (1997)

174. M.A. Weilert, D.L. Whitaker, H.J. Maris, G.M. Seidel, Magnetic levitation of liquid helium. J. Low Temp. Phys. **106**(1–2), 101–131 (1997)

175. D.L. Whitaker, C. Kim, C.L. Vicente, M.A. Weilert, H.J. Maris, G.M. Seidel, Shape oscillations in levitated He II drops. J. Low Temp. Phys. **113**(3–4), 491–499 (1998)

176. J.J. Niemela, Vortex nucleation and the levitation of charged helium II drops. J. Low Temp. Phys. **119**(3), 351–356 (2000)

177. A.J. Leggett, Superfluidity. Rev. Mod. Phys. **71**(2), S318–S323 (1999)

178. O. Gessner, A.F. Vilesov, Imaging quantum vortices in superfluid helium droplets. Annu. Rev. Phys. Chem. **70**, 173–198 (2019)

179. F. Ancilotto, M. Barranco, F. Coppens, J. Eloranta, N. Halberstadt, A. Hernando, D. Mateo, M. Pi, Density functional theory of doped superfluid liquid helium and nanodroplets. Int. Rev. Phys. Chem. **36**(4), 621–707 (2017)

180. F. Ancilotto, M. Pi, M. Barranco, Vortex arrays in nanoscopic superfluid helium droplets. Phys. Rev. B Condens. Matter **91**(10), 100503 (2015)

181. J.M. Escartín, F. Ancilotto, M. Barranco, M. Pi, Vorticity and quantum turbulence in the merging of superfluid helium nanodroplets. Phys. Rev. B **99**(14) (2019)

182. M. Pi, J.M. Escartín, F. Ancilotto, M. Barranco, Coexistence of vortex arrays and surface capillary waves in spinning prolate superfluid $^4$He nanodroplets. Phys. Rev. B **104**(9) (2021)

183. J. Gspann, H. Vollmar, Metastable excitations of large clusters of He-3, He-4 or Ne atoms. J. Chem. Phys. **73**(4), 1657–1664 (1980)

184. J. Gspann, H. Vollmar, Ejection of positive cluster ions from large electron-bombarded He-3 or He-4 clusters. J. Low Temp. Phys. **45**(3–4), 343–355 (1981)

185. W.E. Keller, *Helium-3 and helium-4* (Plenum Press, New York, USA, 1969)

186. D.D. Osheroff, R.C. Richardson, D.M. Lee, Evidence for a new phase of solid He-3. Phys. Rev. Lett. **28**(14), 885–888 (1972)

187. O.V. Lounasmaa, E. Thuneberg, Vortices in rotating superfluid $^3$He. Proc. Natl. Acad. Sci. 96(14), 7760–7767 (1999)
188. J.M. Escartín, F. Ancilotto, M. Barranco, M. Pi, Merging of superfluid helium nanodroplets with vortices. Phys. Rev. B 105(2) (2022)
189. M. Pi, F. Ancilotto, M. Barranco, Rotating $^3$He droplets. J. Chem. Phys. 152(18), 184111 (2020)
190. S.G. Alves, A.F. Vilesov, S.C. Ferreira, Effects of the mean free path and relaxation in a model for the aggregation of particles in superfluid media. J. Chem. Phys. 130(24), 244506 (2009)
191. E. Loginov, L.F. Gomez, N. Chiang, A. Halder, N. Guggemos, V.V. Kresin, A.F. Vilesov, Photoabsorption of Ag-N(N similar to 6–6000) nanoclusters formed in helium droplets: Transition from compact to multicenter aggregation. Phys. Rev. Lett. 106(23), 233401 (2011)
192. K. Nauta, R.E. Miller, Nonequilibrium self-assembly of long chains of polar molecules in superfluid helium. Science 283(5409), 1895–1897 (1999)
193. E. Loginov, L.F. Gomez, A.F. Vilesov, Formation of core-shell silver-ethane clusters in He droplets. J. Phys. Chem. A 117(46), 11774–11782 (2013)
194. E. Loginov, L.F. Gomez, B.G. Sartakov, A.F. Vilesov, Formation of large Ag clusters with shells of methane, ethylene, and acetylene in He droplets. J. Phys. Chem. A 120(34), 6738–6744 (2016)
195. E. Loginov, L.F. Gomez, B.G. Sartakov, A.F. Vilesov, Formation of core-shell ethane-silver clusters in He droplets. J. Phys. Chem. A 121(32), 5978–5982 (2017)
196. A. Przystawik, S. Göde, T. Döppner, J. Tiggesbäumker, K.H. Meiwes-Broer, Light-induced collapse of metastable magnesium complexes formed in helium nanodroplets. Phys. Rev. A 78(2), 021202 (2008)
197. S. Göde, R. Irsig, J. Tiggesbäumker, K.H. Meiwes-Broer, Time-resolved studies on the collapse of magnesium atom foam in helium nanodroplets. New J. Phys. 15, 015026 (2013)
198. K. Nauta, D.T. Moore, R.E. Miller, Molecular orientation in superfluid liquid helium droplets: High resolution infrared spectroscopy as a probe of solvent-solute interactions. Faraday Discuss. 113, 261–278 (1999)
199. A.S. Chatterley, B. Shepperson, H. Stapelfeldt, Three-dimensional molecular alignment inside helium nanodroplets. Phys. Rev. Lett. 119(7), 073202 (2017)
200. J.W. Niman, B.S. Kamerin, D.J. Merthe, L. Kranabetter, V.V. Kresin, Oriented polar molecules trapped in cold helium nanodroplets: Electrostatic deflection, size separation, and charge migration. Phys. Rev. Lett. 123(4), 043203 (2019)
201. H. Saarikoski, S.M. Reimann, A. Harju, M. Manninen, Vortices in quantum droplets: Analogies between boson and fermion systems. Rev. Mod. Phys. 82(3), 2785–2834 (2010)
202. H.C. van de Hulst, *Light scattering by small particles* (Dover Publications Inc., New York City, 1957)
203. S.W. van Sciver, C.F. Barenghi, Visualisation of Quantum turbulence, in *Progress in low temperature physics: Quantum turbulence.* ed. by M. Tsubota, W.P. Halperin (Elsevier, Amsterdam, 2009), pp. 247–303
204. F. Ancilotto, M. Pi, M. Barranco, Vortex arrays in a rotating superfluid He-4 nanocylinder. Phys. Rev. B Condens. Matter 90(17), 174512 (2014)
205. C. Bernando, A.F. Vilesov, Kinematics of the doped quantum vortices in superfluid helium droplets. J. Low Temp. Phys. 191, 242–256 (2018)
206. F. Coppens, F. Ancilotto, M. Barranco, N. Halberstadt, M. Pi, Capture of Xe and Ar atoms by quantized vortices in He-4 nanodroplets. Phys. Chem. Chem. Phys. 19(36), 24805–24818 (2017)
207. G.P. Bewley, D.P. Lathrop, K.R. Sreenivasan, Superfluid helium—visualization of quantized vortices. Nature 441(7093), 588–588 (2006)
208. C. Raman, J.R. Abo-Shaeer, J.M. Vogels, K. Xu, W. Ketterle, Vortex nucleation in a stirred Bose-Einstein condensate. Phys. Rev. Lett. 87(21), 210402 (2001)
209. P. Rosenbusch, V. Bretin, J. Dalibard, Dynamics of a single vortex line in a Bose-Einstein condensate. Phys. Rev. Lett. 89(20), 200403 (2002)

210. A.L. Fetter, Rotating trapped Bose-Einstein condensates. Rev. Mod. Phys. **81**(2), 647–691 (2009)
211. I.A. Pshenichnyuk, N.G. Berloff, Inelastic scattering of xenon atoms by quantized vortices in superfluids. Phys. Rev. B **94**(18), 184505 (2016)
212. D. Kivotides, C.F. Barenghi, Y.A. Sergeev, Interactions between particles and quantized vortices in superfluid helium. Phys. Rev. B **77**(1), 014527 (2008)
213. D. Kivotides, Y.A. Sergeev, C.F. Barenghi, Dynamics of solid particles in a tangle of superfluid vortices at low temperatures. Phys. Fluids **20**(5), 055105 (2008)
214. D.R. Tilley, J. Tilley, *Superfluidity and superconductivity*. Graduate student series in physics, ed. D.F. Brewer. (Institute of Physics Publishing, Bristol, 1990)
215. O.F. Hagena, W. Obert, Cluster formation in expanding supersonic jets—effect of pressure, temperature, nozzle size, and test gas. J. Chem. Phys. **56**(5), 1793–1802 (1972)
216. B. Sheehy, L.F. DiMauro, Atomic and molecular dynamics in intense optical fields. Annu. Rev. Phys. Chem. **47**(1), 463–494 (1996)
217. K. Yamanouchi, The next frontier. Science **295**(5560), 1659 (2002)
218. T. Gorkhover, M. Adolph, D. Rupp, S. Schorb, S.W. Epp, B. Erk, L. Foucar, R. Hartmann, N. Kimmel, K.U. Kuhnel, D. Rolles, B. Rudek, A. Rudenko, R. Andritschke, A. Aquila, J.D. Bozek, N. Coppola, T. Erke, F. Filsinger, H. Gorke, H. Graafsma, L. Gumprecht, G. Hauser, S. Herrmann, H. Hirsemann, A. Homke, P. Holl, C. Kaiser, F. Krasniqi, J.H. Meyer, M. Matysek, M. Messerschmidt, D. Miessner, B. Nilsson, D. Pietschner, G. Potdevin, C. Reich, G. Schaller, C. Schmidt, F. Schopper, C.D. Schroter, J. Schulz, H. Soltau, G. Weidenspointner, I. Schlichting, L. Strueder, J. Ullrich, T. Möller, C. Bostedt, Nanoplasma dynamics of single large xenon clusters irradiated with superintense X-ray pulses from the linac coherent light source free-electron laser. Phys. Rev. Lett. **108**(24), 245005 (2012)
219. T. Nishiyama, Y. Kumagai, A. Niozu, H. Fukuzawa, K. Motomura, M. Bucher, Y. Ito, T. Takanashi, K. Asa, Y. Sato, D. You, Y. Li, T. Ono, E. Kukk, C. Miron, L. Neagu, C. Callegari, M. Di Fraia, G. Rossi, D.E. Galli, T. Pincelli, A. Colombo, T. Kameshima, Y. Joti, T. Hatsui, S. Owada, T. Katayama, T. Togashi, K. Tono, M. Yabashi, K. Matsuda, C. Bostedt, K. Nagaya, K. Ueda, Ultrafast structural dynamics of nanoparticles in intense laser fields. Phys. Rev. Lett. **123**(12), 123201 (2019)
220. C. Peltz, C. Varin, T. Brabec, T. Fennel, Time-resolved X-ray imaging of anisotropic nanoplasma expansion. Phys. Rev. Lett. **113**(13), 133401 (2014)
221. T. Ditmire, T. Donnelly, A.M. Rubenchik, R.W. Falcone, M.D. Perry, Interaction of intense laser pulses with atomic clusters. Phys. Rev. A **53**(5), 3379–3402 (1996)
222. T. Ditmire, J.W.G. Tisch, E. Springate, M.B. Mason, N. Hay, J.P. Marangos, M.H.R. Hutchinson, High energy ion explosion of atomic clusters: Transition from molecular to plasma behavior. Phys. Rev. Lett. **78**(14), 2732–2735 (1997)
223. B. Tabbert, H. Günther, G. zu Putlitz, Optical investigation of impurities in superfluid $^4$He. J. Low Temp. Phys. **109**(5), 653–707 (1997)
224. A. Mauracher, O. Echt, A.M. Ellis, S. Yang, D.K. Bohme, J. Postler, A. Kaiser, S. Denifl, P. Scheier, Cold physics and chemistry: Collisions, ionization and reactions inside helium nanodroplets close to zero K. Phys. Rep. Rev. Sect. Phys. Lett. **751**, 1–90 (2018)
225. B. Schütte, M. Arbeiter, A. Mermillod-Blondin, M.J.J. Vrakking, A. Rouzee, T. Fennel, Ionization avalanching in clusters ignited by extreme-ultraviolet driven seed electrons. Phys. Rev. Lett. **116**(3), 033001 (2016)
226. A. Heidenreich, B. Grüner, M. Rometsch, S.R. Krishnan, F. Stienkemeier, M. Mudrich, Efficiency of dopant-induced ignition of helium nanoplasmas. New J. Phys. **18**(7), 073046 (2016)
227. C.C. Tsao, J.D. Lobo, M. Okumura, S.Y. Lo, Generation of charged droplets by field ionization of liquid helium. J. Phys. D Appl. Phys. **31**(17), 2195–2204 (1998)
228. R.E. Grisenti, J.P. Toennies, Cryogenic microjet source for orthotropic beams of ultralarge superfluid helium droplets. Phys. Rev. Lett. **90**(23), 234501 (2003)
229. R.M.P. Tanyag, A.J. Feinberg, S.M.O. O'Connell, A.F. Vilesov, Disintegration of diminutive liquid helium jets in vacuum. J. Chem. Phys. **152**(23): 234306. 2020 (2020)

230. A. Echelmeier, M. Sonker, A. Ros, Microfluidic sample delivery for serial crystallography using XFELs. Anal. Bioanal. Chem. **411**(25), 6535–6547 (2019)
231. K. Kolatzki, M.L. Schubert, A. Ulmer, T. Möller, D. Rupp, R.M.P. Tanyag, Micrometer-sized droplets from liquid helium jets at low stagnation pressures. Phys. Fluids **34**(1), 012002 (2022)
232. F.-Z. Zhao, B. Zhang, E.-K. Yan, B. Sun, Z.-J. Wang, J.-H. He, D.-C. Yin, A guide to sample delivery systems for serial crystallography. FEBS J. **286**(22), 4402–4417 (2019)
233. M.S. Paoletti, D.P. Lathrop, Quantum turbulence. Ann. Rev. Condens. Matter Phys. **2**(2), 213–234 (2011)
234. W. Guo, M. La Mantia, D.P. Lathrop, S.W. Van Sciver, Visualization of two-fluid flows of superfluid helium-4. Proc. Natl. Acad. Sci. U.S.A. **111**, 4653–4658 (2014)
235. E. Fonda, D.P. Meichle, N.T. Ouellette, S. Hormoz, D.P. Lathrop, Direct observation of Kelvin waves excited by quantized vortex reconnection. Proc. Natl. Acad. Sci. U.S.A. **111**, 4707–4710 (2014)

# Chapter 8
# Electron Diffraction of Molecules and Clusters in Superfluid Helium Droplets

**Jie Zhang, Yunteng He, Lei Lei, Yuzhong Yao, Stephen Bradford, and Wei Kong**

**Abstract** In an effort to solve the crystallization problem in crystallography, we have been engaged in developing a method termed "serial single molecule electron diffraction imaging" (SS-EDI). The unique features of SS-EDI are superfluid helium droplet cooling and field-induced orientation. With two features combined, the process constitutes a molecular goniometer. Unfortunately, the helium atoms surrounding the sample molecule also contribute to a diffraction background. In this chapter, we analyze the properties of a superfluid helium droplet beam and its doping statistics, and demonstrate the feasibility of overcoming the background issue by using the velocity slip phenomenon of a pulsed droplet beam. Electron diffraction profiles and pair correlation functions of monomer-doped droplets, small cluster and nanocluster -doped droplets are presented. The timing of the pulsed electron gun and the effective doping efficiency under different dopant pressures can both be controlled for size selection. This work clears any doubt of the effectiveness of superfluid helium droplets in SS-EDI, thereby advancing the effort in demonstrating the "proof-of-concept" one step further.

J. Zhang (✉) · L. Lei · Y. Yao · S. Bradford · W. Kong
Chemistry Department, Oregon State University, Corvallis, OR 97331, USA
e-mail: zhangji@oregonstate.edu

L. Lei
e-mail: leil@oregonstate.edu

Y. Yao
e-mail: yaoyu@oregonstate.edu

S. Bradford
e-mail: bradfost@oregonstate.edu

W. Kong
e-mail: Wei.Kong@oregonstate.edu

Y. He
Central Community College, Kearney, NE 68848, USA
e-mail: yuntenghe@cccneb.edu

© The Author(s) 2022
A. Slenczka and J. P. Toennies (eds.), *Molecules in Superfluid Helium Nanodroplets*,
Topics in Applied Physics 145, https://doi.org/10.1007/978-3-030-94896-2_8

## 8.1  Introduction

More than 80% of the atomic-resolution structures can be attributed to single-crystal crystallography in the Protein Data Bank. Still, the difficulty and unpredictability of crystallization remain the nemesis of the technique. Recently, several new ideas have been introduced to solve the crystallization problem in crystallography [1–5]. One of the most successful is termed "diffract and destroy" [1], where ultrashort and ultra-intense x-ray photons are used to diffract from a single particle before the particle is destroyed by the radiation. The fundamental premise is that sufficient sampling of a randomly oriented sample population allows complete sampling of all possible orientations. To date, several dozens of new protein structures have already been solved using this method [6, 7]. The method has been adopted to determine the shape of and detect the vortices in superfluid helium droplets [8]. Another method employs electrons because of their much larger diffraction cross sections [9] and easier accessibility in laboratories than ultra-short x-ray photons. In addition, sample alignment in a laser field prior to diffraction has also been demonstrated, simplifying the data interpretation tremendously [2, 3]. The ease in aligning a molecule embedded in superfluid helium droplets has further prompted the idea of using Coulomb explosion to obtain structures of small molecules [4, 10, 11].

Our group has been developing a method called serial single molecule electron diffraction imaging (SS-EDI) as a potential means to solve structures of large biological molecules and nanomaterials [5]. The procedure starts with electrospray ionization to produce ions for doping into superfluid helium droplets. The cooled ions are aligned by an elliptically polarized laser field and subjected to radiation by high energy electrons. The diffraction patterns from the isolated molecules embedded in superfluid helium droplets are accumulated as the sample is refreshed in repetitive pulses for the desired signal-to-noise (S/N) ratio. Three-dimensional information is obtained from diffraction images collected from different orientations of the sample achieved by different polarizations of the alignment laser beam. So far, we have successfully demonstrated the feasibility of doping proteins such as the green fluorescent protein into superfluid helium droplets; [12–14] and performed electron diffraction (ED) of several neutral molecules and clusters embedded in superfluid helium droplets, without laser alignment [15–19].

The role of superfluid helium droplets in SS-EDI is multifaceted. The low temperature of ~ 0.4 K within the droplets is crucial for field-induced alignment and orientation; compared with a room temperature sample, superfluid helium droplet cooling achieves a reduction in the required field strength by nearly four orders of magnitude for the same degree of orientation [20]. In addition, the superfluidity of the droplets allows free rotation of the dopant in response to an external electromagnetic field [21–23]. However, the presence of helium atoms surrounding the embedded molecule inevitably contributes to the diffraction background, particularly if the droplet size reaches the order of $10^6$ helium atoms [24]. Here, we demonstrate the feasibility of solving structural information of molecules and clusters embedded in superfluid droplets.

## 8.2 Theory

### 8.2.1 Theoretical Concept of Gas-Phase Electron Diffraction

A brief review of the fundamental theory about electron diffraction is presented in this section. More details can be found in the references [25–28].

The de Broglie-wavelength $\lambda$ of a high-speed electron including the relativistic correction is calculated as:

$$\lambda = \frac{h}{\sqrt{2m_e \, eV(1 + eV/mc^2)}}, \tag{8.1}$$

with the acceleration voltage $V$, the electron charge $e$, the electron mass $m_e$, the speed of light $c$, and the Planck constant $h$. For an electron of 40 keV, the de Broglie-wavelength is $6.0 \times 10^{-12}$ m. For simplicity, the electron beam is treated as a plane wave, and the wavefunction $\Psi_0$ of the incident electrons traveling along the z-direction is described as:

$$\Psi_0 = e^{i\mathbf{k_0}z}, \tag{8.2}$$

with a wave vector $\mathbf{k_0}$, where $|\mathbf{k_0}| = 2\pi/\lambda$.

The *independent-atom model* treatment, commonly accepted in the community of gas-phase electron diffraction, is used to calculate the molecular diffraction pattern shown in Fig. 8.1. In this approach, the atoms within a molecule are treated as perfect spherical independent diffraction centers. The elastic scattering wavefunction $\Psi_n'$ of the nth atom can be written as:

$$\Psi_n' = K\frac{e^{ik_0R}}{R} f_n(\theta)e^{i(\mathbf{k_0}-\mathbf{k'})\mathbf{r_n}}, \tag{8.3}$$

where $K = \frac{8\pi^2 me^2}{h^2}$, $R$ is the distance between scattering center and observation point, the atomic position vector $\mathbf{r_n}$, the scattered wave vector $\mathbf{k'}$ and the scattering angle $\theta$ with respect to $z$, and the atomic scattering factor $f_n(\theta)$, which includes amplitude and a phase shift term. Then the overall scattering wavefunction for a molecule containing N atoms, is the sum of the scattering wavefunctions from individual atoms:

$$\Psi' = \sum_{n=1}^{N} K\frac{e^{ik_0R}}{R} f_n(\theta)e^{i(\mathbf{k_0}-\mathbf{k'})\mathbf{r_n}}, \tag{8.4}$$

In traditional gas phase electron diffraction (GPED), the magnitude of the momentum transfer $s$ is defined as:

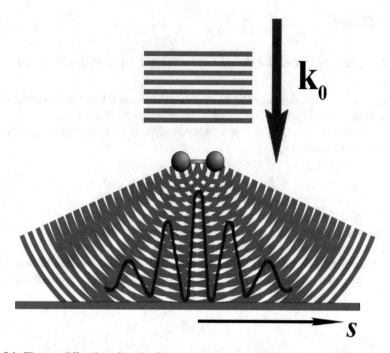

**Fig. 8.1** Electron diffraction of a pair of atoms

$$|\mathbf{s}| = |\mathbf{k_0} - \mathbf{k'}| = 2k_0 \sin\left(\frac{\theta}{2}\right) = \frac{4\pi}{\lambda} \sin\left(\frac{\theta}{2}\right), \tag{8.5}$$

and the total scattering intensity of the molecule can be written as a function of $s$:

$$I(s) = \frac{K^2 I_0}{R^2} \sum_{n=1}^{N} \sum_{m=1}^{N} f_n(s) f_m^*(s) e^{i\mathbf{sr}_{nm}}, \tag{8.6}$$

with $\mathbf{r}_{nm} = \mathbf{r}_n - \mathbf{r}_m$, the distance between the nth atom and the mth atom. Here $r_{nm} \ll R$, $I_0$ is the intensity of the incident electron beam. The scattering factor is rewritten as a function of $s$ with amplitude $|f(s)|$ and the phase term $\eta(s)$ with values tabulated in ref. [29]. The total diffraction intensity $I(s)$ can be split into two parts: a non-structural related contribution from individual atoms, denoted as atomic scattering intensity $I_a(s)$, and a structural related contribution determined by interatomic distances, denote as molecular scattering intensity $I_m(s)$.

$$I(s) = I_m(s) + I_a(s)$$

$$= \frac{K^2 I_0}{R^2} \sum_{n=1}^{N} |f_n(s)|^2$$

$$+ \frac{K^2 I_0}{R^2} \sum_{n=1}^{N} \sum_{\substack{m=1 \\ m \neq n}}^{N} |f_n(s)||f_m(s)| \cos(\eta_n(s) - \eta_m(s)) e^{i s \mathbf{r}_{nm}}, \tag{8.7}$$

For randomly oriented gas-phase molecules, the molecular scattering intensity can be calculated by integrating over all orientations:

$$I_m(s) = \frac{K^2 I_0}{R^2} \sum_{n=1}^{N} \sum_{\substack{m=1 \\ m \neq n}}^{N} |f_n(s)||f_m(s)| \cos(\eta_n(s) - \eta_m(s)) \frac{\sin(s r_{nm})}{s r_{nm}}, \tag{8.8}$$

In addition to elastic scattering, inelastic scattering should be considered for small angle diffractions. Inelastic scattering is considered to only spreads over a small angle, and the intensity of inelastic scattering decreases rapidly when $s$ is increasing. In practice, inelastic scattering is considered incoherent and is often treated as non-structural related background during data analysis.

## 8.2.2   Implementation and Challenges

Although the helium atom is light with a small scattering factor, in the case of electron diffraction of a sample-doped helium droplet, hundreds or more helium atoms in one droplet still generate a considerable background. They can easily overshadow the diffraction pattern of the sample. It is a unique feature of the helium droplet environment that requires additional treatment.

The total diffraction intensity $I(s)$ of a doped droplet is shown as the following:

$$I(s) = I_m(s) + I_a(s) + I_d(s), \tag{8.9}$$

where $I_d(s)$ is the contribution from the diffraction of the helium droplets. More discussion of this term will be covered in Sect. 8.4. Each helium droplet may contain a different number of helium atoms, and the analytical expression for this term is not yet clear. Additionally, the observed diffraction patterns show that no structural information of the helium atoms could be detected for droplets containing less than a few thousand helium atoms, so we treat the helium droplet diffraction as a monotonic non-structural contribution. Doping causes a certain amount of helium atoms to evaporate, but helium's contribution to the diffraction profile remains similar to that prior to doping. To account for this effect, we record the diffraction pattern of a pure helium droplet. We then use a scaling factor to model the contribution of helium in the overall diffraction profile.

In GPED, the modified molecular scattering curves $sM(s)$ are used to magnify the contribution from molecular scattering by removing all contributions from atoms and the helium droplet background and are further magnified by the momentum transfer $s$:

$$sM(s) = \frac{I_m}{I_a} \cdot s, \tag{8.10}$$

The Eq. 8.10 can be further written as:

$$sM(s) = \frac{\alpha_s \cdot I_{total}(s) - \beta \cdot I_{droplet}(s) - \alpha_b \cdot I_{background}(s) - I_{T,at}(s)}{I_{T,at}(s)} \cdot s, \tag{8.11}$$

where $I_{total}(s)$, $I_{droplet}(s)$, and $I_{background}(s)$ are the experimental intensities of the doped droplets, pure droplets, and background; $\alpha_s, \beta$, and $\alpha_b$ are the corresponding fitting parameters; and $I_{T,at}(s)$ is the theoretical atomic diffraction intensity from the dopant molecule. All three fitting parameters contain a common response factor, and the ratio $\frac{\beta}{\alpha_s}$ indicates the ratio of the number of helium atoms after doping to that without doping.

The above method has been used for doped droplets dominated by single dopant doping. When the dopant number increases following the Poisson distribution, a direct fitting of the total diffraction intensity with all the possible candidate structures can be adopted:

$$I_{total}(s) = \beta \cdot I_{droplet}(s) + \alpha_1 \cdot I_1 + \sum_i \alpha_{2_i} I_{2_i} + \sum_i \alpha_{3_i} I_{3_i} + \sum_i \alpha_{4_i} I_{4_i} + ...,$$
$$\tag{8.12}$$

where $I_{2_i}, I_{3_i}$, and $I_{4_i}$ are the theoretical diffraction intensities for a given structure $i$ of dimers, trimers and tetramers, and $\beta, \alpha_1, \alpha_{2_i}, \alpha_{3_i}$, and $\alpha_{4_i}$ are the fitting parameters. Similar to Eq. 8.11, the value $\beta$ represents the remaining helium's contribution after doping relative to that of a neat droplet beam. The value $\alpha_1$ represents the contribution from monomer and the values of $\alpha_{2_i}, \alpha_{3_i}$, and $\alpha_{4_i}$ are the contributions of clusters of dimer, trimer and tetramers with structure $i$: all contain a common scaling factor. Only the ratios of the $\alpha_{n_i}/\alpha_1$ can be compared with the Poisson statistics. When the signal-to-noise level of the experimental data is limited, additional constraints with the pre-assumption of doping statistics need to be considered. Details of the procedure and the statistical methods will be explained in the case studies.

## 8.3 Experiment

The apparatus used for electron diffraction in our laboratory to study samples doped in superfluid helium droplets is illustrated in Fig. 8.2. It combines a superfluid helium droplet source with a GPED system. The apparatus can be divided into three regions: a droplet source chamber, a doping chamber, and a diffraction and detection (D&D) chamber.

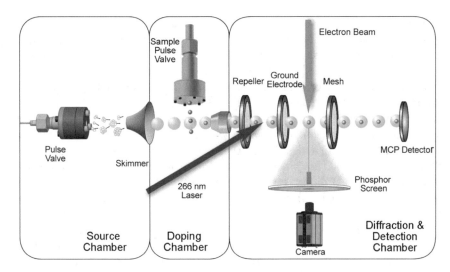

**Fig. 8.2** Experimental setup showing the sample Pulse Valve and in-line Time-of-flight

The droplet source consists of a pulse valve (PV, Cryogenic Copper Even-Lavie Valve) and a skimmer cone of an orifice of 2 mm. High purity helium (99.9995%) is connected to the PV with an effective nozzle diameter of 50 μm and is kept at a temperature between 10 and 20 K with a stagnation pressure between 50 and 70 atm. The skimmer located 11 cm downstream of the PV separates the source chamber from the doping chamber. The droplet source chamber's vacuum level is maintained at $1 \times 10^{-7}$ torr when the PV is off and reaches up to $5 \times 10^{-6}$ torr when the PV is on.

The sample is introduced into the doping chamber by either a heated pickup cell or a sample pulse valve (SPV). For samples requiring heating temperatures above 200 °C, a heated pickup cell located 5 cm downstream from the skimmer base is used. The vapor pressure is controlled by varying the temperature of the cell. For samples with higher vapor pressures, a SPV is mounted 10 cm away from the skimmer base and 5–10 mm from the droplet beam axis. With the use of the SPV, the amount of introduced sample can be significantly reduced, thus resulting in less diffused sample present in the diffraction chamber. The doping chamber is separated from the diffraction chamber by a home-made conical skimmer with a diameter of 4 mm. The doping chamber's vacuum level is $1 \times 10^{-7}$ torr without the sample and increases to $10^{-5}$ torr when the sample is present.

The GPED system located in the D&D chamber includes the electron source, the diffraction system, and the imaging system. The collimated electron beam is pulsed such that it would spatially and temporally overlap with the doped droplets, and the resulting diffracted electrons are detected by a phosphor screen and a camera. The electrons are emitted from a LaB$_6$ filament, biased at 40 keV, then focused onto the detector by a magnetic lens included in the electron gun (Kimball Physics, EGH-6210A). An extra magnetic lens can be mounted below the commercial electron gun to reduce the beam size from a diameter of 3 mm to 0.3 mm at the diffraction spot. The diffracted electrons hit the detector while the undiffracted beam is stopped and measured by a Faraday cup that is 4.8 mm in diameter, 50.8 mm in height, and located 2 mm above the detector. The phosphor detector, as well as the imaging system, can be mounted at different distances from the diffraction spot to collect different ranges of diffraction angles. In most of our experiments, the distance is set to cover up to a diffraction angle of 4.4°, equivalent to $s = 8$ Å$^{-1}$. The imaging system consists of a scintillation screen with a phosphor coating (Beam Imaging Solutions P43, 40 mm) and an EMCCD camera (Andor Technology, iXON Ultra) to record the image. The camera is synchronized with the pulsed electron beam to capture individual diffraction pattern on the phosphor screen. To dynamically remove the background diffraction, a "toggle" method is used. In addition to images from droplets, extra images are measured between consecutive droplet pulses and are accumulated and saved as the background image. When the doping sample is present, the difference between the two images is the net diffraction from the sample doped droplets $I_{total}$. When the sample is not present, the difference corresponds to the net diffraction of pure droplets $I_{doplet}$. Hence, the overall experiment's effective repetition rate is equivalent to half of that of the electron beam. To protect the gun filament and reduce the ambient diffraction, a cold trap is connected to the shielding elements of the electron beam. With the cold trap, the vacuum level of the D&D chamber can be maintained at $8 \times 10^{-9}$ torr.

To assist in the electron source characterization and optimization of the experimental conditions, a rotatable 6-position wheel holding a variety of components is mounted inside the diffraction chamber as illustrated in Fig. 8.3. Each component on the wheel can be positioned in the diffraction spot at the center of the chamber via

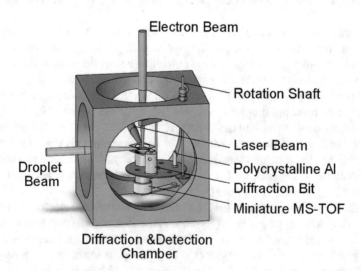

**Fig. 8.3** Rotatable wheel inside D&D chamber

**Table 8.1** List of components on the rotatable wheel

| Name | Function | Note |
|---|---|---|
| Faraday cup | Beam current measurement | |
| Phosphor screen | Electron beam observation | Oriented 45° with electron beam |
| | | Images can be observed through an optical window |
| Polycrystalline aluminum | Camera length and wavelength of the electrons source calibration | A standard TEM calibration sample deposit on a 3 mm copper grid (Electron Microscopy Sciences, Catalog #80044) |
| Sample needle | Gas-phase ED sample delivery | Use for comparison between GPED and diffraction from sample embedded in helium droplets |
| Diffraction bit | Shielding and cleansing the electron beam | Made of brass with an aperture of 6 mm in diameter for droplet path and a graphite aperture of 3 mm located 5 mm above the diffraction spot |
| Mesh detector | Droplet size distribution measurement | |
| Miniature TOF | High-resolution (0.2 u) mass analysis | Drift tube oriented perpendicular with the droplet beam direction |
| Alignment cutter | For the sizes and positions measurement of the electron beam and laser beam | Made of tungsten carbide in an L-shape with a width of 5 mm in both directions |

a rotating shaft located outside the vacuum chamber. The components' details are summarized in Table 8.1, and all of them are not necessarily coexist on the wheel.

Another TOF–MS oriented in-line with the droplet beam consists of a pair of repeller and ground electrodes, and an MCP detector. It is located inside the D&D Chamber for ion detection with high m/z. It can be operated simultaneously with the image collection of electron diffraction. In both TOF–MS setups, the doped droplets are ionized via the 4th harmonic at 266 nm from a pulsed Nd: YAG laser (Quantel, Brilliant). The laser power density on the level of $10^6$ W/cm$^2$ is typically high enough to ionize or fragment the dopants inside the superfluid helium droplets via 2 or 3 multiphoton absorptions while leaving the helium droplets mostly intact. When the miniature TOF–MS is in use, the ionization spot is at the same location as the diffraction spot, whereas in the case of the in-line TOF–MS, the ionization region is 5 cm upstream. In this latter case, the laser timing needs to be adjusted accordingly.

**Fig. 8.4** Arrival time profiles of benzene doped helium droplets recorded at a source temperature of 20 K and a doping pressure of 2.5 × 10⁻⁴ Torr. The acceleration voltages on the repeller electrode are labeled in the figure

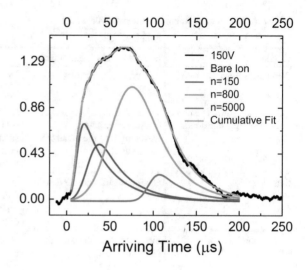

## 8.4 Characterization of Droplet Sizes

Benzene doped droplet was used for the characterization of droplet sizes for different pre-cooled temperatures. The ionization threshold of benzene is only 9.24 eV [30], and 2 photons at 266 nm, with a total energy of 9.35 eV is sufficient for ionization. The 2-photon process is achievable at a power density of $10^5$ to $10^6$ W/cm$^2$ without any focusing lens. The mass spectrum of benzene doped droplets contains no fragments, with only parent ions and clusters of parent ions with helium, ideal for size measurement. Since most of the excess energy (0.1 eV) released upon ionization is taken away by the departing photoelectron, size measurements from benzene doped droplets should be reliable. One example of droplets forming with a stagnation temperature of 20 K is shown in Fig. 8.4. The detected current contains contributions of bare ions and ion-doped droplets, and the latter could contain multiple groups with different sizes, as revealed in our previous measurements [12, 31].

The time profiles are fitted using the exponentially modified Gaussian probability distribution function [32], i.e., a convolution of a Gaussian function with an exponential decay function, and used the nested F-test to avoid overfitting. The center of the fitted Gaussian function is then used to represent the group arriving time. Combined with the acceleration voltage and travel distance, each group's size can then be obtained.

**Fig. 8.5** Diffraction images from $CBr_4$ doped helium droplets. Left: Raw image from $CBr_4$ doped helium droplets, Middle: Wiener filtered regenerated image from left one by refilling the portion inside the dashed box with data from the opposite side of the box and averaging over all four quadrants, Right: image after droplet background removal. Reprinted from "Electron diffraction of CBr4 in superfluid helium droplets: A step towards single molecule diffraction", J. Chem. Phys. 145, 034,307 (2016), with the permission of AIP Publishing

## 8.5    Image and Data Processing

The procedure of image treatment and data process for structure analysis is discussed in this section. One of the examples used here is the diffraction image of $CBr_4$ doped droplets. As mentioned in Sect. 8.2, each unique pair of atoms generates a set of rings, and the diffraction image is highly symmetric with the superposition of all the rings from each pair. However, the Faraday cup and its supporting arm cast a shadow on the phosphor screen as shown in the raw image of $CBr_4$ doped droplet in Fig. 8.5a. The portion inside the red dash box is then removed and replaced by the content in the opposite direction. In this experiment, the diffraction signal is relatively weak and a Wiener filter [33] is used in the image treatment to reduce the noise level of the image. A radial profile of the treated image can be calculated by a MATLAB script and then used in structure fitting as $I_{total}(s)$. For a better visualization effect, the image is then averaged over all four quadrants to smooth the image further. With the fitting parameters, the helium droplet contribution part is then removed to emphasize the dopant contribution as in Fig. 8.5c. Contrasting features can be seen by the naked eye only in this figure in the case of the weak signal.

## 8.6    Case Study

### 8.6.1    Electron Diffraction of Pure Droplets at Different Temperatures

The electron diffraction pattern of pure helium droplets under different temperatures of the helium source was recorded. The diffraction intensity $I_{total}$ is shown in Fig. 8.6

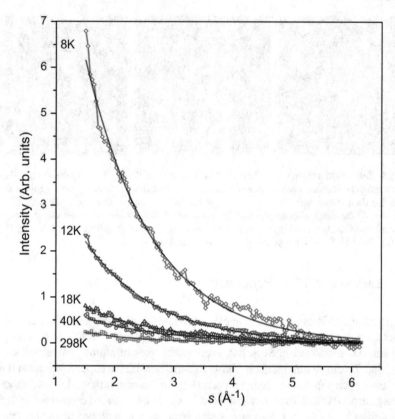

**Fig. 8.6** Total electron diffraction intensity from pure superfluid helium droplets. The smooth line shows single exponential fits to the experimental curve. Reprinted from "Electron Diffraction of Superfluid Helium Droplets" https://pubs.acs.org/doi/abs/10.1021/jz5006829. J. Phys. Chem. Lett. 2014, 5, 11, 1801–1805 with the permission from ACS publishing. Further permission related to the material excerpted should contact ACS

as a function of momentum transfer $s$. The diffraction intensity increases with the decrease in temperature of the droplet source, and the decay rate also exhibits a dependence on the source temperature. A single exponential function can be used to fit the decay of the diffraction profiles, and the fitted width parameters ($w$) are listed in Table 8.2.

The curves of different temperatures can be categorized into two groups: the curves of 8 and 12 K with higher diffraction intensities and narrower distributions and 18 and 40 K with weaker intensities and slower decays. The diffraction profiles of the latter group are essentially identical to that of pure helium gas at 298 K. We have then attempted a biexponential fitting for the lower temperature group ($\leq 12$ K) with one of the exponents fixed at 1.48 Å$^{-1}$ (exponent of the higher temperature group), and the resulting second exponential function has a much faster decay, on the order of 0.3 Å$^{-1}$. We have also performed statistical analysis (F-test) and confirmed the biexponential

**Table 8.2** Width parameters (w) from electron diffraction profiles of superfluid helium droplets[a]

| Source temperature | Exponential $y = y_0 + A \cdot e^{(-s/w)}$ | Droplet size | Surface atoms (%) |
|---|---|---|---|
| 8 K | $1.25 \pm 0.03$ | $10^6$ (220 Å) | 8 |
| 12 K | $1.27 \pm 0.03$ | $10^5$ | 17 |
| 18 K | $1.49 \pm 0.11$ | $10^3$ (22 Å) | 60 |
| 40 K | $1.48 \pm 0.10$ | $<10^3$ | >60 |
| 298 K | $1.68 \pm 0.21$ | 1 | 100 |

[a] Adapted from "Electron Diffraction of Superfluid Helium Droplets" https://pubs.acs.org/doi/abs/10.1021/jz5006829. *J. Phys. Chem. Lett.* 2014, 5, 11, 1801–1805

nature of the decay profiles. For further evidence, we have performed a biexponential analysis for the higher temperature group and confirmed the experimental data's single exponential nature.

The amplitude of the exponential fitting is affected by many factors. Under constant electron fluxes and constant gas fluxes from the nozzle, the amplitude from atomic diffraction should be similar. The amplitude from molecular diffraction, that is, coherent diffraction from a correlated atom pair, should be related to the number of helium pairs within each droplet. For the higher temperature group with negligible coherent molecular diffraction, we therefore expect similar amplitudes of atomic diffraction because the total number of atoms arriving at the diffraction region is similar. This expectation is qualitatively confirmed in Fig. 8.6. On the other hand, from 12 to 8 K, we do expect a rise in diffraction amplitude due to the presence of correlated atom pairs in large droplets, again as evidenced from Fig. 8.6.

Compared with previous reports of neutron diffraction of bulk superfluid helium [34–36], our monotonic decay profiles lack the weak oscillatory portion of the pair correlation function at large $s$ values. We attribute this difference to two reasons: one is the polydispersity in the size of the droplets and the other is the variation in density from the core to the surface of a helium droplet. Theoretical simulations of superfluid helium droplets have revealed a diffuse surface layer and a bulklike interior for a droplet of over 100 atoms.

Although the droplet size distribution varies from setup to setup for pulsed droplet sources, we tentatively use the average droplet size from ref. [37] as a general guide. If we further assume a surface layer of 6 Å, the resulting percent of surface atoms under each source temperature is listed in Table 8.2. Below 12 K, most of the atoms in a droplet are considered interior atoms, whereas above 18 K, diffuse surface atoms dominate. In this sense, biexponential functions should be better representations of the experimental data at the two lowest source temperatures, a point confirmed from our statistical analysis. When the source temperature is above 18 K, diffraction profiles of the droplet beam are essentially the same as that of gas-phase helium atoms, consisting of only incoherent atomic scattering. The negligible pair correlation between atoms of small droplets is essential for analyzing the diffraction pattern of doped droplets: all recorded coherent molecular scattering should be from the doped molecules, whereas coherence of the surrounding helium atoms can be neglected.

One thing to note here is that this work is done at the early stage of our study; thus, the whole apparatus used is a prototype with low sensitivity and limited signal to noise ratio. With this work, we have demonstrated electron diffraction of pure superfluid helium droplets. The diffraction profile of pure helium droplets is affected by the polydispersity of the droplet beam, but it is in qualitative agreement with the size variation of the droplets. Larger droplets with a substantial compact interior component demonstrate stronger diffraction and faster decay with momentum transfer, whereas smaller droplets converge to gas phase isolated molecules when the droplet source temperature reaches 18 K.

## 8.6.2   Single Dopant Case: Ferrocene

Velocity slip [38] can beexploited in order to achieve extensive single molecule doping of small droplets in some favorable cases. Each collision between a dopant molecule and a droplet can result in the evaporation of a certain amount of helium atoms, cool the translational, rotational, and vibrational degrees of freedom, and eliminate the binding energy between the dopant and the droplet. When a large amount of helium atoms are required in this process, for example in the case of large molecules with many low frequency vibrational modes and high binding energy with the helium environment, a large size separation is created between those that can only accommodate one dopant molecule and those that are large enough to accommodate two dopant molecules. By timing the pulsed electron gun at the leading edge of the doped droplets, we can sample only singly doped small droplets, while larger droplets containing more dopant molecules arrive at the diffraction region after the termination of the electron pulse.

Ferrocene powder was placed in a SPV heated to 346 K to provide sufficient vapor pressure for doping. In the toggle mode, the sample PV and the electron gun were operated at 14 Hz, while the droplet PV kept at 18 K ran at half the frequency, and the difference image obtained when the droplet PV was on and off was the net image from the droplets, with or without dopant. A successful example of this approach is illustrated in Fig. 8.6 from the diffraction of ferrocene-doped droplets [19]. Fig. 8.7b shows the radial distributions directly obtained from an image accumulated from 200,000 shots. The electron gun's timing was set to sample droplets with about 2000 atoms/droplet. With these profiles as the intensities of Eq. 8.11 and based on comparisons with the theoretical $sM(s)$ [39], multilinear regression is performed to obtain the coefficients of each component $\alpha_s$, $\beta$, and $\alpha_b$. The left half of the image shown in Fig. 8.7a is the difference image after removing pure droplets' contribution based on the obtained coefficients. The right side of the image is the theoretical calculation.

Due to the transient nature of the doping process, we could not measure the actual pressure in the doping region. However, basedon the Poisson statistics [40] and the fitting results $\beta/\alpha_s$, we can determine the average number of effective collisions between ferrocene and droplets, assuming that trapping of one ferrocene molecule

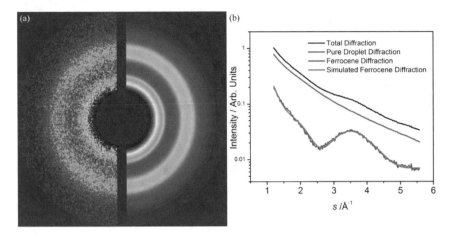

**Fig. 8.7** Electron diffraction of ferrocene doped droplets. **a** The left half of the image is the averaged experimental data after removing the contribution of helium, and right half is the simulation result. **b** is the radial profile with and without dopant. Reprinted from "Communication: Electron diffraction of ferrocene in superfluid helium droplets" *J. Chem. Phys.* 144, 221,101 (2016), with the permission of AIP Publishing

results in a droplet size change from 2000 to 800. We calculate the equivalent pressure for doping based on the experiment setup and the probability of picking zero ($P_0$) and one ($P_1$) ferrocene molecules. Further assuming that no droplets contain two or more ferrocene molecules in the diffraction region, we can then calculate the fraction of singly doped droplets.

Table 8.3 shows that at delays of 1000 and 1200 $\mu$s, more than 80% of the droplets are singly doped. It is worth noting that under these doping conditions, for a droplet of size 2000, only 7% are doped with one ferrocene molecule and 2% are undoped, while the remaining 91% are destroyed because of further collisions with dopant molecules. The above results also imply that regardless of the doping pressure, as long as helium atoms' contribution can be effectively removed according to Eq. 8.11, there is no essential difference between the resulting $sM(s)$ profile. Hence the images obtained from the last three columns were added to improve the signal-to-noise ratio of the final result shown in Fig. 8.7. The data from the first column were not used because of its low ferrocene content. The excellent quality of the fitting procedure

**Table 8.3** Fitting results at different delay times between the two PVs[a]

| Delay ($\mu$s) | 500 | 1000 | 1200 | 4000 |
|---|---|---|---|---|
| $\beta/\alpha_s$ | 0.16 | 0.04 | 0.04 | 0.10 |
| Pressure (Torr) | $6.6 \times 10^{-5}$ | $1.1 \times 10^{-5}$ | $1.1 \times 10^{-4}$ | $8.1 \times 10^{-5}$ |
| $P_1/(P_0 + P_1)$ | 0.72 | 0.81 | 0.81 | 0.76 |

[a] Adapted from "Communication: Electron diffraction of ferrocene in superfluid helium droplets". *J. Chem. Phys.* 144, 221,101 (2016)

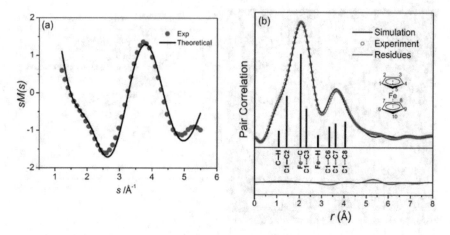

**Fig. 8.8** Modified molecular scattering intensities (**a**) and pair correlation profiles (**b**) of doped ferrocene. The residue is shown in the lower panel of (**b**). Reprinted from "Communication: Electron diffraction of ferrocene in superfluid helium droplets" J. Chem. Phys. 144, 221,101 (2016), with the permission of AIP Publishing

also confirms the hypothesis on the sampling condition: in all delay conditions, only singly doped droplets are sampled in the experiment, with no detectable contributions from ferrocene dimers.

Inverse Fourier transform of the modified molecular scattering intensity results in the pair correlation function and gives the distances of atomic pairs in the molecule. Figure 8.8a shows the experimental and theoretical $sM(s)$ profiles and the comparison between the pair correlation functions obtained from Fig. 8.8a and from the calculation of known gas phase structure is shown in Fig. 8.8b. There is a good agreement between the ferrocene from the doped droplet and the simulation from gas phase ferrocene. However, the high damping factor used in the transform introduces artificial wiggles in the simulation. Gas phase ferrocene is known to be in the eclipse conformation for the two pentacene rings [41]. The resolution is directly related to momentum transfer, which is limited by the detector's size. Unfortunately, with our current setup, we cannot resolve between the eclipse and the staggered structure. To resolve such a difference from the diffraction pattern, the range of s values needs to exceed 12 Å$^{-1}$.

This work demonstrates the feasibility of sampling only singly doped droplets. Several factors determine the degree of size selection via velocity slip, including the physical and chemical properties of the dopant and the dimension and performance of the experimental apparatus. The current approach hence is not "one-size-fits-all". In many cases, statistical analysis and deconvolution procedures are necessary.

## 8.6.3  Small Cluster of the Simple Molecules: CBr₄

With the aid of electron impact ionization (EI) and non-resonant multiphoton ionization (MPI), the most probable size of the droplet beam is determined to be ~2000 helium atoms/droplet in the study of $CBr_4$ doping experiment, in agreement with the general result by Gomez et al. [37]. Doping one $CBr_4$ molecule requires evaporation of ~600 helium atoms [40]. The MPI experiment also confirmed the velocity slip feature in the pulsed droplet beam: the overall droplet beam spans over 200 μs in the diffraction region, but depending on the ionization laser's timing, different sized droplets are sampled in the experiment.

Different from the ferrocene case, the droplets contain not only singly doped but also multiple doped $CBr_4$. There are many possible structures of the $CBr_4$ clusters. To modeling all of the structures are extremely difficult with the limited data resolution. Among contributions of intermolecular atomic pairs from clusters, only the shortest and most prominent Br⋯Br intermolecular pairs are considered in the treatment of the cluster to simplify the fitting process. The Eq. 8.11 is modified to add terms to represent the cluster contribution as:

$$sM(s) = \frac{\alpha_s \cdot I_{total}(s) - \beta \cdot I_{droplet}(s) - \alpha_b \cdot I_{background}(s) - I_{T,at}(s) - C_c \cdot I_{T,c}(s)}{I_{T,at}(s)} \cdot s,$$

(8.13)

$I_{T,c}(s)$ is the theoretical diffraction profile of the shortest Br⋯Br pair with a separation of 4 Å [42, 43] and $C_c$ is the weight of the theoretical contribution of the cluster in the overall diffraction. The radial profiles of different components are shown in the main part of Fig. 8.9. There the total experimental radial distribution ("Total Exp") is compared with the scaled theoretical contribution of clusters in the form of Br⋯Br interference denoted "Br ⋯Br Calculation," and the difference between the experimental data and this theoretical component is labeled as "Exp w/o Br⋯Br." The latter profile should be compared with the theoretical diffraction profile labeled as "CBr₄ Calculation." The halo of the Faraday cup is not included in the profiles. The experimental corrected radial profile contains one clearly observable ring, corresponding to the feature at $s = 2.5$ Å$^{-1}$ and a weaker feature at about 4.3 Å$^{-1}$.

The gas phase electron diffraction images of $CBr_4$ are also recorded by injecting the gaseous sample through a needle positioned directly in place of the droplet beam. The radial profiles and the corresponding pair correlation functions are shown in Fig. 8.10, and they are similar to those from a previous report [44] (not shown in Fig. 8.10). It is important to note that the "doped" trace in Fig. 8.9 is the net difference recorded with and without the droplet beam, while the doping chamber was maintained at a constant pressure. Contributions from the diffused gas in the trace of the "doped" sample are therefore removed. Nevertheless, the doped sample shows a similar diffraction profile as the gaseous sample, and both experimental results are similar to the theoretical calculation. Although the gaseous sample was at room temperature while the doped sample was at 0.38 K, the predominant factor

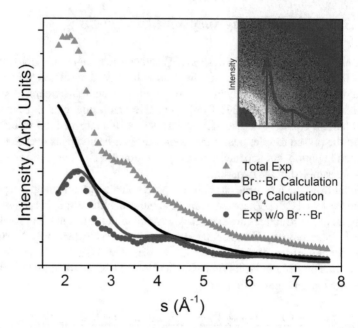

**Fig. 8.9** Radial profiles relevant to diffraction from $CBr_4$ doped superfluid helium droplets. Reprinted from "Electron diffraction of $CBr_4$ in superfluid helium droplets: A step towards single molecule diffraction", J. Chem. Phys. 145, 034,307 (2016), with the permission of AIP Publishing

determining the width of each interference ring is the wave physics, not the vibrational movement of the atoms in a molecule [9]. Evidently, the superfluid helium environment exerts negligible perturbation to the enclosed molecular structure. The weak features near the major features centered at 4.5 and 6.5 Å$^{-1}$ in the $sM(s)$ profile from the doped sample could be due to residual contributions from clusters of $CBr_4$. The same argument applies to the pair correlation profiles, where the profile from the doped sample also contains an extra shoulder near 4 Å.

The diffraction of the droplet beam contains contributions from dopant clusters as well as pure droplets. The contribution of pure droplets from the fitting of Eq. 8.13, $\beta/\alpha_s$, should be proportional to $P_0$, the probability of not picking up any dopant. The contribution of dopant molecular clusters $C_c$ should be proportional to the ratio between the number of intermolecular Br$\cdots$Br pairs with a separation of 4 Å and the average number of dopant molecules in a droplet $\langle k \rangle$

$$\langle k \rangle = \sum_k k \cdot P_k, \qquad (8.14)$$

where $P_k$ is the probability of picking up k molecules by a droplet. If we assume that the electron gun samples the most probable size of the droplet beam of 2000, then there should be at maximum 3 dopant molecules in a droplet. Since there are

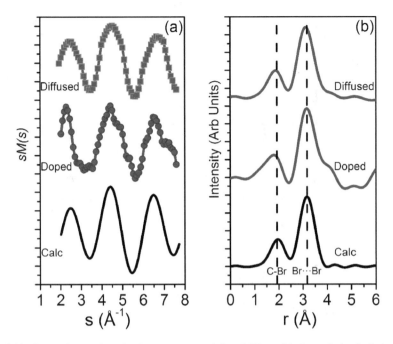

**Fig. 8.10** Comparisons of results from gaseous and doped $CBr_4$ with theoretical calculations: **a** molecular diffraction profiles $sM(s)$, and **b** pair correlation functions. Reprinted from "Electron diffraction of $CBr_4$ in superfluid helium droplets: A step towards single molecule diffraction", J. Chem. Phys. 145, 034,307 (2016), with the permission of AIP Publishing

3 equivalent Br···Br pairs with a separation of 4 Å in a dimer and 4 such pairs in a trimer [45], the value of $C_c$ should be proportional to $(3P_2 + 4P_3)/\langle k \rangle$.

Figure 8.11 compares the experimental fitting results with the Poisson statistics. Figure 8.11a shows the contribution of pure undoped droplets relative to that of doped droplets, and the ratio (vertical axis) is expected to be proportional to the probability of undoped droplets. At a doping pressure of $1 \times 10^{-5}$ Torr, less than 15% of the droplets are undoped, and this value is reproduced from the experimental fitting value of $\beta/\alpha_s$. More and more droplets are doped with increasing doping pressure, and the portion of undoped pure droplets decreases accordingly. At the lowest doping pressure, the error bar is considerably large and a disagreement exists. We attribute this point to the residual gas in the doping chamber since the base pressure in the doping chamber without any doping gas was $1 \times 10^{-6}$ Torr. Figure 8.11b compares the contribution of clusters calculated using $(3P_2+4P_3)/\langle k \rangle$ (designated as "Cluster" in the figure) based on the Poisson statistics [31, 46] with the experimental fitting value for clusters $C_c$. At a doping pressure of $1 \times 10^{-5}$ Torr, contributions of clusters relative to monomers are slightly more than 1:1, and this ratio increases to 1.3:1 at higher doping pressures. This upper limit is attributed to the maximum number of molecules that an average-sized cluster can pick up before it is destroyed completely. The coefficients of $P_2$ and $P_3$ in the expression $(3P_2 + 4P_3)/\langle k \rangle (3P_2 + 4P_3)/\langle k \rangle$ do

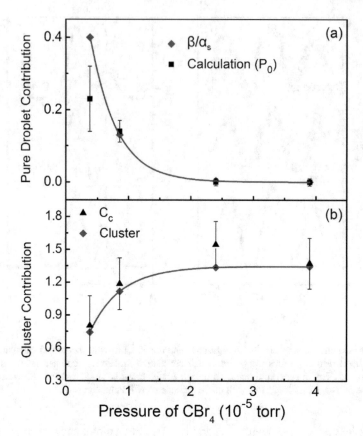

**Fig. 8.11** Comparisons between Poisson statistics and experimental results. **a** Calculated probabilities of undoped droplets $P_0$ in a droplet beam and fitting results $\beta/\alpha_s$ from experiment. **b** Contributions of $CBr_4$ clusters $C_c$ from experiment and calculated ratios of cluster and monomer contributions (designated as "Cluster"). Reprinted from "Electron diffraction of CBr4 in superfluid helium droplets: A step towards single molecule diffraction", *J. Chem. Phys.* 145, 034,307 (2016), with the permission of AIP Publishing

not play a critical role when varied between 3 and 6: they only affect the scaling of the profile, not the general trend.

The above comparison highlights the crucial issue in using superfluid helium droplets as an ultra-cold gentle matrix for electron diffraction from field aligned and/or oriented molecules, i.e., the helium background. The large unwanted diffraction intensity from undoped droplet can potentially overwhelm the detector. This issue should be much more severe for an ion doped droplet beam since the equivalent vapor pressure of ions at the space charge limit is only $10^{-9}$ Torr [13]; in this case, more than 99% of the droplets contain no protein ions at all even in the unrealistically favorable assumption of Poisson statistics. Under low doping conditions, it is therefore the presence of undoped droplets that dominate the background, and elimination of undoped droplets is essential for reducing the background of helium.

The agreement between experiment and analysis alludes to the background problem's solution in electron diffraction of embedded molecules in superfluid helium droplets. By reducing the background from pure undoped droplets via multiple doping, with small corrections for dimers and trimers, clearly resolved diffraction rings of $CBr_4$ similar to those of gas phase molecules can be observed. This condition is achievable for neutral molecules by heavy doping via increased doping pressure or path length. For charged species from an ESI source, fortunately, eliminating neutral undoped droplets from the charged doped droplets is straightforward using electric fields (magnetic fields are generally avoided because of the difficulty in field containment for electron diffraction). An ion-doped droplet beam can be bent from the initial path via an electric field generated by a stack of electrodes in, for example, a reflectron type of design, which also has the benefit of compressing the droplet pulse spatially and temporally. Alternatively, charged droplets can be accelerated or decelerated relative to the neutral undoped droplets, while a pulsed laser and/or electron gun can be synchronized to interact only with the doped beam. Electrostatic steering has the additional benefit of size selection, which could be used to eliminate excessively large droplets due to their high helium content and correspondingly large background contribution.

## 8.6.4   Halogen Bond Case in the Case of $I_2$

By taking advantage of the velocity slip in the pulsed droplet beam, shifting the electron gun's timing relative to the droplet pulse offers a limited degree of control over droplet size and dopant cluster size. We have explored this option in the study of iodine clusters doped in superfluid helium droplets [16]. In this effort, we have chosen two experimental conditions: different source temperatures, doping conditions, and time delays for the electron gun, which resulted in very different diffraction patterns. Figure 8.12 shows the diffraction profiles and the pair correlation functions under the two different conditions: the top panel was obtained under a lower effective doping pressure and by sampling the leading edge of the droplet beam, while the bottom panel was obtained under opposite conditions. The experimental radial profiles (red dots) in the two panels are quite different, indicating significantly different structures sampled under the two different experimental conditions.

The upper panel corresponds to diffraction from smaller sized iodine cluster. With the comparison between the theoretical $sM(s)$ of $I_2$ monomer and the experimental result, the conclusion is that there must be iodine clusters or mixtures of monomers and clusters dominant the diffraction with the possibility of only monomers be eliminated. Based on the timing of the electron gun and the size distribution of the droplet beam, as well as the calibration from previous experiment, the general size of the sampled droplets should be smaller than 1500 atoms/droplet [15, 31]. In general, pickup of one iodine molecule requires removal of 400–600 helium atoms for cooling, and to ensure that the doped droplets continue to travel into the diffraction region, at least 500 helium atoms need to remain with the droplet after doping [31, 40]. With

**Fig. 8.12** Modified
molecular scattering
intensities of smaller iodine
clusters (**a**) and larger iodine
clusters (**b**). The solid black
lines are calculated profiles
based on proposed
structures. Reprint from
"Self-Assembly of Iodine in
Superfluid Helium Droplets:
Halogen Bonds and
Nanocrystals" Angew.
Chem. Int. Ed. Engl. 2017
Mar 20; 56(13): 3541–3545
with the permission of
WILEY

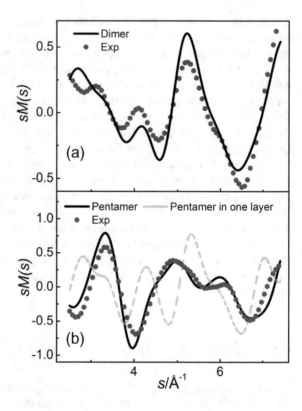

these into consideration, the sampled iodine cluster in the top panel of Fig. 8.12
should contain no more than three iodine molecules.

We were then surprised to discover that no structural information was available
in the literature on iodine dimers or larger clusters. We then fixed the intramolecular
bond length at 2.67 Å, manually varied the relative distance and angle between the
two iodine molecules, calculated the theoretical $sM(s)$, and compared the resulting
diffraction profiles with the experiment by relying on the fitting result of multilinear
regressions of the diffraction profiles from Eq. 8.11. The best fit is a possible "L"
shape structure with an adjusted intermolecular distance of 3.65 Å between the two
nearest iodine atoms. The calculation is not a perfect reproduction of the experimental
result, but it has sufficient merit in reproducing the general trend. We have also
calculated the $sM(s)$ profiles of trimers based on several possible structures, a one-
layer structure from several different cuts of crystalline iodine, and a bi-layer structure
with a dimer and a 3rd molecule in a different plane. None of the profiles can be
considered qualitatively acceptable. We therefore conclude that the diffraction profile
is predominantly due to iodine dimers.

The lower panel in Fig. 8.12 corresponds to diffraction from larger sized iodine
clusters, such as tetramers, pentamers or even hexamers. Among those, the pentamers
is the statistically most likely cluster. We tried to place all five iodine molecules in

one layer according to the crystalline structure [47] (light-blue dashed line), and the result is qualitatively unacceptable. Realizing that the most salient feature in the diffraction profile centered at 3.5 $Å^{-1}$, corresponding to a distance of ~ 4.3 Å, we then considered pseudo-double layer structures with two iodine molecules in each plane and the fifth adjustable out of either plane. We cut fragments from an iodine single crystal, calculated the diffraction profiles, and compared with the experimental result. The black line shows the best result from this adjustment. We can confidently state that the larger iodine clusters sampled in the bottom panel have a bi-layer feature. Unlike the case of the smaller clusters, the calculated $sM(s)$ profiles for tetramers, pentamers, and hexamers are all similar, as long as the iodine molecules form bi-layer structures. The diffraction technique is insensitive to the actual size of the iodine cluster under the current conditions. The pairings from atoms on the outer edges of each cluster do not have repeats, hence their contribution in the overall diffraction profile is overshadowed by those that have many repeats, such as the interlayer distances between corresponding atoms. We choose pentamer as a representation of this cluster group, partly because we have the best success in reproducing the experimental diffraction profile with one particular pentamer structure. We do acknowledge that the experimental data could well be a mixture of clusters with sizes from tetramer to hexamer but with similar structural motifs.

Figure 8.13 shows the pair correlation functions of the two diffraction profiles obtained from Fig. 8.12. The limited range of s values from our image detector requires a large damping factor in the calculation, which not only broadens the profile but also introduces extraneous oscillations in large distances. The estimated uncertainty in the resulting distance is on the order of 0.1 Å. The inset of each panel shows our proposed structures and the numbering schemes. To avoid clumsiness in labeling, all intramolecular distances between the two covalently bonded iodine atoms are labeled "Intra", while only a few intermolecular distances are labeled. The shaded region represents intermolecular pairs such as 3···8, 4···9 and 5···10.

For the case of dimer dominated diffraction, each unique interatomic distance can be more or less resolved under the current conditions, although the fitting is still imperfect. The proposed structure has a distance of 3.65 Å between atoms 1 and 3, and this value is substantially shorter than the sum of the van der Waals distances of two iodine atoms (4.3 Å), but are similar to the in-plane intermolecular distance of 3.5 Å in crystalline iodine [47]. Moreover, all four iodine atoms are in the same plane in the proposed dimer structure. It is the first time that halogen bonds are observed from iodine clusters.

Our experiment is incapable of resolving several unique intermolecular distances for the larger clusters, but a few features are still identifiable. The intramolecular distance at 2.67 Å, which has five repeats, is clearly resolved. The current pentamer structure contains two halogen bonds between atoms 2···3 and 2···5 at ~3.5 Å, while other distances of the parallelepiped of atoms 3···5 and 3···7 are van der Waals in nature. Attempts to rearrange the 1–2 atoms into a more symmetric position stapling the two layers failed to match the current fitting result. An interesting result is the structure of the larger clusters where instead of a single layer, a bi-layer structure seems to dominate when more than three iodine molecules are present. It is possible

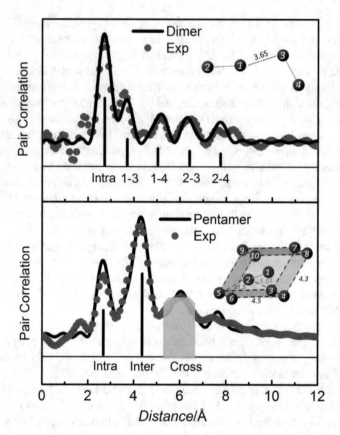

**Fig. 8.13** Pair correlation profiles of doped iodine clusters and their proposed structures. The insets show the structures of the clusters and the numbering of the atoms. Reprint from "Self-Assembly of Iodine in Superfluid Helium Droplets: Halogen Bonds and Nanocrystals" Angew. Chem. Int. Ed. Engl. 2017 Mar 20; 56(13): 3541–3545 with the permission of WILEY

that different from bulk crystals, small clusters are more stable in a bi-layer structure, particularly in a superfluid helium environment. Alternatively, the missing one-layer structure might be related to the limited size range sampled by the electron beam. It is possible that within the size range of the droplet beam sampled by the electron gun, only bi-layer structures can survive the evaporative cooling process and maintain the traveling momentum to the diffraction region. Unfortunately, given the insensitivity of the diffraction technique, further experimental confirmation of this speculation is difficult, if possible at all.

## 8.6.5 CS₂

The experimental results of electron diffraction of $CS_2$ are presented in this section [48]. It demonstrates the feasibility of structural determination of dimers, trimers, tetramers, and clusters containing a large number of monomers embedded in superfluid helium droplets. We can narrow down the range of parameters of the least squares fitting procedure of the diffraction patterns from detailed droplet size measurements and modeling of the doping statistics.

The experiment setup is the same as explained in the previous ferrocene case. Room temperature $CS_2$ with a vapor pressure of 344 Torr is routed to the sample PV via a vacuum feedthrough. Figure 8.14 shows the mass spectrum of $CS_2$ doped in superfluid helium droplets at a stagnation temperature of 10 K. At least three photons at 266 nm are needed for ionization, and the resulting mass spectrum shows extensive fragmentation. The presence of attached helium with the sample ions $C^+$ and $S^+$ manifests the presence of the helium droplet. Although not seen explicitly in the mass spectrum, the presence of $CS_2$ clusters can be deduced from the presence of $CS_2C^+$.

Based on the size analysis mentioned in Sect. 8.4, two groups of doped droplets at the source temperature of 20 K can carry $CS_2$ to the diffraction region. The predominant one has an average size of 800 atoms/droplet and a less abundant group has a much larger average size, with 5000 atoms/droplet. When the source temperature

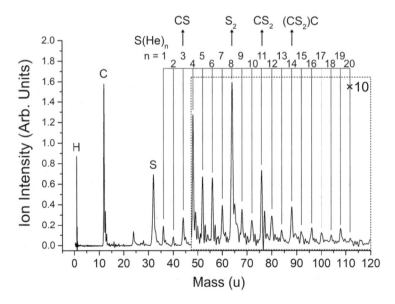

**Fig. 8.14** Mass spectrum of $CS_2$ doped helium droplets recorded at a source temperature of 10 K and a doping pressure of $5 \times 10^{-4}$ Torr. Reprint from "Electron diffraction of CS2 nanoclusters embedded in superfluid helium droplets" *J. Chem. Phys.* 152, 224,306 (2020) with permission from AIP publishing

is reduced to 10 K, the two size groups are $2.2 \times 10^4$ atoms/droplet and $1.4 \times 10^5$ atoms/droplet with similar abundance. It is also worth noting that each size group travels at a unique speed, regardless of the source temperature, and the effect of the source temperature is primarily manifested in the relative abundances of the different groups, while the average size of each group is only mildly dependent on the source temperature [12]. The overall average size of the droplets at a fixed source pressure and temperature, weighted by each group's abundance, is in agreement with literature reports [37, 49]. Due to limited velocity slip [38], however, there is more than one group of droplets with different sizes and velocities at any sampling time within the droplet pulse. Hence, the pulsed electron gun's time setting selects the size composition of the sampled droplets during the diffraction experiment. We note here that these sizes are measured after ionization; hence, the corresponding size of the droplet prior to doping should be larger by a few hundred—on the order of 500—after picking up each dopant molecule [31, 40].

Figure 8.15 shows two scaled radial profiles of the diffraction patterns obtained at two different time settings of the electron gun: the solid squares were recorded at a delay time of 1289 $\mu$s after the droplet PV, and the open squares were at 1269 $\mu$s. Both profiles result from the diffraction of neutral samples without ionization and without sample orientation, and the diffraction profiles are concentric rings due to the orientation average. To emphasize the contribution of molecular interference and to contrast the difference between the two results, the radial profiles are normalized by each profile's exposure time, and the intensity is multiplied by $s$. The difference in the scaled profiles is due to the different doping conditions of the droplet beam: at the earlier timing of 1269 $\mu$s, the droplets are smaller and contain fewer sample molecules.

Figure 8.16 shows the diffraction profile after further shifting the electron gun's timing to 1581 $\mu$s, lowering the source temperature to 10 K, and increasing the effective doping pressure from $2 \times 10^{-4}$ Torr to $5 \times 10^{-4}$ Torr (estimated doping pressure). Compared with Fig. 8.14, the overall diffraction intensity has decreased, but the feature of molecular interference is more prominent, implying more contribution from the sample than from the helium atoms.

Structures of gas phase clusters and single crystals of $CS_2$ have been reported in the literature [50–54]. Experiments in high resolution spectroscopy have concluded that gas phase clusters prefer highly symmetric shapes: the dimer is cross-shaped with a $D_{2d}$ symmetry, [52] as shown in the inset of Fig. 8.14, the trimer forms a pinwheel with a $D_3$ symmetry [50], and the tetramer forms a barrel with a $D_{2d}$ symmetry [51]. A few theoretical calculations have proposed various geometries for small clusters, but the most stable isomer for each cluster size agrees with the experimental results [55, 56]. For clusters containing more than four monomers, convergence to the crystal structure has been predicted [54]. The crystal structure of $CS_2$ at temperatures below 150 K is known to be orthorhombic, containing pairs of $CS_2$. A unit cell contains three monomers at the edges in addition to a central monomer slanted at 55.56° to the c axis, as shown in the inset of Fig. 8.16. The molecular arrangement in crystals is substantially different from those of reported gas phase clusters.

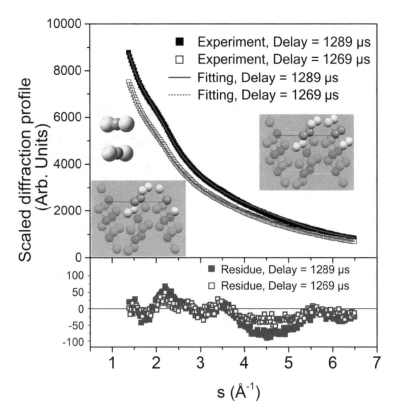

**Fig. 8.15** Scaled radial profiles of diffraction patterns from $CS_2$ doped droplets. The experimental profiles were recorded at a source temperature of 20 K, two different time settings of the electron gun: solid symbols at 1289 µs and open symbols at 1269 µs. All profiles are results of multiplying the radial intensity by s to emphasize the contribution of molecular interference. The insets show the resulting structures for the dimer, trimer, and tetramer. Reprint from "Electron diffraction of CS2 nanoclusters embedded in superfluid helium droplets" *J. Chem. Phys.* 152, 224,306 (2020) with permission from AIP publishing

Equation 8.12 is used for fitting the diffraction profiles, including different cuts from the crystals for dimers, trimers, and tetramers, in addition to their gas phase structures. In this case, $\alpha_{ig}$ ($2 \leq i \leq 4$) as contributions of gas phase clusters containing i monomers, and $\alpha_{ic}$ ($2 \leq i \leq 4$) represents all possible cuts of clusters from the crystal structure. A modified Poisson model using Markovian arrival process with non-identical exponential interarrival times [57] to accommodate the change of cross section reduction with sequential pick up of $CS_2$. The details of the treatment is explained in the reference [48]. The calculated probabilities $P_n$ of pickup n $CS_2$ molecules with a source temperature of 20 K are listed as the first number in each cell of Table 8.4. Even for the larger size group at 1289 µs, the probability of $P_4/P_1$ is only 0.04. About 50% of the droplets contain no sample at all for both delay times, and about 35% of the droplets contain just one $CS_2$. The second number in each

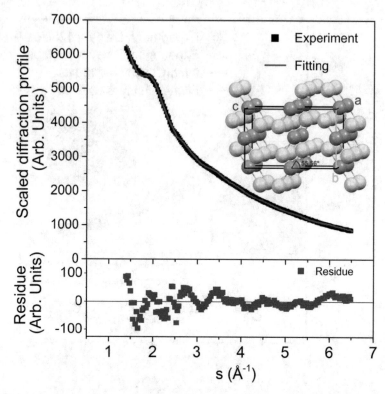

**Fig. 8.16** Scaled radial profile of the diffraction pattern from CS2 doped droplets recorded at a source temperature of 10 K. The profile is scaled by s to emphasize the contribution of molecular interference. The inset shows the structure of the crystal and the axes of the unit cell. Reprint from "Electron diffraction of CS2 nanoclusters embedded in superfluid helium droplets" J. Chem. Phys. 152, 224,306 (2020) with permission from AIP publishing

**Table 8.4** Doping statistics and fitting results of diffraction images from Fig. 15[a]

| Delay time | 1269 $\mu$s | 1289 $\mu$s |
|---|---|---|
| Droplet sizes | 1100 (78%), 5300 (22%) | 1200 (84%), 6500 (16%) |
| Pure droplet ($P_0$, $\beta$) | 0.49, 0.44 | 0.49, 0.47 |
| Monomer ($P_1$, $\alpha_1$) | 0.34, 6.99 | 0.35, 15.0 |
| Dimer ($P_2/P_1$, $\alpha_{2g}/\alpha_1$) | 0.18, 0.18 | 0.13, 0.13 |
| Trimer ($P_3/P_1$, $\alpha_{3c}/\alpha_1$) | 0.10, 0.05 | 0.09, 0.03 |
| Tetramer ($P_4/P_1$, $\alpha_4/\alpha_1$) | 0.04, 0.03 | 0.04, 0.03 |

[a] Adapted from *J. Chem. Phys.* 152, 224,306 (2020)

cell of the above table shows the fitting results of the diffraction profiles based on Eq. 2.4. The values of $\beta$ represent the degree of helium loss due to the pickup process. In both delay times, nearly 50% of helium is lost before the droplets can reach the diffraction region. The absolute values of $\alpha_i$ have no significance since it is affected by the detector's sensitivity, while the ratio of $\alpha_i/\alpha_1$ shows the relative contributions of clusters to monomers.

With the statistical analysis of the fitting result, only the gas phase structure for dimers $\alpha_{2g}$, the closely packed crystal structures for trimers $\alpha_{3c}$ and the tetramer $\alpha_{4c}$ are statistically significant in the diffraction profiles. Furthermore, the contribution of trimers is lower than the calculation results based on the Markovian arrival process. The discrepancy could be related to trimers' rlative instability compared to the other clusters since crystalline $CS_2$ essentially is a composition of dimers [53]. The values on the relative contributions of the other clusters to that of monomers are very consistent between the diffraction fitting results and the calculation based on doping statistics.

The fitting for the diffraction profile of Fig. 8.16 is simpler because we only considered crystalline structures, as suggested in the literature [54]. Both groups of droplets contain at least $2 \times 10^4$ atoms/droplet, and after picking up one $CS_2$, the droplet size changes by less than 3%. Standard Poisson statistics is therefore applicable. At the experiment doping pressure, the most probable number of $CS_2$ monomers for the smaller droplet group is 12 and for the larger group is 40. The continuous red line results from fitting the linear regression, with only $\alpha_c \cdot I_c$ term other than the droplet contributing in Eq. 8.12, with $I_c$ is the theoretical diffraction profile containing numerous monomers based on the crystalline structure of $CS_2$, and $\alpha_c$ is the fitting coefficient. When more than four monomers form a nanocrystal within our detection range, the resulting diffraction profile becomes insensitive to the cluster's size. The calculated diffraction profile results from two unit cells along the a axis, one unit along the b axis, and one unit along the c axis with an overall monomer count of 18. The profile is insensitive to the number of unit cells along each direction, as long as there is more than one monomer (not unit cell) along each direction. However, limiting the structure to two-dimensions spanning only the a and c axis resulted in a substantially lower quality of fitting. It is highly plausible that the nanocrystal is most likely of the same 3D structure as that of a single crystal of $CS_2$.

It is interesting to see that, similar to the iodine case, clusters containing more than two monomeric units deviate from their corresponding gas phase structures and seem to adopt structures of their bulk crystals. The preference for bulk structures in helium droplets has been attributed to the droplet environment's fast cooling effect, even though the helium environment should be more similar to that of vacuum than to that of typical crystal formation. A newly captured dopant can be trapped in a metastable configuration upon entering the droplet, incapable of finding the global minimum.

## 8.6.6 Diffraction of Molecules Only with Light Atoms: Pyrene

All the previous cases mentioned above involve molecular species that contain at least one heavy atom (with an atomic number larger than 20) to help with the contrast between the molecular diffraction and the atomic diffraction from helium. However, biological samples contain mostly carbon atoms, and the contrast issue due to similar diffraction cross sections of carbon [58] and helium has to be addressed. The work with all-light-atom-containing species, pyrene (Py, $C_{16}H_{10}$) demonstrates the feasibility of extracting structural information from the helium background for molecular systems that do not contain any contrasting element [17].

Pyrene's doping condition is characterized by a mass spectrometer via MPI with a 266 nm pulsed laser at a power density of $10^6$ W/cm$^2$ (without focus lens). The miniature TOF located on the wheel shown in Fig. 8.3 resolves the ionized parent, fragment and cluster ions. Figure 8.17 presents the TOF mass spectra of gaseous pyrene, pyrene-doped droplets, and the difference. No fragmentation is observed under this laser power level. The gaseous sample only contains monomeric parent ions, while doped droplets show cluster ions $Py_n^+$ (n = 2–4). Since the degree of fragmentation after ejection from the doped droplet is unknown. the presence of pyrene clusters should be treated as evidence of existence and cannot be used for

**Fig. 8.17** TOF mass spectra of pyrene-related species. Reprinted with permission from "Electron Diffraction of Pyrene Nanoclusters Embedded in Superfluid Helium Droplets" J. Phys. Chem. Lett. 2020, 11, 3, 724–729. Copyright 2020 American Chemical Society

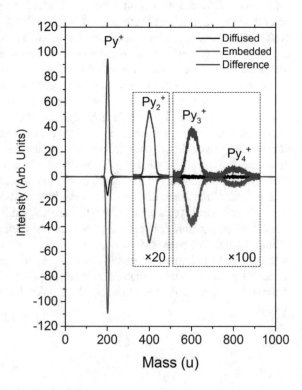

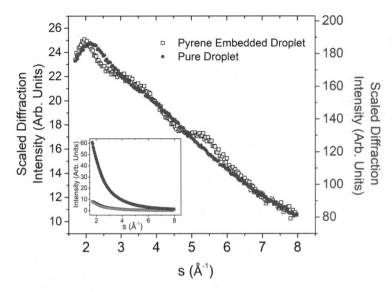

**Fig. 8.18** Radial profiles of diffraction patterns from neat and pyrene doped droplets. The inset shows the relative intensities of the radial profiles. Reprinted with permission from "Electron Diffraction of Pyrene Nanoclusters Embedded in Superfluid Helium Droplets" J. Phys. Chem. Lett. 2020, 11, 3, 724–729. Copyright 2020 American Chemical Society

quantitative analysis of the clusters. The energy of two photons at 266 nm (total energy: 9.35 eV) is more than sufficient to both ionize [59, 60] and dissociate (or dissociate and ionize) a pyrene dimer to produce $Py^+ + Py$ [61, 62].

Figure 8.18 shows the scaled radial profiles of the experimental diffraction patterns, which are obtained with an accumulation of 232,559 shots (12.92 h at a repetition rate of 5 Hz). The insert shows the unscaled radial distribution from the raw experimental data. After a scaling factor of 7.5 for the doped droplet, no difference can be seen between the doped and neat droplets on the linear scale. To contrast the difference between the two results, the radial profiles are scaled by $s^2$, where $s$ is the momentum transfer.

To derive structural information from the diffraction profiles, contributions from the helium background and from all possible pyrene clusters need to be included. Figure 8.19 shows the theoretical diffraction profiles of pyrene clusters based on a few theoretical calculations [61, 63–66] and some representative cuts from crystalline pyrene [67, 68]. The crystal structure and designations of molecular axes are shown in the inset of (a).

Several theoretical calculations on the structures of pyrene clusters have been reported in the literature. The most recent is by Dontot, Spiegelman, and Rapacioli (DSR), reporting a rotation angle of 67° but a slightly nonparallel arrangement between the two molecular planes [61]. However, the authors reported a shallow minimum, with four other structures competitive within 20 meV: they all have parallel molecular planes but are shifted or rotated by different angles, as shown in the inset

**Fig. 8.19** Theoretical diffraction profiles from selected structures of pyrene clusters. The diffraction profile of each structure is color-coded within each panel. Reprinted with permission from "Electron Diffraction of Pyrene Nanoclusters Embedded in Superfluid Helium Droplets" J. Phys. Chem. Lett. 2020, 11, 3, 724–729. Copyright 2020 American Chemical Society

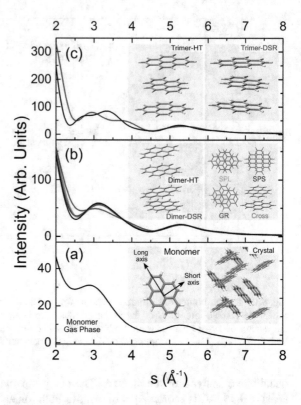

of Fig. 18b. The structure labeled SPL is the global minimum by Gonzales and Lim [66], and it involves a parallel slip between the two monomers along the long axis and an interplanar distance of 3.51 Å, in agreement with the distance in the dimeric unit of crystals. The other three parallel dimers include SPS—slip along the short axis, GR—slip along a C–C bond, and cross—a rotation of 90° [62]. All four structures have very similar diffraction profiles and hence are referred to as the Para dimer in the following discussion. The trimer structure from the DSR calculation is stacked but slightly nonparallel, quite different from a trimeric cut of crystalline pyrene, while the tetramer structure is a 3 + 1 construct, with the fourth pyrene nearly perpendicular to the stacked trimer. An earlier report by Takeuchi (HT structure) contains a parallel dimer [65], a parallel trimer, and a near-cyclic tetramer. Although slightly different from the four parallel dimers, the HT dimer has a very similar diffraction profile to those of the parallel dimers.

If we include all the possible clusters for a global fit, there will be too many independent parameters with limited data points from the diffraction profile. To alleviate model complexity, we chose to fit four sets of structures independently, including the DSR and the HT set, a mix set containing the parallel dimer and the DSR trimer, and a mix set containing the HT dimer and the DSR trimer. The structures of trimers and tetramers derived from crystalline pyrene are eliminated because

**Table 8.5** Constrained least-squares fitting result of embedded $Py_n$ ($n = 1$–3) in superfluid helium droplets from the best model[a]

| Term | Coefficient | Standard error | Coefficient Ratio | Ratio from Doping |
|---|---|---|---|---|
| $\beta$ | 0.06194 | 0.00054 | | |
| $\alpha_0$ | 0.01210 | 0.00170 | | |
| $\alpha_1$ | 0.00798 | 0.00069 | 18 | 99 |
| $\alpha_{2, para}$ | 0.00262 | 0.00072 | 6 | 8 |
| $\alpha_{3, DSR}$ | 0.00044 | 0.00030 | 1 | 1 |

[a] Adapted with permission from "Electron Diffraction of Pyrene Nanoclusters Embedded in Superfluid Helium Droplets" J. Phys. Chem. Lett. 2020, 11, 3, 724–729. Copyright 2020 American Chemical Society

when added to any one of the sets the resulting coefficients for these structures are essentially zero. The Alkaike information criterion and bootstrap resampling method through balanced variable selection is used to confirm the significance of the regression coefficients, and the details are explained in the reference paper. [17] Table 8.5 summarizes the best fitting coefficients and their ratios and uncertainties.

Figure 8.20 compares the experimental data with the fitting results, and the residue is shown in the bottom panel. Similar to Fig. 8.17, both the radial profiles and the

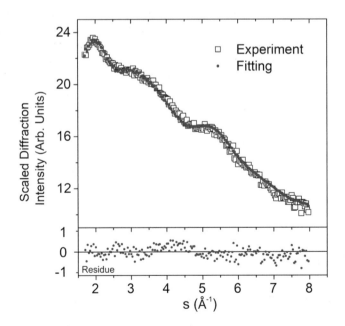

**Fig. 8.20** Comparison of scaled experimental and fitting results. The residue is the difference between the scaled radial profiles. Reprinted with permission from "Electron Diffraction of Pyrene Nanoclusters Embedded in Superfluid Helium Droplets" J. Phys. Chem. Lett. 2020, 11, 3, 724–729. Copyright 2020 American Chemical Society

residues are scaled with $s^2$. The small value of β signifies that more than 90% of the helium atoms could not reach the diffraction region. This level of elimination is on par with our previous work on ferrocene and iodine [16, 19]. The effective high vapor pressure in the doping region destroys most of the small droplets with or without a dopant monomer.

The doping process is modeled using Poisson statistics to further understand the contribution of $Py_n$ in the diffraction pattern. We estimate the number of evaporated helium atoms (2000) upon cooling a pyrene molecule to 0.4 K based on the heat capacity of solid pyrene (229 J/K·mol) [69] and the binding energy of helium (0.6 meV) [70]. After the first collision, 4% of the helium atoms is lost in a droplet of $5 \times 10^4$ atoms/droplet. This size change is negligible and standard Poisson distribution can be used to calculate the probability of doping [46]. The doping pressure is estimated with the empirical formula of supersonic expansion as $1.3 \times 10^{-5}$ Torr. With a doping distance of 7 mm, the probability of doping 0–4 pyrene is 0.71:0.24:0.04:0.004:0.0005 (the ratios of the corresponding $\alpha_i$ values are listed in the last column of Table 8.5). The relative abundance of $Py_2$ and $Py_3$ is in qualitative agreement with that from the fitting. The much larger contribution of monomers from the doping statistics than that from fitting of the diffraction pattern is attributed to contamination in the neat droplet diffraction profile. The doping statistics and the fitting results of the diffraction profile are on par with the abundance of $Py_n^+$ in the mass spectra of Fig. 8.16. We have limited information on the ionization mechanism of $Py_n^+$. However, we speculate that the abundant $Py^+$ is most likely a result of dissociation of $Py_n$ or $Py_n^+$ after desorption from the droplet. The missing contribution from $Py_4$ in the diffraction profile should be a result of low concentration.

In conclusion, the diffraction profile from this experiment of pyrene doped droplets contains mostly contributions from Py and $Py_2$, with indications of a ~10% contribution from $Py_3$. The structure of $Py_2$ contains two parallel pyrene molecules, and that of $Py_3$ appears to be stacked but not completely parallel. This structure of $Py_3$, in our best fitting model, is different from that of the crystalline structure, demonstrating that at least in superfluid helium droplets, the stacking force prevails against the tendency of forming a 3-D closely packed structure. Unlike our previous work, pyrene contains no heavy atoms, and the success of this work offers promise in obtaining molecular parameters from all-light-atom containing species in superfluid helium droplets.

## 8.7  Conclusion

The extreme cooling effect of superfluid helium droplets is attractive in the structural solving of unstable species under normal conditions. Still, the associated background issue in electron diffraction is of practical concern. By taking advantage of the velocity slip of pulsed droplet beams and using Poisson statistics with the Markovian arrival process in sample doping, the background problem for neutral molecule-doped in droplets can be largely minimized. The success of the above variety cases shows the feasibility of this approach even with an all-light-atom system. As for the electron diffraction of protein ions in superfluid helium droplets, the background problem will be easied with manipulating charged particles using electric fields. The space charge limit gives a low charge density of ions will result in a limited diffraction strength. However, the complete separation of undoped droplets and the ability to select small droplets are both beneficial to the ultimate diffraction signal. The next step is to perform diffraction of the ion-doped droplets. The "proof-of-principle" experiment involving the laser-induced alignment of ions embedded in superfluid helium droplets will ensue thereafter.

# References

1. R. Neutze, R. Wouts, d.S.D. van, E. Weckert, J. Hajdu, Nature **406**, 752 (2000)
2. J. Yang, M. Centurion, Struct. Chem. **26**, 1513 (2015)
3. E.T. Karamatskos et al., Nat. Commun. **10**, 3364 (2019)
4. C. Schouder, A.S. Chatterley, F. Calvo, L. Christiansen, H. Stapelfeldt, Struct. Dyn. **6**, 044301 (2019)
5. J. Beckman, W. Kong, V.G. Voinov, W.M. Freund, (Oregon State University, USA, March 8, 2016)
6. R. Neutze, Philos. Trans. R. Soc. Lond. B Biol Sci. **369**, 20130318 (2014)
7. L.C. Johansson, B. Stauch, A. Ishchenko, V. Cherezov, Trends Biochem. Sci. **42**, 749 (2017)
8. O. Gessner, A.F. Vilesov, Annu. Rev. Phys. Chem. **70**, 173 (2019)
9. L.O. Brockway, Rev. Mod. Phys. **8**, 231 (1936)
10. J.D. Pickering, B. Shpperson, L. Christiansen, H. Stapelfeldt, Phys. Rev. A **99**, 043403 (2019)
11. J.D. Pickering, B. Shepperson, L. Christiansen, H. Stapelfeldt, J. Chem. Phys. **149**, 154306 (2018)
12. M. Alghamdi, J. Zhang, A. Oswalt, J.J. Porter, R.A. Mehl, W. Kong, J. Phys. Chem. A **121**, 6671 (2017)
13. L. Chen, J. Zhang, W.M. Freund, W. Kong, J. Chem. Phys. **143**, 044310 (2015)
14. J. Zhang, L. Chen, W.M. Freund, W. Kong, J. Chem. Phys. **143**, 074201 (2015)
15. Y. He, J. Zhang, W. Kong, J. Chem. Phys. **145**, 034307 (2016)
16. Y. He, J. Zhang, L. Lei, W. Kong, Angew. Chem., Int. Ed. **56**, 3541 (2017)
17. L. Lei, Y. Yao, J. Zhang, D. Tronrud, W. Kong, C. Zhang, L. Xue, L. Dontot, M. Rapacioli, J. Phys. Chem. Lett. **11**, 724 (2020)
18. J. Zhang, S. Bradford, W. Kong, J. Chem. Phys. **152**, 224306 (2020)
19. J. Zhang, Y. He, and W. Kong, J. Chem. Phys. **144**, 221101 (2016).
20. W. Kong, L. Pei, J. Zhang, Int. Rev. Phys. Chem. **28**, 33 (2009)
21. J.P. Toennies, A.F. Vilesov, Angew. Chem., Int. Ed. **43**, 2622 (2004)
22. J.P. Toennies, Phys. Scr. **76**, C15 (2007)
23. F. Stienkemeier, K.K. Lehmann, J. Phys. B: At., Mol. Opt. Phys. **39**, R127 (2006)
24. J. Zhang, Y. He, W.M. Freund, W. Kong, J. Phys. Chem. Lett. **5**, 1801 (2014)
25. J. M. Cowley, *Diffraction Physics*, 3rd edn (ELSEVIER, 1995)
26. A. Domenicano, Gas-phase electron diffraction, in *Strength from Weakness: Structural Consequences of Weak Interactions in Molecules, Supermolecules, and Crystals* (Springer, Dordrecht, 2002), Vol. 68, NATO Science Series, Series II: Mathematics, Physics and Chemistry
27. J.C. Williamson, A.H. Zewail, J. Phys. Chem. **98**, 2766 (1994)
28. R. Glauber, V. Schomaker, Phys. Rev. **89**, 667 (1953)
29. E. Prince, *International Tables for Crystallography, Mathematical, Physical and Chemical Tables* (Springer, 2004)

30. G.I. Nemeth, H.L. Selzle, E.W. Schlag, Chem. Phys. Lett. **215**, 151 (1993)
31. Y. He, J. Zhang, W. Kong, J. Chem. Phys. **144**, 084302 (2016)
32. E. Grushka, Anal. Chem. **44**, 1733 (1972)
33. W.K. Pratt, IEEE Trans. Comput. **C-21**, 636 (1972)
34. D.G. Hurst, D.G. Henshaw, Phys. Rev. **100**, 994 (1955)
35. T.R. Sosnick, W.M. Snow, P.E. Sokol, R.N. Silver, Europhys. Lett. **9**, 707 (1989)
36. E.C. Svensson, V.F. Sears, A.D.B. Woods, P. Martel, Phys. Rev. B: Condens. Matter **21**, 3638 (1980)
37. L.F. Gomez, E. Loginov, R. Sliter, A.F. Vilesov, J. Chem. Phys. **135**, 154201 (2011).
38. S. Yang, A.M. Ellis, Rev. Sci. Instrum. **79**, 016106 (2008)
39. C. Park, J. Almlöf, J. Chem. Phys. **95**, 1829 (1991)
40. M. Lewerenz, B. Schilling, J.P. Toennies, J. Chem. Phys. **102**, 8191 (1995)
41. R.K. Bohn, A. Haaland, J. Organomet. Chem. **5**, 470 (1966)
42. R. Mahlanen, J.-P. Jalkanen, T.A. Pakkanen, Chem. Phys. **313**, 271 (2005)
43. M. Capdevila-Cortada, J.J. Novoa, CrystEngComm **17**, 3354 (2015)
44. H. Thomassen, K. Hedberg, J. Mol. Struct. **240**, 151 (1990)
45. M.R. Chowdhury, J.C. Dore, J. Non-Cryst. Solids **46**, 343 (1981)
46. M. Hartmann, R.E. Miller, J.P. Toennies, A.F. Vilesov, Science **272**, 1631 (1996)
47. F. van Bolhuis, P.B. Koster, T. Migchelsen, Acta Crystallogr. A **23**, 90 (1967)
48. J. Zhang, S.D. Bradford, W. Kong, C. Zhang, and L. Xue, J. Chem. Phys. **152**, 224306 (2020)
49. J. Harms, J.P. Toennies, F. Dalfovo, Phys. Rev. B: Condens. Matter Mater. Phys. **58**, 3341 (1998)
50. M. Rezaei, J. Norooz Oliaee, N. Moazzen-Ahmadi, and A. R. W. McKellar, Phys. Chem. Chem. Phys. **13**, 12635 (2011).
51. M. Rezaei, J. Norooz Oliaee, N. Moazzen-Ahmadi, A.R.W. McKellar, Chem. Phys. Lett. **570**, 12 (2013)
52. M. Rezaei, J.N. Oliaee, N. Moazzen-Ahmadi, A.R.W. McKellar, J. Chem. Phys. **134**, 144306 (2011)
53. N.C. Baenziger, W.L. Duax, J. Chem. Phys. **48**, 2974 (1968)
54. G. Singh, R. Verma, S.R. Gadre, J. Phys. Chem. A **119**, 13055 (2015)
55. H. Farrokhpour, Z. Mombeini, M. Namazian, M.L. Coote, J. Comput. Chem. **32**, 797 (2011)
56. G. Singh, S.R. Gadre, Indian J. Chem., Sect. A: Inorg., Bio-inorg., Phys., Theor. Anal. Chem. **53A**, 1019 (2014)
57. M.F. Neuts, J. Appl. Prob. **16**, 764 (1979)
58. A.S. Jablonski, F. Powell, C.J., *NIST electron elastic-scattering cross-section, Database, Version 3.2, SRD 64* (National Institute of Standards and Technology, Gaithersburg, MD, 2010)
59. Z.H. Khan, Acta Phys. Pol., A **82**, 937 (1992)
60. C. Joblin, L. Dontot, G.A. Garcia, F. Spiegelman, M. Rapacioli, L. Nahon, P. Parneix, T. Pino, P. Bréchignac, J. Phys. Chem. Lett. **8**, 3697 (2017)
61. L. Dontot, F. Spiegelman, M. Rapacioli, J. Phys. Chem. A **123**, 9531 (2019)
62. R. Podeszwa, K. Szalewicz, Phys. Chem. Chem. Phys. **10**, 2735 (2008)
63. M. Rapacioli, F. Calvo, F. Spiegelman, C. Joblin, D.J. Wales, J. Phys. Chem. A **109**, 2487 (2005)
64. M. Rapacioli, F. Spiegelman, D. Talbi, T. Mineva, A. Goursot, T. Heine, and G. Seifert, J. Chem. Phys. **130**, 244304 (2009)
65. H. Takeuchi, Comput. Theor. Chem. **1021**, 84 (2013)
66. C. Gonzalez, E.C. Lim, J. Phys. Chem. A **107**, 10105 (2003)
67. J.M. Robertson, J.G. White, J. Chem. Soc. **358** (1947)
68. A.C. Hazell, F.K. Larsen, M.S. Lehmann, Acta Crystallogr., Sect. B **28**, 2977 (1972)
69. N.K. Smith, R.C. Stewart Jr., A.G. Osborn, D.W. Scott, J. Chem. Thermodyn. **12**, 919 (1980)
70. J.P. Toennies, Mol. Phys. **111**, 1879 (2013)

# Chapter 9
# Laser-Induced Alignment of Molecules in Helium Nanodroplets

Jens H. Nielsen, Dominik Pentlehner, Lars Christiansen,
Benjamin Shepperson, Anders A. Søndergaard, Adam S. Chatterley,
James D. Pickering, Constant A. Schouder, Alberto Viñas Muñoz,
Lorenz Kranabetter, and Henrik Stapelfeldt

**Abstract** Moderately intense, nonresonant laser pulses can be used to accurately control how gas phase molecules are oriented in space. This topic, driven by intense experimental and theoretical efforts, has been ever growing and developed for more than 20 years, and laser-induced alignment methods are used routinely in a number of applications in physics and chemistry. Starting in 2013, we have demonstrated that laser-induced alignment also applies to molecules dissolved in helium nanodroplets. Here we present an overview of this new work discussing alignment in both the nonadiabatic (short-pulse) and adiabatic (long-pulse) limit. We show how femtosecond or picosecond pulses can set molecules into coherent rotation that lasts for a long time and reflects the rotational structure of the helium-solvated molecules, provided the pulses are weak or, conversely, results in desolvation of the molecules when the pulses are strong. For long pulses we show that the 0.4 K temperature of the droplets, shared

J. H. Nielsen · A. A. Søndergaard
Department of Physics and Astronomy, University of Aarhus,
Ny Munkegade 120, 8000 Aarhus C, Denmark

D. Pentlehner · L. Christiansen · B. Shepperson · A. S. Chatterley · J. D. Pickering ·
C. A. Schouder · A. V. Muñoz · L. Kranabetter · H. Stapelfeldt (✉)
Department of Chemistry, Aarhus University, Langelandsgade 140, 8000 Aarhus C, Denmark
e-mail: henriks@chem.au.dk

D. Pentlehner
e-mail: Dominik.Pentlehner@th-rosenheim.de

A. S. Chatterley
e-mail: aschatterley@chem.au.dk

J. D. Pickering
e-mail: j.pickering@leicester.ac.uk

C. A. Schouder
e-mail: constantschouder@chem.au.dk

A. V. Muñoz
e-mail: avinas@chem.au.dk

L. Kranabetter
e-mail: kranabetter@chem.au.dk

© The Author(s) 2022
A. Slenczka and J. P. Toennies (eds.), *Molecules in Superfluid Helium Nanodroplets*,
Topics in Applied Physics 145, https://doi.org/10.1007/978-3-030-94896-2_9

with the molecules or molecular complexes, leads to exceptionally high degrees of alignment. Upon rapid truncation of the laser pulse, the strong alignment can be made effectively field-free, lasting for about 10 ps thanks to slowing of molecular rotation by the helium environment. Finally, we discuss how the combination of strongly aligned molecular dimers and laser-induced Coulomb explosion imaging enables determination of the structure of the dimers. As a background and reference point, the first third of the article introduces some of the central concepts of laser-induced alignment for isolated molecules, illustrated by numerical and experimental examples.

## 9.1 Introduction

Firstly, it is necessary to define *what* we mean by alignment of molecules, the central topic of this article. Alignment refers to molecular axes being confined with respect to axes fixed in the laboratory system. The simplest case is 1-dimensional (1D) alignment, where a single molecular axis is being confined. This is illustrated on Fig. 9.1 (b) for the linear carbonyl sulfide (OCS) molecules aligned along a space fixed (vertical) axis, in contrast to the case of randomly oriented molecules depicted in Fig. 9.1 (a). The concept of 1D alignment can also apply to asymmetric top molecules, shown in Fig. 9.1 (d). As explained below, when a laser pulse is employed to induce alignment, the aligned axis is the most polarizable axis, which coincides with the C-I axis of the 3,5-difluoro-iodobenzene molecules, chosen as the illustrative example here. The molecules are, however, free to rotate around the aligned axis. A more complete control of the spatial orientation of the molecule requires that this free rotation is arrested. This case, illustrated in Fig. 9.1 (e), is referred to as 3-dimensional (3D) alignment where the three principal polarizability axes are fixed with respect to a space-fixed, Cartesian coordinate system, often denoted as the (X,Y,Z)-coordinate system. Finally, if the molecules are polar, it is often also relevant to control the direction of the permanent dipole moment, a case termed orientation. Figure 9.1 (f) illustrates the situation where the 3,5-difluoro-iodobenzene are both 3D aligned and oriented, which is jointly called 3D orientation. The concept of orientation applies equally to linear (or symmetric top) molecules. In the case of the OCS molecules this would be the situation where the S-end of the molecules all point in the same direction.

Secondly, we must ask *why* it is interesting and relevant to align molecules. One major reason is that samples of aligned and/or oriented molecules make it possible to study or exploit the ubiquitous orientational dependence of molecules' interaction with other molecules, atoms or polarized light. In fact, the orientational dependence of bimolecular reactions was a primary motivating factor for why researchers started to develop techniques to orient and align molecules over 60 years ago [1]. Molecules can also interact with pulses of light and over the past 10–20 years the study and exploitation of how interactions between laser pulses, in particular short and intense laser pulses, and molecules depend on the orientation of the molecule

**(a) Random orientation**    **(c) Random orientation**    **(e) 3D alignment**

**(b) 1D alignment**    **(d) 1D alignment**    **(f) 3D orientation**

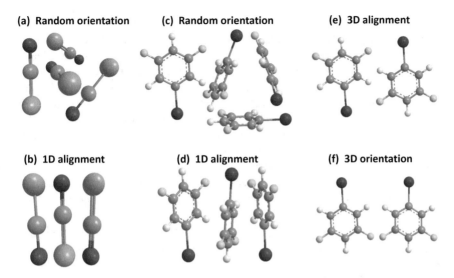

**Fig. 9.1** Illustration of 1D alignment, 3D alignment and 3D orientation using the carbonyl sulfide and 3,5-difluoro-iodobenzene molecules as examples

with respect to the polarization state of the laser pulses has been a topic of intense investigation [2]. A second, major advantage of aligned molecules is the fact that they make it possible to perform molecular frame (MF) measurements, which can significantly increase the information content from experimental observables because the blurring they normally suffer from averaging over randomly oriented molecules is strongly reduced. An illustrative example is MF photoelectron angular distributions from aligned molecules where high-information structures appear that are completely absent in experiments on randomly oriented molecules [3, 4].

Finally, the question is *how* to align molecules. The main techniques developed early on were rotational state-selection of molecules by means of hexapolar electric fields [1, 5], 'brute-force' orientation by a strong static electric field [6–9], collisional alignment in a cold molecular beam [10–12], and photoselection whereby polarized light creates alignment in a vibrationally or electronically excited state [13, 14].

The method discussed here employs nonresonant, moderately intense laser pulses to induce alignment. The fundamental interaction responsible for alignment is the polarizability interaction between the molecule and the electric field of the laser pulse. The laser pulse induces an electric dipole moment in the molecule, which in turn interacts with the electric field of the laser pulse. Almost all molecules have an anisotropic polarizability tensor and, consequently, for these molecules the polarizability interaction depends on their spatial orientation with respect to the polarization of the electric field of the laser pulse. The laser pulse forces the molecules to rotate towards an orientation where the polarizability interaction is optimized and this is what leads to alignment of the molecules. In the simplest case of a linearly polarized laser pulse, the potential energy of the polarizability interaction has a minimum when the most polarizable axis is parallel to the polarization axis. This means that a linearly polarized laser pulse has the potential to induce 1D alignment of molecules. Most of this article concerns 1D alignment of molecules induced by linearly polarized laser pulses. It is, however, also possible to induce 3D alignment using instead an elliptically polarized laser pulse. In this case, the polarizability interaction is optimized when the most polarizable molecular axis is parallel to the major polarization axis and, simultaneously, the second most polarizable molecular axis is parallel to the minor polarization axis. Note that nonresonant means that the laser pulse causes no linear absorption, i.e. the photon energy is not resonant with any transitions in the molecules studied. Moderately intense means that the laser pulse is strong enough that the polarizability interaction induces pronounced alignment, yet weak enough that it does not cause electronic excitation or ionization due to multiphoton absorption.

Laser-induced alignment emerged in the last half of the 1990s [15–21] and beginning of the 2000s [22–30] at the interface between stereochemistry, spectroscopy, strong laser field physics, and wave packet dynamics. Since then the field has undergone a continuous expansion in scope and applications, however with essentially all studies concentrating on gas phase molecules. In 2013 we showed that it is possible to extend laser-induced alignment to molecules embedded inside liquid helium nanodroplets. The purpose of the current article is to provide an overview of laser-induced alignment of molecules embedded in helium nanodroplets, based on work in the period from 2013 to now. The first part of the paper, Sect. 9.2 introduces some of the basic concepts of laser-induced alignment developed for gas phase molecules, exemplified by a few experimental results and calculations. This provides some background and useful reference points for the discussion of molecules in He droplets given in Sect. 9.3. Many more details about laser-induced alignment of gas phase molecules can be found in the existing review articles on the subject [31–35].

## 9.2  Alignment of Isolated Molecules

### 9.2.1  Laser-Induced Alignment: Basics

We start by discussing the rotational dynamics of a linear molecule (in a $\Sigma$ state) induced by a linearly polarized, nonresonant, laser pulse. Theoretically, these dynamics are described by the time-dependent rotational Schrödinger equation with $U(\theta)$ denoting the polarizability interaction:

$$i\hbar \frac{\partial \Psi_{rot}(t)}{\partial t} = \left( B\hat{J}^2 + U(\theta) \right) \Psi_{rot}(t) \tag{9.1}$$

$$= \left( B\hat{J}^2 - \frac{E_0(t)^2}{4} \left( (\alpha_\parallel - \alpha_\perp) \cos^2(\theta) + \alpha_\perp \right) \right) \Psi_{rot}(t) \tag{9.2}$$

$E_0(t)$: the amplitude (envelope) of the electric field of the laser pulse, $\hat{J}^2$: the squared rotational angular momentum operator, $\Psi_{rot}$: the rotational wave function, $B$: the rotational constant, $\alpha_\parallel$ and $\alpha_\perp$: the polarizability parallel and perpendicular to the most polarizable axis, i.e. the molecular axis of a linear molecule, $\theta$: the (polar) angle between the molecular axis and the polarization axis of the laser pulse. We assume that before the laser pulse is turned on, the molecule is populated in a single rotational eigenstate, $|J_k M_k\rangle$, of the field-free Hamiltonian ($B\hat{J}^2$) i.e. $\Psi_{rot}(-\infty) = |J_k M_k\rangle$. Here $J_k$ is the initial quantum number of the rotational angular momentum, $M_k$ its projection on a space-fixed axis parallel to the laser polarization and $|J_k M_k\rangle$ is the rotational eigenfunction given by a spherical harmonic, $Y_{J_k}^{M_k}$. The solution to Eq. 9.2 can be expressed as:

$$\Psi_{rot}^{(k)}(t) = \sum_J d_{JM_k}(t)|JM_k\rangle \tag{9.3}$$

where $d_{JM_k}(t)$ are the expansion coefficients and the index $k$ on $\Psi_{rot}^{(k)}$ indicates the initial state. In other words, the laser pulse creates a superposition of different angular momentum states, $|JM_k\rangle$, but leaves the projection on the laser polarization unchanged. Such a superposition, which is termed a rotational wave packet, can lead to angular confinement, i.e. alignment of the molecule along the polarization axis, provided the phase relationship between the different components$|JM_k\rangle$ in Eq. 9.4 is favorable.

The rotational dynamics depend on the turn-on and turn-off time of the laser pulse. The two most common regimes of alignment are termed adiabatic and nonadiabatic. Here the alignment pulse duration, $\tau_{\text{align}}$, and thus both the turn-on and turn-off time, is either much shorter or much longer than the intrinsic rotational period, $\tau_{\text{rot}}$, of the molecules, respectively.[1] Here $\tau_{\text{rot}}$ is defined as $1/2B$ with $B$ given in units of Hz.

In the adiabatic regime, the initial rotational eigenstate $|J_k M_k\rangle$ evolves adiabatically into the corresponding eigenstate of the complete Hamiltonian (given in Eq. 9.2). Such a state, expressed by Eq. 9.3, is called a pendular state [15, 37] and can lead to strong alignment. Upon turn-off, the pendular state evolves back to the initial rotational quantum state. In the adiabatic regime a sample of molecules is, therefore, only aligned during the laser pulse. By contrast, in the nonadiabatic regime, the laser pulse leaves the molecule in a coherent superposition of field-free rotational eigenstates. This superposition can be expressed as:

$$\Psi_{rot}^{(k)}(t) = \sum_J d_{JM_k} \exp\left(\frac{-iBJ(J+1)t}{\hbar}\right)|JM_k\rangle \tag{9.4}$$

where $d_{JM_k}$ in general are complex numbers. An important consequence of Eq. 9.4 is that in the nonadiabatic regime, alignment can occur a long time after the pulse and, in general, the alignment will continue to evolve after the laser pulse is turned off.

---

[1] Nonadiabatic alignment dynamics can also be induced by turning on the laser pulse slowly and then, typically at the peak, turning it off rapidly [29, 36].

To characterize the degree of alignment, the following measure is normally used:

$$\langle \cos^2 \theta \rangle_k = \langle \Psi_{rot}^{(k)} | \cos^2 \theta | \Psi_{rot}^{(k)} \rangle \tag{9.5}$$

In practice, measurements involve a sample of molecules initially populated in different rotational states, typically assumed to follow a Boltzmann distribution. To account for this, $\langle \cos^2 \theta \rangle_k$ must be averaged over all initial states to give $\langle \cos^2 \theta \rangle$:

$$\langle \cos^2 \theta \rangle = \sum_k f(k) \langle \cos^2 \theta \rangle_k \tag{9.6}$$

where $f(k)$ is the normalized population of state $k$.

### 9.2.2 Nonadiabatic and Adiabatic Alignment: OCS Example

In this and the next section, we illustrate numerically the mechanism and alignment dynamics in the adiabatic and nonadiabatic regimes. We use the OCS molecule as an example and solve Eq. 9.2 for a laser pulse with a duration ($\tau_{align}$) that is either much smaller or larger than $\tau_{rot}$. For OCS, $B = 6.0858$ GHz, so $\tau_{rot} = 82.2$ ps. For the short-pulse or nonadiabatic regime, we use $\tau_{align} = 300$ fs and for the long-pulse or adiabatic regime, we use $\tau_{align} = 600$ ps.

Figure 9.2 shows calculated values of $\langle \cos^2 \theta \rangle$ for OCS molecules as a function of time, $t$, after the peak of the linearly polarized, 300 fs laser pulse with a peak intensity of $6.5 \times 10^{12}$ W/cm$^2$. The rotational temperature of the molecules is set to 1 K to match what can be achieved experimentally in cold molecular beams [38]. The calculations take into account that experimentally the molecules are exposed to a distribution of alignment pulse intensities due to the finite beam waist of the probe laser beam used to measure the time dependent degree of alignment.[2] In practice, this is implemented by averaging the calculations over an intensity distribution determined by the actual beam waists of the laser beams. Here we used $\omega_0 = 35$ and 25 μm that are typical values for the alignment and probe beam, respectively.

Before the arrival of the pulse, $\langle \cos^2 \theta \rangle = 1/3$, the value for a sample of randomly aligned molecules. During and shortly after the pulse, $\langle \cos^2 \theta \rangle$ rises and reaches a first maximum, 0.67, at $t = 0.95$ ps. At this time, some of the angular momentum states in the rotational wave packet (Eq. 9.4) are in phase, which produces alignment. The J-dependence of the complex exponential functions in Eq. 9.4 causes the angular momentum states to fall out of phase at later times, a behavior referred to as wave packet dispersion. On the alignment trace in Fig. 9.2, this manifests as a rapid drop of $\langle \cos^2 \theta \rangle$ for $t > 0.95$ ps. Due to the regular rotational energy structure

---

[2] The probe pulse, sent a time $t$ after the alignment pulse, induces Coulomb explosion of the molecules to probe their spatial orientation. Details are given in Sect. 9.2.3.

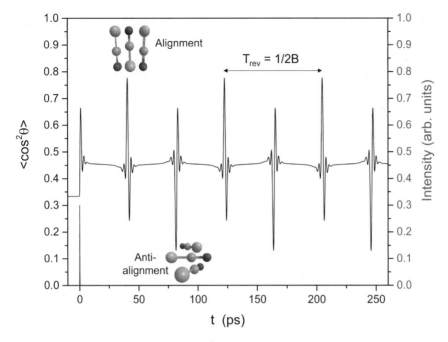

**Fig. 9.2** Nonadiabatic alignment dynamics (black curve) calculated for a sample of OCS molecules with a rotational temperature of 1 K. The intensity profile of the alignment pulse is shown by the red curve. $\tau_{align} = 300$ fs and $I_{align} = 6.5 \times 10^{12}$ W/cm$^2$

of linear molecules, modelled as rigid rotors, i.e. $E_{rot}(J) = BJ(J+1)$, the angular momentum states do come back into phase again at longer times.

In particular, at times separated by integer multiples of $1/2B$, each of the complex exponential functions in Eq. 9.4 accumulates a phase of $2\pi N$, where $N$ is an integer. This means that the wave packets given by Eq. 9.4 are periodic with a period of $1/2B$. The same will hold for any expectation value of an operator, like $\langle \cos^2 \theta \rangle$, i.e. $\langle \cos^2 \theta \rangle$ is also $1/2B$-periodic. Therefore, the initial alignment maximum is repeated every $1/2B$, and this is called the revival period, $T_{rev}$. The $1/2B$ periodicity, which is 82.2 ps for OCS, manifests itself in the alignment trace in Fig. 9.2, where the transients centered at 82.4 ps, 164.6 ps and 246.7 ps are the 1st, 2nd and 3rd full revivals. Here $\langle \cos^2 \theta \rangle$ first dips to a low value corresponding to anti-alignment, where the molecules are confined to the plane perpendicular to the polarization axis, and then rises steeply to a high value where the molecules are aligned along the polarization axis. The prominent transients between the full revivals are termed half revivals. Here, the complex exponential functions are shifted by $\pi$ compared to their values at the full revivals. This reverses the order of alignment and anti-alignment such that the molecules first align and then anti-align. Note that the peak value of $\langle \cos^2 \theta \rangle$ at the half revival, 0.80, exceeds that at the full revivals, 0.67. Higher-order fractional revivals, such as quarter revivals, appear if the OCS molecules are populated initially in a single rotational state rather than in a Boltzmann distribution of states. They also appear for molecules where the population of odd and even $J$-states are influenced by nuclear spin statistics, for instance for $N_2$ [30], $CO_2$ [39] and $I_2$ [23]. The appearance of quarter revivals is illustrated by the experimental results on $I_2$ molecules in Fig. 9.9.

In a classical picture, the alignment pulse exerts a torque on the molecule and thereby sets it into rotation towards the laser polarization. In the limit where the rotation of the molecule during the laser pulse is negligible, the angular velocity, $\omega$, gained by the molecule-laser interaction is given by [40]:

$$\omega = \frac{1}{2} \frac{\Delta\alpha F_{align} \sin(2\theta_0)}{I \varepsilon_0 c}, \tag{9.7}$$

where $\Delta\alpha = \alpha_\parallel - \alpha_\perp$ is the polarizability anisotropy, $I$ is the moment of inertia, $F_{align}$ the fluence of the laser pulse, and $\theta_0$ the initial angle between the molecule and the polarization of the laser pulse. Molecules at different initial angles acquire different angular velocities and therefore do not line up with the polarization vector at exactly the same time. In particular molecules with $\theta_0$ close to $90°$ arrive later than those with $\theta_0$ close to $0°$. This explains why the degree of alignment at the prompt maximum is not necessarily particularly high. In this classical model, often termed the delta-kick model, the molecules will continue to rotate, which explains why the degree of alignment decreases after the prompt peak, but they will never reach the same degree of alignment at later times due to the continuum of classically available angular frequencies. Thus, the presence of revivals is a phenomenon that must be described by the quantum model.

The 300 fs alignment pulse converts each molecule from residing in a single rotational eigenstate into a superposition of eigenstates as expressed by Eq. 9.4. The underlying mechanism is multiple Raman-type transitions between rotational states with $\Delta J = 0, \pm 2$ [32, 41], whereas the vibrational and electronic states are unchanged. Information about which rotational states are populated after the alignment pulse is obtained by Fourier transformation of $\langle \cos^2 \theta \rangle(t)$. The resulting power spectrum, displayed in Fig. 9.3 (a), shows a series of regularly spaced peaks. The spectral peaks reflect the frequencies of the nonzero matrix elements $\langle JM| \cos^2 \theta |J'M \rangle$, i.e. the coherence (coupling) between state $|JM\rangle$ and $|J'M\rangle$. The matrix element is only nonzero if $J - J' = \pm 2, 0$ and thus the frequencies are given by:

$$\nu_{(J-J+2)} = B(4J + 6), \tag{9.8}$$

assuming a rigid rotor model of the molecules.

All the peaks in Fig. 9.3 (a) have been assigned and labelled $(J - J + 2)$. It is seen that states up to $J = 20$ are populated. For comparison, the distribution of J-states prior to the pulse, given by a Boltzmann distribution with a temperature of 1 K, is displayed in Fig. 9.3 (b). Here, essentially all molecules reside in states with $J \leq 4$, showing that the polarizability interaction shifts the molecules to much higher-lying rotational states. Figure 9.3 (a) shows that the weight of $\nu_{(J-J+2)}$ is centered around 300 GHz. This defines the oscillation period of $\langle \cos^2 \theta \rangle$ during the revivals. In fact, the

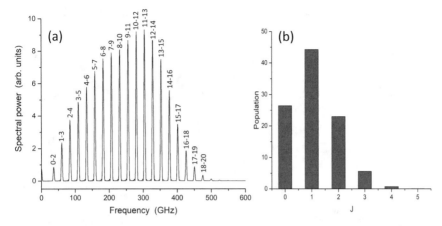

**Fig. 9.3  a** Power spectrum of $\langle \cos^2 \theta \rangle (t)$ shown in Fig. 9.2. **b** Boltzmann distribution of rotational states calculated for OCS molecules at a temperature of 1 K

time difference between the minimum and the maximum of $\langle \cos^2 \theta \rangle$ at both the half and the full revivals is $\sim$1.6 ps—giving an oscillation period of 3.2 ps corresponding to 310 GHz.

Next, we discuss the alignment dynamics in the adiabatic limit. Figure 9.4 (a) shows the time-dependence of $\langle \cos^2 \theta \rangle$ for OCS molecules when a linearly polarized, 600 ps long laser pulse with a peak intensity of $1 \times 10^{12}$ W/cm$^2$ is used to induce alignment. The center of the pulse defines $t = 0$. Again, the rotational temperature of the molecules is set to 1 K and the calculations are averaged over the intensities in the probed focal volume of the alignment pulse. The degree of alignment now closely follows the intensity profile of the laser pulse. It is seen that $\langle \cos^2 \theta \rangle$ rises concurrently with the laser pulse, reaches the maximum value of 0.84 at the peak of the pulse, and returns to the isotropic value of 0.33 when the pulse turns off. This is the characteristic adiabatic behavior of laser-induced alignment and it clearly differs from that of nonadiabatic alignment.

The picture of the adiabatic alignment dynamics is that the laser pulse is turned on sufficiently slowly that each rotational eigenstate of the field-free molecule is transferred into the corresponding eigenstate in the presence of the laser field [15, 19]. These states are called pendular states since they correspond to a molecule librating in the angular potential well created by $U(\theta)$, see Eq. 9.4, a motion similar to the oscillation of a pendulum. If the laser pulse is also turned off sufficiently slowly, the pendular states return to the field-free states. In other words, in the true adiabatic limit, the distribution of rotational states after the pulse is the same as before the pulse as if the molecules never knew that they were aligned for a period of time. This situation stands in stark contrast to that of nonadiabatic alignment where the molecules are left in coherent superpositions of field-free eigenstates after the pulse, which leads to a distinct post-pulse time dependence of $\langle \cos^2 \theta \rangle$ as discussed above.

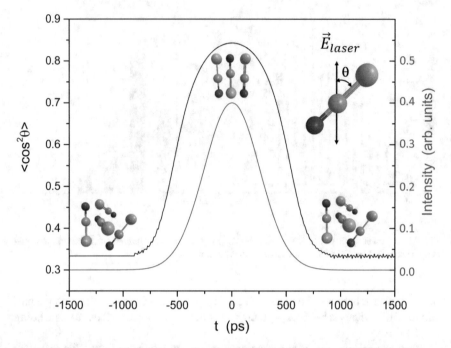

**Fig. 9.4** Adiabatic alignment dynamics (black curve) calculated for a sample of OCS molecules with a rotational temperature of 1 K. The intensity profile of the alignment pulse is shown by the red curve. $\tau_{align} = 600$ ps and $I_{align} = 1.0 \times 10^{12}$ W/cm$^2$. The inset illustrates the polar angle $\theta$ between the inter-atomic axis of a molecule and the laser polarization, represented by the double-headed arrow

In the adiabatic limit, the degree of alignment for a given molecule is only determined by the intensity of the laser pulse and the rotational temperature [36]. The intensity dependence is illustrated experimentally in Sect. 9.2.4. Here we note that if the intensity of the alignment pulse is increased too much, it will start ionizing and/or dissociating the molecules. Consequently, the intensity must be kept sufficiently low to avoid these unwanted processes. In practice, the limit is in the range $10^{12} - 10^{13}$ W/cm$^2$ depending on the molecule and the pulse duration.

One significant advantage of adiabatic alignment is that for the many different molecules that can routinely be brought down to rotational temperatures of a few K in cold, supersonic beams, high degrees of alignment can be achieved, lasting for as long as the pulse is turned on. In the present example, $\langle \cos^2 \theta \rangle > 0.8$ for 250 ps. Thus, adiabatic alignment appears useful for applications such as following reaction dynamics in the molecular frame, where observation times of several tens of picoseconds can be required. The advantage of the long-lasting character of adiabatic alignment was demonstrated in real-time measurements of torsional motion of axially chiral molecules [42–44]. The feasibility of this approach requires that the alignment pulse does not perturb the reaction of the molecule or any other process studied. In Sect. 9.3.6 we discuss how rapid truncation of the laser pulse at its peak can convert the high degree of adiabatic alignment into field-free alignment, lasting tens of ps, for molecules embedded in He nanodroplets. Another advantage of adiabatic alignment is that it is straightforward to extend 1D alignment of an asymmetric top molecule to 3D alignment, see Sect. 9.3.6.

In the nonadiabatic limit, the short duration of the time windows in which strong alignment exists is typically not long enough to enable chemical reaction dynamics to be followed.[3] For instance, in the OCS example with a 300 fs alignment pulse, the strongest alignment, achieved during the half revival, amounts to $\langle \cos^2 \theta \rangle$ staying above 0.8 for only 0.2 ps. The alignment in this time interval occurs, however, long after the alignment pulse is turned off, i.e. under completely field-free conditions. This has proven very useful in many studies of e.g. high-order harmonic generation by intense fs laser pulses [2, 46–50]. For such applications, a high degree of alignment lasting for a few hundred fs is more than enough time to do an experiment.

### 9.2.3 Experimental Setup

We now turn to discussing the practicalities of actually performing alignment experiments on either isolated molecules or molecules in helium droplets. A schematic diagram of a typical experimental setup, depicting the key components, is shown in Fig. 9.5. There are four main parts. (1) A continuous beam of He droplets doped with molecules, propagating along the Z-axis. (2) A pulsed molecular beam of isolated molecules propagating along the X-axis. (3) Pulsed laser beams propagating along the Y-axis. (4) A velocity map imaging (VMI) spectrometer with a flight-axis parallel to the X-axis.[4]

The helium nanodroplets are produced by continuously expanding high purity (99.9999%) helium gas through a cryogenically cooled 5-$\mu$m-diameter aperture into vacuum. The stagnation pressure is typically between 20 and 40 bar while the stagnation temperature is varied between 10 and 18 K. This makes it possible to vary the mean size of the droplets between ∼2000 and 12000 helium atoms [52]. The He droplets pass through a skimmer and enter a pickup cell containing a gas of molecules (or atoms). In most of the experiments described here, the partial pressure of the gas is adjusted to optimize for single molecule doping of each droplet but the partial pressure can also be increased to enable pickup of two molecules by each droplet and subsequent formation of a dimer,[5] see Sects. 9.3.5 and 9.3.7. Hereafter the doped droplets pass through another skimmer (not shown on Fig. 9.5) and enter the 'target chamber' where they interact with the laser pulses. This takes place in the middle

---

[3] One exception is found here [45].

[4] Note that the (X,Y,Z) coordinate systems used to label the experimental setup is rotated for some of the results in this article.

[5] Although not used in the studies presented here, we note that the He droplet instrument is equipped with two in-line pickup cells. This makes it possible to form e.g. a dimer composed of two molecules or larger heterogeneous complexes [53, 54].

(a)

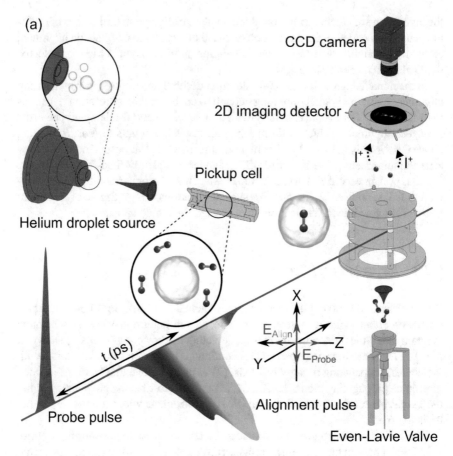

**Fig. 9.5** Schematic diagram of the experimental setup. Depicted are the helium droplet source, the pickup cell, the 2D imaging detector, the CCD camera, and the Even-Lavie valve used in the studies on isolated molecules. The direction of the sketched pulse forms indicates the polarisation direction of the laser pulses used to align (along the z-axis) and probe (along the x-axis) the molecules. Adapted from [51] with permission from American Physical Society (APS). Copyright (2018) by APS

of a velocity-map imaging (VMI) spectrometer. For the alignment experiments, two collinear laser beams, crossing the He droplet beam at 90°, are used.

Both laser beams originate from an amplified Ti-Sapphire femtosecond laser system and thus their central wavelength is 800 nm. The pulses in the first laser beam are used to induce alignment of the molecules. Essentially all molecules studied have negligible absorption at 800 nm, which means that the alignment pulses fulfill the requirement of being nonresonant. For the nonadiabatic alignment measurements, the duration of the alignment pulses is in the range from 300 fs to 15 ps. Such durations are obtained by sending a part of the compressed output from the amplified laser system through a pulse stretcher composed of two transmission gratings in a double-pass geometry [55]. For the adiabatic alignment measurements, a part of the uncompressed output of the laser system is used. The pulses in this beam have a duration of 160 ps.[6]

The pulses in the second laser beam are taken directly from the compressed output of the amplified laser system and their duration is ~40 fs. These probe pulses are used to measure the spatial orientation of the molecules in the following way. The intensity of these pulses, typically a few times $10^{14}$ W/cm$^2$, is high enough to cause rapid multiple ionization of the irradiated molecules. Most of the resulting multiply charged molecular cations break apart into positively charged fragment ions due to internal electrostatic repulsion. This process is termed Coulomb explosion [57]. In many cases, the fragment ions recoil along a molecular axis and thus detection of the emission direction of the fragment ions provides direct information about the spatial orientation of the molecules at the instant that the probe pulse arrives. In the case of 1D alignment, detection of the angular distribution of a single ion species, like I$^+$ ions from an I$_2$ molecule, is sufficient to fully characterize the degree of alignment. For the case of 3D alignment, two ion species are typically needed to characterize how the molecules are aligned. We note there are other ways to measure alignment of molecules including the optical Kerr effect [26, 39], photodissociation [19], four-wave mixing [58] and photoionization yields [59].

The ion detection is implemented by a VMI spectrometer [60, 61]. This means that ions created when the probe pulse interacts with the molecules are projected by a weak electrostatic field onto a 2D microchannel plate (MCP) backed by a phosphor screen. The ion images are recorded by a CCD or a CMOS camera that monitors the phosphor screen and captures the fluorescence created by electrons from the MCP impinging on the phosphor screen. On-line software analysis determines and saves the coordinates of each individual particle hit. The MCP is gated in time by a high voltage switch so that only ions with a certain mass-to-charge ratio are detected at one time.

To achieve the intensities of the alignment pulse and the probe pulse needed for the experiments, both laser beams are focused by a lens with a 30 cm focal length. At the crossing point with the beam of He droplets, the Gaussian beam waist, $\omega_0$ of the alignment (probe) beam is typically 35 μm (25 μm). The laser beams are carefully spatially overlapped, so that the smaller beam waist of the probe beam ensures that only molecules that have been exposed to the alignment beam are being ionized and then detected.

The experimental setup is also equipped with a molecular beam of isolated molecules (or dimers). It is formed by expanding a few mbar of a molecular gas in about 60–80 bar of He gas into vacuum through a pulsed Even-Lavie valve [62]. The molecular beam is skimmed and sent to the target chamber where it intersects the focused laser beams at the same spatial position as the droplet beam. The advantage of this setup is that it becomes possible to conduct experiments under the exact same laser conditions for isolated molecules and for molecules embedded in He nanodroplets. More details on the experimental setup can be found in [63].

---

[6] If needed, an external grating stretcher can increase the duration [56].

## 9.2.4 Experimental Observations of Adiabatic Alignment

In this section, we present experimental results on alignment of gas phase molecules, using $I_2$ as an example. We start in the (quasi-) long-pulse limit. As mentioned in Sect. 9.2.3, 2D ion images constitute the basic experimental observables. From such images, we can extract information about the spatial orientation of the molecules. For $I_2$ molecules, $I^+$ ions are detected. Figure 9.6 (a1) shows an $I^+$ images recorded with the probe pulse only, polarized perpendicular to the detector plane. These ions are produced when the probe pulse ionizes the $I_2$ molecules, and the resulting singly or multiply charged molecular ions fragment into an $I^+$-I, $I^+$-$I^+$ or $I^+$-$I^{2+}$ pair [64]. The key observation for our purpose is that the image is circularly symmetric. This is to be expected since the $I^+$ angular distribution must be symmetric around the polarization axis because the (multiple) ionization rate of the $I_2$ molecules, and thus the emission direction of $I^+$ fragment ions, depends only on the polar angle between the probe pulse polarization and the $I_2$ internuclear axis. The circularly symmetric

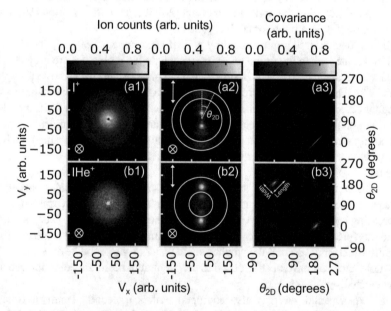

**Fig. 9.6** (a1)–(a2) [(b1)–(b2)]: $I^+$ [$IHe^+$] ion images from Coulomb exploding isolated [He-solvated] $I_2$ molecules with the probe pulse only (1) and with the alignment pulse included (2). The images represent the detection of the ion velocities, $v_x$, $v_y$ in the detector plane. (a3) [(b3)]: Angular covariance map of the image in (a2) [(b2)]. The polarization directions of the alignment pulse (vertical: ↕) and probe pulse (perpendicular to the detector plane: ⊗) are shown on the ion images. White circles indicate the radial ranges used for calculating $\langle \cos^2 \theta_{2D} \rangle$ and the angular covariance maps. For the images in the middle column the probe pulse was sent at $t = 0$ and $I_{align}$ = 0.83 TW/cm$^2$. Adapted from [63] with the permission of American Institute of Physics (AIP) Publishing. Copyright 2017 by AIP

image allows us to conclude that the molecules are randomly oriented in the detector plane and thus serves as a reference for the next measurements where alignment in the detector plane is induced.

Figure 9.6 (a2) shows the $I^+$ images when an alignment pulse is included. The pulse is 160 ps long and is linearly polarized along the Y-axis. The probe pulse is synchronized to the peak of the alignment pulse. Now the $I^+$ ions are tightly confined along the direction of the alignment pulse polarization. In line with many previous works, we interpret this as evidence of 1D alignment by the alignment pulse. To quantify the degree of alignment, we determine $\langle \cos^2 \theta_{2D} \rangle$ from the average of all $I^+$ ions detected between the two white circles. Here $\theta_{2D}$ is the angle between the polarization axis of the alignment pulse and the projection of the recoil vector of an $I^+$ ion on the detector plane [see Fig. 9.6 (a2)]. The reason that we determine $\langle \cos^2 \theta_{2D} \rangle$ and not $\langle \cos^2 \theta \rangle$, is that the detector only records the components of the velocity vector in the detector plane. However, it can be shown that $\langle \cos^2 \theta_{2D} \rangle$ contains the same information as $\langle \cos^2 \theta \rangle$ [65].

The ions between the two circles are chosen because they originate from double ionization of the $I_2$ molecules and subsequent Coulomb explosion into a pair of $I^+$ ions: $I_2^{2+} \rightarrow I^+ + I^+$. In such a Coulomb fragmentation process, each of the $I^+$ ions should recoil back-to-back, precisely along the $I_2$ internuclear axis, meaning that their emission direction is a direct measure of the alignment of the $I_2$-bond-axis at the instant the probe pulse triggers the Coulomb explosion. This makes these $I^+$ ions ideal observables for measuring the degree of alignment. To prove that the ions between the white circles are indeed produced by Coulomb explosion, we determined the angular covariance map of the $I^+$ ions, see Fig. 9.6 (a3). An angular covariance map allows identification of possible correlations in the emission direction of the ions, details are given in Refs. [43, 66–68]. Here two narrow diagonal lines centered at $(0°, 180°)$ and $(180°, 0°)$ stand out and show that the emission direction of an $I^+$ ion is strongly correlated with another $I^+$ ion departing in the opposite direction. This identifies the ions as originating from the $I^+$-$I^+$ channel. As discussed in Ref. [69] the length of the signal in the covariance map (as defined in panel (b3)) is a measure of the distribution of the molecular axes, i.e. the degree of alignment, whereas the width is a measure of the degree of axial recoil. The observation of a width of $\sim 1°$ shows that the axial recoil approximation is indeed fulfilled, i.e., that the two $I^+$ ions fly apart back-to-back along the interatomic axis of their parent molecule.

To explore the alignment dynamics, $\langle \cos^2 \theta_{2D} \rangle$ was measured as a function of time, $t$ measured from the peak of the alignment pulse. We remark that before the pulse, $\langle \cos^2 \theta_{2D} \rangle = 0.50$. This is the value of $\langle \cos^2 \theta_{2D} \rangle$ characterizing randomly oriented molecules that corresponds to 1/3 for $\langle \cos^2 \theta \rangle$. The blue curve in Fig. 9.7 shows that $\langle \cos^2 \theta_{2D} \rangle$ follows the intensity profile of the laser pulse during the rising part and reaches the maximum at the peak of the pulse. We note that this adiabatic behavior occurs despite the fact that the rise time, $\sim 100$ ps is significantly shorter than $\tau_{\text{rot}} = 446$ ps[70, 71]. Upon turn-off of the alignment pulse, $\langle \cos^2 \theta_{2D} \rangle$ does not return all the way to 0.50, indicating that the molecules are left in a superposition of eigenstates rather than in single eigenstates. This is corroborated by measurements out to 750 ps, showing characteristic revival structures in $\langle \cos^2 \theta_{2D} \rangle$ although with a

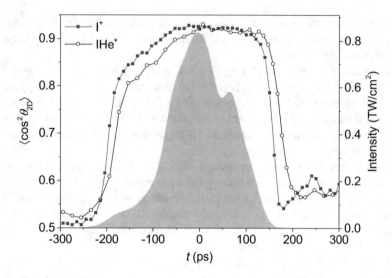

**Fig. 9.7** $\langle \cos^2 \theta_{2D} \rangle(t)$ for both isolated $I_2$ molecules (filled blue squares) and $I_2$ molecules in He droplets (open black circles). The grey shape shows the intensity profile of the alignment pulse (right vertical axis). Reproduced from [63] with the permission of American Institute of Physics (AIP) Publishing. Copyright 2017 by AIP

smaller amplitude than that at $t = 0$ ps. These nonadiabatic effects are not considered here, but details can be found in Ref. [51]. Here, we focus instead on the maximum degree of alignment obtained at the peak of the alignment pulse.

The blue filled squares in Fig. 9.8 shows $\langle \cos^2 \theta_{2D} \rangle$, obtained at $t = 0$ ps, as a function of the intensity of the alignment pulse, $I_{align}$. The curve rises gradually from 0.50 at $I_{align} = 0$ W/cm$^2$, then levels out and ends at $\sim 0.92$ for $I_{align} = 8.3 \times 10^{11}$ W/cm$^2$. The $I_2$ molecules should be able to withstand an intensity of several TW/cm$^2$, so it should be possible to increase the degree of alignment even further by simply increasing $I_{align}$. In Sect. 9.3.5 we show that an alternative and potentially more useful way to increase $\langle \cos^2 \theta_{2D} \rangle$ is to use $I_2$ molecules in He droplets (the data represented by the black lines in Fig. 9.8) because their rotational temperature is lower than that of the gas phase molecules.

### 9.2.5 Experimental Observations of Nonadiabatic Alignment

Next, we turn to alignment in the short-pulse limit induced by 450 fs long, linearly polarized laser pulses. Images of $I^+$ ion images were recorded for a large number of delays, $t$, between the centers of the alignment and the probe pulses. For each image, $\langle \cos^2 \theta_{2D} \rangle$, is determined from the ions produced through the $I^+$-$I^+$ Coulomb explosion channel, as described above. The black traces in Fig. 9.9 show $\langle \cos^2 \theta_{2D} \rangle$ as a function of $t$ for nine different fluences of the alignment pulse, $F_{align}$. We note

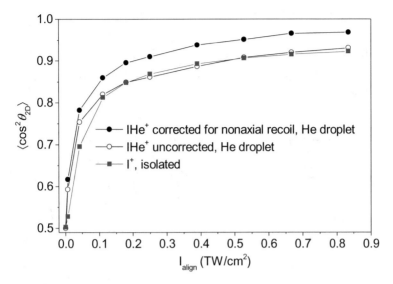

**Fig. 9.8** $\langle \cos^2 \theta_{2D} \rangle$ at the peak of the alignment pulse as a function of $I_{align}$ for isolated $I_2$ molecules (filled blue squares) and for $I_2$ molecules in He droplet without (open black circles) and after (filled black circles) correcting for nonaxial recoil (see text). Reproduced from [63] with the permission of American Institute of Physics (AIP) Publishing. Copyright 2017 by AIP

that the intensity of the alignment pulse, $I_{align}$, is given by $F_{align}/\tau_{align}$, so for e.g. $F_{align} = 0.25 \, J/cm^2$, $I_{align} = 5.5 \times 10^{11} \, W/cm^2$.[7]

The overall structure of the nine alignment traces is similar to that calculated for OCS and shown in Fig. 9.2: a prompt peak shortly after the pulse, a half revival, centered at $\sim$225 ps, and a full revival centered at $\sim$448 ps. In addition, there are transients at $\sim$109 ps and $\sim$333 ps, which are assigned as the quarter and three-quarter revival, respectively. These quarter revivals appear because of the unequal population of rotational states with odd and even $J$, caused by the nuclear spin statistical weight of the odd/even J states = 21/15.

The effect of increasing $F_{align}$ is twofold. First, the prompt peak and the revivals narrow, and their oscillatory structure becomes faster. The zoomed-in region, displayed in the right column of Fig. 9.9 illustrates this effect for the prompt peak. Second, the amplitude of the prompt peak and of the revivals increase as $F_{align}$ is increased up to 3.7 J/cm$^2$. Notably, the global maximum (minimum) of $\langle \cos^2 \theta_{2D} \rangle$, attained at the half (full) revival increases (decreases) from 0.61 (0.42) at $F_{align} = 0.25$ J/cm$^2$ to 0.88 (0.28) at $F_{align} = 3.7$ J/cm$^2$. In other words, both the peak alignment and peak anti-alignment sharpen when $F_{align}$ is increased. Increasing $F_{align}$ beyond 3.7 J/cm$^2$, leads to a gradual weakening of both the global alignment and anti-alignment maxima in the alignment traces.

---

[7] The measurements with $F_{align}$ = 6.4, 7.4 and 8.7 J/cm$^2$ were recorded with an alignment pulse duration of 1300fs to avoid ionization.

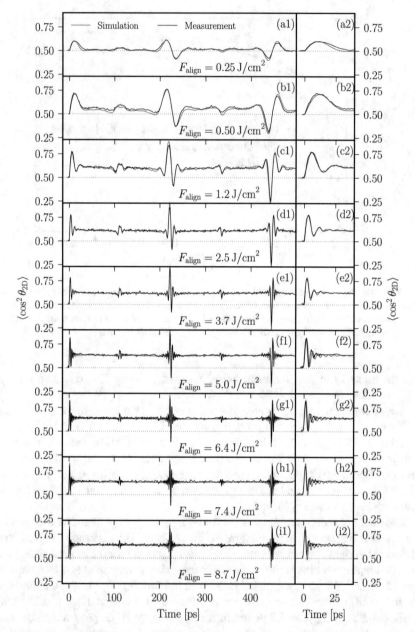

**Fig. 9.9** Time-dependence of $\langle \cos^2 \theta_{2D} \rangle$ for isolated $I_2$ molecules at 9 different fluences of a 450 fs alignment laser pulse. The rightmost panels show a zoom of the first 30 ps

Both effects are caused by the fact that as the intensity of the alignment pulse increases, the $I_2$ molecules are excited to increasingly higher rotational states, which can be seen directly in the spectra of $\langle \cos^2 \theta_{2D} \rangle$, discussed below. This increased width of the wave packet in angular momentum space, enables a tighter angular confinement of the molecular axes but only if the different angular momentum states in the rotational wave packets have a well-defined phase relationship [72]. This is analogous to how broadening a laser pulse in frequency space allows it to become narrower in the time-domain, provided that the frequency components in the pulse are phase-locked. We interpret the weakening of the alignment and the anti-alignment at $F_{align} > 3.7 \, \text{J/cm}^2$ as due to a non-optimal phasing of the angular momentum states in the wave packet. The reason is that the frequencies of the highest angular momentum components in the wave packet are so high that they start evolving during the pulse leading to a phase shift compared to the lower angular momentum components. Expressed classically, the delta-kick model is no longer valid at the highest fluences. Using two alignment pulses, or more generally a shaped alignment pulse, the phase relationship of the components in broad rotational wave packets can be optimized, to increase the degree of alignment further [73, 74]. We note that at the highest fluences the rapid oscillatory structure of the half and full revivals is also influenced by the centrifugal distortion, i.e. deviation from the rigid rotor structure [75]. This dispersive effect is much more pronounced for molecules in He droplets and will be discussed in Sect. 9.3.3.

As mentioned in the discussion of the simulated OCS nonadiabatic alignment dynamics, the spectral content of the wave packets is revealed by Fourier transformation of $\langle \cos^2 \theta_{2D} \rangle (t)$. Figure 9.10 shows the spectrum for each alignment trace. At the lowest fluences, $F_{align} = 0.25 \, \text{J/cm}^2$ and $F_{align} = 0.50 \, \text{J/cm}^2$, only about 8 peaks are observed, see inset on Fig. 9.10 (b). The assignment of the spectral peaks shows that no rotational states higher than $J = 9$ are significantly populated. For comparison, we have plotted the Boltzmann distribution of rotational states for $I_2$ at a temperature of 1 K, the estimated rotational temperature of the molecular beam (see also Sect. 9.3.5). As shown by the inset in Fig. 9.10 (a), the states populated are in the range $J = 0$–10, i.e. almost the same as those derived from the spectrum in Fig. 9.10 (b). We conclude that at $F_{align} = 0.25 \, \text{J/cm}^2$ and $F_{align} = 0.50 \, \text{J/cm}^2$, the probability of a Raman transition between $J$ states is so low that the redistribution of rotational states is very modest.

This is no longer the case when $F_{align}$ is increased. Figure 9.10 shows how the spectrum broadens and shifts to higher frequencies as $F_{align}$ increases. At the highest fluence, $F_{align} = 8.7 \, \text{J/cm}^2$, the spectral width is greater than 200 GHz and there are spectral components beyond 450 GHz. One of the highest-frequency peaks that can be identified is centered at $\sim$454 GHz, which corresponds to the frequency beat between $J = 100$ and $J = 102$. These high-lying $J$ states have been populated by a sequence of $\Delta J = \pm 2$ Raman transitions starting from the initially populated low-lying rotational states. We note that in many of the spectra from $F_{align} = 1.2 \, \text{J/cm}^2$ to $8.7 \, \text{J/cm}^2$ there is an alternation in the amplitude of the peaks, with those corresponding to odd $J$ states being stronger than those corresponding to even $J$ states.

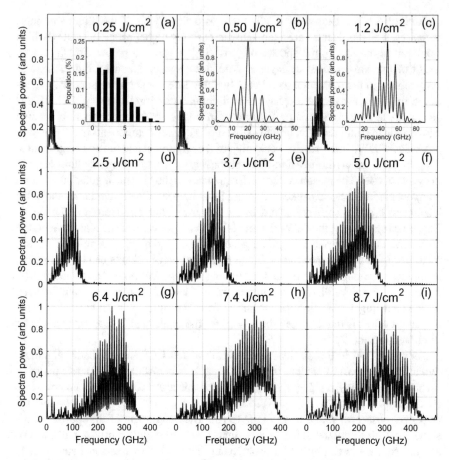

**Fig. 9.10** Power spectra of each of the $\langle \cos^2 \theta \rangle (t)$ traces shown in Fig. 9.9. In (**b**) and (**c**), the inset shows a zoom on the spectral peaks. The inset in (**a**) is a Boltzmann distribution of rotational states calculated for $I_2$ molecules at a temperature of 1 K

This is a consequence of the nuclear spin statistics. The nuclear spin of $^{127}I$ is 5/2, which gives a statistical weight of 21 to odd and 15 to even $J$ states.

To analyze the experimental data we make comparisons to the predictions of both the quantum model and the classical delta-kick model. In the quantum model, the time-dependent Schrödinger equation, Eq. 9.2 is solved numerically for the $I_2$ molecules to give the experimental observable, $\langle \cos^2 \theta_{2D} \rangle$. The calculations were averaged over the focal volume determined by the measured beam waists of the alignment ($\omega_0 = 30$ μm) and probe beams ($\omega_0 = 25$ μm) and with the rotational temperature as a free parameter. As seen in Fig. 9.9 the agreement between the experimental (black curves) and simulated results (red curves), is very good. The minor discrepancies observed may be due to neglect of centrifugal distortion effects (at the highest fluences) in the simulation and the fact that the ionization efficiency

**Table 9.1** Comparison of the maximal $J$ quantum number calculated classically, $J_{max}^{clas}$, (see text) and observed in the experimental spectra, $J_{max}^{obs}$—for four different fluences of the alignment pulse

| $F_{align}$, J/cm$^2$ | $\omega$, $10^{12}$ Hz | $E_{rot}^{clas}$, cm$^{-1}$ | $J_{max}^{clas}$ | $J_{max}^{obs}$ |
|---|---|---|---|---|
| 1.2 | 0.23 | 10 | 16 | 16 |
| 2.5 | 0.49 | 45 | 34 | 34 |
| 5.0 | 0.98 | 179 | 69 | 65 |
| 8.7 | 1.7 | 541 | 120 | 102 |

of the probe pulse depends on the molecular alignment. Overall, the good agreement demonstrates that time-dependent rotational dynamics of gas phase linear molecules, induced by fs or ps laser pulses, is well-understood (Table 9.1).

We also applied the delta-kick model to produce a classical prediction of how much rotational excitation the alignment pulse induces. For this, we used Eq. 9.7 with $\theta_0 = 45°$, to calculate the maximal angular velocity, $\omega$, of a molecule after the interaction with the laser pulse, assuming that the molecule did not rotate before the laser pulse (which is reasonable for a rotational temperature of 1 K). Classically, the rotational energy, $E_{rot}^{clas}$, is given by $E_{rot}^{clas} = \frac{I}{2}\omega^2$, where $I$ is the moment of inertia of an $I_2$ molecule. If we then equate $E_{rot}^{clas}$, in units of cm$^{-1}$, with the quantum expression for the rotational energy, $BJ(J + 1)$, ($B$ in units of cm$^{-1}$) we can determine the maximal $J$ quantum number, $J_{max}^{clas}$. This can be compared to the observed maximal quantum number, $J_{max}^{obs}$ read off from the spectra in Fig. 9.10. Table 9.1 list the values of $\omega$, $E_{rot}^{clas}$, $J_{max}^{clas}$ and $J_{max}^{obs}$ for four different fluences of the experiment. At $F_{align} = 1.2$ J/cm$^2$ and 2.5 J/cm$^2$ the classical model captures the observations essentially spot-on. At the two higher fluences, the classical calculation slightly overestimates the observed maximal rotational quantum state in the wave packets. This corroborate that the rotational dynamics of the highest-lying angular momentum states excited by the alignment pulse are so fast that the delta-kick model is not perfectly valid, i.e. the molecules rotate slightly during the pulse.

Finally, we illustrate the spectroscopic aspect of nonadiabatic alignment, i.e., how it is possible to determine accurate rotational constants from the time-dependent $\langle \cos^2 \theta_{2D} \rangle$ measurements. We do so to introduce a technique applied to molecules in He droplets as discussed in Sect. 9.3.3. We use data for $F_{align} = 0.5$ J/cm$^2$ now recorded in the interval $-10$ ps to 3230 ps. Figure 9.11(a) shows $\langle \cos^2 \theta_{2D} \rangle (t)$ and Fig. 9.11 (b) the corresponding spectrum obtained by Fourier transformation. The alignment trace exhibits a sequence of full and fractional revivals. Their decaying amplitude as a function of time is due to coupling between the rotational angular momenta and the nuclear spins through the electric quadrupole interaction, see Ref. [76] for details. Here we focus on the position of the peaks in the spectrum. Note that these peaks are much narrower than those in Fig. 9.10 (b), due to the >5 times higher spectral resolution resulting from the > 5 times longer time measurement. The black squares in Fig. 9.12 show the central frequencies of the peaks in the spectrum and the red line the fit to B(4J+6), see Eq. 9.8. The best fit is obtained

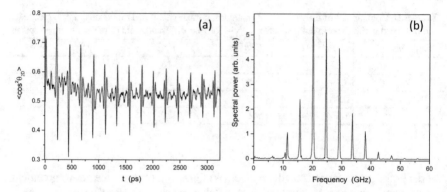

**Fig. 9.11** **a** $\langle \cos^2 \theta_{2D} \rangle$ for isolated $I_2$ molecules with $F_{\text{align}} = 0.50\,\text{J/cm}^2$. **b** Power spectrum of $\langle \cos^2 \theta_{2D} \rangle (t)$ shown in (**a**)

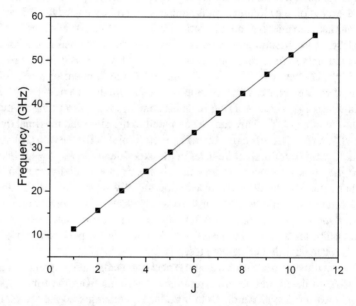

**Fig. 9.12** Central frequencies of the $(J–J+2)$ peaks in the power spectra versus $J$. The red line represent the best fit using Eq. 9.8

for B = $0.03731 \pm 0.00002\,\text{cm}^{-1}$ = $1.119 \pm 0.0001\,\text{GHz}$ , which agrees exactly with the value given on https://webbook.nist.gov/chemistry/. Our results demonstrate the possibility for accurate determination of rotational constants, and thus information on molecular structure, from time-resolved rotational dynamics measurements, a discipline broadly termed rotational coherence spectroscopy [77, 78].

### 9.2.6 Laser-Induced Alignment: A Versatile and Useful Technique

The sections above have demonstrated the ability of moderately intense laser pulses to induce high degrees of alignment in both the nonadiabatic and the adiabatic regime. Space limitation of this article prevents us from providing a full account of laser-induced alignment, so again we refer readers to the aforementioned reviews. We do, however, want to mention a few central points here. First of all, laser-induced alignment applies to all molecules that have an anisotropic polarizability tensor, which includes all molecules expect for spherical tops like methane and sulphur-hexafluoride. In practise, laser-induced alignment has been applied to a variety of molecules ranging from the simplest diatomics like $N_2$ [30] to substituted biphenyls [79]. As shown in the coming sections, using He droplets extends the scope of laser-induced alignment to even larger and more complex systems. In the previous sections, 1D alignment was discussed but we stress that it is also possible to create 3D aligned molecules, in particular by using elliptically polarized pulses in the adiabatic regime [22, 80], but also by using fs/ps pulses [81–83] or combinations of long and short pulses [84, 85]. In Sect. 9.3.6, we discuss 3D alignment of molecules in He droplets. Finally, we would like to point out that in the case of polar molecules, laser pulses in combination with a weak static electric field makes it possible to both create strong alignment and a very high degree of orientation [86–91]. This technique, now referred to as mixed-field orientation, is particularly efficient when the polar molecules are quantum-state selected by an inhomogeneous static electric field prior to the laser-interaction [88, 89, 92]. Mixed-field orientation has been applied to both 1D and 3D orientation [80, 93].

The ability to exert rigorous control over the rotation, alignment and orientation of molecules has proven useful in a number of applications in molecular science and opened new opportunities. Examples include imaging of molecular orbitals by high harmonic generation [46, 94, 95] or by photoelectron angular distributions [3, 96, 97]; alignment-dependent yields of and molecular frame photoelectron distributions from strong-field ionization by intense, polarized fs pulses [98–104]; imaging of static molecular structures by Coulomb explosion, by electron diffraction [105–109] or by x-ray diffraction[110, 111]; time-resolved imaging of molecular structure during intramolecular processes such as torsion [42, 44] and dissociation [45, 112]; high-resolution rotational coherence spectroscopy [113–115]; development of the optical centrifuge and the application of molecular superrotors [116–120]; intramolecular charge migration with sub-fs time resolution using high harmonic generation [50]; and determination of the absolute configuration of chiral molecules [121].

## 9.3  Alignment of Molecules in Helium Nanodroplets

### 9.3.1  Alignment of Molecules in a Dissipative Environment?

An interesting and perhaps obvious question to ask is whether laser-induced alignment techniques can be extended to molecules that are no longer isolated. This question was addressed theoretically around 2005 focusing on nonadiabatic alignment of linear molecules subject to binary collisions in a dense gas of atoms or molecules [122, 123]. It was shown that the collisions caused both dephasing and population relaxation of the energy levels in the rotational wave packets. As a consequence, the amplitudes of the revivals in the $\langle \cos^2 \theta \rangle$ traces were reduced when the revival order increased, and the permanent alignment level decayed gradually. The revival period remained, however, identical to that of the isolated molecules since the collisions did not change the energy of the available rotational states. A few years later, the first experiments, exploring systems such as $CO_2$ molecules in a gas of Ar atoms [124] or a pure gas of $N_2$ molecules [125], were reported. Subsequently, a number of related studies have been published extending measurements and detailed theoretical analysis to other molecular species and mixtures [126–129].

For molecules in classical solvents, like water or ethanol, the situation is qualitatively different. Firstly, a solute molecule is no longer rotating freely due to the high collision rate with the solvent molecules (or atoms) [130]. When the molecules are immersed in a liquid they typically lose the discrete rotational energy level structure characteristic of the gas phase. Secondly, even if a laser pulse could initiate coherent rotation of a molecule, the high collision rate would rapidly perturb the rotation. Thirdly, an alignment pulse would interact not just with the solute molecule but also with the many solvent molecules (or atoms) surrounding it. It seems likely that these three circumstances obstruct the transfer of molecular alignment, based on laser-induced formation of coherent superposition of rotational states, from gas phase molecules to molecules in a solvent.

### 9.3.2  Alignment of Molecules in He Droplets: First Experiments

In 2007 we started to think about using molecules embedded in helium nanodroplets as an alternative system for studying laser-induced alignment of molecules in a dissipative environment. At that time, our motivations were manifold. Firstly, infrared spectra of molecules, such as OCS, $SF_6$, and $N_2O$, in He droplets exhibit a discrete spectral line structure quite similar to that observed for gas phase molecules. This structure had been interpreted as free rotation of the molecules along with a local solvation shell of He atoms, leading to effective rotational constants smaller than those of the corresponding isolated molecules [52, 53, 131]. The rotational coherence lifetimes, estimated from the linewidths of IR and microwave spectra, were on the order of nanoseconds, corresponding to multiple rotational periods. Secondly, IR spectra had shown that the rotational temperature of molecules in He droplets was around $0.4-0.5$ K [132–134]. Such low temperatures would be advantageous for creating a high degree of alignment [135]. Thirdly, the interaction between an alignment pulse and the solvent (the He atoms) should be negligible due to the low, isotropic polarizability and high ionization potential of He atoms. Fourthly, recent laser-based photodissociation experiments [136, 137] had shown that ion detection by velocity map imaging was possible for molecules in He droplets [138]. This strongly indicated that the detection methods, notably femtosecond laser-induced photodissociation and Coulomb explosion, employed to characterize the degree of alignment for gas phase molecules, could also be applied to molecules in He droplets. Finally, helium droplets had been shown to be able to pick up and solvate essentially any molecular species that could be brought into gas phase [52–54, 139–142]. Therefore it should become possible to study alignment for a variety of molecular systems, including some that would otherwise be difficult to explore in cold molecular beams.

The aforementioned considerations built the expectation that both adiabatic and nonadiabatic alignment of molecules in He droplets would be possible. Regarding nonadiabatic alignment, we expected that the alignment dynamics for a given molecule slower be slower than that observed in the gas phase due to the increase of the moment of inertia from the solvation shell [134, 143]. Thus, it seemed likely that the revival periods would be larger compared to that of isolated molecules, and also that the number of revivals might be limited due to a finite rotational coherence lifetime resulting from coupling between the molecule and the surrounding He solvent. We also expected adiabatic alignment to be feasible, and furthermore, that the degree of alignment would benefit from the low rotational temperature [135].

The construction of an experimental setup in our laboratory for exploring laser-induced alignment of molecules was completed in 2010. The plan was to explore alignment in the nonadiabatic limit using a 450 fs kick pulse to initiate alignment and a delayed, intense 30 fs probe pulse to determine the time-dependent degree of alignment via Coulomb explosion. Initially, a pulsed cryogenic valve was employed to produce He droplets [144], to match the pulsed nature of the laser beams. The droplets were doped with either iodobenzene (IB) or methyliodide (MeI) molecules. As a first test, we used only the probe pulse, linearly polarized in the detector plane. The intensity of the probe pulse was $I_{align} = 1.8 \times 10^{14}$ W/cm$^2$, which is strong enough to multiply ionize IB or MeI molecules and create I$^+$ fragments from Coulomb explosion [145, 146]. For isolated molecules the I$^+$ images are angularly anisotropic with most ions detected along the polarization direction. The reason is that the probability for multiple ionization is highest (lowest) for those molecules that happen to have their most polarizable axis (the C-I axis for both species) parallel (perpendicular) to the polarization vector of the probe pulse at the instant that it arrives. Details on this 'geometrical' alignment or enhanced ionization are given in [147]. Our first objective was to detect such angular confinement for the I$^+$ ions created by Coulomb exploding either the IB or MeI molecules in the droplets, as this would prove our ability to detect an anisotropic angular distribution of fragment ions, which was exactly what was needed for the planned alignment experiments.

No such angular anisotropy was, however, detected in the $I^+$ images. Instead, they were circularly symmetric, independent of whether the probe pulse polarization was parallel or perpendicular to the detector. The explanation, we believe, was that each droplet was doped with multiple molecules, which then formed dimers, trimers and even larger clusters. In these oligomers, the C-I axes are not necessarily parallel, and the polarizability tensors could become more isotropic than for the monomers. This would strongly reduce the alignment-dependence of the strong-field ionization. An obvious follow-up question is to ask why the droplets picked up several molecules. We believe that this was because the pulsed nozzle produced very large droplets, with a size of $10^5 - 10^6$ He atoms per droplet [144]. The sheer size of these droplets gives them a very large cross section for picking up molecules or atoms. In principle, the multiple doping can be avoided by lowering the doping pressure so much that each droplet picks up at most one molecule. We tried to enter this regime but at the lowest doping pressure, where the $I^+$ signal was detectable, the $I^+$ images remained isotropic. We believe this was a consequence of the low droplet number density, resulting from their large size. The upshot was that for these large droplets, in the single doping regime there are so few doped droplets that they cannot produce detectable ion signals.

As a consequence of the lack of success with the pulsed valve, the experimental setup was instead equipped with a continuous nozzle of the Göttingen design [149, 150], manufactured by Alkwin Slenczka at Regensburg University. Figure 9.13 shows $IHe^+$ images recorded when He droplets from this new source, doped with either MeI (top row) or IB molecules (bottow row) are irradiated by the probe pulse.[8] When the probe pulse is polarized perpendicular to the detector, Fig. 9.13 (a), (c), the image is circularly symmetric which is to be expected if the molecules are randomly oriented. When the polarization is parallel to the detector plane, Fig. 9.13 (b), (d), the images develop an anisotropy with the ions being confined along the polarization direction. We interpret this as a result of the alignment-dependent strong-field ionization, well-known for isolated molecules, mentioned above. The measurements were repeated for both somewhat higher and somewhat lower doping pressures. No changes were observed in the angular distributions of the $IHe^+$ ions indicating that the experiment took place under single-doping conditions, where each droplet was doped with at most one molecule.

Next, the nonadiabatic alignment experiment was carried out. The experiment was first conducted on isolated MeI molecules provided by the molecular beam. Similar to the $I_2$ experiment described in Sect. 9.2.5, the 450 fs alignment pulse was linearly polarized parallel to the detector while the 30 fs probe was polarized perpendicular to the detector. Examples of $I^+$ images at $t = 0.3$ ps and $t = 33.6$ ps are shown in Fig. 9.14 (a2), (a3). An image obtained with only the probe pulse present is included as a reference, Fig. 9.14 (a1). In the image at $t = 0.3$ ps, the ions are

---

[8] We prefer to detect $IHe^+$ ions because $I^+$ ion images may contain a small contribution from unsolvated molecules drifting from the pickup cell to the target region. Although the experimental setup is designed to make this contribution very small, the detection of $IHe^+$ ions ensures background-free conditions.

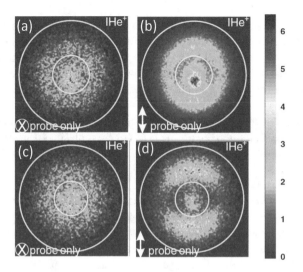

**Fig. 9.13** 2D velocity images of IHe$^+$ resulting from irradiation of He droplets doped with either methyl iodide (top row) or iodobenzene (bottow row) molecules by a 30 fs long probe pulse. The polarization direction of the linearly polarized probe pulse is indicated on each panel. $I_{probe} = 1.8 \times 10^{14}$ W/cm$^2$. No alignment pulse was included. Adapted from [148] with permission from American Physical Society (APS). Copyright (2013) by APS

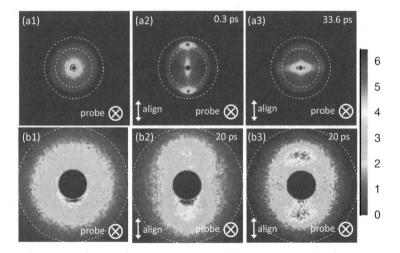

**Fig. 9.14** 2D velocity images of I$^+$ ion images resulting from irradiation of CH$_3$I molecules with the probe pulse [(b3): IHe$^+$ ions]. Top (bottom) row: Isolated (He-solvated) molecules. (a1) and (b1): Probe pulse only. In the other images, the alignment pulse is included, its polarization direction indicated by the double headed arrows. The probe time is also given on the panels. $I_{align} = 1.2 \times 10^{13}$ W/cm$^2$

localized along the polarization of the alignment pulse whereas at $t = 33.6$ ps the localization is perpendicular to the polarization vector. This demonstrates alignment and anti-alignment, respectively. As for the gas phase $I_2$ experiment in Sect. 9.2.5, we determined $\langle \cos^2 \theta_{2D} \rangle$ from the ion hits in the radial range (in between the two yellow circles) corresponding to the directional Coulomb explosion channel, here $I^+$-$CH_3^+$.

The time-dependence of $\langle \cos^2 \theta_{2D} \rangle$, shown by the black curve in Fig. 9.15, has a structure similar to that observed for the $I_2$ gas phase molecules, Fig. 9.9, i.e. narrow, periodically occurring alignment transients identified as the prompt alignment peak ($t = 0.3$ ps), the half revivals ($t = 0.3$ ps $+ (N - \frac{1}{2})T_{rev}$) and the full revivals ($t = 0.3$ ps $+ NT_{rev}$), where $T_{rev} = 1/2B$ and $N = 1, 2, 3, \ldots$ [146]. Hereafter, the molecular beam with the isolated MeI molecules was blocked and instead the doped He droplet beam was let into the target chamber. The laser beams were left untouched to ensure that the parameters of the alignment and the probe pulses were well-suited for inducing and observing laser-induced nonadiabatic alignment.

Figure 9.14 (b2) shows an $I^+$ image, recorded at $t = 20$ ps. The most important observation is that the ions are confined along the alignment pulse polarization showing that the MeI molecules are aligned at that time. A similar confinement is also present in the $IHe^+$ image displayed in Fig. 9.14 (b3). Compared to the gas phase results, the radial (velocity) distribution is broader, which could be caused by the fragments ions scattering on, and thus exchanging energy and momentum with, the He atoms on their way out of the droplets. Also, the probe laser may ionize some He atoms in the vicinity of the MeI molecule. The repulsion between such $He^+$ ions and the $I^+$ fragment ions will add kinetic energy to the latter. To account for the larger velocity range of the $I^+$ (or $IHe^+$) fragment ions, a larger radial range (in between the yellow circles) was employed to determine $\langle \cos^2 \theta_{2D} \rangle$. Note that the ion distributions in the three images, Fig. 9.14 (b1)–(b3) are slightly offset from the image center. This shift results from the fact that the He droplet beam propagates parallel to the detector whereas the molecular beam with the isolated molecules propagates perpendicularly to the detector, see Figs. 9.14 (a) and 9.5.

The time-dependence of $\langle \cos^2 \theta_{2D} \rangle$ for the MeI molecules in He droplets is represented by the red curve in Fig. 9.15. The curve rises to a maximum of 0.60 at $t \sim$ 20 ps, whereupon it decreases to 0.50, the $\langle \cos^2 \theta_{2D} \rangle$ value for randomly oriented molecules, at $t \sim 110$–120 ps. Hereafter $\langle \cos^2 \theta_{2D} \rangle$ remains flat at 0.50 until the end of the recording time, 1000 ps. At the time this experimental result was obtained and published, it seemed at odds with the expectations mentioned above. Most strikingly, there were no revivals observed. Also, the degree of alignment was significantly below that found for isolated molecules with an identical alignment pulse. As we explain in the next section, experiments at much lower intensities and on other molecular species were needed to understand laser-induced rotational dynamics of molecules in He nanodroplets.

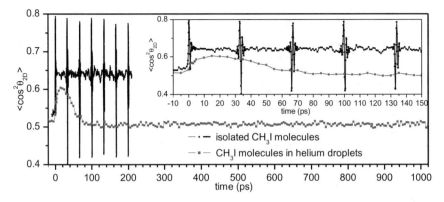

**Fig. 9.15** Time dependence of $\langle \cos^2 \theta_{2D} \rangle$ for isolated $CH_3I$ molecules (black squares) and for $CH_3I$ molecules in He droplets (red squares). The alignment and the probe pulse parameters are identical for the two data series. $I_{align} = 1.2 \times 10^{13}$ W/cm². The inset expands on the first 150 ps. Reproduced from [151] with permission from American Physical Society (APS). Copyright (2013) by APS

### 9.3.3 Nonadiabatic Alignment in the Weak-Field Limit: Free Rotation (Reconciling the Time and the Frequency Domains)

In the first part of this section, we describe nonadiabatic alignment experiments on OCS molecules in He droplets. Their rotational energy structure has been characterized by IR and MW spectroscopy, thereby making OCS an ideal case for comparing, and hopefully reconciling, results from time-resolved alignment dynamics with results from frequency-resolved spectroscopy. Infrared and microwave spectroscopy have shown that the energy, $E_{rot}$, of the five lowest rotational levels are given by:

$$E_{rot} = BJ(J+1) - DJ^2(J+1)^2 \tag{9.9}$$

where $J$ is the rotational angular momentum quantum number and $B = 2.19$ GHz and $D = 11.4$ MHz [152]. The reason that no higher $J$-states are explored is the combination of the fact that at a temperature of 0.4 K only $J = 0, 1, 2, 3$ are significantly populated and that the selection rules for the transitions driven by the IR or MW radiation is $\Delta J = \pm 1$. The gas phase-like rotational energy-level structure given by Eq. 9.9 builds the expectation that a short, nonresonant laser pulse can make a superposition of these states and thus initiate coherent rotational dynamics in a similar way to what has been done with isolated molecules.

The experimental procedure of our OCS experiment was similar to that described for MeI in Sect. 9.3.2 except that the alignment pulses were 15 ps long and the fragment ions detected were $S^+$ ions. The black curves in Fig. 9.16 (a1)–(c1) show the time-dependence of $\langle \cos^2 \theta_{2D} \rangle (t)$ recorded at three different values of the fluence.

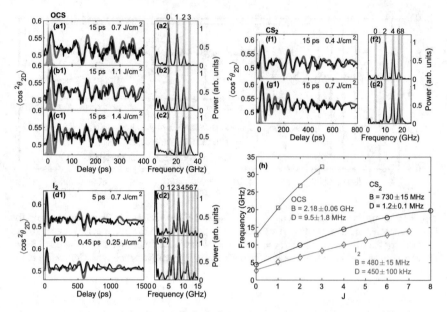

**Fig. 9.16  a**1–**g**1: The time dependence of $\langle\cos^2\theta_{2D}\rangle$ for OCS, CS$_2$ and I$_2$ molecules at different durations and fluences of the alignment pulse, given on each panel. Black (red) curves: experimental (simulated) results. The intensity profile of the alignment pulses are shown by the shaded grey area. **a**2–**g**2: The power spectra of the corresponding $\langle\cos^2\theta_{2D}\rangle$ traces. The spectral peaks, highlighted by the colored vertical bands, are assigned as $(J$–$J+2)$ coherences with $J$ given on top of the panels (blue: even, red: odd). **h**: Central frequencies of the $(J$–$J+2)$ peaks in the power spectra versus $J$. The full lines represent the best fits using Eq. 9.10. The $B$ and $D$ constants from the fits are given for each molecule. Adapted from [153] with permission from American Physical Society (APS). Copyright (2020) by APS

All three alignment traces reach a maximum shortly after the alignment pulse, which is followed by oscillations with a gradually decreasing amplitude. At first glance, the three curves may appear noisy and rather uninformative but this impression changes when one views the corresponding power spectra obtained by Fourier transformation of $\langle\cos^2\theta_{2D}\rangle(t)$. The power spectra contain discrete peaks just as for the spectra of the gas phase I$_2$ molecules presented in Fig. 9.10.

In Sect. 9.2.5, we described how the spectral peaks reflect the frequencies of the nonzero matrix elements $\langle JM|\cos^2\theta_{2D}|J'M\rangle$—the coherence (coupling) between state $|JM\rangle$ and $|J'M\rangle$, where $M$ is the projection of the angular momentum on the polarization axis of the alignment pulse. From Eq. 9.9 the frequencies corresponding to the dominant $\Delta J = J' - J = 2$ coherences [65], labeled $(J$–$J+2)$, are given by:

$$\nu_{(J-J+2)} = B(4J + 6) - D(8J^3 + 36J^2 + 60J + 36). \qquad (9.10)$$

Using Eq. 9.10 the three peaks in Fig. 9.16 (a2) at 12.8, 20.6 and 27.2 GHz are assigned as pertaining to the (0–2), (1–3) and (2–4) coherences, respectively. The spectra for the two higher fluences, Fig. 9.16 (b2)–(c2) contain the same spectral peaks as illustrated by the vertical colored bands. At $F_{align} = 1.4\,J/cm^2$, an extra peak shows up at 32.3 GHz. This peak is assigned as the (3–5) coherence. It can be seen that the weight of the spectral peaks shift to higher frequencies as the fluence is increased. The same observation was made for the nonadiabatic alignment meaurements on $I_2$ molecules, see Fig. 9.10. As mentioned, the classical explanation of the shift is that a stronger alignment pulse imparts more angular momentum to the molecule, which will then lead to higher rotational frequencies. Now the central positions of the four peaks in the spectra can be plotted as a function of $J$ and fitted using Eq. 9.10 with $B$ and $D$ as the fitting parameters. The experimental points along with the best fit, obtained for $B = 2.18 \pm 0.06\,GHz$ and $D = 9.5 \pm 1.8\,MHz$, are shown in Fig. 9.16 (h). These findings agree well with the values from IR spectroscopy, $B = 2.19\,GHz$ and $D = 11.4\,MHz$, where $D$ was the average for the $v=0$ and $v=1$ vibrational states in the IR transition [152].

From this we can draw two conclusions. Firstly, the excellent fitting of Eq. 9.10 to the positions of the spectral peaks strongly indicates that laser-induced rotation of OCS molecules in He droplets is well described by a gas phase model, employing the effective $B$ and $D$ constants, when the rotational excitation is modest, here $J \leq 5$. This point is further discussed in the next paragraph. Secondly, the rotational structure of a He-solvated molecule is encoded in $\langle \cos^2 \theta_{2D} \rangle (t)$ and the rotational constants can be retrieved by Fourier transformation. This introduces nonadiabatic alignment as a rotational coherence spectroscopy method for molecules embedded in He nanodroplets. Examples for molecules other than OCS will be given below.

To test how well a gas phase model describes the observed alignment dynamics, we calculated $\langle \cos^2 \theta_{2D} \rangle (t)$ by solving the time-dependent rotational Schrödinger equation for a linear molecule exposed to the 15 ps alignment pulse, using the experimental pulse shape, and the $B$ and $D$ values from the fit. The calculations were averaged over the initially populated rotational states, given by a Boltzmann distribution with $T = 0.37\,K$, and over the focal volume determined by the measured beam waists of the alignment ($\omega_0 = 30\,\mu m$) and probe beams ($\omega_0 = 25\,\mu m$). Also, the effect of inhomogeneous broadening was implemented by a Gaussian distribution of the $B$ constants with a FWHM, $(\Delta B)$ of 90 MHz and a constant $B/D$ ratio. The distribution of B constants can arise from differences in droplet sizes and shapes, the location of the molecule inside a droplet and coupling of the rotation and centre-of-mass motion of the molecule [154]. The value of $\Delta B$ was taken as half of the width, $W$, of the $J$: $1-0$ transition in a MW spectroscopy experiment on OCS molecules in He droplets [155].

The red curves in Fig. 9.16 (a1)–(c1) show the calculated degree of alignment. The simulated $\langle \cos^2 \theta_{2D} \rangle$ values were scaled by a factor of 0.3, symmetrically centered around $\langle \cos^2 \theta_{2D} \rangle = 0.5$, to account for the non-axial recoil of the $S^+$ ions from the Coulomb explosion of the OCS molecules [63]. The calculated $\langle \cos^2 \theta_{2D} \rangle (t)$ is very close to the measured trace and captures in detail most of the oscillatory patterns observed. The very good agreement between the calculated and experimental data corroborates the previous conclusion that gas phase modelling with effective $B$ and $D$ constants can accurately describe ps laser-induced rotational dynamics of OCS molecules in He droplets. Nevertheless, the observed dynamics appear very different from that of isolated molecules. As discussed below, this is due to the much larger $D$ constant for molecules in He droplets compared to isolated molecules (OCS: $D_{He} \approx 6.5 \times 10^3 D_{gas}$) [152, 156] and the presence of inhomogeneous broadening.

For the OCS molecules aligned with the 15 ps pulse, we calculated $\langle \cos^2 \theta_{2D} \rangle(t)$ for three values of the $D$ constant, with or without the effect of inhomogeneous broadening. When $D = 0$ and all molecules have the same $B$ value, i.e. no inhomogeneous broadening, Fig. 9.17 (a1) shows that $\langle \cos^2 \theta_{2D} \rangle(t)$ is identical to that of isolated OCS molecules except that the revival period is increased by a factor 2.8 due to the effective $B$ constant. The calculation for $D = 5.0\,\text{MHz}$, Fig. 9.17 (b1), shows that the centrifugal term introduces an additional oscillatory structure in $\langle \cos^2 \theta_{2D} \rangle(t)$ and distorts the shape of the revivals. For gas phase molecules, it was already observed and understood that the centrifugal term modulates the shape of rotational revivals [115, 157] but the influence was moderate [9] and the different revivals remained separated from each other. In Fig. 9.17 (b1), the effect of the centrifugal term is so large that, with the exception of the half-revival, there is essentially no longer distinct, separated revivals. This trend is even more pronounced for the calculation with the experimental $D$ value, Fig. 9.17 (c1). The yellow and blue bands provide a rigid rotor reference, Fig. 9.17 (a1), for how the centrifugal term distorts and shifts the rotational revivals.

The panels in the right column of Fig. 9.17 show $\langle \cos^2 \theta_{2D} \rangle(t)$ when inhomogeneous broadening is included by averaging calculated alignment traces over a Gaussian distribution of $B$ constants with a FWHM width of 90 MHz. The main influence is a gradual reduction of the amplitude of the oscillations in the alignment traces while preserving the average value of $\langle \cos^2 \theta_{2D} \rangle$. This dispersion effect produces alignment traces which (for $D = 9.5\,\text{MHz}$) agree very well with the experimental results for all three fluences studied, see Fig. 9.16 (a1)–(c1).

Experiments were carried out for two other linear molecules, $CS_2$ and $I_2$. The alignment traces and the corresponding power spectra are displayed in Fig. 9.16 panels (d)–(g).[10] Again, each spectral peak is assigned to a $(J$–$J + 2)$-coherence, noting that for $CS_2$ only even $J$-states are possible due to the nuclear spin statistics. The center of the spectral peaks are plotted versus $J$ and as for OCS, Eq. 9.10 provides excellent fits to the experimental results, illustrated by the red ($I_2$) and blue ($CS_2$) points/lines in Fig. 9.16 (h). The $B$ and $D$ values extracted from the best fits are given on the figure. No other experimental values from frequency resolved spectroscopy are available in the literature but a recent path integral Monte Carlo (PIMC) simulation gives $B = 756 \pm 9$ MHz, in good agreement with the experimental value. For the case of $CS_2$, IR spectroscopy could have been used to determine B and D, just as it has been used for $CO_2$ [159]. For $I_2$, however, neither IR nor MW would apply because $I_2$ neither has a permanent dipole moment, nor a dipole moment that changes during vibration.

---

[9] Unless extreme rotational states, excited by an optical centrifuge, are populated [158].

[10] The $I_2$ results shown were obtained for alignment pulse durations of 450 fs and 5 ps. Measurements were also carried out at 15 ps showing spectral peaks at the same positions.

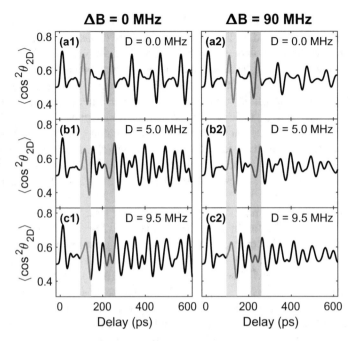

**Fig. 9.17** $\langle \cos^2 \theta_{2D} \rangle$ as a function of time calculated for OCS (B=2.17 GHz) for three different values of the D constant, without (left column) or with (right column) inhomogeneous broadening included. T = 0.37 K and F = 0.7 J/cm². The yellow and blue bands highlight the position of the half-and full-revival for the D=0 case. Reproduced from [153] with permission from American Physical Society (APS). Copyright (2020) by APS

We also calculated the time-dependent degree of alignment for $I_2$ and $CS_2$. As for OCS, it was necessary to scale the simulated data to account for non-axial recoil. The scaling factor was 0.37 for $CS_2$ and 0.75 for $I_2$. The calculated results, shown by the red curves in Fig. 9.16 panels (d1)–(g1) agree very well with the experimental findings. Since the two molecules had not been spectroscopically studied in He droplets before, we had to choose a width for the inhomogeneous distribution of the $B$ constant. The best agreement with the measured $\langle \cos^2 \theta_{2D} \rangle (t)$ was obtained for $\Delta B = 50$ MHz for $CS_2$ and 40 MHz for $I_2$. As for OCS, we also did calculations where the $D$ constant was varied from 0 to the experimental value - 1.2 MHz for $CS_2$ and 450 kHz for $I_2$. The result of these calculations, that are not shown here, identify the valley-peak structure around $t = 200$ ps for $CS_2$ as the quarter revival and the oscillatory structure in the 550–700 ps range for $I_2$ as the half revival.

Based on the excellent agreement between the measured and calculated $\langle \cos^2 \theta_{2D} \rangle (t)$ for the three different molecules, we conclude that for the relatively weak, nonresonant fs or ps pulses used, the mechanism of nonadiabatic alignment is the same for molecules in He droplets as for isolated molecules. This means that first the polarizability interaction between the laser pulse and the He-solvated molecule creates a rotational wave packet. The interaction strength is essentially the same as for an isolated molecule because for a given laser pulse intensity, it is only determined by the polarizability anisotropy of the molecule.[11] The rotational wave packet will, however, be different because the rotational energy levels are not the same for the He-solvated molecules as for the isolated molecules. In fact, the orders-of-magnitude larger $D$ values for the He-solvated molecules cause a strong dispersion of $\langle \cos^2 \theta_{2D} \rangle (t)$ making only the first fractional revivals identifiable. This is in stark contrast to the gas phase case where a large number of revivals are observable. Furthermore, the distribution of $B$ (and $D$) values also contrasts the gas phase case where $B$ and $D$ are defined by the molecular structure only. The inhomogeneous $B$-distribution causes a gradual decay of the amplitude of the oscillations in the $\langle \cos^2 \theta_{2D} \rangle$ trace.

One question remains, namely why is the rotational structure of the He-solvated molecules gas phase-like for the lowest rotational states? We believe this is a consequence of the superfluidity of the He droplets. According to the power spectra, shown in Fig. 9.16, the maximum $J$ values and thus rotational energies, $E_{rot}$( Eq. 9.9), are the following: OCS: $J = 5$, $E = 1.9 \, \text{cm}^{-1}$; CS$_2$: $J = 10$, $E = 2.2 \, \text{cm}^{-1}$; I$_2$: $J = 9$, $E = 1.3 \, \text{cm}^{-1}$. In all three cases, $E_{rot}$ falls in the regime where the density of states in the He droplets is low and thus the coupling between molecular rotation and the phonons (rotons) of the He droplet should be weak [53]. If stronger alignment pulses are applied, we expect that higher rotational states will be excited. Then the free-rotor description is probably no longer valid since the density of states in the He droplet increases strongly, implying that the coupling of rotation to the phonons (rotons) increases, which in turn will make the lifetime of highly excited rotational states (very) short [160]. We note that it is unlikely that Eq. 9.9 will provide accurate or meaningful descriptions of highly-excited rotational states. In the case of OCS, Eq. 9.9 predicts negative energies for $J > 9$, which is unphysical.

### 9.3.4 Nonadiabatic Alignment in the Strong-Field Limit: Breaking Free

To illustrate how increasing the alignment intensity influences the alignment dynamics of molecules in He droplets, we use experimental results for the I$_2$ molecule. The experimental strategy was identical to that described above. An alignment pulse, linearly polarized parallel to the detector and with a duration $\tau_{align}$ = 450 fs, initiates the rotational dynamics, whereupon the delayed probe pulse, $I_{probe}$ = $3.7 \times 10^{14}$ W/cm$^2$, Coulomb explodes the molecules and IHe$^+$ ion images are recorded. The time-dependence of $\langle \cos^2 \theta_{2D} \rangle$ obtained from these images, for nine different fluences, is shown in Fig. 9.18. The alignment trace recorded at the lowest fluence, $F_{align}$ = 0.25 J/cm$^2$, is the same as the one displayed in Fig. 9.16 (e1) and it

---

[11] The He atoms in the shell will also be polarized but the polarizability is isotropic.

shows the characteristic peak shortly after the alignment pulse and the half-revival at around 550–700 ps.

As $F_{align}$ is increased to 0.50 and then to 1.2 J/cm$^2$ the prompt alignment peak grows in amplitude and appears earlier. This is the same behavior seen for OCS in He droplets, Fig. 9.16 panels (a1)–(c1) and for the isolated $I_2$ molecules, Fig. 9.9, and is the result of excitation of higher rotational states, and thus faster rotation and higher degree of alignment. The half-revival structure remains $F_{align}$ = 0.50 and 1.2 J/cm$^2$, although its amplitude decreases somewhat. The spectra, obtained by Fourier transformation (not shown) corroborate that the increase of $F_{align}$ leads to higher frequencies of the rotational wave packet. At $F_{align}$ = 1.2 J/cm$^2$ , the spectrum shows a clustering of the frequencies in a peak around 18–19 GHz. When $F_{align}$ is further increased to 2.5 and 3.7 J/cm$^2$, the revival structure disappears from the time traces, as seen in Fig. 9.18 (d1)–(e1). Unlike the gas phase case, the frequencies in the spectrum do not shift to higher frequencies but rather stay below 20 GHz. We are currently analyzing this phenomenon and its relation to the phonon (roton) excitations of the He droplet. Here, we will instead focus on another non-gas phase phenomenon. Up to $F_{align}$ = 2.5 J/cm$^2$ the rising part of the prompt peak is smooth and increases faster as $F_{align}$ is increased. At $F_{align}$ = 3.7 J/cm$^2$ the rising part changes qualitatively to become much steeper and even develops a small substructure - see the right column of Fig. 9.18 for a zoom-in on the first 100 ps. This trend continues when $F_{align}$ is further increased and the substructure grows to a prominent sharp peak ending with a maximum already at $t \sim 1.3$ ps for $F_{align}$ = 8.7 J/cm$^2$, see Fig. 9.18(i).

This initial dynamics are almost as fast as for an isolated molecule, which can be seen on Fig. 9.19 where the $\langle \cos^2 \theta_{2D} \rangle$ trace during the first 20 ps is shown for both He-solvated molecules and isolated molecules. The maximum of the prompt peak appears at close to the same time in the two cases. We believe the reason for these rapid dynamics of the He-solvated molecules is that the $I_2$ molecules transiently decouples from their He solvation shell. The explanation for the decoupling is that the interaction with the alignment pulse induces a rotational speed of the He-solvated molecules that is so high that the He atoms in the solvation shell detach from the molecule due to the centrifugal force.

To substantiate our explanation, we use the classical delta-kick model, Eq. 9.7, to calculate the angular velocity that a He-dressed $I_2$ molecule gains from the interactions with the laser pulse using the effective moment of inertia, $I_{eff}$, of $I_2$ in the droplets. According to the experimental result in Sect. 9.3.3, $I_{eff}$ is equal to 2.3 times the moment of inertia of a gas phase molecule. In our classical model, a He-dressed molecule is treated as an $I_2$ molecule rigidly attached to eight He atoms placed in the minima of the $I_2$–He potential [152, 162]. This means six He atoms in the central ring around the molecule at a distance, $r_{He}$, of 3.7 Å from the center of the molecule and one He atom at each end, at $r_{He}$ = 5.0 Å, see Fig. 9.20. We note that the value of $I_{eff}$ determined from this structure (Fig. 9.20) gives $I_{eff}$ = 1.8 $I_{gas}$. If two He atoms are placed at each end of the molecule, then $I_{eff}$ = 2.3 $I_{gas}$.

A simple criterion for estimating if a He atom in the solvation shell can be detached is: $E_{rot}(He) > E_{binding}(He)$, where $E_{rot}(He) = \frac{1}{2} m_{He} r_{He}^2 \omega^2$ is the rotational energy of a He atom and $E_{binding}(He) \approx 16$ cm$^{-1}$ is the ground-state binding energy of the

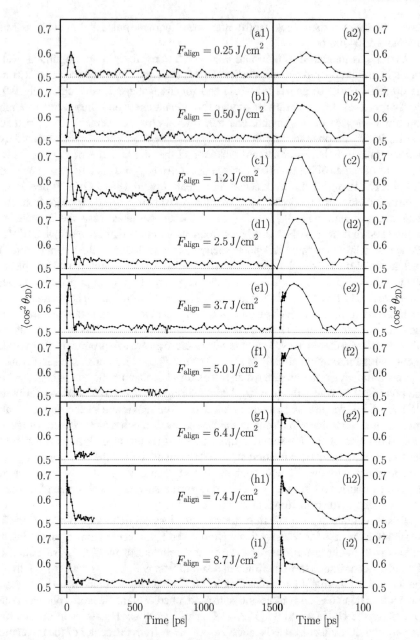

**Fig. 9.18** Time-dependence of $\langle \cos^2 \theta_{2D} \rangle$ for $I_2$ molecules in He droplets at 9 different fluences of a 450 fs alignment laser pulse. The rightmost panels show a zoom of the first 100 ps. Adapted from [161] with permission from American Physical Society (APS). Copyright (2017) by APS

$HeI_2$ complex [162, 163]. Table 9.2 displays $E_{rot}(He)$ calculated for the different fluences. For the value of $r_{He}$ we have taken 5.0 Å. At $F_{kick} \gtrsim 6 \, J/cm^2$ the criterion is met implying that one or several He atoms detach from the molecule (lower panels in Fig. 9.20) since the binding energies of the first few He atoms are similar [164]. If one He atom is detached, the angular velocity of the $I_2$ molecule and its remaining He solvent atoms will increase, which causes detachment of more He atoms and then another increase in $\omega$ etc. It is likely that this self-reinforcing process can lead to rapid detachment of all the He atoms in the solvation shell. A schematic representation of the rotational dynamics in the high fluence limit, where the detachment process occurs, and in the low fluence limit where the molecule and its solvated shell rotates as a concerted system, is shown in Fig. 9.20 (Table 9.2).

Similar rapid, short-time alignment dynamics were also observed for OCS and $CS_2$ molecules in He droplets. Figure 9.21 shows the time-dependent alignment dynamics for OCS and $CS_2$ molecules, in both cases induced by a 450 fs alignment pulse and probed by Coulomb explosion and recording of the $S^+$ fragment ions. The red curves in Fig. 9.21 are obtained for He-solvated molecules. The traces for both OCS and $CS_2$ resemble that of $I_2$ at $F_{align} = 8.7 \, J/cm^2$—a sharp initial spike of $\langle \cos^2 \theta_{2D} \rangle$ followed by a decay over ~50–60 ps to the equilibrium value of 0.5. The $\langle \cos^2 \theta_{2D} \rangle (t)$ traces for isolated molecules in the supersonic beam are shown as references. The right panels in Fig. 9.21 compare the alignment dynamics of He-solvated molecules to

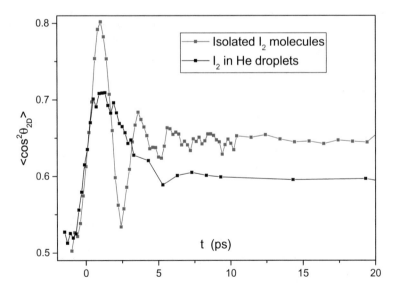

**Fig. 9.19** $\langle \cos^2 \theta_{2D} \rangle$, at early times for isolated $I_2$ molecules (red) and $I_2$ molecules in He droplets (black) recorded for $F_{align} = 8.7 \, J/cm^2$. The parameters of the alignment pulse and the probe pulse were identical for the measurements on the isolated molecules and on the molecules in He droplets. Adapted from [161] with permission from American Physical Society (APS). Copyright (2017) by APS

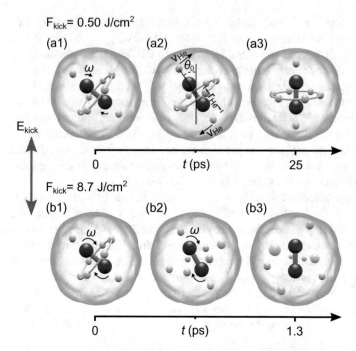

**Fig. 9.20** Schematic illustration of laser-induced rotation of $I_2$ molecules inside He droplets, based on the classical model described in the text, for a weak [(a1)–(a3)] and a strong [(b1)–(b3)] alignment pulse. **a2** illustrates parameters used in the classical model. $\theta_0$: The starting angle between the molecular axis and the alignment pulse polarization, $r_{He}$: Distance from the He atom at the ends to the axis of rotation, $v_{He} = \omega\, r_{He}$: The linear speed of the He atoms at the ends of the molecule gained from the laser-molecule interaction. Reproduced from [161] with permission from American Physical Society (APS). Copyright (2017) by APS

**Table 9.2** Classical calculation of the maximum angular velocity using Eq. (9.7) with $\theta_0 = 45°$ for the nine different fluences used in the experiment. From $\omega$ the linear speed, $v_{He}$, and the rotational energy, $E_{rot}(He)$, of the He atoms at the ends of the molecules are calculated—see text

| $F_{align}$, J/cm$^2$ | $\omega$, $10^{10}$ Hz | $v_{He}$, m/s | $E_{rot}(He)$, cm$^{-1}$ |
|---|---|---|---|
| 0.25 | 2.7 | 13 | 0.029 |
| 0.50 | 5.5 | 26 | 0.12 |
| 1.2 | 14 | 65 | 0.71 |
| 2.5 | 27 | 130 | 2.8 |
| 3.7 | 41 | 195 | 6.4 |
| 5.0 | 54 | 260 | 11 |
| 6.4 | 70 | 338 | 19 |
| 7.4 | 81 | 390 | 26 |
| 8.7 | 95 | 454 | 35 |

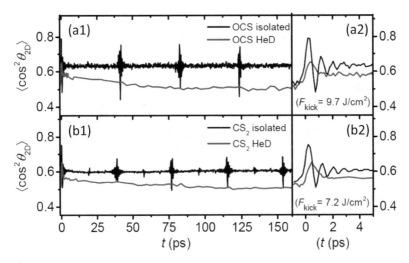

**Fig. 9.21** $\langle\cos^2\theta_{2D}\rangle$(t) for isolated (black curves) and He-solvated molecules (red curves). Upper row: OCS with $F_{align}$ = 9.7 J/cm². Lower row: CS₂ with $F_{align}$ = 7.2 J/cm². The parameters of the alignment pulse and the probe pulse were identical for the measurements on the isolated molecules and on the molecules in He droplets. The rightmost panels show a zoom of the first 5 ps

that of isolated molecules during the first 5 ps. It is seen that $\langle\cos^2\theta_{2D}\rangle$ evolves almost as fast for the molecules in He droplets as for the gas phase molecules. Repeating the classical calculation for OCS and CS₂ with the effective moment of inertia determined in Sect. 9.3.3, shows that the alignment pulse at the fluences used makes the He-solvated molecules rotate so fast that the He atoms will be detached, in an analogous way to the I₂ case just discussed. We note that a quantum model of the laser-induced rotational dynamics of a CS₂ molecule weakly bonded to a single He atom, gives almost the same value for the alignment laser pulse intensity (fluence) needed to dissociate the system [165, 166].

## 9.3.5   Adiabatic Alignment of Molecules in He Nanodroplets

In this section, we discuss alignment induced by laser pulses with a duration of 160 ps. We first describe experiments conducted on I₂ molecules in He droplets. In Sect. 9.2.4, it was already mentioned that although 160 ps is shorter than the rotational period $(1/2B = 446$ ps) for isolated molecules, the degree of alignment exhibited adiabatic behavior, i.e. $\langle\cos^2\theta_{2D}\rangle(t)$ followed the temporal intensity profile of the alignment pulse. Encouraged by those results, we used the same 160 ps pulses to explore alignment of He-solvated I₂ molecules although their moment of inertia is 2.3 times higher than for the isolated molecules.

The experiment was conducted in the same way as for the isolated molecules, but using IHe$^+$ ions as the observables rather than I$^+$ ions. Images of IHe$^+$ with the probe pulse only and with the probe pulse synchronized to the peak of the alignment pulse are shown in Fig. 9.6 (b1) and (b2). In the latter case, the ions are strongly localized around the polarization direction of the alignment pulse, showing that the molecules are strongly 1D aligned. The degree of alignment, $\langle \cos^2 \theta_{2D} \rangle$, is determined from the ions in the radial range between the two white circles. These ions result from double ionization of the I$_2$ molecules and subsequent Coulomb explosion into two I$^+$ ions, each of which then picks up a single He atom. This assignment is based on the angular covariance map, determined from the ions between the two white circles, displayed in Fig. 9.6 (c1). The two covariance signals centered at (0°, 180°) and (180°, 0°) identify the ions as originating from double ionization of the I$_2$ molecules and fragmentation into a pair of correlated IHe$^+$ ions. The length of the covariance signals is small, thus showing that the molecules are well-aligned, but the width, $\sim 11°$, is much larger than in the case of the isolated molecules. We believe the increased width is due to distortion of the ion trajectories from collisions with He atoms on the way out of the droplet. The angular distribution of the IHe$^+$ ions is therefore wider than the distribution of the molecular axes and represents an underestimate of the true degree of alignment. As is discussed below, we can deconvolute the effect of the deviation from axial recoil in the probe process using analysis of the angular covariance maps [69].

The time-dependence of $\langle \cos^2 \theta_{2D} \rangle(t)$ is shown by the black line in Fig. 9.7. It is seen that $\langle \cos^2 \theta_{2D} \rangle(t)$ follows the shape of the laser pulse (shaded grey area). Compared to the gas phase results (blue line) the $\langle \cos^2 \theta_{2D} \rangle$ curve is, however, slightly shifted to later times. Additionally, upon turn-off of the alignment pulse $\langle \cos^2 \theta_{2D} \rangle(t)$ does not return to 0.50. As for the gas phase molecules, this means that the He-solvated molecules are left in a coherent superposition of rotational states after the pulse. In fact, measurements at times out to 1500 ps reveal clear revival structures [51]. Like for the isolated molecules we omit any discussion of these nonadiabatic effects and concentrate on the maximum degree of alignment obtained at the peak of the alignment pulse.

The black open circles in Fig. 9.8 shows $\langle \cos^2 \theta_{2D} \rangle(t)$, obtained at $t = 0$ ps, as a function of the intensity of the alignment pulse, $I_{\text{align}}$. The intensity dependence of $\langle \cos^2 \theta_{2D} \rangle$ for the He-solvated molecules is very similar to that of the gas phase molecules (blue squares). There is a rather steep initial rise from 0.50 at $I_{\text{align}} = 0$ and then a more moderate rise from $I_{\text{align}} = 2.5 \times 10^{11}$ W/cm$^2$ ending at $\langle \cos^2 \theta_{2D} \rangle = 0.92$ at the highest intensity used. We expect the $\langle \cos^2 \theta_{2D} \rangle$ values of the isolated molecules to represent an accurate measurement of the true degree of alignment due to the essentially perfect axial recoil of the I$^+$ fragments. For the molecules in He droplets the deviation of axial recoil, as mentioned above, implies that the true degree of alignment is underestimated. Using a method based on analysis of the correlations between the IHe$^+$ fragment ions employing the angular covariance maps shown in Fig. 9.6 (b3), it is possible to correct the $\langle \cos^2 \theta_{2D} \rangle$ measured for the effect of nonaxial recoil [69]. The result of this correction method is represented by the black filled circles in Fig. 9.8. Comparing this curve to the blue curve, it is clear that $\langle \cos^2 \theta_{2D} \rangle$

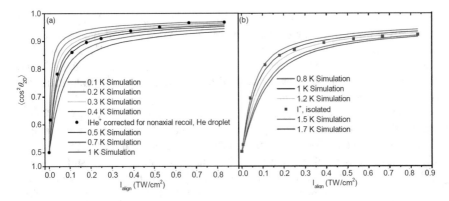

**Fig. 9.22** $\langle\cos^2\theta_{2D}\rangle$ for **a** $I_2$ molecules in He droplets and **b** isolated $I_2$ molecules calculated as a function of the alignment pulse intensity for different rotational temperatures. The black filled circles in (**a**) and the filled blue squares, representing the experimental results, are the same as the black filled circles and filled blue squares in Fig. 9.8. Adapted from Ref. [63] with the permission of American Institute of Physics (AIP) Publishing. Copyright 2017 by AIP

is significantly higher for $I_2$ in He droplets than for isolated $I_2$ molecules at all intensities. In particular, for the highest intensity, the corrected $\langle\cos^2\theta_{2D}\rangle$ value is 0.96. For comparison, a sample of perfectly aligned molecules has $\langle\cos^2\theta_{2D}\rangle = 1$.

To model and thus gain insight into alignment of the He-solvated molecules, we solved the rotational Schödinger equation (compare Eq. 9.2) for the $I_2$ molecules exposed to the experimental 160 ps alignment pulse. The molecules were treated as isolated molecules but with the effective $B$ and $D$ constants determined from the rotational coherence spectroscopy experiment discussed in Sect. 9.3.3. This treatment is motivated by the success of the model to accurately reproduce the $\langle\cos^2\theta_{2D}\rangle(t)$ for the nonadiabatic alignment experiments on $I_2$, OCS and $CS_2$, as shown in Sect. 9.3.3. Also, the intensity of the laser pulse is sufficiently low and varies slowly enough that the He solvation shell is expected to follow adiabatically, as rationalized by angulon theory arguments [63].[12] The result of the calculations, expressed as $\langle\cos^2\theta_{2D}\rangle$, at the peak of the pulse, as a function of the alignment intensity at different rotational temperatures of the $I_2$ molecules are shown in Fig. 9.22 (a).

At all intensities the experimental data points fall within the range confined by the calculated 0.2 K and 0.5 K curves, and on average lie close to the calculated 0.4 K curve. We believe that the agreement between the data points and the 0.4 K curve would be even better if the former could be corrected for the alignment-dependent ionization efficiency of the probe pulse: The probe pulse ionizes, and thus Coulomb explodes, the molecules aligned parallel (perpendicular) to its polarization axis most (least) efficiently. At intermediate degrees of alignment such as for the three data points between 0.1 and 0.3 TW/cm², $\langle\cos^2\theta_{2D}\rangle$ will be underestimated because the best aligned molecules are least efficiently probed. A correction for this probe

---

[12] Details on angulon theory can be found in [167, 168].

pulse selectivity would increase $\langle \cos^2 \theta_{2D} \rangle$ for the three points and bring them closer to the 0.4 K curve. The agreement between the simulated curve for 0.4 K and the experimental results lead us to conclude that in this limit of quasi-adiabatic alignment the He-solvated molecules behave as gas phase molecules at a temperature of 0.4 K. This conclusion is the same as we reached for nonadiabatic alignment, induced by fs or few-ps long alignment pulses.

The reason that the molecules in the He droplets obtain a higher degree of alignment than the molecules in the supersonic beam is that the rotational temperature is lower in the droplets. For the isolated molecules, we also calculated the degree of alignment at the peak of the pulse for different temperatures. The results, together with the experimental data, are displayed in Fig. 9.22 (b). It can be seen that the data points, apart from the points at the highest intensities, fall within the range confined by the calculated 0.8 K and 1.2 K curves. A temperature around 1 K is consistent with previous estimates of the rotational temperature for molecular beams produced by an Even-Lavie valve operated under similar conditions [93].

To complete the argumentation that the degree of alignment for a given molecule that is exposed to a specific laser field strength, $E_0$, is strongly dependent on temperature, we note that in 2001, Tamar Seideman showed that in the limit where $kT/B \gg 1$:

$$\langle \cos^2 \theta \rangle = 1 - \sqrt{\pi kT/\Delta\alpha E_0^2}. \tag{9.11}$$

This expression shows that $\langle \cos^2 \theta \rangle$ is determined by $T$ and independent of $B$. For $I_2$ in He droplets, $kT/B = 8.34\,\text{GHz}/0.48\,\text{GHz} = 17.4$ and for the isolated $I_2$, $kT/B = 20.9\,\text{GHz}/1.12\,\text{GHz} = 18.7$, in both cases the ratio is much larger than 1. Using Eq. 9.11 we obtain $\langle \cos^2 \theta \rangle = 0.94$ for the He-droplet case ($T = 0.4\,\text{K}$) and 0.90 for the gas phase case ($T = 1.0\,\text{K}$), with $\Delta\alpha = 7.0\,\text{Å}^3$ and $E_0 = 2.5\text{e}7$ V/cm, corresponding to the highest intensity of 0.83 TW/cm$^2$. These values are in good agreement with the experimental $\langle \cos^2 \theta_{2D} \rangle$ values of 0.96 and 0.92, noting that for a given distribution of aligned molecules, $\langle \cos^2 \theta_{2D} \rangle$ is always slightly larger than $\langle \cos^2 \theta \rangle$ [65, 145].

The effectiveness of He droplets as a medium in which strong molecular alignment can be created was also demonstrated for 1,4-diiodobenzene and 1,4-dibromobenzene, with direct comparison between He-solvated molecules and isolated molecules in a supersonic beam [63]. For the He-solvated molecules, $\langle \cos^2 \theta_{2D} \rangle$ again exceeded 0.96 at the highest intensity applied, 0.83 TW/cm$^2$. In the following, though, we would like to present another example, which both illustrates the applicability of adiabatic alignment to a molecular dimer and also how the intensity-dependent degree of alignment again matches calculations with a rotational temperature of 0.4 K. The example is the carbon disulfide dimer. In the gas phase, the dimer has a cross-shaped structure with D$_{2d}$ symmetry, and it is a prolate symmetric top, see Fig. 9.23. For now we assume the dimer has the same structure inside He droplets but in Sect. 9.3.7 we actually show that laser-induced Coulomb explosion imaging can be used to determine the structure of the CS$_2$ dimers in He droplets.

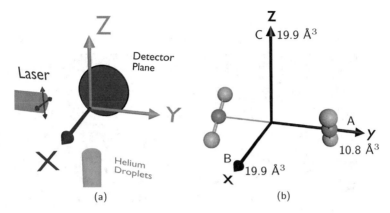

**Fig. 9.23** **a** Sketch of the space-fixed (XYZ) coordinate system. **b** Sketch of the molecular (xyz) coordinate system and the $CS_2$ dimer in its ground state geometry. Polarizability elements of the dimer (in units of $Å^3$) are annotated onto each molecular axis. Adapted from [169] with permission from American Physical Society (APS). Copyright (2019) by APS

Experimentally, the He droplets were sent through a gas of $CS_2$ molecules with a pressure sufficiently high that there is a non-negligible probability that some of the droplets pick up two $CS_2$ molecules. Since the attraction between the two $CS_2$ molecules is much larger than between a $CS_2$ and a He atom, the two molecules are expected to approach each other and can then form a dimer with the binding energy being dissipated into the He solvent and subsequently removed from the droplet by evaporation of ~162 He atoms—evaporation of one He atom removes $5 \, cm^{-1}$ of energy [52] and the binding energy of $(CS_2)_2$ is ~ $810 \, cm^{-1}$ [170]. The doped droplets are irradiated by a 30 fs, 800 nm, $3 \times 10^{14} \, W/cm^2$, probe pulse. Hereby the $CS_2$ are singly ionized, leading to formation of $CS_2^+$ ions, which is the experimental observable considered in this section. Some of the molecules are also multiply ionized, which results in Coulomb explosion and production of $S^+$ fragment ions. The use of these ions for structure determination of the dimer is discussed in Sect. 9.3.7. To induce alignment, the 160 ps pulses are used with the probe pulse synchronized to its peak.

Figure 9.24 (a1) shows a 2D velocity image of $CS_2^+$ ions with the probe pulse only, polarized perpendicular to the detector plane. The image is dominated by an intense signal in the central portion corresponding to low kinetic energy of the $CS_2^+$ ions. We assign this signal to ionization of $CS_2$ monomers in droplets doped with a single molecule. There is also a significant amount of signal detected at larger radii—outside of the yellow ring—corresponding to larger kinetic energies. We assign these ions as originating from double ionization of $(CS_2)_2$ from droplets containing a dimer, and subsequent breakup of the $(CS2)2^{2+}$ dication into two $CS_2^+$ fragment ions. This assignment was substantiated in the same way as we identified the Coulomb explosion for the $I_2$ molecule, either in gas phase or in He droplets, via calculation of the angular covariance map [43, 66], from ions in the radial range outside of the annotated yellow

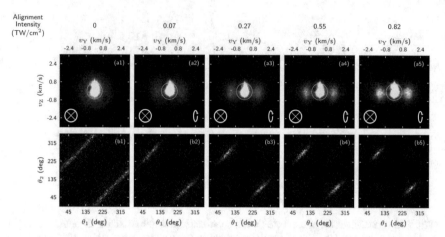

**Fig. 9.24** **a**1–a5: 2D velocity images of $CS_2^+$ ion images at different intensities of the alignment pulse, annotated above each column. **b**1–b5): corresponding angular covariance maps determined from the ions outside the annotated yellow circles in row (**a**). The polarization state of the probe (alignment) laser is shown in the lower left (right) corner of each ion image. The data were recorded under doping conditions where $CS_2$ dimers could be formed. Adapted from [169] with permission from American Physical Society (APS). Copyright (2019) by APS

circle. The result is displayed in Fig. 9.24 (b1). Two distinct diagonal lines, centered at $\theta_2 = \theta_1 - 180°$ and $\theta_2 = \theta_1 + 180°$, stand out and show the correlation of a $CS_2^+$ ion with another $CS_2^+$ ion departing in the opposite direction [$\theta_i$, $i = 1,2$ is the angle between an ion hit and the vertical center line].

The emission direction of the $CS_2^+$ fragment ions from Coulomb explosion of $(CS_2)_2$ is determined by the spatial orientation of the C-C intermolecular axis of the parent dimer at the instant of ionization, similar to how the emission direction of $I^+$ or $IHe^+$ ions is a measure of the spatial orientation of the I-I axis of their parent $I_2$ molecule. Thus, the uniform extent of the covariance signal over 360° shows that the C-C axes are randomly oriented, which is to be expected in the absence of an alignment pulse. When the alignment pulse, circularly polarized in the XZ-plane (Fig. 9.23), is included the $CS_2^+$ ions images develop an asymmetry with the ions becoming confined along the Y-axis. As the intensity of the alignment pulse is increased, the confinement grows stronger as can be seen in the sequence of images on Fig. 9.24. These observations demonstrate that the $CS_2$ dimers are aligned with the C-C axis confined along the Y-axis. The effect of a circularly polarized alignment field is to confine the least polarizable molecular axis perpendicular to the polarization plane [22], which is the same as the propagation direction of the laser pulse (the Y-axis)—see Fig. 9.23 (a). Consequently, the $CS_2$ dimer inside the He droplets must have a structure where the least polarizable axis coincides with the C-C axis. This is the case for a cross-shape, as that of the gas phase dimer, but not for a T-shape, or a slipped-parallel shape [171]. Therefore, the ion images in Fig. 9.24 lead us to conclude that the dimer is cross-shaped and aligned with the C-C axis parallel to

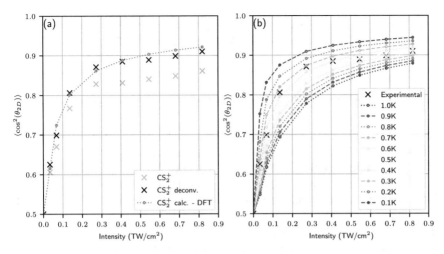

**Fig. 9.25** Experimental (crosses) and simulated (circles and dashed line) degree of alignment of the $CS_2$ dimer as a function of alignment intensity. The alignment was induced adiabatically using a circularly polarized laser pulse. Panel **a** shows a comparison of the experimental degree of alignment with the simulated degree of alignment calculated at 0.4 K. Panel **b** shows a comparison of the experimental degree of alignment with the simulated degree of alignment calculated at a variety of different ensemble temperatures. Adapted from [169] with permission from American Physical Society (APS). Copyright (2019) by APS

the Y-axis. The dihedral angle between the two monomer axes cannot be determined from detecting $CS_2^+$ ions, but in Sect. 9.3.7 we show that this is possible from images of $S^+$ ions.

The quantification of the degree of alignment is done similar to that outlined in the previous sections. The grey crosses in Fig. 9.25 shows $\langle \cos^2 \theta_{2D} \rangle$ as a function of $I_{align}$ based on the five images in Fig. 9.24 and another four at other intensities. Here $\theta_{2D}$ is the angle between an ion hit and the propagation direction of the laser pulse, i.e. the Y axis, see Fig. 9.23. When the $CS_2^+$ ions travel out of the droplet, they are also subject to collisions with He atoms. To avoid underestimating the true degree of alignment, it is necessary to correct for this deviation from nonaxial recoil of the fragment ions, as discussed previously. The corrected data points are shown by the black crosses in Fig. 9.25 (a). They are well-matched by a calculation of $\langle \cos^2 \theta_{2D} \rangle$ at the peak of the alignment pulse for the different intensities used in the experiment. The calculated values of $\langle \cos^2 \theta_{2D} \rangle$ were obtained by solving the time-dependent rotational Schrödinger equation for the dimers exposed to the circularly polarized alignment pulse with the experimental parameters of its duration, intensity and spot size and for a 0.4 K rotational temperature. Finally, the calculations were repeated for a series of different temperatures. As can be seen on Fig. 9.25 (b), the experimental data agrees well with the simulated data using an ensemble temperature between 0.3 and 0.5 K. This agreement implies that the alignment of the $CS_2$ dimers in the He droplets is well-described by alignment of a 0.4 K ensemble of gas phase $CS_2$ dimers.

This is in line with the expected temperature of the dimers inside the He droplets [52] and consistent with the findings for the $I_2$ molecules presented in Sect. 9.3.5.

### 9.3.6 Long-Lasting Field-Free Alignment of Molecules

In the previous section, we demonstrated that in the (quasi-) adiabatic limit of laser-induced alignment, molecules embedded in He nanodroplets can be very strongly aligned thanks to the 0.4 K rotational temperature of the molecules. We focused on 1D alignment, induced by a linearly polarized laser pulse but note that 3D alignment, induced by an elliptically polarized laser pulse, also benefits from the 0.4 K temperature and leads to higher degrees of alignment for He-solvated molecules compared to that of isolated molecules in a cold supersonic beam [172]. One disadvantage of adiabatic alignment is that the strongest alignment occurs at the peak of the pulse when the intensity is highest. For some applications of aligned molecules, this can be a serious obstacle, because the alignment field may perturb the molecules severely. An important example is molecules in electronically excited states, where the alignment field can cause further excitation, dissociation or ionization. Thus, the excited state reactions originally expected may not be observed because they are outmatched by processes induced by the alignment field [173]. In this section, we show how the alignment pulse can be turned off at its peak and how molecules in He droplets keep a very high degree of alignment for 5–20 ps without the presence of the alignment field. Comparison is made to molecules in the gas phase where, in contrast, the degree of alignment disappears rapidly after the pulse is turned off.

The experimental setup is similar to that used for the adiabatic alignment experiments except that the alignment laser beam is now sent through a longpass optical filter just before it is overlapped with the probe laser beam. This filter spectrally truncates the alignment pulses, which leads to a corresponding temporal truncation because the pulses are highly chirped [175]. The intensity profile of the alignment pulse, displayed by the grey shaded area in Fig. 9.26, shows a slow turn-on in $\sim$100 ps followed by a rapid turn-off at the peak, where the intensity drops by more than two orders of magnitude over $\sim$10 ps.

Figure 9.26 (a) shows the alignment dynamics of $I_2$ molecules, both isolated (black curve) and in He droplets (purple curve) exposed to the truncated pulse with linear polarization. All parameters of the alignment and the probe pulses are identical in the measurements on isolated and He-solvated molecules. During the rise of the pulse, the two curves follow the intensity profile, as already shown on Fig. 9.7. After the truncation, both curves drop, but the drop is much slower for the He-solvated molecules. The right panel of Fig. 9.26 (a) shows that $\langle \cos^2 \theta_{2D} \rangle (t)$ retains a value between 0.86 and 0.80 from $t = 5$ ps to $t = 11$ ps, a period in which the alignment pulse intensity is reduced to less than 1 % of the value at the peak. When $\langle \cos^2 \theta_{2D} \rangle (t)$ is corrected for the nonaxial recoil of the $I^+$ ions, the $\langle \cos^2 \theta_{2D} \rangle (t)$ values are between 0.90 and 0.84. By contrast, in the same interval, $\langle \cos^2 \theta_{2D} \rangle (t)$ for the isolated molecules drops from 0.69 to 0.52. These observations show that by rapidly truncating the alignment pulse,

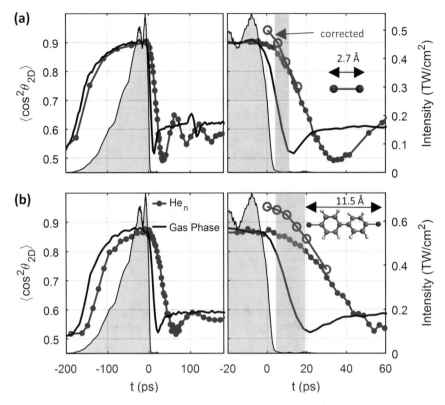

**Fig. 9.26** $\langle \cos^2 \theta_{2D} \rangle (t)$ measured for $I_2$ (panels **a**) and 4,4'-diiodobiphenyl (panels **b**) molecules. Black curves: isolated molecules, purple curves: molecules in He droplets. The intensity profile of the truncated alignment pulse is shown by the grey shaded area and refers to the right vertical axis. The panels on the right show a zoom of the post-truncation region. The green shaded area in each panel marks the interval where the alignment field intensity is < 1 % of its peak and $\langle \cos^2 \theta_{2D} \rangle \geq 0.80$ for molecules solvated in He droplets. The structure of each molecule (shown as an inset), along with a scalebar representing the I–I distance. In droplets, the values for $\langle \cos^2 \theta_{2D} \rangle$ after correction for non-axial recoil at selected times are also shown as blue open circles. Reproduced from [174]

it is possible to create a 6 ps windows in which the $I_2$ molecules remain strongly aligned and where the alignment pulse intensity is reduced by more than a factor of 100, weak enough to be considered field-free for many applications.

We note that $\langle \cos^2 \theta_{2D} \rangle (t)$ does not return to 0.5 after truncation for either the isolated or the He-solvated molecules. In fact, the truncation creates a rotational wave packet in both cases. For gas phase molecules, this has been studied before [29, 36, 41, 176, 177]. In that case, distinct rotational revivals appear, qualitatively similar to the situation when a fs or ps pulse is used to create the rotational wave packet. Here we show the long-time dynamics for the He-solvated molecules to demonstrate that

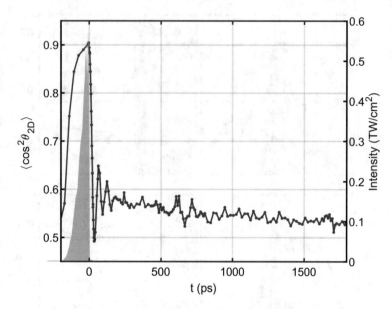

**Fig. 9.27** $\langle\cos^2\theta_{2D}\rangle(t)$ measured for $I_2$ molecules in He droplets. The alignment dynamics is induced by the truncated pulse with an intensity profile similar to that shown in Fig. 9.26

the truncated pulses can also create revivals for molecules in He droplets. Figure 9.27 shows $\langle\cos^2\theta_{2D}\rangle(t)$ in the time interval $-250$ ps to 1800 ps. A distinct revival structure is observed at $\sim 650$ ps after the truncation, a time position fully consistent with that of the half-revival observed for the 450 fs or 5 ps alignment pulses, see Fig. 9.16 and Fig. 9.18. The polarity of the revival is opposite to that of the revivals shown in those figures. This is because there is a different phase relationship between the eigenstates in the wave packet when it is created by truncation, i.e. from molecules that are already adiabatically aligned, compared to when it is created by a short laser pulse, acting on molecules that are initially randomly aligned. Finally, we point out that the average value of $\langle\cos^2\theta_{2D}\rangle(t)$ (the permanent alignment level) decreases over the time measured, which indicates a decay of some of the rotational states in the wave packet.

Measurements were also conducted on a significantly larger molecule, namely 4,4'diiodobiphenyl (DIBP). Its most polarizable axis is along the I-I axis, so experimentally $I^+$ ions are ideal observables for characterizing 1D alignment, just as in the case of the $I_2$ molecules. The time-dependence of $\langle\cos^2\theta_{2D}\rangle(t)$, obtained from $I^+$ ($IHe^+$) images for the isolated (He-solvated) molecules are shown in Fig. 9.26 (b). Despite the much larger moment of inertia for DIBP compared to $I_2$, $\langle\cos^2\theta_{2D}\rangle(t)$ still rises concurrently with the intensity profile of the pulse. After the truncation, the $\langle\cos^2\theta_{2D}\rangle(t)$ curves behave in a similar way to that observed for $I_2$ - a rapid decrease for the isolated molecules and lingering behavior for the He-solvated molecules. After correction for nonaxial recoil, we find that in the interval 5–20 ps, where the

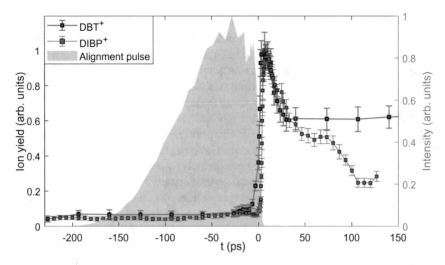

**Fig. 9.28** Time-dependence of the parent ion yields when 1D aligned DBT or DIBP molecules are ionized with the probe pulse at time $t$. $I_{probe} = 2.4 \times 10^{14}$ W/cm$^2$

laser intensity is reduced to $< 1$ % of its maximum value, $\langle \cos^2 \theta_{2D} \rangle (t)$ remains between 0.94 and 0.84 for the He-solvated molecules while it drops from 0.78 to 0.53 for the isolated molecules. Thus, for DIBP in He droplets, there is a 15 ps long field-free alignment window.

To assess the field-free nature of the alignment created we measured the yield of intact parent cations following ionization of either 5,5"-dibromo-2,2':5',2"-terthiophene (DBT) or DIBP molecules with the probe pulse. DBT is an oligomer composed of three thiophene units. If the parent ions are created while the alignment pulse is on, we expect them to be fragmented by the remainder of the pulse, an effect generally occurring for most molecules, see e.g. [173, 178]. This destruction of parent ions precludes their use as experimental observables. Figure 9.28 shows the yield of DBT$^+$ and DIBP$^+$ ions created by ionization of 1D aligned molecules with the probe pulse sent at time $t$. When the probe pulse arrives during the alignment pulse the parent ion signal is almost zero but when it arrives after the pulse both the DBT$^+$ and DIBP$^+$ signal increases sharply and reaches a maximum at $t \sim 7 - 10$ ps, i.e. while the molecules are still strongly aligned, see Figs. 9.26 and 9.29. The sudden increase of the parent signal by almost a factor of 20 shows that the truncation reduces the alignment pulse intensity sufficiently to prevent destruction of the parent ions. Figure 9.28 demonstrates that our method enables ionization experiments on sharply aligned molecules, without the destruction of the fragile parent ions by the alignment field. Without this new ability, structural determination of dimers of large molecules by Coulomb explosion imaging, discussed in Sect. 9.3.7, would not be possible.

Finally, we discuss 3D alignment and how the truncated pulses also enable creation of a time window with 3D aligned molecules under field-free conditions. We focus on the example of DBT. The molecular structure is sketched on the inset in Fig. 9.29 (c). Also, the most polarizable axis (MPA) and the second most polarizable axis (SMPA) are shown. To induce 3D alignment we use again the truncated laser pulse but now elliptically rather than linearly polarized. The idea, developed for gas phase molecules, is that the MPA of the molecule aligns along the major polarization axis of the laser pulse and the SMPA along the minor polarization axis [22, 80]. The last principal polarizability axis of the molecule is then automatically aligned perpendicular to the polarization plane. To characterize the 3D alignment two fragment ion species from the probe-pulse-induced Coulomb explosion are detected: 1) $Br^+$ ions as they are expected to recoil (approximately) along the direction of the MPA. 2) $S^+$ ions as they are expected to recoil (approximately) along the SMPA.

Figure 9.30 (a) shows a $Br^+$ ion image recorded for an alignment pulse linearly polarized along the Y-axis and with the probe pulse sent at $t = 0$, i.e. just before the truncation begins. The ions are sharply confined along the Y-axis showing pronounced 1D alignment of the DBT molecules as expected for the polarization state of the alignment pulse. This is a side view of the molecule. Figure 9.30 (c) shows a $S^+$ image. The alignment pulse is still linearly polarized but now the polarization is rotated to be parallel to the Z-axis. The circularly symmetric image shows that the rotation of the SMPA around the Z-axis is uniform. This is an end view of the molecule. Combining the information from the $Br^+$ and $S^+$ images, we conclude that the alignment pulse confine the MPA but leaves the rotation of the SMPA unrestricted.

When the polarization state of the alignment pulse is changed to elliptical with the major (minor) polarization axis parallel to the Z-axis (Y-axis), intensity ratio $I_Z:I_Y = 3:1$, the $S^+$ image Fig. 9.30 (d) acquires a pronounced angular confinement along the Y-axis. This shows that the SMPA is now aligned along the Y-axis. For the same elliptical polarization but now with the major axis parallel to the Y-axis, the $Br^+$ image remains sharply angularly confined along the Y-axis, Fig. 9.30 (b), i.e. the MPA is still strongly aligned. The simultaneous angular confinement of the $Br^+$ and the $S^+$ ions is the experimental evidence of 3D alignment similar to what has been used for a range of other asymmetric top molecules in the gas phase [22, 43, 80, 83, 172, 179].

To quantify the degree of alignment, we determine $\langle \cos^2 \theta_{2D} \rangle$ from the $Br^+$ images and $\langle \cos^2 \alpha \rangle$ from the $S^+$ images, $\alpha$ is defined in Fig. 9.29 (a). The time-dependence of $\langle \cos^2 \theta_{2D} \rangle$ and $\langle \cos^2 \alpha \rangle$ are shown in Fig. 9.29 (c). They both rise concurrently with the alignment pulse and reaches maximum values at its peak. These values, $\langle \cos^2 \theta_{2D} \rangle \sim 0.80$ and $\langle \cos^2 \alpha \rangle \sim 0.60$ are lower than those found for e.g. 3,5-dichloroiodobenzene molecules in He droplets [172]. We explain the lower values by the fact that the $Br^+$ and $S^+$ fragment ions do not recoil directly along the aligned axes as illustrated in Fig. 9.29 (a), (b). Therefore, $\langle \cos^2 \theta_{2D} \rangle$ and $\langle \cos^2 \alpha \rangle$ can only provide qualitative rather than quantitative measures for the degree of alignment [180]. The important finding is, however, the lingering of the alignment after pulse truncation. From $t = 5$ to 24 ps, where the laser intensity is $< 1\%$ of the peak value, $\langle \cos^2 \theta_{2D} \rangle$ for the $Br^+$ ions drops only from 0.80 to 0.72, and $\langle \cos^2 \alpha \rangle$ remains $> 0.58$ for the $S^+$ ions. This demonstrates a 19 ps-long interval, marked by the green area, where the molecules are 3D aligned under field-free conditions.

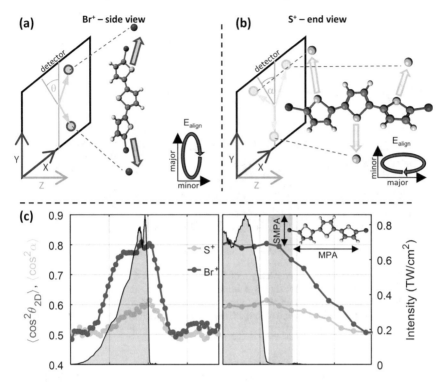

**Fig. 9.29  a-b**: Illustration of how 3D alignment of DBT is characterized. **a**: Side view where the MPA of DBT is confined to the major polarization axis of the alignment pulse, directed along the Y-axis. **b**: End view of DBT, with the MPA confined to the major polarization axis directed on the Z-axis. In the side view, alignment is characterized by $\langle \cos^2 \theta_{2D} \rangle$ where $\theta_{2D}$ is the angle between the emission direction of a $Br^+$ ion and the Y-axis, while in the end view it is characterized by $\langle \cos^2 \alpha \rangle$ where $\alpha$ is the angle between the emission direction of a $S^+$ ion and the Y-axis. **c**: The time dependence of $\langle \cos^2 \theta_{2D} \rangle$ and $\langle \cos^2 \alpha \rangle$ for DBT molecules induced by an elliptically polarized alignment pulse with an intensity ratio of 3:1. The panels on the right show a zoom of the post-truncation region and the green shaded areas mark the interval where the alignment field intensity is <1% of its peak value and $\langle \cos^2 \theta_{2D} \rangle$ has dropped by less than 10%. The MPA and SMPA are overlaid on the molecular structure. Adapted from [174]

## 9.3.7   Structure Determination of Dimers in He Nanodroplets

In this section, we discuss how the combination of laser-induced alignment and Coulomb explosion imaging makes it possible to determine structure of molecular complexes formed inside He nanodroplets. So far, we have introduced Coulomb explosion of molecules and of molecular dimers, triggered by multiple ionization with an intense fs probe pulse, as a technique to determine the spatial orientation of such systems. Coulomb explosion can also serve another useful purpose in determination of molecular structure. This approach is termed Coulomb explosion imaging (CEI)

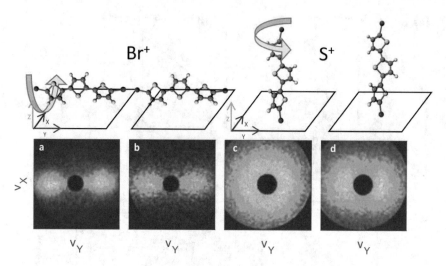

**Fig. 9.30** Two-dimensional velocity images of $Br^+$ ions (**a-b**) and $S^+$ ions (**c-d**) produced from Coulomb explosion of DBT molecules in He droplets at the peak of either a linearly polarized (**a,c**) or an elliptically polarized (**b,d**) alignment pulse with a peak intensity of $9 \times 10^{11}$ W/cm$^2$. The molecular alignment and rotational freedom are depicted above each ion image. Reproduced from the supplementary material of [174]

and relies on the fact that the molecular structure is encoded in the kinetic energy and emission direction of the recoiling fragment ions.

The CEI technique was originally introduced in experiments where a MeV beam of small molecular ions was sent through a few-hundred Å thin metal foil [181]. However most later studies, including many current activities, employ intense fs laser pulses to trigger the Coulomb explosion event [57]. Typically, the pulses are in the near IR region, as in the work described here, but VUV [182] and x-ray pulses [183, 184] are also used. Coulomb explosion imaging has been applied in a variety of different studies, for example determination of the absolute configuration of chiral molecules [121, 185–187], time-resolved imaging of intramolecular processes such as torsion [42, 44], dissociation [188, 189], roaming [190] and proton migration [191], identification of structural isomers [192, 193], and imaging of interatomic or intermolecular wavefunctions of diatomic molecules [194–196] or of weakly bonded dimers [170, 197, 198]. Key to obtaining structural information is the identification of correlations in the angular distributions of the fragment ions, in practice implemented by coincidence or covariance analysis. More recently, it was shown that aligning the molecules in well-defined spatial orientations with respect to the ion imaging detector prior to the Coulomb explosion event increases the structural information available from the recorded fragment ion momenta [43]. Such molecular frame measurement is an example of one of the main advantages offered by aligned molecules as mentioned in Sect. 9.1.

Here we give two examples of structure determination of molecular dimers inside He droplets. The first concerns the CS$_2$ dimer. In Sect. 9.3.5 we already discussed how this dimer can be aligned by a laser pulse and how velocity images of $CS_2^+$ ions, resulting from double ionization of aligned dimers and subsequent Coulomb explosion into two $CS_2^+$ fragment ions, make it possible to infer that the dimer structure is cross-shaped. The dihedral angle, i.e. the angle between the interatomic axes of the $CS_2$ monomers could, however, not be determined. We now discuss how this angle can be determined from images of $S^+$ ions. The pump and the probe laser pulses were the same as those used for the measurement described in Sect. 9.3.5.

Figure 9.31 (a1) shows a velocity map image of $S^+$ ions under doping cell conditions where each droplet picks up at most one molecule. The angular distribution of the $S^+$ ions is localized along the vertical axis, which is parallel to the polarization direction of the alignment pulse. In analogy with the adiabatic alignment experiments on $I_2$, Sect. 9.3.5, we conclude that the $CS_2$ molecules are 1D aligned. The identification of the $S^+$ ions as arising from Coulomb explosion of the $CS_2$ molecules is obtained from the angular covariance map of the $S^+$ ions, Fig. 9.31 (a2), determined from ions outside the yellow circle. Strong correlations of ions with a relative emission of $180°$ are observed similar to the observations for $I_2$. Such a correlation can originate from double ionization of $CS_2$ and fragmentation into ($S^+$, C and $S^+$) or triple ionization of $CS_2$ and fragmentation into ($S^+$, $C^+$ and $S^+$).

The more interesting observations appear when the doping pressure is increased such that a significant number of the droplets become doped with $CS_2$ dimers. The $S^+$ image, shown in Fig. 9.31 (b1) is still confined along the vertical axis. It is wider than the distribution in Fig. 9.31 (a1) but it is not easy to identify a clear signature of the contribution from the dimers in this image alone. However, the contribution from the dimers clearly stands out in the angular covariance map of the image displayed in Fig. 9.31 (b2). In addition to the prominent monomer signals at $(0°, 180°)$ and $(180°, 0°)$ strong correlations are present centered approximately at $(-45°, 45°)$, $(-45°, 135°)$, $(-45°, 225°)$, $(45°, 135°)$, $(45°, 225°)$, $(135°, 225°)$ and, since the covariance map is an autovariance map, at six equivalent positions obtained by mirroring in the central diagonal. Analysis of these extra islands in the covariance map makes it possible to conclude that the dihedral angle is $90°$. The details are given in [171]. The method described here for $CS_2$ was also applied to determine the structure of the OCS dimer [199]. The structure was found to be a slipped-parallel shape similar to the structure found for gas phase dimers [200].

The second example concerns a more complex system namely the dimer of tetracene (Tc), a polycyclic aromatic hydrocarbon composed of four benzene rings fused together in a linear geometry. As a starting point velocity map images of parent ions, $Tc^+$, were recorded. The angular covariance map of the $Tc^+$ ions created with the alignment pulse polarized along the X-axis (perpendicular to the detector, see Fig. 9.5), is depicted in Fig. 9.32($a_e$). The prominent diagonal stripes show correlation between ions departing with a relative angle of $180°$. As in the previous case of $(CS_2)_2$, we conclude that Tc dimers are formed and that they can be identified by double ionization and subsequent Coulomb explosion into a pair of $Tc^+$ ions. Next, we recorded $Tc^+$ images and, in particular, determined the corresponding angular covariance maps for different alignment geometries. In the case of the $CS_2$ dimer this was done by synchronizing the probe pulse to the peak of the 160 ps long alignment

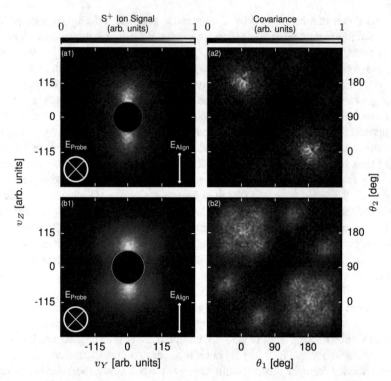

**Fig. 9.31** (a1)-(b1) $S^+$ ion images and (a2)-(b2) corresponding angular covariance maps created from ions outside the yellow circles. The polarization state of the probe (alignment) pulse is given in the lower left (right) corner of each ion image. Image (a1) [(b1)] was recorded under the monomer-[dimer]-doping-condition. Here the central region is removed due to background contaminants occurring at m/z = 32 Da. Reproduced from [171] with permission from American Physical Society (APS). Copyright (2018) by APS

pulse to benefit from the highest degree of alignment. This approach does not work for the Tc case because the $Tc^+$ ion is destroyed by the alignment pulse similar to what happened to the DIBP molecules illustrated in Fig. 9.28. Now we can, however, make use of the ability to create field-free alignment by rapidly turning off the alignment field. Therefore, the measurements on aligned Tc dimers are recorded using the truncated alignment pulse and the probe pulse sent 10 ps after the truncation. In contrast, the $CS_2^+$ ions were not significantly fragmented by the alignment pulse, and so in that case, no truncation was needed.

Column two from the left in Fig. 9.32 shows the angular covariance maps for $Tc^+$ recorded for different alignment geometries. The covariance maps change when the polarization of the alignment pulse is changed from linear to elliptical. The approach to deducing the dimer structure is to compare the experimental covariance maps to covariance maps simulated for the most plausible structures. This was done employing seven different dimer conformations. The angular covariance maps, sub-

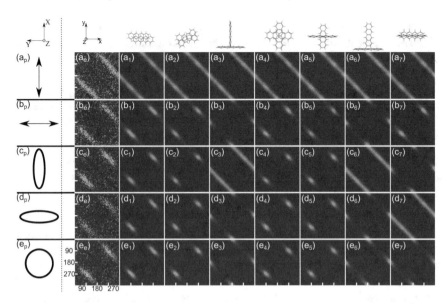

**Fig. 9.32** Angular covariance maps obtained for the Tc dimer. The left column shows the polarization state of the alignment pulse for each row. (the laboratory axes are illustrated at the top). Index e (second column from the left) refers to the experimental results, while indices 1–7 refer to simulated results for different plausible conformations of the tetracene dimer depicted on top of the corresponding column. Each panel axis ranges linearly from 0 to 360°. Reproduced from [201] licensed under a Creative Commons Attribution (CC BY) license

indexed 1–7, are displayed in Fig. 9.32. Details of the numerical simulations are given in [201]. The comparison shows that conformations 1, 2, 4, and 5, produce covariance maps consistent with the experimental data.

To narrow down the possible structures for the dimers we would ideally like to analyze covariance maps for another fragment ion species, just as the $S^+$ ions were analyzed in addition to the $CS_2^+$ ions for the carbon disulfide dimer case. For tetracene, fragmentation will produce $H^+$ or hydrocarbon ions and neither of these will be particularly structure-informative because they can originate from different parts of the Tc molecule. Instead, we performed an alternative type of measurement by recording the alignment-dependent ionization yield of the Tc dimer. In practice, the $Tc^+$ ion yield was measured for the 1D aligned dimers with the probe pulse polarized either parallel or perpendicular to the alignment polarization. For $I_{\text{probe}} = 3 \times 10^{12}$ W/cm$^2$, the $Tc^+$ yield in the parallel geometry was 5.5 times higher than in the perpendicular geometry. Such a large difference in the ionization yield implies that the dimer structure must be highly anisotropic. In particular, the dimer must possesses a 'long' axis leading to the highest ionization rate when the probe pulse is polarized along it. Such an anisotropic structure is compatible with conformations 1 and 2 but not with conformations 4 and 5. Thus, we conclude that the possible conformations of the Tc dimer is a slipped-parallel or parallel-slightly rotated structure.

Recently, we have applied the CEI method to the bromobenzene homodimer, $(BrPh)_2$, and the bromobenzene-iodine, $BrPh–I_2$, heterodimer [202]. As for the $CS_2$ dimer, both $(BrPh)_2$ and $BrPh–I_2$ produce 'tracer-ions', here $Br^+$ and $I^+$, in addition to the parent ions. Analysis of the angular covariance maps for both the parent ions and the atomic fragment ions, lead us to conclude that for $BrPh–I_2$, the $I_2$ molecular axis is approximately perpendicular to the C–Br axis of bromobenzene, whereas $(BrPh)_2$, has a stacked planar structure, with the two bromine atoms pointing in opposite directions.

## 9.4 Conclusion

The work presented in this article has shown that laser-induced alignment techniques, originally developed for isolated molecules in the gas phase, can be transferred to molecules embedded in He nanodroplets. In the (quasi) adiabatic limit, here realized with 160 ps long alignment pulses, the alignment dynamics observed for different molecular species was essentially the same as that of isolated molecules, i.e. the degree of alignment followed the intensity profile of the alignment laser pulse. Furthermore, the intensity dependence of the strongest alignment, reached at the peak of the pulse, agreed well with a gas phase model calculation for a rotational temperature of 0.4 K. These observations indicate that the mechanism of adiabatic laser-induced alignment of molecules in He droplets is the same as that of isolated molecules with a 0.4 K temperature. This temperature, which is lower than the 1–2 K temperature typically reached in cold supersonic beams of isolated molecules, is advantageous for creating very high degrees of alignment, with $\langle \cos^2 \theta_{2D} \rangle$ values in excess of 0.95. By truncating the alignment pulse in time just after its peak, we demonstrated that a 5–20 ps long time window can be created, where the alignment pulse intensity is reduced by several orders of magnitude and yet where the degree of alignment remains high. The reduction of the intensity was sufficient to eliminate the fragmentation of parent ions of large molecules that the remainder of the alignment pulse usually causes. This opens new opportunities for exploring e.g. how the ionization rate of large molecules induced by strong laser pulses or by VUV laser pulses, depends on the molecular alignment with respect to the polarization state of the ionizing laser pulse. In the field-free period, the intensity of the alignment pulse is also expected to be low enough that molecules in electronically excited states will remain intact. This opens unexplored possibilities for femtosecond time-resolved imaging of molecules undergoing fundamental photo-induced intramolecular processes in electronically excited states using Coulomb explosion, or diffraction by ultrashort x-ray [203, 204] or electron pulses [205–208].

One application, discussed here, of adiabatically aligned molecules is structure determination of molecular dimers through Coulomb explosion imaging. We showed that by aligning molecular dimers, either 1-dimensionally or 3-dimensionally, prior to their Coulomb explosion by an intense laser fs laser pulse, made it possible to determine the dimer configuration from the angular covariance maps of the fragment ions. For dimers of small molecules like $CS_2$ or OCS, the structural resolution of the CEI method cannot compete with that available from IR spectroscopy. For larger systems like the tetracene dimer, the IR spectra may, however, be too congested to extract any structural information. The biggest promise of the CEI technique lies, however, in its ability to capture the structure of dimers as they change on the natural atomic time scale. Experimentally, this can be realized by using a fs pump laser pulse to initiate some event of interest like exciplex formation [209, 210] from a weakly-bonded dimer and then following the structural evolution by Coulomb explosion with an intense fs probe pulse sent at a sequence of different times. More generally, it should be possible to trigger bimolecular reactions [211, 212], by initiating the dynamics from prereactive complexes with the fs pump pulse.

It was also shown that molecules in He droplets can be aligned by laser pulses that are much shorter than the intrinsic rotational period of the molecules. In this nonadiabatic limit, two regimes were identified. When the laser pulses are weak, meaning that the rotational energy acquired by the molecules remain below a few $cm^{-1}$, the alignment dynamics can be accurately described by a gas phase model where the molecule is characterized by the effective rotational and centrifugal distortion constants, and the inhomogenoeus broadening of these two parameters is taken into account. Thus, the molecules essentially rotate (quasi-)freely although the time-dependent degree of alignment in practice appear different from that of gas phase molecules because the centrifugal distortion constant, 3–4 orders of magnitude larger than for isolated molecules, causes a strong dispersion of $\langle \cos^2 \theta_{2D} \rangle$, and the inhomogeneous broadening a gradual decay of the amplitude of the oscillations of $\langle \cos^2 \theta_{2D} \rangle$. We believe the quasi-free rotation is a result of the superfluidity of the He droplets.

When the duration of the alignment pulse was kept short (hundreds of fs or a few ps) but its intensity increased significantly, we observed that the rotational dynamics during the first few ps were essentially as fast as for isolated molecules under the same laser conditions. This phenomenon was interpreted as the molecule and its He solvation shell being set into so fast rotation by the strong alignment pulse that the weakly bonded He atoms in the shell were shed off due to the centrifugal force. The interpretation was backed by classical calculations.

Looking forward, there are exciting opportunities to be explored with nonadiabatic alignment of molecules in He droplets. First, the quasi-free rotation observed for low intensities of the alignment pulse is, we believe, a result of superfluidity of the He droplets: the induced rotational energy remains so low that coupling between rotation and the He excitations (rotons, phonons) is very weak. When the intensity of the alignment pulse is increased higher-lying rotational states will be excited and the coupling to the He excitations will increase strongly. It will be interesting to investigate both the structure and dynamics (lifetime) of rotational states in this unexplored regime which is inaccessible to MW and IR spectroscopy. Another opportunity is to use the short, very intense alignment pulse to transiently decouple a molecule from its solvation shell. This will create a situation where the molecule-He system is extremely far away from equilibrium. For instance, we estimate that a $CS_2$ molecule can acquire more than $500 \, cm^{-1}$ of rotational energy by the interaction with an alignment pulse. In comparison the rotational energy at 0.4 K is $\sim 0.3\,\mathrm{cm}^{-1}$. We propose to measure how long it takes for the equilibrium to be restored. In practise this could be done by measuring the rotational dynamics induced by a weak alignment pulse sent after the intense distortion pulse and observing how much the weak pulse must be delayed with respect to the distortion pulse that the alignment trace recorded becomes identical to a reference trace obtained without any distortion pulse. Finally, the studies discussed in this article concerned molecules embedded inside He droplets. Dimers of alkali atoms are, however, residing on the surface of He droplets [213–215]. It will be interesting to apply the nonadiabatic alignment schemes to investigate the rotation of alkali dimers, a topic that is so far unexplored. Such studies, which are now ongoing in our group, may also provide information about their binding to the surface.

**Acknowledgements** The work described here was supported by The Independent Research Fund Denmark (grant no. 10-083041 and no. 8021-00232B), The Lundbeck Foundation (grant no. R45-A4280), The Carlsberg Foundation (grant no. 2009-01-0378), The European Research Council-AdG (Project No. 320459, DropletControl), The European Union Horizon 2020 research and innovation program under the Marie Sklodowska-Curie Grant Agreement No. 641789 MEDEA and No. 674960 ASPIRE, and The Villum Foundation: Villum Experiment (grant no. 23177) and Villum Investigator (grant no. 25886.)

# References

1. P.R. Brooks, Science **193**, 11–16 (1976)
2. M. Lein, J. Phys. B: At. Mol. Opt. Phys. **40**, R135–R173 (2007)
3. L. Holmegaard, J.L. Hansen, L. Kalhoj, S. Louise Kragh, H. Stapelfeldt, F. Filsinger, J. Küpper, G. Meijer, D. Dimitrovski, M. Abu-samha, C.P.J. Martiny, L. Bojer Madsen, Nat. Phys. **6**, 428–432 (2010)
4. J. Arlt, D.P. Singh, J.O.F. Thompson, A.S. Chatterley, P. Hockett, H. Stapelfeldt, K.L. Reid, Mol. Phys. **119**, e1836411 (2021)
5. D.H. Parker, R.B. Bernstein, Annu. Rev. Phys. Chem. **40**, 561 (1989)
6. H.J. Loesch, A. Remscheid, J. Chem. Phys. **93**, 4779–4790 (1990)
7. B. Friedrich, D.R. Herschbach, Nature **353**, 412–414 (1991)
8. J.M. Rost, J.C. Griffin, B. Friedrich, D.R. Herschbach, Phys. Rev. Lett. **68**, 1299–1302 (1992)
9. P.A. Block, E.J. Bohac, R.E. Miller, Phys. Rev. Lett. **68**, 1303–1306 (1992)
10. B. Friedrich, D.P. Pullman, D.R. Herschbach, J. Phys. Chem. **95**, 8118–8129 (1991)
11. V. Aquilanti, D. Ascenzi, D. Cappelletti, F. Pirani, Nature **371**, 399–402 (1994)
12. V. Aquilanti, M. Bartolomei, F. Pirani, D. Cappelletti, F. Vecchiocattivi, Y. Shimizu, T. Kasai, Phys. Chem. Chem. Phys. **7**, 291 (2005)
13. R.E. Drullinger, R.N. Zare, J. Chem. Phys. **51**, 5532–5542 (1969)
14. M.J. Weida, C.S. Parmenter, J. Chem. Phys. **107**, 7138–7147 (1997)
15. B. Friedrich, D. Herschbach, Phys. Rev. Lett. **74**, 4623–4626 (1995)
16. T. Seideman, J. Chem. Phys. **103**, 7887–7896 (1995)
17. W. Kim, P.M. Felker, J. Chem. Phys. **104**, 1147–1150 (1996)
18. T. Seideman, Phys. Rev. Lett. **83**, 4971 (1999)
19. H. Sakai, C.P. Safvan, J.J. Larsen, K.M. Hilligsøe, K. Hald, H. Stapelfeldt, J. Chem. Phys. **110**, 10235–10238 (1999)
20. N.E. Henriksen, Chem. Phys. Lett. **312**, 196–202 (1999)

21. J. Ortigoso, M. Rodríguez, M. Gupta, B. Friedrich, J. Chem. Phys. **110**, 3870–3875 (1999)
22. J.J. Larsen, K. Hald, N. Bjerre, H. Stapelfeldt, T. Seideman, Phys. Rev. Lett. **85**, 2470 (2000)
23. F. Rosca-Pruna, M.J.J. Vrakking, Phys. Rev. Lett. **87**, 153902 (2001)
24. I.S. Averbukh, R. Arvieu, Phys. Rev. Lett. **87**, 163601 (2001)
25. M. Machholm, N.E. Henriksen, Phys. Rev. Lett. **87**, 193001 (2001)
26. V. Renard, M. Renard, S. Guérin, Y.T. Pashayan, B. Lavorel, O. Faucher, H.R. Jauslin, Phys. Rev. Lett. **90**, 153601 (2003)
27. E. Péronne, M.D. Poulsen, C.Z. Bisgaard, H. Stapelfeldt, T. Seideman, Phys. Rev. Lett. **91**, 043003 (2003)
28. M. Leibscher, I.S. Averbukh, H. Rabitz, Phys. Rev. Lett. **90**, 213001 (2003)
29. J.G. Underwood, M. Spanner, M.Y. Ivanov, J. Mottershead, B.J. Sussman, A. Stolow, Phys. Rev. Lett. **90**, 223001 (2003)
30. P.W. Dooley, I.V. Litvinyuk, K.F. Lee, D.M. Rayner, M. Spanner, D.M. Villeneuve, P.B. Corkum, Phys. Rev. A **68**, 023406 (2003)
31. H. Stapelfeldt, T. Seideman, Rev. Mod. Phys. **75**, 543 (2003)
32. T. Seideman, E. Hamilton, Adv. At. Mol. Opt. Phys **52**, 289–329 (2005)
33. Y. Ohshima, H. Hasegawa, Int. Rev. Phys. Chem. **29**, 619–663 (2010)
34. S. Fleischer, Y. Khodorkovsky, E. Gershnabel, Y. Prior, I.S. Averbukh, Isr. J. Chem. **52**, 414–437 (2012)
35. C.P. Koch, M. Lemeshko, D. Sugny, Rev. Mod. Phys. **91**, 035005 (2019)
36. T. Seideman, J. Chem. Phys. **115**, 5965–5973 (2001)
37. B. Friedrich, D. Herschbach, J. Phys. Chem. **99**, 15686–15693 (1995)
38. J.H. Nielsen, P. Simesen, C.Z. Bisgaard, H. Stapelfeldt, F. Filsinger, B. Friedrich, G. Meijer, J. Küpper, Phys. Chem. Chem. Phys. **13**, 18971–18975 (2011)
39. V. Renard, M. Renard, A. Rouzée, S. Guérin, H.R. Jauslin, B. Lavorel, O. Faucher, Phys. Rev. A **70**, 033420 (2004)
40. M. Leibscher, I.S. Averbukh, H. Rabitz, Phys. Rev. A **69**, 013402 (2004)
41. B.J. Sussman, J.G. Underwood, R. Lausten, M.Y. Ivanov, A. Stolow, Phys. Rev. A **73**, 053403 (2006)
42. C.B. Madsen, L.B. Madsen, S.S. Viftrup, M.P. Johansson, T.B. Poulsen, L. Holmegaard, V. Kumarappan, K.A. Jørgensen, H. Stapelfeldt, Phys. Rev. Lett. **102**, 073007–4 (2009)
43. J.L. Hansen, J.H. Nielsen, C.B. Madsen, A.T. Lindhardt, M.P. Johansson, T. Skrydstrup, L.B. Madsen, H. Stapelfeldt, J. Chem. Phys. **136**, 204310–204310–10 (2012)
44. L. Christensen, J.H. Nielsen, C.B. Brandt, C.B. Madsen, L.B. Madsen, C.S. Slater, A. Lauer, M. Brouard, M.P. Johansson, B. Shepperson, H. Stapelfeldt, Phys. Rev. Lett. **113**, 073005 (2014)
45. C.Z. Bisgaard, O.J. Clarkin, G. Wu, A.M.D. Lee, O. Gessner, C.C. Hayden, A. Stolow, Science **323**, 1464–1468 (2009)
46. J. Itatani, J. Levesque, D. Zeidler, H. Niikura, H. Pépin, J.C. Kieffer, P.B. Corkum, D.M. Villeneuve, Nature **432**, 867–871 (2004)
47. T. Kanai, S. Minemoto, H. Sakai, Nature **435**, 470–474 (2005)
48. J.P. Marangos, S. Baker, N. Kajumba, J.S. Robinson, J.W.G. Tisch, R. Torres, Phys. Chem. Chem. Phys. **10**, 35–48 (2008)
49. B.K. McFarland, J.P. Farrell, P.H. Bucksbaum, M. Gühr, Science **322**, 1232–1235 (2008)
50. P.M. Kraus, B. Mignolet, D. Baykusheva, A. Rupenyan, L. Horný, E.F. Penka, G. Grassi, O.I. Tolstikhin, J. Schneider, F. Jensen, L.B. Madsen, A.D. Bandrauk, F. Remacle, H.J. Wörner, Science **350**, 790–795 (2015)
51. B. Shepperson, A.S. Chatterley, L. Christiansen, A.A. Søndergaard, H. Stapelfeldt, Phys. Rev. A **97**, 013427 (2018)
52. J.P. Toennies, A.F. Vilesov, Angew. Chem. Int. Ed. **43**, 2622–2648 (2004)
53. M.Y. Choi, G.E. Douberly, T.M. Falconer, W.K. Lewis, C.M. Lindsay, J.M. Merritt, P.L. Stiles, R.E. Miller, Int. Rev. Phys. Chem. **25**, 15 (2006)
54. S. Yang, A.M. Ellis, Chem. Soc. Rev. **42**, 472–484 (2012)

55. J.L. Hansen, Imaging molecular frame dynamics using spatially oriented molecules. Ph.D. thesis, Aarhus University, Aarhus (2012)

56. S. Trippel, T.G. Mullins, N.L. Müller, J.S. Kienitz, K. Dlugolecki, J. Küpper, Mol. Phys. **111**, 1738–1743 (2013)

57. T. Yatsuhashi, N. Nakashima, J. Photochem. Photobiol. C: Photochem. Rev. **34**, 52–84 (2018)

58. X. Ren, V. Makhija, V. Kumarappan, Phys. Rev. A **85**, 033405 (2012)

59. J. Mikosch, C.Z. Bisgaard, A.E. Boguslavskiy, I. Wilkinson, A. Stolow, The J. Chem. Phys. **139**, 024304 (2013)

60. D.W. Chandler, P.L. Houston, The J. Chem. Phys. **87**, 1445–1447 (1987)

61. A.T.J.B. Eppink, D.H. Parker, Rev. Sci. Instrum. **68**, 3477–3484 (1997)

62. U. Even, J. Jortner, D. Noy, N. Lavie, C. Cossart-Magos, J. Chem. Phys. **112**, 8068–8071 (2000)

63. B. Shepperson, A.S. Chatterley, A.A. Søndergaard, L. Christiansen, M. Lemeshko, H. Stapelfeldt, The J. Chem. Phys. **147**, 013946 (2017)

64. J.H. Posthumus, A.J. Giles, M.R. Thompson, K. Codling, J. Phys. B: At. Mol. Opt. Phys. **29**, 5811–5829 (1996)

65. A.A. Søndergaard, B. Shepperson, H. Stapelfeldt, J. Chem. Phys. **147**, 013905 (2017)

66. C.S. Slater, S. Blake, M. Brouard, A. Lauer, C. Vallance, J.J. John, R. Turchetta, A. Nomerotski, L. Christensen, J.H. Nielsen, M.P. Johansson, H. Stapelfeldt, Phys. Rev. A **89**, 011401 (2014)

67. L.J. Frasinski, J. Phys. B: At. Mol. Opt. Phys. **49**, 152004 (2016)

68. C. Vallance, D. Heathcote, J.W.L. Lee, J. Phys. Chem. A **125**, 1117–1133 (2021)

69. L. Christensen, L. Christiansen, B. Shepperson, H. Stapelfeldt, Phys. Rev. A **94**, 023410 (2016)

70. R. Torres, R. de Nalda, J.P. Marangos, Phys. Rev. A **72**, 023420 (2005)

71. S. Trippel, T. Mullins, N.L.M. Müller, J.S. Kienitz, J.J. Omiste, H. Stapelfeldt, R. González-Férez, J. Küpper, Phys. Rev. A **89**, 051401 (2014)

72. S. Guérin, J. Mod. Opt. **55**, 3193–3201 (2008)

73. S. Guérin, A. Rouzée, E. Hertz, Phys. Rev. A **77**, 041404 (2008)

74. E.T. Karamatskos, S. Raabe, T. Mullins, A. Trabattoni, P. Stammer, G. Goldsztejn, R.R. Johansen, K. Długołecki, H. Stapelfeldt, M.J.J. Vrakking, S. Trippel, A. Rouzée, J. Küpper, Nat. Commun. **10**, 3364 (2019)

75. A.A. Søndergaard, *Understanding Laser-Induced Alignment and Rotation of Molecules Embedded in Helium Nanodroplets* (Aarhus University, Thesis, 2016)

76. E.F. Thomas, A.A. Søndergaard, B. Shepperson, N.E. Henriksen, H. Stapelfeldt, Phys. Rev. Lett. **120**, 163202 (2018)

77. P.M. Felker, J. Phys. Chem. **96**, 7844–7857 (1992)

78. C. Riehn, Chem. Phys. **283**, 297–329 (2002)

79. C.B. Madsen, L.B. Madsen, S.S. Viftrup, M.P. Johansson, T.B. Poulsen, L. Holmegaard, V. Kumarappan, K.A. Jørgensen, H. Stapelfeldt, J. Chem. Phys. **130**, 234310–9 (2009)

80. I. Nevo, L. Holmegaard, J.H. Nielsen, J.L. Hansen, H. Stapelfeldt, F. Filsinger, G. Meijer, J. Küpper, Phys. Chem. Chem. Phys. **11**, 9912–9918 (2009)

81. J.G. Underwood, B.J. Sussman, A. Stolow, Phys. Rev. Lett. **94**, 143002 (2005)

82. K.F. Lee, D.M. Villeneuve, P.B. Corkum, A. Stolow, J.G. Underwood, Phys. Rev. Lett. **97**, 173001 (2006)

83. X. Ren, V. Makhija, V. Kumarappan, Phys. Rev. Lett. **112**, 173602 (2014)

84. S.S. Viftrup, V. Kumarappan, S. Trippel, H. Stapelfeldt, E. Hamilton, T. Seideman, Phys. Rev. Lett. **99**, 143602–4 (2007)

85. J.L. Hansen, H. Stapelfeldt, D. Dimitrovski, M. Abu-samha, C.P.J. Martiny, L.B. Madsen, Phys. Rev. Lett. **106**, 073001 (2011)

86. B. Friedrich, D. Herschbach, J. Phys. Chem. A **103**, 10280–10288 (1999)

87. B. Friedrich, D. Herschbach, J. Chem. Phys. **111**, 6157–6160 (1999)

88. L. Holmegaard, J.H. Nielsen, I. Nevo, H. Stapelfeldt, F. Filsinger, J. Küpper, G. Meijer, Phys. Rev. Lett. **102**, 023001 (2009)

89. O. Ghafur, A. Rouzee, A. Gijsbertsen, W.K. Siu, S. Stolte, M.J.J. Vrakking, Nat Phys **5**, 289–293 (2009)
90. J.H. Nielsen, H. Stapelfeldt, J. Küpper, B. Friedrich, J.J. Omiste, R. González-Férez, Phys. Rev. Lett. **108**, 193001 (2012)
91. J.J. Omiste, R. González-Férez, Phys. Rev. A **94**, 063408 (2016)
92. S. Trippel, T. Mullins, N.L. Müller, J.S. Kienitz, R. González-Férez, J. Küpper, Phys. Rev. Lett. **114**, 103003 (2015)
93. F. Filsinger, J. Küpper, G. Meijer, L. Holmegaard, J.H. Nielsen, I. Nevo, J.L. Hansen, H. Stapelfeldt, J. Chem. Phys. **131**, 064309–13 (2009)
94. S.Haessler, J. Caillat, W. Boutu, C. Giovanetti-Teixeira, T. Ruchon, T. Auguste, Z. Diveki, P. Breger, A. Maquet, B. Carré, R. Taïeb, P. Saliéres, Nat. Phys. **6**, 200–206 (2010)
95. C. Vozzi, M. Negro, F. Calegari, G. Sansone, M. Nisoli, S. De Silvestri, S. Stagira, Nat. Phys. **7**, 822–826 (2011)
96. J.L. Hansen, L. Holmegaard, L. Kalhøj, S.L. Kragh, H. Stapelfeldt, F. Filsinger, G. Meijer, J. Küpper, D. Dimitrovski, M. Abu-samha, C.P.J. Martiny, L.B. Madsen, Phys. Rev. A **83**, 023406 (2011)
97. D. Dimitrovski, M. Abu-samha, L.B. Madsen, F. Filsinger, G. Meijer, J. Küpper, L. Holmegaard, L. Kalhøj, J.H. Nielsen, H. Stapelfeldt, Phys. Rev. A **83**, 023405 (2011)
98. I.V. Litvinyuk, K.F. Lee, P.W. Dooley, D.M. Rayner, D.M. Villeneuve, P.B. Corkum, Phys. Rev. Lett. **90**, 233003 (2003)
99. D. Pavicic, K.F. Lee, D.M. Rayner, P.B. Corkum, D.M. Villeneuve, Phys. Rev. Lett. **98**, 243001 (2007)
100. J.L. Hansen, L. Holmegaard, J.H. Nielsen, H. Stapelfeldt, D. Dimitrovski, L.B. Madsen, J. Phys. B: At. Mol. Opt. Phys. **45**, 015101 (2012)
101. J. Maurer, D. Dimitrovski, L. Christensen, L.B. Madsen, H. Stapelfeldt, Phys. Rev. Lett. **109**, 123001 (2012)
102. D. Dimitrovski, J. Maurer, H. Stapelfeldt, L. Madsen, Phys. Rev. Lett. **113**, 103005 (2014)
103. R. Johansen, K.G. Bay, L. Christensen, J. Thøgersen, D. Dimitrovski, L.B. Madsen, H. Stapelfeldt, J. Phys. B: At. Mol. Opt. Phys. **49**, 205601 (2016)
104. P. Sándor, A. Sissay, F. Mauger, M.W. Gordon, T.T. Gorman, T.D. Scarborough, M.B. Gaarde, K. Lopata, K.J. Schafer, R.R. Jones, J. Chem. Phys. **151**, 194308 (2019)
105. C.J. Hensley, J. Yang, M. Centurion, Phys. Rev. Lett. **109**, 133202 (2012)
106. R. Boll, D. Anielski, C. Bostedt, J.D. Bozek, L. Christensen, R. Coffee, S. De, P. Decleva, S.W. Epp, B. Erk, L. Foucar, F. Krasniqi, J. Küpper, A. Rouzée, B. Rudek, A. Rudenko, S. Schorb, H. Stapelfeldt, M. Stener, S. Stern, S. Techert, S. Trippel, M.J.J. Vrakking, J. Ullrich, D. Rolles, Phys. Rev. A **88**, 061402 (2013)
107. M.G. Pullen, B. Wolter, A.T. Le, M. Baudisch, M. Hemmer, A. Senftleben, C.D. Schröter, J. Ullrich, R. Moshammer, C.D. Lin, J. Biegert, Nat. Commun. **6** (2015)
108. J. Yang, M. Guehr, T. Vecchione, M.S. Robinson, R. Li, N. Hartmann, X. Shen, R. Coffee, J. Corbett, A. Fry, K. Gaffney, T. Gorkhover, C. Hast, K. Jobe, I. Makasyuk, A. Reid, J. Robinson, S. Vetter, F. Wang, S. Weathersby, C. Yoneda, M. Centurion, X. Wang, Nat. Commun. **7**, 1–9 (2016)
109. Wolter, B., Pullen, M.G., Le, A.T., Baudisch, M., Doblhoff-Dier, K., Senftleben, A., Hemmer, M., Schröter, C.D., Ullrich, J., Pfeifer, T., Moshammer, R., Gräfe, S., Vendrell, O., Lin, C.D., Biegert, J., Science **354**, 308–312 (2016)
110. J. Küpper, S. Stern, L. Holmegaard, F. Filsinger, A. Rouzée, A. Rudenko, P. Johnsson, A.V. Martin, M. Adolph, A. Aquila, S. Bajt, A. Barty, C. Bostedt, J. Bozek, C. Caleman, R. Coffee, N. Coppola, T. Delmas, S. Epp, B. Erk, L. Foucar, T. Gorkhover, L. Gumprecht, A. Hartmann, R. Hartmann, G. Hauser, P. Holl, A. Hömke, N. Kimmel, F. Krasniqi, K.U. Kühnel, J. Maurer, M. Messerschmidt, R. Moshammer, C. Reich, B. Rudek, R. Santra, I. Schlichting, C. Schmidt, S. Schorb, J. Schulz, H. Soltau, J.C.H. Spence, D. Starodub, L. Strüder, J. Thøgersen, M.J.J. Vrakking, G. Weidenspointner, T.A. White, C. Wunderer, G. Meijer, J. Ullrich, H. Stapelfeldt, D. Rolles, H.N. Chapman, Phys. Rev. Lett. **112**, 083002 (2014)

111. T. Kierspel, A. Morgan, J. Wiese, T. Mullins, A. Aquila, A. Barty, R. Bean, R. Boll, S. Boutet, P. Bucksbaum, H.N. Chapman, L. Christensen, A. Fry, M. Hunter, J.E. Koglin, M. Liang, V. Mariani, A. Natan, J. Robinson, D. Rolles, A. Rudenko, K. Schnorr, H. Stapelfeldt, S. Stern, J. Thøgersen, C.H. Yoon, F. Wang, J. Küpper, J. Chem. Phys. **152**, 084307 (2020)
112. P. Hockett, C.Z. Bisgaard, O.J. Clarkin, A. Stolow, Nat. Phys. **7**, 612–615 (2011)
113. C. Schröter, K. Kosma, T. Schultz, Science **333**, 1011–1015 (2011)
114. C. Schröter, J.C. Lee, T. Schultz, PNAS **115**, 5072–5076 (2018)
115. A.S. Chatterley, M.O. Baatrup, C.A. Schouder, H. Stapelfeldt, Phys. Chem. Chem. Phys. **22**, 3245–3253 (2020)
116. D.M. Villeneuve, S.A. Aseyev, P. Dietrich, M. Spanner, M.Y. Ivanov, P.B. Corkum, Phys. Rev. Lett. **85**, 542–545 (2000)
117. A. Korobenko, A.A. Milner, V. Milner, Phys. Rev. Lett. **112**, 113004 (2014)
118. A.A. Milner, A. Korobenko, V. Milner, Phys. Rev. A **93**, 053408 (2016)
119. A. Korobenko, V. Milner, Phys. Rev. Lett. **116**, 183001 (2016)
120. A.A. Milner, J.A. Fordyce, I. MacPhail-Bartley, W. Wasserman, V. Milner, I. Tutunnikov, I.S. Averbukh, Phys. Rev. Lett. **122**, 223201 (2019)
121. L. Christensen, J.H. Nielsen, C.S. Slater, A. Lauer, M. Brouard, H. Stapelfeldt, Phys. Rev. A **92**, 033411 (2015)
122. S. Ramakrishna, T. Seideman, Phys. Rev. Lett. **95**, 113001 (2005)
123. S. Ramakrishna, T. Seideman, J. Chem. Phys. **124**, 034101–034111 (2006)
124. T. Vieillard, F. Chaussard, D. Sugny, B. Lavorel, O. Faucher, J. Raman Spectrosc. **39**, 694–699 (2008)
125. N. Owschimikow, F. Königsmann, J. Maurer, P. Giese, A. Ott, B. Schmidt, N. Schwentner, J. Chem. Phys. **133**, 044311–044311–13 (2010)
126. J.M. Hartmann, C. Boulet, J. Chem. Phys. **136**, 184302–184302–17 (2012)
127. F. Chaussard, T. Vieillard, F. Billard, O. Faucher, J.M. Hartmann, C. Boulet, B. Lavorel, J. Raman Spectrosc. **46**, 691–694 (2015)
128. J. Ma, H. Zhang, B. Lavorel, F. Billard, E. Hertz, J. Wu, C. Boulet, J.M. Hartmann, O. Faucher, Nat. Commun. **10**, 1–7 (2019)
129. J. Ma, H. Zhang, B. Lavorel, F. Billard, J. Wu, C. Boulet, J.M. Hartmann, O. Faucher, Phys. Rev. A **101**, 043417 (2020)
130. J. Jang, R.M. Stratt, J. Chem. Phys. **113**, 5901–5916 (2000)
131. F. Stienkemeier, K.K. Lehmann, J. Phys. B: At. Mol. Opt. Phys. **39**, R127 (2006)
132. M. Hartmann, R.E. Miller, J.P. Toennies, A. Vilesov, Phys. Rev. Lett. **75**, 1566 (1995)
133. C. Callegari, A. Conjusteau, I. Reinhard, K.K. Lehmann, G. Scoles, J. Chem. Phys. **113**, 10535–10550 (2000)
134. G.E. Douberly, Molecules in helium nanodroplets, in *Topics in Applied Physics*, ed. by A. Slenczka, J.P. Toennies (2021)
135. V. Kumarappan, C.Z. Bisgaard, S.S. Viftrup, L. Holmegaard, H. Stapelfeldt, J. Chem. Phys. **125**, 194309–7 (2006)
136. A. Braun, M. Drabbels, Phys. Rev. Lett. **93**, 253401 (2004)
137. A. Braun, M. Drabbels, J. Chem. Phys. **127**, 114304 (2007)
138. M. Mudrich, F. Stienkemeier, Int. Rev. Phys. Chem. **33**, 301–339 (2014)
139. F. Bierau, P. Kupser, G. Meijer, G. von Helden, Phys. Rev. Lett. **105**, 133402 (2010)
140. A.I.G. Florez, D.S. Ahn, S. Gewinner, W. Schoellkopf, G. von Helden, Phys. Chem. Chem. Phys. **17**, 21902–21911 (2015)
141. A. Mauracher, O. Echt, A.M. Ellis, S. Yang, D.K. Bohme, J. Postler, A. Kaiser, S. Denifl, P. Scheier, Phys. Rep. **751**, 1–90 (2018)
142. G. von Helden, Molecules in helium nanodroplets, in *Topics in Applied Physics*, ed. by A. Slenczka, J.P. Toennies (2021)
143. A. Slenczka, Molecules in helium nanodroplets, in *Topics in Applied Physics*, ed. by A. Slenczka, J.P. Toennies (2021)
144. D. Pentlehner, R. Riechers, B. Dick, A. Slenczka, U. Even, N. Lavie, R. Brown, K. Luria, Rev. Sci. Instrum. **80**, 043302 (2009)

145. J.J. Larsen, H. Sakai, C.P. Safvan, I. Wendt-Larsen, H. Stapelfeldt, J. Chem. Phys. **111**, 7774–7781 (1999)
146. E. Hamilton, T. Seideman, T. Ejdrup, M.D. Poulsen, C.Z. Bisgaard, S.S. Viftrup, H. Stapelfeldt, Phys. Rev. A **72**, 043402–12 (2005)
147. J.H. Posthumus, Rep. Prog. Phys. **67**, 623 (2004)
148. D. Pentlehner, J.H. Nielsen, L. Christiansen, A. Slenczka, H. Stapelfeldt, Phys. Rev. A **87**, 063401 (2013)
149. A. Scheidemann, B. Schilling, J.P. Toennies, J. Phys. Chem. **97**, 2128–2138 (1993)
150. M. Lewerenz, B. Schilling, J.P. Toennies, Chem. Phys. Lett. **206**, 381–387 (1993)
151. D. Pentlehner, J.H. Nielsen, A. Slenczka, K. Mølmer, H. Stapelfeldt, Phys. Rev. Lett. **110**, 093002 (2013)
152. S. Grebenev, M. Hartmann, M. Havenith, B. Sartakov, J.P. Toennies, A.F. Vilesov, J. Chem. Phys. **112**, 4485 (2000)
153. A.S. Chatterley, L. Christiansen, C.A. Schouder, A.V. Jørgensen, B. Shepperson, I.N. Cherepanov, G. Bighin, R.E. Zillich, M. Lemeshko, H. Stapelfeldt, Phys. Rev. Lett. **125**, 013001 (2020)
154. K.K. Lehmann, Mol. Phys. **97**, 645–666 (1999)
155. R. Lehnig, P.L. Raston, W. Jäger, Faraday Discuss. **142**, 297–309 (2009)
156. K.K. Lehmann, J. Chem. Phys. **114**, 4643–4648 (2001)
157. D. Rosenberg, R. Damari, S. Kallush, S. Fleischer, J. Phys. Chem. Lett. **8**, 5128–5135 (2017)
158. A.A. Milner, A. Korobenko, J.W. Hepburn, V. Milner, The J. Chem. Phys. **147**, 124202 (2017)
159. K. Nauta, R.E. Miller, J. Chem. Phys. **115**, 10254–10260 (2001)
160. R. Zillich, Y. Kwon, K. Whaley, Phys. Rev. Lett. **93**, 250401 (2004)
161. B. Shepperson, A.A. Søndergaard, L. Christiansen, J. Kaczmarczyk, R.E. Zillich, M. Lemeshko, H. Stapelfeldt, Phys. Rev. Lett. **118**, 203203 (2017)
162. L. García-Gutierrez, L. Delgado-Tellez, l. Valdés, R. Prosmiti, P. Villarreal, G. Delgado-Barrio, J. Phys. Chem. A **113**, 5754–5762 (2009)
163. S.E. Ray, A.B. McCoy, J.J. Glennon, J.P. Darr, E.J. Fesser, J.R. Lancaster, R.A. Loomis, The J. Chem. Phys. **125**, 164314 (2006)
164. F. Paesani, K.B. Whaley, J. Chem. Phys. **121**, 4180–4192 (2004)
165. A.A. Søndergaard, R.E. Zillich, H. Stapelfeldt, J. Chem. Phys. **147**, 074304 (2017)
166. P. Vindel-Zandbergen, J. Jiang, M. Lewerenz, C. Meier, M. Barranco, M. Pi, N. Halberstadt, J. Chem. Phys. **149**, 124301 (2018)
167. R. Schmidt, M. Lemeshko, Phys. Rev. Lett. **114**, 203001 (2015)
168. M. Lemeshko, Phys. Rev. Lett. **118**, 095301 (2017)
169. J.D. Pickering, B. Shepperson, L. Christiansen, H. Stapelfeldt, Phys. Rev. A **99**, 043403 (2019)
170. C.A. Schouder, A.S. Chatterley, L.B. Madsen, F. Jensen, H. Stapelfeldt, Phys. Rev. A **102**, 063125 (2020)
171. J.D. Pickering, B. Shepperson, B.A. Hübschmann, F. Thorning, H. Stapelfeldt, Phys. Rev. Lett. **120**, 113202 (2018)
172. A.S. Chatterley, B. Shepperson, H. Stapelfeldt, Phys. Rev. Lett. **119**, 073202 (2017)
173. R. Boll, A. Rouzée, M. Adolph, D. Anielski, A. Aquila, S. Bari, C. Bomme, C. Bostedt, J.D. Bozek, H.N. Chapman, L. Christensen, R. Coffee, N. Coppola, S. De, P. Decleva, S.W. Epp, B. Erk, F. Filsinger, L. Foucar, T. Gorkhover, L. Gumprecht, A. Hömke, L. Holmegaard, P. Johnsson, J.S. Kienitz, T. Kierspel, F. Krasniqi, K.U. Kühnel, J. Maurer, J. Messerschmidt, R. Moshammer, N.L.M. Müller, B. Rudek, E. Savelyev, I. Schlichting, C. Schmidt, F. Scholz, S. Schorb, J. Schulz, J. Seltmann, M. Stener, S. Stern, S. Techert, J. Thøgersen, S. Trippel, J. Viefhaus, M. Vrakking, H. Stapelfeldt, J. Küpper, J. Ullrich, A. Rudenko, D. Rolles, Faraday Discuss. (2014)
174. A.S. Chatterley, C. Schouder, L. Christiansen, B. Shepperson, M.H. Rasmussen, H. Stapelfeldt, Nat. Commun. **10**, 133 (2019)
175. A.S. Chatterley, E.T. Karamatskos, C. Schouder, L. Christiansen, A.V. Jørgensen, T. Mullins, J. Küpper, H. Stapelfeldt, J. Chem. Phys. **148**, 221105 (2018)
176. Z.C. Yan, T. Seideman, The J. Chem. Phys. **111**, 4113–4120 (1999)

177. A. Goban, S. Minemoto, H. Sakai, Phys. Rev. Lett. **101**, 013001 (2008)
178. C.S. Slater, S. Blake, M. Brouard, A. Lauer, C. Vallance, C.S. Bohun, L. Christensen, J.H. Nielsen, M.P. Johansson, H. Stapelfeldt, Phys. Rev. A **91**, 053424 (2015)
179. H. Tanji, S. Minemoto, H. Sakai, Phys. Rev. A **72**, 063401 (2005)
180. V. Makhija, X. Ren, V. Kumarappan, Phys. Rev. A **85**, 033425 (2012)
181. Z. Vager, R. Naaman, E.P. Kanter, Science **244**, 426–431 (1989)
182. I. Luzon, K. Jagtap, E. Livshits, O. Lioubashevski, R. Baer, D. Strasser, Phys. Chem. Chem. Phys. **19**, 13488–13495 (2017)
183. T. Takanashi, K. Nakamura, E. Kukk, K. Motomura, H. Fukuzawa, K. Nagaya, S.I. Wada, Y. Kumagai, D. Iablonskyi, Y. Ito, Y. Sakakibara, D. You, T. Nishiyama, K. Asa, Y. Sato, T. Umemoto, K. Kariyazono, K. Ochiai, M. Kanno, K. Yamazaki, K. Kooser, C. Nicolas, C. Miron, T. Asavei, L. Neagu, M. Schöffler, G. Kastirke, X.J. Liu, A. Rudenko, S. Owada, T. Katayama, T. Togashi, K. Tono, M. Yabashi, H. Kono, K. Ueda, Phys. Chem. Chem. Phys. **19**, 19707–19721 (2017)
184. A. Rudenko, L. Inhester, K. Hanasaki, X. Li, S.J. Robatjazi, B. Erk, R. Boll, K. Toyota, Y. Hao, O. Vendrell, C. Bomme, E. Savelyev, B. Rudek, L. Foucar, S.H. Southworth, C.S. Lehmann, B. Kraessig, T. Marchenko, M. Simon, K. Ueda, K.R. Ferguson, M. Bucher, T. Gorkhover, S. Carron, R. Alonso-Mori, J.E. Koglin, J. Correa, G.J. Williams, S. Boutet, L. Young, C. Bostedt, S.K. Son, R. Santra, D. Rolles, Nature **546**, 129–132 (2017)
185. M. Pitzer, M. Kunitski, A.S. Johnson, T. Jahnke, H. Sann, F. Sturm, L.P.H. Schmidt, H. Schmidt-Böcking, R. Dörner, J. Stohner, J. Kiedrowski, M. Reggelin, S. Marquardt, A. Schießer, R. Berger, M.S. Schöffler, Science **341**, 1096–1100 (2013)
186. P. Herwig, K. Zawatzky, M. Grieser, O. Heber, B. Jordon-Thaden, C. Krantz, O. Novotný, R. Repnow, V. Schurig, D. Schwalm, Z. Vager, A. Wolf, O. Trapp, H. Kreckel, Science **342**, 1084–1086 (2013)
187. E.F. Thomas, N.E. Henriksen, J. Chem. Phys. **150**, 024301 (2019)
188. H. Stapelfeldt, E. Constant, P.B. Corkum, Phys. Rev. Lett. **74**, 3780–3783 (1995)
189. M. Burt, R. Boll, J.W.L. Lee, K. Amini, H. Köckert, C. Vallance, A.S. Gentleman, S.R. Mackenzie, S. Bari, C. Bomme, S. Düsterer, B. Erk, B. Manschwetus, E. Müller, D. Rompotis, E. Savelyev, N. Schirmel, S. Techert, R. Treusch, J. Küpper, S. Trippel, J. Wiese, H. Stapelfeldt, B.C. de Miranda, R. Guillemin, I. Ismail, L. Journel, T. Marchenko, J. Palaudoux, F. Penent, M.N. Piancastelli, M. Simon, O. Travnikova, F. Brausse, G. Goldsztejn, A. Rouzée, M. Géléoc, R. Geneaux, T. Ruchon, J. Underwood, D.M.P. Holland, A.S. Mereshchenko, P.K. Olshin, P. Johnsson, S. Maclot, J. Lahl, A. Rudenko, F. Ziaee, M. Brouard, D. Rolles, Phys. Rev. A **96**, 043415 (2017)
190. T. Endo, S.P. Neville, V. Wanie, S. Beaulieu, C. Qu, J. Deschamps, P. Lassonde, B.E. Schmidt, H. Fujise, M. Fushitani, A. Hishikawa, P.L. Houston, J.M. Bowman, M.S. Schuurman, F. Légaré, H. Ibrahim, Science **370**, 1072–1077 (2020)
191. H. Ibrahim, B. Wales, S. Beaulieu, B.E. Schmidt, N. Thiré, E.P. Fowe, R. Bisson, C.T. Hebeisen, V. Wanie, M. Giguére, J.C. Kieffer, M. Spanner, A.D. Bandrauk, J. Sanderson, M.S. Schuurman, F. Légaré, Nat. Commun. **5**, 4422 (2014)
192. U. Ablikim, C. Bomme, H. Xiong, E. Savelyev, R. Obaid, B. Kaderiya, S. Augustin, K. Schnorr, I. Dumitriu, T. Osipov, R. Bilodeau, D. Kilcoyne, V. Kumarappan, A. Rudenko, N. Berrah, D. Rolles, Sci. Rep. **6**, srep38202 (2016)
193. M. Burt, K. Amini, J.W.L. Lee, L. Christiansen, R.R. Johansen, Y. Kobayashi, J.D. Pickering, C. Vallance, M. Brouard, H. Stapelfeldt, The J. Chem. Physi. **148**, 091102 (2018)
194. E. Skovsen, M. Machholm, T. Ejdrup, J. Thøgersen, H. Stapelfeldt, Phys. Rev. Lett. **89**, 133004 (2002)
195. C. Petersen, E. Péronne, J. Thøgersen, H. Stapelfeldt, M. Machholm, Phys. Rev. A **70**, 033404 (2004)
196. T. Ergler, A. Rudenko, B. Feuerstein, K. Zrost, C.D. Schröter, R. Moshammer, J. Ullrich, Phys. Rev. Lett. **97**, 193001 (2006)
197. S. Zeller, M. Kunitski, J. Voigtsberger, A. Kalinin, A. Schottelius, C. Schober, M. Waitz, H. Sann, A. Hartung, T. Bauer, M. Pitzer, F. Trinter, C. Goihl, C. Janke, M. Richter, G. Kastirke,

M. Weller, A. Czasch, M. Kitzler, M. Braune, R.E. Grisenti, W. Schöllkopf, L.P.H. Schmidt, M.S. Schöffler, J.B. Williams, T. Jahnke, R. Dörner, PNAS **113**, 14651–14655 (2016)
198. M. Kunitski, Molecules in helium nanodroplets, in *Topics in Applied Physics*, ed. by A. Slenczka, J.P. Toennies (2021)
199. J.D. Pickering, B. Shepperson, L. Christiansen, H. Stapelfeldt, J. Chem. Phys. **149**, 154306 (2018)
200. N. Moazzen-Ahmadi, A.R.W. McKellar, Int. Rev. Phys. Chem. **32**, 611–650 (2013)
201. C. Schouder, A.S. Chatterley, F. Calvo, L. Christiansen, H. Stapelfeldt, Struct. Dyn. **6**, 044301 (2019)
202. Schouder, C., Chatterley, A.S., Johny, M., Hübschmann, F., Al-Refaie, A.F., Calvo, F., Küpper, J., Stapelfeldt, H., To be submitted (2021)
203. L.F. Gomez, K.R. Ferguson, J.P. Cryan, C. Bacellar, R.M.P. Tanyag, C. Jones, S. Schorb, D. Anielski, A. Belkacem, C. Bernando, R. Boll, J. Bozek, S. Carron, G. Chen, T. Delmas, L. Englert, S.W. Epp, B. Erk, L. Foucar, R. Hartmann, A. Hexemer, M. Huth, J. Kwok, S.R. Leone, J.H.S. Ma, F.R.N.C. Maia, E. Malmerberg, S. Marchesini, D.M. Neumark, B. Poon, J. Prell, D. Rolles, B. Rudek, A. Rudenko, M. Seifrid, K.R. Siefermann, F.P. Sturm, M. Swiggers, J. Ullrich, F. Weise, P. Zwart, C. Bostedt, O. Gessner, A.F. Vilesov, Science **345**, 906–909 (2014)
204. R.M.P. Tanyag, B. Langbehn, D. Rupp, T. Möller, Molecules in helium nanodroplets, in *Topics in Applied Physics*, ed. by A. Slenczka, J.P. Toennies (2021)
205. Y. He, J. Zhang, W. Kong, J. Chem. Phys. **145**, 034307 (2016)
206. L. Lei, Y. Yao, J. Zhang, D. Tronrud, W. Kong, C. Zhang, L. Xue, L. Dontot, M. Rapacioli, J. Phys. Chem. Lett. **11**, 724–729 (2020)
207. J. Zhang, S.D. Bradford, W. Kong, C. Zhang, L. Xue, J. Chem. Phys. **152**, 224306 (2020)
208. J. Zhang, Y. He, L. Lei, Y. Yao, S. Bradford, W. Kong, Molecules in helium nanodroplets, in *Topics in Applied Physics*, ed. by A. Slenczka, J.P. Toennies (2021)
209. H. Saigusa, E.C. Lim, Chem. Phys. Lett. **336**, 65–70 (2001)
210. M. Miyazaki, M. Fujii, Phys. Chem. Chem. Phys. **17**, 25989–25997 (2015)
211. D. Zhong, P.Y. Cheng, A.H. Zewail, The J. Chem. Phys. **105**, 7864–7867 (1996)
212. M.D. Wheeler, D.T. Anderson, M.I. Lester, Int. Rev. Phys. Chem. **19**, 501–529 (2000)
213. F. Stienkemeier, J. Higgins, W.E. Ernst, G. Scoles, Phys. Rev. Lett. **74**, 3592–3595 (1995)
214. F. Ancilotto, G. DeToffol, F. Toigo, Phys. Rev. B **52**, 16125–16129 (1995)
215. L. Bruder, M. Koch, M. Mudrich, F. Stienkemeier, Molecules in helium nanodroplets, in *Topics in Applied Physics*, ed. by A. Slenczka, J.P. Toennies (2021)

# Chapter 10
# Ultrafast Dynamics in Helium Droplets

Lukas Bruder, Markus Koch, Marcel Mudrich, and Frank Stienkemeier

**Abstract**  Helium nanodroplets are peculiar systems, as condensed superfluid entities on the nanoscale, and as vessels for studies of molecules and molecular aggregates and their quantum properties at very low temperature. For both aspects, the dynamics upon the interaction with light is fundamental for understanding the properties of the systems. In this chapter we focus on time-resolved experiments in order to study ultrafast dynamics in neat as well as doped helium nanodroplets. Recent experimental approaches are reviewed, ranging from time-correlated photon detection to femtosecond pump-probe photoelectron and photoion spectroscopy, coherent multidimensional spectroscopy as well as applications of strong laser fields and novel, extreme ultraviolet light sources. The experiments examined in more detail investigate the dynamics of atomic and molecular dopants, including coherent wave packet dynamics and long-lived vibrational coherences of molecules attached to and immersed inside helium droplets. Furthermore, the dynamics of highly-excited helium droplets including interatomic Coulombic decay and nanoplasma states are discussed. Finally, an outlook concludes on the perspectives of time-resolved experiments with helium droplets, including recent options provided by new radiation sources of femto- or even attosecond laser pulses up to the soft X-ray range.

L. Bruder · F. Stienkemeier (✉)
Institute of Physics, University of Freiburg, Hermann-Herder-Str. 3, 79104 Freiburg, Germany
e-mail: stienkemeier@uni-freiburg.de

L. Bruder
e-mail: lukas.bruder@physik.uni-freiburg.de

M. Koch
Institute of Experimental Physics, Graz University of Technology, Petersgasse 16, 8010 Graz, Austria
e-mail: markus.koch@tugraz.at

M. Mudrich
Institute of Physics and Astronomy, Aarhus University, Ny Munkegade 120, 8000 Aarhus C, Denmark
e-mail: mudrich@phys.au.dk

Indian Institute of Technology Madras, Chennai 600036, India

© The Author(s) 2022
A. Slenczka and J. P. Toennies (eds.), *Molecules in Superfluid Helium Nanodroplets*,
Topics in Applied Physics 145, https://doi.org/10.1007/978-3-030-94896-2_10

## 10.1   Introduction

Modern time-resolved experimental techniques using ultrafast lasers offer a fascinating approach to unraveling the intriguing dynamics of helium nanodroplet systems. Although it is tempting to separate the dynamical properties from the static picture of a molecular system, such an effort in many respects is not viable at all, because statics and dynamics are directly linked from fundamental principles of interactions. Moreover, key static aspects like e.g. the structure of a complex, may even only be understood by probing its dynamics. For instance, characteristics of interaction potentials determining structural properties may not be accessible by spectroscopy but may be determined from relaxation schemes and the motion of vibrational wavepackets therein. Therefore, experimental methods employing short-pulse lasers play a key role in gaining insight both into the structure and dynamics on the atomic and molecular level.

A fundamental connection of measurements using high-resolution continuous-wave lasers in the frequency domain and measurements with pulsed lasers in the time domain can be understood from the line width of a transition which is associated to the lifetime by Fourier transformation. In this way, even quantitative aspects of decay mechanisms are already encoded in the homogeneous and inhomogeneous broadenings of measured spectra. The other way around, from time-domain "Fourier-transform" spectroscopy, detailed spectroscopic information in the frequency domain can be gained with very high resolution even when using spectrally very broad, femtosecond laser pulses. However, many important dynamical aspects can only be characterized when measuring in the time domain; in particular, if energy decay paths of a system can furcate into many degrees of freedom or corresponding states, or if a series of secondary processes can be triggered by the laser excitation processes. Typical examples include structural dynamics like e.g. desorption or fragmentation processes on the one hand, and, on the other hand, electron dynamics from non-adiabatic couplings or the interaction in a many-body system. Finally, by applying coherent multidimensional spectroscopy methods, simultaneous high-frequency and high-time resolution down to the Fourier limit is achieved. First recent results in this direction will be discussed at the end of the chapter.

With respect to helium droplets or helium in general, the superfluid properties are strongly related to dynamical processes. E.g., frictionless flow and vorticity, which are key peculiarities of superfluid behavior, are inherently interwoven with motion and dynamical aspects of the material. The key experiments probing bulk superfluid properties like the Andronikashvili experiment of rotating disks [1], or the observation of superfluid film flow, the fountain effect or heat transport (zero sound) [2], directly examine motion of the system or a motional behavior of a probe interacting with the superfluid.

When probing the dynamics on the nanoscale down to atomic/molecular dimensions, due to the shorter lengthscales on the one hand, and the lower involved masses on the other hand, the corresponding time scales shorten down to the picosecond (ps) or femtosecond (fs) time range. Electronic processes have typical time scales in the femtosecond range; vibrations and rotations of individual molecules extend their dynamics into the picosecond range; motions involving larger structures or/and weak interactions may reach nanosecond time duration. As a consequence, in order to access dynamics in nanodroplets, it is almost always inevitable to have femtosecond time resolution of the laser pulses triggering and probing the dynamics.

The combination of ultrafast methods with helium droplets is exceptionally interesting. On the one hand, the role of helium droplets as an ideal spectroscopic matrix allows for the study of nuclear motion without strong perturbation of a strongly-interacting environment. On the other hand, doped droplets serve as unique model system for the relaxation dynamics of heterogeneous nanosystems. In this direction, rare-gas clusters also gained much attention in combination with new radiation sources providing extreme high-field or/and short-wavelength laser pulses. Here, molecular and cluster beam samples in vacuum are required to keep the complexity of the condensed systems on a level that is still tractable by theoretical modelling. Finally, nanoscopic superfluidity still bears fundamental open questions, in particular, when it comes to the relevance to short-time dynamics.

A variety of time-resolved techniques have been developed over the years and applied to specific experiments involving helium nanodroplets. In overview articles, time-resolved experiments on pure and doped helium nanodroplets have already been in the focus of reviewed work [3–5], however, not including the recent prominent developments in ultrafast laser techniques.

Before discussing in detail specific topics on the dynamics in helium droplets, in the following chapter, time-resolved experimental techniques will be introduced and the applications to helium nanodroplets will be summarized.

## 10.2   Time-Resolved Techniques Applied to Helium Nanodroplets

### 10.2.1   Time-Resolved Photon Detection

In order to perform a time-resolved measurement, a start and stop event has to be registered with defined timing. Because of the statistical nature of spontaneous events, it is practical to start timing with a laser-induced preparation of the system employing an ultrashort pulse which provides accurate timing. A straight-forward approach is time-resolved detection of signals like e.g. the arrival of emitted photons by standard electronics. A prominent variant is the so-called time-correlated single photon counting (TCSPC), i.e. detecting single photons from fast multi-channel-plate (MCP)-amplified photon signals combined with fast digitizing of arrival times. In this approach, the timing resolution is typically limited to a few tens of picoseconds. In helium droplets experiments, TCSPC served as the initial approach to studying the dynamics of photo-induced processes of dopants.

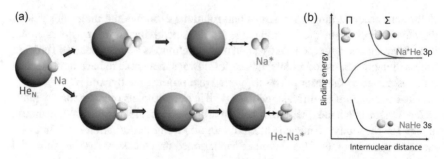

**Fig. 10.1** (a) Upon electronic excitation of an alkali-doped helium nanodroplet, desorption of the alkali atom (upper branch) or the formation of an exciplex molecule (lower branch) is induced, depending on the alignment of the *p*-orbital of the excited state ($\Sigma$ or $\Pi$ configuration, denoting the projection of the orbital angular momentum with respect to the internuclear axis). (**b**) Schematic potential diagram of diatomic states

In terms of systems that feature interesting dynamics, alkali atoms play a peculiar role because they do not submerge into helium nanodroplets but are located at the surface. The reason originates in the, compared to other atoms and molecules, large volume occupied by the valence electron, and the repulsive interaction of condensed helium to additional electrons, leading to bubble-like structures around alkali atoms. From simple arguments of surface tension and volume energy contributions, a binding motive at the surface without evolving a full bubble is energetically more stable when compared to the interior state. In other words, the alkali containing bubbles float, forming dimple-like textures on the surface of droplets. At the beginning of time-resolved studies, alkali-doped helium nanodroplets were in the focus because frequency-domain studies had manifested the surface location of dopants, desorption of dopants, the formation of exciplex molecules [6–10], fragmentation of dopant molecules, as well as spin flips [11, 12]. All of these aspects raised interesting questions concerning their dynamics.

The first TCSPC studies were performed on Na-doped helium nanodroplets [13] excited on the prominent $3p \leftarrow 3s$ transition ($D_1$ and $D_2$ lines). Depending on the orientation of the excited *p*-orbital perpendicular or parallel to the surface of the helium droplet, strong repulsive forces and consequently desorption of the excited atom, or attractive forces leading to NaHe* exciplex formation, respectively, set in upon laser interaction (Fig. 10.1). Exciplexes are complexes of a metal atom with one or a few He atoms which are stable only in an electronically excited state.

The time-resolved fluorescence measured when exciting the repulsive $\Sigma$-configuration of the NaHe$_N$ absorption band had an appearance time of 50–70 ps, significantly longer when compared to the instrument resolution of 20 ps. The latter was determined as the onset of fluorescence of gas-phase sodium atoms excited at the same transition. Shifting the laser in wavelength for the formation of NaHe exciplexes, and only collecting their red-shifted fluorescence, revealed a biexponential rise with 50–70 ps and 700 ps, which was assigned in comparison with theory to the two excited fine-structure states $^2\Pi_{1/2}$ and $^2\Pi_{1/3}$, respectively. These studies were extended both on the theory side, including the helium droplet interaction, and experimentally comparing different alkali metals (K, Rb) [14]. Interestingly, when going down the periodic table, the formation times of exciplexes along the $J = 1/2$ asymptote ($n\,^2P_{1/2} \leftarrow n\,^2S_{1/2}$) were measured to scale with the spin-orbit interaction strength, i.e. increasing into the nanosecond range. Opposed to that, upon excitation along the $J = 3/2$ path ($n\,^2P_{3/2} \leftarrow n\,^2S_{1/2}$), the formation times decreased, which was attributed to an enhanced long-range dispersion interaction for the heavier alkalis.

Experiments on Al atoms residing inside helium droplets provided first results on electronic relaxation dynamics [15]. A fast nonradiative quenching of the excited $3^2D$ state into the $4^2S$ state, releasing about $7000\,\mathrm{cm}^{-1}$ in energy, was measured to take place within 50 ps, which unfortunately matched the time-resolution in that measurement. Here, TCSPC was only able to provide an upper bound for the relaxation time.

Studies on photo-induced nonadiabatic dynamics of alkali trimers were also commenced with the technique of TCSPC. The peculiar properties of helium droplets to isolate van der Waals-bound high-spin quartet states [11, 16, 17] enabled to observe spin dynamics, i.e. forming covalently-bound alkali molecules upon intersystem crossing [18]. E.g., in the case of sodium, an intersystem-crossing time of 1.4 ns was determined, which significantly decreased for higher excitation energies approaching the access point to the doublet manifold. At the same time, vibronically-resolved data gave insight into the vibrational cooling of the trimers, which appeared to be on the same time scale as the spin-flip dynamics.

The influence of the helium droplet environment on spin-flip dynamics and predissociation was extended in detailed studies on the excitation of alkali dimers in triplet states ($1\,^3\Pi_g \leftarrow 1\,^3\Sigma_u^+$) [19]. The appearance of both molecular and atomic fragment emission was measured having a rise time < 80 ps, independent of the addressed vibrational excitation of the upper state. Predissociation and intersystem crossing appear to be in competition and on the same time scale. The intersystem crossing time in this case was deduced to be of the order of 10 ps which is surprisingly fast, considering the weak interaction to the helium surface which induces the process.

## 10.2.2   Pump-Probe Fluorescence Detection

To overcome the limitations of time resolution given by electronics, the femtosecond pump-probe technique was introduced many years ago in order to study details of molecular dynamics. For his pioneering work, Ahmed H. Zewail was awarded the Nobel Prize in 1999. In pump-probe studies, two or more ultrashort laser pulses are employed. In the simplest variant the process to be studied is triggered by an ultrashort laser pulse and the dynamics is probed by triggering a detection process, again with an ultrashort pulse delayed in time. The time resolution is only given by the properties of the laser pulses and the precision of setting a delay between the two pulses; for both, attosecond timing can be reached, covering the range needed for molecular processes and even electronic dynamics. An obvious extension of the TCSPC approach would be a pump-probe laser-induced fluorescence scheme. When using high-intensity femtosecond pulses, high photoionization rates can be achieved even in multi-photon processes. Alternatively to fluorescence detection, photoion or photoelectron detection can be advantageous because of the high detection efficiency of charged particles and mass and/or kinetic energy selectivity of detected particles can be obtained. For this reason, most of the results discussed in the following include ionization of the sample in the probe step.

## 10.2.3 Time-Resolved Spectroscopy by Photoion Detection

The first femtosecond pump-probe studies of doped He nanodroplets were carried out by the groups of Stienkemeier and Schulz at the Max-Born-Institut in Berlin in the late 1990s. Using the output of a mode-locked Ti:Sa laser in combination with mass-resolved ion detection using a quadrupole mass spectrometer, the yields of photoions where measured as a function of the delay between pairs of near-infrared (NIR) laser pulses. Owing to their large resonant absorption cross sections and extremely low ionization energies, alkali atoms, molecules, and clusters are well suited for this photoionization scheme and have been studied in detail in a series of such experiments [4, 10, 20–26]. Most importantly, due to their weak binding to the surface of He droplets, alkali atoms and molecules ionized by a resonant multiphoton process tend to detach from the droplets. This facilitates their detection as bare ions or as complexes with one or a few attached He atoms.

In photoionization experiments, the dynamics are determined by the interaction of both the neutral and ionic species with the He droplet which can qualitatively differ. Ionized dopants experience a much stronger attractive interaction towards the helium density than neutrals in their ground or excited states. This is in contrast to the above discussed fluorescence experiments, which solely focus on neutral species. For the case of an alkali-atom dopant, a schematic representation of the pump-probe photoionization scheme is shown in Fig. 10.2. Resonant excitation by the pump pulse induces the desorption of the atom from the droplet surface. Upon photoionization by the probe pulse, the dopant-droplet interaction suddenly changes from repulsive to attractive (Fig. 10.2b). Subsequently, the photoion either falls back into the droplet where it is solvated by forming a dense He shell around it, or it continues to move away from the droplet to be detected as a bare ion. In a series of femtosecond pump-probe experiments supported by TDDFT simulations, Mudrich and coworkers have obtained detailed insights into the competing dynamics between desorption and re-absorption of surface-attached alkali dopants [27–30] (Sect. 10.3.1). A similar behavior is also observed for other species immersed inside helium droplets by Koch and coworkers [31–33] (Sect. 10.3.2). From these experiments as well as theoretical predictions, it seems that the interplay between the de-solvation of excited

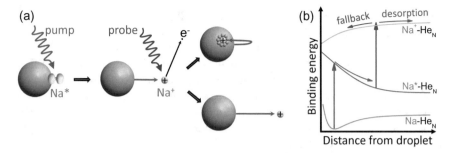

**Fig. 10.2** Femtosecond pump-probe photoionization scheme for the example of alkali-atom dopants. (**a**) Resonant dopant excitation by the pump pulse leads to a dopant-droplet repulsion, initiating the desorption of the dopant from the droplet surface. Photoionization of the dopant by the probe pulse causes the ion either to fall back into the droplet where it is solvated by forming high-density He solvation layers around it (upper branch), or to detach as a free ion (lower branch). (**b**) Generic pseudodiatomic alkali-He$_N$ potential curves illustrating the energetics of the involved states

neutrals and the solvation of the ionized species is a general trend in pump-probe photoionization experiments of alkalis and other dopants.

In addition, photoinduced processes on the intra and inter-dopant level overlay and interplay with the dopant-droplet interaction dynamics, to which pump-probe photoionization experiments provide access as well. Examples are the femtosecond and picosecond dynamics of exciplex formation for potassium (K) and rubidium (Rb) atoms attached to He droplets [10, 21]. Also, highly regular sub-fs oscillations were observed in the pump-probe traces due to electronic wavepacket interference. This phenomenon and a derived new spectroscopic scheme will be discussed in Sect. 10.6. Subsequent measurements on alkali dimers (Na$_2$, K$_2$, Rb$_2$) and trimers (K$_3$, Rb$_3$ and K-Rb heterotrimers) revealed essentially unperturbed vibrational wavepacket dynamics of the free molecules after their detachment from the droplets [22, 23, 25, 34], see Sect. 10.4.1. In a few instances, indications for the interaction of the He droplet with the vibrating molecules, causing dephasing and relaxation, were found and discussed in the framework of a quantum mechanical oscillator coupled to a superfluid bath [35, 36]. Most importantly, long-lasting vibrational coherences were measured, facilitated by the ultracold He droplet environment that prepares the molecules in their vibrational ground state prior to excitation. In this way, it was possible to measure highly resolved vibrational spectra of alkali dimers, trimers, and alkali-He exciplexes [22, 24, 25, 34, 37]. In these early experiments, a high laser pulse repetition rate (80 MHz) was used, which, in principle, introduces some ambiguity due to the possibility to excite and probe each droplet multiple times as it moves through the interaction region. As such, signals from species attached to the He droplet surface as well as from atoms and molecules already desorbed off the droplets into the gas phase may both have contributed in these experiments. Therefore, later experiments were carried out using amplified pulses at a repetition rate in the range 5-200 kHz where this ambiguity can be excluded. These experiments have mainly focused on the desorption dynamics of excited alkali atoms, molecules [38], and alkali-He exciplexes [27–30].

Another line of fs pump-probe experiments was pursued by the group of Tigges-bäumker and Meiwes-Broer in Rostock, based on their vast experience in strong-field ionization of metal clusters. When being exposed to intense NIR pulses, metal clus-ters (free or embedded in He nanodroplets) are multiply ionized and charge states of the exploding ions as high as $Z = 10$ for silver (Ag), $Z = 12$ for lead (Pb), and $Z = 13$ for cadmium (Cd) clusters were observed [39]. In addition to highly charged atomic and fragment clusters ions, mass spectra of strong-field ionized metal-doped He nanodroplets display regular progressions due to complexes of metal ions with attached He atoms [39–41]. For magnesium (Mg) ions, up to 150 attached He atoms were detected [40]. These are indicative for the formation of so-called 'snowballs', stable ion-He complexes first observed in bulk liquid He [42]. The term derives from the fact that for some species, the density of the local He shell around the cation adopts a regular structure and surpasses that of solid He. The pump-probe dynamics of snowball complexes of Ag were found to be opposite to that of the bare metal ions, indicating that He droplets can feature a cage effect causing the re-aggregation of fragments [40].

Initially, He nanodroplets were mostly regarded as an alternative method for form-ing metal clusters and the focus had been on the charging dynamics and Coulomb explosion of the dopant metal atoms, see Sect. 10.5.3 [43–46]. Later, the important role of the He shell in the ionization dynamics was recognized [40, 45, 47, 48] and the focus shifted more towards the nanoplasma dynamics of the He nanodroplets themselves [49–52]. In particular, resonant heating and charging of the nanoplasma manifests itself by enhanced yields of singly and even doubly charged He ions at a pump-probe delay around 0.5 ps, see Sect. 10.5.3 [51–53].

Particularly peculiar photoionization dynamics was observed by the Rostock group for Mg-doped He nanodroplets [54, 55]. Based on linear absorption spec-tra and on fs pump-probe resonant ionization traces of multiply doped He droplets, it was concluded that Mg atoms aggregate in He nanodroplets in an unusual way to form a foam-like structure where the metal atoms arrange themselves in a regular 10 Å-spaced network separated by He atoms. This structure collapses upon elec-tronic excitation to form metallic clusters. Thus, the transient mass spectra reveal a sharp drop of the yield of $Mg^+$ and small $Mg_n^+$ cluster ions within 350 fs due to the decreased ionisation cross-section of Mg as the electronic properties evolve from the atomic to a bulk-like state. Subsequent slow recovery of the Mg ion signals within $\approx 50$ ps was associated with the escape dynamics out of the He droplets. The for-mation of Mg foam in He droplets was essentially confirmed by theoretical model calculations [56].

The group of Gessner and Neumark in Berkeley performed seminal studies on the ultrafast dynamics of pure He nanodroplets using resonant excitation by XUV pulses. Using electron and ion imaging detection, intricate relaxation dynamics of highly excited He droplets were observed, including the emission of Rydberg atoms, small He clusters, and very low-energy electrons on various time scales. These studies have recently been reviewed [5] and will be discussed in more detail in Sect. 10.5.1.

More recently, the group of Koch in Graz succeeded in measuring excited-state dynamics of indium (In) atoms and dimers embedded inside He nanodroplets, see Sect. 10.3.2 [31–33]. Similarly to surface-bound alkali atoms, the delay-dependent ion yield revealed the ejection dynamics of the excited atom out of the He droplet. Indium dimers featured long-lasting vibrational coherences similarly to alkali dimers, despite their initial state of solvation inside the He droplets.

## 10.2.4 Time-Resolved Photoelectron Spectroscopy

Photoelectrons (PE) are an important observable for tracking ultrafast processes in $He_N$ with time-resolved pump–probe photoionization. In contrast, photo-ions are in many situations hard to detect and/or add additional dynamics due to their strong attractive interaction with the droplet, as discussed in the previous section. Time-resolved photoelectron spectroscopy (TRPES) is a well established pump–probe photoionization technique for measuring the temporal evolution of the kinetic energy and the yield of the generated PEs [57–59]. The probe pulse couples (or projects) the excited state wavefunction onto the ionic state, a process that is governed by electronic selection rules and the Franck–Condon overlap of the involved vibrational states. The evolution of the excited state energy and its ionization probability provides insight into the dynamics of electrons and nuclei, which are often coupled non-adiabatically. The interpretation of transient signals therefore relies on quantum-chemical simulations. Photophysical and photochemical processes that can be observed include, among others, intra- and intermolecular electron and proton transfer, non-adiabatic energy relaxation, quantum beats and wave packets of electronic, vibrational and rotational degrees of freedom, or photodissociation and -association.

The applicability of TRPES for the observation of ultrafast molecular processes inside $He_N$ stands or falls with the influence of the He environment on the PE observable. While this influence is moderate at picosecond timescales after photoexcation, in agreement with early frequency-domain PE studies [60–62], it can be significantly stronger within the first picosecond after photoexcitation, especially for atomic and small molecular chromophores. However, numerous femtosecond time-resolved experiments have shown that the coupled electronic and nuclear dynamics of chromophores in $He_N$ can be observed with TRPES, as demonstrated for bare droplets [5, 63–66] (Sect. 10.5.1), as well as with surface-located [29, 30] (Sect. 10.3.1) and fully solvated dopants [31–33] (Sect. 10.3.2 & 10.4.2). Even if the photoexcitation process drives the chromophore–droplet system strongly out of equilibrium [31–33, 66], accurate insight can be obtained.

In the following we discuss the increased He influence on PE spectra within the first picosecond after photoexcitation for the In–$He_N$ system [31], which is shown in Fig. 10.3. The corresponding potential energy curves can be seen in Fig. 10.8b. Within the first picosecond the PE peak energy decreases by 290 meV (from 610 meV to 320 meV) due to significant rearrangement of the He solvation shell around the In atom in response to photoexcitation. These dynamics are accompanied by the

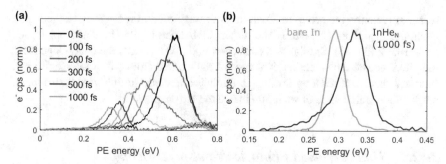

**Fig. 10.3** The He influence on photoelectron spectra in femtosecond pump—probe photoionization for different time delays [31]. The spectra are obtained with In atoms located inside He$_N$ (pump pulse for In-He$_N$: 376 nm, 3,30 eV; pump pulse for bare In: 410 nm, 3.02 eV; probe pulse: 405 nm, 3.06 eV; In ionization energy: 5.79 eV, pump–probe cross correlation time: 150 fs). (**a**) Evolution of a PE peak due to dynamics of the solvation shell. (**b**) Comparison of PE peaks for solvated atoms with equilibrated solvation shell (1000 fs pump-probe time delay, red trace) and bare atoms (red trace)

transfer of electronic energy of the dopant to kinetic energy of the surrounding He atoms, as discussed in Sect. 10.3.2. Additionally, larger linewidths and increased peak areas are observed for short delays (Fig. 10.3a). During the pump–probe cross correlation time (150 fs for this experiment) the simultaneous presence of pump and probe photons leads to saturation and peak distortion. Afterwards, up to $\approx$ 500 fs, the line width is increased because of two reasons: (i) the combination of transient peak shift ($\approx$ 1 meV/fs) and 150 fs cross correlation and, (ii) the Franck–Condon overlap of the excited and ionic states, which are distorted inside the He$_N$ (see Fig. 10.8).

The He-related influence on the PE spectrum within the first picosecond will be superimposed on the TRPES signal of intrinsic nuclear and electronic dynamics of embedded molecules. The magnitude of this influence corresponds to the distortion of the excitation band, as observed by frequency domain spectroscopy (see also Sect.10.3.2). Accordingly, sharp electronic transitions (zero-phonon lines) frequently observed for larger molecules [67, 68] indicate that this initial influence might be less severe for such systems.

After $\approx$ 1 ps, when the He solvation shell has equilibrated, the PE signal (Fig. 10.3b, red curve) peaks at slightly higher energies (30 meV increase) compared to the bare atom peak (red), representing the reduced ionization potential inside the droplet [61]. The increased width of the in-droplet peak compared to the bare-atom peak (62 meV versus 35 meV) is again related to the Franck–Condon overlap of the distorted excited and ionic states inside the droplet (see Fig. 10.8). The PE peak also exhibits a wing extending to lower energies, even below the bare-atom band, which is a signature of energy relaxation of the photoelectrons due to binary collisions with individual He atoms on the way out of the droplet, as previously observed in PE spectroscopy experiments with nanosecond laser pulses [61].

On the droplet surface, the PE peak-shift is slower by a factor of 2–3 [29, 30] due to the lower He density (Sect. 10.3.1). In pump-probe photoionization of nanoplasmas inside $He_N$, the photoelectron kinetic energy was recently used as observable for the temporal evolution of the plasma (Sect. 10.5.3). After plasma ignition with a strong-field pulse, the kinetic energy of electrons released by the probe pulse corresponds to the electron temperature in the plasma [52]. Strong-field probing revealed the appearance of photoelectron spectra characteristic for above-threshold ionization, indicating electron recombination into high-lying Rydberg states [53].

## 10.2.5  Time-Resolved Correlation Spectroscopy

The detection of photoelectrons or -ions can be extended by establishing correlation between the ionization products, such as ion–electron or ion–ion correlations. The assignment of electron spectra to ion fragments, for example, allows for the identification of different pathways in strong-field ionization of molecules [69]. In femtosecond pump-probe photofragmentation experiments bond breaking may occur in the electronically excited state (after pump pulse excitation) or in the ionic state (after probe pulse ionization); two ionization channels that can be disentangled through electron–ion correlation but remain indistinguishable by sole detection of electrons or ions [70–72]. Correlation can be established either by coincidence detection [73, 74], where pairs of charged particles are detected for single ionization events, or by covariance-mapping, where correlations are revealed based on statistical fluctuations [75]. While coincidence detection requires disadvantageously low count rates (typ. $< 0.3$ ionization events per laser shot) in order to avoid so-called false coincidences [76], covariance-mapping allows for much larger signal rates. Recently, the analysis of photoelectron-photoion coincidence measurements with Bayesian probability theory was demonstrated to enable high count rates and provides additional advantages, such as an increased signal-to-noise ratio, exclusion of false coincidences, proper pump-only signal subtraction, and confidence intervals of the spectrum [77, 78].

The hurdle for correlated detection in combination with $He_N$ is the trapping mechanism of ions inside the droplets due to strong attractive forces (Sect.10.2.3). Ion trapping can be overcome if the ions are provided with sufficient kinetic energy to escape the droplet, as it is the case in Coulomb explosion after double ionization [79, 80], or for vibrational IR excitation of molecular ions [81]. In addition, dopant ions are ejected from the droplets to some extent when being indirectly created through Penning or charge-transfer ionization via excited or ionized He, respectively [82–84]. Ion–ion and ion–electron coincidence detection enabled the identification of inter-atomic autoionization processes inside $He_N$ such as interatomic Coulombic decay (ICD) in pure $He_N$ [79] and double ionization of alkali dimers through ICD or through electron-transfer mediated decay (ETMD) [80, 85, 86] (see Sect. 10.5.2)

A very recent example of time-resolved electron–ion covariance spectroscopy of the $In_2$–$He_N$ system is shown in Fig. 10.4. Pump–probe photoionization of $In_2$ inside $He_N$ leads to an unexpected strong $In-He_n^+$ ($n = 0, ..., \approx 30$) ion signal within the first tens of picoseconds [33], whereas ion yields are typically zero within $\approx 50\,ps$

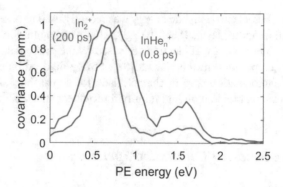

**Fig. 10.4** Covariance detection of electrons and ions after pump-probe photoionization of $In_2-$ $He_N$. The PE spectrum at 0.8 ps (blue) is correlated to the detection of $InHe_n^+$ ($n = 0, ..., \approx 30$), while the 200 ps PE spectrum (red) is correlated to $In_2^+$. Both spectra are normalized. The signal peaking at 1.5 eV in both spectra is likely due to a erroneous correlation of electrons from pump-only ionization and the respective ions

after pump excitation due to trapping (see Fig. 10.13). Comparison of electron–ion covariance measurements at short (0.8 ps) and long (200 ps) time delays sheds light on the process (Fig. 10.4): At long delays (red trace) the prominent PE peak at $\approx 0.65$ eV is correlated to $In_2^+$, identifying photoionization of excited $In_2^*$ after ejection from the droplet. The $In–He_n^+$ PE peak in question at short delays (blue trace) appears at slightly higher energies ($\approx 0.80$ eV), whereas the PE peak of $In^*$ would be expected at lower energies ($\approx 0.32$ eV, see Fig. 10.3), and should also be sharper. This indicates that the $In–He_n^+$ signal originates actually from unfragmented $In_2^*$ molecules, and that fragmentation occurs after ionization in the $In_2^+$ ionic state. A strong increase of the $In–He_n^+$ yield with probe power supports this assumption and suggests that ion ejection is caused by probe-pulse induced $In_2^+$ excitation to a repulsive $In_2^{+*}$ state. Note that a sole TRPES measurement would not be able to screen out the photoelectrons associated to $In–He_n^+$ since they have the same energy as those leading to $In_2^+$.

Having discussed the major experimental tools suitable for time-resolved spectroscopy of pure and doped He droplets, we will highlight some recent application examples in the next section. Before we come to that, we conclude this section by giving an overview of the theoretical work on the dynamics of pure and doped He nanodroplets, as many of the time-resolved experiments have greatly benefited from model calculations.

## 10.2.6 Time-Dependent Density-Functional Theory Simulations

The structure of pure and doped He nanodroplets has been studied theoretically by various methods, the most accurate being Quantum Monte Carlo (QMC) [87]. However, the computational cost quickly exceeds currently available computer resources when it comes to simulating experimentally relevant nanodroplet systems. Furthermore, QMC cannot describe the dynamic evolution of superfluid He in real time. These limitations can be overcome by time-dependent density functional theory (TDDFT) which can be applied to much larger systems than QMC and allows for a time-dependent formulation. A promising recent development is a hybrid path-integral molecular dynamics/bosonic path-integral Monte Carlo method [88]. This method provides the theoretical foundation of simulating fluxional molecules and reactive complexes in He environments seamlessly from one He atom up to bulk He at the accuracy level of coupled cluster electronic structure calculations.

TDDFT is the only method to date that can successfully reproduce results from a wide range of time-resolved experiments in superfluid He on the atomic scale. The great benefit of these simulations is that detailed insights into the structural dynamics of the entire system is obtained, including density modulations of the He such as surface or bulk density waves and solvation shells, which are not directly accessible experimentally. Likewise, the He dynamics induced by ions such as the formation of snowballs and the nucleation of vortices, which have so far eluded experimental detection, are still amenable to TDDFT simulations. Thus, during the last decade, TDDFT has emerged as a powerful tool to describe the structure and dynamics of doped He droplets, thereby valuably complementing the experimental advances. This work has been summarized in two review articles [7, 89]. The method has mostly been developed and promoted by M. Barranco and his group in Barcelona, and the code is freely available [90].

Inspired by experiments, a variety of metal atoms and ions have been studied in view of the structure and dynamics of their complexes with $He_N$ [7, 89]. Upon electronic excitation of either surface-bound alkali atoms [91–93] or initially submerged atoms (silver, Ag) [94], in most cases the excited atoms were ejected from the droplets within a few ps or a few tens of ps, respectively. Depending on the excited state, either bare atoms were ejected or exciplexes were formed, which in turn either desorbed from the droplets or remained attached to them [29, 94]. Ag atoms in the lowest excited state were ejected from the droplet with a speed consistent with the Landau velocity $v_L = 58$ m/s, which was measured experimentally for excited Ag and other molecular dopants [95]. It was concluded that excited dopants interacting repulsively with the He droplets are expelled towards the droplet surface while repeatedly exciting pairs of rotons such that their speed stays below $v_L$.

The microscopic dynamics of metal ions located near the He droplet surface have so far only become accessible through simulations as ions tend to remain tightly bound in the He droplet interior and therefore elude detection. For the $Ba^+He_N$ system, it was found that due to the relatively strong attractive ion-He interaction,

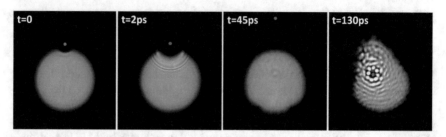

**Fig. 10.5** TDDFT simulations of the dynamics of a Rb atom on a He nanodroplet consisting of 1000 He atoms. At $t = 0$, the Rb atom is excited into the lowest excited state $(5p\Pi_{1/2})$ and at $t = 20$ ps it is ionized. Based on results reported in [28]

the velocity of the $Ba^+$ cation during the solvation process temporarily exceeds $v_L$, leading to the nucleation of a quantized ring vortex [96]. When formed at the He droplet surface, the $Ba^+$ ion moves towards the center of the droplet. After about 8 ps, the $Ba^+$ is fully surrounded by He and a few ps later a dense solvation layer of He forms with transiently appearing spots where He localizes; but eventually the He shell remains smooth. Thereafter, the solvated ion keeps oscillating inside the droplet without friction at a velocity $< v_L$. Due to their large masses and stronger attractive interactions with the He, $Rb^+$ and $Cs^+$ ions initially located at the droplet surface form snowballs at the droplet surface within 10 ps [97]. At longer times the snowballs become solvated by the He droplet which rearranges itself around the stationary ion. Large density fluctuations induced by the cation solvation process lead to the nucleation of quantized vortices. In the case of $Cs^+$, the initial phase of snowball formation prevents the ion from penetrating into the He droplet. The snowball therefore forms at the surface of the droplet in $\approx 30$ ps. Due to the effective shielding of the $Cs^+$ charge by the surrounding He atoms, it is only weakly bound to the droplet. For relatively small He droplets consisting of 1000 atoms, the $Cs^+$ snowball even detached after 90 ps from the droplet due to He density fluctuations.

To compare with recent fs pump-probe experiments, TDDFT calculations were performed that simulated the sequence of pump-probe excitation and ionization at a variable delay. In this way it was possible to reproduce the combined process of desorption of the excited atom and the subsequent fallback and solvation of the ion [28, 98]. Figure 10.5 shows snapshots of the simulated evolution of a Rb atom excited into its lowest excited state $5p\Pi_{1/2}$ at $t = 0$ (green dot turning blue). At $t = 20$ ps it is ionized (red dot). At $t = 2$ ps, the excited Rb atom departs from the droplet leaving behind He density waves traveling through the droplet. At $t = 45$ ps, the $Rb^+$ is at its largest elongation away from the droplet, before falling back into the droplet to form a snowball ($t = 130$ ps). The time constants obtained from the simulation are in good agreement with the experimental results [28]. Similar simulations were performed to complement recent XUV pump-probe studies of the photodynamics of pure He nanodroplets [66, 99]. Here, one or two excited He atoms (He*) in a He droplet take the role of the dopants. Surprisingly, the response of the He droplet strongly resembles that of excited metal atoms in the sense that a bubble forms around the

He* within $\lesssim 0.5$ ps, followed by the ejection of He* from the droplet. In the case two He* are located near each other, the two bubbles merge, which causes the He* to decay by ICD, see Sect. 10.5.2.

More recently, the structure and dynamics of rotating He nanodroplets has been a focus of TDDFT simulations [100–102]. In particular, the formation of quantized vortices in $^4$He nanodroplets and their ability to capture dopant atoms was investigated in detail [103, 104]. Furthermore, collisions of atoms with He droplets as they occur in the experiments during the pick-up process where addressed [105]. Even the merging of two He nanodroplets was simulated, with particular focus on vorticity and quantum turbulence [106].

The Gonzalez group recently developed a hybrid method using DFT to describe the He droplet and a quantum wave packet treatment of the dopant [107] that allows to investigate the He influence on intramolecular processes. With this approach, they obtained predictions of the femtosecond time-resolved dynamics of dimer molecules inside He droplets, including photodissociation of $Cl_2$ [107–110], molecule formation of $Ne_2$ [111, 112], vibrational energy relaxation of $I_2$ [113], and rotational energy relaxation of $H_2$ [114].

## 10.3   Dynamics of Atomic Dopants

The weak influence of the He environment on dopants often results in a negligible perturbation of the ground-state structure of dopants [115], as well as in minor influence on their vibrational and rotational degrees of freedom [67]. Photoexcitation of electronic transitions, in contrast, can lead to a considerable rearrangement of the He solvation shell triggered by a change of the repulsive interaction between the chromophore dopant and the He atoms. The solvent-related response to photoexcitation can best be investigated with atomic dopants in order to avoid complications related to internal degrees of freedom. Since fully solvated dopants experience a stronger but symmetric He interaction, compared to the weaker and asymmetric interaction of surface-located dopants, these situations will be discussed separately.

### 10.3.1   Surface-Located Atoms

While most atoms and molecules are submerged in the interior of He nanodroplets, alkali atoms and small alkali clusters reside in weakly bound dimple states at the droplet surface [6, 7]. Upon electronic excitation, all alkali atoms promptly detach from the He droplet surface due to enhanced Pauli repulsion acting between the diffuse excited valence electron and the He. The only exceptions are Rb and Cs in their lowest excited states where small photon excess energies are insufficient to induce direct desorption [116] and indirect desorption through M*-He exciplex formation is prevented by a barrier along the M*-He potential [9, 117]. In contrast, alkali ions tend to form strongly bound snowball complexes in the bulk of the He droplets as a result of attractive polarization forces [118, 119].

The kinematics of the desorption of atoms induced by optical excitation was first studied experimentally by nanosecond (ns) electron and ion imaging spectroscopy and theoretically by TDDFT [91, 92, 120–122]. Likewise, the dynamics of solvation of alkali ions formed by photoionization was treated by TDDFT and experimentally using ns ion imaging and mass spectrometry [121, 123, 124]. The observed linear dependence of the mean kinetic energy of the desorbed excited atoms on the laser frequency points at an impulsive desorption process [92, 120, 122]. This process is well described by one-dimensional pseudo-diatomic potential curves which quantify the effective interaction between the dopant and the He droplet as a whole [6, 117, 125]. Even the angular distributions of ion images agreed very well with the description of the alkali-droplet complex in terms of a pseudo-diatomic molecule. In some cases, in particular for highly excited states, electronic relaxation occurred in the course of desorption due to curve crossings induced by the interaction with the He droplet [92, 121, 126]. The energy partitioning between the He and the desorbing atom depends on the alkali species and on the quantum state, and appears to be related to the size and shape of the electron orbital [92, 122]. Alkali-He exciplexes were formed either directly by laser-excitation into bound molecular states [120, 122], or by a tunneling process [14, 29, 126]. The desorption of the exciplexes occurred either promptly as for excited atoms, or by a thermal process driven by vibrational or spin relaxation [29, 120, 126].

The early fs pump-probe experiments with He droplets doped with alkali metals mostly focused on the formation of alkali-He exciplexes [10, 21, 24, 127] and on electronic and vibrational coherences of alkali atoms and molecules, respectively [22–25, 34–36]. As dual fs pulses at high repetition rate (80 MHz) were used, the exact location of the dopants, attached to the droplets or in the vacuum, has remained somewhat uncertain; resonant absorption from multiple laser pulses may have induced the desorption prior to the pump-probe process.

Time-resolved measurements of the desorption dynamics of excited alkali atoms and exciplexes have so far only been reported for Rb and RbHe exciplexes. Using amplified fs pulses of the Ti:Sa laser and harmonics thereof, yields, kinetic energies, and angular distributions of electrons and ions for various excited states of the RbHe$_N$ complex have been traced [27–30]. This is achieved by the method of velocity-map imaging (VMI), where the velocity distribution of electrons or ions is mapped onto a two-dimensional spatial distribution in the plane of a position-sensitive detector [128]. Given cylindrical symmetry with respect to an axis perpendicular to the spectrometer axis (usually the laser polarization), the measured two-dimensional distribution can be converted into the three-dimensional velocity distribution by inverse Abel transformation. The radial part reflects the kinetic energy spectrum and the angular part contains information about the symmetry of the state that is photoionized.

As an example, Fig. 10.6 shows typical velocity-map ion images recorded upon excitation of the $6p\Pi$ state [panel (a)] and the $6p\Sigma$ state [panel b)] [28, 30]. In the $6p\Pi$ state, a large fraction of Rb atoms form RbHe exciplexes prior to desorption.

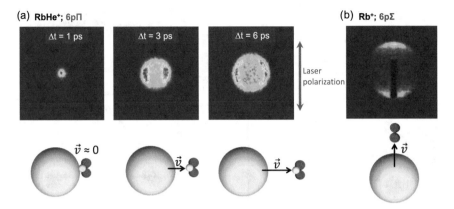

**Fig. 10.6** Velocity-map ion images of $[RbHe]^+$ and $Rb^+$ created by fs pump-probe photoionization of Rb-doped He nanodroplets at a laser wavelength of 415 nm [excitation of the $6p\Pi$ state, (**a**)] at various delays, and at 403 nm [$6p\Sigma$ state, (**b**)] for a delay of 4.8 ps. Based on results reported in [30]. The bottom panel schematically illustrates the corresponding photoinduced processes

As it is expected for promptly desorbing atoms, the $[RbHe]^+$ ions feature a typical ring-like intensity distribution $I_{RbHe}$ with an angular dependence $I_{RbHe} \propto \sin^2 \theta$ with respect to the polarization of the laser pulses which is characteristic of a $\Sigma \rightarrow \Pi$ perpendicular dissociative transition. The increase of the mean radius of this distribution as a function of delay clearly shows that desorption of RbHe exciplexes proceeds as a prompt, pseudo-diatomic dissociation process. A schematic representation is shown at the bottom of Fig. 10.6. Excitation of the $RbHe_N$ complex into the $6p\Sigma$ state results in a $I_{Rb} \propto \cos^2 \theta$-angular intensity distribution according to a $\Sigma \rightarrow \Sigma$ parallel transition causing prompt dissociation, see panel (b).

Figure 10.7a depicts the pseudo-diatomic potential curves involved in the two processes. Owing to the repulsive character of the excited states, the Rb atoms promptly desorb as free atoms ($6p\Sigma$ state) or as a mixture of Rb atoms and RbHe exciplexes ($6p\Pi$ state). In contrast, the ionic potential curve is attractive, causing $Rb^+$ or $[RbHe]^+$ to fall back into the droplet when created near the droplet surface, i.e. at short pump-probe delay. The condition for the ion to fall back into the droplet or to move away is given by the balance of the kinetic energy gained by repulsion in the neutral excited state on the one hand, and the potential energy barrier in the ionic state on the other. Indeed, the yield of detected $Rb^+$ at a laser wavelength of 403 nm ($6p\Sigma$ excitation) nearly vanishes at short delay and steeply rises around 0.5 ps, see Fig. 10.7b. The yield of large $[RbHe_n]^+$, $n \approx 5000$, complexes (not shown) features the opposite behavior [27]. This confirms the concept that ions fall back into the droplet when created near the surface. The yield of photoelectrons (not shown) displays no significant pump-probe dependence, indicating that electrons are emitted from the dopants irrespective of the dopants' position with respect to the droplet surface. A transient maximum of the $[RbHe]^+$ yield around 1 ps was interpreted in

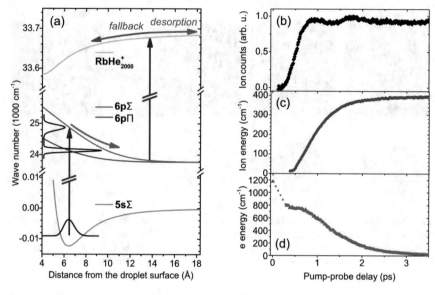

**Fig. 10.7** (a) Illustration of the pump-probe scheme for probing the desorption dynamics of Rb atoms excited to $6p$-correlated states, based on the RbHe$_N$ pseudo-diatomic potential curves. Based on results reported in [27]. (b) Detected Rb$^+$ ion yield at a laser wavelength of 403 nm ($6p\Sigma$ excitation); (c) Rb$^+$ ion kinetic energies; (d) electron energy. Based on results reported in [30]

terms of associative photoionization of [RbHe]$^+$, i.e. the direct optical excitation of a bound cationic state [27].

The Rb$^+$ ion kinetic energy monotonously rises within $\approx 1$ ps [Fig. 10.7c] due to the acceleration of the excited Rb atom away from the droplet surface. Concurrently, the photoelectron energy drops by about 1200 cm$^{-1}$ [panel (d)] due to the increase of the potential-energy difference between the pseudo-diatomic excited state and the ionized state as the Rb atom moves away from the droplet.

Similar delay-dependent electron and ion signals were observed in a two-color pump-probe experiment where Rb atoms were excited to the $5p$-correlated states. The main difference was a slower dynamics by a factor of nearly 100. Ion yields and kinetic energies continuously rose on the 100 ps time scale. This is due to less repulsive pseudo-diatomic potential curves in these states compared to the $6p$-correlated states [117]. Both the dynamics in the $6p$ and $5p$ states were well reproduced by TDDFT simulations [28]. A detailed ion and electron-imaging study of RbHe exciplexes formed in the $5p\Pi$ state, combined with TDDFT simulations, revealed that the desorption of the RbHe exciplexes, proceeding within $\approx 700$ ps, is induced by $^2\Pi_{3/2} \rightarrow ^2\Pi_{1/2}$ spin relaxation. The formation time of the RbHe exciplex was found to range between 20 and 50 ps [29].

These studies were further extended to Rb$_2$ dimers formed on He nanodroplets [38]. Similarly to alkali atoms, Rb$_2$ excited to intermediate states were found to promptly desorb off the droplets. However, both angular and energy distributions of detected $Rb_2^+$ ions appear to be most crucially determined by the $Rb_2$ intramolecular symmetries rather than by the symmetries of the $Rb_2He_N$ pseudo-diatomic complex. The pump-probe dynamics of $Rb_2^+$ was found to be slower than that of $Rb^+$ in the same wavelength range of the pump pulse, pointing at a weaker effective guest-host repulsion for excited molecules than for atoms.

To summarize this section, as general trends, an excited alkali atom or molecule tends to promptly desorb off the He droplet surface, in good agreement with a pseudodiatomic dissociation model. In contrast, an ion, formed at the droplet surface, sinks into the bulk of the droplet where a dense He shell form around it. Pump-probe photoionization signals manifest the competing dynamics of desorption of the excited neutral and the falling back of the ion into the droplet. Another trend is that a resonantly excited metal atom tends to form a metal-He exciplex with variable abundance depending on the symmetry of the excited state. These metal-He exciplexes tend to promptly desorb as well; exceptions are the lowest excited states where the repulsion between the excited alkali and the He droplet is weak. The combination of fs pump-probe photoionization spectroscopy with VMI of electrons and ions reveals detailed information about the dynamics and kinematics of the desorption process for specific excited states. In future experiments, it would be interesting to extend these studies to larger alkali oligomers and other types of metals which are initially located deeper within the He droplet surface (Mg, Ca), or in the droplet interior (Ag, Al, Cr, Cu,…). Likewise, direct measurements of the dynamics of ejection of excited ions [81, 124] would by highly desirable.

## 10.3.2   Solvated Atoms—Solvation Dynamics

In their electronic ground state, atomic dopants inside $He_N$ are surrounded by a He solvation shell that forms through equilibration of attractive dopant–He forces and repulsive Pauli interactions between the dopant's valence electrons and the closed-shell helium. Electronic excitation of the dopant often causes its valence electron to expand radially, inducing a strong interaction with the solvation shell. The energy related to this expansion process has to be provided as excess photon energy, represented by photoexcitation bands that are typically blue-shifted and broadened by several hundred wave numbers with respect to the bare-atom transition (see Fig. 10.8a), as observed in frequency-domain experiments for atomic dopants such as Al [15], Ag [62], or Cr [129]. Additionally, these experiments have revealed dopant ejection after photoexcitation, indicating the heliophobic character of the excited state. Electronic relaxation through nonradiative population transfer to lower states induced by curve crossings due to interaction with surrounding He atoms was observed [15, 60, 62, 129–131], as well as excipex formation. These frequency-domain results raise a number of questions about the nature of dopant photoexcitation inside a $He_N$: (i) Which primary solvent-related processes are triggered by photoexcitation, (ii) what are their characteristic time scales, (iii) how is the excess energy dispersed

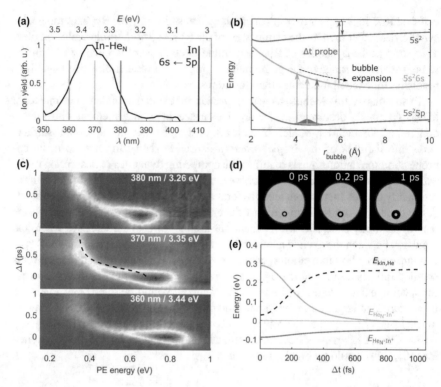

**Fig. 10.8** Fast solvent response to photoexcitation of In–He$_N$: Expansion of the solvation shell and energy dissipation. (**a**) Photoexcitation spectra of In–He$_N$ and bare In in the range of the $6s \leftarrow 5p$ transition [132]. Vertical bars indicate phoexcitation energies corresponding to the spectrograms in (c). (**b**) Sketch of the In–He$_N$ potential energy surfaces as function of the bubble radius for In in its ground [$5s^2 5p$ ($^2P_{1/2}$), blue], lowest excited [$5s^2 6s$ ($^2S_{1/2}$), green] and ionic ground state [$5s^2$ ($^1S_0$), red]. Pump excitation at different photon energies (c.f., (a)) and probe ionization at 405 nm are indicated. Red arrows correspond to the PE kinetic energy measured by TRPES. (**c**) PE spectrograms showing the initial bubble expansion obtained with different photoexcitation energies [32], as indicated in (a) and (b). The simulated PE peak shift ($E_{\text{He}_N\text{-In*}} - E_{\text{He}_N\text{-In}^+}$ from (e)) is shown as dashed line. (**d**) He density distributions of In–He$_{4000}$ at selected times after photoexcitation, as calculated with TDDFT [31]. (**e**) Interaction energy $E_{\text{He}_N\text{-In*}}$ of the $5s^2 6s$ excited state (green curve) and interaction energy $E_{\text{He}_N\text{-In}^+}$ of the $5s^2$ ionic state (red curve) [31]. Additionally, the kinetic energy of the He atoms, $E_{\text{kin, He}}$, is plotted as dashed line.

into the system, and (iv) what is the dependence of these processes on experimental parameters such as droplet size or excitation energy?

In the following we discuss answers to these questions that were obtained with TRPES of indium (In) atoms inside He$_N$ [31, 32]. In TRPES inside He$_N$ the valence electron, which is electronically excited by the pump pulse, is exploited to sense the temporal evolution of the He environment by retrieving the ionization energy with the probe pulse (see Sect.10.2.4). Rearrangement of the solvation shell around the dopant through nuclear relaxation can be followed as transient PE peak shift as the potential energies of the excited and the ionic state depend on the dopant distance to neighboring He atoms. The observed processes can be distinguished by their time scale into a fast expansion of the solvation shell to form a He bubble ($\approx 500$ fs), as well as a slower bubble oscillation and dopant ejection from the droplet ($\approx 50$ ps), which will be discussed separately.

### 10.3.2.1   Fast Solvent Response: Expansion of the Solvation Shell

The electronic photoexcitation process of In atoms inside $He_N$ is depicted in Fig. 10.8b, which shows the In–$He_N$ pseudo-diatomic potential energy curves as function of the bubble radius: Photoexcitation of the In atom in its ground-state solvation shell (4.5 Å radius) leads initially to an increased excited state energy, which is represented by the broadened and blue-shifted excitation spectrum (Fig. 10.8a) [132]. Relaxation towards an expanded solvation shell causes energetical shifts in both the excited and the ionic state, which can be tracked by probe-pulse ionization at increasing time delays and observation of the resulting PE energy.

Corresponding PE spectra for different excitation wavelengths are shown in Fig. 10.8c [31]. Each spectrum shows a rapid initial shift of the PE peak to lower energies, that levels off in all three cases within 1 ps to the same value of 320 meV. The 1 ps PE-energy is approximately 0.03 eV above the gas phase value due to the reduced ionization potential of solvated atoms.

With increasing photoexcitation energy, the PE peak maximum at time zero shifts to higher values, from 0.65 eV for 380 nm to 0.79 eV, for 360 nm [32]. Additionally, the PE peak shift proceeds faster for higher excitation energies, as becomes evident by comparison the initial slopes for the three spectra in Fig. 10.8c. This trend is expected from the increasing steepness of the excited state potential energy curve in Fig. 10.8b, suggesting a stronger acceleration of the excited state wave packet for higher excitation energies. In combination, these results show that the excitation excess energy is fully transferred to kinetic energy of the solvation shell within 1 ps. Concerning variation of the droplet size, no influence is found on the PE spectra ($\bar{N} = 2600 - 40000$), showing that bubble expansion is a purely local process that only depends on the its environmental fluid density [32].

Additional insight into the ultrafast In–$He_N$ photoexcitation dynamics is obtained from TDDFT simulations using the BCN-TLS-HeDFT computing package [90], based on In–He pair potentials of the ground, excited and ionic state [31, 132]. He density plots at selected times after photoexcitation (Fig. 10.8d) show that the solvation shell almost doubles in radius within the first picosecond from 4.5 to 8.1 Å. Based on this time evolution of the He density the time-dependencies of the excited and ionic states can be computed by integrating the corresponding pair potentials over the corresponding droplet densities. Figure 10.8e shows that the He environment increases the excited state energy ($E_{He_N In*}$), while it decreases the ionic state energy ($E_{He_N In^+}$). The difference of both energy deviations ($E_{He_N In*} - E_{He_N In^+}$) corresponds to the transient PE shift and is in excellent agreement with the TRPES measurement (Fig. 10.8c, dashed line). The TDDFT simulation thus sheds light onto the dissipation

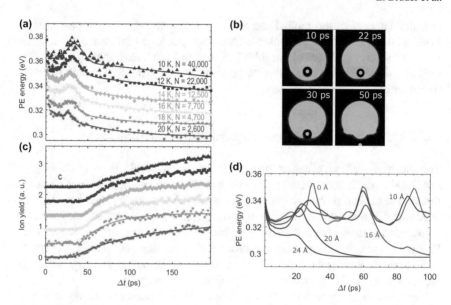

**Fig. 10.9** Slow solvent response to photoexcitation of In–He$_N$: Bubble oscillation and dopant ejection [32]. (**a**) Transient PE energies for different droplet sizes, showing an overall decrease due to dopant ejection with a local maximum due to bubble contraction (all curves are vertically offset by the same amount). (**b**) He density distributions of In–He$_{4000}$ at selected times after photoexcitation, as calculated with TDDFT. (**c**) Transient ion yield for different droplet sizes representing dopant ejection (all curves are vertically offset by the same amount). (**d**) Simulated PE transients for trajectories originating at different distances to the droplet center (droplet size $N = 4000$, 36 Å radius)

process of the excess energy, as it allows to quantify the individual contributions to the ionization energy measured by TRPES. Only with the TDDFT results it becomes clear that the excess energy of the photoexcitation process is initially stored as potential energy of the excited state and subsequently converted into kinetic energy of the surrounding He atoms ($E_{kin,He}$ in Fig. 10.8e) within the first picosecond, leading to He density waves propagating through the droplet (see Fig. 10.9b).

### 10.3.2.2 Slower Solvent Response: Bubble Oscillation and Dopant Ejection

After adaption of the He environment to the electronically excited dopant within the first picosecond, the heliophobic character of the excited state leads to dopant ejection on a $10 - 100$ ps timescale. Additionally, the impulsive stimulation of the He solvation layer initiates a collective oscillation of the He bubble. Although the two processes overlap in time, they can be sensed and distinguished with TRPES, underlining its sensitivity.

Figure 10.9a shows the transient shift of the PE peak position up to 200 ps, exhibiting a gradual $\approx 20$ meV decrease to the bare-atom PE energy. This PE peak shift represents dopant ejection from the droplet and is influenced by the distributions of both droplet sizes and starting locations within the observed ensemble. Superimposed on the PE energy reduction, a temporary increase is observed at ≈ 30 ps that is caused by the increase of He density around the dopant in consequence of the first contraction of the bubble oscillation. Importantly, neither the PE peak shift nor the temporal increase exhibit a dependence on the droplet size (Fig. 10.9a), which indicates that the In atoms are not equally distributed within the droplet but rather are confined within a small spherical shell beneath the surface. Also, the excitation energy has no influence on ejection signal and only weakly increases the bubble oscillation period (not shown) [32], which is in line with the interpretation that the excitation excess energy is fully transferred to kinetic energy of the solvation shell within 1 ps. The momentum of the dopant is thus not changed upon photoexcitation inside the droplet due to the symmetry of the bubble expansion, in contrast to surface-located dopants [133].

Complementary information on the ejection dynamics is obtained from photoions, which can only be detected if ionization by the probe pulse takes place outside the droplet at sufficient distance to its surface [133] (see Sect.10.2.3). The transient ion yield (Fig. 10.9c) remains essentially zero up to ≈ 50 ps as this observable is insensitive to dynamics inside the droplet (bubble expansion and oscillation). The signal onset at 50 ps and its rise up to 200 ps represents In ejection and support the TRPES results (Fig. 10.9a); in particular, the ion transients are also nearly independent of the droplet size.

Further insight into the dynamics on the 10-100 ps time scale is obtained from TDDFT simulations, which predict correct time-scales for both the bubble oscillation and ejection from the droplet, as can be seen from the corresponding He density distributions in Fig. 10.9b). Computed PE transients (Fig. 10.9d) strongly depend on the location within the droplet where the photoexcitation takes place: center-located atoms experience a periodic PE increase of ≈ 30 meV, representing the bubble contraction, while off-center locations show a limited number of bubble oscillations followed by gradual ≈ 20 meV energy decrease due to ejection. In particular, the single bubble contraction predicted for the trajectory originating at 20 Å distance from the droplet center (16 Å beneath the droplet surface) supports the assumption that the In atoms are contained within a small region beneath the droplet surface.

These results on the In–He$_N$ system are in agreement with a recent study on pure He$_N$, triggered by XUV pulses from a free electron laser [66], see Sect. 10.5.1. There, the TRPES results show signatures for creation of a He bubble around a localized excitation, He*, on a very similar timescale (≈ 500 fs) and much faster ejection of the He* (2.5 ps), which might be related to near-surface excitation and stronger acceleration of He atoms compared to the heavier In.

In summary, the In–He$_N$ experiments provide the first real-time study of the solvation dynamics triggered by photoexcitation of an impurity embedded inside a helium droplet, in contrast to the surface-dopants discussed above (Sect. 10.3.1). The atomic impurities used do not show any internal dynamics on the relevant time scales and are thus ideal probes for the response of the superfluid helium solvent. As a remarkable result, similar solvation and ejection dynamics are found for impurities embedded inside the droplets, impurities attached to the surface or even for single, directly excited helium atoms in the helium droplet. For molecular dopants, the described solvent-related dynamics—solvation shell expansion, bubble oscillation and dopant ejection—will be superimposed on intramolecular dynamics because electronic excitation is the primary process in photochemical reactions. A mechanistic description of these processes will thus be key to the conception and interpretation of ultrafast photochemical studies inside $He_N$. For larger molecules one can expect less pronounced solvation shell dynamics since excited molecular orbitals may experience less contact to the He surrounding, as indicated by sharp electronic transitions (zero-phonon lines) that are frequently observed for these dopants [67, 68]. A systematic characterization of these processes for different classes of molecules in future experiments will be essential.

### 10.3.3 Dynamics of Superfluid Droplets Compared to Normalfluid $^3$He Droplets

A fascinating property of helium is its superfluid nature and related dynamics, posing fundamental questions on the existence of such phenomena in confined, nanoscale droplets. Non-dissipative flow has been observed measuring ro-vibrational spectra, confirming superfluidity in nanodroplets [134]. Even in molecular complexes containing only a few helium atoms non-classical inertia has been verified [135]. The experimental results triggered significant interest from theory, leading to a deeper understanding of nanoscopic superfluidity [136–138]. Ultrafast time-resolved studies came into play measuring the vibrational relaxation of attached molecules, indicating evidence for a Landau-critical velocity on the molecular level connected to vibrational motion [35]. Indeed, the Landau velocity was then confirmed from measuring the velocity of ejected dopants [95].

For all such studies, one ideally compares the Bose-Einstein-condensed superfluid $^4$He droplets with $^3$He droplets representing a Fermi fluid. The latter can readily be formed in droplet beam sources (see [3] and references therein). Comparing studies were seminal confirming superfluidity in helium droplets with infrared spectroscopy [134], furthermore, in electronic spectra connected to the structure of zero-phonon lines and phonon wings [139, 140], as well in recent X-ray diffraction imaging experiments [141] and corresponding theory [7, 89, 102]. Femtosecond time-resolve experiments comparing $^3$He droplets have been studied with alkalies as probe species. In laser-induced fluorescence (LIF) spectra of Na atoms, the effect on changing the helium isotope becomes apparent as a significant shift of the spectra [142]. However, in comparison with theory this is well understood from the decreased density of $^3$He and respective binding energies to the helium surface dimple. Differences are even more pronounced in excitation spectra of alkaline earth dopants, providing a sensitive probe of their location [143]. Nevertheless, superfluid dynamics did not show up in

LIF data since observing the electronic excitation probes the environment "frozen" in the ground state configuration.

In time-resolved measurements the formation of RbHe exciplexes has been probed with femtosecond time resolution [21] revealing a faster formation of $^4$HeRb (8.5 ps) compared to $^3$HeRb (11.6 ps). This was a surprising result because intuitively the lighter $^3$He is expected to evolve a faster dynamic. Also from the interaction potentials and a theoretical model, based on the helium tunneling into the bound exciplex configuration, a 40-fold acceleration of forming a $^3$HeRb was calculated [21]. Apparently, the tunneling model does not give the right picture. It was speculated that a difference in the vibrational relaxation when entering the bound molecular potential could be responsible for the different formation times.

## 10.4   Vibrational Dynamics of Molecular Dopants

Vibrational wave packets (WP) were among the first dynamical processes investigated by time-domain spectroscopy. A vibrational WP in a molecule can be seen as quantum mechanical analogue to classical vibration that, due to its coherent nature, provides insight into both the intrinsic structure of the molecule and its interaction with the environment. In a pump-probe experiment, the pump pulse can launch a WP in the excited electronic state (or the ground state) by simultaneously populating several vibrational levels with a well-defined initial phase. The pulse thus creates a coherent superposition of nuclear eigenstates (Fig. 10.10) [144] that evolves in time and can be tracked by photoionization with the probe pulse, projecting the WP onto the ion continuum. Dependence of the ionization probability and ionization energy on the reaction coordinate (e.g., internuclear distance) yields a periodic modulation of the PE yield and energy, as well as the photoion yield, all at characteristic frequencies that correspond to the energetic distances between excited vibrational

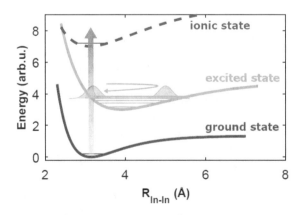

**Fig. 10.10** Generation and observation of vibrational wave packets in a pump-probe experiment

states. Dispersion in an anharmonic potential energy curve leads to dephasing and rephasing of the WP at characteristic revival times.

If the molecule is embedded in a dissipative environment, collisions with solvent molecules may cause vibrational relaxations and decoherence (deterioration of the phase relation within the observed ensemble of molecules), resulting in an irreversible loss of modulation contrast. Information about decoherence, energy relaxation or even deformation of the potential energy curve by the solvation shell, can be retrieved from a spectrogram, as obtained from Fourier transformation of the oscillating signal within a sliding time window. Vibrational WPs of small molecules (mostly dimers) in He$_N$ have been used to probe the He influence on nuclear structure and dynamics, both at the droplet surface and in its interior, which will be discussed separately in the following.

### 10.4.1 Vibrational Wavepackets in Alkali Dimers and Trimers

Alkali diatomic molecules were among the first molecules to be studied by time-resolved laser spectroscopy due to their strong electronic transitions in the NIR and visible (VIS) ranges of the spectrum. Low ionization potentials make alkali dimers accessible to photoionization spectroscopy using comparatively low photon energies and laser intensities. Besides, potential-energy curves can be calculated with high precision, thus facilitating the interpretation of spectroscopic data.

A number of interesting phenomena have been studied using these simple molecules, e.g., wavepacket propagation in spin-orbit-coupled states [145, 146], fractional revivals of vibrational wavepackets [147, 148], the competition of different ionization pathways [149, 150], and isotope-selective ionization [151]. Detailed insights into the vibrational dynamics have been obtained by applying new experimental techniques such as photoelectron spectroscopy [152] and optimal control schemes using shaped laser pulses [153].

Alkali dimers have newly attracted interest due to the recent advances in the formation of ultracold molecules out of ultracold atomic ensembles by means of Feshbach resonances [154] and photoassociation [155]. These studies require the knowledge of molecular spectra with great precision. However, conventional molecular spectroscopy usually probes molecules in their singlet ground state; triplet states, which play important roles in the physics of ultracold molecule physics, are more difficult to access experimentally.

Alkali dimers are efficiently formed by aggregation of atoms picked up by He nanodroplets. Due to the lower binding energy of alkali dimers in their lowest metastable triplet state ($\approx 300\,\text{cm}^{-1}$) compared to the singlet ground state ($\approx 3000\,\text{cm}^{-1}$), preferentially triplet dimers remain attached to the surface of the droplets [12, 156]. While triplet dimers are oriented parallel to the He surface, singlet states tend to adopt a more erect configuration [157–159]. For Li$_2$ in its triplet state, drastically enhanced vibrational quenching rates were predicted for the triplet state as compared to the singlet ground state [158, 160].

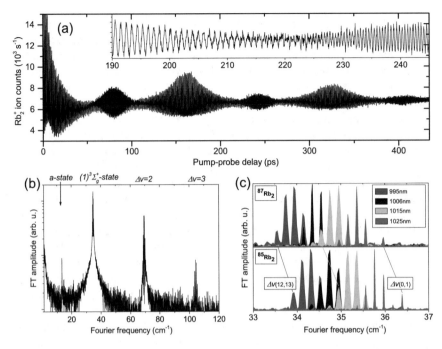

**Fig. 10.11** (a) Pump-probe trace of vibrational WPs in high-spin $Rb_2$ molecules formed on the surface of He droplets recorded at an excitation wavelength of $\lambda = 1060$ nm. (b), (c) Fourier spectra inferred from (a) and other traces. Based on results reported in [25]

Vibrational WP dynamics in triplet states of an alkali dimer attached to a He nanodroplet was observed for the first time using $Na_2$ [23]. No influence of the He droplet on the WP dynamics was observed, likely due to the desorption of the dimer off the droplet prior to the actual pump-probe process. For $K_2$ in singlet states, indications for the influence of the He droplet on the vibrational dynamics [22] were found. Transient modulations of both amplitudes and frequencies of vibrational frequency components were observed, from which the time constant for the desorption dynamics was estimated to range between 3 and 8 ps.

For $Rb_2$ in the triplet states $a^3\Sigma_u^+$ and $(1)^3\Sigma_g^+$, long-lived vibrational coherences were observed up to pump-probe delays $\gtrsim 1.5$ ns, see e.g. Fig. 10.11a [25]. Likely, the fast desorption of the excited $Rb_2$ and its low internal temperature facilitates the detection of WP interferences with high contrast, including full and fractional revivals. Fourier analysis of the time traces provides high-resolution vibrational spectra (see Fig. 10.11b, resolution about 900 MHz), which are in excellent agreement with *ab initio* calculations and of interest for ultracold molecules experiments [161]. Even individual beat frequencies for the two isotopologs, $^{85}Rb_2$ and $^{87}Rb_2$ were resolved [Fig. 10.11c]. This shows that high-resolution spectroscopic data can be extracted from fs pump-probe experiments on doped He nanodroplets.

By comparing the measured data with theoretical results based on dissipative quantum dynamics calculations, it was found that the most important effect of the He environment is vibrational relaxation causing dephasing and energy dissipation [35, 36]. Alternatively, rotational wavepacket dynamics was considered as a cause of the observed decay of the oscillation amplitude [162]. However, unphysically high rotational temperatures would have to be assumed. Besides, no rotational revival structure was observed, which would show up at delay times around 0.6 ns [36]. The strong dependence of the measured dephasing time on the laser wavelength cannot be rationalized by rotational dynamics, either. However, contributions of rotational dynamics to the fast decay observed at short delays $\lesssim 0.3$ ps cannot be excluded.

In the $K_2$ case, the best agreement between theory and experiment was achieved when damping of the WP motion was neglected for slowly moving WPs [35]. Likewise, the WP dynamics of $Rb_2$ was best described by low damping rates for the WP motion in the lower vibrational states $v \lesssim 15$ of the $(1)^3\Sigma_g^+$ state, whereas higher vibrations were more strongly damped [36]. It is tempting to relate these findings to the critical Landau velocity $v_L$ for frictionless motion in superfluid He [95]. However, more systematic measurements, in particular for molecules immersed in the bulk of the droplets, are needed to unambiguously assess the role of superfluidity in the damping of molecular vibrations in or on He nanodroplets.

One great advantage of the He nanodroplet isolation technique is that heterogeneous, polyatomic complexes can be formed rather easily and with some degree of control [67]. By multiply doping He droplets with different species, small heterogeneous clusters [34, 163–165] up to core-shell nanoparticles can be aggregated [166, 167]. In this way, the fs spectroscopy studies of alkali dimers were extended to alkali trimers, specifically $Rb_3$, $Rb_2K$, $RbK_2$, and $K_3$ in quartet states [34]. Similarly to the alkali dimers, long-lived vibrational coherences were observed in certain regions of the spectral range accessible by the Ti:Sa laser. Thus, vibrational spectra with a spectral resolution of the order of 0.2 cm$^{-1}$ were measured. A typical sliding window Fourier spectrum, or spectrogram, recorded on the mass of $Rb_3$, is shown in Fig. 10.12a. The power spectrum of the full pump-probe scan is shown in panel (b). Clearly, distinct frequencies are visible which persist over delay times ranging between $\approx 10$ and $\approx 100$ ps.

In contrast to the straight-forward assignment of pump–probe power spectra recorded for alkali dimers, the interpretation of the trimer spectra turns out to be much more involved. This is due to the concurrent effect of Jahn-Teller and spin-orbit-couplings in these rather heavy species. Nevertheless, the measured frequencies were assigned to beats of vibrational normal modes in the $(1)^4A_2'$ and $(2)^4E'$ states by comparing to *ab initio* calculations. The most prominent lines are the asymmetric stretch and bending modes $Q_{x/y}$ of the lowest quartet state excited by impulsive Raman scattering, followed by the $Q_{x/y}$-modes of the excited electronic states. The symmetric stretch mode $Q_s$ is only visible in the spectrum of $Rb_3$, whereas it is significantly broadened due to fast dephasing within a few ps. Intramolecular vibrational relaxation, intersystem-crossing or vibrational relaxation by coupling to the He droplet may be the cause of this fast damping.

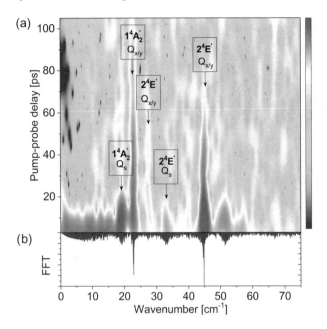

**Fig. 10.12** (**a**) Sliding window Fourier spectrum of Rb$_3$ measured by fs pump-probe photoionization spectroscopy at a laser wavelength of 850 nm. The corresponding electronic states as well as vibrational modes are indicated. (**b**) Fourier spectrum including the whole delay range. Based on results reported in [34]

In summary, very long-lived vibrational coherences over hundreds of picoseconds up to nanoseconds were observed for alkali-molecules attached to the surface of helium nanodroplets. This is in stark contrast to molecules in solution or thin films were decoherence is generally much stronger. Nevertheless, signs of dephasing and decoherence induced by the molecule-droplet interactions were found. Due to the general tendency of excited alkali molecules to desorb from the droplet surface, it is hard to pinpoint decoherence effects based on vibrational interference spectra, though. The high spectral resolution achievable by coherent wavepacket interference spectroscopy combined with the synthesis advantages of helium nanodroplets provide a valuable tool for high-resolution spectroscopy of rare molecular species, as has been shown for high-spin alkali molecules and heterogeneous alkali complexes which are not accessible by other methods.

## 10.4.2  Vibrational Wave Packets in Solvated Dimers

While vibrational dynamics of alkali molecules on the droplet surface were found to be only weakly influenced by the He environment, as discussed in the previous chapter, the perturbation of fully solvated molecules can be assumed to

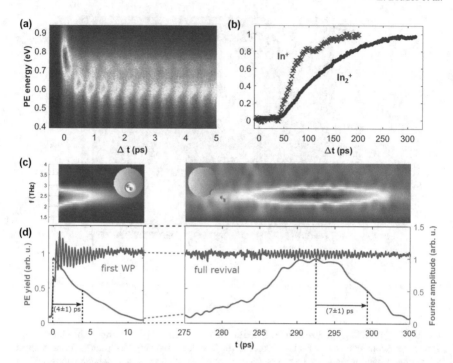

**Fig. 10.13** In$_2$ vibrational wave packet dynamics inside He$_N$ [33]. (**a**) PE spectrum showing the initial oscillations after photoexcitation. (**b**) Comparison of the transient In$^+$ and In$_2^+$ ion yields. (**c**) Spectrogram showing the oscillation frequency obtained from sliding-window Fourier transformation of the initial WP signal (left) and the full revival (right). (**d**) Integrated PE yield (blue) and oscillation amplitue from sliding-window Fourier transformation (red)

be stronger, which might lead to complete suppression of intramolecular dynamics [168]. Recently it could be demonstrated, however, that the He influence on vibrational dynamics of solvated dimer molecules can also be low. Figure 10.13 shows the time-resolved pump-probe photoionization spectra of In$_2$–He$_N$, which contains characteristic signatures of both solvent-related and intramolecular dynamics. By comparison with the In atom dynamics (Sect. 10.3.2), the initial PE peak shift from 0.75 to 0.60 eV within the first picosecond in Fig. 10.13a can be identified as bubble expansion in response to valence electron expansion due to photoexcitation. Also, the ion transients (Fig. 10.13b) show that the electronically excited In$_2^*$ is ejected from the droplet. The slower ion rise compared to the atom, is related to a reduced interaction of the molecular valence electron with the He environment and the larger mass of the molecule.

Intramolecular vibrational WP dynamics are represented as modulation of the PE signal with a periodicity of 0.42 ps (2.42 THz oscillation frequency), as can be seen in the PE spectrum (Fig. 10.13a). The integrated PE yield (Fig. 10.13d, blue curve) shows that the initial WP modulation decays within 10 ps and reappears as full revival between 280 and 300 ps, although with reduced amplitude. The time-dependency of the oscillating signal is shaped by a combination of WP dispersion in the anharmonic potential and decoherence due to He interaction and can be analyzed by sliding-window Fourier transformation. The corresponding spectrograms are shown in Fig. 10.13c and the (normalized) time-dependent amplitude in Fig. 10.13d (red trace).

Since the full revival at 290 ps is observed at times where the excited dimers have left the droplets (see Fig. 10.13b), direct comparison of WP oscillation of solvated $In_2$ inside $He_N$ and bare $In_2$ in gas phase is possible and reveals the decoherence imposed by the He environment. In the He environment, both dispersion and decoherence are active so that the WP oscillation decay proceeds twice as fast [50% decrease within $(4 \pm 1)$ ps], as compared to the bare $In_2$ [$(7 \pm 1)$ ps] where solely dispersion is active. This comparison allows to estimate a lower limit of the decoherence half-life of about 10 ps. The oscillation amplitude of the revival, however, suggests a longer half-life since the delayed and slow ion yield increase (Fig. 10.13b) indicates average $In_2$– He interaction of several tens of picoseconds, while the oscillation amplitude of the revival has decayed to only 20% of the initial value. The fact that the same oscillation frequencies are observed for solvated and bare $In_2$ indicates that the distortion of the excited potential energy curve is not significant for this excitation energy, and that no phase-conserving vibrational relaxation takes place. Further experiments are required in order to test for potential energy distortions at higher energies and to identify the contributions of elastic depahsing and vibrational energy relaxation [113].

Comparison of these long decoherence times (tens of picoseconds) to the much faster decoherence in conventional solvents, typically hundreds of femtoseconds to few picoseconds in special cases [169], underlines the superiority of superfluid He as a solvent. This low interference with nuclear motion can be rationalized by considering the $In_2$ molecule and its surrounding He bubble as coupled oscillators that interact only weakly due to their different oscillation periods ($In_2$: 0.42 ps, bubble: $\approx$ 30 ps, see Sect. 10.3.2). Also, the $In_2$ oscillation energy of 80 cm$^{-1}$ significantly exceeds the elementary excitations of $He_N$ (phonons and ripplons) [67]. These results indicate that molecular systems that exhibit a slow decay of nuclear coherence in gas-phase can now be investigated in a thermal bath environment.

In summary, the $In_2$–$He_N$ experiments provide an impressive example for the preservation of vibrational coherence in a fully solvated molecule despite the disturbance of the helium bath. The vibrational coherence is preserved even while the molecule propagates through the medium and gets ejected. This enabled a direct comparison of the coherent WP motion for the solvated and free gas-phase molecule. In future experiments it will be interesting to examine the influence of additional mediator atoms or molecules on the WP properties of the $In_2$-$He_N$ system. It will be straight forward to monitor alterations of the decoherence time and probably oscillation frequency as function of amount and interaction strengths of mediator particles.

## 10.5  Dynamics of Highly Excited Helium Droplets

He nanodroplets are particularly attractive as model systems for studying the photo-dynamics of finite-size condensed-phase systems, both experimentally and theoret-ically [5, 66, 89]. (i) He atoms have a simple electronic structure which simplifies electron spectra and model calculations; (ii) interatomic binding is extremely weak which allows one to neglect chemical binding; He droplets can essentially be treated as assemblies of unperturbed atoms; (iii) the structure of He nanodroplets is homo-geneous and nearly size-independent due to their superfluid nature; this avoids the congestion of spectra by multiple phases and structures as it is the case for solid clusters [170, 171]. Furthermore, exploring transient phenomena associated with superfluidity is a particularly fascinating aspect of time-resolved He nanodroplet spectroscopy [5, 31, 66, 95, 172].

There are essentially two experimental approaches to studying the dynamics of excited and ionized He nanodroplets: Electron bombardment and resonant absorp-tion or ionization by extreme ultraviolet radiation (XUV). Electron bombardment in combination with mass spectrometry is a well-established technique; positive and negative ions of both He and dopants can be created relatively easily [173–176]. However, the energy resolution is rather limited in such experiments and so far no time-resolved measurements have been reported.

The first XUV experiments were performed by the group of Toennies *et al.* in Berlin using synchrotron radiation (BESSY I) [177]. This mass spectrometric study established the key aspects of photoionization dynamics of He droplets: Ionisation occurs not only by direct electron emission at photon energies exceeding the ioniza-tion energy of He atoms ($E_i = 24.6\,\text{eV}$) but also by autoionization at photon energies in the range $23\,\text{eV} \lesssim h\nu \le E_i$. The dominant ionisation products in this regime are $He_2^+$ ions and small $He_n^+$ clusters as well as large cationic clusters with $n \gtrsim 10^3$. In doped droplets, the dopants are ionized indirectly by a Penning ionization-like process through He* 'excitons' whereas no evidence for direct photoionization of dopants was found.

At $h\nu \le E_i$, $He^+$ ions are formed in the droplets which first undergo resonant charge hopping (19-35 fs per hop) over a distance of $3.1\,\text{Å}$ before localizing by forming $He_2^+$ [178, 179]. Alternatively, when a dopant is present in the droplet, the $He^+$ can localize by charge transfer ionization of the dopant [180, 181]. The excess energy is carried away by emission of a photon (radiative charge transfer, RCT [182]). In many cases, this reaction leads to the ejection of the dopant ion or a complex of the dopant ion with He atoms. A more intricate variant of charge transfer ionization is electron-transfer mediated decay (ETMD) [183], where the excess energy is transferred back to the electron-donating dopant or to a third particle nearby. RCT and ETMD have been studied in doped He nanodroplets for various alkali and alkaline earth dopants [82, 84, 86, 184].

These studies were refined in a series of synchrotron experiments carried out by D. Neumark and coworkers in Berkeley. By applying photoion and electron imaging detection, further insights into the relaxation of photoexcited or ionized

He droplets were obtained [185–187]. In particular, extremely low-energy electrons were observed in the regime of He droplet autoionization, whereas for $h\nu > E_i$, electrons had as much as 0.5 eV higher kinetic energy than those from atomic He at the same photon energy. By implementing photoelectron–photoion coincidence (PEPICO) imaging detection at the synchrotron facility ELETTRA in Trieste, these studies were further extended in the direction of interatomic Coulombic decay (ICD) processes [79, 80, 82–84, 86, 184, 188–190], see Sect. 10.5.2 as well as photoelectron spectroscopy [189–191]. In these studies, the role of the He droplets was to serve as an inert substrate that prepares a molecular complex in its vibronic groundstate; the reaction was then initiated by the interaction with an energetic He$^*$ or He$^+$ created in the droplet.

Complementary synchrotron experiments were carried out by the group of Möller *et al.* in Hamburg using fluorescence detection of excited pure He nanodroplets. Sharp atomic and molecular lines in the emission spectra indicated the localization of droplet excitation on an excited He atom (He$^*$) or He$_2^*$ excimer, and the formation of void bubbles around them [192–194]. Relaxation into lower-lying singlet and even triplet states was observed; the latter were induced by electron-ion recombination in the droplet autoionization regime [192]. Upon excitation of high-lying electronically excited states, unusual Rydberg states located at the surface of He nanodroplets were evidenced by time-correlated fluorescence spectroscopy [195].

## 10.5.1   Time-Resolved XUV Spectroscopy of Pure He Nanodroplets

Time-resolved spectroscopy of He nanodroplets first became possible thanks to the development of XUV light sources based on high-harmonics generation of intense NIR laser pulses. In a series of XUV-pump and NIR-probe experiments, the Berkeley group succeeded in tracing the relaxation dynamics of resonantly excited pure He nanodroplets [5]. By imaging the emitted electrons [63–65] and ions [65, 196, 197] following excitation of the $1s4p$-correlated state of the droplet ($h\nu = 23.6 \pm 0.2$ eV), *intra*-band and *inter*-band relaxation into lower lying droplet states was inferred, leading to the expulsion of free He$^*$ Rydberg atoms. Specifically, He atoms in orbitally aligned $1s4p$-states and in unaligned $1s3d$ states were found to be the dominant fragments. The ejection timescales of atoms in $1s4p$ and $1s3d$-states were $\lesssim 120$ fs and $\approx 220$ fs, respectively. The component of very low-energy electrons associated with droplet autoionization was observed with a rise time of 2–3 ps.

With the advent of intense XUV and X-ray radiation sources provided by free-electron lasers (FELs), it has become possible to excite, ionize, and even directly image He nanodroplets by intense XUV or X-ray pulses [101, 198–200]. In the following we discuss recent experiments that systematically studied the relaxation of He nanodroplets resonantly excited into the lowest absorption band using tunable XUV-FEL pulses [66]. Owing to the use of UV probe pulses ($3^{rd}$ harmonic of the

Ti:Sa laser), all final states of the relaxation could be detected. The He droplets in the $1s2p$-correlated state was found to undergo ultrafast interband relaxation to the $1s2s$ state. Subsequently, the excitation localizes by forming a bubble around a localized excited atom, He*, in either of the two metastable $1s2s$ $^1S$ or $^3S$ states. Eventually, the He* is transported to the droplet surface where it is ejected or where it resides in a dimple prior to forming a $He_2^*$ excimer [173]. The interpretation of the measured high-resolution TRPES is supported by TDDFT simulations carried out by the group of M. Barranco. They essentially confirm a three-step relaxation process: Ultrafast electron localization, electronic relaxation into metastable states, and the formation of a bubble, which eventually bursts at the droplet surface, thereby ejecting a single excited He atom.

Figure 10.14a and b schematically depict the pump-probe scheme used to excite pure He nanodroplets into the $1s4p$ $^1P$ state at $h\nu = 23.8$ eV. The tunable XUV pulse is generated by the seeded FEL FERMI, Trieste [202]. The relaxation dynamics is probed by TRPES using a time-delayed near-UV pulse with a photon energy $h\nu' = 3.1$ eV. The grey shaded area in (a) depicts the absorption spectrum of medium-sized He nanodroplets measured by fluorescence emission [203]. The pump and probe photons are represented as red and blue vertical arrows, respectively. The dotted curved arrows indicate the relaxation into lower excited states of the droplet or free atoms.

Typical photoelectron spectra recorded for pump-probe delays up to 150 ps are shown in Fig. 10.14c [201]. The bright red dot at zero delay and electron energy $\approx 2.3$ eV is due to 1+1 resonant two-photon ionization of the droplet $1s4p$ excitation, whereas all features at delays $> 0.3$ ps are due to 1+2 three-photon ionization as the excited-state population relaxes into $1s2s, p$ states within $\lesssim 1$ ps. Thereafter, the population accumulates in the metastable $1s2s$ $^{1,3}S$ atomic states. The $1s2s$ $^3S$ metastable triplet state is most likely formed by electron-ion recombination following droplet autoionization [192, 195]. The low-energy component in the electron spectrum is due to He-droplet autoionization [185]. The subsequent relaxation of the $1s2s$ states proceeds as in the aforementioned case of direct optical excitation [66]. Thus, despite the extremely weak binding of the He atoms in the droplets and the superfluid nature thereof, energy dissipation is very efficient; up to 4 eV of electron energy is dissipated within 1 ps by intraband and interband relaxations, *i. e.* the coupling of electronic and nanofluid nuclear degrees of freedom [66]. Note that the electron spectra contain replicas of the discussed features at multiples of the probe photon energy (not shown). These peaks are due to above-threshold ionization (ATI) induced by the absorption of multiple probe photons by helium nanodroplets containing several electronically excited states. When comparing to ATI electron spectra of excited helium atoms, ATI is found to be drastically enhanced in excited helium nanodroplets. The enhancement of ATI in multiply excited helium nanodroplets is attributed to laser-assisted electron scattering and a collective coupling between the excited helium atoms within the droplets [204].

In future experiments and model calculations, the initial step of the localization of the droplet excitation on one individual atomic center should be studied in more detail. At present, neither the range of delocalization, nor the time scale of the collapse

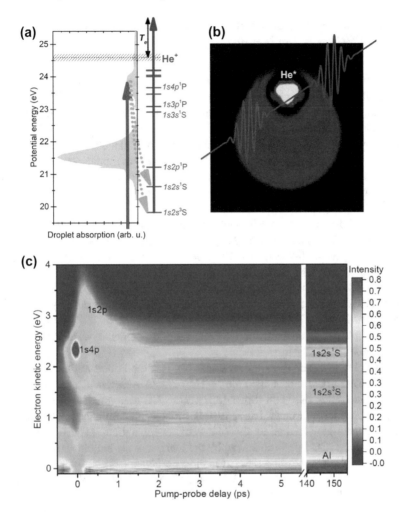

**Fig. 10.14** (a) XUV pump—UV probe scheme used for TRPES of 1s4p-excited pure He nanodroplets. (b) Snapshot of the 2D He density profile obtained from time-dependent density-function simulations. Adapted from [66], published under a CC BY 4.0 license. (c) Transient photoelectron spectra. All features at delay times > 0.3 ps are due to two-photon ionization by the 3.1 eV-probe pulse or autoionization (AI). Based on results reported in [201]

of the initial delocalized state is known. Only for small He clusters has the vibronic relaxation dynamics been addressed theoretically [205]. Shorter pulses, as provided by XUV attosecond sources, will be instrumental for resolving this type of excitonic dynamics in He nanodroplets and in other nanoclusters [206].

## 10.5.2 Interatomic Coulombic Decay Processes in Doped Helium Nanodroplets

A number of indirect ionization processes have been evidenced in recent synchrotron studies, all of which are related to interatomic Coulombic decay (ICD) [207–209]. This term, first introduced by Cederbaum *et al.* in 1997 [210], subsumes various autoionization channels occurring in weakly bonded matter, in which not only the initially excited state, but also neighbouring atoms or molecules take part. ICD has mostly been studied using rare-gas dimers and clusters as model systems, but more relevant condensed phase systems such as liquid water comes more and more to the fore, in particular in view of the possible relevance of ICD for radiation damage in biological matter [211–213].

In He nanodroplets, these processes rely on the interatomic transfer of charge or energy within the droplet [79, 179, 214] or between the excited or ionized He droplet and a dopant particle [82–84, 187–189]. Some of these processes lead to efficient double ionization of the He droplet or the dopant following absorption of a single photon [80, 86, 184]. Recent experiments at the XUV FEL FERMI, Trieste, further extended these works to collective autoionization processes of multiply excited He nanodroplets irradiated by intense, resonant XUV pulses [215–218]. Pure and doped He nanodroplets have also attracted the interest of theoreticians as testbeds for ICD investigations [219–221].

When only a few excitations are present in one He droplet, autoionization proceeds according to the ICD mechanism $He^* + He^* \rightarrow He + He^+ + e^-_{ICD}$, first proposed by Kuleff *et al.* for the neon dimer [222]. The ICD electrons are created with a characteristic energy around 16 eV, as this process predominantly involves pairs of $1s2s$-relaxed $He^*$ located at short interatomic separation. TDDFT simulations indicate that the ICD rate is enhanced by the merger of the bubbles around the two adjacent $He^*$, thereby pushing them even closer together [99].

Indeed, electron spectra recorded for multiply excited He nanodroplets display a clear peak at 16.6 eV, see Fig. 10.15a. Note that for increasing XUV intensity, this peak broadens and shifts towards lower energy. In addition, a low-energy component gains in intensity, which is indicative for thermal electron emission. Both features mark the transition of the multiply-excited He nanodroplet into a nanoplasma, which in this case is induced by collective autoionization [215, 218]. The collective Coulomb potential of the evolving nanoplasma tends to down shift the energy of emitted electrons [53].

To directly measure the ICD rate, the same pump-probe scheme as described before was applied for slightly higher pulse energies [Fig. 10.14a]. By photoionizing the $He^*$ with the probe pulse, the population of $He^*$ pairs was depleted, thereby interrupting the ICD process. Indeed, the yield of ICD electrons becomes minimal around a delay of 200 fs, while the yield of electrons emitted by photoionizing $He^*$ reaches a maximum, see Fig. 10.15b. From the subsequent rise of the ICD signal the time scale of ICD in this system was deduced by comparing with a simple Monte-Carlo simulations [solid lines in Fig. 10.15b]. ICD was found to be surprisingly fast

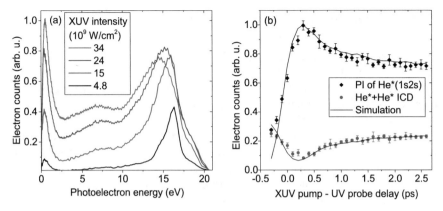

**Fig. 10.15** (**a**) Electron spectra recorded for He nanodroplets irradated by resonant XUV pulses at variable intensity. The peak around 16 eV indicates ICD of pairs of He*. Peak broadening at increasing intensity is due to the formation of a collective Coulomb potential as the droplet evolves into a nanoplasma. Based on results reported in [218]. (**b**) Yields of electrons created by photoionization (PI) of He* and of ICD electrons as a function of XUV-pump and UV-probe delay. The minimum in the ICD trace is due to the transient depletion of He*, thereby suppressing ICD. Based on results reported in [99]

(400-900 fs) and only weakly depended on the initial number of excitations per He droplets, which were controlled by the XUV intensity and the He droplet size [99]. This counterintuitive result is rationalized by the relatively strong interatomic attraction acting between two He* and the merger of bubbles around two interacting He*. The strong distance dependencies of these two effects makes the ICD rate extremely sensitive to the initial separation of the He* pairs; those pairs with small initial separations $\lesssim 10$ Å decay very effectively, whereas all other He* are likely to be ejected out of the droplets before getting sufficiently close to one another to decay by ICD.

In summary, He nanodroplets are attractive targets for studying ICD processes as they bridge the gap between van der Waals molecules and condensed phase systems. Both homogeneous and heterogeneous ICD processes have been evidenced in pure and doped He droplets, respectively. Using intense ultrashort XUV-FEL pulses, the dynamics of a resonant ICD process was measured for the first time in a condensed-phase system using He nanodroplets. The unexpectedly short ICD time was rationalized by the peculiar quantum fluid dynamics of He droplets. In future experiments it appears promising to apply the technique of pump-probe depletion of excited-state populations for elucidating the dynamics of other types of ICD processes, including those that involve dopant atoms and molecules.

### 10.5.3  Dynamics of Helium Nanoplasmas

Helium nanodroplets are mostly used for isolating molecules at low temperature in a transparent and extremely inert environment. The particularly favourable properties of He droplets originate in the extremely high excitation and ionisation energy in combination with the extremely low droplet temperature and the resulting superfluid state. These properties make He droplets the 'ideal spectroscopic matrix' [3, 67]. In contrast, doped He nanodroplets can turn into a highly reactive medium, a so-called nanoplasma, when illuminated by intense ($\gtrsim 10^{15}$ Wcm$^{-2}$) NIR laser pulses. Alternatively, multiple excitation by intense resonant XUV pulses also leads to the formation of a nanoplasma due to collective autoionization—a combination of ICD and inelastic scattering processes [215–218].

At first glance, He droplets appear less suited for studying nanoplasmas due to the high threshold intensity needed for singly ionizing He at 800 nm wavelength ($1.5 \times 10^{15}$ Wcm$^{-2}$ [223]), and due to the small number of two electrons each He atom can at most contribute to building up a nanoplasma. However, He droplets doped with heavier species have recently revealed a diverse strong-field ionization dynamics resulting from the extremely large differences in ionization energies of the dopants and the He host medium. The controlled location of dopants inside or at the surface of the droplets adds another control parameter for nanoplasma ignition [224].

Initiated by tunnel-ionization of the dopant atoms acting as seeds, a He nanodroplet evolves into a nanoplasma, a highly ionized collective state, which can greatly enhance ionisation and fragmentation of the embedded dopants [47, 49, 50, 52, 225]. Alternatively, an ionizing XUV pump pulse preceding the driving NIR pulse can be used to ignite the nanoplasma, as recently demonstrated for pure argon clusters [226]. The intense NIR probe pulse acts as a strong time-varying external driving field which forces the seed electrons into oscillatory motion within the cluster. The resulting electron-impact ionization in the field of the created ionic cores enhances the build-up of a confined plasma-like state. During this laser-driven ionization process, a large fraction of electrons released from their parent atoms remain trapped in the space charge potential of the cluster (inner ionisation). The critical phase of light-matter interaction sets in as the plasma expands and the dipolar eigenfrequency of the plasma ('plasmon resonance') meets the frequency of the driving laser field [227, 228]. Under such resonance conditions, the nanoplasma becomes highly light absorbing. Consequently, the nanoplasma heats up dramatically and emits electrons and ions (outer ionisation). Very high ion charge states, electron energies up to multi-keV and ion kinetic energies up to MeV, and even XUV and X-ray photons have been detected [227, 228].

The first strong-field ionization experiments were carried out by the Rostock group using metal cluster-doped He nanodroplets. The focus was mainly on the charging of the embedded metal cluster rather than on the dynamics of the He nanoplasma. Single intense laser pulses of variable duration as well as dual pulses were used, and soon a significant influence of the He environment on the ionization dynamics of the metal core was realized [45, 229]. The condition for resonant charging was reached earlier

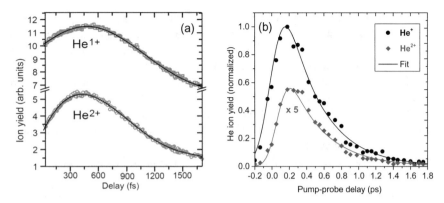

**Fig. 10.16** (a) Yields of He ions as a function of delay between two intense NIR pulses of $\approx 10$ fs duration. The mean droplet size was $15,000$ He atoms, and the mean number of Xe dopants is 15; based on results reported in [51]. (b) He ion yields for He droplets doped with about 30 Ar atoms as a function of the delay between a soft X-ray (250 eV) pump pulse and an intense NIR probe pulse

in time than for the bare metal cluster, due to more efficient initial non-resonant charging of the metal core in the presence of the He environment that supplied additional electrons generated by electron-impact ionization of He shell. This lead to a faster expansion and thus to an earlier resonant matching of the plasmon and the photon energies. As a further consequence of the metal–He interaction in large droplets, caging of fragments was observed, which induced the reaggregation of the metal clusters [45].

The active role of the He shell in the strong-field ionization process of doped He nanodroplets was confirmed by classical molecular dynamics (MD) simulations [47–49, 51, 224, 225]. A direct manifestation of the strong dopant-He coupling in the nanoplasma state is the observed efficient charging of Xe dopants up to $Xe^{21+}$, by far exceeding the charge states reached for free Xe atoms or Xe clusters of the size of the dopant cluster at the used NIR intensities $\approx 10^{15}$ Wcm$^{-2}$ [52, 225]. The delay-dependent nanoplasma absorption and the electron energies were predicted to feature two maxima due to distinct resonance conditions of the dopant and the He nanoplasma components which are met at different times in the course of the expansion [47, 48].

In NIR pump-probe experiments, a plasmon resonance feature was clearly visible in the transient $He^+$ and $He^{2+}$ ion yields [51], see Fig. 10.16a, and in TR-PES [52]. The maximum shifted from a delay time of 100 fs for small He nanodroplets, He$_N$, $N \approx 6000$, to about 500 fs for $N \approx 15,000$, in good agreement with MD simulations [51]. The efficiency of the dual pulse scheme for igniting and driving a nanoplasma in doped He nanodroplets was confirmed by applying an optimal control scheme to enhance the strong-field induced emission of highly charged atomic ions from embedded silver clusters [46].

More detailed insight into the dynamics of a NIR-ignited He nanoplasma was recently obtained by following in time the energy of Auger electrons emitted by a correlated electronic decay process akin to ICD inside the nanoplasma [53]. Similar correlated decay processes have recently been observed for nanoplasmas induced in heavier rare gas clusters [230, 231]. The delay-dependent shifting of Auger electron energies and above-threshold ionization (ATI) peaks by more than 15 eV reflects the evolution of the collective Coulomb potential created by the He nanoplasma through electron emission on the time scale of tens of ps. Single-shot electron velocity-map images of He nanoplasmas display large variety of signal types, most crucially depending on the cluster size [232]. The common feature is a two-component distribution for each single-cluster event: a bright inner part with nearly circular shape corresponding to electron energies up to a few eV, surrounded by an extended background of more energetic electrons.

In a recent experiment carried out using the FEL FLASH at DESY in Hamburg, a nanoplasma was ignited by irradiating doped He nanodroplets with soft x-ray ($h\nu = 250$ eV) pump pulses and time-delayed NIR probe pulses [233]. Fig. 10.16b shows the measured $He^+$ and $He^{2+}$ ion yields. In contrast to the experiments that employ NIR dual pulses, here the X-ray pump pulse selectively inner-shell ionized the dopant cluster owing to the much larger absorption cross section of the dopants (Ar, Kr, Xe) compared to He. Classical MD simulations indicated that the pronounced maximum of the He ion yields at a delay of about 200 fs was partly due to the plasmon resonance, and partly to electron migration from the He shell to the highly charged dopant-cluster core leading to a transient increase of the total number of quasi-free electrons present in the cluster volume due to electron-He collisions. The MD simulations, which reproduced the experimental pump-probe curves, also showed that the expansion of an X-ray-ionized Ar cluster embedded in a He nanodroplet is strongly damped compared to a free Ar cluster of the same size. Thus, He droplets act as efficient tampers that slow down the explosion of embedded nanostructures, a property that could be exploited for improving coherent diffraction images [234].

In another recent FEL-based experiment, the dynamics of strong-field induced nanoplasmas in He droplets were probed using single-shot, single-particle fs time-resolved X-ray coherent diffractive imaging (CDI) at the Linac Coherent Light Source (LCLS) [235]. NIR-induced nanoplasma formation and subsequent droplet evolution were probed by delayed X-rays pulses ($\approx 100$ fs, $h\nu' = 600$ eV). Delay-dependent CDI patterns revealed distinct dynamics evolving on multiple timescales.

To summarize this section, a He nanodroplet can be turned from a weakly-interacting cryo-matrix into a highly charged, highly reactive nanoplasma by irradiation by intense NIR or XUV pulses. The strong-field ionization dynamics turns out to be extremely non-linear and highly sensitive to the presence of dopants owing to the large difference in ionization energies between dopants and the He host droplet. The characteristic pump-probe dynamics is a time-delayed absorption resonance associated with the evolution of a collective plasmon resonance. Still open questions pertain to the mechanisms and dynamics of the early phase of nanoplasma ignition and the late stage of recombination of electrons and ions during the expansion of the nanoplasma. In the latter phase, highly excited atoms and ions are populated which can in turn interact and decay by ICD-like processes. Furthermore, the enhanced emission of highly directional energetic electrons by plasmonic enhancement effects [236, 237], as recently observed with heavier rare-gas clusters, might be efficient for He nanodroplets as well. The property of He nanodroplets to act as a tamper that protects embedded molecules and nanostructures against ultrafast charging and fragmentation makes them interesting for single-shot X-ray coherent-diffraction imaging, a new technique that bears enormous potential for bio-molecular imaging and nanoscience [238, 239].

## 10.6  Coherent Multidimensional Spectroscopy in Helium Nanodroplets

Regarding the ultrafast spectroscopy concepts applied to helium nanodroplets, we have so far discussed time-resolved photoion and photoelectron spectroscopy. In these experiments, the attainable time and frequency resolution is directly given by the duration and spectral width of the pump and probe pulses. Furthermore, a well-defined phase relation between pump and probe pulses is not required which simplifies the demands on the optical setup.

This is in contrast to ultrafast coherent control and quantum interference spectroscopy methods [240, 241]. Here, phase-locked pulse sequences are applied and the quantum interference between different excitation pathways is probed or controlled. While coherent control is a topic of high interest in many fields [242], the focus of this contribution lies on spectroscopic applications. Quantum interference spectroscopy bears the advantages of a high time-frequency resolution as well as the capability to selectively probe specific signal contributions by appropriate design of the pulse sequences. Examples are the detection of multiple-quantum coherences which provide a highly sensitive probe for inter-particle interactions [37, 243] or photon-echos giving insight into ensemble inhomogeneities [244]. Established methods involve WP interferometry and coherent multidimensional spectroscopy (CMDS).

### 10.6.1  Spectroscopic Concepts of Wave Packet Interferometry and Coherent Multidimensional Spectroscopy

The concept of WP interferometry has been applied in many different experiments to probe and control the dynamics of various quantum systems, as discussed in two review articles [241, 245]. The terms WP interferometry and quantum interference spectroscopy are often used equivalently. Depending on the investigated system, the experiment may be more intuitively described by the interference of WPs or quantum pathways excited in the system. In the following we will apply the quantum pathway picture. Figure 10.17 shows the basic concept. Pump and probe pulses each excite a specific pathway in the system (Fig. 10.17a) leading to the same final state population. Since both pathways propagate along different states during the pump-probe delay

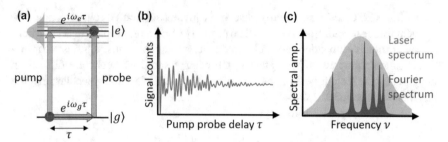

**Fig. 10.17** Wave packet interferometry scheme. (**a**) Pump-probe pulses excite two different quantum pathways leading to the same final state. The interference between the pathways depends on the different phase factors accumulated during the pump-probe delay $\tau$. (**b**) Schematic time-domain signal for the case of six excited states, as in (**a**). Oscillations reflect the constructive/destructive pathway interference. (**c**) Fourier transform of the signal, yielding the absorption spectrum of the system (blue) along with the laser spectrum (gray)

$\tau$, they accumulate different phases, giving rise to an alternating constructive and destructive interference pattern in the signal with a periodicity of $2\pi/(\omega_e - \omega_g)$ (Fig. 10.17b). At the same time the overall decay of the signal amplitude reflects dephasing and decoherence effects. Hence, the fringe pattern contains the information of an absorption spectrum, which is obtained by a Fourier transform of the signal (Fig. 10.17c). With this approach, the spectral resolution is given by the length of the time-domain signal and is thus decoupled from the spectral width of the laser spectrum. As such, state-resolved information can be gained even for broadband pulse spectra covering many resonances in the system.

There is a conceptual similarity to the detection of coherent vibrational WP oscillations discussed in several examples in Sect. 10.4. In these experiments, the propagation of vibrational wave packets along an electronic potential energy surface is probed. A Fourier analysis of the WP beating provides in analogy spectral information beyond the frequency resolution given by the femtosecond pulses. This scheme is primarily sensitive to the system's vibrational and rotational degrees of freedom, whereas WP interferometry also maps the electronic properties of the system including vibrational-electronic couplings.

In terms of dynamics, WP interferometry provides limited information. The method can only monitor changes of WPs within the Frank-Condon window between the ground and excited state. Processes such as the decay of WPs into new states or the transient change of the potential energy surface due to chemical reactions or perturbations by the environment may be hidden. In contrast, non-interferometric pump-probe experiments as discussed further above can offer a much enhanced observation window and dynamics can be monitored over a large parameter space. The situation is different if additional probe laser pulses are added to the WP interferometry scheme. These pulses may then probe the state of the system outside of the ground-excited state Frank-Condon window and thus extend the observation window for the system dynamics.

CMDS is an example for a particular powerful nonlinear extension of WP interferometry. This method greatly improves the information content deductible from ultrafast spectroscopy experiments [246–248]. CMDS combines the resolution advantage of WP interferometry with the extended sensitivity to dynamics known from classical pump-probe spectroscopy. The result is a nonlinear spectroscopy scheme which features several spectroscopic advantages not simultaneously present in any other technique. These include the direct spectroscopic access to couplings and relaxation pathways in the probed sample, the high time-frequency resolution as well as the capability to reveal system-bath interactions and inhomogeneities in real-time.

There are many variants of CMDS in terms of spectral range, detection scheme and number of excitation pulses [249]. A detailed description of all aspects is beyond the scope of this book. Here, we will restrict our discussion on population-detected two-dimensional (2D) spectroscopy in the VIS spectral domain, which probes electronic transitions. This is the only variant so far applied to helium nanodroplet samples [250] and can be readily explained in the framework of WP interference.

In the 2D version of CMDS, the method basically correlates two WP interferometry measurements, each performed by a phase-locked pulse pair (Fig. 10.18a). Performing a 2D Fourier transform of the data with respect to the time delays $\tau$ (between pulse 1 and 2) and $t$ (between 3, 4), yields a 2D frequency-correlation spectrum (Fig. 10.18c), hence the name *multidimensional* spectroscopy. These spectra show the frequency-resolved absorption (x-axis) directly correlated to the frequency-resolved/detection (y-axis) of the sample. In addition, the time delay $T$ in between the two WP interferometry experiments probes the time evolution of the system. Due to the underlying interferometric measurement scheme a high time-frequency resolution is achieved which automatically adapts to the time scales and spectral linewidths of the system [251].

The interaction of the quantum system with the four-pulse sequence gives rise to a multitude of nonlinaer signals (examples shown in Fig. 10.18b). To categorize the signals, it is convenient to adapt the common terminology from transient absorption spectroscopy: stimulated emission (SE) and excited state absorption (ESA) signals both probe the excited state, whereas ground state bleach (GSB) signals probe the ground state properties. ESA pathways involve transitions to higher-lying states and contribute with negative amplitude to the spectra, while SE and GSB contribute both with positive amplitude[1].

The 2D spectra can be interpreted as follows: (i) spectral peaks on the diagonal reflect the linear absorption spectrum, however with the additional information of 2D line shapes, directly dissecting the inhomogeneous (along diagonal) and homogeneous (along anti-diagonal) broadening in the system. Hence, time-resolved information about the system-bath interactions can be directly gained from the line shape analysis [244, 254]. (ii) Off-diagonal peaks (termed *cross peaks*), indicate couplings between different states and energy relaxation (peaks AB and BA in Fig. 10.18c). This

---

[1] While this is the case in the photoionization 2D spectroscopy experiments presented here, the different sign of ESA and SE/GSB amplitudes is not strictly given in all detection schemes [252, 253]

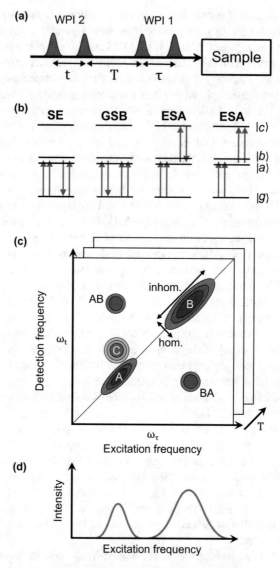

**Fig. 10.18** The principle of 2D spectroscopy. (**a**) Pulse sequence exciting the sample. Pulses 1, 2 and 3, 4 form phase-locked pulse pairs to perform two correlated WP interferometry experiments (WPI 1,2), whereas the time evolution of the system is probed in between the pulse pairs. (**b**) Model energy-level system along with a selection of possible nonlinear signals induced by the four-pulse sequence. SE: stimulated emission, GSB: ground-state bleach, ESA: excited-state absorption. (**c**) 2D frequency spectrum obtained from a 2D Fourier transform of the signal. Peaks A and B on the diagonal represent the $|g\rangle \leftrightarrow |a\rangle$, $|b\rangle$ resonances. Their 2D lineshapes reflect the inhomogeneous and homogeneous linewidth along the diagonal and antidiagonal, respectively. Peak C denotes an excited state absorption from $|a\rangle$ to the higher-lying state $|c\rangle$. AB and BA denote cross peaks which reflect couplings between states $|a\rangle$ and $|b\rangle$. (**d**) Projection of the 2D spectrum onto the x-axis, resulting in a one-dimensional spectrum as it would be measured with conventional absorption spectroscopy

greatly simplifies the identification of relaxation pathways and allows to follow the relaxation dynamics in real-time [255–257]. (iii) Transitions to higher-lying states (ESA) contribute with negative amplitude and are thus readily identified (peak C in Fig. 10.18c). All this information is difficult to deduce from one-dimensional spectroscopy, where 2D lineshape information is not available and cross peaks overlap spectrally with diagonal features. This is schematically expressed by a projection of the 2D spectrum onto a single axis (Fig. 10.18d).

CMDS has been so far mainly applied in the condensed phase where the method has achieved considerable success, as highlighted in several review articles [246–248, 254, 258, 259]. Certainly, the advantages of CMDS are also very beneficial in the gas phase, in particular in more complex gas phase samples, such as species embedded in helium nanodroplets offering the study of intra/inter-molecular dynamics and peculiar system-bath interactions. However, an extension of CMDS to the gas phase has been so far vastly impeded by the difficult signal-to-noise challenge due to the low sample densities predominant in gas phase experiments. Over the last years, Bruder and Stienkemeier et al. have developed a specialized experimental approach [250] to solve this issue, which is outlined below.

## 10.6.2   Resolving the Experimental Challenges

The implementation of CMDS experiments faces two major signal-to-noise challenges. On the one hand, CMDS is a nonlinear spectroscopy scheme. Thus, a high dynamic range is required to uncover the weak nonlinear signals from dominant linear signals and general background noise. To give some numbers, in CMDS experiments of doped helium nanodroplet species, background signals are typically one to three orders of magnitude larger than the CMDS signal. On the other hand, the underlying interferometric measurement scheme adds an additional noise source stemming from phase jitter between the optical pulses. Sub-cycle phase stability is demanded, which implies a reduction of optical pathlength fluctuations to $< \lambda/50$ [260], a value that is very hard to achieve with conventional interferometers. This requirement also applies to coherent control and WP interferometry experiments, however, in CMDS it scales with higher order due to the nonlinear WP interferometry scheme.

Several techniques were developed which solve these issues in the condensed phase, as summarized in Ref. [249]. Out of these methods, the phase modulation technique developed by Marcus and coworkers [261, 262] is most suitable for the application in helium droplet beam experiments for several reasons. It provides efficient phase stabilization to reduce phase jitter, extraordinary sensitivity by incorporating lock-in detection and it can be combined with efficient photoionization detection schemes [37, 243]. It is also compatible with high power, high repetition rate ($> 100$ kHz) laser systems which improve statistics while avoiding saturation of optical transitions and detectors.

The phase modulation technique is shown in Fig. 10.19. Precise phase beatings are imprinted in the optical pulse sequence which transfers to a characteristic ampli-

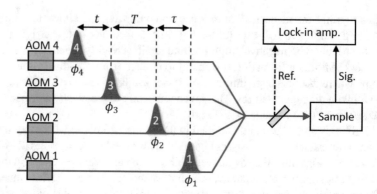

**Fig. 10.19** Phase modulation scheme to improve phase stability and sensitivity in 2D spectroscopy experiments. A sequence of four laser pulses excites the sample. Four acousto-optical modulators (AOMs) shift the carrier envelope phase $\phi_i$ of the individual laser pulses in each laser cycle by a well-defined value, which results in a continuous modulation of the quantum interference signals. A lock-in amplifier is used for demodulation. An optical interference signal is coupled-out from the optical setup for referencing the lock-in demodulation process

tude modulation of the interference signals in the time domain. Background signals are not modulated or appear at other modulation frequencies, enabling selective, highly efficient lock-in amplification of the interference signals. Moreover, hetero-dyne detection with an optical interference signal is implemented which leads to a cancellation of the optical phase jitter and rotating frame detection. The latter results in a downshift of quantum interference frequencies by several orders of magnitude.

The advantages of the phase modulation technique are demonstrated in Fig. 10.20, showing a comparison between conventional and phase-modulated WP interferometry of a rubidium-doped helium nanodroplet sample [37]. With the spectral bandwidth of the femtosecond laser, the $5S_{1/2} \rightarrow 5P_{3/2}$ (D$_2$ line) as well as the $5P_{3/2} \rightarrow 5D_{5/2,3/2}$ atomic transitions in Rb are resonantly excited and detected by 1+2 REMPI combined with mass-resolved ion detection. The quantum interference signals in the time domain exhibit rapid oscillations corresponding to the constructive and destructive interference of excitation pathways induced as a function of the pump-probe delay. As a striking feature, the oscillation period in the phase-modulated WP interferometry measurements is more than a factor of 100 larger compared to the conventional technique (Fig. 10.20a,b), which is due to the rotating frame detection. Hence, much sparser sampling of the signal is possible while deducing the same amount of information.

A Fourier transform yields the absorption spectrum revealing a drastic difference in the signal-to-noise performance of both experiments. In the conventional WP interferometry, the atomic resonances can only be qualitatively identified due to the strong phase jitter on the signal. In contrast, the phase modulation approach delivers a highly resolved spectrum with excellent signal-to-noise ratio. Obviously, the frequency spectrum precisely resembles the signature of free gas-phase Rb atoms without any sign of droplet-induced perturbations, implying that the current

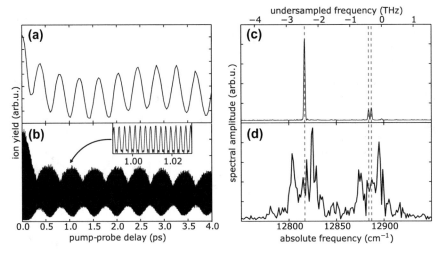

**Fig. 10.20** Performance comparison between phase-modulated and conventional WP interferometry. Time domain WP interferometry signals obtained with the phase modulation technique (**a**) and without (**b**) showing the first 4 ps of the signal. (**c**), (**d**) Respective Fourier transforms of the full data set (50 ps length). In (c), the scales on the top/bottom show the rotating frame and the up-shifted frequency axis, respectively. Dashed vertical lines indicate the atomic resonances: $5S_{1/2} \rightarrow 5P_{3/2}$ and $5P_{3/2} \rightarrow 5D_{3/2,5/2}$. Adapted from [37]—Published by the PCCP Owner Societies. Licensed under CC BY 3.0

experiment is predominantly sensitive to already desorbed atoms. As mentioned above, this is explained by the high laser repetition rate (80 MHz), supporting the desorption of the atoms and subsequent probing in the gas phase in the same experiment. Quantum interference experiments with lower repetition rate are presented further below in Sect. 10.6.4. In summary, the phase modulation experiment in Fig. 10.20 featuring a great signal-to-noise improvement for highly dilute helium droplet samples marks an important milestone and opened-up the door for CMDS experiments of helium nanodroplet samples.

### 10.6.3 High Resolution Wave Packet Interferometry

The previous example indicates the prospective of using the phase modulation technique for high resolution spectroscopy. In WP interferometry, the frequency-resolution limit $\Delta \nu$ of the experimental apparatus is directly connected with the scanned pump-probe delay range $\Delta \tau_{\text{range}}$ by $\Delta \nu = 1/\Delta \tau_{\text{range}}$. With mechanical delay stages scanning ranges of $< 2$ ns are realistic [36] which corresponds to a resolution limit of 500 MHz. Frequency-comb-based approaches can in principle extend the scanning range to reach a frequency resolution of $\approx 100$ MHz [263]. This option however rises considerably the demands on the laser source and is not further discussed

**Fig. 10.21** Rb and RbHe
energy levels along with the
pump probe scheme used in
the WP interferometry
experiments. The shape of
atomic Rb (yellow) and He
(blue) orbitals is sketched.
Adapted
from [37]—Published by the
PCCP Owner Societies.
Licensed under CC BY 3.0

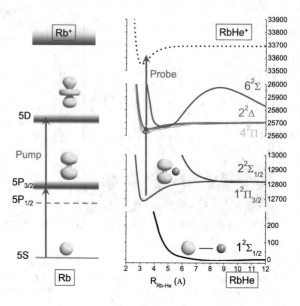

here. In general, much higher spectral resolution is achieved with continuous-wave
lasers. However, the advantage of WP interferometry is a flexible time-frequency
resolution to study spectral and temporal aspects of the target system. Moreover,
the intense femtosecond pulses improve the signal strength in multiphoton probing
schemes. This applies to photoionization schemes but also to some exotic molecules
such as alkali-helium (AkHe) exciplexes.

As already discussed above (Sect. 10.2.1, 10.3.1), AkHe molecules exhibit an
anti-bonding ground state whereas some excited electronic states support bound configurations. As an example, the Rb atomic levels along with the RbHe pair potentials
are given in Fig. 10.21. Upon electronic excitation of the AkHe$_N$ system, exciplexes
may form and desorb from the droplet. Desorbed metastable AkHe$_n$ complexes up
to $n \leq 4$ have been observed with mass spectrometry [10]. The formation, probing
and detection thus requires a multiphoton experiment which is highly favorable with
femtosecond laser sources, making WP interferometry the ideal spectroscopic tool
to study the level structure of these systems.

Despite many experimental and theoretical studies devoted to these peculiar
molecules [8–10, 13, 14, 21, 24, 126, 127, 264, 265] some questions about the formation mechanism and associated formation times remain unsolved[14, 21, 127].
Moreover, until recently high resolution spectral data has not been available, partly
due to the difficult accessibility by standard absorption/emission spectroscopy [8, 9]
and the limited resolution given by other techniques [24, 126, 127]. The first highly
resolved vibronic spectrum of an AkHe molecule has been obtained with the phase
modulation technique [37].

In the RbHe molecule, WPs between the electronic states correlated to the 5P
and 5D atomic asymptotes of rubidium were induced and probed via subsequent

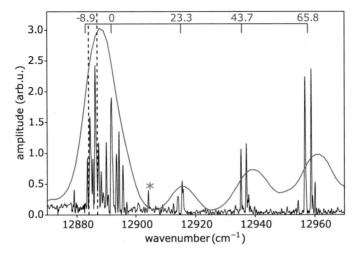

**Fig. 10.22** High resolution Rb*He spectrum recorded with phase-modulated WP interferometry (black)[37] along with femtosecond-pump, picosecond-probe photoionization measurements (red)[127]. Beat frequencies deduced from conventional WP interferometry (blue)[24] are shown on the top. Vertical dashed lines indicate the atomic $5P_{3/2} \rightarrow 5D_{3/2,5/2}$ transitions. The asterix marks an artificial peak coming from low frequency noise. Adapted from[37]—Published by the PCCP Owner Societies. Licensed under CC BY 3.0

photoionization (Fig. 10.21). The resulting Fourier transform spectrum is shown in Fig. 10.22, revealing a clean, highly resolved vibronic spectrum with a resolution of $0.3\,\text{cm}^{-1}$. In contrast to the vibrational WP studies (Sect. 10.4) probing purely vibrational modes, the experiment here detects vibronic resonances between different electronic states. The strong spin-oribit coupling in the RbHe system results in a complex manifold of many closely spaced electronic states which explains the highly structured spectrum in Fig. 10.22. Reasonable good agreement with a theoretical model is found[37] which is remarkable considering the high degree of spectral details and the experimental resolution being clearly beyond the precision of current models. The spectroscopic potential of the novel WP interferometry technique is shown by a comparison with previous experiments based on picosecond pulse shaping and conventional WP interferometry. While good agreement is found between the different experimental techniques, the new method provides a factor of $\geq 10$ higher resolution. This example hence underlines the advantage of femtosecond spectroscopy techniques in the spectral study of metastable molecules and provides new benchmark spectroscopic data for the development of ab-initio methods.

### 10.6.4 Ultrafast Droplet-Induced Coherence Decay in Alkali Dopants

The peculiarities of the alkali-helium droplet interaction have been already discussed. The repulsive character of most excited $AkHe_N$ states induces a line-broadening in the order of 10–100 wavenumbers [125]. Accordingly, ground-excited state coherences are expected to decay on the order of 0.1–1 ps. Quantum interference spectroscopy should provide a real-time analysis of this process. In fact, time resolved quantum interference studies were among the first femtosecond experiments of doped droplet species [20]. In these early attempts, insufficient phase stability prohibited a Fourier analysis of the transient interference signals. Instead, indications about the droplet interaction were directly deduced from the coherence decay times which were in the order of few hundred femtoseconds for low excited states in $KHe_N$.

With the novel phase-modulated WP interferometry technique a high resolution study of the decoherence process becomes possible. While the experiment in Fig. 10.20 had probed already desorbed Rb atoms and thus renders insensitive to droplet-induced dynamics, Fig. 10.23 shows data from the same target system using a modified experimental setup, now clearly revealing the ultrafast decoherence induced by the guest-host interaction [266]. The experimental modifications comprise of a lower laser repetition rate (80 MHz $\rightarrow$ 200 kHz) and a separate, delayed ionization laser ($\lambda = 520$ nm, delay $\sim$ns) to ionize the species after full desorption. Moreover, very broad bandwidth femtosecond pulses (FWHM = $1600$ cm$^{-1}$) are used to simultaneously cover the absorption lines of Rb atoms and $Rb_2$, $Rb_3$ molecules adsorbed to the droplet surface.

The time domain signal reveals an extremely fast coherence decay within $\approx 150$ fs, followed by a much weaker but persistent interference signal extending beyond

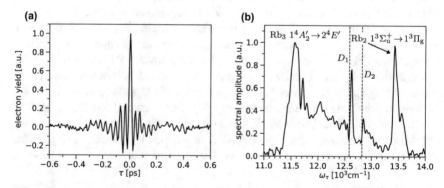

**Fig. 10.23** Droplet-induced decoherence in Rb atoms and $Rb_2$, $Rb_3$ molecules tracked by phase-modulated WP interferometry. (a) Pump-probe transient in the time domain, effusive atomic background is subtracted. The pronounced spike at 0 fs stems from the optical pump-probe cross-correlation mapped to the continuum by three-photon ionization. (b) Fourier transform. Labels indicate the excited resonances in the Rb molecules. Dashed vertical lines mark the atomic $D_{1,2}$ transitions

delays of 1.5 ps. A Fourier transform uncovers a rich absorption spectrum showing the absorption bands of the Rb trimer: $1\,^4A'_2 \rightarrow 2\,^4E'$ at $11600\,\text{cm}^{-1}$, the dimer: $a\,^3\Sigma_u^+ \rightarrow 1\,^3\Pi_g$ at $13500\,\text{cm}^{-1}$ and monomer: $5s\,^2\Sigma_{1/2} \rightarrow 5p\,^2\Pi_{1/2,3/2}$ at $12600\,\text{cm}^{-1}$ and $12830\,\text{cm}^{-1}$, $5s\,^2\Sigma_{1/2} \rightarrow 5p\,^2\Sigma_{1/2}$ at $12850\,\text{cm}^{-1}$. These features are in good agreement with previous steady-state absorption spectroscopy [9, 267, 268]. In particular, the monomer response now resembles the blue-shifted strongly broadened absorption profile characteristic for the pseudo-diatomic model [125]. The broad blue shoulder of the $Rb_3$ resonance ($11800$–$12600\,\text{cm}^{-1}$) was also observed in femtosecond absorption mass-spectrometry, where a clear correlation to the $Rb_3^+$ ion yield was determined [25].

This example shows the sensitivity of WP interferometry to the guest-host interaction in doped droplet beam experiments and the capability to probe complex spectra extending over a broad spectral range and many resonances from different species. The experiment serves as precursor study for 2D spectroscopy on these samples. An extension to 2D spectroscopy facilitates the direct correlation of absorption and emission of each spectral feature/dopant species and allows to follow their dynamical evolution in real time as discussed below.

### 10.6.5  Coherent Multidimensional Spectroscopy of Doped Helium Nanodroplets

The unique properties of 2D spectroscopy render it a powerful tool for the study of ultrafast dynamics and guest-host interactions in doped helium droplet samples. The latter effect is particularly pronounced for alkalis, which hence provide an ideal test system for 2D spectroscopy experiments. Figure 10.24 shows 2D spectroscopy data for rubidium-doped helium droplets. In these experiments, a four-pulse sequence induces the 2D signal (cf. Figs. 10.18, 10.19) which is detected via photoionization. The ionization step is performed either by an additional interaction with the fourth laser pulse or by a delayed fifth pulse and is combined either with electron (Fig. 10.24a,b) or ion detection (c). The different ionization and detection schemes explain the different appearance of peak amplitudes in the spectra. The 2D maps show clear, pronounced peaks well separated from the noise floor, which is remarkable considering the challenging signal-to-noise conditions in these experiments. These measurements constitute the first 2D spectroscopy study of isolated cold molecules [269].

The 2D spectra directly disclose the correlations between the absorption and emission of the $Rb_2$ and $Rb_3$ molecules, revealing various cross peaks and ESA signals, which were not observed in previous experiments. For the $Rb_2$ molecule, exemplary the excitation and probing scheme is shown in Fig. 10.24d. The $Rb_2$ data exhibits two strong ESA features (labeled $ESA_1$, $ESA_2$) which extend into the complex $Rb_2$ Rydberg manifold, featuring a high density of electronic states (not shown in Fig. 10.24d). Despite the complex level structure, some clear conclusions can be drawn with the help of 2D spectroscopy. The position of the ESA peaks along the

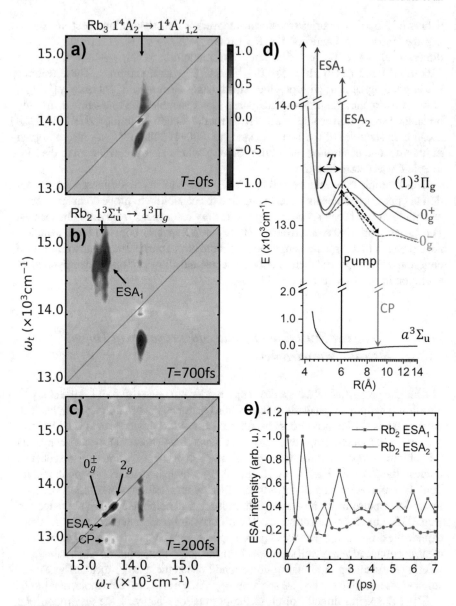

**Fig. 10.24** CMDS results for $Rb_2$ and $Rb_3$ molecules formed on the surface of helium nanodroplets. 2D spectra detected via photoelectrons (**a**), (**b**) and photoions (**c**) for different evolution times $T = 0, 200, 700\,fs$ as labeled. X-axis: absorption, y-axis: detection frequency. Labels indicate the molecular transitions and spin-orbit splittings. Two distinct ESA features ($ESA_{1,2}$) and a cross peak (CP) are marked, as well. (**d**) Ab-initio $Rb_2$ potential energy curves and concluded photodynamics. Labels of probe-transitions correspond to the ones in (a-c). The helium perturbation on the $0_g^+$ state is schematically drawn as dashed curve. (**e**) Time evolution of the ESA amplitudes reveal coherent vibrational WP oscillations with a phase shift of $\pi$ between the traces. Adapted from [269]— Published by Springer Nature, licensed under CC BY 4.0

detection axis show the spectral position of the Frank-Condon windows to the Rydberg states. At the same time, the coherent vibrational WP oscillations reflected in the ESA peaks (Fig. 10.24e) pin down the location of the Frank-Condon windows. From the clear $\pi$-phase shift between both traces the existence of two Frank-Condon windows located at the inner and outer turning point of the excited state potential, respectively, becomes apparent (see sketch Fig. 10.24d).

The 2D data also permits a refined interpretation of a Stokes shift observed in the $Rb_2$ emission (cross peak red-shifted by $600\,cm^{-1}$, labeled CP in Fig. 10.24c). This feature was previously interpreted as the emission from vibrationally relaxed free gas-phase $Rb_2$ molecules [268]. In contrast, the high time-frequency resolution in the 2D spectroscopy experiment uncovers an ultrafast intra-molecular relaxation within $< 100\,fs$ (not shown) into the outer potential well of the $1\,^3\Pi_g$ state, catalyzed by the helium perturbation (sketched in Fig. 10.24d) [269].

While the $Rb_2$ molecule offered rich intra-molecular dynamics on femtosecond time scales, the $Rb_3$ molecule serves as a sensitive probe for the dynamics of the quantum fluid droplet. Many of the above discussed ultrafast dynamics studies investigated guest-host interactions in doped droplets with the goal to deduce properties about the quantum fluid itself. To avoid less available femtosecond XUV light sources, often impurities are embedded as probes that are optically accessible. In time-resolved photoionization studies of impurities, the droplet response is inferred from the transient energy shift between the neutral excited and ionic state of the guest-host interaction potential (Fig. 10.8). 2D spectroscopy offers an alternative approach which probes the matrix-shift along the ground and excited states of the purely neutral interaction potential (Fig. 10.25a).

For the large inertia of mass of the $Rb_3$ molecule, any short-time dynamics along the interaction coordinate may be solely attributed to the response of the helium density. Hence the molecule provides an ideal probe for the dynamics of the quantum fluid at the droplet surface. Upon impulsive excitation of the molecule with the femtosecond laser pulses, the helium density will repel and the system will relax along the interaction coordinate (Fig. 10.25a). The process can be followed in real time in the 2D spectroscopy data (Fig. 10.25b). Here, a pronounced dynamic Stokes shift is observable, which stems from the SE signal probing the system's relaxation along the excited state of the interaction potential. The energy shift reaches an asymptotic value of $(150 \pm 19)\,cm^{-1}$ within $\approx 2.5\,ps$ which marks the time-scale for the ultrafast rearrangement of the helium density to reach equilibrium. In comparison, the desorption of Rb atoms and molecules for the lowest excitations commonly takes place on much longer time scales (Sect. 10.3.1). A similar dynamic is expected for the $Rb_2$ molecule, which is, however, covered by the persistent $ESA_2$ peak.

These 2D spectroscopy experiments have demonstrated the power and added value of applying CMDS to doped helium droplet species. Ak molecules attached to He droplets have been extensively studied in recent years, both with high-resolution steady-state spectroscopy and time-resolved pump-probe experiments. Yet, the high time-frequency resolution and the ability to directly correlate absorption and emission spectra in 2D spectroscopy has still brought new insight into these systems which shows great promise for future studies on other systems. In particular, the

**(a)**

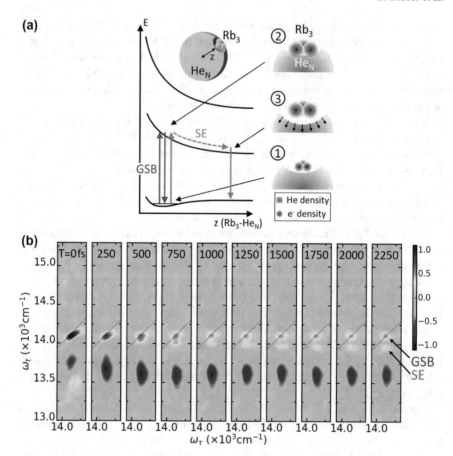

**Fig. 10.25** Real-time observation of the helium surface repulsion. (**a**) Sketch of the Rb$_3$–He$_N$ interaction potentials. Steps 1–3 show the repulsion of the helium density following the impulsive molecular excitation. The excited state relaxation process is traced by the SE signal. The GSB probes the ground state where no system-bath dynamics occur. (**b**) Time-evolution of the spectra showing the cut-out of the Rb$_3$ $1\,^4A_2' \rightarrow 1\,^4A_{1,2}''$ excitation. A clear dynamic Stokes shift (spectral splitting of SE and GSB signals) is visible, which converges to a constant red-shift of $150\,cm^{-1}$ within $\approx 2.5\,ps$. Adapted from [269]—Published by Springer Nature, licensed under CC BY 4.0

spectra of many organic molecules dissolved inside helium nanodroplets show only weak perturbation [270, 271]. Hence, high-resolution 2D spectroscopy of organic compounds are at hand which would provide invaluable complementary information to condensed phase studies. Very recently, Bruder and Stienkemeier et al. have performed the first CMDS experiments of an organic molecule fully dissolved in helium droplets [272] which marks the highest resolution so far achieved in a molecular 2D spectrum and thus underlines this potential.

## 10.7   Conclusions and Outlook

The various ultrafast spectroscopy methods presented in this chapter have uncovered a rich variety of structural and electronic dynamics in pure and doped helium nanodroplets with unique features that are not found in other quantum systems. On the one hand, helium nanodroplets offer the possibility for high resolution studies of species dissolved in a weakly-perturbing environment which is in stark contrast to studies in the condensed phase. Naively, high spectral resolution is associated with continuous-wave laser spectroscopy. However, as has been discussed in this chapter, femtosecond laser pulses open-up the preparation and study of coherent WPs which can provide high spectral information through Fourier analysis with comparable resolution to steady-state methods. Moreover, the direct WP analysis in the time-domain provides access to phase information and decay processes which may not be accessible in steady-state spectroscopy. These aspects have been well demonstrated in the observation of extremely long-lived vibrational WPs in alkali molecules attached to the helium droplet surface (Sect. 10.4.1), the recurrence of vibrational revivals in ejected $In_2$ molecules (Sect. 10.4.2) or in the identification of multiple ionization pathways from phase shifts in the WP motion (Sect. 10.6.5). Intense femtosecond pulses also open-up nonlinear multiphoton experiments which provide access to strong-field effects, to the study of higher-lying states, and the properties of metastable species. Examples were the formation and high-resolution study of metastable exciplex molecules (Sect. 10.6.3), or nanoplasmas ignited inside helium nanodroplets with remarkable efficiency (Sect. 10.5.3). These studies, however, also demonstrate that the interaction with high-intensity or/and high photon energy laser pulses in most cases leads to the deposition of large amounts of energy into almost all degrees of freedom, in this way wiping out the low-temperature quantum properties of superfluid droplets.

On the other hand, time-resolved experiments have provided insight into the dynamics of the droplets themselves. Novel coherent XUV light sources have for the first time enabled the direct study of the multifaceted relaxation dynamics inside helium nanodroplets in real-time (Sect. 10.5.1). Ultrafast electronic relaxation, bubble formation and ejection of metastable He atoms from the droplets were directly measured. Furthermore, insight into structural dynamics of the helium density have been gained from time-resolved studies with impurities embedded inside the droplets or attached to their surface. The experiments have uncovered a general behavior of the quantum fluid which can be categorized into a fast change of the helium solvation shell and a somewhat slower transport dynamics. The primary, fast response of the helium density in the local environment of the impurity (solvation shell) takes place on a few hundred femtoseconds to a few picoseconds as a response to the impulsive electronic excitation of the impurity. While this process is accurately predicted by theory (Sect. 10.2.6), only very recently TRPES (Sect. 10.3.2) and 2D spectroscopy experiments (Sect. 10.6.5) have provided the first experimental access to these dynamics. The experiments permit a comparison of the time scales for the helium repulsion inside and at the surface of the droplets, revealing a slower equilibration at the surface. While a different behavior of the quantum fluid inside the droplet compared to surface states is expected, the complex aspects of the dynamics with respect to the dopant properties and density modes of the droplet will require further experiments in this direction for more general conclusions.

The secondary transport dynamics often occur on a longer time scale in the picosecond to nanosecond domain. As a typical feature of helium nanodroplets, electronic excitation of impurities almost always triggers a propagation along the impurity-droplet interaction potential leading to the ejection of the dopant from the droplet or to the trapping of the dopant inside snowball structures in the droplet core. These dynamics directly manifest themselves in pump-probe measurements of alkali atoms and molecules which first desorb from the droplet surface when being resonantly excited, and then fall back into the droplet when being ionized (Sect. 10.3.1). The impurity transport and detachment may be accompanied by helium density oscillations which are observable in TRPES measurements (Sect. 10.3.2). Time-dependent density functional simulations nicely visualize the full evolution of the dopant-droplet system, and even reproduce the experimentally observed dynamics quantitatively (Sect. 10.2.6). In addition, the dynamics of the quantum fluid superimpose and interplay with the intra-molecular and inter-molecular dynamics of the impurities which may comprise of intra-molecular vibrational energy redistribution (IVR), spin and electronic relaxation as well as dissociation and charge transfer processes (Sect. 10.4.2, 10.3.1, 10.5.2, 10.6.5) and often take place on a comparable time scale as the dynamics of the helium density.

This interplay of dopant and helium bath dynamics makes helium nanodroplets a challenging, however at the same time, a fascinating target system for spectroscopic studies. While pure and doped helium droplets form an enclosed nanosystem which is still amenable to theoretical models, these systems feature intriguing aspects of fundamental molecular dynamics, system-bath interactions and unique quantum fluid properties as it is not found in any other system. The here discussed time-resolved studies have provided a first glimpse into the ultrafast dynamics of these systems and have founded the experimental and theoretical basis for further exploration. In this view, two major routes are identified which concern the vast synthesis abilities provided by helium nanodroplets as well as the rapid technological developments in XUV light sources and XUV spectroscopy methods.

First, the ability of He droplets to generate microsolvation environments [273, 274] will provide new options to the field of femtochemistry. Flexible pickup possibilities allow to design the environment around a molecule of interest in terms of number of solvent molecules and their interaction strength (polarizability, dipole moment, hydrogen bond). The advantages provided by time-domain spectroscopy, including coherence phenomena and phase information, can now be applied to these systems. In particular, building up the solvation shell piece-by-piece will allow to track ultrafast dynamics as the environmental conditions bridge from isolation to full solvation, shedding light on the gap between accurate gas-phase studies and real-world systems in solution. Furthermore, the repertoire of time-domain techniques ranging from pump-probe to multidimensional spectroscopy, can be applied to droplet-specific systems that have previously evaded time-domain investigations, including tailor-made complexes [275, 276], fragile agglomerates [11, 277], or highly reactive species [278]. For example, diverse combinations of donor-acceptor pairs can be prepared for charge-transfer studies, where the time-domain approach will be essential to disentangle nuclear and electronic dynamics. Another field of interest would be the investigation of exciton dynamics in molecular aggregates [279], including migration, fission and annihilation mechanisms. In view of fundamental photophysical reactions and the development of efficient photo-switches, the real-time study of isomerization inside the quantum fluid [280] and inside microsolvation environments would contribute new insights.

Second, the novel developments in ultrafast XUV light sources open up fascinating perspectives in the time-resolved study of pure and doped helium nanodroplets as they offer direct access to the optical properties of superfluid helium. The static and dynamics properties of helium droplets have been studied in recent years using all types of XUV light sources [64, 66, 184, 192, 215]. Still, questions remain, in particular, about collective phenomena such as correlated electronic decay processes observed in these systems. As one example, a variety of highly efficient ICD processes in doped and pure droplets have been evidenced in recent years. To capture the full kinematics and dynamics of such processes, covariance and coincidence detection methods are instrumental. With the advent of high-repetition-rate intense femtosecond lasers and XUV radiation sources, combining coincidence detection with femtosecond time-resolved spectroscopy of helium nanodroplets is in reach.

Furthermore, CMDS and related coherent nonlinear spectroscopy methods provide selective probes for collective properties [243] and variations in the local environment of many-body quantum systems [244, 254]. As such, XUV-CMDS experiments would shed new light on the inhomogeneity and many-body nature of the helium droplet absorption spectrum. Furthermore, with a transfer of coherent nonlinear methods to the X-ray domain, localized core resonances would be accessible, thus facilitating the study of dopant complexes inside helium droplets with unprecedented atomic sensitivity. Recently, XUV-WP interferometry of helium atoms was demonstrated which shows that the phase stability issue in XUV quantum interference experiments can be solved [281]. Moreover, coherent XUV and X-ray wave mixing experiments were established detecting nonlinear mixing signals with nanometer resolution and site-specificity [282–284]. These developments open-up the perspective for XUV and X-ray coherent nonlinear and even multidimensional spectroscopy experiments.

Yet, the most direct probing of the structural dynamics of nanoparticles is achieved through the new technique of X-ray single-shot coherent diffraction imaging (CDI) [198]. Currently, this technique relies on radiation provided by one of the few existing XUV and X-ray FEL facilities. However, tremendous progress is being made in generating intense and femtosecond and even attosecond pulses in the XUV and X-ray ranges [285]. New radiation sources such as high-harmonic generation based on high-power table-top femtosecond lasers will make it possible to directly visualize the structural dynamics of helium nanodroplets and other nanoparticles using pump-probe and possibly more sophisticated CDI schemes [235, 286].

**Acknowledgements** M.K. acknowledges financial support by the Austrian Science Fund (FWF) under Grants P29369-N36 and P33166-N, as well as support from NAWI Graz. M.M. gratefully acknowledges funding from the Carlsberg Foundation and from the SPARC Programme, MHRD, India. F.S. and L.B. gratefully acknowledge financial support by the European Research Council (ERC) with the Advanced Grant "COCONIS" (694965) and by the Deutsche Forschungsgemeinschaft (DFG) IRTG CoCo (2079). F.S. acknowledges funding within the DFG Priority Program QUTIF (STI 125/22).

# References

1. E.L. Andronikashvili, J. Phys. U.S.S.R. **10**, 201 (1946)
2. C. Enss, S. Hunklinger, *Low-Temperature Physics* (Springer-Verlag, Berlin Heidelberg, 2005)
3. F. Stienkemeier, K. Lehmann, J. Phys. B **39**, R127 (2006)
4. M. Mudrich, F. Stienkemeier, Int. Rev. Phys. Chem. **33**, 301 (2014)
5. M.P. Ziemkiewicz, D.M. Neumark, O. Gessner, Int. Rev. Phys. Chem. **34**, 239 (2015)
6. F. Stienkemeier, J. Higgins, C. Callegari, S.I. Kanorsky, W.E. Ernst, G. Scoles, Z. Phys. D. **38**, 253 (1996)
7. M. Barranco, R. Guardiola, S. Hernández, R. Mayol, J. Navarro, M. Pi, J. Low Temp. Phys. **142**, 1 (2006)
8. J. Reho, J. Higgins, C. Callegari, K.K. Lehmann, G. Scoles, J. Chem. Phys. **113**, 9686 (2000)
9. F.R. Brühl, R.A. Trasca, W.E. Ernst, J. Chem. Phys. **115**, 10220 (2001)
10. C.P. Schulz, P. Claas, F. Stienkemeier, Phys. Rev. Lett. **87**, 153401 (2001)
11. J. Higgins, C. Callegari, J. Reho, F. Stienkemeier, W.E. Ernst, K.K. Lehmann, M. Gutowski, G. Scoles, Science **273**, 629 (1996)
12. J. Higgins, C. Callegari, J. Reho, F. Stienkemeier, W.E. Ernst, M. Gutowski, G. Scoles, J. Phys. Chem. A **102**, 4952 (1998)
13. J. Reho, C. Callegari, J. Higgins, W.E. Ernst, K.K. Lehmann, G. Scoles, Faraday Discussion **108**, 161 (1997)
14. J. Reho, J. Higgins, C. Callegari, K.K. Lehmann, G. Scoles, J. Chem. Phys. **113**, 9694 (2000)
15. J.H. Reho, U. Merker, M.R. Radcliff, K.K. Lehmann, G. Scoles, J. Phys. Chem. A **104**, 3620 (2000)
16. J. Nagl, G. Auböck, A. Hauser, O. Allard, C. Callegari, W. Ernst, J. Chem. Phys. **128**, 154320 (2008)
17. O. Bünermann, F. Stienkemeier, Eur. Phys. J. D **61**, 645 (2011)
18. J. Reho, J. Higgins, M. Nooijen, K.K. Lehmann, G. Scoles, M. Gutowski, J. Chem. Phys. **115**, 10265 (2001)
19. J. Reho, J. Higgins, K.K. Lehmann, Faraday Discussion **118**, 33 (2001)
20. F. Stienkemeier, F. Meier, A. Hägele, H.O. Lutz, E. Schreiber, C.P. Schulz, I.V. Hertel, Phys. Rev. Lett. **83**, 2320 (1999)
21. G. Droppelmann, O. Bünermann, C.P. Schulz, F. Stienkemeier, Phys. Rev. Lett. **93**, 0233402 (2004)
22. P. Claas, G. Droppelmann, C.P. Schulz, M. Mudrich, F. Stienkemeier, J. Phys. B **39**, S1151 (2006)
23. P. Claas, G. Droppelmann, C.P. Schulz, M. Mudrich, F. Stienkemeier, J. Phys. Chem. A **111**, 7537 (2007)
24. M. Mudrich, G. Droppelmann, P. Claas, C. Schulz, F. Stienkemeier, Phys. Rev. Lett. **100**, 023401 (2008)

25. M. Mudrich, P. Heister, T. Hippler, C. Giese, O. Dulieu, F. Stienkemeier, Phys. Rev. A **80**, 042512 (2009)
26. M. Schlesinger, W.T. Strunz, Phys. Rev. A **77**, 012111 (2008)
27. J. von Vangerow, O. John, F. Stienkemeier, M. Mudrich, J. Chem. Phys. **143**, 034302 (2015)
28. J. Von Vangerow, F. Coppens, A. Leal, M. Pi, M. Barranco, N. Halberstadt, F. Stienkemeier, M. Mudrich, J. Phys. Chem. Lett. **8**, 307 (2017)
29. F. Coppens, J. Von Vangerow, M. Barranco, N. Halberstadt, F. Stienkemeier, M. Pi, M. Mudrich, Phys. Chem. Chem. Phys. **20**, 9309 (2018)
30. N.V. Dozmorov, A.V. Baklanov, J. von Vangerow, F. Stienkemeier, J.A.M. Fordyce, M. Mudrich, Phys. Rev. A **98**, 043403 (2018)
31. B. Thaler, S. Ranftl, P. Heim, S. Cesnik, L. Treiber, R. Meyer, A.W. Hauser, W.E. Ernst, M. Koch, Nat. Commun. **9**, 4006 (2018)
32. B. Thaler, P. Heim, L. Treiber, M. Koch, J. Chem. Phys. **152**, 014307 (2020)
33. B. Thaler, M. Meyer, P. Heim, M. Koch, Phys. Rev. Lett. **124**, 115301 (2020)
34. C. Giese, F. Stienkemeier, M. Mudrich, A.W. Hauser, W.E. Ernst, Phys. Chem. Chem. Phys. **13**, 18769 (2011)
35. M. Schlesinger, M. Mudrich, F. Stienkemeier, W.T. Strunz, Chem. Phys. Lett. **490**, 245 (2010)
36. B. Grüner, M. Schlesinger, P. Heister, W.T. Strunz, F. Stienkemeier, M. Mudrich, Phys. Chem. Chem. Phys. **13**, 6816 (2011)
37. L. Bruder, M. Mudrich, F. Stienkemeier, Phys. Chem. Chem. Phys. **17**, 23877 (2015)
38. A. Sieg, J. von Vangerow, F. Stienkemeier, O. Dulieu, M. Mudrich, J. Phys. Chem. A. **120**, 7641 (2016)
39. J. Tiggesbäumker, F. Stienkemeier, Phys. Chem. Chem. Phys. **9**, 4748 (2007)
40. T. Döppner, T. Diederich, S. Göde, A. Przystawik, J. Tiggesbäumker, K.-H. Meiwes-Broer, J. Chem. Phys. **126**, 244513 (2007)
41. S. Müller, M. Mudrich, F. Stienkemeier, J. Chem. Phys. **131**, 044319 (2009)
42. K. Atkins, Phys. Rev. **116**, 1339 (1959)
43. T. Döppner, S. Teuber, M. Schumacher, J. Tiggesbäumker, K. Meiwes-Broer, Appl. Phys. B. **71**, 357 (2000)
44. T. Döppner, T. Diederich, J. Tiggesbäumker, K.-H. Meiwes-Broer, Eur. Phys. J. D **16**, 13 (2001)
45. T. Döppner, T. Diederich, A. Przystawik, N.X. Truong, T. Fennel, J. Tiggesbäumker, K.H. Meiwes-Broer, Phys. Chem. Chem. Phys. **9**, 4639 (2007)
46. N. Truong et al., Phys. Rev. A **81**, 013201 (2010)
47. A. Mikaberidze, U. Saalmann, J.M. Rost, Phys. Rev. A **77**, 041201 (2008)
48. C. Peltz, T. Fennel, Eur. Phys. J. D **63**, 281 (2011)
49. A. Mikaberidze, U. Saalmann, J.M. Rost, Phys. Rev. Lett. **102**, 128102 (2009)
50. S. Krishnan et al., Phys. Rev. Lett. **107**, 173402 (2011)
51. S.R. Krishnan et al., New J. Phys. **14**, 075016 (2012)
52. M. Kelbg, A. Heidenreich, L. Kazak, M. Zabel, B. Krebs, K.-H. Meiwes-Broer, J. Tiggesbäumker, J. Phys. Chem. A **122**, 8107 (2018)
53. M. Kelbg, M. Zabel, B. Krebs, L. Kazak, K.-H. Meiwes-Broer, J. Tiggesbäumker, Phys. Rev. Lett. **125**, 093202 (2020)
54. A. Przystawik, S. Göde, T. Döppner, J. Tiggesbäumker, K.-H. Meiwes-Broer, Phys. Rev. A **78**, 021202 (2008)
55. S. Göde, R. Irsig, J. Tiggesbäumker, K.-H. Meiwes-Broer, New J. Phys. **15**, 015026 (2013)
56. A. Hernando, M. Barranco, R. Mayol, M. Pi, F. Ancilotto, Phys. Rev. B **78**, 184515 (2008)
57. A. Stolow, A.E. Bragg, D.M. Neumark, Chem. Rev. **104**, 1719 (2004)
58. I.V. Hertel, W. Radloff, Rep. Prog. Phys. **69**, 1897 (2006)
59. T. Weinacht, B.J. Pearson, *Time-Resolved Spectroscopy: An Experimental Perspective* (CRC Press, Taylor & Francis Group, Boca Raton, FL 33487–2742, 2019)
60. P. Radcliffe, A. Przystawik, T. Diederich, T. Döppner, J. Tiggesbäumker, K.-H. Meiwes-Broer, Phys. Rev. Lett. **92**, 173403 (2004)
61. E. Loginov, D. Rossi, M. Drabbels, Phys. Rev. Lett. **95**, 163401 (2005)

62. E. Loginov, M. Drabbels, J. Phys. Chem. A. **111**, 7504 (2007)
63. O. Kornilov, C.C. Wang, O. Bünermann, A.T. Healy, M. Leonard, C. Peng, S.R. Leone, D.M. Neumark, O. Gessner, J. Phys. Chem. **114**, 1437 (2010)
64. O. Kornilov, O. Bünermann, D.J. Haxton, S.R. Leone, D.M. Neumark, O. Gessner, J. Phys. Chem. **115**, 7891 (2011)
65. M.P. Ziemkiewicz, C. Bacellar, K.R. Siefermann, S.R. Leone, D.M. Neumark, O. Gessner, J. Chem. Phy. **141**, (2014)
66. M. Mudrich et al., Nat. Commun. **11** (2020)
67. J.P. Toennies, A.F. Vilesov, Angew. Chem. Int. Ed. **43**, 2622 (2004)
68. C. Callegari, W.E. Ernst, in *Handbook of High Resolution Spectroscopy*, edited by F. Merkt, M. Quack (John Wiley & Sons, Chichester, 2011)
69. A.E. Boguslavskiy, J. Mikosch, A. Gijsbertsen, M. Spanner, S. Patchkovskii, N. Gador, M.J.J. Vrakking, A. Stolow, Science **335**, 1336 (2012)
70. P. Maierhofer, M. Bainschab, B. Thaler, P. Heim, W.E. Ernst, M. Koch, J. Phys. Chem. A **120**, 6418 (2016)
71. M. Koch, B. Thaler, P. Heim, W.E. Ernst, J. Phys. Chem. A **121**, 6398 (2017)
72. M. Koch, P. Heim, B. Thaler, M. Kitzler, W.E. Ernst, J. Phys. B: At., Mol. Opt. Phys. **50**, 125102 (2017)
73. R.E. Continetti, Ann. Rev. Phy. Chem. **52**, 165 (2001)
74. T. Arion, U. Hergenhahn, J. Elec. Spectros. Relat. Phen. **200**, 222 (2015)
75. L.J. Frasinski, K. Codling, P.A. Hatherly, Science **246**, 1029 (1989)
76. V. Stert, W. Radloff, C. Schulz, I. Hertel, Eur. Phys. J. D **5**, 97 (1999)
77. M. Rumetshofer, P. Heim, B. Thaler, W.E. Ernst, M. Koch, W. von der Linden, Phys. Rev. A **97**, 062503 (2018)
78. P. Heim, M. Rumetshofer, S. Ranftl, B. Thaler, W.E. Ernst, M. Koch, W. von der Linden, Entropy **21**, 93 (2019)
79. M. Shcherbinin, A.C. LaForge, V. Sharma, M. Devetta, R. Richter, R. Moshammer, T. Pfeifer, M. Mudrich, Phys. Rev. A **96**, 013407 (2017)
80. A. LaForge, M. Shcherbinin, F. Stienkemeier, R. Richter, R. Moshammer, T. Pfeifer, M. Mudrich, Nat. Phys. **15**, 247 (2019)
81. S. Smolarek, N.B. Brauer, W.J. Buma, M. Drabbels, J. Am. Chem. Soc. **132**, 14086 (2010)
82. D. Buchta et al., J. Phys. Chem. A **117**, 4394 (2013)
83. M. Shcherbinin, A.C. LaForge, M. Hanif, R. Richter, M. Mudrich, J. Phys. Chem. A **122**, 1855 (2018)
84. L. Ben Ltaief et al., J. Phys. Chem. Lett. **10**, 6904 (2019)
85. A.C. LaForge et al., Phys. Rev. Lett. **116**, 203001 (2016)
86. L.B. Ltaief et al., Phys. Chem. Chem. Phys. **22**, 8557 (2020)
87. *Microscopic Approaches to Quantum Liquids in Confined Geometries, Series on Advances in Quantum Many-Body Theory*, edited by E. Krotscheck, J. Navarro (World Scientific Pub Co, Singapore, 2002), p. 450
88. F. Brieuc, C. Schran, F. Uhl, H. Forbert, D. Marx, J. Chem. Phys. **152**, 210901 (2020)
89. F. Ancilotto, M. Barranco, F. Coppens, J. Eloranta, N. Halberstadt, A. Hernando, D. Mateo, M. Pi, Int. Rev. Phys. Chem. **36**, 621 (2017)
90. M. Pi, F. Ancilotto, F. Coppens, N. Halberstadt, A. Hernando, A. Leal, D. Mateo, R. Mayol, M. Barranco, $^4$He-DFT BCN-TLS: a computer package for simulating structural properties and dynamics of doped liquid helium-4 systems. https://github.com/bcntls2016/ (2016)
91. A. Hernando, M. Barranco, M. Pi, E. Loginov, M. Langlet, M. Drabbels, Phys. Chem. Chem. Phys. **14**, 3996 (2012)
92. J. von Vangerow, A. Sieg, F. Stienkemeier, M. Mudrich, A. Leal, D. Mateo, A. Hernando, M. Barranco, M. Pi, J. Phys. Chem. A **118**, 6604 (2014)
93. M. Martinez, F. Coppens, M. Barranco, N. Halberstadt, M. Pi, Phy. Chem. Chem. Phy. **21**, 3626 (2019)
94. D. Mateo, A. Hernando, M. Barranco, E. Loginov, M. Drabbels, M. Pi, Phys. Chem. Chem. Phys. **15**, 18388 (2013)

95. N.B. Brauer, S. Smolarek, E. Loginov, D. Mateo, A. Hernando, M. Pi, M. Barranco, W.J. Buma, M. Drabbels, Phys. Rev. Lett. **111**, 153002 (2013)
96. D. Mateo, A. Leal, A. Hernando, M. Barranco, M. Pi, F. Cargnoni, M. Mella, X. Zhang, M. Drabbels, J. Chem. Phys. **140**, 131101 (2014)
97. A. Leal, D. Mateo, A. Hernando, M. Pi, M. Barranco, A. Ponti, F. Cargnoni, M. Drabbels, Phys. Rev. B **90**, 224518 (2014)
98. F. Coppens, J. von Vangerow, A. Leal, M. Barranco, N. Halberstadt, M. Mudrich, M. Pi, F. Stienkemeier, Eur. Phys. J. D **73**, 94 (2019)
99. A.C. LaForge et al., Phys. Rev. X **11**, 021011 (2021)
100. F. Ancilotto, M. Barranco, M. Pi, Phy. Rev. B **97**, 184515 (2018)
101. S.M. O'Connell et al., Phys. Rev. Lett. **124**, 215301 (2020)
102. M. Pi, F. Ancilotto, J.M. Escartín, R. Mayol, M. Barranco, Phys. Rev. B **102**, 060502 (2020)
103. A. Leal, D. Mateo, A. Hernando, M. Pi, M. Barranco, Phys. Chem. Chem. Phys. **16**, 23206 (2014)
104. F. Coppens, F. Ancilotto, M. Barranco, N. Halberstadt, M. Pi, Phys. Chem. Chem. Phys. **19**, 24805 (2017)
105. F. Coppens, A. Leal, M. Barranco, N. Halberstadt, M. Pi, J. Low Temp. Phys. **187**, 439 (2017)
106. J.M. Escartín, F. Ancilotto, M. Barranco, M. Pi, Phys. Rev. B **99**, 140505 (2019)
107. A. Vilà, M. González, R. Mayol, J. Chem. Theory Comput. **11**, 899 (2015)
108. A. Vilà, A., M. González, R. Mayol, Phys. Chem. Chem. Phys. **17**, 32241 (2015)
109. A. Vilà, M. González, R. Mayol, Phys. Chem. Chem. Phys. **18**, 2409 (2016)
110. A. Vilà, M. González, Phys. Chem. Chem. Phys. **18**, 27630 (2016)
111. A. Vilà, M. González, Phys. Chem. Chem. Phys. **18**, 31869 (2016)
112. M. Blancafort-Jorquera, A. Vilà, M. González, Phys. Chem. Chem. Phys. **21**, 24218 (2019)
113. A. Vilà, M. Paniagua, M. González, Phys. Chem. Chem. Phys. **20**, 118 (2018)
114. M. Blancafort-Jorquera, A. Vilà, M. González, Phys. Chem. Chem. Phys. **21**, 21007 (2019)
115. M. Koch, G. Auböck, C. Callegari, W.E. Ernst, Phys. Rev. Lett. **103**, 035302 (2009)
116. G. Auböck, J. Nagl, C. Callegari, W.E. Ernst, Phys. Rev. Lett. **101**, 035301 (2008)
117. C. Callegari, F. Ancilotto, J. Phys. Chem. A **115**, 6789 (2011)
118. D.E. Galli, M. Buzzacchi, L. Reatto, J. Chem. Phys. **115**, 10239 (2001)
119. M. Rossi, M. Verona, D. Galli, L. Reatto, Phys. Rev. B **69**, 212510 (2004)
120. L. Fechner, B. Grüner, A. Sieg, C. Callegari, F. Ancilotto, F. Stienkemeier, M. Mudrich, Phys. Chem. Chem. Phys. **14**, 3843 (2012)
121. E. Loginov, C. Callegari, F. Ancilotto, M. Drabbels, J. Phys. Chem. A **115**, 6779 (2011)
122. E. Loginov, M. Drabbels, J. Phys. Chem. A **118**, 2738 (2014)
123. M. Theisen, F. Lackner, W.E. Ernst, J. Phys. Chem. A **115**, 7005 (2011)
124. X. Zhang, M. Drabbels, J. Chem. Phys. **137**, 051102 (2012)
125. O. Bünermann, G. Droppelmann, A. Hernando, R. Mayol, F. Stienkemeier, J. Phys. Chem. A **111**, 12684 (2007)
126. E. Loginov, A. Hernando, J.A. Beswick, N. Halberstadt, M. Drabbels, J. Phys. Chem. A **119**, 6033 (2015)
127. C. Giese, T. Mullins, B. Grüner, M. Weidemüller, F. Stienkemeier, M. Mudrich, J. Chem. Phys. **137**, (2012)
128. A.T.J.B. Eppink, D.H. Parker, Rev. Sci. Instrum. **68**, 3477 (1997)
129. A. Kautsch, M. Hasewend, M. Koch, W.E. Ernst, Phys. Rev. A **86**, 033428 (2012)
130. A. Kautsch, M. Koch, W.E. Ernst, J. Phys. Chem. A **117**, 9621 (2013)
131. M. Koch, A. Kautsch, F. Lackner, W.E. Ernst, J. Phys. Chem. A **118**, 8373 (2014)
132. B. Thaler, R. Meyer, P. Heim, S. Ranftl, J.V. Pototschnig, A.W. Hauser, M. Koch, W.E. Ernst, J. Phys. Chem. A **123**, 3977 (2019)
133. J. von Vangerow, F. Coppens, A. Leal, M. Pi, M. Barranco, N. Halberstadt, F. Stienkemeier, M. Mudrich, J. Phy. Chem. Lett. **8**, 307 (2017)
134. S. Grebenev, J.P. Toennies, A.F. Vilesov, Science **279**, 2083 (1998)
135. A.R.W. McKellar, Y. Xu, W. Jäger, Phy. Rev. Lett. **97**, 183401 (2006)
136. P. Sindzingre, M.L. Klein, D.M. Ceperley, Phys. Rev. Lett. **63**, 1601 (1989)

137. F. Paesani, A. Viel, F.A. Gianturco, K.B. Whaley, Phys. Rev. Lett. **90**, 073401 (2003)
138. S. Moroni, A. Sarsa, S. Fantoni, K.E. Schmidt, S. Baroni, Phys. Rev. Lett. **90**, 143401 (2003)
139. S. Grebenev, B. Sartakov, J.P. Toennies, A.F. Vilesov, Science **289**, 1532 (2000)
140. N. Poertner, J.P. Toennies, A.F. Vilesov, G. Benedek, V. Hizhnyakov, EPL (Europhysics Letters) **88**, 26007 (2009)
141. D. Verma et al., Phys. Rev. B **102**, 014504 (2020)
142. F. Stienkemeier, O. Bünermann, R. Mayol, F. Ancilotto, M. Barranco, M. Pi, Phys. Rev. B **70**, 214509 (2004)
143. A. Hernando, R. Mayol, M. Pi, M. Barranco, F. Ancilotto, O. Bünermann, F. Stienkemeier, J. Phys. Chem. A **111**, 7303 (2007)
144. G. Wu, P. Hockett, A. Stolow, Phys. Chem. Chem. Phys. **13**, 18447 (2011)
145. S. Rutz, R. de Vivie-Riedle, E. Schreiber, Phys. Rev. A **54**, 306 (1996)
146. B. Zhang, N. Gador, T. Hansson, Phys. Rev. Lett. **91**, 173006 (2003)
147. S. Rutz, E. Schreiber, Chem. Phys. Lett. **269**, 9 (1997)
148. M.J.J. Vrakking, D.M. Villeneuve, A. Stolow, Phys. Rev. A **54**, R37 (1996)
149. R. de Vivie-Riedle, K. Kobe, J. Manz, W. Meyer, B. Reischl, S. Rutz, E. Schreiber, L. Wöste, J. Phys. Chem. **100**, 7789 (1996)
150. C. Nicole, M.A. Bouchène, C. Meier, S. Magnier, E. Schreiber, B. Girard, J. Chem. Phys. **111**, 7857 (1999)
151. A. Lindinger, J.P. Toennies, A.F. Vilesov, J. Chem. Phys. **121**, 12282 (2004)
152. M. Wollenhaupt, V. Engel, T. Baumert, Ann. Rev. Phys. Chem. **56**, 25 (2005)
153. T. Baumert, G. Gerber, in *Femtosecond Laser Spectroscopy*, edited by P. Hannaford (Springer Verlag, , 2005), p. chap. 9
154. C. Chin, R. Grimm, P. Julienne, E. Tiesinga, Rev. Mod. Phys. **82**, 1225 (2010)
155. K.M. Jones, E. Tiesinga, P.D. Lett, P.S. Julienne, Rev. Mod. Phys. **78**, 483 (2006)
156. F. Stienkemeier, W.E. Ernst, J. Higgins, G. Scoles, J. Chem. Phys. **102**, 615 (1995)
157. F. Ancilotto, G. DeToffol, F. Toigo, Phys. Rev. B **52**, 16125 (1995)
158. S. Bovino, E. Coccia, E. Bodo, D. Lopez-Durán, F. A. Gianturco, J. Chem. Phys. **130**, (2009)
159. G. Guillon, A. Zanchet, M. Leino, A. Viel, R.E. Zillich, J. Phys. Chem. A **115**, 6918 (2011)
160. S. Bovino, E. Bodo, E. Yurtsever, F.A. Gianturco, J. Chem. Phys. **128**, 224312 (2008)
161. J.G. Danzl, E. Haller, M. Gustavsson, M.J. Mark, R. Hart, N. Bouloufa, O. Dulieu, H. Ritsch, H.-C. Nägerl, Science **321**, 1062 (2008)
162. M. Gruebele, A.H. Zewail, J. Chem. Phys. **98**, 883 (1993)
163. G. Droppelmann, M. Mudrich, C.P. Schulz, F. Stienkemeier, Eur. Phys. J. D **52**, 67 (2009)
164. S. Müller, S. Krapf, T. Koslowski, M. Mudrich, F. Stienkemeier, Phys. Rev. Lett. **102**, 183401 (2009)
165. M. Dvorak, M. Müller, O. Bünermann, F. Stienkemeier, J. Chem. Phys. **140**, 144301 (2014)
166. A. Boatwright, C. Feng, D. Spence, E. Latimer, C. Binns, A.M. Ellis, S. Yang, Faraday discussions **162**, 113 (2013)
167. G. Haberfehlner, P. Thaler, D. Knez, A. Volk, F. Hofer, W.E. Ernst, G. Kothleitner, Nat. commun. **6**, 1 (2015)
168. H. Schmidt, J. von Vangerow, F. Stienkemeier, A.S. Bogomolov, A.V. Baklanov, D.M. Reich, W. Skomorowski, C.P. Koch, M. Mudrich, J. Chem. Phy. **142**, 044303 (2015)
169. R. Monni, G. Auböck, D. Kinschel, K.M. Aziz-Lange, H.B. Gray, A. Vlcek, M. Chergui, Chem. Phy. Lett. **683**, 112 (2017)
170. M. Lundwall, M. Tchaplyguine, G. Öhrwall, A. Lindblad, S. Peredkov, T. Rander, S. Svensson, O. Björneholm, Surface science **594**, 12 (2005)
171. A. Masson, L. Poisson, M.-A. Gaveau, B. Soep, J.-M. Mestdagh, V. Mazet, F. Spiegelman, J. Chem. Phys. **133**, 054307 (2010)
172. A.V. Benderskii, J. Eloranta, R. Zadoyan, V.A. Apkarian, J. Chem. Phys. **117**, 1201 (2002)
173. H. Buchenau, E.L. Knuth, J. Northby, J.P. Toennies, C. Winkler, J. Chem. Phys. **92**, 6875 (1990)
174. A. Mauracher, M. Daxner, J. Postler, S.E. Huber, S. Denifl, P. Scheier, J.P. Toennies, J. Phys. Chem. Lett. **5**, 2444 (2014)

175. A. Mauracher, O. Echt, A. Ellis, S. Yang, D. Bohme, J. Postler, A. Kaiser, S. Denifl, P. Scheier, Phys. Rep. **751**, 1 (2018)
176. T. González-Lezana, O. Echt, M. Gatchell, M. Bartolomei, J. Campos-Martínez, P. Scheier, International Reviews in Physical Chemistry **39**, 465 (2020)
177. R. Fröchtenicht, J.P. Toennies, A. Vilesov, Chem. Phys. Lett. **229**, 1 (1994)
178. N. Halberstadt, K.C. Janda, Chem. Phys. Lett. **282**, 409 (1998)
179. D. Buchta et al., J. Chem. Phys. **139**, 084301 (2013)
180. T. Ruchti, B.E. Callicoatt, K.C. Janda, Phys. Chem. Chem. Phys. **2**, 4075 (2000)
181. W. Lewis, C. Lindsay, R. Bemish, R. Miller, J. Am. Chem. Soc. **127**, 7235 (2005)
182. N. Saito et al., Chem. Phys. Lett. **441**, 16 (2007)
183. J. Zobeley, R. Santra, L.S. Cederbaum, J. Chem. Phys. **115**, 5076 (2001)
184. A. LaForge et al., Phy. Rev. Lett. **116**, 203001 (2016)
185. D.S. Peterka, A. Lindinger, L. Poisson, M. Ahmed, D.M. Neumark, Phys. Rev. Lett. **91**, 043401 (2003)
186. D.S. Peterka, J.H. Kim, C.C. Wang, L. Poisson, D.M. Neumark, J. Phys. Chem. A **111**, 7449 (2007)
187. C.C. Wang, O. Kornilov, O. Gessner, J.H. Kim, D.S. Peterka, D.M. Neumark, J. Phys. Chem. **112**, 9356 (2008)
188. S. Mandal et al., Phys. Chem. Chem. Phys. **22**, 10149 (2020)
189. L.B. Ltaief, M. Shcherbinin, S. Mandal, S.R. Krishnan, R. Richter, T. Pfeifer, M. Mudrich, J. Phy. B: Atomic. Molecul. Optic. Phys. **53**, 204001 (2020)
190. L.B. Ltaief, M. Shcherbinin, S. Mandal, S. Krishnan, R. Richter, S. Turchini, N. Zema, M. Mudrich, J. Low. Temp. Phys. (2021)
191. M. Shcherbinin, F.V. Westergaard, M. Hanif, S.R. Krishnan, A. LaForge, R. Richter, T. Pfeifer, M. Mudrich, J. Chem. Phys. **150**, 044304 (2019)
192. K. von Haeften, A.R. B.d. Castro, M. Joppien, L. Moussavizadeh, R.V. Pietrowski, T.Möller, Phys. Rev. Lett. **78**, 4371 (1997)
193. K. von Haeften, T. Laarmann, H. Wabnitz, T. Möller, Phys. Rev. Lett. **87**, 153403 (2001)
194. K. von Haeften, T. Laarmann, H. Wabnitz, T. Möller, Phys. Rev. Lett. **88**, 233401 (2002)
195. K. von Haeften, T. Laarmann, H. Wabnitz, T. Möller, J. Phys. B **38**, 373 (2005)
196. O. Bünermann, O. Kornilov, S.R. Leone, D.M. Neumark, O. Gessner, IEEE. J. Sel. Top. Quantum Electron. **18**, 308 (2012)
197. O. Bünermann, O. Kornilov, D.J. Haxton, S.R. Leone, D.M. Neumark, O. Gessner, J. Chem. Phys. **137**, 214302 (2012)
198. L.F. Gomez et al., Science **345**, 906 (2014)
199. B. Langbehn et al., Phys. Rev. Lett. **121**, 255301 (2018)
200. O. Gessner, A.F. Vilesov, Ann. Rev. Phy. Chem. **70**, 173 (2019)
201. J.D. Asmussen et al., Phys. Chem. Chem. Phys. **23**, 15138 (2021)
202. V. Lyamayev et al., J. Phys. B **46**, 164007 (2013)
203. M. Joppien, R. Karnbach, T. Möller, Phys. Rev. Lett. **71**, 2654 (1993)
204. R. Michiels et al., Phys. Rev. Lett. **127**, 093201 (2021)
205. K.D. Closser, O. Gessner, M. Head-Gordon, J. Chem. Phys. **140**, 134306 (2014)
206. C.-Z. Gao, P.M. Dinh, P.-G. Reinhard, E. Suraud, Phys. Chem. Chem. Phys. **19**, 19784 (2017)
207. U. Hergenhahn, J. Electron. Spectrosc. Relat. Phenom. **184**, 78 (2011)
208. T. Jahnke, J. Phys. B **48**, 082001 (2015)
209. T. Jahnke et al., Chem. Rev. **120**, 11295 (2020)
210. L.S. Cederbaum, J. Zobeley, F. Tarantelli, Phys. Rev. Lett. **79**, 4778 (1997)
211. M. Mucke et al., Nat. Phys. **6**, 143 (2010)
212. V. Stumpf, K. Gokhberg, L.S. Cederbaum, Nat. Chem. **8**, 237 (2016)
213. X. Ren, E. Wang, A.D. Skitnevskaya, A.B. Trofimov, K. Gokhberg, A. Dorn, Nat. Phys. **14**, 1062–1066 (2018)
214. F. Wiegandt et al., Phys. Rev. A **100**, 022707 (2019)
215. Y. Ovcharenko et al., Phys. Rev. Lett. **112**, 073401 (2014)
216. A.C. LaForge et al., Sci. Rep. **4**, 3621 (2014)

217. R. Katzy, M. Singer, S. Izadnia, A.C. LaForge, F. Stienkemeier, Rev. Sci. Instr. **87**, 013105 (2016)
218. Y. Ovcharenko et al., New J. Phys. **22**, 083043 (2020)
219. N.V. Kryzhevoi, V. Averbukh, L.S. Cederbaum, Phys. Rev. B **76**, 094513 (2007)
220. N.V. Kryzhevoi, D. Mateo, M. Pi, M. Barranco, L.S. Cederbaum, Phy. Chem. Chem. Phy. **15**, 18167 (2013)
221. S. Kazandjian et al., Phys. Rev. A **98**, 050701 (2018)
222. A.I. Kuleff, K. Gokhberg, S. Kopelke, L.S. Cederbaum, Phys. Rev. Lett. **105**, 043004 (2010)
223. S. Augst, D. Strickland, D.D. Meyerhofer, S.L. Chin, J.H. Eberly, Phys. Rev. Lett. **63**, 2212 (1989)
224. A. Heidenreich, B. Grüner, M. Rometsch, S. Krishnan, F. Stienkemeier, M. Mudrich, New J. Phys. **18**, 073046 (2016)
225. A. Heidenreich, B. Grüner, D. Schomas, F. Stienkemeier, S.R. Krishnan, M. Mudrich, J. Modern Opt. **64**, 1061 (2017)
226. B. Schütte, M. Arbeiter, A. Mermillod-Blondin, M.J. Vrakking, A. Rouzée, T. Fennel, Phys. Rev. Lett. **116**, 033001 (2016)
227. U. Saalmann, C. Siedschlag, J.M. Rost, J. Phys. B **39**, R39 (2006)
228. T. Fennel, K.-H. Meiwes-Broer, J. Tiggesbäumker, P.-G. Reinhard, P.M. Dinh, E. Suraud, Rev. Mod. Phys. **82**, 1793 (2010)
229. T. Döppner, S. Teuber, T. Diederich, T. Fennel, P. Radcliffe, J. Tiggesbäumker, K.-H. Meiwes-Broer, Eur. Phys. J. D **24**, 157 (2003)
230. B. Schütte, M. Arbeiter, T. Fennel, G. Jabbari, A. Kuleff, M. Vrakking, A. Rouzée, Nat. Phys. **6**, 8596 (2015)
231. T. Oelze et al., Sci. Rep. **7**, 40736 (2017)
232. C. Medina et al., New J. Phys. **23**, 053011 (2021)
233. D. Schomas *et al.*, arXiv preprint arXiv:2005.02944 (2020)
234. C. Gnodtke, U. Saalmann, J.M. Rost, Phys. Rev. A **79**, 041201 (2009)
235. C. Bacellar *et al.*, submitted (2021)
236. B. Schütte et al., Sci. Rep. **6**, 39664 (2016)
237. J. Passig et al., Nat. commun. **8**, 1 (2017)
238. J. Miao, T. Ishikawa, I.K. Robinson, M.M. Murnane, Science **348**, 530 (2015)
239. M.J. Bogan et al., Nano Lett. **8**, 310 (2008)
240. M. Dantus, V.V. Lozovoy, Chem. Rev. **104**, 1813 (2004), publisher: American Chemical Society
241. K. Ohmori, Annu. Rev. Phys. Chem. **60**, 487 (2009)
242. M. Shapiro, P. Brumer, Phy. Rep. **425**, 195 (2006)
243. L. Bruder, A. Eisfeld, U. Bangert, M. Binz, M. Jakob, D. Uhl, M. Schulz-Weiling, E.R. Grant, F. Stienkemeier, Phys. Chem. Chem. Phys. **21**, 2276 (2019)
244. K. Lazonder, M.S. Pshenichnikov, D.A. Wiersma, Opt. Lett. **31**, 3354 (2006)
245. M. Fushitani, Annu. Rep. Prog. Chem., Sect. C: Phys. Chem. **104**, 272 (2008)
246. R.M. Hochstrasser, PNAS **104**, 14190 (2007)
247. M. Cho, Chem. Rev. **108**, 1331 (2008)
248. P. Nuernberger, S. Ruetzel, T. Brixner, Angew. Chem. Int. Ed. **54**, 11368 (2015)
249. F.D. Fuller, J.P. Ogilvie, Annu. Rev. Phys. Chem. **66**, 667 (2015)
250. L. Bruder, U. Bangert, M. Binz, D. Uhl, F. Stienkemeier, J. Phys. B: At. Mol. Opt. Phys. **52**, 183501 (2019)
251. D.M. Jonas, Annu. Rev. Phys. Chem. **54**, 425 (2003)
252. P. Malý, T. Mančal, J. Phys. Chem. Lett. **9**, 5654 (2018)
253. O. Kühn, T. Mančal, T. Pullerits, J. Phys. Chem. Lett. **11**, 838 (2020)
254. M. Maiuri, J. Brazard, Top Curr Chem (Z) **376**, 10 (2018)
255. T. Brixner, J. Stenger, H.M. Vaswani, M. Cho, R.E. Blankenship, G.R. Fleming, Nature **434**, 625 (2005)
256. J. Dostál, J. Pšenčík, D. Zigmantas, Nat. Chem. **8**, 705 (2016)
257. K.L.M. Lewis, J.P. Ogilvie, J. Phys. Chem. Lett. **3**, 503 (2012)

258. G. Moody, S.T. Cundiff, Adv. Phys. **2**, 641 (2017)
259. M.K. Petti, J.P. Lomont, M. Maj, M.T. Zanni, J. Phys. Chem. B **122**, 1771 (2018)
260. P. Hamm, M. Zanni, *Concepts and Methods of 2D Infrared Spectroscopy* (Cambridge University Press, 2011)
261. P.F. Tekavec, T.R. Dyke, A.H. Marcus, J. Chem. Phys. **125**, 194303 (2006)
262. P.F. Tekavec, G.A. Lott, A.H. Marcus, J. Chem. Phys. **127**, 214307 (2007)
263. B. Lomsadze, B.C. Smith, S.T. Cundiff, Nat. Photon. **12**, 676 (2018)
264. K. Hirano, K. Enomoto, M. Kumakura, Y. Takahashi, T. Yabuzaki, Phys. Rev. A **68**, (2003)
265. M. Zbiri, C. Daul, J. Chem. Phys. **121**, 11625 (2004)
266. M. Binz, Ph.D. thesis, University of Freiburg, 2021
267. J. Nagl, G. Auböck, A.W. Hauser, O. Allard, C. Callegari, W.E. Ernst, Phy. Rev. Lett. **100**, 063001 (2008)
268. O. Allard, J. Nagl, G. Auböck, C. Callegari, W.E. Ernst, Journal of Physics B-Atomic Molecular And Optical. Physics **39**, S1169 (2006)
269. L. Bruder, U. Bangert, M. Binz, D. Uhl, R. Vexiau, N. Bouloufa-Maafa, O. Dulieu, F. Stienkemeier, Nat. Commun. **9**, 4823 (2018)
270. M. Wewer, F. Stienkemeier, J. Chem. Phys. **120**, 1239 (2004)
271. R. Lehnig, A. Slenczka, J. Chem. Phys. **118**, 8256 (2003)
272. U. Bangert, F. Stienkemeier, L. Bruder, arXiv:2112.05418 (2021)
273. D. Mani *et al.*, Sci. Adv. **5**, eaav8179 (2019)
274. A. Gutberlet, G. Schwaab, O. Birer, M. Masia, A. Kaczmarek, H. Forbert, M. Havenith, D. Marx, Science **324**, 1545 (2009)
275. K. Nauta, R.E. Miller, Science **287**, 293 (2000)
276. G. Haberfehlner, P. Thaler, D. Knez, A. Volk, F. Hofer, W.E. Ernst, G. Kothleitner, Nat. Commun. **6**, 8779 (2015)
277. K. Nauta, R.E. Miller, Phys. Rev. Lett. **82**, 4480 (1999)
278. J. Küpper, J.M. Merritt, Int. Rev. Phys. Chem. **26**, 249 (2007)
279. J. Roden, A. Eisfeld, M. Dvořák, O. Bünermann, F. Stienkemeier, J. Chem. Phys. **134**, 054907 (2011)
280. G.E. Douberly, J.M. Merritt, R.E. Miller, Phys. Chem. Chem. Phys. **7**, 463 (2005)
281. A. Wituschek et al., Nat. Commun. **11**, 1 (2020)
282. F. Bencivenga *et al.*, Sci. Adv. **5**, eaaw5805 (2019)
283. C. Svetina et al., Opt. Lett. **44**, 574 (2019)
284. R. Mincigrucci *et al.*, arXiv preprint arXiv:2010.04860 (2020)
285. D. Rupp et al., Nat. Commun. **8**, 1 (2017)
286. L. Flückiger et al., New J. Phys. **18**, 043017 (2016)

# Chapter 11
# Synthesis of Metallic Nanoparticles in Helium Droplets

**Florian Lackner**

**Abstract** Helium droplets provide a unique cold and inert synthesis environment for the formation of nanoparticles. Over the past decade, the method has evolved into a versatile tool, ready to be used for the creation of new nanomaterials. Species with different characteristics can be combined in a core@shell configuration, allowing for the formation of nanoparticles with tailored properties. The realm of structures that can be formed extends from clusters, comprising only a few atoms, to spherical sub-10 nm particles and nanowires with a length on the order of a few hundred nanometers. The formed nanoparticles can be deposited on any desired substrate under soft-landing conditions. This chapter is concerned with the formation of metal and metal oxide nanoparticles with helium droplets. The synthesis process is explained in detail, covering aspects that range from the doping of helium droplets to the behavior of deposited particles on a surface. Different metal particle systems are reviewed and methods for the creation of metal oxide particles are discussed. Selected experiments related to optical properties as well as the structure and stability of synthesized nanoparticles are presented.

## 11.1 Introduction

During the 1990s helium nanodroplets emerged as an outstanding matrix for spectroscopy experiments [1, 2]. It was readily realized that these droplets of liquid helium can pick up atoms and molecules [3, 4], which subsequently form complexes in their interior or at their surface [5]. The assembling of molecules and small clusters became a major topic in the field and helium droplet isolation spectroscopy evolved into an excellent method for the characterization of many elusive and hitherto unknown species. Examples of molecules and complexes that could be isolated in helium droplets encompass high-spin alkali dimers and trimers [6–8], cyclic water hexamers [9] and chains of polar molecules [10] combined with small metal clusters [11].

F. Lackner (✉)
Institute of Experimental Physics, Graz University of Technology,
Petersgasse 16, 8010 Graz, Austria

© The Author(s) 2022
A. Slenczka and J. P. Toennies (eds.), *Molecules in Superfluid Helium Nanodroplets*,
Topics in Applied Physics 145, https://doi.org/10.1007/978-3-030-94896-2_11

Major research efforts have been directed towards metal dopants. Helium droplets act as an efficient cryostat, which enabled the preparation of very cold metal clusters [12, 13]. Early experiments focused on the optical and electronic properties of small clusters, studied in-situ using mass spectrometry based on laser or electron impact ionization [14]. The first surface deposition experiments with large metal clusters in the nanometer size regime (i.e. small nanoparticles [15]) were reported in 2007 [16]. Soon thereafter, in 2011, transmission electron microscopy images of deposited particles were presented [17], which sparked the advent of the method as a new tool for the formation of nanoparticles.

Since then, many research groups embarked on the investigation of the possibilities offered by this unique synthesis approach [18–25]. This chapter provides a review over this particular research branch in helium droplet science and focusses on the formation of metal and metal oxide nanoparticles. A thorough description of the doping process as well as the agglomeration and deposition of nanoparticles grown in helium droplets is presented. Selected experiments with metal and metal oxide nanoparticles are discussed, covering aspects such as nanoscale oxidation and thermal stability. Special attention is given to plasmonic metals in helium droplets, core@shell structures and recent experiments that may open up new research directions for the next decade of nanoparticle synthesis with helium droplets.

## 11.2 Nanoparticle Synthesis with Helium Droplets

The expansion of helium under high pressure through a small, cold nozzle into vacuum leads to the generation of a helium droplet beam [1, 2]. Continuous helium droplet sources are typically equipped with a $5 \mu m$ nozzle. Helium stagnation pressures between 20 and 80 bar are employed and the nozzle is cooled to temperatures below about 25 K. The helium droplet beam is skimmed and guided into a chamber with one or more pickup regions where the desired dopant materials are provided in the gas phase. Upon collision, atoms and molecules are picked up by the helium droplets and cool down rapidly to a temperature of 0.4 K [26]. Dopants are typically located inside the droplet where they are free to move, which results in the agglomeration of larger complexes [5]. Depending on the initial helium droplet size and the doping rate, nanoparticles with different sizes and shapes are formed. Subsequently, the synthesized particles can be deposited on any desired substrate that is placed into the beam. In the course of the deposition process, the helium evaporates and only the plain particles remain on the surface. The deposited nanoparticles can then be investigated or employed in other experiments outside the helium droplet apparatus.

The following section describes the nanoparticle synthesis process in detail, from doping, agglomeration and particle growth to possible particle sizes and shapes that can be obtained and the deposition on different substrates.

## 11.2.1 Doping of Helium Nanodroplets

Helium droplets are doped at a pickup region where the desired dopant material is provided in form of gas phase atoms or molecules. The required dopant pressure in such a pickup zone, which is typically a few centimeters long, is on the order of about $10^{-4}$ mbar. For many metals that have been deposited in helium droplets in the past, the required pressure is reached at temperatures around the melting point. Currently, the standard approach for the synthesis of nanoparticles in helium droplets is based on the use of resistively heated pickup ovens that provide the desired dopants.

### 11.2.1.1 Doping of Helium Droplets with Resistively Heated Pickup Ovens

Temperatures up to about 1500 °C are typically reached with resistively heated ovens. Such pickup sources have been successfully employed to synthesize and deposit nanoparticles consisting of many different metals, including Ag [21, 27], Au, Ni [22, 28, 29], Fe [30], Co [31], Cr [28, 32], Pd [25], Al [33], Cu, Mg [34] and Zn [35].

A sketch of a pickup oven arrangement with two doping zones is shown in Fig. 11.1. The pickup ovens, highlighted in red and blue, consist of alumina coated tungsten evaporation baskets, which can maintain a maximum temperature of about 1500 °C [22]. The additional basket oriented upside down helps to confine the evaporated dopants and avoids a deposition of large amounts of material in the region above the lower basket, which is important for preventing clogging issues. In operation, the baskets are surrounded by cylindrical water-cooled jackets, which are placed on-top of the water-cooled base plate. Water-cooling is crucial for high temperature pickup

**Fig. 11.1** Sketch of a helium droplet ($He_N$) source and a pickup region with two resistively heated ovens

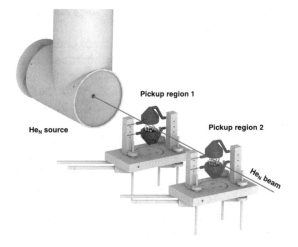

sources and avoids the melting of nearby copper wires and a heating of the vacuum chamber. The latter could result in a doping of the droplets with unwanted species such as water, an effect that can further be reduced by using liquid nitrogen cold traps.

Oven temperatures may be measured using a pyrometer or thermocouples. However, during the operation at high temperatures a direct temperature measurement can be difficult. Alternatively, the oven temperature can be estimated based on the heating power, considering that the resistivity of the employed tungsten heating wires is a function of temperature [36]. A more accurate pickup oven temperature can be obtained from the measurement of the solid-liquid transition of a doping material. An example of such a temperature calibration has recently been demonstrated based on the laser induced fluorescence collected from Au nanoparticles functionalized with rhodamine B molecules in helium droplets [37]. With this method, it was possible to observe the melting and freezing plateau during heat-up and cool-down of the Au pickup oven, providing a known temperature reference point.

### 11.2.1.2 Alternative Pickup Sources

Beyond the standard resistively heated pickup sources there is a large variety of alternative approaches that have been employed in the past for the doping of helium droplets. Some of these methods are capable of reaching the necessary pickup pressures for materials with very high melting points.

Temperatures up to about 1700 °C have been reached using an electron beam bombardment source [38], which has been successfully employed for spectroscopy experiments on Cr doped helium droplets. [39, 40] In this approach, thermionic emission from a filament provides electrons that are accelerated by a high voltage towards a crucible that holds the doping material. The kinetic energy of the electrons is transformed into heat at the target. The increased metal pressure, which is established in the region above the crucible, enables the pickup of atoms and the formation of clusters in helium droplets.

Another approach is the direct ohmic heating of the doping material. In this case, a sample is placed between two electrodes and directly heated by an electric current. This leads to the generation of an increased dopant pressure close by the sample surface. The method has been employed for the doping of He droplets with Si [41] as well as with C and Ta atoms [42, 43]. Temperatures up to 2200 °C and beyond have been reported.

The required pickup pressure for materials with very high melting temperatures can be reached with laser ablation [44–47]. With this technique, species such as Mo and Ta as well as Ti have been successfully trapped in helium droplets [44].

However, while it has been shown that metals with a very high melting point can be isolated in helium droplets, most of the results are limited to small clusters, consisting of only a few atoms. An application of such methods for the preparation of larger nanoparticles could unlock the full potential of the approach and provide access to a large variety of metal and metal oxide dopants.

Finally, it is noted that large organic molecules can be trapped in helium droplets employing an electrospray ionization (ESI) pickup source [48, 49]. Furthermore, highly reactive and elusive chemical species can be isolated in helium droplets by using a pyrolysis source [50, 51]. The preparation of helium droplets doped with gases and liquids is possible by using a gas pickup cell that is connected via a variable leak valve to a reservoir.

Different types of pickup sources can be combined in a helium droplet setup in order to realize sequential doping schemes that allow for the combination of a large variety of materials.

## 11.2.2 Aggregation of Nanoparticles

After pickup, atoms and molecules roam around inside the superfluid host environment. Eventually, the dopants will collide and agglomerate to larger nanostructures, provided that a sufficiently large number of dopants is present. The size of the nanoparticles that can be created depends on the initial size of the helium droplets: For each dopant atom or molecule that is added to the droplet a certain amount of He atoms evaporates. The excess energy that is introduced to the system is dissipated as long as He atoms are available. Thereby, each He atom is carrying away an energy of approximately $5\,cm^{-1}$ ($0.62\,meV$) [2]. The initial velocity of a dopant typically exceeds the critical Landau velocity, which has been experimentally determined as $\sim 56\,ms^{-1}$ for Ag [52]. Kinetic energy is dissipated fast by the creation of elementary excitations, such as phonons, rotons and ripplons, inside and on the surface of the droplet [23]. More important, however, is the energy that is liberated by the formation of bonds during the aggregation process. In fact, the transferred kinetic energy is about two orders of magnitude smaller than the released binding energy and can therefore be neglected [53].

The vast majority of dopants agglomerate inside the helium droplet. A notable example are alkali and alkaline earth atoms, which reside at the droplet surface after pickup [54, 55] where they can form clusters [56–58]. For the alkali metals, it has been found that if a certain size is exceeded, the clusters migrate from the surface to the interior of the droplet [59, 60]. For potassium clusters $K_n$, for example, the critical size corresponds to about $n = 80$ [61]. Larger alkali nanoparticles, except $Cs_n$ [59], thus, can be assumed to reside inside the helium droplet.

For a metal cluster, the binding energy per atom, $E_b$, depends on the total number of atoms $n$ and can be estimated with the following equation [53]:

$$E_b(n) = E_{bulk} + 2^{\frac{1}{3}} \left( \frac{1}{2} D_e - E_{bulk} \right) n^{-\frac{1}{3}} \qquad (11.1)$$

In the case of the aggregation of Ag and Au particles, for example, the bulk binding energies per atom, $E_{bulk}$, correspond to $2.95\,eV$ and $3.81\,eV$ [62], respectively, with dimer dissociation energies, $D_e$, of $1.65\,eV$ and $2.29\,eV$ [63]. The number of atoms that can form a particle in a helium droplet is, in principle, limited by the complete evaporation of all He atoms. According to (11.1), the maximum number of atoms that can be deposited in a helium droplet consisting of $10^6$ He atoms corresponds to 266 for Ag and 209 for Au. Considering the example of Ta, which has a very high bulk binding energy per atom ($E_{bulk} = 8.1$ eV and $D_e = 4$ eV) [62, 63], the number of atoms that can be packed into such a droplet is estimated with only 109, about half the number obtained for Au. Note that compared to metals, the number of noble gas atoms that can be accommodated by helium droplets of a particular size is about an order of magnitude higher due to the weaker binding energies. This simple estimation shows that the agglomeration process depends on the properties of the dopant species, which has important consequences for adjusting doping ratios in an experiment, as discussed further down.

Collision time scales for pairs of Cu, Ag and Au dopants have been calculated and range from about 10 ns in 20 nm diameter droplets to 10 μs in droplets with 200 nm diameter [64]. Considering a helium droplet beam velocity of about 200 ms$^{-1}$ (i.e. 200 μm/μs), it takes about 50 μs for a droplet to travel 1 cm. The pickup region is typically separated by a few 10 cm or more from the place where the droplets are investigated or where the particles are deposited. It becomes evident that collision time scales are much smaller than the time it takes for a droplet to travel between both places and it can be assumed that the particle formation process is completed shortly after the droplet has left the pickup region [64].

### 11.2.2.1   Monitoring the Doping and Aggregation Process

The titration method offers an experimental approach for the measurement of helium droplet sizes. Reliable numbers for a droplet source stagnation pressure of 20 bar (5 μm nozzle) have been reported for the droplet size regime of $N_{He} = 10^4 - 10^{10}$ [65]. This is the size range relevant for nanoparticle synthesis experiments for which [65] provides an excellent reference. The technique is based on the measurement of the attenuation of the helium droplet beam upon doping with rare gas atoms. The aggregation of these atoms goes along with the evaporation of He atoms and, consequently, an attenuation of the beam flux. The attenuation can be measured either by a pressure gauge or a mass spectrometer, both are typically available at a state-of-the-art setup.

The attenuation of the beam provides information on the number of dopants that are deposited in a droplet and can therefore be used to adjust a desired doping level and the resulting particle size. A simple formula for the estimation of the mean number of dopants per droplet $\langle N_{dopant} \rangle$ is given by [17, 53, 66]:

$$\langle N_{dopant} \rangle = A \langle N_{He} \rangle \frac{E_{He}}{E_{bulk}} \tag{11.2}$$

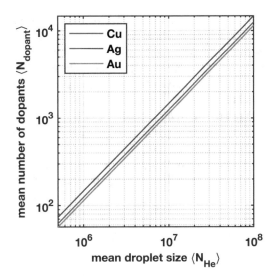

**Fig. 11.2** Mean number of dopant atoms $\langle N_{dopant} \rangle$ for Cu, Ag and Au as a function of the mean droplet size $\langle N_{He} \rangle$ calculated from (11.2) for an attenuation factor of $A = 0.7$

The factor $A$ corresponds to the measured beam attenuation $A = \frac{\Delta P}{P_{He}}$ obtained from the initial beam intensity $P_{He}$ and the difference between the initial and final beam intensity after doping $\Delta P$. As the parameter $A$ corresponds to a ratio, absolute numbers are not needed and the beam intensity can, for example, be measured with a quadrupole mass spectrometer set to the He or $He_2$ mass window. Here, $E_{He} = 0.62$ meV is considered as dissociation energy for $He_N \rightarrow He_{N-1} + He$ [2, 25]. Note that other sources use 0.76 meV, the He vaporization enthalpy at 0.62 K [66, 67]. $\langle N_{He} \rangle$ is the mean number of He atoms per droplet and $E_{bulk}$ is the binding energy per dopant using the bulk binding energy as an estimation. Equation (11.2) may be refined considering (11.1) above. Figure 11.2 shows an example of the calculated number of dopant atoms for the three coinage metals Cu, Ag and Au as a function of the droplet size for an attenuation factor of 0.7. It can be seen that for each species slightly different results are obtained. The determination of absolute numbers with this method should be treated with care, ideally the obtained particle sizes are checked by scanning transmission electron microscopy after deposition.

However, the method is well suited to adjust the ratio between the number of atoms from each dopant material in the helium droplets. Here, it is very important that binding energy differences are considered. The attenuation of the beam by each dopant species that is added at a different pickup zone can be measured. Upon deposition, dopant ratios may be investigated by scanning transmission electron microscopy (STEM) or X-ray photoelectron spectroscopy (XPS).

An alternative approach used to monitor the doping level and the agglomeration process makes use of a quartz crystal microbalance (QMB), which is available in many state-of-the-art helium droplet machines [16, 20, 22]. A QMB enables the direct measurement of the amount of material that is deposited. Such devices measure the deposition rate based on the mass-dependent change of the resonance frequency of a quartz crystal. However, the deposition rates obtained with standard helium droplet machines are rather low, typically in the ng/s regime, such that micro balances are operated close to their detection limit. This implies that care has to be taken with respect to electromagnetic stray-fields and temperature drifts. Furthermore, as a microbalance measures the amount of deposited material, the sticking coefficient, which expresses the probability with which a particle remains bound to the surface upon deposition, will have an impact on the result. Typically, the sticking coefficients are considered to be very high, but they are dependent on the dopant—substrate combination and possible backscattering effects have not yet been experimentally investigated. However, microbalances can be very valuable in order to adjust deposition rates at a particular apparatus in a reproducible manner.

### 11.2.3   Nanoparticle Growth

The nanoparticle growth process is dependent on the initial size of the employed helium droplets. Two different processes can be distinguished [68]:

(i) Single center growth is the dominating process in small helium droplets. In this case, the required time for the recombination of two dopants ($t_{rec}$) is shorter than the average time between two successive pickup events ($t_{n,n+1}$). Thus, a single nucleus is formed inside the droplet to which additional dopants are continuously added.

(ii) In large helium droplets, the doping rate can be so high that the dopants nucleate to clusters at different sites inside the droplet ($t_{n,n+1} < t_{rec}$). Subsequently, these clusters will recombine with each other and form larger cluster-cluster aggregates inside the droplet. Figure 11.3, taken from [68], shows $t_{rec}$ and $t_{n,n+1}$ for the case of Ag dopants and an attenuation factor $A = 0.7$ as a function of the initial helium droplet size $N_{He}$. The single- and multi-center growth regimes are indicated.

The transition from single- to multi-center aggregation is also reflected by structural differences in core@shell nanoparticles formed in droplets of different sizes, which has been studied in detail for Ag@Au [69]. In these experiments, double-core particles are only observed if the initial droplet size exceeds $N_{He} = 5 \times 10^5$ He atoms (corresponding to a droplet radius of 18 nm), in agreement with the crossover in Fig. 11.3. In sufficiently large droplets, the particles grown at multiple aggregation centers can still be separated when the second pickup zone is reached. Subsequently, these core particles are covered by a shell layer. The result of such a process is presented in Fig. 11.4b, which shows a Au particle with two Ag cores. An example of a small spherical Ag@Au particle with a single core is shown in Fig. 11.4a [69]. At an initial droplet size of about $5 \times 10^7$ He atoms ($\sim$80 nm radius) the study suggests a

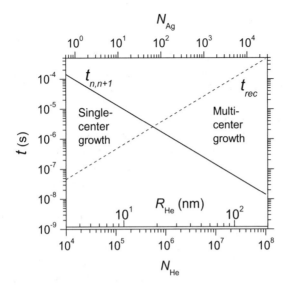

**Fig. 11.3** Dopant recombination time ($t_{rec}$) and time between two successive pickup events ($t_{n,n+1}$) as a function of the helium droplet size. Single center growth is dominant for small helium droplets whereas multi-center growth is prevailing in large droplets, as shown here for doping with Ag atoms and an attenuation factor of $A = 0.7$. Reprinted figure with permission from [68]. Copyright (2011) by the American Physical Society

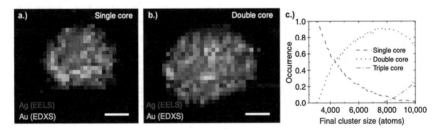

**Fig. 11.4** Panel **a** and **b** show elemental maps of deposited core@shell particles with a single and double Ag core (red), surrounded by a Au shell (green). The scale bars represent 2 nm. Panel **c** shows the calculated occurrence probability of single-, double- and triple-core clusters as a function of final cluster size, i.e. the total number of Ag and Au atoms, for droplets consisting of $5 \times 10^7$ He atoms. Figure adapted from [69], licensed under CC-BY 4.0

crossover from particles with a single core to particles which exhibit a double core at about 4000–5000 added dopants, as shown in Fig. 11.4c. At high doping rates, even triple-core particles can be formed. The results demonstrate that not only the helium droplet size plays an important role in this context but also the adjusted doping level.

### 11.2.3.1 Quantum Vortices and Nanoparticle Growth

Quantum vortices in rotating superfluid helium droplets carry angular momentum [70]. Transmission electron microscopy images of deposited wire-like nanoparticles provided the first indication for the presence of quantum vortices in helium droplets [27]. Direct evidence for vortices in helium droplets has later been presented and is based on diffraction images of individual droplets recorded by X-ray scattering at a free electron laser facility [70–72], see chapter *X-ray and XUV Imaging of Helium Nanodroplets* by Tanyag, Langbehn, Rupp and Möller in this volume [73].

Vortices attract dopants through hydrodynamic forces, which results in an agglomeration of particles along the vortex cores [53]. Large helium droplets can host many vortices in array-like structures [74]. The presence of such vortex structures in helium droplets has a huge impact on the shape of the formed particles because they serve as a scaffold during particle growth. The capture cross section of vortices hosted by large droplets can be three orders of magnitude larger than the collision cross section of individual dopants [23]. Dopant–vortex recombination time scales can be found in [23, 53]. For helium droplets with a diameter of about $1\,\mu m$, for example, the time it takes for a metal particle to get trapped by a vortex has been calculated to lie within 10 and $100\,\mu s$ for different species and doping levels [53]. Individual particles can move along the vortex core and fuse together [23]. Depending on the initial helium droplet size, the number of hosted vortices and the doping level, these particles may be synthesized by multi-center growth in the droplet volume or at nucleation centers directly pinned to vortices. If the doping level is high enough, continuous filament-like structures are formed.

This process has been explored for different doping level regimes by [53], Fig. 11.5 shows selected results obtained for Au. For weak doping ($A = 0.04$), segmented structures aligned along a track are observed. The doping level, in this case, is not high enough for the formation of continuous filaments, indicating that there are indeed

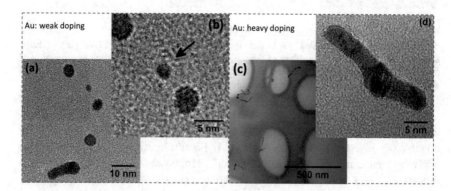

**Fig. 11.5** Deposited Au nanoparticles synthesized in helium droplets. For weak doping ($A = 0.04$), individual, well separated nanoparticles are observed whereas heavy doping ($A = 0.75$) leads to the formation of elongated nanowire structures. Figure adapted from [53]. Published by the PCCP Owner Societies

separated aggregation centers along vortices. For heavy doping ($A = 0.75$), continuous nanowire-structures are observed. In this case, the growth process has continued, individual aggregated particles are fused together and elongated filaments are formed. Note that in the course of this process, particles are not expected to melt completely [53]. Thus, the morphology of the coagulated particles is largely preserved, as revealed by scanning transmission electron microscopy images of deposited nanoparticles for many different metals, see, for example, Fig. 11.5d or Fig. 11.6d.

At the moment, the minimum size of a helium droplet that is required in order to host a vortex is not known. X-ray scattering experiments have confirmed the presence of vortices in droplets larger than about 100 nm [27, 72]. A hint for a possible minimum size may be deduced from the fact that elongated particles, which require vortices as a scaffold during the growth process, begin to emerge in droplets with diameters larger than about 50 nm. However, vortices are theoretically predicted even for very small droplets [70, 75, 76]. Clever experiments capable of detecting vortex signatures in smaller helium droplets will have to be developed in order to fully answer this question [77].

### 11.2.3.2 The Foam Hypothesis

The foam hypothesis describes a scenario according to which dopants do not agglomerate to clusters or nanoparticles but remain separated from each other, resulting in a metastable, foam-like super-structure inside the helium droplet. Different observations gave rise to speculations about such structures in the past, with Mg as the most famous example [78, 79].

The presence of dopants that do not form a bond upon doping has long been considered as a possible scenario, used, for example, to explain the observed mismatch between coagulation and pickup cross sections [5]. Evidence for separated dopants has been found during experiments with alkali and alkaline earth metals [80–82] and some indications for foam-structures have been found for Al in helium droplets [83, 84]. A similar situation has been predicted by DFT calculations for Ne atoms in superfluid (bulk) helium [85].

The best studied example in this context is Mg, for which experimental evidence for the formation of foam-like structures has been presented [78, 79, 86]. The stability of such structures is explained by local minima that emerge due to the modulation of the long-range van der Waals part of the Mg dimer potential energy curve by the surrounding helium. This causes a potential barrier at larger interatomic separations, which hinders the formation of compact clusters. Spectroscopic evidence [78] for a foam-configuration is based on the recording of atom-like transition at mass windows that correspond to larger Mg clusters using resonant multi-photon ionization spectroscopy. Note that similar spectra have been observed for other species, such as Al, [83, 84] Au [87] and Cr [39].

However, all experimental indications for foam-like structures have in common that they were observed for small complexes formed in helium droplets with $N_{He} < 10^5$. In particular, for Mg it has been shown that beyond a mean number of 70 Mg atoms per droplet, the spectral signatures associated with foam structures disappear [86]. For the two dopants for which foam-like structures have been suggested, Mg and Al, it has been shown that compact nanoparticles are observed upon deposition [33, 34, 88–90]. Thus, it seems as if such metastable structures only play a role for small clusters and it is still questionable if the scenario applies for other materials beyond Mg.

### 11.2.4 Core@shell Nanoparticles

A speciality of the helium droplet synthesis approach is the formation of core@shell nanoparticles [18, 91]. In particular, with the method it is possible to synthesize spherical sub-5 nm core@shell particles and core@shell nanowires with diameters below 10 nm. The core and shell material can be selected independent from each other, which allows for the design of nanoparticles with tailored properties. For spherical particles, the core diameter and the shell thickness can be well controlled [92]. Core@shell particles may also be formed with other methods, using, for example, wet chemical approaches or cluster beam techniques [93–95]. Compared to these methods, however, the helium droplet approach is extremely flexible and provides an inert and cold synthesis environment that enables the combination of a large variety of different materials in a configuration only determined by the pickup sequence. Without adaption of the experimental setup, species that are very different in nature can be combined in nanoparticles, from metals and metal oxides to organic molecules, gases and even highly reactive species such as alkali metals.

Bimetallic Au–Ag nanoparticles with core@shell structure were among the first that have been formed and deposited using the helium droplet approach. With this material combination, it was demonstrated that the pickup sequence dictates the core and shell materials [91]. Modern scanning transmission electron microscopy (STEM) allows for the creation of elemental maps for selected nanoparticles, providing an important tool for the characterization of core@shell structures. Depending on the element of interest, electron energy loss spectroscopy (EELS) or energy dispersive X-ray (EDX) spectroscopy is the method of choice, both provide element sensitivity. EELS is typically used for lighter elements whereas EDX is more sensitive for heavier elements. By scanning the electron beam across a particle it is possible to record EELS and EDX spectra for each pixel of an image. An evaluation of the counts within a certain interval in the EELS or EDX spectrum around a spectral feature associated with a particular element allows for the creation of a map, which reflects the spatial distribution of this element. A selected result from such an analysis of a nanoparticle deposited on a TEM substrate is shown in Fig. 11.6 [91], with elemental maps for Ag (a) and Au (b). The combination of both images in Fig. 11.6c, created using color-codes for Ag (green) and Au (red), reveals the core@shell structure of the particle, which corresponds, in this case, to a Au core surrounded by a Ag shell. Figure 11.6d shows a high-angle annular dark-field (HAADF) image of a similar Au@Ag nanoparticle. Even though the lattice structure is resolved, in this case a

**Fig. 11.6** Images obtained by scanning transmission electron microscopy of core@shell Au@Ag nanowires [91]. Panel **a** shows an EELS map for Ag, panel **b** corresponds to an EDX map for Au. The color-coded image **c** is created by combining both elemental maps (green for Ag and red for Au). Panel **d** shows a high-resolution HAADF image of a different Au@Ag particle comparable in size. Reprinted with permission from [91]. Copyright (2014) by the American Physical Society

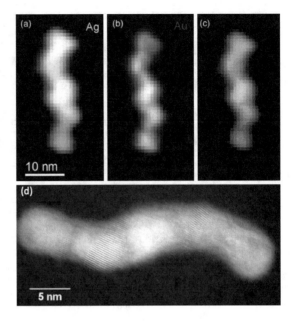

core@shell contrast is not obvious, demonstrating the need for element sensitive tools for the characterization of core@shell nanoparticles. In this image it can also be seen that the particle exhibits differently oriented facets, originating from individual clusters that are fused together in the course of the synthesis process along a vortex line.

## 11.2.5  Deposition of Nanoparticles

A particular advantage associated with the helium droplet technique is the soft deposition process [16, 96–98], enabled by the cushioning of the impact by the liquid helium that surrounds the nanoparticles. Furthermore, the droplets provide a very cold environment with a velocity that corresponds to only about $200-300\,\mathrm{ms}^{-1}$. During the deposition process, the kinetic energy is lower than the binding energy per atom, which is characteristic for particle deposition in the soft-landing regime [96, 99]. Ab-initio calculations for the deposition of a Au atom solvated in a $\mathrm{He}_{300}$ droplet on a $\mathrm{TiO}_2$ surface resulted in a landing energy below 0.15 eV [97]. This value is much smaller than the binding energy per (bulk) Au atom of 3.81 eV, supporting the experimentally observed soft-landing scenario. These considerations also hold for larger particles, as shown by the calculation of the deposition of $\mathrm{Ag}_{5000}$ particles solvated in $\mathrm{He}_{100000}$ droplets on an amorphous carbon substrate [98]. Only for velocities beyond about 1000 m/s a melting of the Ag particle and its subsequent spreading on the substrate is predicted. An example of a deposited $\mathrm{Ag}_{5000}$

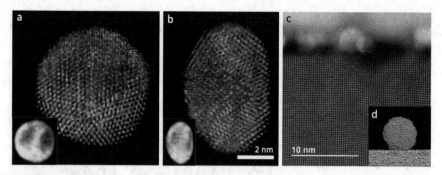

**Fig. 11.7** The left two panels correspond to an electron tomography reconstruction of the structure of a deposited Ag@Au core@shell particle, top (**a**) and side (**b**) view. The insets on the bottom left show the elemental distribution of Ag (dark) and Au (bright). The size of the lenticular shaped particle, which comprises a double Ag core, corresponds to $8 \times 7 \times 5\,nm^3$. Panel **c** shows a side-view STEM image of deposited Au nanoparticles on a crystalline $TiO_2$ substrate. Focused ion beam milling was employed to prepare a thin sample slice suitable for STEM imaging. Panel **d** shows the result from an ab-initio calculation of a deposited $Ag_{5000}$ particle. Image **a** and **b** adapted from [69], licensed under CC-BY 4.0. The image in panel **c** is reprinted with permission from [102]. Copyright (2016) American Chemical Society. The image in panel **d** is reprinted with permission from [98]. Copyright (2017) American Chemical Society

nanoparticle from these calculations is shown in Fig. 11.7d. It should be noted, however, that even though the deposition process is very soft, structural rearrangements as a consequence of the impact can occur. Together with experiments that explored the structure and morphology of deposited Ag particles [21], molecular dynamics simulations for $Ag_n$ with $n = 100-2000$ revealed that deposited particles can adopt an energetically more favorable structure [96]. In particular, it was found that larger nanoparticles tend to retain their original morphology, while smaller ones undergo structural rearrangements. Furthermore, the presented theoretical results highlight the dependence of the landing process on the interaction between atoms in the particle and the substrate, among other parameters [96].

The fact that the deposition process is very soft is also evidenced by the ability of the approach to synthesize and deposit core@shell nanoparticles: It has not been observed that the energy released during the impact influences the internal core@shell structure of the formed nanoparticles, which could manifest, for example, in the formation of alloys. Core@shell particles have been formed using various different materials and substrates and, in the absence of oxidation effects, their structure always reflects the pickup sequence.

The soft deposition process has been exploited in previous experiments and enabled the decoration of ultra-thin substrates. Examples encompass few atomic layer thick hexagonal boron-nitride (hBN) and $0.5 \times 0.5\,mm^2$ free-standing, 10 nm thick SiN substrates. Ultra-thin hBN substrates have been employed for plasmon spectroscopy experiments with deposited particles using scanning transmission electron microscopy (STEM) in combination with electron energy loss spectroscopy (EELS) [100]. Free-standing SiN substrates provide a sufficiently large unobstructed area to allow for the use of advanced XUV absorption spectroscopy methods for the investigation of nanoparticles [101].

Electron tomography enables the 3D reconstruction of deposited nanoparticles [69]. A selected result is shown in Fig. 11.7, composed from scanning transmission electron microscopy images recorded from different perspectives. Images (a) and (b) in Fig. 11.7 show a Ag@Au core@shell particle, top and side view, respectively. The elemental distribution of Ag (dark) and Au (bright) within the particle can be seen in the inset panels. The size of this lenticular shaped particle, which exhibits a double Ag core, corresponds to $8 \times 7 \times 5$ nm$^3$. It can be seen that the deposited particle is flattened due to the interaction between the particle and the amorphous carbon substrate. The smooth shape is explained by surface diffusion processes that proceed upon deposition and during the investigation in the scanning transmission electron microscope. However, the internal icosahedral morphology is preserved as well as the core@shell structure. Alloying processes are not observed, which entails that the deposition process is indeed very soft [69].

Similar results have been obtained for Au nanoparticles on a TiO$_2$ substrate. In this case, focused ion beam (FIB) milling has been employed to prepare thin sample slices that can be studied by STEM, allowing for a view on the deposited particles from the side. A selected image is presented in Fig. 11.7c [102], which shows the interface between Au particles and a TiO$_2$ substrate. Two particles can be identified, with $\sim$2.5 nm and $\sim$3.7 nm diameter, for which the Au lattice structure is resolved. The TiO$_2$ crystal structure is observed in the region below the particles, the white area on top corresponds to a ZnO capping layer needed for FIB-sample preparation purposes. The shape of the nanoparticle can clearly be recognized and appears only slightly flattened, similar to the results obtained from electron tomography.

However, it has to be kept in mind that the form and shape of the deposited particles does not necessarily reflect the actual situation in the helium droplet. For Xe dopants in large helium droplets, it has been directly shown that the dopant material is pinned to vortices and multiple separated filaments are formed [70]. It is generally assumed that metal dopants form similar filaments. However, images recorded from deposited particles always show single elongated, branched structures. This indicates that the particles rearrange during or after deposition. On the surface, particles from different vortices may be fused together and surface diffusion processes can affect their final form.

### 11.2.5.1  Nanoparticles at the Surface

With the helium droplet synthesis technique it is possible to deposit nanoparticles on every substrate that is placed into the droplet beam. However, while the soft deposition process preserves the integrity of the particle and the substrate, the particles may be affected by dynamic processes that can influence the obtained structures. This has to be taken into account for subsequent experiments.

In the majority of previous experiments, nanoparticles were deposited on corrugated surfaces which are not ideally flat. Typical examples are amorphous carbon or

**Fig. 11.8** TEM images of Ag nanoparticles deposited on amorphous carbon substrates ($N_{He} \sim 10^{10}$, $\sim 2 \times 10^6$ Ag atoms per droplet) for three different deposition times of **a** 4 s, **b** 2 min and **c** 30 min. Images reprinted with permission from [98]. Copyright (2017) American Chemical Society

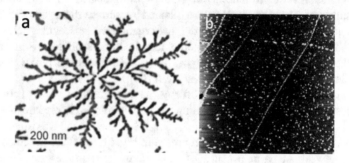

**Fig. 11.9** Panel **a** shows agglomerated Ag nanoparticles on a graphite surface, recorded by scanning electron microscopy (SEM). Panel **b** shows a scanning tunneling microscopy (STM) image (871.1 nm × 871.1 nm) of Ag nanoparticles on sputtered highly oriented pyrolytic graphite (HOPG) with a low number of defects. Note that the nanoparticles in both images have been synthesized with alternative cluster sources and not with helium droplets, see [104, 106] for details. The image in panel **a** is reprinted with permission from [104]. Copyright (2002) by the American Physical Society. The image in panel **b** is reprinted from [106]. Copyright (2005), with permission from Elsevier

SiN substrates as used for transmission electron microscopy. On such substrates, the nanoparticles are typically not mobile and can be expected to remain at the position where they have been deposited. However, this only holds if the surface coverage is low: In the case of the deposition of large (segmented) Ag filaments, synthesized in $N_{He} \sim 10^{10}$ droplets that contain on average $\sim 2 \times 10^6$ Ag atoms, it has been observed that large Ag free areas appear between the particles with increasing surface coverage, best seen in Fig. 11.8b [98]. Such large void areas between the structures are not expected based on the statistic nature of the deposition process, which would result in a random distribution. These areas are explained by the impact of large helium droplets, which can push around the particles that are already present at the surface. For comparison, panel a in Fig. 11.8 shows an image recorded for low surface coverage, achieved by exposing the substrate to the droplet beam for only about 4 s. In this case the Ag filaments are well separated. For very high surface coverage, as shown in Fig. 11.8c, the observed pattern changes again and larger structures are formed by the agglomeration of deposited particles on the surface.

Another effect that can influence the final structure of nanoparticles on surfaces is observed on extremely flat substrates. If the interaction between substrate and nanoparticle is weak, the deposited particles are mobile, which can lead to the formation of islands [103]. Under certain conditions, this causes the formation of well separated, ramified structures. An example of the result of such a process for $Ag_{150}$ particles deposited under soft-landing conditions (thus, comparable to helium droplet synthesis) on graphite using a gas-aggregation cluster source is shown in Fig. 11.9a [104]. Note that in the presence of substrate defects the mobility of the deposited clusters is lowered [105, 106]. Thus, particles agglomerate preferentially at steps on the surface, as can be seen in Fig. 11.9b, which shows a scanning tunneling microscopy (STM) image of Ag nanoparticles on highly oriented pyrolytic graphite (HOPG) [106].

These examples show that the resulting structures at the surface can be very different compared to the original form of the deposited particles. Many processes can affect the final size, shape and structure, which has to be considered when working with nanoparticles on surfaces.

## 11.2.6  Size and Shape of Nanoparticles Synthesized with Helium Droplets

The size and shape of the formed nanoparticles is determined by the initial helium droplet diameter and the adjusted doping level [28, 91]. For droplets with average diameters increasing from 25 to 1700 nm, the evolution of the particle size and shape can be followed in Fig. 11.10 [28]:

(i) In helium droplets with average diameters below 50 nm, predominantly small and spherical particles are formed. The obtained nanoparticle diameters are typically below 5 nm. Assuming a continuous helium droplet source with a

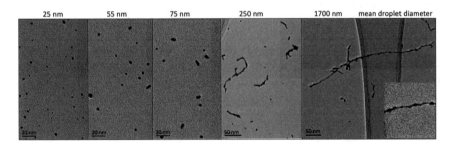

**Fig. 11.10** Evolution of the size and shape of deposited nanoparticles with increasing helium droplet size. At a mean droplet diameter of 25 nm the resulting particles are spherical with diameters well below 10 nm. Between 50 and 100 nm the particles become more and more elongated. Deposited particles formed in larger helium droplets exhibit nanowire-like shapes with a length up to a few 100 nm. Adapted with permission from [28]. Copyright (2014) American Chemical Society

5 μm nozzle and 20 bar He stagnation pressure, such droplets are generated at
nozzle temperatures higher than 9 K [65].

(ii) In the droplet diameter range from 50 nm (20 bar, 9 K) to about 100 nm (20 bar,
7 K), small spherical nanoparticles are still present, however, the number of
elongated, rod-like particles increases with the droplet size for high doping
rates. It is important to note that also droplets beyond 50 nm diameter can be
used to produce small spherical particles if lower doping rates are chosen [69].

(iii) He droplets with diameters beyond 100 nm enable the synthesis of long
nanowire structures. For example, in droplets with diameters of about 1 μm
(20 bar, ∼5.4 K), the formed structures can have a length up to a few 100 nm.
These structures typically exhibit multiple branches and do not correspond to
ideal one-dimensional wires.

The fact that the transition from one regime to another proceeds gradually is not
surprising considering the broad helium droplet size distribution [1, 2, 107, 108].
The droplet size distribution translates into a nanoparticle size distribution because
the pickup probability scales with the geometric cross-section of the droplet [5], i.e.
larger droplets will collect more dopants than smaller ones.

Interestingly, in previous helium droplet synthesis experiments, nanoparticles with
diameters beyond 10 nm have not been observed. This seems to be a limit that cannot be overcome by conventional helium droplet setups. Considering small helium
droplets, the particle diameter is limited by the maximum number of dopants that can
be added to the droplet until all He atoms are evaporated. At a certain helium droplet
size, the presence of vortices leads to the formation of elongated wire-structures,
which sets an upper limit to the diameter of spherical particles. However, also the
diameter of wire-structures is limited and does typically also not exceed 10 nm.

### 11.2.6.1   Size Distribution of Deposited Nanoparticles

An important parameter in the nanoparticle synthesis business is the obtained size
distribution. For many applications, narrow size distributions are desirable, in particular, for small particles in the sub-10 nm regime where many properties can vary
substantially with size [109, 110]. As discussed above, the helium droplets that are
present in the beam have different diameters. The width of the helium droplet size
distribution is typically on the same order of magnitude as the mean value [1, 2, 107,
108]. Consequently, also nanoparticles are produced with a certain size distribution
because the number of dopants that are picked up depends on the size of the helium
droplet.

For small spherical particles with a few nanometer diameter, the full-width-at-
half-maximum (FWHM) is typically on the order of about 1–2 nm [21, 92, 102]. An
example of particles formed in helium droplets with a diameter of approximately
60 nm (50 bar, 11.5 K) is shown in Fig. 11.11. The mean diameter of the particles
formed at these conditions and the adjusted doping level is about 2.5 nm. The width
of the nanoparticle size distribution, shown in the right panel in Fig. 11.11, has a

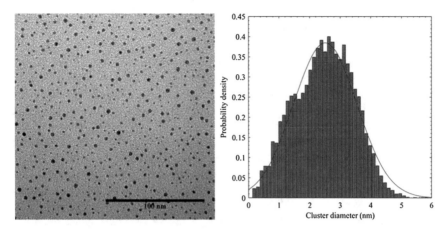

**Fig. 11.11** Typical transmission electron microscopy image obtained for Ag nanoparticles, deposited for 10 min onto an amorphous carbon substrate. The distribution of nanoparticles created under the selected conditions (50 bar, 11.5 K) is shown on the right. The mean particle diameter corresponds to 2.5 nm, with a full-width-at-half-maximum (FWHM) of the distribution of 2.5 nm. Reprinted from [21], with the permission of AIP Publishing

**Fig. 11.12** Distribution of particle diameters and lengths obtained for short oxidized Co nanowires. The mean value $\mu$ and standard deviation $\sigma$ of the Gaussian function fitted to the data, plotted in red, are listed in the figure. Image reprinted from the Supplementary Information of [101], with the permission of AIP Publishing

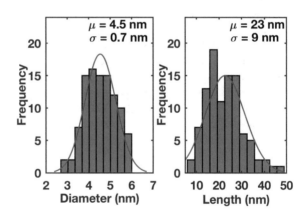

FWHM of 2.5 nm. Note that in this particular case there is a weak shoulder observed for smaller diameters, which is explained by the presence of particles with different morphologies on the substrate, as analyzed in detail in [21].

Compared to spherical particles, the size distributions obtained for nanorods and nanowires are similar with respect to the diameter, however, their length distribution is much broader. An example is shown in Fig. 11.12 for oxidized Co nanowires with a diameter of 4.5 nm and a FWHM of the distribution of about 1.7 nm. Note that in this case, the particles are oxidized after they have been deposited and the distribution may have changed during this process. The mean length of the particles in this example is 23 nm but the width of the length distribution extends over a few 10 nm [101].

## 11.3 Metal Nanoparticles

Already in the early days of helium droplet research, metal clusters consisting of up to about 100 atoms have been formed and investigated [12, 111–113]. Back then, however, these clusters were analyzed in-situ using laser spectroscopy and mass spectroscopic techniques [14]. Even though considered already earlier [1], the first helium droplet deposition experiments were reported in 2007 [16]. This pioneering work made use of a micro balance setup to measure deposition rates for Ag, Au and bimetallic Ag–Au structures, demonstrating that nanoparticles isolated in helium droplets can be deposited onto a surface. Shortly thereafter, the first images recorded by transmission electron microscopy were presented [17], with surprising results for metal nanoparticles formed in large helium droplets [27]: It was discovered that the deposited particles exhibit an elongated, wire-like structure, which provided indirect evidence for the existence of vortices in helium nanodroplets. These initial experiments sparked the advent of helium droplet based nanoparticle synthesis and many groups embarked on an investigation of the possibilities offered by this new approach [17, 19–21].

Since then, a large variety of different metals has been employed for the synthesis and deposition of nanoparticles, encompassing Cu, Ag, Au, Cr, Fe, Co, Ni, Al, Pd and Zn [22, 25, 27, 28, 31, 33–35, 91, 114], all of which show very similar structures: Spherical sub-10 nm particles are formed in smaller helium droplets, as discussed above, whereas nanorods and nanowire structures are formed with large droplets. The fact that all these materials exhibit similar structures shows that properties such as size and shape are determined by the synthesis environment. Thus, the approach allows for the formation of nanoparticles that are all very similar in size and shape, independent of the material. Upon deposition, however, these materials can behave very different, which has been explored in many experiments on deposited nanoparticles.

### 11.3.1 Thermal Stability of Metal Particles and Nanoscale Alloying Processes

The use of heatable substrates for scanning transmission electron microscopy enabled the investigation of temperature dependent processes in deposited nanoparticles [25]. The surface of these substrates typically consists of amorphous carbon or SiN, temperatures up to about 1300 °C can be reached. Recent heating experiments have addressed the thermal stability of nanowires and investigated nanoscale alloying processes.

### 11.3.1.1 Thermal Stability of Deposited Nanowires

Initially, the results obtained for Ag nanoparticles were puzzling as they did not show a continuous wire-like shape when synthesized with large helium droplets. Moreover, they appeared segmented, consisting of individual particles aligned along a track (see e.g. Fig. 11.8a or Fig. 11.13 (Ag b)). [27, 114] This issue was resolved by temperature dependent studies with deposited Ag nanoparticles. In these experiments it has been discovered that the segmented form of the particles can be explained by Rayleigh breakup, a phenomenon that involves diffusion processes that proceed at the surface of deposited nanoparticles [115]. Figure 11.13 shows a selected set of scanning transmission electron microscopy images recorded for Au and Ag nanowires at two different temperatures. The results demonstrate that if Ag particles are kept at low temperatures (-15 °C) they are not segmented, in contrast to their room temperature counterparts. Rayleigh breakup is also observed for Au nanowires, for which higher temperatures are necessary to induce the segmentation process. The process has been simulated using a cellular automaton approach [117], also shown in Fig. 11.13. Depending on the local particle curvature, the chemical potential is driving the diffusion of atoms at the surface, an effect that enhances with increasing temperature. As a consequence, the particle edges become smoother and, eventually, the nanowires

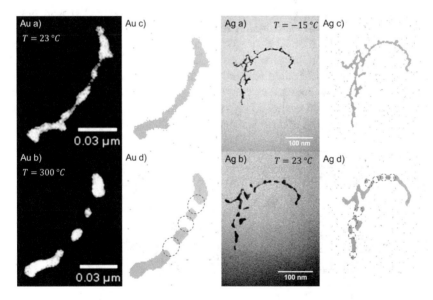

**Fig. 11.13** Comparison between measured surface diffusion processes and cellular automaton prediction of Au and Ag nanowires. In **a** and **c** one can see the initial structure and the converted structure used in the cellular automaton approach (see [117] for details on the approach), respectively. Those structures evolve to the ones shown in (**d**), when heating is applied, in very good agreement with the final structures observed experimentally (**b**). Note that for Ag, Rayleigh breakup is already observed at room temperature. Reprinted from [117]. Published by the PCCP Owner Societies, licensed under CC-BY 3.0

are segmented. This process has also been investigated for Ni and Cu nanowires [116, 117]. Note that surface diffusion processes can also be enhanced by irradiation with an electron beam during the recording of transmission electron microscopy images [118].

### 11.3.1.2 Nanoscale Alloying Experiments

For small spherical nanoparticles, formed by single-center aggregation, the core diameter and the shell thickness can be controlled. This has been exploited in a series of experiments dedicated to the investigation of nanoscale alloying processes in core@shell nanoparticles. An important advantage of the helium droplet approach in this type of experiments, thereby, is that it allows for the formation of clean and residual-free particles.

Nanoscale alloying processes have been investigated for Ag@Au as well as Au@Ag nanoparticles with a diameter of about 4 nm [119]. Figure 11.14 shows selected results for Ag@Au. At room temperature, 23 °C (296 K), the scanning transmission electron microscopy (STEM) image and the corresponding radial intensity profile both show a core@shell contrast. In the high-angle annular dark-field (HAADF) image, the Ag appears as a dark core surrounded by bright Au because the contrast scales with the number of atoms (sample thickness) and the square of the mean atomic number Z of the scattering material ("Z-contrast") [120]. Interestingly, upon heating to 300 °C (573 K) a core@shell contrast is no longer observable, indicating that the particle is completely alloyed. The diffusion dynamics that proceed between the initial and terminal temperature can be analyzed based on the radial intensity profile. Using constant heating time steps, a diffusion constant $D(T)$ can be obtained from such an analysis. For both investigated systems in [119], Ag@Au and Au@Ag, the alloying process is completed at 300 °C, considerably lower than

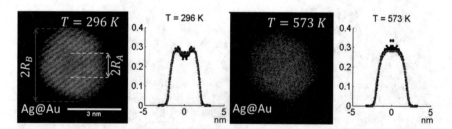

**Fig. 11.14** Ag@Au core@shell nanoparticle at room temperature, 23 °C (296 K) and at 300 °C (573 K) on a heatable carbon substrate. The scanning transmission electron microscopy (STEM) images are accompanied by radial intensity profiles. The dip in the room temperature profile originates from the Ag core that appears as dark region in the STEM image. At 300 °C, the alloying process is completed and a core@shell contrast is no longer observed. Figure adapted from [119], published by the Royal Society of Chemistry, licensed under CC-BY 3.0

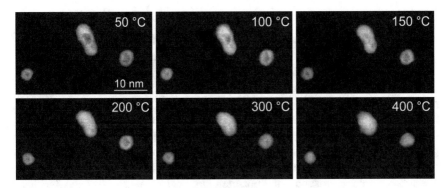

**Fig. 11.15** High-angle annular dark-field (HAADF) images of Ni@Au core@shell clusters for a series of temperatures recorded by scanning transmission electron microscopy. The alloying process is completed at 400 °C. At this temperature also the modification of the particle shape becomes evident. Reprinted with permission from [29]. Copyright 2018 American Chemical Society

compared to the bulk. This is attributed to surface size effects that become important in the sub-10 nm particle size regime.

Temperature dependent effects have also been studied for the iron-triade elements, Fe, Co and Ni, in combination with Au [29, 31, 121]. Figure 11.15 shows Ni@Au nanoparticles with different sizes and shapes on a heatable carbon substrate [29]. In these images, the Ni cores appear as dark regions inside the bright Au shells. For the nanorod visible in the top part of the figure, a weakening of the contrast is already observed around 100–150 °C. However, only at a temperature of 400 °C the contrast vanishes for all particles. It becomes evident that the stability of the particles and, hence, the alloying process depends on the position of the Ni core within the particle. It has been found that a decentralized Ni core is more stable than a centric Ni core [121]. Heating to 300 °C for 30 min and subsequent cooling under ultra-high vacuum (UHV) conditions, in the absence of any oxygen, results in the formation of multiple Ni cores inside the Au host matrix, a process referred to as spinodal decomposition [121].

## 11.3.2   Plasmonic Metals in Helium Droplets

In bulk material, the plasma frequency dictates the collective oscillations of conduction band electrons in a metal. At the surface, the translational invariance is broken and lower energy modes are supported, known as surface plasmons. If the plasmonic structures approach sizes that are commensurate with or smaller than the wavelength of an incoming electromagnetic field, surface plasmons become localized and can no longer propagate [122, 123]. Localized surface plasmon resonances (LSPR) give rise to strong optical absorption and enable the concentration and enhancement of the electromagnetic field in the subwavelength regime. The spectral position of the localized surface plasmon resonance depends on the size, shape and material of a nanoparticle as well as on the environment. The LSPR provides the foundation of many applications of plasmonic nanoparticles, from surface enhanced Raman spectroscopy (SERS) [124] to sensor technologies [125].

Matrix isolation spectroscopy has contributed to the research on localized surface plasmons [126]. Helium droplet isolation spectroscopy, in particular, has been used to investigate surface plasmons in small sub-10 nm sized nanoparticles. This is an interesting regime where quantum size effects emerge [127] and where a transition from discrete, quantized molecule-like transitions to broadband plasmon excitations occurs [123, 128–130]. The typical plasmonic materials, Ag and Au, behave very differently when approaching the quantum size regime below 10 nm: A localized surface plasmon resonance in Au is only observed for particle diameters beyond ~2 nm [130]. In contrast, for Ag, transitions associated with a plasmon resonance have been reported for clusters consisting of only a few atoms (e.g. for $Ag_9^+$ and $Ag_{11}^+$) [129]. Furthermore, differences in the electronic band structure and the relative position of the LSPR between these materials result in a stronger damping of the plasmon intensity for Au than for Ag [123, 131].

The classic plasmonic materials, Cu, Ag, and Au, have been isolated in helium droplets and have been investigated by means of laser spectroscopy in-situ, covering the range from atoms and small clusters [13, 87, 132–135] to nanoparticles [68, 136]. The majority of helium droplet-based experiments with plasmonic materials has been carried out with Ag dopants. Using resonant two-photon ionization spectroscopy, a feature associated with a plasmon resonance has been observed for $Ag_8$ clusters in helium droplets [13, 132]. A very large range of Ag particle sizes, from clusters consisting of a few atoms on average up to the nanowire regime, has been investigated using beam depletion spectroscopy [68]. By setting a mass spectrometer at the end of the helium droplet apparatus to the $He_2^+$ mass window ($m = 8$), the loss of He atoms upon laser excitation of metal electrons and the concomitant shrinking of the droplet by the dissipation of electronic energy can be followed as a function of the photon energy of a tunable laser source. The resulting absorption spectra, taken from [68], are shown in Fig. 11.16, recorded for $Ag_n$ clusters embedded in helium nanodroplets with mean sizes starting from $n = 6$ a) up to $n = 6000$ e). Calculated spectra are plotted using dashed lines, while experimental spectra are represented by solid grey lines. The localized surface plasmon resonance can be identified as strong peak in the spectrum, its position shifts from 3.8 eV for small clusters to 3.6 eV for larger particles. Furthermore, for larger Ag nanoparticles an additional feature emerges in the infrared region, attributed to the structural change that occurs when the multi-center growth regime is reached and particles aggregate at vortex cores. The coupling of plasmon modes as well as the emerging of transversal modes in elongated nanostructures both can explain this additional feature. This observation provided an in-situ spectroscopy based evidence for the transition from single to multi-center aggregation [68]. Similar results have been obtained for Cu particles in helium droplets [136].

The excitation of a localized surface plasmon resonance allows for an efficient transfer of energy to the metal nanoparticle in the helium droplet. The energy stored

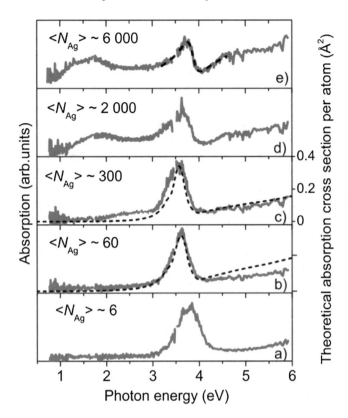

**Fig. 11.16** Beam depletion spectra recorded for Ag$_n$ clusters embedded in helium nanodroplets with mean sizes starting from $n = 6$ a) up to $n = 6000$ e). Theoretical spectra are plotted as dashed lines, while experimental spectra are represented by solid grey lines. The localized surface plasmon resonance (LSPR) is observed around 3.6 eV, depending on the average Ag particle size. Reprinted with permission from [68]. Copyright (2011) by the American Physical Society

in the plasmon leads, via non-radiative electronic relaxation, to the generation of heat in the Ag particle, which is dissipated by the helium droplet. Using pulsed laser systems that can provide high intensities in the few mJ/cm$^2$ regime, depletion spectra have been recorded, which indicate that large Ag clusters can reconstruct inside the helium droplet if exposed to high intensity laser pulses [66, 67]. This observation has been explained by the formation of a bubble around the Ag, which isolates the particle from the He bath. The local temperatures, thereby, can exceed the melting point.

### 11.3.2.1 Deposited Plasmonic Nanoparticles

If nanoparticles are continuously deposited, surface coverages on the order of a few 10% can be obtained. This requires deposition time scales of a few minutes for nanowires and several hours for small spherical nanoparticles. Plasmonic nanoparticles have large absorption cross sections, which enables an ex-situ investigation of samples with low surface coverage using different methods. In fact, the areas where particles are deposited can be seen by eye, as shown in the photograph in Fig. 11.17, which corresponds to a fused silica glass coverslip that holds spherical ∼5 nm nanoparticles with about 25% surface coverage. Two deposition areas, marked by arrows, are visible with a size of about 5 × 5 mm².

From particles as shown in Fig. 11.17, extinction spectra can be recorded using UV/vis spectrophotometry [92]. This enabled the investigation of the dependence of the localized surface plasmon resonance in Ag@Au core@shell particles on the Ag:Au ratio. Results from this study, carried out for particles with an average diameter of about 5 nm and a surface coverage of about 25%, are shown in Fig. 11.18. The absorbance of plain Ag (blue) and Au (red) nanoparticles is peaking at 447 nm and 555 nm, respectively. Due to the presence of the fused silica substrate, the peak position appears red-shifted compared to the spectra of Ag particles isolated in helium droplets. An important requirement for the investigation of deposited nanoparticles is that the surface coverage is kept low in order to avoid an inter-particle coupling of plasmon modes, which typically causes the emerging of absorption features in the infra-red for the studied materials. Furthermore, high surface coverages can give rise to particle agglomeration, which would also influence the spectra. However, this is not the case for the spectra in Fig. 11.18. The Ag:Au ratio can be well controlled by the temperatures of the two ovens that hold the Ag and Au dopants. By adjusting the pickup levels, the Ag:Au ratio has been set to 2:1 (green), 1:1 (black) and 1:2 (yellow). It can be seen that the LSPR shifts from its position for bare Ag nanoparticles towards the Au resonance with increasing Au contend.

An important application of plasmonic metal nanoparticles is surface enhanced Raman spectroscopy (SERS) [124]. This technique exploits the enhancement of the

**Fig. 11.17** Photograph of two nanoparticle deposition areas, marked by arrows, with a surface coverage of about 25%

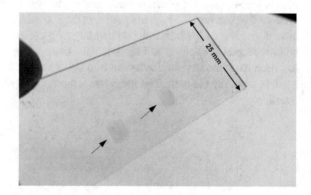

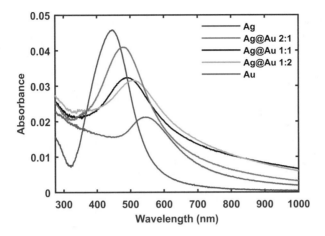

**Fig. 11.18** Extinction spectra of spherical Ag@Au nanoparticles on a glass coverslip with a surface coverage of about 25%. Nanoparticles with different Ag:Au ratios of 1:2 (orange), 1:1 (black) and 2:1 (green) have been deposited, spectra of bare Ag and Au particles are shown in blue and red, respectively. The LSPR peak maximum shifts from 447 nm (Ag) to 555 nm (Au), depending on the Au content. Reprinted from [92], licensed under CC BY 4.0

electromagnetic field close by a nanoparticle or, in particular, at gaps between adjacent particles. Raman scattering is typically very weak, however, the field enhancement effect can boost the local field intensity by orders of magnitude such that also the intensity of the Raman scattered light is increased. It has been shown that this effect also applies for the small 5 nm diameter nanoparticles deposited on glass slides with the helium droplet synthesis technique [92]. An example of SERS spectra recorded for such nanoparticles using a 532 nm laser, functionalized ex-situ with 4-metyhlbenzenethiol (4-MBT) molecules, is presented in Fig. 11.19. Ag nanoparticles, which exhibit a very strong plasmon resonance, give rise to the strongest SERS signal. For the other Ag:Au ratios, the SERS intensity decreases with increasing Au content.

Electron energy loss spectroscopy (EELS) provides an alternative and very powerful approach for the study of plasmons in deposited nanoparticles using a scanning transmission electron microscope [100]. In this technique, high energy electrons are accelerated towards a target. The majority of electrons are elastically scattered and may be collected, for example, by a high-angle annular dark field (HAADF) detector, which can create images with atomic resolution. However, some electrons are inelastically scattered, whereby different type of processes can give rise to an energy-loss. The excitation of plasmon modes in a nanoparticle is an example of an inelastic scattering interaction, which is revealed by electron-energy loss spectra in the low energy-loss regime [137]. In addition to optically allowed dipole modes, as, for example, seen in Fig. 11.18, EELS provides also access to dark modes such as quadrupole or bulk plasmon modes [138]. An important criterium that has to be

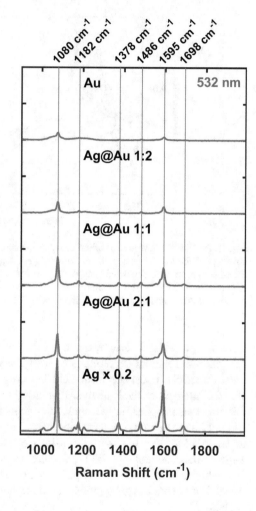

**Fig. 11.19** Surface enhanced Raman spectra recorded using Ag, Au and Ag@Au nanoparticles on a glass coverslip, functionalized with 4-MBT molecules. Different Ag:Au ratios have been adjusted, the spectra have been recorded using a 532 nm laser. Reprinted from [92] licensed under CC BY 4.0

fulfilled by the substrate is that it has to be transparent for electrons in the low-loss region where plasmon excitations are typically observed. Thus, ultra-thin substrates are favorable for this technique. The deposition of nanoparticles on very thin substrates, however, is a specialty of the helium droplet method due to the soft deposition process. Consequently, the method has been employed to deposit nanoparticles on hexagonal boron-nitride (hBN) substrates with a thickness of only a few atomic layers, which has been found to be an excellent substrate for plasmon spectroscopy with a scanning transmission electron microscope [100]. By scanning the electron beam across a selected nanoparticle, EELS spectra can be recorded for each pixel of the image and a plasmon map can be created. An example of such an EELS map is presented in Fig. 11.20, recorded for a Ag@Au core@shell nanorod (7 × 20 nm) [100]. In this case, the spatial distribution of the intensity within a selected interval (1.95 eV ± 0.09 eV) of the EELS spectrum, which corresponds to the longitudinal

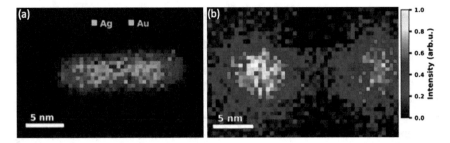

**Fig. 11.20** Elemental map (**a**) and plasmon map (**b**) of a Ag@Au ($7 \times 20\,nm$) nanoparticle, created from electron energy loss spectra (EELS) recorded in a scanning transmission electron microscope. The plasmon map shows the spatial EELS intensity distribution for a selected interval ($1.95\,eV \pm 0.09\,eV$) around the dipolar plasmon mode of the particle in panel (**a**). Images from [100], reprinted with the permission of AIP Publishing

dipolar plasmon mode of this nanoparticle, has been plotted. EELS spectra also provide element sensitive information in the high-loss regime. With this information the elemental map shown in Fig. 11.20a has been created.

### 11.3.3  Metal Nanoparticles and Molecules

Infrared spectroscopy experiments have been performed in order to explore the interaction between dopant molecules and metal nanoparticles. These experiments have been carried out for Ag@ethane clusters [139, 140] as well as Ag clusters surrounded by methane, ethylene and acetylene shells [141]. The position of the band frequencies resembles the spectra of crystalline samples rather than gaseous molecules. Consequently, at the droplet temperature of 0.4 K, these dopants are considered to be solid. By inspecting the C–H stretch modes of these hydrocarbon molecules, it can be distinguished between molecules attached to a Ag surface and molecules in the dopant molecule volume. The formation of a shell layer around the Ag particle can be followed with this approach: If only a few molecules are added, features corresponding to Ag–molecule aggregates are observed. Once the first shell layer has been established, the addition of molecules gives rise to volume-like features, slightly shifted to features originating from the first layer. Interestingly, it has been shown that the helium droplet approach enables also the synthesis of ethane@Ag [140, 142], i.e. an ethane core surrounded by a Ag shell, a system which can be stabilized in the low-temperature He droplet environment.

### 11.3.4  Beyond Two-Component Core@shell Nanoparticles

The synthesis of core@shell particles is often claimed as one of the key advantages of the helium droplet synthesis approach. The formation of core@shell nanoparticles in helium droplets has, so far, focused on the combination of two different materials. However, recently it has been demonstrated that also more than two species can be combined in a core@shell@shell configuration [37].

In these experiments the structure of the formed particles has been probed using laser induced fluorescence (LIF) spectroscopy, employing rhodamine B (RB) fluorophores as reporter molecules. The experiment is sketched in Fig. 11.21: Large helium nanodroplets, consisting of about $10^{10}$ He atoms per droplet on average, enter the pickup region where they first pass a Au pickup oven. Au nanoparticles nucleate at vortices were they form filament structures at high doping levels. Subsequently, the droplet beam passes a gas pickup cell used to form an intermediate layer of dopants that surrounds the Au particles. Different species, which are expected to be solid in the cold droplet, have been tested, including hexane, Ar and isopropyl alcohol. In a third pickup region, RB molecules are deposited in the droplets using a resistively heated pickup oven. The molecules are then excited with a 532 nm laser, resonant to the $\pi$-$\pi^*$ transition of the dye molecule and the laser induced fluorescence signal is recorded with a spectrometer.

Fig. 11.22a shows a compilation of recorded LIF spectra. A prominent fluorescence peak with a maximum at about 590 nm can be identified. The blue spectrum provides a reference for bare RB complexes in helium droplets. The red spectrum in Fig. 11.22a is obtained for Au@RB particles without an intermediate shell layer. It is evident that the fluorescence signal is quenched when Au is added to the droplets, the shape of the feature is not affected. However, if a hexane layer is inserted between the Au core and the RB shell, the fluorescence signal increases again, which can

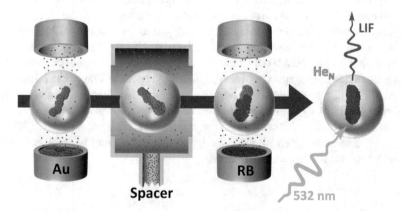

**Fig. 11.21** Sketch of a setup with three pickup zones used for the fabrication of core@shell@shell particles in helium droplets. Adapted from [37], published under CC-BY license

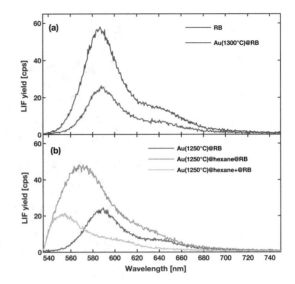

be seen in Fig. 11.22b, orange spectrum. This is explained by the formation of an isolating hexane layer around the Au core, which inhibits direct contact between RB molecules and the Au metal. The integrated fluorescence yield obtained for the orange Au@hexane@RB spectrum is a factor of two higher than for the blue Au@RB spectrum. With increasing hexane pickup level, the fluorescence decreases again (green curve), accompanied by a red-shift of the peak.

The enhancement of the fluorescence signal upon the addition of the intermediate layer demonstrates that the hexane molecules indeed form a shell around the Au core and that core@shell@shell nanostructures are formed.

## 11.4  Metal Oxide Nanoparticles

Metal oxide nanoparticles possess interesting properties and many species are technologically relevant with widespread industrial applications [143]. For the production of metal oxide nanoparticles with helium droplets, two different recipes have been followed in previous experiments.

The first and obvious approach is based on the direct evaporation of metal oxides in a pickup oven. However, this can be difficult as many metal oxides have very high melting points and require, thus, extremely high pickup temperatures. For $V_2O_5$ it has been shown that helium droplets can be directly doped with metal oxides. The required pickup oven temperature is on the order of about 900°C [144]. An advantage of the method is, that in helium droplets the ionization process is typically less destructive than compared to direct electron impact ionization of complexes in the gas phase [24, 145, 146]. This is explained by the indirect ionization mechanism, which

544                                                                          F. Lackner

**Fig. 11.23** High-angle
annular dark-field (HAADF)
image of $V_2O_5$ nanoparticles
recorded by scanning
transmission electron
microscopy. Reprinted from
[147]. Published by the
PCCP Owner Societies,
licensed under CC-BY 3.0

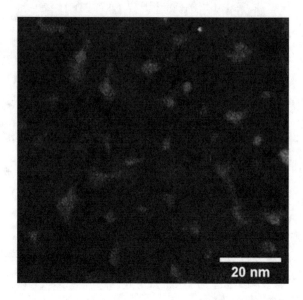

20 nm

proceeds via charge hopping, and the high cooling rate provided by the liquid helium
matrix. Thus, the time-of-flight mass spectra recorded for $(V_2O_5)_n$ oligomers were
not dominated by fragments. The results revealed that $(V_2O_5)_n$ oligomers sublimate
preferentially in form of complexes with even $n$, which represent the dominant species
observed in the mass spectra [144].

In addition to the study of small $(V_2O_5)_n$ oligomers, it has also been possible to
form $V_2O_5$ nanoparticles in helium droplets, which have been deposited and analyzed
by scanning transmission electron microscopy [147]. Figure 11.23 shows an example
of the formed vanadium oxide nanoparticles. Visual inspection suggests that the
particles are larger than typical bare metal nanoparticles formed at these experimental
conditions even though the surface coverage is low such that coagulation effects are
not expected. This may be a hint for a strong interaction between substrate and
particles in this case.

A second approach for the synthesis of metal oxide nanoparticles is the controlled
exposure of bare metal particles to oxygen after deposition [35]. The exposure of
nanoparticles to ambient air is often unavoidable, for example, in order to transport samples to an electron microscope or other experimental setups. Note that the
oxidation of deposited nanoparticles can influence their structure, which has to be
considered for subsequent experiments [29, 30].

In a recent study, nanoparticles formed in helium droplets have been deliberately oxidized by exposure to oxygen after deposition in order to fabricate Ag@ZnO
core@shell particles [35]. In these experiments, the droplets were first doped with
Ag and, subsequently, with Zn in order to form Ag@Zn nanoparticles, which have
been deposited on indium tin oxide (ITO) substrates. Ultraviolet photoelectron spectroscopy (UPS) has been employed to trace the Zn oxidation process at the Zn 3d
core-level peak around 10 eV binding energy, which is accessible with standard He

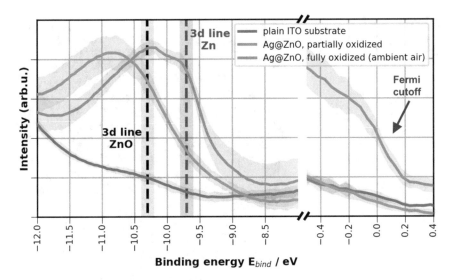

**Fig. 11.24** UPS spectra of Ag@ZnO nanoparticles on ITO substrates. Partially oxidized particles (orange) are compared to particles exposed to air for 1 h (green) and the plain ITO substrate (blue). A shift of the Zn 3d core level peak is observed, accompanied by a decrease of the signal at the Fermi cutoff. Adapted from [35] licensed under CC BY 4.0

discharge light sources. Figure 11.24 shows a comparison between UPS spectra of slightly oxidized nanoparticles (orange) and nanoparticles exposed to ambient air for about 1 h (green). The shift of the 3d core level peak reflects the transition from Ag@Zn with a partially oxidized Zn shell to Ag@ZnO particles with a fully oxidized shell layer. This is accompanied by the disappearance of the signal at the Fermi cutoff in the spectrum.

## 11.4.1 Determination of Oxidation States

Many different approaches have been applied in the past to analyze the oxidation state of metal oxide nanoparticles synthesized with the helium droplet technique.

Scanning transmission electron microscopy (STEM) is an excellent approach for this purpose. From STEM images with atomic resolution, lattice constants can be obtained, which allows for a determination of the oxidation state of a material based on a comparison to tabulated literature values. An example is shown in Fig. 11.25a, which corresponds to a high-resolution STEM high-angle annular dark-field (HAADF) image of a Ag@ZnO nanoparticle for which the hexagonal lattice structure of the ZnO is well resolved [35]. The corresponding ZnO wurtzite unit cell is shown in Fig. 11.25c. Two strategies can be followed for the determination of lattice constants. If many lattices are identified by visual inspection, an outline of the

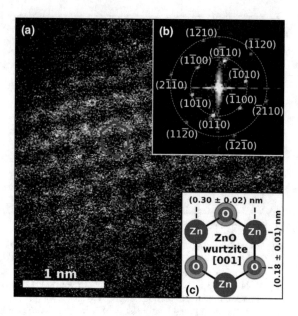

**Fig. 11.25** Real space scanning transmission electron microscopy image of the ZnO shell of a Ag@ZnO nanoparticle (**a**). The wurtzite structure can be identified in the image, the corresponding unit cell with lattice constants is shown in panel (**c**). The 2D Fourier transform of the image is shown in the inset panel (**b**), the characteristic lattice reflexes, which reflect the hexagonal ZnO structure, are visible. Adapted from [35], licensed under CC BY 4.0

image contrast enables a direct measurement of the lattice spacing from the image [147]. However, an analysis of lattice reflexes in Fourier space is often advisable, in particular, in case of a lower image quality [101]. A 2D Fourier transform of the image that shows the ZnO shell of a particle in Fig. 11.25a is presented in the inset in panel (b). Many lattice reflexes can be identified and assigned based on comparison to bulk values. Note, however, that the lattice constants obtained for nanoscale objects can deviate from their macroscopic counterparts. A method related to the analysis of Fourier transformed STEM images is X-ray powder diffraction (XRD), which may also be employed in future for an analysis of nanoparticles formed and deposited with helium droplets. Furthermore, EELS spectra can provide insight into the oxidation state of deposited nanoparticles [30, 147].

In addition to methods based on transmission electron microscopy, UV/vis extinction spectra provide information on the oxidation state of deposited nanoparticles. This approach has, for example, been used for deposited $V_2O_5$ particles such as shown in Fig. 11.23 [147]. For vanadium oxides, an identification of the oxidation state is complicated due to the large number of possibilities. However, only the vanadium pentoxide is transparent in the visible regime. Thus, the absence of absorption features in the visible supports an interpretation of $V_2O_5$ as dominant species on the substrate.

Ultraviolet photoelectron spectroscopy (UPS) has already been introduced above (Fig. 11.24) as method sensitive to oxidation processes [35]. X-ray photoelectron spectroscopy can also be used to characterize the oxidation state of a material, as has been done, for example, for deposited Ni [18] and Al [33] containing nanoparticles synthesized in helium droplets. X-ray absorption near edge structures (XANES) can also provide insight into the presence of oxides but the method requires access to synchrotron facilities [32].

An aspect that has to be considered in this context is that mixtures of different oxides and bare metals may be obtained by helium droplet synthesis. This is particularly relevant in the case of an oxidation of particles by exposure to ambient air after deposition. For Co nanoparticles, for example, this resulted in the formation of partially oxidized CoO particles that contain areas with bare Co metal [101].

## 11.4.2 Oxidation Experiments with Deposited Metal Nanoparticles

Helium droplet synthesis enables the formation and deposition of small sub-10 nm nanoparticles under ultrahigh vacuum (UHV) conditions. The subsequent exposure of the deposited particles to oxygen or ambient air allows for an investigation of oxidation processes at the nanoscale.

The mechanism that drives the oxidation can be different in such small nanoparticles than compared to larger structures or the bulk material. This has been subject to an interesting experiment on Al and Au@Al nanoparticles [33]. The results indicate that the reaction between small Al clusters and oxygen involves an etching process whereas the oxidation of larger Al nanoparticles proceeds via heterogeneous oxidation. For small particles, this leads to a rapid generation of heat and an ejection of $Al_2O$ products, which goes along with a destruction of the particles. Consequently, small aluminum oxide nanoparticles, with a diameter below 4 nm, are not found in scanning transmission electron microscopy (STEM) images. The same phenomenon is observed for small Au@Al nanoparticles. An example of such Au@Al particles, initially equipped with a 1 nm Au core, after deposition and exposure to ambient air, is shown in Fig. 11.26b. Only residues of the deposited particles are found. Intact Au@Al structures with an oxidized Al shell have only been observed for particles with Au core sizes exceeding 4 nm diameter. An example can be seen in Fig. 11.26a, which shows intact Au@Al nanoparticles after exposure to ambient air with a 5 nm Au core and a 1 nm shell layer. Note that in this case the number of Al atoms is smaller than required to build a plain Al particle with 4 nm diameter. The observed transition between the two regimes at around 4 nm coincides with the particle size where a drastic change of the coordination number occurs. Consequently, beyond 4 nm particle diameter the released energy is dispersed among a larger number of atoms and the particles can survive the oxidation process.

For the Ni@Au system it has been demonstrated that oxidation can cause a structural inversion. This results in the formation of Au@NiO particles [29, 121], triggered by the diffusion of Ni atoms to the surface in the presence of oxygen, which lowers the barrier for diffusion. Note that a similar inversion process can be induced by the electron beam in an electron microscope [148].

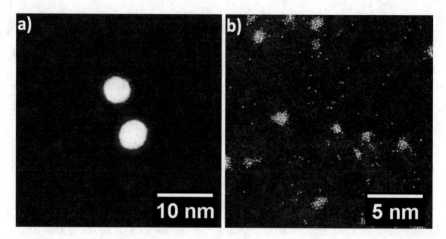

**Fig. 11.26** Au@Al core@shell nanoparticles with different Au core sizes after deposition and exposure to ambient air. While particles with large Au cores (5 nm) surrounded by a 1 nm Al shell, as shown in panel (**a**), retain their integrity, the oxidation process leaves behind scattered Au complexes for small Au cores (1 nm), as seen in panel (**b**). Adapted with permission from [33]. Copyright 2019 American Chemical Society

A study of deposited Fe@Au core@shell particles [30], which have been exposed to ambient air, revealed that in this case both the structure as dictated by the pickup sequence and the inverted structure are observed at room-temperature in scanning transmission electron microscopy images. Even stable Fe@Au@Fe-oxide particles are present on the substrate. Furthermore, indications for Janus-like particles, for which areas with Fe oxide and Au coexist next to each other, have been found. Examples for small spherical Fe@Au and Fe@Au@Fe-oxide particles are presented in Fig. 11.27. These experiments revealed that a critical Au shell thickness of only 2–3 layers Au is required to protect the Fe core from oxidation.

Another material for which a structural inversion has been observed is the Cu-Mg system [34, 90]. In these experiments, the Mg pickup cell was passed by the helium droplets prior to the Cu pickup oven, however, the resulting elemental maps showed structures with Cu core particles surrounded by a shell layer of MgO. Interestingly, the Cu cores did not oxidize in these experiments.

These findings have to be kept in mind when designing nanoparticle synthesis experiments. Helium droplet synthesis is typically considered as a unique tool that allows for a combination of a sheer unlimited amount of materials simply by changing the dopant material in the pickup cells. While this may hold as long as particles are in the ultra-high vacuum (UHV), upon exposure to air, oxidation effects can influence the structure of the deposited particles or even cause their complete disintegration.

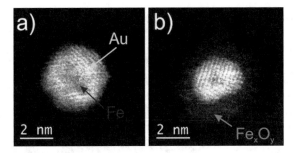

**Fig. 11.27** Example of Fe@Au particles deposited on an amorphous carbon TEM grid. Panel **a** shows a particle with an intact Fe core protected from oxidation by the Au layer. Panel **b** shows a particle that was subject to incomplete oxidation, which resulted in a Fe@Au@Fe-oxide structure. Adapted from [30], licensed under CC BY 3.0. Published by the Royal Society of Chemistry

## 11.4.3 Metal Core—Transition Metal Oxide Shell Nanoparticles

The ability to select the core and shell species from a wide variety of materials opens the door for the creation of novel nanostructures with tailored properties. A recent example is the combination of plasmonic Ag core particles with ZnO shells [35]. Figure 11.28a shows a high-angle annular dark-field (HAADF) image of spherical Ag@ZnO nanoparticles with a diameter of about 5.8 nm and Ag core diameters around 3 nm. The Zn shell has been oxidized after the particles were deposited by exposure to ambient air for about 1 h. Interestingly, this procedure resulted in the formation of ZnO shells with a very uniform layer thickness of 1.3 nm, which fully cover the Ag cores. Similar results have been obtained for Ag@ZnO nanowire structures, for which also a very uniform ZnO shell was observed, with a thickness of 1.6 nm [35]. The presence of Ag and ZnO in form of core@shell structure is confirmed by elemental maps created from energy dispersive X-ray (EDX) spectra acquired using a scanning transmission electron microscope (STEM). Figure 11.28b shows the Ag distribution, which correlates very well with the bright cores in the corresponding HAADF image (a). Panels (c) and (d) show the Zn and O distribution, respectively. The Zn rich areas are clearly larger than the Ag cores, indicating that Zn is present in a shell layer. The distribution of oxygen is less well defined, pointing at a typical issue of such elemental maps: If the sample has been exposed to air, a small amount of oxygen can be found everywhere, on the nanoparticles as well as on the substrate. Another typical contaminant is carbon in form of organic molecules, which is often encountered in scanning transmission electron microscopy experiments. Thus, care has to be taken when the oxygen or carbon distribution within such images is analyzed.

An intriguing characteristic of particles that comprise a small spherical Ag core and a ZnO shell layer is that such a structure combines a plasmonic material with a localized surface plasmon resonance (LSPR) and a semiconducting material. The

**Fig. 11.28** High-angle
annular dark-field (HAADF)
images (**a**) of Ag core @
ZnO shell particles recorded
by scanning transmission
electron microscopy
(STEM). The Ag appears as
a bright core in the center of
the particles, surrounded by
a darker ZnO shell layer.
Spatially resolved EDX
maps recorded for the same
sample area as shown in (**a**)
reveal the distribution of Ag
(**b**), Zn (**c**) and O (**d**).
Reprinted from [35],
licensed under CC BY 4.0

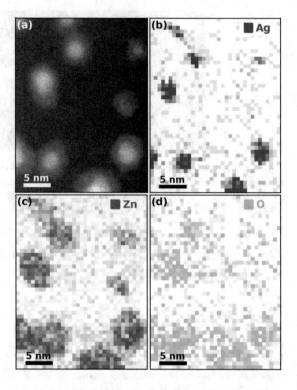

combination of plasmonic metal nanoparticles with transition metal oxides may
bear great potential for the enhancement of the efficiency of devices that harvest
solar energy [149–151]. In order to investigate if such a nanoscale Ag core preserves its plasmon resonance while inside the ZnO shell, two-photon photoelectron
(2PPE) spectroscopy experiments have been carried out [35]. These experiments
made use of a NanoESCA energy-filtered photoemission electron microscope (EF-
PEEM) [152]. The results are shown in Fig. 11.29, obtained using a 3.02 eV (410 nm)
p-polarized laser resonant to the LSPR of the Ag particles. Plain Ag nanoparticles
(orange spectrum) show a strong enhancement, giving rise to the emergence of electrons with high kinetic energies. Furthermore, a well defined Fermi level cutoff at
about $E_{kin} = 1.9$ eV can be identified in the 2PPE spectrum, which is characteristic
for metals. Plain ZnO particles, on the other hand, only show a strong secondary
electron peak at low kinetic energies, with only little high energy electrons and no
signal at the Fermi cutoff (blue spectrum). If the ZnO particles are equipped with
a Ag core, the spectrum (green) resembles the plain Ag spectrum, indicating that
the resulting Ag@ZnO particles combine the plasmonic enhancement properties of
plain Ag particles with the semiconducting ZnO material.

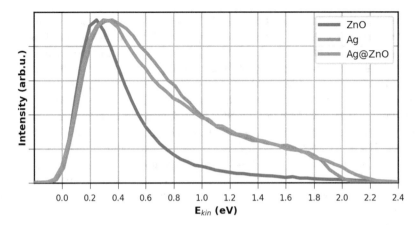

**Fig. 11.29** Two-photon photoelectron (2PPE) spectra of ZnO (blue), Ag (orange), and Ag@ZnO (green) nanoparticles deposited on an ITO substrate. The significant increase of the electron yield at higher kinetic energies is attributed to an enhancement by the excitation of Ag surface plasmons. Reprinted from [35], licensed under CC BY 4.0

## 11.5  Outlook

Within the past decade, the helium droplet synthesis approach has evolved into a versatile tool for the production of small nanoparticles. Today, helium droplets are routinely doped with metals and experimental strategies for the production of metal oxide particles have been presented. The advantage of the method is that it provides an inert synthesis environment, which enables the fabrication of very clean nanoparticles. The assembling of nanostructures by the doping of helium droplets in an atom-by-atom manner allows for a control of the particle size, ranging from small clusters that comprise only a few atoms to spherical particles and elongated nanowires. The deposition of particles is very gentle due to the soft-landing conditions ensured by the cushioning effect due to the liquid helium droplet that evaporates completely in the course of the process. A major advantage is the combination of different materials in form of core@shell particles with sizes in the sub-10 nm regime. Dopants can be easily exchanged and the core and shell layer material are defined by the pickup sequence. On the downside, the method is not scaleable and can only produce nanomaterials in the sub-milligram mass regime. The amount of deposited material, however, is sufficient for the study of fundamental properties of nanoparticles.

Future experiments will exploit the unique advantages offered by the helium droplet synthesis approach. The sub-10 nm particle size regime, where the method excels, is very interesting for applications, for example, in catalysis [29, 102]. For such small nanoparticles, size effects become important and their properties can be very different from the bulk material [109]. A versatile method such as helium droplet synthesis could be used to rapidly explore different dopant combinations in order to search for new nanomaterials. Small nanoparticles synthesized with helium droplets can also have special magnetic properties [32]. Reactive materials, such as aluminum or alkali metals, possess interesting optical properties but are difficult to handle by conventional synthesis methods [153]. However, these materials can be isolated and investigated very well in the inert liquid helium droplet environment, the deposition takes place under ultra-high vacuum (UHV) conditions. The synthesis of metal core—transition metal oxide shell particles has just recently been demonstrated with the approach [35], combining materials of interest for plasmonics and photo-catalysis [149–151]. A deposition of such nanoparticles on ultra-thin substrates may enable the investigation of mechanisms related to plasmon decay or charge-carrier dynamics with advanced time-resolved XUV spectroscopy methods [101, 154]. Furthermore, the possibility of combining multiple materials with selected properties [37] in a helium droplet opens new perspectives for the creation of unique tailored nanomaterials.

**Acknowledgements** The author thanks Wolfgang E. Ernst for his support over the past decade and the years of joint research. Stimulating discussions with Wolfgang E. Ernst and Roman Messner are gratefully acknowledged. This work has been supported by the Austrian Science Fund (FWF) grant P30940 and NAWI Graz.

# References

1. J.P. Toennies, A.F. Vilesov, Superfluid helium droplets: a uniquely cold nanomatrix for molecules and molecular complexes. Angew. Chem. Int. **43**(20), 2622–2648 (2004)
2. C. Callegari, W.E. Ernst, Helium droplets as nanocryostats for molecular spectroscopy—from the vacuum ultraviolet to the microwave regime, in *Handbook of High-Resolution Spectroscopy*, ed. by M. Quack, F. Merkt (Wiley, Chichester, 2011), pp. 1551–1594
3. A. Scheidemann, J.P. Toennies, J.A. Northby, Capture of neon atoms by $^4$He clusters. Phys. Rev. Lett. **64**(16), 1899 (1990)
4. S. Goyal, D.L. Schutt, G. Scoles, Vibrational spectroscopy of sulfur hexafluoride attached to helium clusters. Phys. Rev. Lett. **69**(6), 933 (1992)
5. M. Lewerenz, B. Schilling, J.P. Toennies, Successive capture and coagulation of atoms and molecules to small clusters in large liquid helium clusters. J. Chem. Phys. **102**, 8191–8207 (1995)
6. F. Stienkemeier, J. Higgins, W.E. Ernst, G. Scoles, Spectroscopy of alkali atoms and molecules attached to liquid He clusters. Z. Phys. B **98**(3), 413–416 (1995)
7. J. Higgins, C. Callegari, J. Reho, F. Stienkemeier, W.E. Ernst, K.K. Lehmann, M. Gutowski, G. Scoles, Photoinduced chemical dynamics of high-spin alkali trimers. Science **273**(5275), 629–631 (1996)
8. J. Higgins, W.E. Ernst, C. Callegari, J. Reho, K.K. Lehmann, G. Scoles, M. Gutowski, Spin polarized alkali clusters: observation of quartet states of the sodium trimer. Phys. Rev. Lett. **77**(22), 4532 (1996)
9. K. Nauta, R.E. Miller, Formation of cyclic water hexamer in liquid helium: the smallest piece of ice. Science **287**(5451), 293–295 (2000)

10. K. Nauta, R.E. Miller, Nonequilibrium self-assembly of long chains of polar molecules in superfluid helium. Science **283**(5409), 1895–1897 (1999)
11. K. Nauta, D.T. Moore, P.L. Stiles, R.E. Miller, Probing the structure of metal cluster-adsorbate systems with high-resolution infrared spectroscopy. Science **292**(5516), 481–484 (2001)
12. A. Bartelt, J.D. Close, F. Federmann, N. Quaas, J.P. Toennies, Cold metal clusters: helium droplets as a nanoscale cryostat. Phys. Rev. Lett. **77**(17), 3525 (1996)
13. F. Federmann, K. Hoffmann, N. Quaas, J.P. Toennies, Spectroscopy of extremely cold silver clusters in helium droplets. Eur. Phys. J. D 11–14 (1999)
14. J. Tiggesbäumker, F. Stienkemeier, Formation and properties of metal clusters isolated in helium droplets. Phys. Chem. Chem. Phys. **9**(34), 4748–4770 (2007)
15. W.G. Kreyling, M. Semmler-Behnke, Q. Chaudhry, A complementary definition of nanomaterial. Nano Today **5**(3), 165–168 (2010)
16. V. Mozhayskiy, M.N. Slipchenko, V.K. Adamchuk, A.F. Vilesov, Use of helium nanodroplets for assembly, transport, and surface deposition of large molecular and atomic clusters. J. Chem. Phys. **127**(9), 094701 (2007)
17. E. Loginov, L.F. Gomez, A.F. Vilesov, Surface deposition and imaging of large Ag clusters formed in He droplets. J. Phys. Chem. A **115**(25), 7199–7204 (2011)
18. A. Boatwright, C. Feng, D. Spence, E. Latimer, C. Binns, A.M. Ellis, S. Yang, Helium droplets: a new route to nanoparticles. Faraday Discuss. **162**, 113–124 (2013)
19. S. Yang, A.M. Ellis, D. Spence, C. Feng, A. Boatwright, E. Latimer, C. Binns, Growing metal nanoparticles in superfluid helium. Nanoscale **5**(23), 11545–11553 (2013)
20. S.B. Emery, K.B. Rider, B.K. Little, C.M. Lindsay, Helium droplet assembled nanocluster films: cluster formation and deposition rates. J. Phys. Chem. C **117**(5), 2358–2368 (2013)
21. A. Volk, P. Thaler, M. Koch, E. Fisslthaler, W. Grogger, W.E. Ernst, High resolution electron microscopy of Ag-clusters in crystalline and non-crystalline morphologies grown inside superfluid helium nanodroplets. J. Chem. Phys. **138**(21), 214312 (2013)
22. P. Thaler, A. Volk, D. Knez, F. Lackner, G. Haberfehlner, J. Steurer, M. Schnedlitz, W.E. Ernst, Synthesis of nanoparticles in helium droplets: a characterization comparing mass-spectra and electron microscopy data. J. Chem. Phys. **143**(13), 134201 (2015)
23. R.M.P. Tanyag, C.F. Jones, C. Bernando, S.M. O'Connell, D. Verma, A.F. Vilesov, Experiments with large superfluid helium nanodroplets, in *Cold Chemistry: Molecular Scattering and Reactivity Near Absolute Zero*, ed. by O. Dulicu, A. Osterwalder. Theoretical and Computational Chemistry Series (Royal Society of Chemistry, 2017), pp. 389–443
24. A. Mauracher, O. Echt, A.M. Ellis, S. Yang, D.K. Bohme, J. Postler, A. Kaiser, S. Denifl, P. Scheier, Cold physics and chemistry: collisions, ionization and reactions inside helium nanodroplets close to zero K. Phys. Rep. **751**, 1–90 (2018)
25. W.E. Ernst, A.W. Hauser, Metal clusters synthesized in helium droplets: structure and dynamics from experiment and theory. Phys. Chem. Chem. Phys. **23**, 7553–7574 (2021). https://doi.org/10.1039/D0CP04349D
26. M. Hartmann, R.E. Miller, J.P. Toennies, A. Vilesov, Rotationally resolved spectroscopy of $SF_6$ in liquid helium clusters: a molecular probe of cluster temperature. Phys. Rev. Lett. **75**(8), 1566 (1995)
27. L.F. Gomez, E. Loginov, A.F. Vilesov, Traces of vortices in superfluid helium droplets. Phys. Rev. Lett. **108**(15), 155302 (2012)
28. E. Latimer, D. Spence, C. Feng, A. Boatwright, A.M. Ellis, S. Yang, Preparation of ultrathin nanowires using superfluid helium droplets. Nano Lett. **14**(5), 2902–2906 (2014)
29. M. Schnedlitz, M. Lasserus, R. Meyer, D. Knez, F. Hofer, W.E. Ernst, A.W. Hauser, Stability of core@shell nanoparticles for catalysis at elevated temperatures: structural inversion in the Ni@Au system observed at atomic resolution. Chem. Mater. **30**(3), 1113–1120 (2018)
30. M. Lasserus, D. Knez, M. Schnedlitz, A.W. Hauser, F. Hofer, W.E. Ernst, On the passivation of iron particles at the nanoscale. Nanoscale Adv. **1**(6), 2276–2283 (2019)
31. M. Schnedlitz, D. Knez, M. Lasserus, F. Hofer, R. Fernández-Perea, A.W. Hauser, M.P. de Lara-Castells, W.E. Ernst, Thermally induced diffusion and restructuring of iron triade (Fe Co, Ni) nanoparticles passivated by several layers of gold. J. Phys. Chem. C **124**(30), 16680–16688 (2020)

32. S. Yang, C. Feng, D. Spence, A.M. Al Hindawi, E. Latimer, A.M. Ellis, C. Binns, D. Peddis, S.S. Dhesi, L. Zhang, Y. Zhang, K.N. Trohidou, M. Vasilakaki, N. Ntallis, I. MacLaren, M.F. de Groot, Robust ferromagnetism of chromium nanoparticles formed in superfluid helium. Adv. Mater. **29**(1), 1604277 (2017)

33. K.R. Overdeep, C.J. Ridge, Y. Xin, T.N. Jensen, S.L. Anderson, C.M. Lindsay, Oxidation of aluminum particles from 1 to 10 nm in diameter: the transition from clusters to nanoparticles. J. Phys. Chem. C **123**(38), 23721–23731 (2019)

34. S.B. Emery, Y. Xin, C.J. Ridge, R.J. Buszek, J.A. Boatz, J.M. Boyle, C.M. Lindsay, Unusual behavior in magnesium-copper cluster matter produced by helium droplet mediated deposition. J. Chem. Phys. **142**(8), 084307 (2015)

35. A. Schiffmann, T. Jauk, D. Knez, H. Fitzek, F. Hofer, F. Lackner, W.E. Ernst, Helium droplet assisted synthesis of plasmonic Ag@ZnO core@shell nanoparticles. Nano Res. **13**(11), 2979–2986 (2020)

36. Section 12: Properties of Solids; Electrical Resistivity of Pure Metals, in *Handbook of Chemistry and Physics*, 84th edn., ed. by David R. Lide (CRC Press, Boca Raton, Florida, 2003)

37. R. Messner, W.E. Ernst, F. Lackner, Shell-isolated Au nanoparticles functionalized with rhodamine B fluorophores in helium nanodroplets. J. Phys. Chem. Lett. **12**(1), 145–150 (2021)

38. M. Ratschek, M. Koch, W.E. Ernst, Doping helium nanodroplets with high temperature metals: formation of chromium clusters. J. Chem. Phys. **136**(10), 104201 (2012)

39. A. Kautsch, M. Koch, W.E. Ernst, Photoinduced molecular dissociation and photoinduced recombination mediated by superfluid helium nanodroplets. Phys. Chem. Chem. Phys. **17**(18), 12310–12316 (2015)

40. M. Koch, A. Kautsch, F. Lackner, W.E. Ernst, One-and two-color resonant photoionization spectroscopy of chromium-doped helium nanodroplets. J. Phys. Chem. A **118**(37), 8373–8379 (2014)

41. S.A. Krasnokutski, F. Huisken, Oxidative reactions of silicon atoms and clusters at ultralow temperature in helium droplets. J. Phys. Chem. A **114**(50), 13045–13049 (2010)

42. S.A. Krasnokutski, F. Huisken, A simple and clean source of low-energy atomic carbon. Appl. Phys. Lett. **105**(11), 113506 (2014)

43. S.A. Krasnokutski, M. Kuhn, A. Kaiser, A. Mauracher, M. Renzler, D.K. Bohme, P. Scheier, Building carbon bridges on and between fullerenes in helium nanodroplets. J. Phys. Chem. Lett. **7**(8), 1440–1445 (2016)

44. M. Mudrich, B. Forkl, S. Müller, M. Dvorak, O. Bünermann, F. Stienkemeier, Kilohertz laser ablation for doping helium nanodroplets. Rev. Sci. Inst. **78**(10), 103106 (2007)

45. W.K. Lewis, B.A. Harruff-Miller, P. Leatherman, M.A. Gord, C.E. Bunker, Helium droplet calorimetry of strongly bound species: carbon clusters from $C_2$ to $C_{12}$. Rev. Sci. Inst. **85**(9), 094102 (2014)

46. J. Jeffs, N.A. Besley, A.J. Stace, G. Sarma, E.M. Cunningham, A. Boatwright, S. Yang, A.M. Ellis, Metastable aluminum atoms floating on the surface of helium nanodroplets. Phys. Rev. Lett. **114**(23), 233401 (2015)

47. R. Katzy, M. Singer, S. Izadnia, A.C. LaForge, F. Stienkemeier, Doping He droplets by laser ablation with a pulsed supersonic jet source. Rev. Sci. Inst. **87**(1), 013105 (2016)

48. F. Bierau, P. Kupser, G. Meijer, G. von Helden, Catching proteins in liquid helium droplets. Phys. Rev. Lett. **105**(13), 133402 (2010)

49. M. Alghamdi, J. Zhang, A. Oswalt, J.J. Porter, R.A. Mehl, W. Kong, Doping of green fluorescent protein into superfluid helium droplets: size and velocity of doped droplets. J. Phys. Chem. A **121**(36), 6671–6678 (2017)

50. J. Küpper, J.M. Merritt, R.E. Miller, Free radicals in superfluid liquid helium nanodroplets: a pyrolysis source for the production of propargyl radical. J. Chem. Phys. **117**(2), 647–652 (2002)

51. A.M. Morrison, J. Agarwal, H.F. Schaefer III., G.E. Douberly, Infrared laser spectroscopy of the $CH_3OO$ radical formed from the reaction of $CH_3$ and $O_2$ within a helium nanodroplet. J. Phys. Chem. A **116**(22), 5299–5304 (2012)

52. N.B. Brauer, S. Smolarek, E. Loginov, D. Mateo, A. Hernando, M. Pi, M. Barranco, W.J. Buma, M. Drabbels, Critical Landau velocity in helium nanodroplets. Phys. Rev. Lett. **111**(15), 153002 (2013)
53. A. Volk, P. Thaler, D. Knez, A.W. Hauser, J. Steurer, W. Grogger, F. Hofer, W.E. Ernst, The impact of doping rates on the morphologies of silver and gold nanowires grown in helium nanodroplets. Phys. Chem. Chem. Phys. **18**(3), 1451–1459 (2016)
54. O. Bünermann, G. Droppelmann, A. Hernando, R. Mayol, F. Stienkemeier, Unraveling the absorption spectra of alkali metal atoms attached to helium nanodroplets. J. Phys. Chem. A **111**(49), 12684–12694 (2007)
55. Y. Ren, V.V. Kresin, Surface location of alkaline-earth-metal-atom impurities on helium nanodroplets. Phys. Rev. A **76**(4), 043204 (2007)
56. S. Vongehr, A.A. Scheidemann, C. Wittig, V.V. Kresin, Growing ultracold sodium clusters by using helium nanodroplets. Chem. Phys. Lett. **353**(1–2), 89–94 (2002)
57. S. Müller, S. Krapf, T. Koslowski, M. Mudrich, F. Stienkemeier, Cold reactions of alkali-metal and water clusters inside helium nanodroplets. Phys. Rev. Lett. **102**(18), 183401 (2009)
58. M. Theisen, F. Lackner, W.E. Ernst, Rb and Cs oligomers in different spin configurations on helium nanodroplets. J. Phys. Chem. A **115**(25), 7005–7009 (2011)
59. C. Stark, V.V. Kresin, Critical sizes for the submersion of alkali clusters into liquid helium. Phys. Rev. B **81**(8), 085401 (2010)
60. L. An der Lan, P. Bartl, C. Leidlmair, H. Schöbel, R. Jochum, S. Denifl, T.D. Märk, A.M. Ellis, P. Scheier, The submersion of sodium clusters in helium nanodroplets: identification of the surface → interior transition. J. Chem. Phys. **135**(4), 044309 (2011)
61. L.A. der Lan, P. Bartl, C. Leidlmair, H. Schöbel, S. Denifl, T.D. Märk, A.M. Ellis, P. Scheier, Submersion of potassium clusters in helium nanodroplets. Phys. Rev. B **85**(11), 115414 (2012)
62. C. Kittel, Einführung in die Festkörperphysik (6. Auflage), R. Oldenbourg Verlag München Wien, 1983)
63. M.D. Morse, Clusters of transition-metal atoms. Chem. Rev. **86**(6), 1049–1109 (1986)
64. A.W. Hauser, A. Volk, P. Thaler, W.E. Ernst, Atomic collisions in suprafluid helium-nanodroplets: timescales for metal-cluster formation derived from He-density functional theory. Phys. Chem. Chem. Phys. **17**(16), 10805–10812 (2015)
65. L.F. Gomez, E. Loginov, R. Sliter, A.F. Vilesov, Sizes of large He droplets. J. Chem. Phys. **135**(15), 154201 (2011)
66. C.F. Jones, C. Bernando, S. Erukala, A.F. Vilesov, Evaporation dynamics from Ag-doped He droplets upon laser excitation. J. Phys. Chem. A **123**(28), 5859–5865 (2019)
67. L.F. Gomez, S.M. O'Connell, C.F. Jones, J. Kwok, A.F. Vilesov, Laser-induced reconstruction of Ag clusters in helium droplets. J. Chem. Phys. **145**(11), 114304 (2016)
68. E. Loginov, L.F. Gomez, N. Chiang, A. Halder, N. Guggemos, V.V. Kresin, A.F. Vilesov, Photoabsorption of $Ag_N$ (N~6-6000) nanoclusters formed in helium droplets: transition from compact to multicenter aggregation. Phys. Rev. Lett. **106**(23), 233401 (2011)
69. G. Haberfehlner, P. Thaler, D. Knez, A. Volk, F. Hofer, W.E. Ernst, G. Kothleitner, Formation of bimetallic clusters in superfluid helium nanodroplets analysed by atomic resolution electron tomography. Nat. Commun. **6**(1), 1–6 (2015)
70. O. Gessner, A.F. Vilesov, Imaging quantum vortices in superfluid helium droplets. Annu. Rev. Phys. Chem. **70**, 173–198 (2019)
71. L.F. Gomez, K.R. Ferguson, J.P. Cryan, C. Bacellar, R.M.P. Tanyag, C.J. Jones, S. Schorb, D. Anielski, A. Ali Belkacem, C. Bernando, R. Boll, J. Bozek, S. Carron, G. Chen, T. Delmas, L. Englert, S.W. Epp, B. Erk, L. Foucar, R. Hartmann, A. Hexemer, M. Huth, J. Kwok, S.R. Leone, J.H.S. Ma, F.R.N.C. Maia, E. Malmerberg, S. Marchesini, D.M. Neumark, B. Poon, J. Prell, D. Rolles, B. Rudek, A. Rudenko, M. Seifrid, K.R. Siefermann, F.P. Sturm, M. Swiggers, J. Ullrich, F. Weise, P. Zwart, C. Bostedt, O. Gessner, A.F. Vilesov, Shapes and vorticities of superfluid helium nanodroplets. Science **345**(6199), 906–909 (2014)
72. R.M.P. Tanyag, C. Bernando, C.F. Jones, C. Bacellar, K.R. Ferguson, D. Anielski, R. Boll, S. Carron, J.P. Cryan, L. Englert, S.W. Epp, B. Erk, L. Foucar, L.F. Gomez, R. Hartmann, D.M. Neumark, D. Rolles, B. Rudek, A. Rudenko, K.R. Siefermann, J. Ullrich, F. Weise, C.

Bostedt, O. Gessner, A.F. Vilesov, Communication: X-ray coherent diffractive imaging by immersion in nanodroplets. Struct. Dyn. **2**, 051102 (2015)

73. R.M.P. Tanyag, B. Langbehn, D. Rupp, T. Möller, X-ray and XUV imaging of helium nanodroplets, in *Molecules in Superfluid Helium Nanodroplets: Spectroscopy, Structures and Dynamics*, ed. by A. Slenczka, J.P. Toennies. Topics in Applied Physics 145 (2022)

74. S.M. O'Connell, R.M.P. Tanyag, D. Verma, C. Bernando, W. Pang, C. Bacellar, C.A. Saladrigas, J. Mahl, B.W. Toulson, Y. Kumagai, P. Walter, Angular momentum in rotating superfluid droplets. Phys. Rev. Lett. **124**(21), 215301 (2020)

75. F. Ancilotto, M. Pi, M. Barranco, Vortex arrays in nanoscopic superfluid helium droplets. Phys. Rev. B **91**(10), 100503 (2015)

76. F. Coppens, F. Ancilotto, M. Barranco, N. Halberstadt, M. Pi, Capture of Xe and Ar atoms by quantized vortices in $^4$He nanodroplets. Phys. Chem. Chem. Phys. **19**(36), 24805–24818 (2017)

77. E. García-Alfonso, F. Coppens, M. Barranco, M. Pi, F. Stienkemeier, N. Halberstadt, Alkali atoms attached to vortex-hosting helium nanodroplets. J. Chem. Phys. **152**(19), 194109 (2020)

78. A. Przystawik, S. Göde, T. Döppner, J. Tiggesbäumker, K.-H. Meiwes–Broer, Light–induced collapse of metastable magnesium complexes formed in helium nanodroplets. Phys. Rev. A **78**, 021202 (2008)

79. A. Hernando, M. Barranco, R. Mayol, M. Pi, F. Ancilotto, Density functional theory of the structure of magnesium-doped helium nanodroplets. Phys. Rev. B **78**(18), 184515 (2008)

80. J. Higgins, C. Callegari, J. Reho, F. Stienkemeier, W.E. Ernst, M. Gutowski, G. Scoles, Helium cluster isolation spectroscopy of alkali dimers in the triplet manifold. J. Phys. Chem. A **102**(26), 4952–4965 (1998)

81. G.E. Douberly, P.L. Stiles, R.E. Miller, R. Schmied, K.K. Lehmann, $(HCN)_m$-$M_n$ (M = K, Ca, Sr): Vibrational excitation induced solvation and desolvation of dopants in and on helium nanodroplets. J. Phys. Chem. A **114**(10), 3391–3402 (2010)

82. F. Lackner, W.E. Ernst, Photoinduced molecule formation of spatially separated atoms on helium nanodroplets. J. Phys. Chem. Lett. **9**(13), 3561–3566 (2018)

83. S.A. Krasnokutski, F. Huisken, Low-temperature chemistry in helium droplets: reactions of aluminum atoms with $O_2$ and $H_2O$. J. Phys. Chem. A **115**(25), 7120–7126 (2011)

84. S.A. Krasnokutski, F. Huisken, Resonant two-photon ionization spectroscopy of Al atoms and dimers solvated in helium nanodroplets. J. Chem. Phys. **142**(8), 084311 (2015)

85. J. Eloranta, Self-assembly of neon into a quantum gel with crystalline structure in superfluid $^4$He: prediction from density functional theory. Phys. Rev. B **77**(13), 134301 (2008)

86. L. Kazak, S. Göde, K.H. Meiwes-Broer, J. Tiggesbäumker, Photoelectron spectroscopy on magnesium ensembles in helium nanodroplets. J. Phys. Chem. A **123**(28), 5951–5956 (2019)

87. R. Messner, A. Schiffmann, J.V. Pototschnig, M. Lasserus, M. Schnedlitz, F. Lackner, W.E. Ernst, Spectroscopy of gold atoms and gold oligomers in helium nanodroplets. J. Chem. Phys. **149**(2), 024305 (2018)

88. S.B. Emery, K.B. Rider, B.K. Little, A.M. Schrand, C.M. Lindsay, Magnesium cluster film synthesis by helium nanodroplets. J. Chem. Phys. **139**(5), 054307 (2013)

89. S.B. Emery, K.B. Rider, C.M. Lindsay, Stabilized magnesium/perfluoropolyether nanocomposite films by helium droplet cluster assembly. Propellants Explos. Pyrotechn. **39**(2), 161–165 (2014)

90. R.J. Buszek, C.J. Ridge, S.B. Emery, C.M. Lindsay, J.A. Boatz, Theoretical study of Cu/Mg core@shell nanocluster formation. J. Phys. Chem. A **120**(48), 9612–9617 (2016)

91. P. Thaler, A. Volk, F. Lackner, J. Steurer, D. Knez, W. Grogger, F. Hofer, W.E. Ernst, Formation of bimetallic core-shell nanowires along vortices in superfluid He nanodroplets. Phys. Rev. B **90**(15), 155442 (2014)

92. F. Lackner, A. Schiffmann, M. Lasserus, R. Messner, M. Schnedlitz, H. Fitzek, P. Pölt, D. Knez, G. Kothleitner, W.E. Ernst, Helium nanodroplet assisted synthesis of bimetallic Ag@Au nanoparticles with tunable localized surface plasmon resonance. Eur. Phys. J. D **73**(5), 104 (2019)

93. G. Schmid, L.F. Chi, Metal clusters and colloids. Adv. Mater. **10**(7), 515–526 (1998)

94. M.B. Cortie, A.M. McDonagh, Synthesis and optical properties of hybrid and alloy plasmonic nanoparticles. Chem. Rev. **111**(6), 3713–3735 (2011)
95. R.E. Palmer, R. Cai, J. Vernieres, Synthesis without solvents: the cluster (nanoparticle) beam route to catalysts and sensors. Acc. Chem. Res. **51**(9), 2296–2304 (2018)
96. P. Thaler, A. Volk, M. Ratschek, M. Koch, W.E. Ernst, Molecular dynamics simulation of the deposition process of cold Ag-clusters under different landing conditions. J. Chem. Phys. **140**(4), 044326 (2014)
97. M.P. de Lara-Castells, N.F. Aguirre, H. Stoll, A.O. Mitrushchenkov, D. Mateo, M. Pi, Communication: unraveling the $^4$He droplet-mediated soft-landing from ab initio-assisted and time-resolved density functional simulations: Au@$^4$He$_{300}$/TiO$_2$ (110). J. Chem. Phys. **142**(13), p131101 (2015)
98. R. Fernández-Perea, L.F. Gómez, C. Cabrillo, M. Pi, A.O. Mitrushchenkov, A.F. Vilesov, M.P. de Lara-Castells, Helium droplet-mediated deposition and aggregation of nanoscale silver clusters on carbon surfaces. J. Phys. Chem. C **121**(40), 22248–22257 (2017)
99. C. Binns, Nanoclusters deposited on surfaces. Surf. Sci. Rep. **44**(1–2), 1–49 (2001)
100. A. Schiffmann, D. Knez, F. Lackner, M. Lasserus, R. Messner, M. Schnedlitz, G. Kothleitner, F. Hofer, W.E. Ernst, Ultra-thin h-BN substrates for nanoscale plasmon spectroscopy. J. Appl. Phys. **125**(2), 023104 (2019)
101. A. Schiffmann, B.W. Toulson, D. Knez, R. Messner, M. Schnedlitz, M. Lasserus, F. Hofer, W.E. Ernst, O. Gessner, F. Lackner, Ultrashort XUV pulse absorption spectroscopy of partially oxidized cobalt nanoparticles. J. Appl. Phys. **127**(18), 184303 (2020)
102. Q. Wu, C.J. Ridge, S. Zhao, D. Zakharov, J. Cen, X. Tong, E. Connors, D. Su, E.A. Stach, C.M. Lindsay, A. Orlov, Development of a new generation of stable, tunable, and catalytically active nanoparticles produced by the helium nanodroplet deposition method. J. Phys. Chem. Lett. **7**(15), 2910–2914 (2016)
103. D. Appy, H. Lei, C.Z. Wang, M.C. Tringides, D.J. Liu, J.W. Evans, P.A. Thiel, Transition metals on the (0 0 0 1) surface of graphite: fundamental aspects of adsorption, diffusion, and morphology. Prog. Surf. Sci. **89**(3–4), 219–238 (2014)
104. C. Bréchignac, P. Cahuzac, F. Carlier, C. Colliex, J. Leroux, A. Masson, B. Yoon, U. Landman, Instability driven fragmentation of nanoscale fractal islands. Phys. Rev. Lett. **88**(19), 196103 (2002)
105. C. Bréchignac, P. Cahuzac, F. Carlier, C. Colliex, M. de Frutos, N. Kebaili, J. Le Roux, A. Masson, B. Yoon, Control of island morphology by dynamic coalescence of soft-landed clusters. Eur. Phys. J. D **16**(1), 265–269 (2001)
106. I. Lopez-Salido, D.C. Lim, Y.D. Kim, Ag nanoparticles on highly ordered pyrolytic graphite (HOPG) surfaces studied using STM and XPS. Surf. Sci. **588**(1–3), 6–18 (2005)
107. U. Henne, J.P. Toennies, Electron capture by large helium droplets. J. Chem. Phys. **108**(22), 9327–9338 (1998)
108. R. Sliter, L.F. Gomez, J. Kwok, A. Vilesov, Sizes distributions of large He droplets. Chem. Phys. Lett. **600**, 29–33 (2014)
109. E. Roduner, Size matters: why nanomaterials are different. Chem. Soc. Rev. **35**(7), 583–592 (2006)
110. E.C. Tyo, S. Vajda, Catalysis by clusters with precise numbers of atoms. Nat. Nanotechnol. **10**(7), 577–588 (2015)
111. T. Diederich, T. Döppner, J. Braune, J. Tiggesbäumker, K.H. Meiwes-Broer, Electron delocalization in magnesium clusters grown in supercold helium droplets. Phys. Rev. Lett. **86**(21), 4807 (2001)
112. T. Diederich, J. Tiggesbäumker, K.H. Meiwes-Broer, Spectroscopy on rare gas-doped silver clusters in helium droplets. J. Chem. Phys. **116**(8), 3263–3269 (2002)
113. T. Döppner, S. Teuber, T. Diederich, T. Fennel, P. Radcliffe, J. Tiggesbäumker, K.H. Meiwes-Broer, Dynamics of free and embedded lead clusters in intense laser fields. Eur. Phys. J. D **24**(1), 157–160 (2003)
114. D. Spence, E. Latimer, C. Feng, A. Boatwright, A.M. Ellis, S. Yang, Vortex-induced aggregation in superfluid helium droplets. Phys. Chem. Chem. Phys. **16**(15), 6903–6906 (2014)

115. A. Volk, D. Knez, P. Thaler, A.W. Hauser, W. Grogger, F. Hofer, W.E. Ernst, Thermal instabilities and Rayleigh breakup of ultrathin silver nanowires grown in helium nanodroplets. Phys. Chem. Chem. Phys. **17**(38), 24570–24575 (2015)
116. A.W. Hauser, M. Schnedlitz, W.E. Ernst, A coarse-grained Monte Carlo approach to diffusion processes in metallic nanoparticles. Eur. Phys. J. D **71**(6), 1–8 (2017)
117. M. Schnedlitz, M. Lasserus, D. Knez, A.W. Hauser, F. Hofer, W.E. Ernst, Thermally induced breakup of metallic nanowires: experiment and theory. Phys. Chem. Chem. Phys. **19**(14), 9402–9408 (2017)
118. D. Knez, M. Schnedlitz, M. Lasserus, A. Schiffmann, W.E. Ernst, F. Hofer, Modelling electron beam induced dynamics in metallic nanoclusters. Ultramicroscopy **192**, 69–79 (2018)
119. M. Lasserus, M. Schnedlitz, D. Knez, R. Messner, A. Schiffmann, F. Lackner, A.W. Hauser, F. Hofer, W.E. Ernst, Thermally induced alloying processes in a bimetallic system at the nanoscale: AgAu sub-5 nm core-shell particles studied at atomic resolution. Nanoscale **10**(4), 2017–2024 (2018)
120. T. Walther, C. Humphreys, A quantitative study of compositional profiles of chemical vapour-deposited strained silicon-germanium/silicon layers by transmission electron microscopy. J. Cryst. Growth **197**, 113–128 (1999)
121. M. Schnedlitz, R. Fernandez-Perea, D. Knez, M. Lasserus, A. Schiffmann, F. Hofer, A.W. Hauser, M.P. de Lara-Castells, W.E. Ernst, Effects of the core location on the structural stability of Ni-Au core-shell nanoparticles. J. Phys. Chem. C **123**(32), 20037–20043 (2019)
122. S.A. Maier, *Plasmonics: Fundamentals and Applications* (Springer Science & Business Media, 2007)
123. U. Kreibig, M. Vollmer, *Optical Properties of Metal Clusters*. Springer Series in Material Science, Vol. 25 (Springer-Verlag, Berlin, Heidelberg, 1995)
124. K. Kneipp, M. Moskovits, H. Kneipp, Surface-enhanced Raman scattering. Phys. Today **60**(11), 40 (2007)
125. M.E. Stewart, C.R. Anderton, L.B. Thompson, J. Maria, S.K. Gray, J.A. Rogers, R.G. Nuzzo, Nanostructured plasmonic sensors. Chem. Rev. **108**(2), 494–521 (2008)
126. H. Haberland, Looking from both sides. Nature **494**(7435), E1–E2 (2013)
127. J.M. Fitzgerald, P. Narang, R.V. Craster, S.A. Maier, V. Giannini, Quantum plasmonics. Proc. IEEE **104**(12), 2307–2322 (2016)
128. K. Selby, V. Kresin, J. Masui, M. Vollmer, W.A. de Heer, A. Scheidemann, W.D. Knight, Photoabsorption spectra of sodium clusters. Phys. Rev. B **43**(6), 4565 (1991)
129. W.A. De Heer, The physics of simple metal clusters: experimental aspects and simple models. Rev. Mod. Phys. **65**(3), 611 (1993)
130. M. Zhou, C. Zeng, Y. Chen, S. Zhao, M.Y. Sfeir, M. Zhu, R. Jin, Evolution from the plasmon to exciton state in ligand-protected atomically precise gold nanoparticles. Nat. Commun. **7**(1), 1–7 (2016)
131. V. Amendola, R. Pilot, M. Frasconi, O.M. Maragó, M.A. Iatì, Surface plasmon resonance in gold nanoparticles: A review. J. Phys. Condens. Matter **29**(20), 203002 (2017)
132. P. Radcliffe, A. Przystawik, T. Diederich, T. Döppner, J. Tiggesbäumker, K.H. Meiwes-Broer, Excited-state relaxation of $Ag_8$ clusters embedded in helium droplets. Phys. Rev. Lett. **92**(17), 173403 (2004)
133. A. Przystawik, P. Radcliffe, S. Göde, K.H. Meiwes-Broer, J. Tiggesbäumker, Spectroscopy of silver dimers in triplet states. J. Phys. B: Atom. Mol. Opt. Phys. **39**(19), S1183 (2006)
134. E. Loginov, M. Drabbels, Excited state dynamics of Ag atoms in helium nanodroplets. J. Phys. Chem. A **111**(31), 7504–7515 (2007)
135. F. Lindebner, A. Kautsch, M. Koch, W.E. Ernst, Laser ionization and spectroscopy of Cu in superfluid helium nanodroplets. Int. J. Mass Spectrom. **365**, 255–259 (2014)
136. L.F. Gomez, E. Loginov, A. Halder, V.V. Kresin, A.F. Vilesov, Formation of unusual copper clusters in helium nanodroplets. Int. J. Nanosci. **12**(02), 1350014 (2013)
137. J.A. Scholl, A.L. Koh, J.A. Dionne, Quantum plasmon resonances of individual metallic nanoparticles. Nature **483**(7390), 421–427 (2012)

138. A.L. Koh, K. Bao, I. Khan, W.E. Smith, G. Kothleitner, P. Nordlander, S.A. Maier, D.W. McComb, Electron energy-loss spectroscopy (EELS) of surface plasmons in single silver nanoparticles and dimers: influence of beam damage and mapping of dark modes. ACS Nano **3**(10), 3015–3022 (2009)

139. E. Loginov, L.F. Gomez, A.F. Vilesov, Formation of core-shell silver-ethane clusters in He droplets. J. Phys. Chem. A **117**(46), 11774–11782 (2013)

140. D. Verma, R.M.P. Tanyag, S.M. O'Connell, A.F. Vilesov, Infrared spectroscopy in superfluid helium droplets. Adv. Phys. X **4**(1), 1553569 (2019)

141. E. Loginov, L.F. Gomez, B.G. Sartakov, A.F. Vilesov, Formation of large Ag clusters with shells of methane, ethylene, and acetylene in He droplets. J. Phys. Chem. A **120**(34), 6738–6744 (2016)

142. E. Loginov, L.F. Gomez, B.G. Sartakov, A.F. Vilesov, Formation of core-shell ethane-silver clusters in he droplets. J. Phys. Chem. A **121**(32), 5978–5982 (2017)

143. W.J. Stark, P.R. Stoessel, W. Wohlleben, A.J.C.S.R. Hafner, Industrial applications of nanoparticles. Chem. Soc. Rev. **44**(16), 5793–5805 (2015)

144. M. Lasserus, M. Schnedlitz, R. Messner, F. Lackner, W.E. Ernst, A.W. Hauser, Vanadium (V) oxide clusters synthesized by sublimation from bulk under fully inert conditions. Chem. Sci. **10**(12), 3473–3480 (2019)

145. H. Schöbel, M. Dampc, F.F. Da Silva, A. Mauracher, F. Zappa, S. Denifl, T.D. Märk, P. Scheier, Electron impact ionization of CCl$_4$ and SF$_6$ embedded in superfluid helium droplets. Int. J. Mass Spectrom. **280**(1–3), 26–31 (2009)

146. P. Bartl, K. Tanzer, C. Mitterdorfer, S. Karolczak, E. Illenberger, S. Denifl, P. Scheier, Electron ionization of different large perfluoroethers embedded in ultracold helium droplets: effective freezing of short-lived decomposition intermediates. Rapid Commun. Mass Spectrom. **27**(2), 298–304 (2013)

147. M. Lasserus, D. Knez, F. Lackner, M. Schnedlitz, R. Messner, D. Schennach, G. Kothleitner, F. Hofer, A.W. Hauser, W.E. Ernst, Synthesis of nanosized vanadium (V) oxide clusters below 10 nm. Phys. Chem. Chem. Phys. **21**(37), 21104–21108 (2019)

148. D. Knez, M. Schnedlitz, M. Lasserus, A.W. Hauser, W.E. Ernst, F. Hofer, G. Kothleitner, The impact of swift electrons on the segregation of Ni-Au nanoalloys. Appl. Phys. Lett. **115**(12), 123103 (2019)

149. M. Xiao, R. Jiang, F. Wang, C. Fang, J. Wang, C.Y. Jimmy, Plasmon-enhanced chemical reactions. J. Mater. Chem. A **1**(19), 5790–5805 (2013)

150. C. Clavero, Plasmon-induced hot-electron generation at nanoparticle/metal-oxide interfaces for photovoltaic and photocatalytic devices. Nat. Photon. **8**(2), 95–103 (2014)

151. D.C. Ratchford, Plasmon-induced charge transfer: challenges and outlook. ACS Nano **13**(12), 13610–13614 (2019)

152. M. Escher, N. Weber, M. Merkel, C. Ziethen, P. Bernhard, G. Schönhense, S. Schmidt, F. Forster, F. Reinert, B. Krömker, D. Funnemann, NanoESCA: a novel energy filter for imaging X-ray photoemission spectroscopy. J. Phys. Condens. Matter **17**(16), S1329 (2005)

153. M.G. Blaber, M.D. Arnold, M.J. Ford, Search for the ideal plasmonic nanoshell: the effects of surface scattering and alternatives to gold and silver. J. Phys. Chem. C **113**(8), 3041–3045 (2009)

154. F. Lackner, J.A. Gessner, F. Siegrist, A. Schiffmann, R. Messner, M. Lasserus, M. Schnedlitz, B.W. Toulson, D. Knez, F. Hofer, O. Gessner, W.E. Ernst, M. Schultze, Attosecond spectroscopy of ultrafast carrier dynamics in nanoparticles, in *International Conference on Ultrafast Phenomena*, ed. F. Kärtner, M. Khalil, R. Li, F. Légaré, T. Tahara. OSA Technical Digest (Optical Society of America) (2020), pp. M4A-13

# Appendix
# Helium Cluster and Droplet Spectroscopy Reviews

[1] S. Albertini, E. Gruber, F. Zappa, S. Krasnokutski, F. Laimer, and P. Scheier: *Chemistry and physics of dopants embedded in helium droplets,* Mass Spec Rev. (2021), 1–39. circa 385 references. https://doi.org/10.1002/mas.21699.

*This review describes recent experiments on the capture of atoms and ions by neutral and charged helium droplets. Ion reactions in droplets and the spectroscopy of ions in droplets are dealt with. Recent studies of chemical reactions are surveyed.*

[2] T. Gonzalez-Lezana, O. Echt, and M. Gatchell: *Solvation of ions in helium,* Int. Rev. Phys. Chem. **39** (4) (2020): 465–516. 350 references. https://doi.org/10.1080/0144235X.2020.1794585.

*Experiments on the solvation of atomic, molecular and cluster ions in helium nanodroplets are reviewed. The latest results on negative and positive and highly charged pure and doped droplets are described.*

[3] O. Gessner, A.F. Vilesov: *Imaging quantum vortices in superfluid helium droplets,* Ann. Rev. Phys. Chem. **70** (2019): 173–198. 151 references. https://doi.org/10.1146/annurev-physchem-042018-052744.

*This review provides an introduction to quantum vorticity in helium droplets and reviews advances in X-ray and extreme ultraviolet scattering for* in situ *detection of droplet shapes and the imaging of vortex structures inside individual, isolated droplets and understanding the rotational motion of isolated, nano- to micrometer-scale superfluid helium droplets. New applications of helium droplets ranging from studies of quantum phase separations to mechanisms of low-temperature aggregation are discussed.*

[4] D. Verma, R. M. P. Tanyag, S. M. O. O'Connell, A. F. Vilesov: *Infrared spectroscopy in superfluid helium droplets,* Adv Phys-X, **4** (2019), **4**, (1):149–179. 206 references. https://doi.org/10.1080/23746149.2018.1553569.

*This review provides an overview of many infrared spectroscopic investigations of small molecules and clusters in helium droplets with special emphasize on organic molecules and radicals and reactive complexes.*

[5] M. Lemeshko and R. Schmidt: *Molecular impurities interacting with a many-particle environment: from helium droplets to ultracold gases*, Chapter 9 in: Theoretical and Computational Chemistry Series No. 11, *Cold Chemistry: molecular scattering and reactivity near zero*, Ed O. Dulieu and A. Osterwalder, The Royal Society of Chemistry, London, UK. (2018): 444–495. 162 references. EPUB ISBN: 978-1-78801-355-0.

*Discusses the rotation of molecules in helium nanodroplets and in Bose–Einstein condensates in terms of the recently developed angulon quasiparticle theory.*

[6] R. M. P. Tanyag, C. F. Jones, C. Bernando, S. M. O. O'Connell, D. Verma and A. F. Vilesov: *Experiments with large superfluid helium nanodroplets*, Chapter 8 in: Theoretical and Computational Chemistry Series No. 11, *Cold Chemistry: molecular scattering and reactivity near zero*, Ed O. Dulieu and A. Osterwalder, The Royal Society of Chemistry, London, UK. (2018): 389–443. 189 references. EPUB ISBN: 978-1-78801-355-0.

*This comprehensive review focuses on experiments on large helium droplets carried out in the group of Vilesov at the University of Southern California. It contains a discussion of elementary excitations (phonons and rotons) and calculated cooling rates under different conditions. Contains images of droplets and of metal wires grown inside helium droplets and a discussion of X-ray diffraction patterns of pure and Xe-doped droplets.*

[7] A. Mauracher, O. Echt, A. M. Ellis, S. Yang, D. K. Bohme, J. Postler, A. Kaiser, S. Denifl, and P. Scheier: *Cold physics and chemistry: collisions, ionization and reactions inside helium nanodroplets close to zero Kelvin*, Phys. Rep. **751** (2018):1–90. 623 references. https://doi.org/10.1016/j.physrep.2018.05.001.

*This is a comprehensive well illustrated review with 62 mostly colored figures which covers most aspects of pick-up and ionization of droplets, cations and anions in droplets, chemical reactions and spectroscopy of highly reactive molecules.*

[8] F. Ancilotto, M. Barranco. F. Coppens, J. Eloranta, N. Halberstadt, A. Hernando, D. Mateo, and M. Pi: *Density functional theory of doped superfluid liquid helium and nanodroplets*, Int. Rev. Phys. Chem. **36**, (4) (2017): 621–707. 295 references. https://doi.org/10.1080/0144235X.2017.1351672.

*As implied by the title the review summarizes the density functional theory of liquid helium. The theory is applied to many current experimental phenomena involving mostly metal atoms and ions doped in pure and mixed $^3He/^4He$ droplets. Electrons, cations and vortices in droplets and their dynamics are illustrated by many colored visualizations of snapshot images. Calculations of rotational superfluidity of molecules in small ($\leq 40$) droplets are compared with experiment. Liquid helium on surfaces and soft landing of droplets on surfaces are also discussed.*

[9] R. Rodriguez-Cantano, T. Gonzalez-Lezana and P. Villarreal: *Path integral Monte Carlo investigations on doped helium clusters*, Int. Rev. Phys. Chem. **35**, (1) (2016): 37–68. 143 references. https://doi.org/10.1080/0144235X.2015.1132595.

*Path integral Monte Carlo calculations are applied to visualize the interaction of the exemplary systems Ca, $Rb_2$, and the anions $He^{*-}$, and $He_2^{*-}$ with small $l(\leq 40)$ helium droplets.*

[10] M. P. Ziemkiewicz, D. M. Neumark and O. Gessner: *Ultrafast electronic dynamics in helium nanodroplets*, Int. Rev. Phys. Chem. **34**, (2) (2015): 239–267. 101 references. https://doi.org/10.1080/0144235X.2015.1051353.

*Femtosecond time-domain experiments in the extreme ultraviolet range are used to probe interband relaxation dynamics, dynamics of highly excited states and photoassociation within the droplets. Some of these channels lead to the ejection of excited Rydberg atoms and molecules of helium.*

[11] M. Mudrich and F. Stienkemeier: *Photoionisation of pure and doped helium nanodroplets*, Int. Rev. Phys. Chem. **33**, (3) (2014): 301–339. 219 references. https://doi.org/10.1080/0144235X.2014.937188.

*Previous studies of laser-induced fluorescence spectroscopy of alkali atoms in droplets are extended to study photoelectron and photoionization with velocity map electron and ion imaging. Femtosecond pump-probe photoionization experiments are also reviewed.*

[12] S. Yang and A. M. Ellis: *Helium droplets: a chemistry perspective* Chem Soc Rev **42**, (2013), 472–484. 62 references. https://doi.org/10.1039/C2CS35277J.

*This tutorial review provides an easy to understand overview of the field of helium droplet experiments with special emphasize on chemical complexes and clusters formed inside droplets and their spectroscopy.*

[13] J. P. Toennies: *Helium clusters and droplets: Microscopic superfluidity and other quantum effects.* Mol. Phys. **111** (2013): 1879–1891. 118 references. https://doi.org/10.1080/00268976.2013.802039.

*This review covers topics ranging from matter-wave experiments on the helium dimer and small $^4He$ clusters and mixed $^4He$-$^3He$ clusters up to spectroscopy of large organic clusters in large droplets with emphasize on evidence for microscopic superfluidity.*

[14] S. Yang and A. M. Ellis: *Clusters and nanoparticles in superfluid helium droplets: Fundamentals, challenges and perspectives*, Nanodroplets, Lect. Notes in Nanoscale Sci. and Techn. **18**, Chapter 10 (2013): 237–264. 106 references. https://doi.org/10.1007/978-1-4614-9472-0_10.

*This article presents a basic overview of several key experiments on molecular spectra, metal atoms on the surface and in the interior of droplets and soft landing on solid surfaces.*

[15] C. Callegari, W. Jäger, and F. Stienkemeier: *Helium nanodroplets*, CRC Handbook of Nanophysics Vol. 3, Nanoparticles and Quantum Dots (K. Sattler, Ed. in Chief), Taylor & Francis, 4/1-4/28 (2011). Ebook ISBN: 978-1-4200-7545-8.

*General review of droplet spectroscopy with an emphasize on spectra of small water clusters and a wide size range of ammonia clusters.*

[16] O. Echt, T. D. Märk, and P. Scheier, *Molecules and clusters embedded in helium nanodroplets*, CRC Handbook of Nanophysics vol. 2, Clusters and Fullerenes (K. Sattler, Ed. in Chief), Taylor & Francis, (2011): 20/1-20/24. ISBN: 9781420075540, 1420075543.

*Starts with background of basic properties of atomic He and the bulk liquid. Ionization, fragmentation and cluster cooling of rare gas clusters are discussed. Discusses experiments on ionization of hydrogen clusters and the electron attachment to droplets doped with water clusters.*

[17] K. Kuyanov-Prozument, D. Skvortsov, M. N. Slipchenko, B. G. Sartakov and A. Vilesov: *Matrix isolation spectroscopy in helium droplets*, Chapter 7 in book "Physics and Chemistry at Low Temperatures" Edited by L. Khriachtchev, Jenny Stanford Publishing, Boca Raton (2011): 203–230. ISBN 13: 978-981-4267-51-9 (hbk).

*This review surveys recent results of the Vilesov group of infrared spectra of small water and ammonia clusters and the relative energies of conformers of 2-chloroethanol (2-CLE) molecules.*

[18] Callegari and W. E. Ernst: *Helium nanodroplets as nanocryostats for molecular spectroscopy-from the vacuum ultraviolet to the microwave regime* in Handbook of High-resolution Spectroscopy; Edited by Martin Quack and Frederic Merkt, John Wiley and Sons, Ltd. (2011): 1551–1594. 173 references. http://onlinelibrary.wiley.com/book/10.1002/9780470749593.

*Contains a comprehensive list of 23 active experimental groups, their techniques, and molecules investigated.*

[19] F. Paesani: *Superfluidity of clusters* in Handbook of Nanophysics Vol. 2; Edited by K. D. Sattler Taylor and Francis, 2010: 12-1-12-16. 162 references. ISBN: 9781420075540, 1420075543.

*A review of the theory in connection with recent spectroscopic studies of doped small He clusters with 2–70 atoms.*

[20] Wei Kong, L. Pei and J. Zhang: *Linear dichroism spectroscopy of gas phase biological molecules embedded in superfluid helium droplets,* Int. Rev. in Phys. Chem. **28** (2009) 33–52. 124 references. https://doi.org/10.1080/014423508025 73678.

*This review is devoted largely to gas phase linear dichroism spectroscopy, including the theoretical and experimental background and a few examples. The advantage and the procedures of aligning biological molecules in superfluid helium droplets are discussed at length.*

[21] A. Slenczka and J. P Toennies: *Chemical dynamics inside superfluid helium nanaodroplets at 0.37 K* in "Low Temperatures and Cold Molecules" (editor: I.W.M. Smith) Imperial College Press, World Scientific, Singapore 2008: 345–392. 57 references. ISBN-13: 978-1-84816-209-9.

*Survey of laser emission spectroscopy, pump-probe photodissociation, photoionization experiments. Photo-induced tautomerization, ion–molecule reactions, photoinduced isomerization and neutral molecule reactive reactions in helium droplets are treated.*

[22] K. Szalewicz: *Interplay between theory and experiment in investigations of molecules embedded in superfluid helium nanodroplets,* Int. Rev. Phys. Chem. **27** (2) (2008): 273–316. 296 references. https://doi.org/10.1080/01442350801933485.

*Written by a well-known quantum chemist with special emphasize on the potential energy surfaces of small molecules with up to 10 attached He atoms. Contains an extensive discussion of theories of the $^4He$- dimer potential and pure bulk helium and an overview of quantum chemical calculations of the potentials of several molecules interacting with up to 100 attached He atoms.*

[23] J. Tiggesbäumker and F. Stienkemeier: *Formation and properties of metal clusters isolated in helium droplets,* Phys. Chem. Chem. Phys. **9** (2007): 4748–4770. 270 references. https://doi.org/10.1039/B703575F.

*This comprehensive review is about the formation and mass spectra of metal clusters on the surface and in the interior of helium droplets.*

[24] J. Küpper, and J. M. Merritt: *Spectroscopy of free radicals and radical containing entrance-channel complexes in superfluid helium nanodroplets*: Int. Rev. Phys. Chem. **26 (2)** (2007): 249–287. 262 references. https://doi.org/10.1080/014423506 01087664.

*Complements the work described in the review No. 27. Describes experiments in which droplets are doped with propargyl radicals, Na and an entire series of hydrocarbon radicals and complexes.*

[25] G. N. Makarov: *On the possibility of selecting molecules embedded in superfluid helium nanodroplets (clusters),* Physics—Uspekhi **49** (2006) 1131–1150. https://doi.org/10.1070/PU2006v049n11ABEH005941.

*This review discusses the state of molecular spectroscopy in droplets and proposes how laser resonance excited isotope molecules can be selected by deflecting them by a secondary beam.*

[26] B. S. Dumesh and L. A Surin: *Unusual rotations in helium and hydrogen nanoclusters and "nanoscopic" superfluidity,* Physics-Uspekhi **49** (11) (2006):1113–1129. https://doi.org/10.1070/PU2006v049n11ABEH006073.

*This review is devoted to the spectroscopy of small He clusters (N < 20) and $H_2$ clusters (N < 17) with attached OCS, $N_2O$, $CO_2$ and CO. Theoretical and experimental rotational spectra are compared and evidence for superfluidity in these small clusters is analyzed.*

[27] F. Stienkemeier, and K. K. Lehmann: *Spectroscopy and dynamics in helium nanodroplets,* J. Phys. B: At. Mol. Opt. Phys. **39** (2006): R127–R166. 269 references. https://doi.org/10.1088/0953-4075/39/8/R01.

*Time-resolved techniques for studying the dynamics mostly between alkali metal atoms and molecules and with the helium environment are reviewed. Theoretical results on the energetics and dynamics of helium droplets are also discussed.*

[28] M. Y. Choi, G. E. Douberly, T. M. Falconer, W. K. Lewis, C. M. Lindsay, J. M. Merritt, P. L. Stiles, and R. E. Miller: *Infrared spectroscopy of helium nanodroplets: novel methods for physics and chemistry,* Int. Rev. Phys. Chem. **25** (1–2) (2006): 15–75. 357 references. http://dx.doi.org/10.1080/01442350600625092.

*A very detailed review reporting the work from the group of the late Roger Miller based on the pendular spectroscopy IR method. Describes first experiments with radicals, metal containing clusters and amino acids. 357 references.*

[29] R. Guardiola: *Drops made of helium atoms: A fascinating many-body system,* In: Condensed Matter Theories, Vol. **20**, Editors: J. W. Clark, R. M. Panoff and H. Li, Nova Science Publishers Inc. (2006): 119–130. ISBN-13: 978-1594549892.

*A compact easily understandable review written by a many-body theoretician about the magic numbers of small $^4He$, $^3He$ and mixed $^4He$-$^3He$ clusters.*

[30] M. Barranco, R. Guardiola, S. Hernández, R. Mayol, J. Navarro, and M. Pi: *Helium nanodroplets: An overview,* J. Low Temp. Phys. **142** (2006):1–81. 283 references. https://doi.org/10.1007/s10909-005-9267-0.

*Provides an extensive survey of theoretical droplet properties such as evaporative cooling, density distributions and energy levels also of small and large $^3He$-$^4He$ mixed droplets. Theoretical density profiles and energies of doped droplets are surveyed. Also describes calculations of vortices inside droplets and droplets on surfaces. A section is also devoted to small helium droplets on adsorbing substrates.*

[31] J. P. Toennies and A. F. Vilesov: *Superfluid helium droplets: A Uniquely cold nanomatrix for molecules and molecular complexes,* Angew. Chem. Int. Ed. **43** (2004):2622–2648. 269 references. https://doi.org/10.1002/anie.200300611.

*A comprehensive highly cited review covering the spectroscopic experiments at the time.*

[32] G. N. Makarov: *Spectroscopy of single molecules and clusters inside helium nanodroplets. Microscopic manifestation of 4He superfluidity,* Physics-Usbekhi **47**(3) (2004): 217–247. 249 references. https://doi.org/10.1070/PU2004v047n03AB EH001698.

*An extensive and comprehensive review with 5 sections on production of droplets, rovibrational spectroscopy, electronic spectroscopy, other applications, and helium droplets and Bose–Einstein condensate in traps.*

[33] J. P. Toennies: *Microscopic superfluidity of small $^4He$ and Para-$H_2$ clusters inside helium droplets*, in Advances in Quantum Many-Body Physics, Vol. 4 (editors: E. Krotschek and J. Navarro), World Scientific, Singapore 2002. 1–39. 166 references. *An update of the 1998 Annu. Rev. Phys. Chem. review No. 42.*

[34] Special issue: *Helium Nanodroplets: A Novel Medium for Chemistry and Physics*, J. Chem. Phys. **115**, (22) (2000).

*Contains five excellent review articles, three of which (No. 35, 37 and 38) are listed below and 17 research articles on the theory of pure droplets and mostly on various aspects of spectra of molecules in droplets.*

[35] C. Calligari, K. K. Lehmann, R. Schmid and G. Scoles: *Helium nanodroplet isolation spectroscopy: methods and recent results*, J. Chem. Phys. **115**, (22), (2000): 10090–10110. 151 references. https://doi.org/10.1063/1.1418746.

*A comprehensive review including theoretical aspects with detailed analysis of several infrared, microwave, double resonance and Stark spectra of simple molecules and clusters in droplets. Contains an extensive table of rotational constants of 52 molecules and clusters in free space and in helium droplets and a table of vibrational lifetimes.*

[36] J. P. Toennies, A. F. Vilesov, and K. B. Whaley: *Superfluid helium droplets: An ultracold nanolaboratory*, Physics Today, February (2001). 18 references. http://sci tation.aip.org/content/aip/magazine/physicstoday/54/2?ver=pdfcov.

*A short concise easy to read review on the state of the art at that time.*

[37] F. Stienkemeier and A. F. Vilesov: *Electronic spectroscopy in He droplets*, J. Chem. Phys. **115** (22) (2001): 10119–10137. 207 references. https://doi.org/10.1063/1.1415433.

*Summary of the results of the Princeton group on the electronic spectra of alkali and other metal atoms attached to He droplets and the spectra of organic molecules mostly from the Göttingen group.*

[38] J A. Northby: *Experimental studies of helium droplets*, J. Chem. Phys. **115**, (22) (2001):10065–10077. 144 references. https://doi.org/10.1063/1.1418249.

*One of the early pioneers describes the state of the art in producing and investigating pure droplets from the smallest ($He_2$) to super large drops magnetically levitated. Excited metastable, negative and positively charged droplets are discussed.*

[39] J. P. Higgins, J. Reho, F. Stienkemeier, W. E. Ernst, K. K. Lehmann, and G. Scoles: *Spectroscopy in, on, and off a beam of superfluid helium nanodroplets*, Atomic and molecular beams, R. Campargue (ed.), Springer (2001): 723–734: 137 references. https://doi.org/10.1007/978-3-642-56800-8_51.

*Reviews the research of the Princeton group of G. Scoles on the electronic spectra of $^1\Sigma$ and $^3\Sigma$ states of $Na_2$ and of $Na_2He$. Contains an interesting discussion of the angular distribution of the beam of desorbed alkali atoms and molecules.*

[40] Y. Kwon, P. Huang, M. V. Patel, D. Blume, and B. Whaley: *Quantum solvation and molecular rotations in superfluid helium clusters*, J. Chem. Phys. **113** (16) (2000): 6489–6501. 137 references. https://doi.org/10.1063/1.1310608.

*An important review of the quantum theory of the free rotation of molecules in helium droplets and the role of superfluidity. The theory addresses the increase of the rotational constants in terms of the non-superfluid density in the first solvation shell.*

[41] S. Grebenev, M. Hartmann, A. Lindinger, N. Pörtner. B. Sartakov, J. P. Toennies, and A.F. Vilesov: *Spectroscopy of molecules in helium droplets*, Physica B **280** (2000) 65–72. 43 references. https://doi.org/10.1016/S0921-4526(99)01451-9.

*Early review from the Göttingen group with emphasize on the evidence for superfluidity from the observation of $S_1 \leftarrow S_0$ transition in glyoxal and rotational levels in the IR spectrum of OCS.*

[42] J. P. Toennies and A. F. Vilesov: *Spectroscopy of atoms and molecules in liquid helium*, Annu. Rev. Phys. Chem. **49** (1998): 1–41. 296 references. https://doi.org/10.1146/annurev.physchem.49.1.1.

*First review on spectroscopy of molecules in doped helium droplets. Contains account of early history of helium matrices and the major pioneering experiments with helium droplets.*

[43] K. B. Whaley: *Spectroscopy and microscopic theory of doped helium clusters*. Adv. Molecular Vibrations and Collision Dynamics. **3**, (1998): 397–451. 248 references.

*The first review devoted largely to the theory of molecules in the superfluid environment of droplets. Contains a discussion of Variational, Diffusion and Path Integral Monte Carlo theories and several examples in which early experiments are analyzed. Still worth reading as an introduction.*

# Index

© The Editor(s) (if applicable) and The Author(s) 2022
A. Slenczka and J. P. Toennies (eds.), *Molecules in Superfluid Helium Nanodroplets*,
Topics in Applied Physics 145, https://doi.org/10.1007/978-3-030-94896-2

Printed in the United States
by Baker & Taylor Publisher Services